Springer Optimization and Its Applications

Volume 161

Aims and Scope

Optimization has continued to expand in all directions at an astonishing rate. New algorithmic and theoretical techniques are continually developing and the diffusion into other disciplines is proceeding at a rapid pace, with a spot light on machine learning, artificial intelligence, and quantum computing. Our knowledge of all aspects of the field has grown even more profound. At the same time, one of the most striking trends in optimization is the constantly increasing emphasis on the interdisciplinary nature of the field. Optimization has been a basic tool in areas not limited to applied mathematics, engineering, medicine, economics, computer science, operations research, and other sciences.

The series **Springer Optimization and Its Applications (SOIA)** aims to publish state-of-the-art expository works (monographs, contributed volumes, textbooks, handbooks) that focus on theory, methods, and applications of optimization. Topics covered include, but are not limited to, nonlinear optimization, combinatorial optimization, continuous optimization, stochastic optimization, Bayesian optimization, optimal control, discrete optimization, multi-objective optimization, and more. New to the series portfolio include Works at the intersection of optimization and machine learning, artificial intelligence, and quantum computing.

Volumes from this series are indexed by Web of Science, zbMATH, Mathematical Reviews, and SCOPUS.

More information about this series at http://www.springer.com/series/7393

Stephan Dempe • Alain Zemkoho
Editors

Bilevel Optimization

Advances and Next Challenges

Springer

Editors
Stephan Dempe
Institute of Numerical Mathematics
and Optimization
TU Bergakademie Freiberg
Freiberg, Germany

Alain Zemkoho
School of Mathematical Sciences
University of Southampton
Southampton, UK

ISSN 1931-6828 ISSN 1931-6836 (electronic)
Springer Optimization and Its Applications
ISBN 978-3-030-52118-9 ISBN 978-3-030-52119-6 (eBook)
https://doi.org/10.1007/978-3-030-52119-6

This Springer imprint is published by the registered company Springer Nature Switzerland AG.
The registered company address is: Gewerbestrasse 11, 6330 Cham, Switzerland

Preface

Bilevel optimization refers to the area of optimization dealing with problems having a hierarchical structure, involving two decision-makers: a leader and a follower. This problem originated from the field of economic game theory and was introduced in the habilitation thesis of Heinrich Freiherr von Stackelberg (October 31, 1905, Moscow to October 12, 1946, Madrid) completed in 1934. This thesis, written in Cologne, on market structure and equilibrium (in German language: Marktform und Gleichgewicht) was published in the same year by Julius Springer, Berlin and Wien [8]. An English translation of the thesis was published in 2011 by Springer [9].

One of the central topics in von Stackelberg's habilitation thesis is a model of duopoly, now known as Stackelberg game. About 50 years later, mathematicians renamed the model into the bilevel optimization (or as synonym "bilevel programming" [1]) problem and its quick development within mathematical optimization started into different directions. One of the initial points of attention was the realization that the problem is not well-posed if the follower's decision is not uniquely defined. Another issue resulted from different possibilities to transform the bilevel problem into single-level problems, which are not necessarily equivalent to the original one. It might be worth to note that later on, two-level (as a synonym for "bilevel") optimization was one of the initial sparks of non-differentiable optimization.

Nowadays, bilevel optimization has further developed into a wide number of different directions (finite and infinite dimensional problems, instances with one or many objective functions in the lower- and/or upper-level problem, as well as problems with discrete or continuous variables in one or both levels). Although we can find deterministic algorithms as well as metaheuristics suggested to solve those problems, the bilevel optimization problem itself is NP-hard. The problem has a huge number of applications, and there is now a strong interaction between bilevel optimization theory and related applications.

This volume was motivated by the 85th anniversary (in 2019) of the aforementioned book "Marktform und Gleichgewicht" by von Stackelberg, which led the way to the creation of the field of bilevel optimization. In order to appreciate how far we have gone in the development of the area and look ahead for where there is still much work to be done, leading experts in bilevel optimization have contributed to the realization of the book. Therefore, unlike other books on the subject, the particularity of this volume is that it is essentially made of surveys of different topics on bilevel optimization. Hence, it can serve as initial contact point for up to date developments in the field. All chapters in the volume are peer-reviewed by experts in mathematical optimization and have been carefully revised before publication.

We would like to take this opportunity to thank all the contributors for helping to put this volume together. We are also very grateful to all anonymous referees for carefully reading the chapters to help enhance the quality of the book. The support provided by the editorial team at Springer is highly appreciated as well.

The book is organized into four parts based on the concentration of the content of the corresponding chapters. Next, we provide a brief guided tour of each part.

Part I: Bilevel Optimization, Game Theory, and Applications

Considering the origin of bilevel optimization, the aim of this part is to present results from the game theory perspective of the problem and some related applications. The emphasis here is on Stackelberg games, Nash games, solution concepts, and the interconnections between them. Chapter 1 considers the optimistic and pessimistic version of bilevel optimization problem and discusses the possible one-level optimization approximations. For each of these approximations, relationships with the original problems are analyzed and closely related generalized Nash equilibrium problems are also crafted. Overall, the implications of these transformations in solving the optimistic and pessimistic bilevel programs are studied.

Chapter 2 introduces the concept of Stackelberg–Nash games and related properties, including existence and uniqueness results, as well as welfare properties. Specifications of these results are given for linear and quadratic "payoff functions" and extensions to multiobjective functions are also discussed.

A survey of multi-leader-follower bilevel optimization problems, which are two-level problems with one or more Nash games at each level, is given in Chap. 3. As the interaction between players at each level is governed by Nash games, this distinguishes this class of problem from multiobjective optimization problems studied later in Chap. 15. More generally, Chap. 3 provides a general introduction to multi-leader-follower bilevel programs and provides an overview of this complex class of the problem, with emphasis on existence results, reformulations, and solution algorithms, as well as a flavor of related applications.

Chapter 4 provides an overview of the solution concepts in bilevel optimization, including the standard ones (for the optimistic and pessimistic formulations) and the intermediate solution concepts, as well as the subgame perfect Nash equilibrium. Relationships between these concepts are discussed with illustrative examples. The other main focus of the chapter is on approximation schemes to compute these equilibria, as well as stability analysis based on various types of perturbations.

The applications of bilevel programming have grown exponentially in the last three decades. Chapter 5 presents one major area of application of bilevel optimization, i.e., in energy and electricity markets. In this chapter, an overview of various practical applications and related challenges is presented, as well as illustrative examples in the area of strategic investment in energy generation and storage.

To conclude this part, Chap. 6 discusses an application of bilevel optimization in the fast-growing area of machine learning; the focus of the chapter is on hyperparameter computation. Hyperparameters represent a key piece of information in the selection process of machine learning models and their computation can naturally be modeled by bilevel optimization; this chapter presents an overview of this topic while introducing new algorithms, which can deal with the learning problem with nonsmooth and/or nonconvex objective functions.

Various other applications of bilevel optimization can be found in Chap. 20.

Part II: Theory and Methods for Linear and Nonlinear Bilevel Optimization Considering the nonsmooth and nonconvex nature of bilevel programs, variational analysis has turned out to be a key tool for the analysis of the problem. The first two chapters of this part provide overviews of techniques to derive necessary optimality conditions for bilevel optimization problems. Chapter 7 focuses on the use of the generalized differentiation tools [7] to derive optimality conditions for the value function reformulation of nonsmooth bilevel programs. In the case where the problem data is smooth, Chap. 8 explores the derivation of optimality conditions via the Karush–Kuhn–Tucker reformulation, when the lower-level problem is convex. In the case where the lower-level problem is nonconvex, the latter chapter combines the Karush–Kuhn–Tucker and value function reformulation to formulate necessary optimality conditions for smooth bilevel programs.

The second main focus of this part is on solution algorithms for bilevel optimization problems. Considering the large number of publications on this subject and the variations in the specific models and available tools to develop efficient methods, we have organized the chapters based on the category of the problem class. Hence, Chap. 9 presents solution algorithms for simple bilevel programs. Note that a simple bilevel program is an optimization problem over the solution set of another optimization problem [2]; the main difference here is that the follower's problem is not a parametric optimization problem as in the context of the general bilevel optimization problem. After a historical review of simple bilevel programs and some classical examples, this chapter focuses on the convergence analysis of various solution methods dedicated to the problem.

Chapter 10 then covers solution methods tailored to bilevel programs with linear upper- and lower-level problems. The chapter takes the reader back in time, with a wide range of techniques introduced to solve this class of problem since optimization experts got into contact with the field of bilevel optimization. It might be useful to recall here that initial works on solution algorithms for bilevel optimization essentially focused on problems of this category [1]. This chapter covers enumerative techniques, methods based on the Karush–Kuhn–Tucker and lower-level value function reformulations, as well as heuristic-based approaches.

Chapter 11 considers bilevel programs with quadratic upper- and lower-level problems. For this class of problems, the authors introduce new techniques to find local and global optimal solutions. The main tool of the analysis here is the global search theory, a new technique dedicated to nonconvex optimization problems. It might be useful to recall that bilevel programs are nonconvex, even in the case where all the functions involved in the problem are linear.

Next, Chap. 12 focuses on general bilevel programs, which are not necessarily linear nor quadratic. To proceed with the analysis, the single-level reformulation considered is based on the mathematical programs with equilibrium constraints (MPEC); this is indeed a synonym for the Karush–Kuhn–Tucker reformulation of the bilevel optimization problem. It is important to mention here that the MPEC represents a large field of optimization in its own right [6], which provides huge opportunities to solve bilevel programs. However, both problems are not necessarily equivalent [3]. This chapter covers the basic MPEC theory, including constraint qualifications and stationarity concepts, as well as MPEC-based solution methods for bilevel optimization and related software environments.

Chapter 13 is dedicated to heuristic-type algorithms to solve bilevel programs. Motivated, in part, by the inherent complexity of the bilevel optimization problem, the development of this class of methods for bilevel optimization has significantly increased over the last decade. The rise in the popularity might also be attributed to the overall growth of such techniques in the wider field of optimization driven in part by the no-free lunch theorem on the need for specifically tailored algorithms for different classes of problem to enhance efficiency [10]. This chapter gives an overview of various classes of methods, including non-surrogate and surrogate-based evolutionary algorithms.

To end this part, Chap. 14 gives an overview of solution algorithms for pessimistic bilevel programs. It is worth mentioning that this class of problem is the most difficult one in the bilevel optimization world, as it entails minimizing a typically upper-semicontinuous function [4]; hence, it very often happens that the pessimistic bilevel program does not have a solution, while the corresponding optimistic version of the problem is solvable. For this reason, work on methods dedicated to solve such problems is still at its infancy. This chapter presents penalty and Kth-Best methods, as well as approximation and reduction approaches that have so far been proposed to tackle the pessimistic bilevel optimization problem.

Part III: Extensions and Uncertainty in Bilevel Optimization

Various extensions of bilevel optimization are possible and a few of them have been briefly touched on in Parts I and II. Recall, for instance, that bilevel programs involving Nash games are studied in Chap. 3 and multiobjective two-level optimization problems are briefly discussed in Chap. 13.

Chapter 15 is uniquely dedicated to solution methods for the latter class of problems. It presents a wide range of methods to compute solutions for multiobjective bilevel optimization problems, after presenting various scalarization and optimality condition-based characterizations of the problem. It might be worth mentioning here that considering the nature of a bilevel optimization problem, with at least two objective functions, there has also been some interest in the literature to establish a connection of this class of problem to multiobjective optimization. The first successful such attempt was made in [5], in the context of linear bilevel optimization. This chapter also provides an overview of this topic with further references.

Chapter 16 introduces bilevel optimal control problems, which are bilevel programs where the upper- or lower-level problem is an optimal control problem of ordinary or partial differential equations. Such problems extend the applicability of bilevel optimization to another wide range of real-world problems. This also gives rise to new challenges on top of the already difficult structure of bilevel programs in finite dimensions. The level of difficulty typically depends on which player is solving the optimal control problem; this chapter navigates through these scenarios, providing the state of the literature in this relatively new area of research while presenting some insights into practical applications and mathematical theory involved.

Another emerging consideration in bilevel optimization is uncertainty, which is drawing more and more attention. Chapter 17 overviews the current state of research in stochastic bilevel optimization, with a focus on linear bilevel programs, where uncertainty is on the right-hand side of the lower-level problems. For this class of the problem, the theoretical details, including stability analysis and the impact of different types of distributions, are investigated. Solution methods for stochastic linear bilevel programs, as well as related challenges, are also discussed.

To conclude this part, Chap. 18 considers linear integer multistage stochastic optimization problems and shows they are intertwined with linear integer multilevel optimization. The chapter provides a thorough critical review of current works around linear multistage optimization while proposing some novel lines of analysis. Since the mathematical description of the problem, as well as the corresponding linear multilevel optimization model, naturally involves value functions, the authors propose a detailed analysis of such functions with related approximation techniques relevant to algorithm development.

Part IV: Numerical and Research Tools for Bilevel Optimization
The last part of this volume presents two technical documents, which can be very helpful for research in bilevel optimization. To support the development of numerical methods to solve bilevel optimization problems with continuous variables, Chap. 19 presents a collection of 173 examples with related codes prepared in MATLAB. This *B*ilevel *O*ptimization *LIB*rary (BOLIB) also contains the currently known best solution for the problems, wherever possible.

Finally, Chap. 20 provides an extensive bibliography on the subject, with more than 1500 references. This chapter covers a wide range of topics that have been the subject of investigation of bilevel optimization in the last four decades, with an abundance of related works. It includes an overview of theoretical properties, solution algorithms, and applications.

Freiberg, Germany Stephan Dempe
Southampton, UK Alain Zemkoho

References

1. W. Candler, R. Norton, Multilevel programming. Technical Report 20, World Bank Development Research Center, Washington (1977)
2. S. Dempe, N. Dinh, J. Dutta, Optimality conditions for a simple convex bilevel programming problem, in *Variational Analysis and generalized Differentiation in Optimization and Control*, ed. by R.S. Burachik, J.-C. Yao (Springer, Berlin, 2010), pp. 149–161
3. S. Dempe, J. Dutta, Is bilevel programming a special case of a mathematical program with complementarity constraints? Math. Program. **131**(1–2), 37–48 (2012)
4. S. Dempe, B.S. Mordukhovich, A.B. Zemkoho, Necessary optimality conditions in pessimistic bilevel programming. Optimization **63**(4), 505–533 (2014)
5. J. Fülöp, On the equivalency between a linear bilevel programming problem and linear optimization over the efficient set. Technical Report, Hungarian Academy of Sciences (1993)
6. Z.-Q. Luo, J.-S. Pang, D. Ralph, *Mathematical Programs with Equilibrium Constraints* (Cambridge University Press, Cambridge, 1996)
7. B.S. Mordukhovich, *Variational Analysis and Applications* (Springer, Berlin, 2018)
8. H.F. Von Stackelberg, *Marktform und Gleichgewicht* (Springer, Wien, 1934)
9. H.F. Von Stackelberg, *Market Structure and Equilibrium*. Translation of "Marktform und Gleichgewicht" by Bazin, Damien, Hill, Rowland, Urch, Lynn (Springer, Berlin, 2011)
10. D.H. Wolpert, W.G. Macready, No free lunch theorems for optimization. IEEE Trans. Evol. Comput. **1**(1), 67–82 (1997)

Contents

Contributors

Didier Aussel Université de Perpignan, Perpignan, France

Damien Bazin Côte d'Azur University, CNRS, GREDEG, Nice, France

Suresh Bolusani Department of Industrial and Systems Engineering, Lehigh University, Bethlehem, PA, USA

Johanna Burtscheidt University of Duisburg-Essen, Faculty of Mathematics, Essen, Germany

Herminia I. Calvete Statistical Methods Department, IUMA, University of Zaragoza, Zaragoza, Spain

Francesco Caruso Department of Economics and Statistics, University of Naples Federico II, Naples, Italy

Zhong Chen School of Information and Mathematics, Yangtze University, Hubei, P.R. China

Matthias Claus University of Duisburg-Essen, Faculty of Mathematics, Essen, Germany

Stefano Coniglio Department of Mathematical Sciences, University of Southampton, Southampton, UK

Kalyanmoy Deb Michigan State University, East Lansing, MI, USA

Stephan Dempe TU Bergakademie Freiberg, Institute of Numerical Mathematics and Optimization, Freiberg, Germany

Joydeep Dutta Indian Institute of Technology, Kanpur, India

Gabriele Eichfelder Institute for Mathematics, TU Ilmenau, Ilmenau, Germany

Yuxin Fan Huazhong University of Science and Technology, Wuhan, P.R. China

Carmen Galé Statistical Methods Department, IUMA, University of Zaragoza, Zaragoza, Spain

Ludovic Julien EconomiX, UPL, University Paris Nanterre, CNRS, Nanterre, France

Youngdae Kim Argonne National Laboratory, Lemont, IL, USA

Lorenzo Lampariello Roma Tre University, Rome, Italy

Sven Leyffer Argonne National Laboratory, Lemont, IL, USA

M. Beatrice Lignola Department of Mathematics and Applications R. Caccioppoli, University of Naples Federico II, Naples, Italy

June Liu School of Management, Huaibei Normal University, Huaibei, Anhui, P. R. China

Zhichao Lu Michigan State University, East Lansing, MI, USA

Pekka Malo Aalto University School of Business, Helsinki, Finland

Patrick Mehlitz Brandenburgische Technische Universität Cottbus Senftenberg, Cottbus, Germany

Boris S. Mordukhovich Department of Mathematics, Wayne State University, Detroit, MI, USA

Jacqueline Morgan Department of Economics and Statistics and Center for Studies in Economics and Finance, University of Naples Federico II, Naples, Italy

Todd Munson Argonne National Laboratory, Lemont, IL, USA

Olivier Musy CRED, University Paris Panthéon Assas, Paris, France

Takayuki Okuno RIKEN AIP, Chuo, Japan

Andrei V. Orlov Matrosov Institute for System Dynamics and Control Theory SB RAS, Irkutsk, Russia

Tanushree Pandit Indian Institute of Technology, Kanpur, India

Salvador Pineda University of Malaga, Malaga, Spain

Ted K. Ralphs Department of Industrial and Systems Engineering, Lehigh University, Bethlehem, PA, USA

Simone Sagratella Sapienza University of Rome, Rome, Italy

Vladimir Shikhman Chemnitz Technical University, Chemnitz, Germany

Ankur Sinha Indian Institute of Management, Ahmedabad, India

Oliver Stein Institute of Operations Research, Karlsruhe Institute of Technology (KIT), Karlsruhe, Germany

Alexander S. Strekalovsky Matrosov Institute for System Dynamics and Control Theory SB RAS, Irkutsk, Russia

Anton Svensson Universidad de Chile, Santiago, Chile
Université de Perpignan, Perpignan, France

Sahar Tahernejad Department of Industrial and Systems Engineering, Lehigh University, Bethlehem, PA, USA

Akiko Takeda Department of Creative Informatics, Graduate School of Information Science and Technology, The University of Tokyo, Bunkyo, Japan
Riken Aip, Chuo, Japan

Diego A. Tejada-Arango Comillas Pontifical University, Madrid, Spain

Andrey Tin School of Mathematics, University of Southampton, Southampton, UK

Gerd Wachsmuth Brandenburgische Technische Universität Cottbus Senftenberg, Cottbus, Germany

Sonja Wogrin Comillas Pontifical University, Madrid, Spain

Jane J. Ye Department of Mathematics and StatisticsUniversity of Victoria Victoria, BC, Canada

Alain B. Zemkoho School of Mathematics, University of Southampton, Southampton, UK

Yue Zheng School of Management, Huaibei Normal University, Huaibei, Anhui, P.R. China

Shenglong Zhou School of Mathematics, University of Southampton, Southampton, UK

Part I
Bilevel Optimization, Game Theory, and Applications

Chapter 1
Interactions Between Bilevel Optimization and Nash Games

Lorenzo Lampariello, Simone Sagratella, Vladimir Shikhman, and Oliver Stein

Abstract We aim at building a bridge between bilevel programming and generalized Nash equilibrium problems. First, we present two Nash games that turn out to be linked to the (approximated) optimistic version of the bilevel problem. Specifically, on the one hand we establish relations between the equilibrium set of a Nash game and global optima of the (approximated) optimistic bilevel problem. On the other hand, correspondences between equilibria of another Nash game and stationary points of the (approximated) optimistic bilevel problem are obtained. Then, building on these ideas, we also propose different Nash-like models that are related to the (approximated) pessimistic version of the bilevel problem. This analysis, being of independent theoretical interest, leads also to algorithmic developments. Finally, we discuss the intrinsic complexity characterizing both the optimistic bilevel and the Nash game models.

Keywords Optimistic bilevel problem · Pessimistic bilevel problem · Generalized Nash equilibrium problem · Approximation techniques · Constraint qualifications · Degeneracies

---

L. Lampariello (✉)
Roma Tre University, Rome, Italy
e-mail: lorenzo.lampariello@uniroma3.it

S. Sagratella
Sapienza University of Rome, Rome, Italy
e-mail: sagratella@diag.uniroma1.it

V. Shikhman
Chemnitz Technical University, Chemnitz, Germany
e-mail: vladimir.shikhman@mathematik.tu-chemnitz.de

O. Stein
Institute of Operations Research, Karlsruhe Institute of Technology (KIT), Karlsruhe, Germany
e-mail: stein@kit.edu

S. Dempe, A. Zemkoho (eds.), *Bilevel Optimization*, Springer Optimization and Its Applications 161, https://doi.org/10.1007/978-3-030-52119-6_1

1.1 Introduction

The aim of this chapter is to provide a simple, high-level analysis of the interactions between two fundamental paradigms in multi-agent optimization, namely bilevel programming and Generalized Nash Equilibrium Problems (GNEP). In particular, we point out differences and similarities between two-level hierarchical optimization and one-level game models. Besides being of independent theoretical interest, this study paves the way to possible algorithmic developments. While we discuss in detail the basics of bilevel programming in this very section, in Sects. 1.2 and 1.3 we introduce different GNEP models that are proven to be tightly related to bilevel problems. Finally, in Sect. 1.4 we investigate further aspects concerning the intrinsic complexity of the models that are here analyzed. We stress that, in order to convey the main ideas, we refrain from giving too many technical details: the interested reader can refer to [7, 9–11], which most of the material in this chapter comes from.

Bilevel problems have a hierarchical structure involving two decision, upper and lower, levels. At the lower level, the so-called follower problem consists in minimizing the objective function f, parametric in the variables x, over the feasible set Y:

$$\begin{aligned} \underset{v}{\text{minimize}} \quad & f(x, v) \\ \text{s.t.} \quad & v \in Y. \end{aligned} \tag{1.1.1}$$

Note that, for the sake of presentation, we consider the case in which the follower problem feasible set does not depend on x, see Sects. 1.2 and 1.3. Nevertheless, many of the results that we describe can be generalized also to the case of a parametric feasible set (see, e.g., [9, 11]). In Sect. 1.4 we focus, in particular, on the follower's feasible set $Y(x)$ given by a finite number of smooth inequality constraints, i.e. it is defined as $\{v \mid g_j(x, v) \geq 0,\ j \in J\}$. Assuming problem (1.1.1) to be solvable for every $x \in X$, let φ be the follower optimal value function

$$\varphi(x) \triangleq \min_{v}\{f(x, v) \mid v \in Y\},$$

and let $S(x)$ be the set of the follower optimal points.

Further, denoting by ε a positive tolerance on the optimal value of problem (1.1.1), the set of follower ε-optimal solutions is given, for every x, by $\{y \in Y \mid f(x, y) \leq \varphi(x) + \varepsilon\}$, which, in general is not a singleton. As a consequence, whenever, among the follower's problem solutions, exclusively those that best suit the leader objective F are chosen, at the upper level one obtains the so-called original optimistic bilevel problem:

$$\begin{aligned} \underset{x}{\text{minimize}} \quad & \min_{y}\{F(x, y) \mid y \in Y,\ f(x, y) \leq \varphi(x) + \varepsilon\} \\ \text{s.t.} \quad & x \in X. \end{aligned} \tag{1.1.2}$$

But other possible points of view could be taken: in fact, "less optimistic" approaches have been considered in the literature (see [16] for a recent survey) up to the original pessimistic case

$$\begin{aligned} &\underset{x}{\text{minimize}} \max_{y} \{F(x, y) \mid y \in Y,\ f(x, y) \leq \varphi(x) + \varepsilon\} \\ &\text{s.t.} \quad x \in X. \end{aligned} \tag{1.1.3}$$

Problem (1.1.3) is the robust version of a bilevel program since, among the follower's optimal points, exclusively the worst ones (with respect to the leader objective F) are selected. We observe that problems (1.1.2) and (1.1.3) are solvable under standard assumptions such as continuity of functions f and F, compactness and nonemptiness of sets X and Y, and convexity of the follower problem (1.1.1). Hence, we can introduce

$$\psi_\varepsilon^o(x) \triangleq \min_y \{F(x, y) \mid y \in Y,\ f(x, y) \leq \varphi(x) + \varepsilon\}$$

and

$$\psi_\varepsilon^p(x) \triangleq \max_y \{F(x, y) \mid y \in Y,\ f(x, y) \leq \varphi(x) + \varepsilon\}$$

as the leader optimistic and pessimistic optimal value functions, respectively. Here we briefly recall that, as testified by the literature, the need for the introduction of the positive tolerance ε emerges for solvability, regularity and stability reasons. Moreover, "in practice, it is usually too much to ask for exact optimal solutions. The follower may be satisfied with an almost optimal solution" [14]. Hence, in the wake of a well-established tradition that dates back at least to [13] and [17], and for ease of discussion, we consider directly the approximated problems (1.1.2) and (1.1.3) rather than their exact counterparts (which are readily obtained by taking $\varepsilon = 0$ in (1.1.2) and (1.1.3)). Although the solution of the problems (1.1.2) and (1.1.3) with a strictly positive perturbation parameter ε is significant and meaningful per se both in a theoretical and in a practical perspectives, the question arises quite naturally on what happens when ε vanishes (see e.g. [12–15] and the rather exhaustive and recent survey [1]).

Concerning the optimistic problem, the following proposition sheds a light on what happens to optimal points when ε goes to 0.

Proposition 1.1.1 (See e.g. [14, Theorem 4.1]) *Let $\varepsilon \searrow 0$ and x^ε be the corresponding sequence of optimal points for the original optimistic problem* (1.1.2)*. Then, any accumulation point x^0 of the sequence $\{x^\varepsilon\}$ is an optimal point of* (1.1.2) *with $\varepsilon = 0$, i.e.,*

$$x^0 \in \arg\min_{x \in X} \psi_0^o(x). \qquad \triangle$$

When dealing with the pessimistic version (1.1.3), the picture becomes more complicated. In fact, even under our initial assumptions, the pessimistic problem (1.1.3) with $\varepsilon = 0$ might not be solvable (see e.g. [11, Example 3.1]). However, also in this unfavourable case, when ε goes to 0 the infimum of the pessimistic problem (1.1.3) can be reached, as detailed next.

Proposition 1.1.2 (See e.g. [11, Theorem 3.6]) *Let $\varepsilon \searrow 0$ and x^ε be the corresponding sequence of optimal points for the original pessimistic problem* (1.1.3)*. Then, at any accumulation point x^0 of the sequence $\{x^\varepsilon\}$ the infimum of* (1.1.3) *with $\varepsilon = 0$ is attained, in the sense that*

$$\text{cl}\, \psi_0^p(x^0) = \inf_{x \in X} \psi_0^p(x),$$

where $\text{cl}\, \psi_0^p$ stands for the closure of ψ_0^p on X. △

As customary in the literature, in order to devise practical solution methods, one can think to move the difficulties in the objective functions for (1.1.2) and (1.1.3) to the constraints level, obtaining the so-called standard optimistic and pessimistic versions of the bilevel problem. Namely, regarding the standard optimistic case,

$$\begin{aligned} \underset{x,y}{\text{minimize}}\ & F(x, y) \\ \text{s.t.}\quad & x \in X,\ y \in Y \\ & f(x, y) \le \varphi(x) + \varepsilon, \end{aligned} \tag{1.1.4}$$

where the constraint $F(x, y) \le \psi_\varepsilon^o(x)$, which should be present in principle, is omitted being redundant. Note that the optimal value function formulation (1.1.4) with $\varepsilon = 0$ has been introduced in [18] and further employed in [24, 25]. As for the standard pessimistic version that has been proposed for the first time in [11],

$$\begin{aligned} \underset{x,y}{\text{minimize}}\ & F(x, y) \\ \text{s.t.}\quad & x \in X,\ y \in Y \\ & f(x, y) \le \varphi(x) + \varepsilon \\ & F(x, y) \ge \psi_\varepsilon^p(x). \end{aligned} \tag{1.1.5}$$

Passing from the original to the standard versions of the problems comes at a price. In fact, local optimal points for (1.1.4) and (1.1.5) might happen not to lead to corresponding local optimal points for (1.1.2) and (1.1.3), respectively. On the other hand, there is a complete correspondence between global optimal points (for a more detailed discussion on these aspects see [3, 11]).

From now on, we assume: F and f to be continuously differentiable, f convex with respect to y for every x, and X and Y nonempty compact convex sets.

We employ standard notation. However, for the reader's convenience, we remark that, considering a real function h depending on two variables blocks, we denote by

$\nabla_1 h(x, y)$ the gradient of $h(\bullet, y)$ evaluated at x, while by $\nabla_2 h(x, y)$ the gradient of $h(x, \bullet)$ evaluated at y. Furthermore, with C convex set and $z \in C$, $N_C(z)$ is the classical normal cone (to C at z) of convex analysis (see e.g. [19, Chapter 6]). As for the definitions of semicontinuity and other properties of single and set-valued mappings such as lower semicontinuity, outer and inner semicontinuity, and local boundedness, we refer the reader to [19].

1.2 Standard Optimistic Bilevel and Equilibrium Problems

We remark that problem (1.1.4) is difficult to treat since φ in the value function constraint is implicitly defined; moreover, it is a nonconvex program even assuming convexity of the defining functions and sets. In order to deal with the implicit nature of φ, we propose two different ways to suitably reformulate the value function constraint: these approaches result in two different GNEP models that are tightly related to the standard optimistic bilevel problem (1.1.4) and where no implicitly defined functions appear (see [9, 10]). Here, for the sake of brevity, we just recall that GNEPs share with bilevel problems a parametric nature but, unlike them, they are problems in which all agents act at the same level. In particular, we call equilibrium of a GNEP a feasible solution from which no player has incentives to deviate unilaterally. GNEPs have been extensively studied in the literature and, as for fundamental definitions, tools and relevant bibliography, we refer the interested reader to the survey paper [6].

The first model reads as:

$$\begin{array}{llcll} \underset{x,y}{\text{minimize}} & F(x,y) & \qquad & \underset{v}{\text{minimize}} & f(x,v) \\ \text{s.t.} & x \in X,\ y \in Y & & \text{s.t.} & v \in Y. \\ & f(x,y) \le f(x,v) + \varepsilon & & & \end{array} \tag{1.2.1}$$

The second GNEP is as follows:

$$\begin{array}{llcll} \underset{x,y}{\text{minimize}} & F(x,y) & \qquad & \underset{u,v}{\text{minimize}} & f(x,v) \\ \text{s.t.} & x \in X,\ y \in Y & & \text{s.t.} & v \in Y \\ & f(x,y) \le f(u,v) + \nabla_1 f(u,v)^T(x-u) + \varepsilon & & & u = x. \end{array} \tag{1.2.2}$$

We say that, for both games, the player controlling x and y is the leader, while the other player is the follower. Note that, in the leader's problem, only the feasible set, specifically the modified value function constraint, depends on the follower's variables. We observe that the follower's problem in (1.2.1) and (1.2.2) is exactly the same as in (1.1.1). The presence of the modified value function constraint, introducing in both GNEPs some degree of hierarchy in an horizontal framework, keeps memory of the original imbalance of power between leader and follower in the standard optimistic bilevel problem (1.1.4). Concerning the GNEP model (1.2.1), one can show that any equilibrium of (1.2.1) leads to a corresponding optimal solution for (1.1.4).

Theorem 1.2.1 ([9, Theorem 3.1 and Corollary 3.1]) *If* (x^*, y^*, v^*) *is an equilibrium of GNEP* (1.2.1)*, then* (x^*, y^*) *is a global solution of the standard optimistic bilevel problem* (1.1.4). △

When dealing with GNEP (1.2.2), to make this game tightly related to the original bilevel problem (1.1.4), it is convenient to assume the convexity of f. We remark that this requirement, while nonstandard, can be obtained under mild assumptions (see [10, Example 2.2]). The main consequences of this condition are summarized in the following result which builds upon the material in [22].

Proposition 1.2.2 ([9, Proposition 2.4]) *If* f *is convex, then the function* φ *is convex and continuously differentiable. Moreover, for any* x*, the set* $\{\nabla_1 f(x, v) \mid v \in S(x)\}$ *is a singleton and, for any* $v \in S(x)$, $\nabla\varphi(x) = \nabla_1 f(x, v)$. △

In view of Proposition 1.2.2, letting (x, y) be feasible for (1.1.4), we say that (x, y) is a stationary point for (1.1.4) if a scalar λ exists such that:

$$\begin{aligned} &0 \in \nabla_1 F(x, y) + \lambda\nabla_1 f(x, y) - \lambda\nabla\varphi(x) + N_X(x) \\ &0 \in \nabla_2 F(x, y) + \lambda\nabla_2 f(x, y) + N_Y(y) \\ &\lambda \in N_{\mathbb{R}_-}(f(x, y) - \varphi(x) - \varepsilon). \end{aligned} \tag{1.2.3}$$

We remark that, even if f is not convex, nevertheless, thanks to the Danskin's Theorem (see, e.g., [2, Theorem 9.13]), φ enjoys some useful continuity properties along with an estimate for its set of subgradients that can be exploited in order to generalize stationarity conditions (1.2.3) in a nonsmooth sense. We also note that, thanks to the presence of ε, standard constraint qualifications hold for (1.1.4) (see [10]), making relations (1.2.3) necessary conditions for optimality of (1.1.4).

Proposition 1.2.3 *The Mangasarian–Fromovitz Constraint Qualification (MFCQ) is satisfied for standard optimistic bilevel problem* (1.1.4). △

Proof In order to prove the claim, suffice it to observe that for the multiplier of the value function constraint to be positive, we must have $f(x, y) - \varphi(x) - \varepsilon = 0$, and thus $f(x, y) = \varphi(x) + \varepsilon > \varphi(x)$. In turn, the convexity of the lower level problem implies that y is not a lower level solution, given x. Hence, the multiplier must be zero at any feasible point and the MFCQ is trivially satisfied. □

As for the GNEP models, one can easily show that the properties in Proposition 1.2.4 hold.

Proposition 1.2.4 (See [10, Remark 3.2]) *If F and f are convex, then GNEP* (1.2.2) *is a convex game whose equilibrium set is nonempty; furthermore, the MFCQ holds.* △

The following theorem, claiming that stationary points of (1.1.4) lead to equilibria of (1.2.2) and vice versa, shows the deep connection between (1.1.4) and (1.2.2).

Theorem 1.2.5 ([10, Theorem 3.3]) *Let F and f be convex. The following claims hold:*

(i) any equilibrium (x^*, y^*, u^*, v^*) of GNEP (1.2.2) *is such that (x^*, y^*) is stationary for the standard optimistic bilevel problem* (1.1.4)*;*
(ii) any stationary point (x^*, y^*) for the standard optimistic bilevel problem (1.1.4) *is such that (x^*, y^*, x^*, v^*) is an equilibrium of GNEP* (1.2.2) *for any $v^* \in S(x^*)$.*

△

The following example, completing the picture, emphasizes the different relations between the two GNEP models and (1.1.4).

Example
Consider the following standard optimistic bilevel problem with $\varepsilon > 0$ such that $\sqrt{\varepsilon} < \frac{1}{6}$:

$$\begin{aligned} \underset{x,y}{\text{minimize}}\ & x^2 + y^2 \\ \text{s.t.}\quad & x \in [-1, 1],\ y \in [-1, 1] \\ & (x + y - 1)^2 \le \varphi(x) + \varepsilon, \end{aligned} \tag{1.2.4}$$

where $\varphi(x) = \min_v\{(x + v - 1)^2 : v \in [-1, 1]\}$. We remark that, while the objective functions of both the leader F and the follower f are convex, nevertheless (1.2.4) is a nonconvex optimization problem with an implicitly defined constraint. The unique solution of (1.2.4) is $(x^*, y^*) = \left(\frac{1-\sqrt{\varepsilon}}{2}, \frac{1-\sqrt{\varepsilon}}{2}\right)$. In this case, GNEP (1.2.1) reads as:

$$\begin{aligned} \underset{x,y}{\text{minimize}}\ & x^2 + y^2 & \qquad \underset{v}{\text{minimize}}\ & (x + v - 1)^2 \\ \text{s.t.}\quad & x \in [-1, 1],\ y \in [-1, 1] & \text{s.t.}\quad & v \in [-1, 1], \\ & (x + y - 1)^2 \le (x + v - 1)^2 + \varepsilon & & \end{aligned} \tag{1.2.5}$$

which, clearly is nonconvex due to the presence of the nonconvex value function constraint in the leader's problem. The point $(x^*, y^*, v^*) =$

(continued)

$\left(\frac{1-\sqrt{\varepsilon}}{2}, \frac{1-\sqrt{\varepsilon}}{2}, \frac{1+\sqrt{\varepsilon}}{2}\right)$, where $v^* = \frac{1+\sqrt{\varepsilon}}{2}$ is the unique solution of the follower's problem when $x = x^*$, is not an equilibrium for (1.2.5): suffice it to observe that the feasible point $\left(-\frac{\sqrt{\varepsilon}}{2}, \frac{1+\sqrt{\varepsilon}}{2}, \frac{1+\sqrt{\varepsilon}}{2}\right)$ entails a strictly smaller value for the leader's objective, given $\sqrt{\varepsilon} < \frac{1}{6}$. Since the unique solution of problem (1.2.4) (which is a fortiori a stationary point) does not lead to an equilibrium of (1.2.5), by Theorem 1.2.1, its set of equilibria is empty.

On the contrary, GNEP (1.2.2), that is,

$$\begin{array}{lll} \underset{x,y}{\text{minimize}} & x^2+y^2 & \qquad \underset{u,v}{\text{minimize}} \ (x+v-1)^2 \\ \text{s.t.} & x \in [-1,1],\ y \in [-1,1] & \qquad \text{s.t.} \quad v \in [-1,1] \\ & (x+y-1)^2 \le (u+v-1)^2 & \qquad \qquad u = x, \\ & \quad +2(u+v-1)(x-u)+\varepsilon & \end{array}$$

in view of Proposition 1.2.4, is convex and, by Theorem 1.2.5, the point (x^*, y^*, x^*, v^*) is one of its equilibria.

Here we summarize the distinctive properties of (1.2.1) compared to those of (1.2.2), pointing out their different relations with (1.1.4):

- under the full convexity of F and f, the GNEP (1.2.2) is a convex game that certainly has an equilibrium, while the GNEP (1.2.1) is a non necessarily convex game with a possibly empty equilibrium set;
- under the full convexity of F and f, equilibria of the GNEP (1.2.2) lead to stationary points of the standard optimistic bilevel problem (1.1.4) and vice versa, while stationary points of the standard optimistic bilevel problem (1.1.4) may not lead to equilibria of the GNEP (1.2.1);
- any equilibrium of the GNEP (1.2.1) leads to a global optimal point of the standard optimistic bilevel problem (1.1.4), while, under the full convexity of F and f, any optimal point of the standard optimistic bilevel problem (1.1.4) leads to an equilibrium of the GNEP (1.2.2), see the scheme below.

Global solutions

GNEP (2.1)	Th.2.1	problem (1.4)	Pr. 2.3, Th.2.5	GNEP (2.2)
x, y, v	$\Rightarrow$	x, y	$\Rightarrow$	x, y, x, v

The scheme above simply means that, on the one hand, the (x, y)-part of the equilibrium set of GNEP (1.2.1) is a subset of the set of global solutions of the standard optimistic bilevel problem (1.1.4), while, on the other hand, the (x, y)-part of the equilibrium set of GNEP (1.2.2) turns out to be a superset of the set of global solutions of (1.1.4).

It is worth specifying, in view of the developments to follow, the case in which the follower's objective f does not depend on x. The resulting model is referred to, in the literature, as *simple bilevel* or *pure hierarchical* problem (see [1] and the references therein). In this favourable situation, we obtain the following strengthened properties:

- the follower's objective f, optimal value function φ, and solution set map S are trivially convex, constant, and fixed, respectively;
- assuming the full convexity of F, the standard optimistic bilevel problem (1.1.4) turns out to be a convex program;
- both GNEPs (1.2.1) and (1.2.2) reduce to the following game:

$$\begin{array}{llcll} \underset{x,y}{\text{minimize}} & F(x, y) & \qquad & \underset{v}{\text{minimize}} & f(v) \\ \text{s.t.} & x \in X,\ y \in Y & & \text{s.t.} & v \in Y. \\ & f(y) \le f(v) + \varepsilon & & & \end{array} \tag{1.2.6}$$

As a straightforward consequence, we get the following strong result.

Corollary 1.2.6 *The following claims hold:*

(i) any equilibrium (x^*, y^*, v^*) *of GNEP* (1.2.6) *is such that* (x^*, y^*) *is global optimal for the standard optimistic bilevel problem* (1.1.4)*;*
(ii) any global optimal point (x^*, y^*) *for the standard optimistic bilevel problem* (1.1.4) *is such that* (x^*, y^*, v^*) *is an equilibrium of GNEP* (1.2.6) *for any* $v^* \in S$.

△

In the light of the considerations above, in this case, the (x, y)-part of the equilibrium sets of GNEP (1.2.1) and (1.2.2) coincide and, in turn, are equivalent to the set of global solutions of the standard optimistic bilevel problem (1.1.4).

In summary, GNEPs (1.2.1) and (1.2.2) are complementary: in fact, (1.2.1) is intended to provide relations between standard optimistic bilevel problems and Nash games in terms of their global optimal solutions, while the peculiar properties of (1.2.2), which is tailored to address stationary solutions, pave the way to algorithmic developments for (1.1.4). In fact, the procedure that simply consists in the alternating solution of the leader's (plus a proximal term) and the follower's convex problems in GNEP (1.2.2), as detailed in the scheme below, eventually provides with a point satisfying relations (1.2.3), i.e., a stationary point for the standard optimistic bilevel problem (1.1.4).

Solution Procedure for the Standard Optimistic Bilevel Problem (1.1.4)
Starting from $(x^0, y^0, v^0) \in X \times Y \times Y$ such that $y^0 = v^0 \in S(x^0)$, with $\tau > 0$, iteratively compute

$$(x^{k+1}, y^{k+1}) = \arg \min_{(x,y)\in X\times Y} \Big\{ F(x, y) + \tfrac{\tau}{2}\|(x, y) - (x^k, y^k)\|^2 : \\ f(x, y) \le f(x^k, v^k) + \nabla_1 f(x^k, v^k)^T (x - x^k) + \varepsilon \Big\},$$

and $v^{k+1} \in S(x^{k+1})$.

We remark that, thanks to the convexity of f, any accumulation point of the sequence produced by the procedure above is shown to be stationary for (1.1.4), see [10, Theorem 4.1]. We emphasize that the solution method above can be suitably modified in order to cope with a nonconvex objective function F and a possibly unbounded set X, also allowing for the possible inexact solution of the leader's problem (see [10, Appendix B]).

1.3 Standard Pessimistic Bilevel and Equilibrium Problems

When passing from optimistic to pessimistic bilevel programs, two main additional difficulties appear. Firstly, while the original pessimistic problem (1.1.3) with $\varepsilon = 0$ possesses a finite infimum under standard assumptions, the existence of corresponding optimal points may only be guaranteed under requirements which usually fail in practical problems. As explained above, this leads to the difference in the statements of Proposition 1.1.1 and Proposition 1.1.2. On the other hand, for $\varepsilon > 0$ the solvability of the perturbed pessimistic program is not an issue under standard assumptions, neither in its original form (1.1.3) nor in standard form (1.1.5).

However, for any $\varepsilon \ge 0$ the second main difficulty lies in the intrinsic three level structure of pessimistic bilevel programs, as opposed to the two level structure of optimistic bilevel programs in standard form. Note that both the original optimistic problem (1.1.2) and the original pessimistic problem (1.1.3) not only possess an upper level player (the leader) who minimizes her objective function in the variable x over the set X, and a lower level player (the follower) who minimizes her objective function in the variable v over the set Y, but also an intermediate player who minimizes or maximizes, respectively, the function $F(x, y)$ in y over the ε-optimal solutions $S_\varepsilon(x)$ of the lower level player.

When passing from the original form (1.1.2) to the standard form (1.1.4) of an optimistic bilevel problem, the constraint $F(x, y) \le \psi_\varepsilon^o(x)$ which models minimality with respect to the intermediate player turns out to be redundant, so that at this point (1.1.4) becomes a two level problem. On the contrary, in the pessimistic bilevel program in standard form (1.1.5) the constraint $F(x, y) \geq \psi_\varepsilon^p(x)$ which models the intermediate player's maximality requirement in general is not redundant, so that the three level structure from the original form (1.1.3) prevails.

The standard pessimistic form (1.1.5) turns out to share with its optimistic counterpart (1.1.4) the same vertical structure, but with an additional level. In the light of this simple observation, the present section will focus on possibilities for the algorithmic treatment of this three level structure. In Sect. 1.2 we have seen that the vertical structure of a two level problem may be reformulated into the horizontal structure of an adequately constructed equilibrium problem. In the same spirit, as for the treatment of three level problems, three possible approaches come to mind:

- Reformulate the vertical three level problem into a horizontal three player equilibrium problem.
- Reformulate the upper and the intermediate level into a two player equilibrium problem, but keep the vertical structure with respect to the lower level problem.
- Reformulate the intermediate and the lower level into a two player equilibrium problem, but keep the vertical structure with respect to the upper level problem. This results in a two level program whose lower level is an equilibrium problem, that is a multi follower game with two followers.

The first two reformulations of a pessimistic bilevel program, as outlined above, are stated explicitly in the appendix of [11]. The additional assumption to obtain a one-to-one correspondence between global solutions of the pessimistic bilevel program and the resulting reformulations essentially turns out to be the x-independence of the lower level ε-optimal solutions $S_\varepsilon(x)$. This requirement may only be expected to hold in special applications. Fortunately the third approach, that is, the reformulation of the pessimistic bilevel problem as a two level problem with an equilibrium problem in the lower level, does not need any additional assumptions. This may be seen as follows.

Subsequently we describe the reformulation of the standard pessimistic bilevel problem (1.1.5) as a multi follower game (see [11] for details). The intermediate and the lower level problem form the two level problem

$$\underset{y}{\text{maximize}}\ F(x, y) \quad \text{s.t.} \quad y \in Y,\ f(x, y) \leq \varphi(x) + \varepsilon \tag{1.3.1}$$

with $\varphi(x) = \min_v \{f(x, v) \mid v \in Y\}$. The crucial observation is that the variable x only serves as a parameter in (1.3.1), so that with respect to its upper and lower level decision variables y and v, respectively, we face a pure hierarchical problem. This allows us to evoke Corollary 1.2.6 and state a strong relationship between the global solutions of (1.3.1) and the equilibria of the GNEP

$$\begin{array}{lll} \underset{y}{\text{minimize}} & -F(x, y) \\ \text{s.t.} & y \in Y \\ & f(x, y) \leq f(x, v) + \varepsilon \end{array} \qquad \begin{array}{ll} \underset{v}{\text{minimize}} & f(x, v) \\ \text{s.t.} & v \in Y \\ & \end{array} \tag{1.3.2}$$

where, again, x serves as a parameter. In fact, given $x \in X$, for any equilibrium (y^*, v^*) of (1.3.2), the point y^* is a global solution of (1.3.1), and for any global solution y^* of (1.3.1) and any $v^* \in S(x)$, the point (y^*, v^*) is an equilibrium of (1.3.2).

Hence, if we denote the set of equilibria of (1.3.2) by $E_\varepsilon(x)$, the pessimistic bilevel program in standard form (1.1.5) and the multi follower game

$$\underset{x,y,v}{\text{minimize}}\ F(x, y) \quad \text{s.t.} \quad x \in X,\ (y, v) \in E_\varepsilon(x) \tag{1.3.3}$$

are equivalent in the following sense.

Theorem 1.3.1 ([11, Proposition 5.1]) *The following relations hold:*

(i) If (x^*, y^*, v^*) is global solution of (1.3.3)*, then (x^*, y^*) is a global solution of* (1.1.5)*.*
(ii) If (x^*, y^*) is a global solution of (1.1.5) *then for any $v^* \in S(x^*)$ the point (x^*, y^*, v^*) is global solution of* (1.3.3)*.* △

Since the global solutions of (1.1.5) correspond to the global solutions of the original pessimistic problem (1.1.3) in a straightforward manner, one may treat the multi follower game (1.3.3) algorithmically to solve (1.1.3). We remark that, as it already was the case when passing from the original to the standard version, also in passing from (1.1.5) to (1.3.3) the relation between local solutions is more intricate [11, Prop. 5.2]. Let us also point out that the problems (1.2.1) and (1.3.2) are structurally different since in the former x is a decision variable, while in the latter it serves as a parameter. This partly explains why an analogue of Theorem 1.3.1(ii) is missing in Theorem 1.2.1.

As a next step, let us briefly discuss a promising algorithmic approach for the solution of the multi follower game (1.3.3). One can replace the equilibrium constraint in (1.3.3) by the equivalent Karush–Kuhn–Tucker conditions. This can be done provided that, in addition to the initial assumptions, F is concave in y for each x, and the set Y has a functional description as system of inequalities $h \leq 0$, where the components of the vector h are continuously differentiable and convex functions. Furthermore, we assume the Slater condition for Y, that is, the existence of some $\bar{v}$ with $h(\bar{v}) < 0$. By the Karush–Kuhn–Tucker theorem, v is a global solution of the right hand side player's problem in (1.3.2) if and only if there exists a multiplier vector μ with

$$\nabla_2 f(x, v) + \nabla h(v)\mu = 0$$
$$0 \leq \mu \perp -h(v) \geq 0.$$

For given (x, v) also the left hand side player's problem in (1.3.2) possesses a Slater point, namely a sufficiently small shift of the point v towards $\bar{v}$. Hence, y is a global solution of this problem if and only if there exist multipliers λ and ξ with

$$\begin{aligned} -\nabla_2 F(x, y) + \nabla h(y)\lambda + \nabla_2 f(x, y)\xi &= 0 \\ 0 \leq \lambda \perp -h(y) &\geq 0 \\ 0 \leq \xi \perp f(x, v) - f(x, y) + \varepsilon &\geq 0. \end{aligned}$$

Consequently there is a strong connection between the optimal solutions of the multi follower game (1.3.3) and the mathematical program with complementarity constraints

$$\begin{aligned} \underset{x,y,v,\lambda,\xi,\mu}{\text{minimize}}\ & F(x, y) \\ \text{s.t.}\quad & x \in X \\ & -\nabla_2 F(x, y) + \nabla h(y)\lambda + \nabla_2 f(x, y)\xi = 0 \\ & 0 \leq \lambda \perp -h(y) \geq 0 \\ & 0 \leq \xi \perp f(x, v) - f(x, y) + \varepsilon \geq 0 \\ & \nabla_2 f(x, v) + \nabla h(v)\mu = 0 \\ & 0 \leq \mu \perp -h(v) \geq 0. \end{aligned} \tag{1.3.4}$$

Since passing from (1.3.3) to (1.3.4) involves lifting the decision variables into a higher dimensional space, the connection between the local solutions of both problems is intricate [11], while the connection between global solutions reads as follows.

Proposition 1.3.2 ([11, Proposition 5.4]) *The following claims hold:*

(i) If $(x^*, y^*, v^*, \lambda^*, \xi^*, \mu^*)$ is a global solution of (1.3.4) *then (x^*, y^*, v^*) is globally optimal for* (1.3.3).
(ii) If (x^*, y^*, v^*) is a global solution of (1.3.3) *then for all corresponding Karush–Kuhn–Tucker multipliers $(\lambda^*, \xi^*, \mu^*)$ the point $(x^*, y^*, v^*, \lambda^*, \xi^*, \mu^*)$ is globally optimal for* (1.3.4).

△

The following scheme summarizes by which results the standard pessimistic bilevel problem (1.1.5) may be solved via its reformulation as the mathematical program with complementarity constraints (1.3.4).

Global solutions

problem (1.5)	Cor.2.6,Th.3.1	problem (3.3)	Pr. 3.2	problem (3.4)
x, y	$\Leftrightarrow$	x, y, v	$\Leftrightarrow$	$x, y, v, \lambda, \xi, \mu$

In turn, problem (1.3.4) can be equivalently reformulated as a mixed integer nonlinear programming problem for which practical solution methods are available. For an example we refer to [11], where the pessimistic version of a regulator problem from economics is solved via the reformulation (1.3.4).

In addition, let us introduce two novel alternative approaches aimed at the solution of the multi follower game (1.3.3). Their common feature is, as for the KKT reformulation, that they replace the equilibrium constraint in (1.3.3) by different characterizations.

For the equilibrium characterization by the Nikaido–Isoda approach [23], let us define the optimal value function of the left hand side player in (1.3.2), given the parameter x and the right hand side player's decision v,

$$\widehat{\psi}_{\varepsilon}^{p}(x, v) \triangleq \min_{y}\{-F(x, y) \mid y \in Y_{\varepsilon}(x, v)\}$$

with the left hand side player's feasible set $Y_{\varepsilon}(x, v) \triangleq \{z \in Y \mid f(x, z) \leq f(x, v) + \varepsilon\}$. The corresponding right hand side player's optimal value function, given the parameter x (and the left hand side player's decision y), reduces to the lower level optimal value function $\varphi(x)$, due to the hierarchical structure of (1.3.1). The gap function of (1.3.2) is thus defined as

$$V_{\varepsilon}(x, y, v) \triangleq -F(x, y) - \widehat{\psi}_{\varepsilon}^{p}(x, v) + f(x, v) - \varphi(x).$$

It may be shown to be nonnegative on the joint feasible set $W_{\varepsilon}(x) \triangleq \{(y, v) \in Y \times Y \mid f(x, y) \leq f(x, v) + \varepsilon\}$ of both players, and the equilibria of (1.3.2) are exactly the global solutions of

$$\underset{y, v}{\text{minimize}}\; V_{\varepsilon}(x, y, v) \quad \text{s.t.} \quad (y, v) \in W_{\varepsilon}(x)$$

with minimal value zero [5, 23]. At first glance, this seems to lead to a two level reformulation of (1.3.3), i.e.,

$$\underset{x,y,v}{\text{minimize}}\ F(x,y) \quad \text{s.t.} \quad x \in X,\ (y,v) \in \underset{(\eta,z)\in W_\varepsilon(x)}{\text{argmin}}\ V_\varepsilon(x,\eta,z),\ \ V_\varepsilon(x,y,v) \le 0.$$

However, as the definition of V_ε comprises the optimal value functions $\widehat{\psi}_\varepsilon^p$ and φ, this actually still is a three level problem. On the other hand, the above considerations also imply

$$E_\varepsilon(x) = \{(y,z) \in W_\varepsilon(x) \mid V_\varepsilon(x,y,v) \le 0\}$$

so that (1.3.3) may be rewritten as the two level problem

$$\underset{x,y,v}{\text{minimize}}\ F(x,y) \quad \text{s.t.} \quad x \in X,\ y,v \in Y,\ f(x,y) \le f(x,v)+\varepsilon,\ V_\varepsilon(x,y,v) \le 0.$$

Moreover, the constraint $V_\varepsilon(x,y,v) \le 0$ may be reformulated as

$$\max_{\eta\in Y_\varepsilon(x,v)} (F(x,\eta) - F(x,y)) \le \min_{z\in Y}(f(x,z) - f(x,v))$$

so that we arrive at the following straightforward result.

Proposition 1.3.3 *The multi follower game* (1.3.3) *possesses the same global as well as local solutions as the generalized semi-infinite program*

$$\begin{aligned} &\underset{x,y,v}{\textit{minimize}}\ F(x,y) \\ &\quad \textit{s.t.} \quad x \in X,\ y,v \in Y,\ f(x,y) \le f(x,v)+\varepsilon, \qquad (1.3.5)\\ &\qquad\quad F(x,\eta) - F(x,y) \le f(x,z) - f(x,v) \quad \forall \eta \in Y_\varepsilon(x,v),\ z \in Y. \end{aligned}$$

△

For introductions to generalized semi-infinite programming we refer to [20, 21].

While convexity assumptions on the player problems in (1.3.2) are not necessary in the Nikaido–Isoda approach above, they are crucial for the following equilibrium characterization. Recall that our blanket assumptions from Sect. 1.1 state the continuous differentiability of F and f, the convexity of f in y for every x, and compactness, convexity as well as nonemptiness of the sets X and Y. This implies the convexity of the right hand side player's problem in (1.3.2) as well as the convexity of the left hand side player's feasible set $Y_\varepsilon(x,v)$. In the following we shall additionally assume, as done for the KKT approach, the concavity of F in y for each x, which makes also the left hand side player's problem convex.

The variational inequality reformulation of (1.3.2) bases on the variational argument that, given (x, v), the point $y \in Y_\varepsilon(x, v)$ is a global solution of the left hand side player's problem if and only if the inequalities

$$\nabla_2(-F(x, y))^T(\eta - y) \geq 0 \quad \forall \eta \in Y_\varepsilon(x, v)$$

hold and that, analogously, $v \in Y$ is a global solution of the right hand side player's problem if and only if the following inequalities are satisfied:

$$\nabla_2 f(x, v)^T(z - v) \geq 0 \quad \forall z \in Y.$$

Hence, the simultaneous validity of all these inequalities characterizes the condition $(y, v) \in E_\varepsilon(x)$, and the following result is easily seen to hold.

Proposition 1.3.4 *The multi follower game* (1.3.3) *possesses the same global as well as local solutions as the generalized semi-infinite program*

$$\begin{aligned} &\mathop{minimize}_{x,y,v} \; F(x, y) \\ &\text{s.t.} \quad x \in X, \; y, v \in Y, \; f(x, y) \leq f(x, v) + \varepsilon, \\ &\begin{pmatrix} -\nabla_2 F(x, y) \\ \nabla_2 f(x, v) \end{pmatrix}^T \left(\begin{pmatrix} \eta \\ z \end{pmatrix} - \begin{pmatrix} y \\ v \end{pmatrix} \right) \geq 0, \; \forall (\eta, z) \in Y_\varepsilon(x, v) \times Y. \end{aligned} \tag{1.3.6}$$

△

We remark that in the generalized semi-infinite program (1.3.6) the index variables η and z enter linearly in the constraint function, as opposed to their role in (1.3.5).

1.4 Intrinsic Complexity of Bilevel and Equilibrium Problems

In this section we illustrate the intrinsic complexity of the bilevel problem (1.4.2) and the corresponding GNEP (1.4.3) as originating from parametric optimization. By taking the study of singularities in parametric optimization into account, the structural analysis of bilevel feasible set and Nash equilibria is performed. It turns out that degeneracies cannot be avoided for both problem types. In order to highlight this phenomenon, we consider the follower problem with the feasible set $Y(x)$ depending on the leader variable x:

$$\begin{aligned} &\mathop{\text{minimize}}_{v} \; f(x, v) \\ &\text{s.t.} \quad v \in Y(x). \end{aligned} \tag{1.4.1}$$

Recall that it is given by a finite number of smooth inequality constraints:

$$Y(x) \triangleq \left\{v \,\middle|\, g_j(x, v) \geq 0,\ j \in J\right\}.$$

The bilevel feasible set is

$$M \triangleq \{(x, v) \mid v \in S(x)\},$$

where $S(x)$ is the set of the follower optimal points. Then, the optimistic bilevel problem can be stated as follows:

$$\underset{(x,v)\in M}{\text{minimize}}\ F(x, v). \tag{1.4.2}$$

The latter is equivalent to the standard optimistic bilevel problem (1.1.4) with

$$\varphi(x) \triangleq \min_v\{f(x, v) \mid v \in Y(x)\},$$

and the vanishing tolerance $\varepsilon = 0$.

We focus here on the analysis of the bilevel feasible set M for the one-parametric case of $\dim(x) = 1$. In [7] a generic classification of minimizers for (1.4.1) has been established. Namely, they differ in a possible violation of nondegeneracy occurring at the local minimizer:

- Linear Independence Constraint Qualification (LICQ),
- Strict Complementarity (SC),
- Second Order Sufficient Condition (SOSC).

Let us denote by $|J_0(x^*, v^*)|$ the number of active constraints at the point of interest $v^* \in S\left(x^*\right)$. As shown in [7], the minimizer v^* can be of one of the following five types :

Type 1: LICQ, SC, and SOSC are satisfied.
Type 2.1: SC is violated, and exactly one active Lagrange multiplier vanishes.
Type 4.1: LICQ is violated, and the rank of active gradients is $|J_0(x^*, v^*)| - 1$.
Type 5.1: LICQ is violated with $|J_0(x^*, v^*)| = \dim(v) + 1$, and MFCQ is violated.
Type 5.2: LICQ is violated with $|J_0(x^*, v^*)| = \dim(v) + 1$, but MFCQ is satisfied.

The corresponding typical branches of (possibly local) minimizers are depicted in Fig. 1.1. Type 1 is referred to as nondegenerate, other Types 2.1, 4.1, 5.1, and 5.2 describe singularities and are referred to as degenerate cases. The derivation of the five types is based on the classification of generalized critical points in one-parametric programming as given in [8].

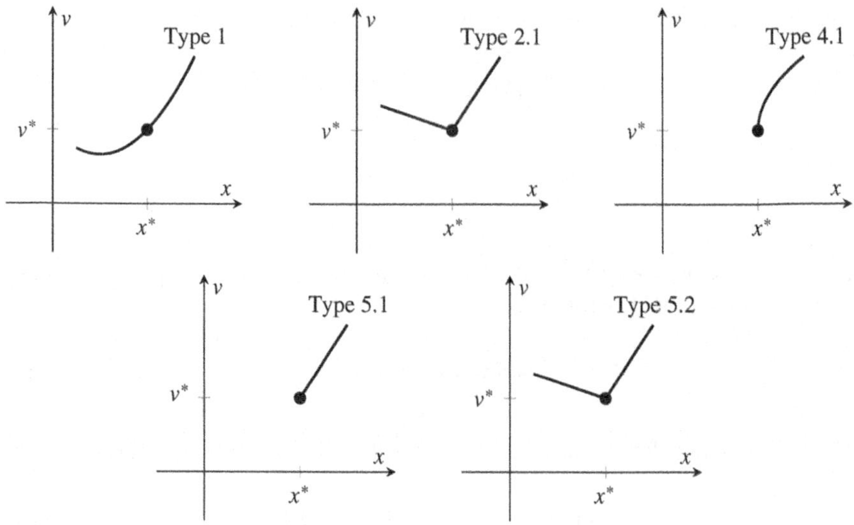

Fig. 1.1 Five types

Now, we turn our attention to the global structure of the bilevel feasible set just around a leader's choice x^*, i.e. to the set

$$M_{x^*} \triangleq \left\{(x, v) \,\middle|\, v \in S(x) \ \text{for } x \text{ close to } x^* \right\}.$$

From the practical point of view, the description of M_{x^*} is actually crucial for the leader. Indeed, assuming that (1.4.1) is solvable in a neighborhood of x^* it might happen that, changing x^*, the leader forces the follower to switch to another branch of global minimizers. This occurs if the previous branch of follower's global minimizers eventually ends up. Such a situation introduces a kind of a shock or catastrophic behavior in bilevel optimization, and leads to discontinuity effects. Moreover, it is even possible that the feasible set ends abruptly.

In what follows, we introduce several cases for the description of M_{x^*}. In fact, as we shall see in Theorem 1.4.1, they turn out to be the only ones in general. First of all, the situations as in Type 1, 2.1, 4.1, 5.1 or 5.2 may occur. Secondly, we present composite cases, where M_{x^*} contains more than one component (see Fig. 1.2).

Let $S(x^*) = \{v_1^*\}$ consist of just one follower optimal point and v_2^* is a local minimizer, such that locally around x^*

$$M_{x^*} = \left\{(x, v(x)) \,\middle|\, v(x) = \begin{cases} v_1(x), \ x \le x^*, \\ v_2(x), \ x > x^*. \end{cases} \right\},$$

or

$$M_{x^*} = \left\{(x, v(x)) \,\middle|\, v(x) = \begin{cases} v_2^*(x), \ x < x^*, \\ v_1^*(x), \ x \ge x^*. \end{cases} \right\}.$$

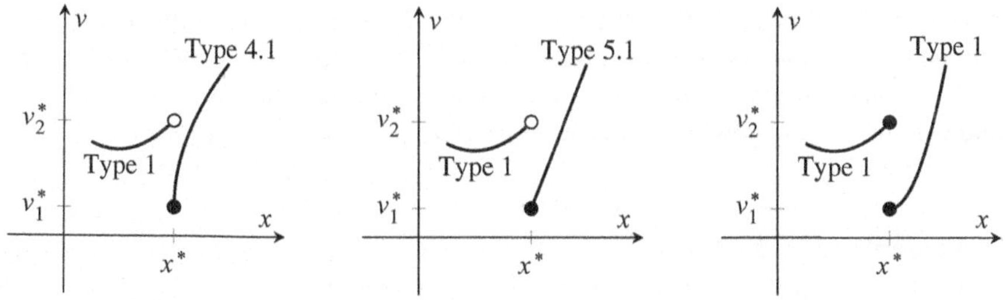

Fig. 1.2 Composite types

Here, $v_1(x)$, $v_2(x)$ are unique global minimizers for x in a neighborhood of x^* with $v_1(x^*) = v_1^*$, $v_2(x^*) = v_2^*$, according to the following cases:

Type 4.1/1: v_1^* is of Type 4.1, and v_2^* is of Type 1.
Type 5.1/1: v_1^* is of Type 5.1, and v_2^* is of Type 1.

Finally, let $S(x^*) = \{v_1^*, v_2^*\}$ consist of two follower optimal points, such that

$$M_{x^*} = \left\{ (x, v(x)) \;\middle|\; v(x) = \begin{cases} v_1(x), & x \le x^*, \\ v_2(x), & x \ge x^*. \end{cases} \right\}.$$

Here, $v_1(x)$, $v_2(x)$ are unique global minimizers for x in a neighborhood of x^* with $v_1(x^*) = v_1^*$, $v_2(x^*) = v_2^*$, according to the following case:

Type 1/1: v_1^* and v_2^* are both of Type 1. Additionally, it holds:

$$\left.\frac{\mathrm{d}\left[f(x, v_2(x)) - f(x, v_1(x))\right]}{\mathrm{d}x}\right|_{x=x^*} \neq 0.$$

As we mentioned before, only the cases just introduced occur in general [4].

Theorem 1.4.1 ([4, Theorem 4.5]) *Let* $\dim(x) = 1$. *Generically, the bilevel feasible set* M_{x^*} *is either empty or given as in one of the cases Type 1, 2.1, 4.1, 5.1, 5.2, Type 4.1/1, Type 5.1/1, Type 1/1. Additionally, it holds:*

$$F(x^*, v_1^*) \neq F(x^*, v_2^*),$$

where v_1^*, v_2^* *correspond to the cases Type 4.1/1, Type 5.1/1 and Type 1/1.* △

Let us now apply the GNEP approach from Sect. 1.2 for Types 2.1, 4.1, 5.1, 5.2 of bilevel problem. For that, we recall the original GNEP model from [9], where the leader feasible set depends on the follower variable:

$$\begin{aligned} \min_{x,y}\ & F(x, y) \\ \text{s.t.}\ & (x, y) \in X \times Y(x), \\ & f(x, y) \le f(x, v) + \varepsilon \end{aligned} \qquad\qquad \begin{aligned} \min_{v}\ & f(x, v) \\ \text{s.t.}\ & v \in Y(x). \end{aligned} \tag{1.4.3}$$

Note that (1.4.3) is a generalization of (1.2.1) with $Y(x) \equiv Y$. The crucial questions will be whether the degeneracies of the follower problem as given in Types 2.1, 4.1, 5.1, 5.2 can be overcome at least at the corresponding equilibrium of GNEP (1.4.3), and whether new degeneracies eventually show up in the leader problem.

Example Type 2.1
Let the data of the bilevel problem be given as follows:

$$F(x, v) = -x + 2v, \quad f(x, v) = (x - v)^2, \quad g_1(x, v) = v.$$

Its feasible set is

$$M = \{(x, \max\{x, 0\}) \mid x \in \mathbb{R}\},$$

and $(0, 0)$ solves the bilevel problem (1.4.2). Note that the feasible set $Y(x) = \{v \mid v \geq 0\}$ does not depend on the leader variable x. Due to Type 2.1, SC is violated at $v = 0$ for the follower problem (1.4.1) with $x = 0$. More precisely, the active Lagrange multiplier vanishes. The corresponding GNEP (1.4.3) reads as:

$$\begin{array}{lll} \underset{x,y}{\text{minimize}} & -x + 2y & \qquad \underset{v}{\text{minimize}} \ (x - v)^2 \\ \text{s.t.} & y \geq 0, & \qquad \text{s.t.} \quad v \geq 0. \\ & (x - y)^2 \leq (x - v)^2 + \varepsilon & \end{array} \tag{1.4.4}$$

The leader problem in (1.4.4) is unsolvable for every ν. Hence, though the solution of the follower problem in (1.4.4) is

$$\nu(x) = \begin{cases} x, \text{ if } x \geq 0, \\ 0, \text{ if } x < 0, \end{cases}$$

the set of equilibria of (1.4.4) is empty.

Example Type 4.1
Let the data of the bilevel problem be given as follows:

$$F(x, v) = x + v, \quad f(x, v) = -v, \quad g_1(x, v) = x - v^2.$$

Its feasible set is

$$M = \left\{ \left(x, \sqrt{x}\right) \mid x \geq 0 \right\},$$

and $(0, 0)$ solves the bilevel problem (1.4.2). Due to Type 4.1, LICQ is violated here at $v = 0$ for the follower problem (1.4.1) with $x = 0$. More precisely, $|J_0(0, 0)| = 1$, and the rank of active gradients is zero. The corresponding GNEP (1.4.3) reads as:

$$\begin{aligned} &\min_{x,y} \text{imize} \; x + y & \qquad & \min_{v} \text{imize} \; -v \\ &\text{s.t.} \quad x - y^2 \geq 0, & & \text{s.t.} \quad x - v^2 \geq 0. \\ &\qquad -y \leq -v + \varepsilon & & \end{aligned} \tag{1.4.5}$$

The solution of the leader problem in (1.4.5) is

$$(x(\nu), y(\nu)) = \begin{cases} \left(\frac{1}{4}, \frac{1}{2}\right), & \text{if } \nu \leq \varepsilon - \frac{1}{2}, \\ \left((\nu - \varepsilon)^2, \nu - \varepsilon\right), & \text{if } \nu > \varepsilon - \frac{1}{2}. \end{cases}$$

The solution of the follower problem in (1.4.5) is

$$\nu(x) = \begin{cases} \sqrt{x}, & \text{if } x \geq 0, \\ \emptyset, & \text{if } x < 0. \end{cases}$$

Hence, the unique equilibrium of (1.4.5) is $(x^*, y^*, v^*) = \left(\frac{\varepsilon^2}{4}, -\frac{\varepsilon}{2}, \frac{\varepsilon}{2}\right)$ for all sufficiently small $\varepsilon > 0$. Moreover, the minimizer $(x^*, y^*) = \left(\frac{\varepsilon^2}{4}, -\frac{\varepsilon}{2}\right)$ of the leader problem is nondegenerate, as well as the minimizer $v^* = \frac{\varepsilon}{2}$ of the follower problem.

Example Type 5.1
Let the data of the bilevel problem be given as follows:

$$F(x,v) = x + v, \quad f(x,v) = v, \quad g_1(x,v) = v, \quad g_2(x,v) = x - v.$$

Its feasible set is

$$M = \{(x,0) \mid x \geq 0\},$$

and $(0,0)$ solves the bilevel problem (1.4.2). Due to Type 5.1, LICQ is violated here at $v = 0$ for the follower problem (1.4.1) with $x = 0$. More precisely, $|J_0(0,0)| = 2$, and MFCQ is violated. The corresponding GNEP (1.4.3) reads as:

$$\begin{aligned} &\underset{x,y}{\text{minimize}} \; x + y && \underset{v}{\text{minimize}} \; v \\ &\text{s.t.} \quad y \geq 0,\; x - y \geq 0, && \text{s.t.} \quad v \geq 0,\; x - v \geq 0. \\ &\qquad\; y \leq v + \varepsilon && \end{aligned} \tag{1.4.6}$$

The solution of the leader problem in (1.4.6) is

$$(x(\nu), y(\nu)) = \begin{cases} (0,0), \text{ if } \nu + \varepsilon \geq 0, \\ \emptyset, \text{ if } \nu + \varepsilon < 0. \end{cases}$$

The solution of the follower problem in (1.4.6) is

$$\nu(x) = \begin{cases} 0, \text{ if } x \geq 0, \\ \emptyset, \text{ if } x < 0. \end{cases}$$

Hence, the unique equilibrium of (1.4.6) is $(x^*, y^*, v^*) = (0,0,0)$ for $\varepsilon > 0$. Moreover, the minimizer $(x^*, y^*) = (0,0)$ of the leader problem is nondegenerate, but the minimizer $v^* = 0$ of the follower problem remains degenerate of Type 5.1.

Example Type 5.2
Let the data of the bilevel problem be given as follows:

$$F(x, v) = -x + 2v, \quad f(x, v) = v, \quad g_1(x, v) = v, \quad g_2(x, v) = -x + v.$$

Its feasible set is

$$M = \{(x, \max\{x, 0\}) \mid x \in \mathbb{R}\},$$

and $(0, 0)$ solves the bilevel problem (1.4.2). Due to Type 5.2, LICQ is violated here at $v = 0$ for the follower problem (1.4.1) with $x = 0$. More precisely, $|J_0(0, 0)| = 2$, and MFCQ is satisfied. The corresponding GNEP (1.4.3) reads as:

$$\begin{array}{llll} \underset{x,y}{\text{minimize}} & -x + 2y & \qquad \underset{v}{\text{minimize}} & v \\ \text{s.t.} & y \geq 0, \ -x + y \geq 0, & \qquad \text{s.t.} & v \geq 0, \ -x + v \geq 0. \\ & y \leq v + \varepsilon & & \end{array} \tag{1.4.7}$$

The solution of the leader problem in (1.4.7) is

$$(x(v), y(v)) = \begin{cases} (0, 0), & \text{if } v + \varepsilon \geq 0, \\ \emptyset, & \text{if } v + \varepsilon < 0. \end{cases}$$

The solution of the follower problem in (1.4.7) is

$$v(x) = \begin{cases} x, & \text{if } x \geq 0, \\ 0, & \text{if } x < 0. \end{cases}$$

Hence, the unique equilibrium of (1.4.7) is $(x^*, y^*, v^*) = (0, 0, 0)$ for $\varepsilon > 0$. Moreover, the minimizer $(x^*, y^*) = (0, 0)$ of the leader problem is nondegenerate, but the minimizer $v^* = 0$ of the follower problem remains degenerate of Type 5.2.

We conclude that the degeneracies of Type 2.1 in the lower level hamper the solvability of the leader problem in GNEP (1.4.3). Degeneracies of Type 4.1 disappear by applying the proposed GNEP approach. However, those of Types 5.1 and 5.2 remain persistent. This phenomenon is stable w.r.t. sufficiently small perturbations of data, and, hence, may unavoidably cause numerical difficulties while solving the corresponding GNEP (1.4.3).

References

1. F. Caruso, M.B. Lignola, J. Morgan, Regularization and approximation methods in Stackelberg games and bilevel optimization. Technical report, Centre for Studies in Economics and Finance (CSEF), University of Naples, Italy, 2019
2. F.H. Clarke, Y.S. Ledyaev, R.J. Stern, P.R. Wolenski, *Nonsmooth Analysis and Control Theory*, vol. 178 (Springer, Berlin, 2008)
3. S. Dempe, B.S. Mordukhovich, A.B. Zemkoho, Sensitivity analysis for two-level value functions with applications to bilevel programming. SIAM J. Optim. **22**(4), 1309–1343 (2012)
4. D. Dorsch, H. Th. Jongen, V. Shikhman, On intrinsic complexity of Nash equilibrium problems and bilevel optimization. J. Optim. Theory Appl. **159**, 606–634 (2013)
5. A. Dreves, C. Kanzow, O. Stein, Nonsmooth optimization reformulations of player convex generalized Nash equilibrium problems. J. Global Optim. **53**, 587–614 (2012)
6. F. Facchinei, C. Kanzow, Generalized Nash equilibrium problems. Ann. Oper. Res. **175**(1), 177–211 (2010)
7. H. Th. Jongen, V. Shikhman, Bilevel optimization: on the structure of the feasible set. Math. Program. **136**, 65–89 (2012)
8. H. Th. Jongen, P. Jonker, F. Twilt, Critical sets in parametric optimization. Math. Program. **34**, 333–353 (1986)
9. L. Lampariello, S. Sagratella, A bridge between bilevel programs and Nash games. J. Optim. Theory Appl. **174**(2), 613–635 (2017)
10. L. Lampariello, S. Sagratella, Numerically tractable optimistic bilevel problems. Comput. Optim. Appl. **76**, 277–303 (2020)
11. L. Lampariello, S. Sagratella, O. Stein, The standard pessimistic bilevel problem. SIAM J. Optim. **29**, 1634–1656 (2019)
12. M.B. Lignola, J. Morgan, Topological existence and stability for Stackelberg problems. J. Optim. Theory Appl. **84**(1), 145–169 (1995)
13. M.B. Lignola, J. Morgan, Stability of regularized bilevel programming problems. J. Optim. Theory Appl. **93**(3), 575–596 (1997)
14. G-H. Lin, M. Xu, J.J. Ye, On solving simple bilevel programs with a nonconvex lower level program. Math. Program. **144**(1–2), 277–305 (2014)
15. P. Loridan, J. Morgan, A theoretical approximation scheme for Stackelberg problems. J. Optim. Theory Appl. **61**(1), 95–110 (1989)
16. L. Mallozzi, R. Messalli, S. Patrì, A. Sacco, *Some Aspects of the Stackelberg Leader/Follower Model* (Springer, Berlin, 2018), pp. 171–181
17. D.A. Molodtsov, V.V., Fedorov, Approximation of two-person games with information exchange. USSR Comput. Math. Math. Phys. **13**(6), 123–142 (1973)
18. J.V. Outrata, On the numerical solution of a class of Stackelberg problems. Z. Oper. Res. **34**(4), 255–277 (1990)
19. R.T. Rockafellar, J.B. Wets, *Variational Analysis* (Springer, Berlin, 1998)
20. O. Stein, *Bi-Level Strategies in Semi-Infinite Programming* (Kluwer Academic Publishers, Boston, 2003)
21. O. Stein, How to solve a semi-infinite optimization problem. Eur. J. Oper. Res. **23**, 312–320 (2012)
22. T. Tanino, T. Ogawa, An algorithm for solving two-level convex optimization problems. Int. J. Syst. Sci. **15**(2), 163–174 (1984)
23. A. von Heusinger, C. Kanzow, Optimization reformulations of the generalized Nash equilibrium problem using Nikaido–Isoda-type functions. Comput. Optim. Appl. **43**, 353–377 (2009)
24. J.J. Ye, D.L. Zhu, Optimality conditions for bilevel programming problems. Optimization **33**(1), 9–27 (1995)
25. J.J. Ye, D.L. Zhu, New necessary optimality conditions for bilevel programs by combining the MPEC and value function approaches. SIAM J. Optim. **20**(4), 1885–1905 (2010)

Chapter 2
On Stackelberg–Nash Equilibria in Bilevel Optimization Games

Damien Bazin, Ludovic Julien, and Olivier Musy

Abstract Hierarchical games with strategic interactions such as the Stackelberg two-stage game epitomize a standard economic application of bilevel optimization problems. In this paper, we survey certain properties of multiple leader–follower noncooperative games, which enable the basic Stackelberg duopoly game to encompass a larger number of decision makers at each level. We focus notably on the existence, uniqueness and welfare properties of these multiple leader–follower games. We also study how this particular bilevel optimization game can be extended to a multi-level decision setting.

Keywords Multiple leader–follower game · Stackelberg–Nash equilibrium

2.1 Introduction

Hierarchical optimization problems concern environments in which groups of individuals decide in a sequential way. The strategic context of the agent is then extended because the decision of each agent becomes influenced by the decisions made by other agents in the past. The agent will also have to take into account the consequences on these decisions of the choices that other individuals will make in the future. In this context, two-level optimization problems correspond to games which have two stages of interconnected decisions—the most common category for such problems (Shi et al. [35], Dempe [12, 13], Sinha et al. [36]). In

D. Bazin
Côte d'Azur University, CNRS, GREDEG, Nice, France
e-mail: damien.bazin@gredeg.cnrs.fr

L. Julien (✉)
EconomiX, UPL, University Paris Nanterre, CNRS, Nanterre, France
e-mail: ludovic.julien@parisnanterre.fr

O. Musy
CRED, University Paris 2 Panthéon Assas, Paris, France
e-mail: olivier.musy@u-paris2.fr

S. Dempe, A. Zemkoho (eds.), *Bilevel Optimization*, Springer Optimization and Its Applications 161, https://doi.org/10.1007/978-3-030-52119-6_2

such environments there are at least two decision makers for which the convex set mapping solution for the lower level problem becomes the feasible set for the upper level problem (Bard [5]). Since this is a common feature of strategic interactions, there are numerous applications of such optimization problems in recent literature, for instance, in the fields of electricity markets (Hu and Ralph [19], Aussel et al. [2, 3]), and transportation (see Dempe [13] or Dempe and Kalashnikov [14]). Economics is the oldest field of application, as the first use of this strategic context was proposed by Stackelberg in 1934, in his book on the study of oligopolies and market structures (Stackelberg [38][1]).

In the current paper, we use this initial application of bilevel optimization problems in the study of industrial organization and market structures. More specifically, we focus on the multiple leader–follower game, which extends the initial Stackelberg duopoly game (restricted to one leader and one follower) to a two-stage quantity setting noncooperative game.[2] The first version of this model was introduced by Sherali [33], and explored by Daughety [9], Ehrenmann [15], Pang and Fukushima [31], Yu and Wang [40], DeMiguel and Xu [11], Julien [22], and Aussel et al. [4]. This nontrivial extension to the basic duopoly game provides a richer set of strategic interactions between several decision makers, notably because the sequential decision making process introduces heterogeneity among firms. Strategic interactions are more complex to handle because the game itself consists of two Cournot simultaneous move games embedded in a Stackelberg sequential competition game. The decision makers who interact simultaneously belong to the same cohort, while those who interact sequentially belong to two distinct cohorts. Decision makers are firms, and these firms are either leaders or followers. Indeed, this model comprises strategic interactions at two levels of decisions as well as strategic interactions at the same level of decisions.

Bearing in mind that this framework implies both simultaneous and sequential interactions, we can define the corresponding strategic equilibrium concept as a Stackelberg–Nash equilibrium (SNE). In this paper, we focus on the existence, uniqueness and welfare properties of this noncooperative equilibrium, which is still actively researched, especially in mathematical economics. We highlight three points: first, the existence of an equilibrium is not trivial in the presence of several followers. Second, the uniqueness of an equilibrium is based on strong technical assumptions regarding the strict concavity of payments. Third, several properties relating to market power and its consequences cannot be captured by the simple duopoly model. By using examples, we also illustrate some of the main features in

[1]The book was published in 1934 in German, but was translated into English in 1952 by Oxford University Press and 2011 by Springer. We refer to the 2011 version, as it corresponds to the original 1934 book.

[2]To the best of our knowledge, the first extension of the Stackelberg duopoly was introduced by Leitmann [29], who considered a model with one leader and several followers. This was further developed by Murphy et al. [34]. It is worth noting that Stackelberg [38] had already envisaged the possibility of several market participants (see Chap. 3).

terms of welfare for this noncooperative equilibrium, which we then compare to the Cournot–Nash equilibrium (CNE) and the competitive equilibrium (CE).

The remainder of the paper is structured as follows. In Sect. 2.2, we consider the standard bilevel multiple leader–follower game and state a number of assumptions. In Sect. 2.3, we define the Stackelberg–Nash equilibrium. Section 2.4 is devoted to the existence and uniqueness of the Stackelberg–Nash equilibrium. Section 2.5 examines two important examples. In Sect. 2.6, we investigate some welfare properties of the Stackelberg–Nash equilibrium. In Sect. 2.7, we consider the challenging extension to a multilevel decision setting, and in Sect. 2.8, we conclude.

2.2 The Model

We adopt the following notational conventions. Let $\mathbf{x} \in \mathbb{R}^n_+$. Then, $\mathbf{x} \geq \mathbf{0}$ means $x_i \geqslant 0, i = 1, \ldots, n$; $\mathbf{x} > \mathbf{0}$ means there is some i such that $x_i > 0$, with $\mathbf{x} \neq \mathbf{0}$, and $\mathbf{x} >> \mathbf{0}$ means $x_i > 0$ for all i, $i = 1, \ldots, n$. The notation $f \in C^s(\mathbb{R}^n)$ is used to indicate that the function f has first through s-th continuous partial derivatives on $\mathbb{R}^n$. So, $f \in C^2(\mathbb{R}^n)$ means f is twice-continuously differentiable. A m dimensional vector function $\mathbf{F}$ is defined by $\mathbf{F} : A \subseteq \mathbb{R}^n \to B \subseteq \mathbb{R}^m$, with $\mathbf{F}(\mathbf{x}) = (f_1(\mathbf{x}), \ldots, f_2(\mathbf{x}), \ldots, f_m(\mathbf{x}))$. The Jacobian matrix of $\mathbf{F}(\mathbf{x})$ with respect to $\mathbf{x}$ at $\bar{\mathbf{x}}$ is denoted by $\mathcal{J}_{\mathbf{F}_\mathbf{x}}(\bar{\mathbf{x}})$, with $\mathcal{J}_{\mathbf{F}_\mathbf{x}}(\bar{\mathbf{x}}) = \left[\frac{\partial(f_1,\ldots,f_j,\ldots,f_m)}{\partial(x_1,\ldots,x_i,\ldots,x_n)}(\bar{\mathbf{x}})\right]$. Its corresponding determinant at $\bar{\mathbf{x}}$ is denoted by $\left|\mathcal{J}_{\mathbf{F}_\mathbf{x}}(\bar{\mathbf{x}})\right|$.

Let us consider a market with one divisible homogeneous product. On the demand side, there is a large number of consumers (a continuum), whose behavior is synthesized using a continuous market demand function, namely $d : \mathbb{R}_+ \to \mathbb{R}_+$, with $p \longmapsto d(p)$, where p is the unit price of the good expressed in a *numéraire*. Indeed, let $X \longmapsto p(X) = d^{-1}(X)$ be the market inverse demand function. This function represents the maximum price consumers are willing to pay to buy the quantity X. On the supply side, there is a finite number of decision makers, i.e., risk-neutral firms, whose finite set is $\mathcal{F}$. The set of firms can be divided into two subsets $\mathcal{F}_L = \{1, \ldots, n_L\}$ and $\mathcal{F}_F = \{1, \ldots, n_F\}$, where $\mathcal{F}_L$ is the subset of leaders and $\mathcal{F}_F$ the subset of followers, with $\mathcal{F}_L \cup \mathcal{F}_F = \mathcal{F}$ and $\mathcal{F}_L \cap \mathcal{F}_F = \varnothing$. We consider $|\mathcal{F}_L| \geqslant 1$ and $|\mathcal{F}_F| \geqslant 1$, where $|A|$ denotes the cardinality of set A. Leaders are indexed by i, $i \in \mathcal{F}_L$, and followers are indexed by j, $j \in \mathcal{F}_F$. Firm i (resp. j) produces x^i_L (resp. x^j_F) units of the good. Likewise, x^i_L and x^j_F represent respectively the supply for leader $i \in \mathcal{F}_L$, and follower $j \in \mathcal{F}_F$. Each firm bears some costs. Let $C^i_L : \mathbb{R}_+ \to \mathbb{R}_+$, with $x^i_L \longrightarrow C^i_L(x^i_L)$ be the cost function of leader $i \in \mathcal{F}_L$. Likewise, for each $j \in \mathcal{F}_F$, we let $C^j_F(x^j_F)$. Thus, there is a market clearing condition which stipulates that the demand balances the aggregate supply X, with $X \equiv \sum_i x^i_L + \sum_j x^j_F$.

We make the following set of assumptions regarding $p(X)$. This we designate as Assumption 2.2.1.

Assumption 2.2.1 *The price function $p(X)$ satisfies:*

(1a) $p(X) \geqslant 0$ *for all* $X \geqslant 0$, *with* $p(X) \in C^2(\mathbb{R}_{++})$;
(1b) $\frac{dp(X)}{dX} < 0$ *for* $X \geqslant 0$;
(1c) $\forall x \geqslant 0$, $\frac{dp(X)}{dX} + kx\frac{d^2p(X)}{(dX)^2} \leqslant 0$, *where* $k > 0$. △

(1a) indicates that the inverse demand function $p(X)$ is positively valued, and that it may or may not intersect the quantity axis and/or the price axis. Therefore, (1a) does not impose too stringent a property on the demand function: it may be strictly concave (convex) or linear, without imposing certain boundary conditions. (1a) also indicates that $p(X)$ is well-behaved: it is twice continuously differentiable on the open set $\mathbb{R}_{++}$.
(1b) indicates that the market demand is strictly decreasing.
(1c) stipulates that marginal revenue for any single firm is a decreasing function of total industry output. This formulation deserves two comments. First, we do not impose that the price function be a concave function, i.e., $\frac{d^2p(X)}{(dX)^2} \leqslant 0$, so we do not preclude (strictly) convex market demand functions. Second, our formulation of the decreasing marginal revenue hypothesis embodies the term k. For any leader firm the term k satisfies $k \neq 1$ unless leaders behave as followers (as in the Cournot model for which $k = 1$).

Likewise, we designate as Assumption 2.2.2 the set of assumptions made concerning the cost functions.

Assumption 2.2.2 *The cost function $C^h(x^h)$, $h \in \mathcal{F}$, satisfies:*

(2a) $\forall h \in \mathcal{F}$, $C^h(x^h) \geqslant 0$ *for all* $x \geqslant 0$, *with* $C^h(x^h) \in C^2(\mathbb{R}_{++})$;
(2b) $\forall h \in \mathcal{F}$, $\frac{dC^h(x^h)}{dx^h} > 0$ *and* $\frac{d^2C^h(x^h)}{(dx^h)^2} \geqslant 0$. △

(2a) stipulates that the cost functions are positive and twice continuously differentiable on the open set $\mathbb{R}_{++}$.
(2b) requires that costs are increasing and convex for all firms (for a discussion on this assumption, which may be weakened, see Julien [22]). When the costs are concave functions, multiple optima may exist.

Let us consider now the noncooperative bilevel optimization game Γ associated with this market. Let $S_L^i = [0, \infty)$ be the strategy set of leader $i \in \mathcal{F}_L$, where the supply x_L^i represents the pure strategy of leader $i \in \mathcal{F}_L$. Similarly, let $S_F^j = [0, \infty)$, where x_F^j is the pure strategy of follower $j \in \mathcal{F}_F$. Let $\mathbf{x}_L = (x_L^1, \ldots, x_L^i, \ldots, x_L^{n_L})$ be a strategy profile for all the leaders. Likewise, $\mathbf{x}_F = (x_F^1, \ldots, x_F^j, \ldots, x_F^{n_F})$ is a strategy profile for all the followers. A strategy profile will be represented by the vector $(\mathbf{x}_L, \mathbf{x}_F)$, with $(\mathbf{x}_L, \mathbf{x}_F) \in \prod_{i \in \mathcal{F}_L} S_L^i \times \prod_{j \in \mathcal{F}_F} S_F^j$. In addition, let $\mathbf{x}_L^{-i} = (x_L^1, \ldots, x_L^{i-1}, x_L^{i+1}, \ldots, x_L^{n_L})$ and $\mathbf{x}_F^{-j} = (x_F^1, \ldots, x_F^{j-1}, x_F^{j+1}, \ldots, x_F^{n_F})$.

Therefore, the profits $\Pi : \prod_{i\in\mathcal{F}_L} S_L^i \times \prod_{j\in\mathcal{F}_F} S_F^j \to \mathbb{R}_+$ of each firm at the lower and upper levels may be written in terms of payoffs as:

$$\Pi_L^i(x_L^i, X^{-i}) = p(x_L^i + X^{-i})x_L^i - C_L^i(x_L^i), i \in \mathcal{F}_L \tag{2.2.1}$$

$$\Pi_F^j(x_F^j, X^{-j}) = p(x_F^j + X^{-j})x_F^j - C_F^j(x_F^j), j \in \mathcal{F}_F, \tag{2.2.2}$$

where $X^{-i} \equiv X - x_L^i$ and $X^{-j} \equiv X - x_F^j$. It is worth noting that under Assumptions 2.2.1 and 2.2.2, the functions (2.2.1) and (2.2.2) are strictly concave.

The sequential game $\boldsymbol{\Gamma}$ displays two levels of decisions, namely 1 and 2, and no discounting. We also assume that the timing of positions is given.[3] Each leader first chooses a quantity to sell, and each follower determines their supply based on the residual demand. Information is again assumed to be complete. Information is imperfect because at level 1 (resp. level 2) a leader (resp. a follower) cannot observe what the other leaders (resp. other followers) decide: the multiple leader–follower model is thus described by a two-stage game which embodies two simultaneous move partial games. Indeed, the leaders play a two-stage game with the followers, but the leaders (the followers) play a simultaneous move game together.

2.3 Stackelberg–Nash Equilibrium: A Definition

The main purpose of this section is to define the SNE. To this end, we study the optimal behavior in each stage of the bilevel game. In this framework, strategic interactions occur within each partial game but also between the partial games through sequential decisions. It is worth noting that the critical difference from the usual two-player games stem from the fact that the optimal decision made by a follower does not necessarily coincide with their best response.[4]

Let us consider the second stage of the game $\boldsymbol{\Gamma}$. Given any strategy profile for leaders $\mathbf{x}_L \in \prod_i S_L^i$ and for all strategy profiles $\mathbf{x}_F^{-j} \in \prod_{-j} S_F^{-j}$ for all followers but j, we can define $\phi^j : \prod_{-j} S_F^{-j} \times \prod_i S_L^i \to S_F^j$, with $x_F^j = \phi^j(\mathbf{x}_F^{-j}, \mathbf{x}_L)$, $j \in \mathcal{F}_F$, as follower j's optimal decision mapping. Thus, the lower level optimization problem for follower j may be written:[5]

---

[3] Hamilton and Slutsky [16] provide theoretical foundations for endogenous timing in duopoly games and for the Stackelberg market outcome.

[4] One difficulty stems from the fact the followers' optimal decision mappings may be mutually inconsistent (Julien [22]).

[5] The same problem could be rewritten as follows. Let the objective of each firm be written as $-\Pi_F^j(x_F^j, X^{-j}) = C_F^j(x_F^j) - p(x_F^j + X^{-j})x_F^j$, for $j \in \mathcal{F}_F$. Then, the follower's problem might be written as $\phi^j(\mathbf{x}_F^{-j}, \mathbf{x}_L) := \min\{-\Pi_L^i(x_L^i, X^{-i}) : (\mathbf{x}_F^{-j}, \mathbf{x}_L) \in \prod_{-j} S_F^{-j} \times \prod_i S_L^i, x_F^j \in S_F^j\}$. A Nash equilibrium has to be sought out between followers.

$$\phi^j(\mathbf{x}_F^{-j}, \mathbf{x}_L) := \max_{\{x_F^j(.)\}} \{\Pi_F^j(x_F^j, \mathbf{x}_F^{-j}, \mathbf{x}_L) : (\mathbf{x}_F^{-j}, \mathbf{x}_L) \in \prod_{-j} \mathcal{S}_F^{-j} \times \prod_i \mathcal{S}_L^i, x_F^j \in \mathcal{S}_F^j\}. \tag{2.3.1}$$

Let $\mathcal{L}(x_F^j, \mathbf{x}_F^{-j}, \mathbf{x}_L, \lambda) := \Pi_F^j(x_F^j, \mathbf{x}_F^{-j}, \mathbf{x}_L) + \lambda x_F^j$ be the Lagrangian, where $\lambda \geqslant 0$ is the Kuhn–Tucker multiplier. By using Assumptions 2.2.1 and 2.2.2, the first-order sufficient condition may be written:

$$\frac{\partial \mathcal{L}(x_F^j, \mathbf{x}_F^{-j}, \mathbf{x}_L, \lambda)}{\partial x_F^j} = p(X) + \frac{dp(X)}{dX} x_F^j - \frac{dC_F^j(x_F^j)}{dx_F^j} + \lambda = 0 \tag{2.3.2}$$

$$\lambda \geqslant 0, x_F^j \geqslant 0, \text{ with } \lambda x_F^j = 0.$$

With Assumptions 2.2.1 and 2.2.2, the optimal decision mapping $\phi^j(\mathbf{x}_F^{-j}, \mathbf{x}_L)$ exists and is unique.[6] Indeed, we have either $\phi^j(\mathbf{x}_F^{-j}, \mathbf{x}_L) = 0$ or $\phi^j(\mathbf{x}_F^{-j}, \mathbf{x}_L) > 0$. Therefore, if $x_F^j > 0$, then $\lambda = 0$, where x_F^j is the solution to the equation $p(X) + x_F^j \frac{dp(X)}{dX} - \frac{dC_F^j(x_F^j)}{dx_F^j} = 0$, which yields $\phi^j(\mathbf{x}_F^{-j}, \mathbf{x}_L) > 0$. Now, if $\lambda > 0$, then $x_F^j = 0$, which means that $\phi^j(\mathbf{x}_F^{-j}, \mathbf{x}_L) = 0$. Then, $\phi^j(\mathbf{x}_F^{-j}, \mathbf{x}_L) \geqslant 0$, $j \in \mathcal{F}_F$. In addition, as for Assumptions 2.2.1 and 2.2.2, Π_F^j is strictly concave in x_F^j, then, according to Berge Maximum Theorem, $\phi^j(\mathbf{x}_F^{-j}, \mathbf{x}_L)$ is continuously differentiable.

This function is not a best response function since it also depends on the decisions of the other followers who make their decision at the lower level. By using the implicit function theorem, we have that:

$$\frac{\partial \phi^j(\mathbf{x}_F^{-j}, \mathbf{x}_L)}{\partial x_F^{-j}} = -\frac{\frac{dp(X)}{dX} + x_F^j \frac{d^2p(X)}{(dX)^2}}{2\frac{dp(X)}{dX} + x_F^j \frac{d^2p(X)}{(dX)^2} - \frac{d^2C_F^j(x_F^j)}{(dx_F^j)^2}}, \tag{2.3.3}$$

as $\frac{\partial \phi^j(\mathbf{x}_F^{-j}, \mathbf{x}_L)}{\partial x_F^{-j}} = -\frac{\frac{\partial^2 \Pi_F^j(.)}{\partial x_F^j \partial x_F^{-j}}}{\frac{\partial^2 \Pi_F^j(.)}{(\partial x_F^j)^2}}$. We have that $\frac{\partial \phi^j(.)}{\partial x_F^{-j}} \in (-1, 0)$, when $\phi^j(.) > 0$, and $\frac{\partial \phi^j(.)}{\partial x_F^{-j}} = 0$ when $\phi^j(.) = 0$. Then, $\frac{\partial \phi^j(.)}{\partial x_F^{-j}} \in (-1, 0]$, $-j, j \in \mathcal{F}_F$. In addition, it is possible to show that $\frac{\partial \phi^j(.)}{\partial x_L^i} \in (-1, 0]$, $i \in \mathcal{F}_L$, $j \in \mathcal{F}_F$.

[6] The payoff function is strictly concave and the strategy set is compact and convex.

We assume that the followers' optimal behaviors as studied in stage 1 of the bilevel optimization process are consistent (see Sect. 2.4).[7] Then, the system of equations which determines such best responses has a unique solution, so we can define the best response for follower j as $\varphi^j : \prod_{i \in \mathcal{F}_L} S_L^i \to S_F^j$, with $x_F^j = \varphi^j(\mathbf{x}_L)$, $j \in \mathcal{F}_F$.[8] Let $\boldsymbol{\varphi} : \prod_{i \in \mathcal{F}_L} S_L^i \to \prod_{i \in \mathcal{F}_L} S_F^j$, with $\boldsymbol{\varphi} = \left(\varphi^1(\mathbf{x}_L), \ldots, \varphi^{n_F}(\mathbf{x}_L)\right)$, be the vector of best responses. The vector function $\boldsymbol{\varphi}(\mathbf{x}_L)$ constitutes a constraint for the decision maker at the upper level as we now have $p(X) = p(x_L^i + X_L^{-i} + \sum_j \varphi^j(x_L^i + X_L^{-i}))$, where $X_L^{-i} \equiv \sum_{-i,-i \neq i} x_L^{-i}$.

Therefore, at the upper level of the game, leader i's optimal decision, which is defined by $\psi^i : \prod_{-i \in \mathcal{F}_L} S_L^{-i} \to S_L^i$, with $x_L^i = \psi^i(\mathbf{x}_L^{-i})$, is the solution to the problem:

$$\psi^i(\mathbf{x}_L^{-i}) := \max_{\{x_L^i(.)\}} \{\Pi_L^i(x_L^i, \mathbf{x}_L^{-i}, \boldsymbol{\varphi}(\mathbf{x}_L)) : \mathbf{x}_L^{-i} \in \prod_{-i} S_L^{-i}, x_L^i \in S_L^i\}. \qquad (2.3.4)$$

Let $\mathcal{L}(x_L^i, \mathbf{x}_L^{-i}, \mu) := \Pi_L^i(x_L^i, \mathbf{x}_L^{-i}, \varphi(\mathbf{x}_L)) + \mu x_L^i$ be the Lagrangian, where $\mu \geqslant 0$ is the Kuhn–Tucker multiplier. As $p(X) = p(x_L^i + X_L^{-i} + \sum_j \varphi^j(x_L^i + X_L^{-i}))$, which is continuous (see Julien [22]), the Kuhn–Tucker conditions may be written:

$$\frac{\partial \mathcal{L}(x_L^i, \mathbf{x}_L^{-i}, \mu)}{\partial x_L^i} = p(X) + \left(1 + \nu^i\right) x_L^i \frac{dp(X)}{dX} - \frac{dC_L^i(x_L^i)}{dx_L^i} + \mu = 0 \quad (2.3.5)$$

$$\mu \geqslant 0, x_L^i \geqslant 0, \text{ with } \mu x_L^i - 0,$$

The term $\nu^i = \frac{\partial \sum_j \varphi^j(\mathbf{x}_L)}{\partial x_L^i}$, with $\nu^i \geqslant -1$, represents the reaction of all followers to leader i's strategy, i.e., the slope of the aggregate best response to i, $i \in \mathcal{F}_L$. By construction, $\nu^i = \nu^{-i} = \nu$ for all $i, -i \in \mathcal{F}_L$. Let $k = (1 + \nu)$. We may have either $\psi^i(\mathbf{x}_L^{-i}) = 0$ or $\psi^i(\mathbf{x}_L^{-i}) > 0$.

[7]This is one critical difference with the standard duopoly game in which the optimal decision of the follower coincides with their best response. Julien [22] provides a consistency condition which helps determine each optimal decision as a function of the strategy profile for the leaders. Indeed, we give a sufficient nondegeneracy condition on the determinant of the Jacobian matrix associated with the set of equations that allows us to implicitly define the best response mappings. Under this condition, the set of equations which implicitly determines the best responses is a variety of the required dimension, that is, the corresponding vector mapping which defines this set of equations is a C^1-diffeomorphism. Here this criterion is satisfied as long as Assumptions 2.2.1 and 2.2.2 both hold. These assumptions can be weakened. It is worth noting that our notion of consistency differs from the notion of price consistency in Leyffer and Munson [30] that results in a square nonlinear complementarity problem.

[8]It is possible to show that the best responses are not increasing, so the game displays actions which are strategic substitutes. Please note that the condition is sufficient, so strategic complementarities could exist provided they are not too strong.

By using Assumptions 2.2.1 and 2.2.2, it is possible to show that, for each $i \in \mathcal{F}_L$, the second-order sufficient condition holds:

$$\frac{\partial^2 \Pi_L^i(x_L^i, \mathbf{x}_L^{-i})}{(\partial x_L^i)^2} = k\left(kx_L^i \frac{d^2 p(X)}{(dX)^2} + 2\frac{dp(X)}{dX}\right) - \frac{d^2 C_L^i(x_L^i)}{(dx_L^i)^2} < 0. \qquad (2.3.6)$$

Finally, it is worth noting that $\frac{\partial^2 \Pi_L^i(.)}{\partial x_L^i \partial x_L^{-i}} = k\left(k\frac{dp(X)}{dX} + x_L^i \frac{d^2 p(X)}{(dX)^2}\right) \leqslant 0$, for each $i \in \mathcal{F}_L$; and, by using the implicit function theorem, we have that

$$\frac{\partial \psi^i(.)}{\partial x_L^{-i}} = -\frac{\frac{\partial^2 \Pi_L^i(.)}{\partial x_L^i \partial x_L^{-i}}}{\frac{\partial^2 \Pi_L^i(.)}{\partial (x_L^i)^2}} = -\frac{k\frac{dp(X)}{dX} + k^2 x_L^i \frac{d^2 p(X)}{(dX)^2}}{2k\frac{dp(X)}{dX} + k^2 x_L^i \frac{d^2 p(X)}{(dX)^2} - \frac{d^2 C_L^i(x_L^i)}{(dx_L^i)^2}}, \qquad (2.3.7)$$

so we can deduce that $\frac{\partial \psi^j(.)}{\partial x_F^{-j}} \in (-1, 0]$, for all $-i \neq i, -i, i \in \mathcal{F}_L$.

The solution to the n_L equations such as (2.3.5) yields the strategy profile for the leaders $\tilde{\mathbf{x}}_L = (\tilde{x}_L^1, \ldots, \tilde{x}_L^i, \ldots, \tilde{x}_L^{n_L})$. From the set of the best responses, i.e., $(\varphi^1(\mathbf{x}_L), \ldots, \varphi^{n_F}(\mathbf{x}_L))$, it is possible to deduce the strategy profile for followers $\tilde{\mathbf{x}}_F = (\tilde{x}_F^1, \ldots, \tilde{x}_F^j, \ldots, \tilde{x}_F^{n_F})$.

We are now able to provide a definition of an SNE for this bilevel game.

Definition 2.3.1 (SNE) A Stackelberg–Nash equilibrium of $\mathbf{\Gamma}$ is given by a strategy profile $(\tilde{\mathbf{x}}_L, \boldsymbol{\varphi}(\tilde{\mathbf{x}}_L)) \in \prod_{i \in \mathcal{F}_L} \mathcal{S}_L^i \times \prod_{j \in \mathcal{F}_F} \mathcal{S}_F^j$, with $\tilde{\mathbf{x}}_F = \boldsymbol{\varphi}(\tilde{\mathbf{x}}_L)$, where $\boldsymbol{\varphi} : \prod_{i \in \mathcal{F}_L} \mathcal{S}_L^i \to \prod_{j \in \mathcal{F}_F} \mathcal{S}_F^j$, such that conditions C1 and C2 hold:

C1 $\forall i \in \mathcal{F}_L$ $\Pi_L^i\left(\tilde{x}_L^i, \tilde{\mathbf{x}}_L^{-i}, \boldsymbol{\varphi}(\tilde{x}_L^i, \tilde{\mathbf{x}}_L^{-i})\right) \geqslant \Pi_L^i\left(x_L^i, \tilde{\mathbf{x}}_L^{-i}, \boldsymbol{\varphi}(x_L^i, \tilde{\mathbf{x}}_L^{-i})\right)$, $\forall \boldsymbol{\varphi}(\mathbf{x}_L) \in \prod_{j \in \mathcal{F}_F} \mathcal{S}_F^j$, $\forall \mathbf{x}_L^{-i} \in \prod_{-i \in \mathcal{F}_L} \mathcal{S}_F^{-i}$ and $\forall x_L^i \in \mathcal{S}_L^i$;

C2 $\forall j \in \mathcal{F}_F$ $\Pi_F^j(\tilde{x}_F^j, \tilde{\mathbf{x}}_F^{-j}, \tilde{\mathbf{x}}_L) \geqslant \Pi_F^j(x_F^j, \tilde{\mathbf{x}}_F^{-j}, \tilde{\mathbf{x}}_L)$, $\forall x_F^j \in \mathcal{S}^j$. △

2.4 Stackelberg–Nash Equilibrium: Existence and Uniqueness

Existence and uniqueness problems are complex in this framework as there are several decision makers at each level: strategic interactions occur within levels but also between the two levels through sequential decisions. Indeed, the n_L leaders play a two-stage game with the n_F followers, but the leaders (the followers) play a simultaneous move game together. Therefore, the bilevel game $\mathbf{\Gamma}$ displays two partial games, namely the lower level game $\mathbf{\Gamma}_F$ and the upper level game $\mathbf{\Gamma}_L$.

The equilibrium of the entire game $\boldsymbol{\Gamma}$ is a pure strategy subgame perfect Nash equilibrium (SPNE), while the equilibria in each partial game are Nash equilibria. We state two results which pertain to existence and uniqueness. Then, we discuss existence and uniqueness within the literature.

The following Theorem may be stated for the bilevel game $\boldsymbol{\Gamma}$ under consideration.

Theorem 2.4.1 (Existence of SNE) *Let us consider the game* Γ, *and let Assumptions 2.2.1 and 2.2.2 be satisfied. Then, there exists a Stackelberg–Nash equilibrium.* △

Proof Here we provide heuristic proof (for more details, see notably Julien [22] with weaker assumptions on costs). As we have many decision makers at the lower and upper levels, we show that there exists a Nash equilibrium at each level of the game, i.e., there exists a strategy profile $(\tilde{\mathbf{x}}_L, \tilde{\mathbf{x}}_F) \in \prod_i \mathcal{S}_L^i \times \prod_j \mathcal{S}_F^j$ such that the leaders and followers strategic optimal plans are mutually consistent. We define the function $\mathbf{\Lambda}_L : \prod_i \mathcal{S}_L^i \to \prod_i \mathcal{S}_L^i$, with $\mathbf{\Lambda}_L(\mathbf{x}_L) = \times_{i=1}^{n_L} \psi^i$. The function $\mathbf{\Lambda}(\mathbf{x}_L)$ is continuous (as each ψ^i given by the solution for (2.3.5) is continuous under Assumptions 2.2.1 and 2.2.2 in $\mathbf{x}_L$ on $\prod_i \mathcal{S}_L^i$, a compact and convex subset of Euclidean space (as the product of compact and convex strategy sets $\mathcal{S}_L^i$, $i \in \mathcal{F}_L$). Then, according to the Brouwer Fixed Point Theorem, the function $\mathbf{\Lambda}(\mathbf{x}_L)$ has a fixed point $\tilde{\mathbf{x}}_L \in \prod_i \mathcal{S}_L^i$, with components $\tilde{x}_L^i$, where $\tilde{x}_L^i \in \mathcal{S}_L^i$, for each $i \in \mathcal{F}_L$. This fixed point is a pure strategy Nash equilibrium of the subgame $\mathbf{\Gamma}_L$. Now let us define $\mathbf{\Lambda}_F : \prod_j \mathcal{S}_F^j \times \prod_i \mathcal{S}_L^i \to \prod_j \mathcal{S}_F^j \times \prod_i \mathcal{S}_L^i$, with $\mathbf{\Lambda}_F(\mathbf{x}_F, \mathbf{x}_L) = \times_{j=1}^{n_F} \phi^j$, where, for each j, ϕ^j is the solution to (2.3.2). Given that $\tilde{\mathbf{x}}_L \in \prod_i \mathcal{S}_L^i$, we have that $\mathbf{\Lambda}_F(\mathbf{x}_F, \tilde{\mathbf{x}}_L) = \times_{j=1}^{n_F} \phi^j(\mathbf{x}_F^{-j}, \tilde{\mathbf{x}}_L)$. A similar argument as the one made for the leaders shows that the function $\mathbf{\Lambda}_F(\mathbf{x}_F, \tilde{\mathbf{x}}_L)$ has a fixed point $\tilde{\mathbf{x}}_F \in \prod_j \mathcal{S}_F^j$, with components $\tilde{x}_F^j$, where $\tilde{x}_F^j \in \mathcal{S}_F^j$, for all $j \in \mathcal{F}_F$. This fixed point is a pure strategy Nash equilibrium of the subgame $\mathbf{\Gamma}_F$. But then, the point $(\tilde{\mathbf{x}}_L, \tilde{\mathbf{x}}_F)$, with $(\tilde{\mathbf{x}}_L, \tilde{\mathbf{x}}_F) \in \prod_i \mathcal{S}_L^i \times \prod_j \mathcal{S}_F^j$ exists, which constitutes a SPNE of $\mathbf{\Gamma}$.

The existence of an equilibrium is obtained here under mild conditions for market demand and costs. Some of these conditions could be relaxed provided the remaining conditions are completed with additional restrictions. For instance, convexity of costs for all firms is not necessary.

The next theorem relies on the uniqueness of the SNE (see Julien [22]).

Theorem 2.4.2 (Uniqueness of SNE) *Let Assumptions 1 and 2 be satisfied. Then, if a Stackelberg–Nash equilibrium exists, it is unique.* △

Proof To show uniqueness, we consider $\boldsymbol{\pi}_L = \left(\frac{\partial \Pi_L^1}{\partial x_L^1}, \ldots, \frac{\partial \Pi_L^i}{\partial x_L^i}, \ldots, \frac{\partial \Pi_L^{n_L}}{\partial x_L^{n_L}}\right)$ (see Julien [22] for more details). Let $\left|J_{-\boldsymbol{\pi}_L}(\tilde{\mathbf{x}}_L, \tilde{\mathbf{x}}_F)\right|$, with $J_{-\boldsymbol{\pi}_L} = -\left(\frac{\partial^2 \Pi_L^i}{\partial x_L^i \partial x_L^{-i}}\right)$, where $\frac{\partial \Pi_L^i}{\partial x_L^i} = p(X) + k x_L^i \frac{dp(X)}{dX} - \frac{dC_L^i(x_L^i)}{dx_L^i}$. By using Corollary 2.1 in Kolstad and Mathiesen [26]), as leaders in the partial game $\mathbf{\Gamma}_L$ behave like Cournot firms, we show this criterion is satisfied, so the SNPE in $\mathbf{\Gamma}_L$ is unique. It is possible to show that:

$$\left|J_{-\pi_L}\right| = \left(1 - k \sum_{i \in \mathcal{F}_L} \frac{\frac{dp(X)}{dX} + kx_L^i \frac{d^2p(X)}{(dX)^2}}{\frac{d^2C_L^i(x_L^i)}{(dx_L^i)^2} - k\frac{dp(X)}{dX}}\right) \prod_{i \in \mathcal{F}_L} \left(\frac{d^2C_L^i(x_L^i)}{(dx_L^i)^2} - k\frac{dp(X)}{dX}\right). \tag{2.4.1}$$

Then, as $sign\left|J_{-\pi_L}\right| = sign(1 - k \sum_{i \in \mathcal{F}_L} \frac{\frac{dp(X)}{dX} + kx_L^i \frac{d^2p(X)}{(dX)^2}}{\frac{d^2C_L^i(x_L^i)}{(dx_L^i)^2} - k\frac{dp(X)}{dX}})$, by using the assumptions on costs and demand, we deduce:

$$\left|J_{-\pi_L}(\tilde{\mathbf{x}}_L, \tilde{\mathbf{x}}_F)\right| > 0. \tag{2.4.2}$$

As $\left|J_{-\pi_L}(\tilde{\mathbf{x}}_L, \tilde{\mathbf{x}}_F)\right| > 0$ there exists a unique Nash equilibrium in the subgame $\mathbf{\Gamma}_L$. Now, given a unique point $\tilde{\mathbf{x}}_L$, and by using a similar argument as the one made previously for the upper level, it is possible to show that $\left|J_{-\pi_F}(\tilde{\mathbf{x}}_L, \tilde{\mathbf{x}}_F)\right| > 0$ at the lower level, with $k = 1$, in (2.4.1). Then, there is a unique pure strategy Nash equilibrium in the subgame $\mathbf{\Gamma}_F$. Then, the SPNE of $\mathbf{\Gamma}$ is unique, which proves the uniqueness of the SNE. □

Remark 2.4.3 If we assume symmetry, the condition for the sign for $\left|J_{-\pi_L}((\tilde{\mathbf{x}}_L, \tilde{\mathbf{x}}_F)\right|$ may be rewritten as $\frac{dp(X)}{dX} + kx_L^i \frac{d^2p(X)}{(dX)^2} < \frac{1}{kn_L}\left(\frac{d^2C_L^i(x_L^i)}{(dx_L^i)^2} - k\frac{dp(X)}{dX}\right)$, which would indicate that "on average" leaders' marginal revenues could be increased but not too much. In addition, $\frac{d^2C_L^i(x_L^i)}{(dx_L^i)^2} - k\frac{dp(X)}{dX} + n_L\left(\frac{dp(X)}{dX} + kx_L^i \frac{d^2p(X)}{(dX)^2}\right) = \frac{\partial^2\Pi_L^1}{(\partial x_L^1)^2} + (n_L - 1)\frac{\partial^2\Pi_L^i}{\partial x_L^i \partial x_L^{-i}} < 0$: the effect of a change in x_L^i on i's marginal profit dominates the sum of the cross effects of similar changes for the supply of other leaders. △

The uniqueness of an SNE holds under strong assumptions. It can happen that multiple Nash equilibria exist at both levels. At the lower level as well as the upper level, multiplicity of equilibria can be generated by strong strategic complementarities caused either by nonconvex costs or market demand functions which do not intersect the axis. The multiplicity of Nash equilibria can lead to coordination failures problems.

Existence and uniqueness have already been explored in the multiple leader–follower model. Sherali [33] shows existence and uniqueness with identical convex costs for leaders, and states some results under the assumptions of linear demand with either linear or quadratic costs (Ehrenmann [15]). Sherali's model is an extension of the seminal paper by Murphy et al. [34] which covers the case of many followers who interact with one leader. In their model the authors provide a characterization of the SNE, along with an algorithm to compute it. They state a Theorem 1 which gives the properties of the *aggregate* best response for the followers expressed as a function of the leader's strategy. This determination stems from a family of optimization programs for the followers based on a price function which is affected by the supply of the leader. They show that this aggregate function is convex, and then, study the problem faced by the leader. Nevertheless, they do not study the conditions under which the followers' optimal decisions are mutually consistent. In the same vein, Tobin [37] provides an efficient algorithm to find a unique SNE by parameterizing the price function by the leader's strategy. Some strong assumptions are made on the thrice-differentiability of the price function and the cost to the leader.

More recently, in line with De Wolf and Smeers [10] and DeMiguel and Xu [11] extend the work by Sherali [33] to include uncertainty with stochastic market demand. Unlike Sherali [33] they allow costs to differ across leaders. Nevertheless, to show that the expected profit of any leader is concave, they assume that the aggregate best response of the followers is convex. However as this assumption does not always hold, these authors must resort to a linear demand. Pang and Fukushima [31], Yu and Wang [40], and Jia et al. [20] prove the existence of an equilibrium point of a finite game with two leaders and several followers without specifying the assumptions made on demand and costs. Kurkarni and Shanbhag [27] show that when the leaders' objectives admit a quasi-potential function, the global and local minimizers of the leaders' optimization problems are global and local equilibria of the game. Finally, Aussel et al. [2] study the existence of an equilibrium in the electricity markets.

2.5 The Linear and the Quadratic Bilevel Optimization Games

In this section, we consider two standard bilevel optimization games: the linear model with asymmetric costs and the quadratic model with symmetric costs. The following specification holds in both models. There are $n_L \geqslant 1$ leader(s) and $n_F \geqslant 1$ follower(s), with $n_L + n_F = n$. Let $p(X) = a - bX$, $a, b > 0$, where $X = X_L + X_F$, with $X_L = \sum_{i=1}^{n_L} x_L^i$ and $X_F = \sum_{j=1}^{n_F} x_F^j$.

2.5.1 The Linear Bilevel Optimization Game

The costs functions are given by $C_L^i(x_L^i) = c_L^i x_L^i, i = 1, \ldots, n_L$, and by $C_F^j(x_F^j) = c_F^j x_F^j$, $j = 1, \ldots, n_F$, with $c_L^i, c_F^j < a$, for all i and all j. The strategy sets are given by $\mathcal{S}_L^i = [0, \frac{a}{b} - c_L^i], i \in \mathcal{F}_L$, and$\mathcal{S}_F^j = [0, \frac{a}{b} - c_F^j], j \in \mathcal{F}_F$.

As a point of reference, when each firm is a price-taker and does not behave strategically, the competitive equilibrium (CE), is such that the market price and the aggregate supply are given by $p^* = \min\{c_L^1, \ldots, c_L^{n_L}, c_F^1, \ldots, c_F^{n_F}\}$ and $X^* = \frac{a-c^*}{b}$, where $c^* = \min\{c_L^1, \ldots, c_L^{n_L}, c_F^1, \ldots, c_F^{n_F}\}$. The corresponding payoffs are given by $(\Pi_L^i)^* = 0$ (resp. $(\Pi_F^j)^* = 0$) when $c_L^i = c^*$ (resp. $c_F^j = c^*$). In addition, the Cournot–Nash equilibrium (CNE), in which all firms play simultaneously, is given by $\hat{x}_L^i = \frac{a+\sum_{-i\neq i} c_L^{-i} + \sum_j c_F^j - (n_L+n_F)c_L^i}{b(n_L+n_F+1)}$, $i \in \mathcal{F}_L$, $\hat{x}_F^j = \frac{a+\sum_i c_L^i + \sum_{-j\neq j} c_F^{-j} - (n_L+n_F)c_F^j}{b(n_L+n_F+1)}$, $j \in \mathcal{F}_F$, $\hat{X} = \frac{a(n_L+n_F)-n_L c_L^i - n_F c_F^j}{b(n_L+n_F+1)}$, and $\tilde{p} = \frac{a+n_L c_L^i + n_F c_F^j}{n_L+n_F+1}$, with corresponding payoffs

$$\hat{\Pi}_L^i = \frac{(a-(n_L+1)c_L^i + n_F c_F^j)(a + \sum_{-i\neq i} c_L^{-i} + \sum_j c_F^j - (n_L+n_F)c_L^i)}{b(n_L+n_F+1)^2}, \; i \in \mathcal{F}_L,$$

$$\hat{\Pi}_F^j = \frac{(a+n_L c_L^i - (n_F+1)c_F^j)(a + \sum_i c_L^i + \sum_{-j\neq j} c_F^{-j} - (n_L+n_F)c_F^j)}{b(n_L+n_F+1)^2}, \; j \in \mathcal{F}_F.$$

At the lower level, follower j's problem may be written as follows:

$$\phi^j(\mathbf{x}_F^{-j}, \mathbf{x}_L) : \max\{[a - b(x_F^j + X_F^{-j} + X_L) - c_F^j x_F^j]x_F^j : \mathcal{S}_F^j = [0, \frac{a}{b} - c_F^j]\}. \tag{2.5.1}$$

The optimal decision mapping for follower j corresponding to the solution to Eq. (2.3.2) is given by:

$$\phi^j(\mathbf{x}_F^{-j}, \mathbf{x}_L) = \frac{a - c_F^j}{2b} - \frac{1}{2}(X_F^{-j} + X_L), \tag{2.5.2}$$

where $X_F^{-j} \equiv \sum_{-j\neq j} x_F^{-j}$. The best response for follower j, $j \in \mathcal{F}_F$, is given by the convex linear function:

$$\varphi^j(\mathbf{x}_L) = \frac{a + \sum_{-j\neq j} c_F^{-j} - n_F c_F^j}{b(n_F+1)} - \frac{1}{n_F+1} X_L. \tag{2.5.3}$$

At the upper level, as the price function may be written as $p(X_L) = \frac{a+\sum_j c_F^j}{n_F+1} - \frac{b}{n_F+1} X_L$, leader i's optimal decision mapping $\psi^i(\mathbf{x}_L^{-i})$ is the solution to the upper level optimization problem, which may be written as follows:

$$\psi^i(\mathbf{x}_L^{-i}) : \max\left\{\left[\frac{a+\sum_j c_F^j}{n_F+1} - \frac{b(x_L^i + X_L^{-i})}{n_F+1} - c_L^i\right] x_L^i : \mathcal{S}_L^i = [0, \frac{a}{b} - c_L^i]\right\}. \tag{2.5.4}$$

The optimal decision mapping of leader i is given by:

$$\psi^i(\mathbf{x}_L^{-i}) = \frac{a-(n_F+1)c_L^i+\sum_j c_F^j}{2b} - \frac{1}{2}X_L^{-i}. \tag{2.5.5}$$

We deduce the equilibrium strategy for leader i:

$$\tilde{x}_L^i = \frac{a-(n_F+1)c_L^i+\sum_j c_F^j}{b(n_L+1)}, i \in \mathcal{F}_L. \tag{2.5.6}$$

Then, as $\tilde{X}_L \equiv \sum_i \tilde{x}_L^i = \frac{an_L+n_L\sum_j c_F^j-(n_F+1)\sum_i c_L^i}{b(n_L+1)}$, by using (2.5.2), we can deduce the equilibrium strategy for follower j:

$$\tilde{x}_F^j = \frac{a-(n_F+1)c_F^j-n_L\sum_j c_F^j+(n_F+1)\sum_i c_L^i}{b(n_L+1)(n_F+1)}, j \in \mathcal{F}_F. \tag{2.5.7}$$

Therefore as $\tilde{X}_F \equiv \sum_j \tilde{x}_F^j = \frac{an_F-(n_Fn_L+n_L+1)\sum_j c_F^j+n_F(n_F+1)\sum_i c_L^i}{b(n_L+1)(n_F+1)}$, so the market price is given by:

$$\tilde{p} = \frac{a+(n_F+1)\sum_i c_L^i+\sum_j c_F^j}{(n_L+1)(n_F+1)}. \tag{2.5.8}$$

The payoffs are then given by:

$$\tilde{\Pi}_L^i = \frac{A[a-(n_F+1)c_L^i+\sum_j c_F^j]}{b(n_L+1)^2(n_F+1)}, i \in \mathcal{F}_L; \tag{2.5.9}$$

$$\tilde{\Pi}_F^j = \frac{B[a-(n_F+1)c_F^j-n_L\sum_j c_F^j+(n_F+1)\sum_i c_L^i]}{b[(n_L+1)(n_F+1)]^2}, j \in \mathcal{F}_F, \tag{2.5.10}$$

where $A \equiv a+(n_F+1)\sum_i c_L^i+\sum_j c_F^j-(n_L+1)(n_F+1)c_L^i$ and $B = a+(n_F+1)\sum_i c_L^i+\sum_j c_F^j-(n_L+1)(n_F+1)c_F^j$.

Finally, consider the particular case where $C_L^i(x_L^i) = cx_L^i, i = 1, \ldots, n_L$, and by $C_F^j(x_F^j) = cx_F^j$, $j = 1, \ldots, n_F$, with $c < a$. The CE, is such that aggregate supply and market price are given respectively by $X^* = \frac{a-c}{b}$, $p^* = c$, and $(\Pi^i)^* = 0$, $i = 1, \ldots, n$.[9] The CNE is given by $\hat{x}_L^i = \hat{x}_F^j = \frac{a-c}{b(n_L+n_F+1)}$, $\hat{X} = \frac{(a-c)(n_L+n_F)}{b(n_L+n_F+1)}X^*$,

[9]The CE supplies are given by $((x_L^i)^*, (x_F^j)^*) = (\alpha X^*, (1-\alpha)X^*)$, with $\alpha \in (0, 1)$. In what follows, we consider the symmetric outcome for which $\alpha = \frac{1}{2}$.

$\tilde{p} = \frac{a+c(n_L+n_F)}{n_L+n_F+1}$, and $\hat{\Pi}^i_L = \hat{\Pi}^j_F = \frac{(a-c)^2}{b(n_L+n_F+1)^2}$, $i \in \mathcal{F}_L$, $j \in \mathcal{F}_F$. The SNE is given by $\tilde{x}^i_L = \frac{a-c}{b(n_L+1)}$, $i \in \mathcal{F}_L$, $\tilde{x}^j_F = \frac{a-c}{b(n_L+1)(n_F+1)}$, $j \in \mathcal{F}_F$, $\tilde{p} = \frac{a+c[n_L(n_F+1)+n_F]}{(n_L+1)(n_F+1)}$, $\tilde{\Pi}^i_L = \frac{(a-c)^2}{b(n_L+1)^2(n_F+1)}$, $i \in \mathcal{F}_L$, and $\tilde{\Pi}^j_F = \frac{(a-c)^2}{b[(n_L+1)(n_F+1)]^2}$, $j \in \mathcal{F}_F$.

We can observe that for each $i \in \mathcal{F}_L$, we have $\tilde{\Pi}^i_L \geqslant \hat{\Pi}^i_L$ whenever $n_L \leqslant \sqrt{n_F+1}$: any leader will achieve a higher payoff provided the number of leaders is not too high.[10] It is worth noting that $\lim_{n_L \to \infty} \tilde{p} = c$ (resp. $\lim_{(n_L,n_F)\to(\infty,\infty)} \tilde{p} = c$): so when the number of leaders (resp. leaders and followers) becomes arbitrarily large the SNE market price coincides with the CE price p^*. This result holds with the CNE in case either the number of leaders or followers goes to infinity (in the inclusive sense!).

2.5.2 The Quadratic Bilevel Optimization Game

The costs functions are given by $C^i_L(x^i_L) = \frac{c}{2}(x^i_L)^2$, $i = 1, \ldots, n_L$, and by $C^j_F(x^j_F) = \frac{c}{2}(x^j_F)^2$, $j = 1, \ldots, n_F$, with $c < a$, for all i and all j. The strategy sets are given by $\mathcal{S}^i_L = [0, \frac{a}{b} - \frac{c}{2}(x^i_L)^2]$, $i \in \mathcal{F}_L$, and$\mathcal{S}^j_F = [0, \frac{a}{b} - \frac{c}{2}(x^i_L)^2]$, $j \in \mathcal{F}_F$.

The CNE is given by $\hat{x}^i_L = \frac{a}{b(n_L+n_F+1)+c}$, for all $i \in \mathcal{F}_L$, $\hat{x}^j_F = \frac{a}{b(n_L+n_F+1)+c}$, for all $j \in \mathcal{F}_F$, $\hat{p} = \frac{a(b+c)}{b(n_L+n_F+1)+c}$, and $\hat{\Pi}^i_L = \frac{a^2(2b+c)}{[b(n_L+n_F+1)+c]^2}$, for all $i \in \mathcal{F}_L$, $\hat{\Pi}^j_F = \frac{a^2(2b+c)}{[b(n_L+n_F+1)+c]^2}$, for all $j \in \mathcal{F}_F$.

Consider now the SNE. By following the same procedure as for the linear bilevel game, the SNE equilibrium supplies are given by:

$$\tilde{x}^i_L = \frac{a}{b(n_L+1) + c(\frac{b}{b+c}n_F + 1)}, \; i \in \mathcal{F}_L; \tag{2.5.11}$$

$$\tilde{x}^j_F = \frac{a[b + c(1 + \frac{b}{b+c}n_F)]}{[c + b(n_F+1)][b(n_L+1) + c(\frac{b}{b+c}n_F + 1)]}, \; j \in \mathcal{F}_F. \tag{2.5.12}$$

Therefore, we deduce the market price

$$\tilde{p} = \frac{a(b+c)[b + c(1 + \frac{b}{b+c}n_F)]}{[c + b(n_F+1)][b(n_L+1) + c(\frac{b}{b+c}n_F + 1)]}. \tag{2.5.13}$$

[10] The welfare properties of the bilevel optimization linear game with symmetric costs are explored in Daughety [9], Julien et al. [24, 25], and in Julien [23].

The corresponding payoffs are given by:

$$\tilde{\Pi}_L^i = \frac{a^2(2b^2 + 3bc + c^2 + bcn_F)}{2[c + b(n_F + 1)][b(n_L + 1) + c(\frac{b}{b+c}n_F + 1)]^2}, i \in \mathcal{F}_L, \tag{2.5.14}$$

$$\tilde{\Pi}_F^j = \frac{a^2(2b + c)[b + c(1 + \frac{b}{b+c}n_F)]^2}{2[c + b(n_F + 1)]^2[b(n_L + 1) + c(\frac{b}{b+c}n_F + 1)]^2}, j \in \mathcal{F}_F. \tag{2.5.15}$$

It is easy to check that, as the production of any leader is higher than the production of any follower, the payoff of any leader is higher.

2.6 Stackelberg–Nash Equilibrium: Welfare Properties

We now turn to the nonoptimality of the SNE and some of its welfare properties. To this end, we compare the SNE market outcome with the CNE, and with the CE. Next, we consider the relation between market concentration and surplus, and also the relation between individual market power, payoffs and mergers.

2.6.1 *The SNE, CNE and CE Aggregate Market Outcomes*

We can state the following proposition, which represents a well-known result.

Proposition 2.6.1 *Let $\tilde{X}$, $\hat{X}$, and X^* be respectively the SNE, the CNE, and the CE aggregate supplies; and $\tilde{p}$, $\hat{p}$, and p^* the corresponding market prices. Then, $\hat{X} < \tilde{X} < X^*$, and $p^* < \tilde{p} < \hat{p}$.* △

In the bilevel optimization game, the leaders can set a higher supply. In addition, the increment in the aggregate supply of leaders more than compensates for the decrease in the aggregate supply of followers when the aggregate best response is negatively sloped, whereas it goes in the same direction when the aggregate best response increases, i.e., when strategies are complements.[11] Therefore, the aggregate supply (market price) is higher (lower) in the SNE than in the CNE, both when strategies are substitutes and when they are complements. The following example illustrates that the noncooperative sequential game leads to higher traded output than in the noncooperative simultaneous game (see Daughety [9]).

[11] When the slope of the aggregate best response is zero, then the SNE can coincide with the CNE (see notably Julien [21]).

Noncooperative Sequential Game Leads to Higher Traded Output
Consider the linear bilevel game given by (2.5.1)–(2.5.10), where $C_L^i(x_L^i) = cx_L^i$, $i = 1, \ldots, n_L$, and by $C_F^j(x_F^j) = cx_F^j$, $j = 1, \ldots, n_F$, with $c < a$. From (2.5.6) and (2.5.7), we can deduce $\tilde{X}_L = \frac{n_L}{n_L+1}X^*$ and $\tilde{X}_F = \frac{n_F}{(n_L+1)(n_F+1)}X^*$. Then, the aggregate supply is $\tilde{X} = \frac{n_L n_F + n_L + n_F}{(n_L+1)(n_F+1)}X^*$, which may be written as $\tilde{X}(n_L, n) = \frac{n + nn_L - n_L^2}{(n_L+1)(n-n_L+1)}X^*$. We see that $\tilde{X} < X^*$. Then, we obtain $p^* < \tilde{p} = \frac{a+c[n_L(n_F+1)+n_F]}{(n_L+1)(n_F+1)}$. We can observe that $\tilde{X}(0, n) = \tilde{X}(n, n) = \frac{n}{n+1}X^*$, which corresponds to the Cournot–Nash equilibria, and $\tilde{X}(2, n) = \frac{4n-9}{4(n-2)} < \tilde{X}(2, n) = \frac{3n-4}{3(n-1)} > X(1, n) = \frac{2n-1}{2n} > X(0, n)$. Then, for fixed n, the aggregate supply is concave in n_L, i.e., $\frac{\partial^2 X(n_L, n)}{(\partial n_L)^2} = -\frac{2X^*}{(n_L+1)^3(n-n_L+1)} < 0$. Indeed, the Cournot–Nash aggregate supply is given by $\hat{X}(n_L, n_F) = \frac{n_L+n_F}{n_L+n_F+1}X^*$. Then, we have $\hat{X}(n_L, n_F) < \tilde{X}(n_L, n_F)$.

Remark 2.6.2 When the aggregate best response for followers has a zero slope in equilibrium, the leaders rationally expect that each strategic decision they undertake should entail no reactions from the followers (Julien [21]). △

2.6.2 *Welfare and Market Power*

If we are to define welfare when studying the variation in aggregate supply for this framework, we must take into consideration the shares of aggregate supply of leaders and followers. In accordance with Julien [23], let $\vartheta_L \equiv \frac{X_L}{X}$, with $0 \leqslant \vartheta_L \leqslant 1$, and $\vartheta_F \equiv \frac{X_F}{X}$, with $0 \leqslant \vartheta_F \leqslant 1$, and where $\vartheta_L + \vartheta_F = 1$. Therefore, the social surplus may be defined as:

$$\mathbf{S}(X) := \int_0^X p(z)dz - (\sum_{i \in \mathcal{F}_L} C_L^i(s_L^i \vartheta_L X) + \sum_{j \in \mathcal{F}_F} C_F^j(s_F^j \vartheta_F X)), \text{ with } X \leqslant X^*, \tag{2.6.1}$$

where $s_L^i \equiv \frac{x_L^i}{X_L}$ is leader i's market share, and $s_F^j \equiv \frac{x_F^j}{X_F}$ is follower j's market share. Differentiating partially with respect to X and decomposing $p(X)$ leads to:

$$\frac{\partial \mathbf{S}(X)}{\partial X} = \sum_{i \in \mathcal{F}_L} s_L^i \vartheta_L (p(X) - \frac{dC_L^i(x_L^i)}{dX}) + \sum_{j \in \mathcal{F}_F} s_F^j \vartheta_F (p(X) - \frac{dC_F^j(x_F^j)}{dX}) \geqslant 0, \tag{2.6.2}$$

as we have that $\sum_{i\in\mathcal{F}_L} s_L^i\vartheta_L + \sum_{j\in\mathcal{F}_F} s_F^j\vartheta_F = 1$, and for fixed s_L^i, s_F^j, ϑ_L and ϑ_F, with $\frac{\partial \mathbf{S}(X)}{\partial X}_{|X=X^*} = 0$.

The social surplus is hence higher at the SNE than at the CNE, and reaches its maximum value at the CE.[12] Therefore, one essential feature of the SNE bilevel game is that the strategic interactions between leaders and followers may be welfare enhancing.

Remark 2.6.3 Daughety [9] shows that, if the aggregate supply is used as a measure of welfare, welfare may be maximized when there is considerable asymmetry in the market, whereas symmetric (Cournot) equilibria for which $n_L = 0$ and $n_L = n$ minimize welfare. Thus, the concentration index may no longer be appropriate for measuring welfare. △

2.6.3 *Market Power and Payoffs*

We now compare the SNE payoffs with the CNE payoffs. To this end, the optimal conditions (2.3.2) and (2.3.5) may be expressed respectively as:

$$p(X) = (1 + m_L^i)\frac{dC_L^i(x_L^i)}{dx_L^i}, \text{ with } m_L^i = \frac{1}{1 + \frac{1+\nu}{\epsilon}\vartheta_L s_L^i} - 1, i \in \mathcal{F}_L; \tag{2.6.3}$$

$$p(X) = (1 + m_F^j)\frac{dC_F^j(x_F^j)}{dx_F^j}, \text{ with } m_F^j = \frac{1}{1 + \frac{1}{\epsilon}\vartheta_F s_F^j} - 1, j \in \mathcal{F}_F, \tag{2.6.4}$$

where m_L^i and m_F^j are leader i's and follower j's markups, and ϵ is the price elasticity of demand, that is, $\epsilon \equiv \frac{dp(X)}{dX}\frac{p}{X}$.

To analyze the relation between market power and individual payoffs, let us consider:

$$L_L^i = -\frac{1+\nu}{\epsilon}\vartheta_L s_L^i, i \in \mathcal{F}_L; \tag{2.6.5}$$

$$L_F^j = -\frac{1}{\epsilon}\vartheta_F s_F^j, j \in \mathcal{F}_F. \tag{2.6.6}$$

[12] Indeed, $\frac{\partial S_C(X)}{\partial X} = -X\frac{dp(X)}{dX} > 0$, with $S_C(X) := \int_0^X p(z)dz - p(X)X$. In addition, if we let $S_P(X) := p(X)(\vartheta_L \sum_{i=1}^{n_L} s_L^i + \vartheta_F \sum_{j=1}^{n_F} s_F^j)X - \sum_{i=1}^{n_L} C_L^i(s_L^i\vartheta_L X) - \sum_{j=1}^{n_F} C_F^j(s_F^j\vartheta_F X)$, then $\frac{dS_P(X)}{dX} = p(X) + X\frac{dp(X)}{dX} - [\vartheta_L \sum_{i=1}^{n_L} s_L^i \frac{dC_L^i(s_L^i\vartheta_L X)}{dX} + \vartheta_F \sum_{j=1}^{n_F} s_F^j \frac{dC_F^j(s_F^j\vartheta_F X)}{dX}] < 0$ (from Assumption 2.2.2b).

where L_L^i and L_F^j are the Lerner indexes for follower j and for leader i respectively.[13]

Proposition 2.6.4 *If $L_L^i > L_F^j$, then $\tilde{\Pi}_L^i > \tilde{\Pi}_F^j$, $i \in \mathcal{F}_F$, $j \in \mathcal{F}_F$. In addition, if $L_L^i = L_F^j$ for all $i \in \mathcal{F}_F$ and $j \in \mathcal{F}_F$, then, $\tilde{\Pi}_L^i \gtrless \tilde{\Pi}_F^j$ if and only if $v \gtrless 0$, $i \in \mathcal{F}_F$, $j \in \mathcal{F}_F$.* △

Proof Immediate from the definition of the Lerner index and by using (2.6.3) and (2.6.4). □

It is worth pointing out that there are certain differences in leaders' (resp. followers') payoffs caused by asymmetries in costs. As this bilevel game embodies strategic interactions among several leaders and followers, we now explore the possibility of merging.

2.6.4 *Welfare and Mergers*

The strategic effects of merging on welfare depend on the noncooperative strategic behavior which prevails in the SNE. The following example illustrates the welfare effects of merging (see Daughety [9]).

Welfare Effects

Consider the linear bilevel game given by (2.5.1)–(2.5.10), where $C_L^i(x_L^i) = cx_L^i$, $i = 1, \ldots, n_L$, and by $C_F^j(x_F^j) = cx_F^j$, $j = 1, \ldots, n_F$, with $c < a$. Let $\tilde{X}(n_L, n) = \frac{n+nn_L-n_L^2}{(n_L+1)(n-n_L+1)} X^*$. First, a merger means that one firm disappears from the market. Consider the following three cases:

1. The merger of two leaders so that the post merger market has $n_L - 1$ leaders but still $n - n_L$ followers;
2. The merger of two followers, so that there are n_L leaders but $n - n_L - 1$ followers; and
3. The merger of one leader and one follower, so that there are n_L leaders but $n - n_L - 1$ followers.

(continued)

[13] The Lerner index for any decision maker is defined in an SNE as the ratio between the excess of the price over the marginal cost and the price, that is, $L := \frac{p(X) - \frac{dc(x)}{dx}}{p(X)}$.

Therefore, in case 1, calculations yield

$$\tilde{X}(n_L - 1, n - 1) - \tilde{X}(n_L, n) = -\frac{(n-1)n_L(n_L+3) - 2n_L + 1}{n_L(n_L+1)(n-n_L+1)} X^* < 0.$$

In cases 2 and 3, we obtain $\tilde{X}(n_L, n-1) - \tilde{X}(n_L, n) = -\frac{1}{(n_L+1)(n-n_L)(n-n_L+1)} X^* < 0$. Thus, welfare is always reduced. Second, if we now consider that the number of leaders increases, the comparative statics yields:

$$\frac{\partial \tilde{X}(n_L, n)}{\partial n_L} = \frac{n - 2n_L}{(n_L+1)^2(n-n_L+1)^2} X^* \lesseqqgtr 0 \text{ for } n \lesseqqgtr 2n_L,$$

$$\frac{\partial \tilde{X}(n_L, n)}{\partial n} = \frac{1}{(n_L+1)(n-n_L+1)^2} X^* > 0,$$

and

$$\frac{\partial^2 \tilde{X}(n_L, n)}{\partial n_L \partial n} = \frac{n - 2n_L}{(n_L+1)^2(n-n_L+1)^2} X^* \lesseqqgtr \text{ for } 3n_L + 1 \lesseqqgtr n.$$

The last effect captures the effect on welfare of changes in industry structure. If we now consider that two followers merge and behave as a leader firm, there are $n-1$ firms with n_L+1 leaders and $n-n_L-2$. Using algebra leads to $\tilde{X}(n_L+1, n-1) - \tilde{X}(n_L, n) = \frac{n-3(n_L+1)}{(n_L+1)(n_L+2)(n-n_L-1)(n-n_L+1)} X^* > 0$ whenever $n_L < \frac{n}{3} - 1$: so, when there are few leaders, merging can increase aggregate supply. More asymmetry is beneficial; it is socially desirable as it enhances welfare. However when $n_L > \frac{n}{2}$, fewer leaders and more followers could increase welfare.

The difference between the two cases can be explained by the fact that, in the second case, the reduction of the number of followers is associated with an increase in the number of leaders.

Remark 2.6.5 It can be shown that two firms which belong to the same cohort and have the same market power rarely have an incentive to merge, whereas a merger between two firms which belong to two distinct cohorts and have different levels of market power is always profitable as the leader firm incorporates the follower firm regardless of the number of rivals. In the SNE the merger better internalizes the effect of the increase in price on payoffs than in the CNE: the decrease in supply is lower than under Cournot quantity competition. △

2.7 Extension to Multilevel Optimization

Bilevel optimization models have been extended to three-level optimization environments (see Bard and Falk [6], Benson [7], Han et al. [17, 18], among others), and to T-level optimization with one decision maker at each level (Boyer and Moreaux [8], Robson [32]). The three level optimization game has been studied in depth by Alguacil et al. [1] and Han et al. [17]. The existence of a noncooperative equilibrium in the multilevel optimization with several decision makers at each level remains an open problem. Nevertheless, the multiple leader–follower game may be extended to cover a T-stage decision setting in the case of the linear model (Watt [39], Lafay [28], and Julien et al. [24, 25]). The extended game should represent a free entry finite horizon hierarchical game. We will focus on the computation and on certain welfare properties. To this end, and for the sake of simplicity, we consider an extended version of the linear model studied in Sect. 2.5, where $C_L^i(x_L^i) = cx_L^i$, $i = 1, \ldots, n_L$, and by $C_F^j(x_F^j) = cx_F^j$, $j = 1, \ldots, n_F$, with $c < a$.

There are now T levels of decisions indexed by t, $t = 1, 2, \ldots, T$. Each level embodies n_t decision makers, with $\sum_{t=1}^T n_t = n$. The full set of sequential levels represents a hierarchy. The supply of firm i in level t is denoted by x_t^i. The aggregate supply in level t is given by $X_t \equiv \sum_{i=1}^{n_t} x_t^i$. The n_t firms behave as leaders with respect to all firms at levels $\tau > t$, and as followers with respect to all firms at levels $\tau < t$. The price function may be written as $p = p(\sum_t X_t)$. Let $p(X) = a - bX$, $a, b > 0$, where $X \equiv \sum_t X_t$. The costs functions are given by $C_t^i(x_t^i) = cx_t^i$, $i = 1, \ldots, n_t, t = 1, \ldots, T$, with $c < a$. The strategy sets are given by $\mathcal{S}_t^i = [0, \frac{a}{b} - c]$, $i = 1, \ldots, n_t, t = 1, \ldots, T$.

Bearing in mind this framework, if firms compete as price-takers, the CE is still given by $X^* = \frac{a-c}{b}$, $p^* = c$, and $(\Pi_t^i)^* = 0$, $i = 1, \ldots, n_t$, $t = 1, \ldots, T$. The CNE is given by $\hat{x}_t^i = \frac{1}{\sum_{t=1}^T n_t + 1} X^*$, $\hat{X} = \frac{\sum_{t=1}^T n_t}{\sum_{t=1}^T n_t + 1} X^*$, $\hat{p} = \frac{a + c\sum_{t=1}^T n_t}{\sum_{t=1}^T n_t + 1}$, and $\hat{\Pi}_t^i = \frac{(a-c)^2}{b} \frac{1}{(\sum_{t=1}^T n_t + 1)^2}$, $i = 1, \ldots, n_t, t = 1, \ldots, T$.

At level t, firm i's profit is given by:

$$\Pi_t^i(x_t^i, X_t^{-i}, \sum_{\tau,\tau\neq t}^{T} X_\tau) = [a - b(x_t^i + X_t^{-i} + \sum_{\tau,\tau\neq t}^{T} X_\tau)]x_F^j - cx_F^j. \qquad (2.7.1)$$

Therefore, the problem of firm i at level t may be written as follows:

$$\max_{\{x_t^i\}} \Pi_t^i(x_t^i, X_t^{-i}, \sum_{\tau=1}^{t-1} X_{t-\tau}, \sum_{\tau=1}^{T-t} X_{t+\tau}) := [a - c - b(x_t^i + X_t^{-i} + \sum_{\tau=1}^{t-1} X_{t-\tau} + \sum_{\tau=1}^{T-t} X_{t+\tau})]x_t^i, \qquad (2.7.2)$$

where $X_t^{-i} \equiv X_t - x_t^i$, and $\sum_{\tau-1}^{t-1} X_{t-\tau}$ and $\sum_{\tau-1}^{T-t} X_{t+\tau}$ denote respectively the aggregate supply of all leaders at level $t-\tau$ for $\tau \in \{1, \ldots, t-1\}$ and the aggregate supply of all followers at level $t+\tau$ for $\tau \in \{1, \ldots, T-t\}$.

The solution to this program yields the optimal decision for firm i at stage t, i.e., $x_t^i = \phi_t^i(X_t^{-i}, \sum_{\tau,\tau\neq t}^{T} X_\tau)$. By solving recursively from the last level T to the first level 1, it is possible to deduce the equilibrium strategy for any firm at any stage (see Watt [39]). Indeed, the SNE strategy of firm i at stage t may be written as follows:[14]

$$\tilde{x}_t^i = \left(\prod_{\tau=1}^{t} \frac{1}{n_\tau + 1}\right) X^*, t = 1, \ldots, T. \tag{2.7.3}$$

Therefore, the aggregate supply is given by:

$$\tilde{X} = \left(\sum_{t=1}^{T} n_t \prod\nolimits_{\tau=1}^{t} \frac{1}{n_\tau + 1}\right) X^*. \tag{2.7.4}$$

Then, we deduce the market price:

$$\tilde{p} = c + (a-c)\left(\prod_{\tau=1}^{t} \frac{1}{n_\tau + 1}\right). \tag{2.7.5}$$

Then, the payoffs are given by:

$$\tilde{\Pi}_t^i = \frac{(a-c)^2}{b}\left(\prod_{\tau=1}^{t} \frac{1}{(n_\tau + 1)^2} \prod_{\tau=t+1}^{T} \frac{1}{n_\tau + 1}\right), t = 1, \ldots, T-1; \tag{2.7.6}$$

$$\tilde{\Pi}_T^i = \frac{(a-c)^2}{b}\left(\prod_{\tau=1}^{T} \frac{1}{(n_\tau + 1)^2}\right). \tag{2.7.7}$$

It is worth noting that the specification $T = 2, n_t = 1, t = 1, 2$, corresponds to the standard bilevel duopoly game. The specification $T = 2, n_1 = n_L$ and $n_2 = n_F$, corresponds to the linear bilevel game from Sect. 2.5.

Proposition 2.7.1 *Consider a market with linear demand and identical constant marginal costs, then the T-level Stackelberg game coincides with a multilevel Cournot game in which firms compete oligopolistically on the residual demands.*

△

[14]For computations of the equilibrium values, see notably Watt [39], Lafay [28], Julien and Musy [24], and Julien et al. [24, 25].

Proof See Julien et al. [25] who show that the assumptions of linear demand and identical (strictly positive) constant marginal costs are necessary and sufficient conditions for Proposition 2.7.1 to hold. □

Therefore each firm within a given stage behaves as if there were no subsequent stages, i.e., it is as if the direct followers for firm i in stage t do not matter. This generalizes the t-stage monopoly property of Boyer and Moreaux [8].

To explore the welfare properties of the linear T-level model, let us define ω, the index of social welfare $\tilde{X}$, as:

$$\omega = \prod_{\tau=1}^{T} \kappa_t n_t = 1 - \prod_{\tau=1}^{T} \frac{1}{n_\tau + 1} = 1 + \kappa_{1,T}. \tag{2.7.8}$$

Then, we are able to state the following two propositions (see Julien et al. [24]).

Proposition 2.7.2 *When the number of firms becomes arbitrarily large, either by arbitrarily increasing the number of firms at each stage by keeping the number of stages T constant, i.e., $\forall t$, $n_t \to \infty$, given $T < \infty$, or by increasing the number of stages without limit, i.e., $T \to \infty$, the T-level SNE aggregate supply converges to the CE aggregate supply.* △

Proof Immediate from the two limits given by $\lim_{T\to\infty}(\prod_{\tau=1}^{T} \kappa_t n_t) = 1$ and $\lim_{n_t\to\infty} (\prod_{\tau=1}^{T} \kappa_t n_t) = 1$. □

Proposition 2.7.3 *In the T-level linear economy, social welfare can be maximized by enlarging the hierarchy or by changing the size of existing stages through the reallocation of firms from the most populated stage until the size of all stages is equalized.* △

Proof See Julien et al. [24]. The relocation reflects the merger analysis provided in Daughety [9] (see preceding subsection). When the number of levels is fixed, the relocation is welfare improving until there is the same number of firms at each level. □

Remark 2.7.4 A sequential market structure with one firm per stage Pareto dominates any other market structure, including the CNE (see Watt [39]). △

Remark 2.7.5 It can be verified that the firms' surplus in the SNE may be inferior to the firms' surplus in the CNE when $T \geqslant 3$, so the firm which chooses to be at the upper level may be better off if the other firms are supplying simultaneously. △

The results contained in Propositions 2.7.2 and 2.7.3 may be used to analyze how an increase in the number of decision makers affects welfare. Indeed, when a new firm enters at level t it causes a decrease in market price as we have

$$\tilde{p}(\sum_{\tau=1}^{t} \tilde{X}_\tau + \tilde{x}_t^{n_t+1}) - \tilde{p}(\sum_{\tau=1}^{t} \tilde{X}_\tau) = \frac{a-c}{n_t \prod_{\tau=1}^{T}(n_\tau + 1)} < 0, \tag{2.7.9}$$

where $\tilde{x}_t^{n_t+1}$, with $\tilde{x}_t^{n_t+1} > 0$, represents the supply of the additional firm. In addition, the maximization of welfare implies the most asymmetric distribution of market power. Nevertheless, these results are valid in a linear economy with identical costs. Indeed, if costs are different, entry is affected by some relocations or extensions. Lafay [28] uses a T-level game in which firms enter at different times or have different commitment abilities. Here firms bear different constant marginal costs. *The linear T-level optimization game confirms the positive effect of an increase in the number of decision makers on welfare*. However, the salient feature is that firms must now forecast future entries in the market. Indeed, asymmetric costs could make entry inefficient. If the firm reasons backwards, and the price is lower when there is further entry, the firm enters the market provided its costs do not exceed the resulting market price.[15]

2.8 Conclusion

We have proposed a short synthesis of the application of bilevel optimization to some simple economic problems related to oligopolistic competition. Using standard assumptions in economics relative to the differentiability of objective functions, we have presented some elements to characterize the Stackelberg–Nash equilibrium of the noncooperative two-stage game, where the game itself consists of two Cournot simultaneous move games embedded in a Stackelberg sequential competition game.

The tractability of the model, especially when assuming linearity of costs and demand, makes it possible to derive certain welfare implications from this bi-level optimization structure, and to compare it with standard alternatives in terms of market structures. Indeed, the T-level optimization game represents a challenge to the modeling of strategic interactions.

Acknowledgements The authors acknowledge an anonymous referee for her/his helpful comments, remarks and suggestions on an earlier version. Any remaining deficiencies are ours.

References

1. N. Alguacil, A. Delgadillo, J.M. Arroyo, A trilevel programming approach for electric grid defense planning. Comput. Oper. Res. **41**, 282–290 (2014)
2. D. Aussel, P. Bendotti, M. Pištěk, Nash equilibrium in a pay-as-bid electricity market: part 1—existence and characterization. Optimization **66**(6), 1013–1025 (2017)

[15]Lafay [28] shows that when constant marginal costs differ among firms, the price contribution by an additional entrant may not be negative since the strategies of all firms are modified when a firm no longer enters the market.

3. D. Aussel, P. Bendotti, M. Pištěk, Nash equilibrium in a pay-as-bid electricity market part 2-best response of a producer. Optimization **66**(6), 1027–1053 (2017)
4. D. Aussel, G. Bouza, S. Dempe, S. Lepaul, Multi-leader disjoint-follower game: formulation as a bilevel optimization problem (2018). Preprint 2018-10, TU Bergakademie Freiberg, Fakultät für Mathematik und Informatik
5. J.F. Bard, Convex two-level optimization. Math. Program. **40**(1–3), 15–27 (1988)
6. J.F. Bard, J.E. Falk, An explicit solution to the multi-level programming problem. Comput. Oper. Res. **9**(1), 77–100 (1982)
7. H.P. Benson, On the structure and properties of a linear multilevel programming problem. J. Optim. Theory Appl. **60**(3), 353–373 (1989)
8. M. Boyer, M. Moreaux, Perfect competition as the limit of a hierarchical market game. Econ. Lett. **22**(2–3), 115–118 (1986)
9. A. F. Daughety, *Beneficial concentration*, The American Economic Review **80** (1990), no. 5, 1231–1237.
10. D. De Wolf, Y. Smeers, A stochastic version of a Stackelberg–Nash–Cournot equilibrium model. Manag. Sci. **43**(2), 190–197 (1997)
11. V. DeMiguel, H. Xu, A stochastic multiple-leader Stackelberg model: analysis, computation, and application. Oper. Res. **57**(5), 1220–1235 (2009)
12. S. Dempe, *Foundations of Bilevel Programming* (Springer, Berlin, 2002)
13. S. Dempe, Bilevel optimization: theory, algorithms and applications, in *Bilevel Optimization: Advances and Next Challenges*, ed. by S. Dempe, A.B. Zemkoho (Springer, Berlin, 2019)
14. S. Dempe, V. Kalashnikov, *Optimization with Multivalued Mappings: Theory, Applications and Algorithms*, vol. 2 (Springer, Berlin, 2006)
15. A. Ehrenmann, Manifolds of multi-leader Cournot equilibria. Oper. Res. Lett. **32**(2), 121–125 (2004)
16. J.H. Hamilton, S.M. Slutsky, Endogenous timing in duopoly games: Stackelberg or Cournot equilibria. Games Econ. Behav. **2**(1), 29–46 (1990)
17. J. Han, J. Lu, Y. Hu, G. Zhang, Tri-level decision-making with multiple followers: model, algorithm and case study. Inform. Sci. **311**, 182–204 (2015)
18. J. Han, G. Zhang, J. Lu, Y. Hu, S. Ma, Model and algorithm for multi-follower tri-level hierarchical decision-making, in *International Conference on Neural Information Processing* (Springer, Berlin, 2014), pp. 398–406
19. X. Hu, D. Ralph, Using EPECs to model bilevel games in restructured electricity markets with locational prices. Oper. Res. **55**(5), 809–827 (2007)
20. W. Jia, S. Xiang, J. He, Y. Yang, *Existence and stability of weakly Pareto–Nash equilibrium for generalized multiobjective multi-leader–follower games*. J. Global Optim. **61**(2), 397–405 (2015)
21. L.A. Julien, A note on Stackelberg competition. J. Econ. **103**(2), 171–187 (2011)
22. L.A. Julien, On noncooperative oligopoly equilibrium in the multiple leader–follower game. Eur. J. Oper. Res. **256**(2), 650–662 (2017)
23. L.A. Julien, Stackelberg games, in *Handbook of Game Theory and Industrial Organization*, ed. by L. Corchon, M. Marini, vol. 1 (Edward Elgar Publishing, Cheltenham, 2018), pp. 261–311
24. L. Julien, O. Musy, A. Saïdi, Do followers really matter in Stackelberg competition? Lect. Econ. **75**(2), 11–27 (2011)
25. L.A. Julien, O. Musy, A.W. Saïdi, On hierarchical competition in oligopoly. J. Econ. **107**(3), 217–237 (2012)
26. C.D. Kolstad, L. Mathiesen, Necessary and sufficient conditions for uniqueness of a Cournot equilibrium. Rev. Econ. Stud. **54**(4), 681–690 (1987)
27. A.A. Kulkarni, U.V. Shanbhag UV, An existence result for hierarchical stackelberg v/s stackelberg games. IEEE Trans. Automat. Contr. **60**(12), 3379–3384 (2015)
28. T. Lafay, A linear generalization of Stackelberg's model. Theory Decis. **69** (2010), no. 2, 317–326.
29. G. Leitmann, On generalized Stackelberg strategies. J. Optim. Theory Appl. **26**(4), 637–643 (1978)

30. S. Leyffer, T. Munson, Solving multi-leader-common-follower games. Optimization Methods & Software **25**(4), 601–623 (2000)
31. J.-S. Pang M. Fukushima, Quasi-variational inequalities, generalized Nash equilibria, and multi-leader–follower games. Comput. Manag. Sci. **2**(1), 21–56 (2005)
32. A.J. Robson, Stackelberg and Marshall. Am. Econ. Rev. 69–82 (1990)
33. H.D. Sherali, A multiple leader Stackelberg model and analysis. Oper. Res. **32**(2), 390–404 (1984)
34. H.D. Sherali, A.L. Soyster, F.H. Murphy, Stackelberg–Nash–Cournot equilibria: characterizations and computations. Oper. Res. **31**(2), 253–276 (1983)
35. C. Shi, G. Zhang, J. Lu, On the definition of linear bilevel programming solution. Appl. Math. Comput. **160**(1), 169–176 (2005)
36. A. Sinha, P. Malo, K. Deb, A review on bilevel optimization: from classical to evolutionary approaches and applications. IEEE Trans. Evol. Comput. **22**(2), 276–295 (2017)
37. R.L. Tobin, Uniqueness results and algorithm for Stackelberg–Cournot–Nash equilibria. Ann. Oper. Res. **34**(1), 21–36 (1992)
38. H. Von Stackelberg, *Market Structure and Equilibrium* (Springer, Berlin, 2011). Translation from the German language edition: *Marktform und Gleichgewicht* (New York, 1934)
39. R. Watt, A generalized oligopoly model. Metroeconomica **53**(1), 46–55 (2002)
40. J. Yu, H.L. Wang, An existence theorem for equilibrium points for multi-leader–follower games. Nonlinear Anal. Theory Methods Appl. **69**(5–6), 1775–1777 (2008)

Chapter 3
A Short State of the Art on Multi-Leader-Follower Games

Didier Aussel and Anton Svensson

Abstract Multi-Leader-Follower games are complex optimization problems that mix a bilevel structure with one or more Nash games. Such kinds of models have been already described in the seminal book of H. von Stackelberg ((1934)Marktform und Gleichgewicht. Springer, Berlin); von Stackelberg et al. ((2011) Market structure and equilibrium. Springer, Heidelberg) and are known to perfectly fit to a lot of applications involving non cooperative situations with hierarchical interactions. Nevertheless it is only recently that theoretical and numerical developments for Multi-Leader-Follower problems have been made. This chapter aims to propose a state of the art of this field of research at the frontier between optimization and economics.

Keywords Multi-Leader-Follower games · Generalized Nash games · Existence · Variational reformulation · Algorithms

3.1 Introduction

A Multi-Leader-Follower Game is a model that describes quite complex interactions between non cooperative players/decision makers and which includes a hierarchy in their decision process. The set of players is divided into two categories, the leaders, or upper level players, and the followers, or the lower level players.

The splitting of the set of players into these two categories is based on a dependence criterion. The followers' strategy is somehow passive in the sense that

---

D. Aussel (✉)
Université de Perpignan, Perpignan, France
e-mail: aussel@univ-perp.fr

A. Svensson
Université de Perpignan, Perpignan, France

Universidad de Chile, Santiago, Chile
e-mail: asvensson@dim.uchile.cl

S. Dempe, A. Zemkoho (eds.), *Bilevel Optimization*, Springer Optimization and Its Applications 161, https://doi.org/10.1007/978-3-030-52119-6_3

they react to what the leaders decide. Contrarily, the leaders' strategy is active, meaning that the leaders take into account the "future" reaction of the followers. More precisely, a Multi-Leader-Follower problem corresponds to the problem of the leaders and consists in choosing optimal strategies/equilibrium strategies (of the leaders) considering a conjectured reaction of the followers.

Remark 3.1.1 In real life applications, the evaluation of these optimal strategies/ equilibrium strategies through the resolution of the associated Multi-Leader-Follower game corresponds actually to the first step of a two steps process. Indeed these optimal strategies/equilibrium strategies are determined based on a conjectured reaction of the followers but as soon as these optimal strategies/equilibrium strategies are revealed by the leaders, then, in a second step, the followers compute the optimal response/Nash equilibrium of their parameterized optimization problem/Nash game and this response can differ from the leader's conjecture in case of multiplicity of the solutions of the followers' problem. △

The two-level structure of a Multi-Leader-Follower game has led some authors to refer the model as a *bilevel game*. In particular, it is an extension of the bilevel programming problem, where the interaction between only one leader and one follower are considered. But in general, in a Multi-Leader-Follower game each level have more than one player and thus under the non cooperative premise we will assume that within the upper level, as well as within the lower level, the players' behavior is modeled by Nash games.

The special case of leaders without followers (or equivalently absence of leaders) trivially reduces to (Generalized Nash game). Nevertheless in this chapter we will only concentrate our attention to cases including a bilevel structure. Now the concept of Multi-Leader-Follower game can be separated into three classes/structures of model:

(SLMFG) *Single-Leader-Multi-Follower game*,
(MLSFG) *Multi-Leader-Single-Follower game*, and
(MLMFG) *Multi-Leader-Multi-Follower game*.

Applications of these models can be found in industrial eco-parks and electricity markets, as we shall shortly describe but also in other areas such as transmission and generation expansion planning, water resource optimal allocation, demand side management, among others.

Let us note that in these models some subtle ambiguities, which are particularly serious in the case of MLSFG and MLMFG, can easily occur in the formulation of the interactions between the players (see Sect. 3.2).

Nevertheless, even if the Multi-Leader-Follower models have been considered a long time ago (see Historical comment below), theory and algorithms have been developed only recently, say during the last decades. This is partially due to the fact that the reformulation of these difficult problems required modern tools of non-smooth and variational analysis like quasi-variational inequalities and/or coderivative calculus. Our aim in this chapter is therefore to propose a short state of art of the domain.

In Sect. 3.2, we introduce the notations as well as two motivating examples, one in Industrial eco-parks and another for the modelling of deregulated electricity markets. Section 3.3 is devoted to recent developments for Single-Leader-Multi-Follower problems while Sect. 3.4 deals with Multi-Leader-Multi-Follower models. In both cases, existence results and reformulations are discussed as well as specific algorithms that have been developed to solve those difficult problems. Finally, in Sect. 3.5, we conclude and suggest possible extensions.

Historical Comments

It is not so known but the Multi-Leader-Follower concepts were actually already considered in Chapter 2 of Stackelberg's famous book [1] (English translation [2]). Indeed in 1934 and considering the case of a family of players interacting in a non cooperative way, H. von Stackelberg discussed essentially two possible relative attitudes of two of the players (let us call them A and B) with regard to the game:

(i) *"Assuming that player A views the behaviour of player B as being independent of his (A's) behaviour, in this case A regards B's supply as a given variable and he orientates himself to it. Thus the behaviour of A is dependent on the behaviour of B. If B therefore trades based on assumptions that match reality, he thus sees the behaviour of his rival A as being dependent on his (B's) behaviour."*

(ii) *"Assuming that A views the behaviour of B as being dependent on his (A's) behaviour, if that matches reality, then it means that B is orientating himself to A's behaviour."*

It is easy to see that some "incompatible" situations may occur. It will be the case when there is at least, in the family of players, a couple (A,B) of them such that A is following an (i) policy but B does not see the behaviour of his rival A as being dependent on his (B's) behaviour, or vice versa.

Now in the case where there is at least one player, say B, that is following at the same time policy (i) with regard to a player A and a positioning (ii) with regard to another player C then the resulting interaction will lead to a (at least) trilevel model. But this case, also considered in Stackelberg's book, is beyond the scope of this chapter.

Then, using the two possible interrelations (i) and (ii), Stackelberg describes, in a systematic way, complex situations from Generalized Nash games to Multi-Leader-Follower games. If the positioning of the players are "compatible" (not leading to incompatible situations) and not leading to a (at least) trilevel model then the family of players can be split into two categories: the ones considering that a subgroup of players' behaviour depend on their behaviour (the so-called *leaders*) and a subfamily of players that position themselves as having their decision/behaviour depending on the decisions of a subgroup of other players' behaviour (the so-called *followers*).

3.2 Notations and Examples of Applications

As emphasized in the previous section, Multi-Leader-Follower games are bilevel models mixing the Nash-equilibrium structure of usual non cooperative game theory within each level, and a hierarchical feature between the two levels.

So let us first recall that a generalized Nash game corresponds to a non cooperative interaction between the players, each of them solving an optimization problem in which the objective function and/or the constraint set depend upon the variables of the other players. The generalized Nash equilibrium problem (GNEP, for short) is to find a strategic combination, called a Nash equilibrium of the game, with the property that no player can gain by deviating unilaterally from it (Nash, J., 1951). The Nash equilibrium can be characterized by being, for each player, a best answer, given the strategies chosen by the others. More precisely, let us assume that we have p players and each player $\nu \in \{1, 2, \ldots, p\}$ has a strategy $x_\nu \in K_\nu \subset \mathbb{R}^{n_\nu}$. We use the classical notation $x = (x_\nu, x_{-\nu})$ in $\mathbb{R}^n$ $(n = n_1 + \cdots, n_p)$ where $x_{-\nu}$ denotes the vector of strategies of all the players but of ν, that is, $x_{-\nu} := (x_1, \ldots, x_{\nu-1}, x_{\nu+1}, \ldots, x_p)$. Given the strategy $x_{-\nu}$, the player ν chooses a strategy x_ν such that it solves the following optimization problem

$$(P(x_{-\nu})) \qquad \min_{x_\nu} \theta_\nu(x_\nu, x_{-\nu}), \quad \text{subject to } x_\nu \in K_\nu(x_{-\nu}),$$

where $\theta_\nu : \mathbb{R}^n \to \mathbb{R}$ is a the ν-player's objective function and $\theta_\nu(x_\nu, x_{-\nu})$ denotes the loss that player ν suffers when the rival players have chosen the strategy $x_{-\nu}$. A *solution* $\bar{x}$ of the (GNEP) is a vector of $\prod_\nu K_\nu(\bar{x}_{-\nu})$ such that, for any ν, $\bar{x}_\nu$ solves $(P(\bar{x}_{-\nu}))$. There is a huge literature on (GNEP) and many problems can be modelled as a Nash equilibrium problem, see e.g. [3] (Fig. 3.1).

Now let us go back to bilevel structures, that is, when hierarchical interrelations link some of the players. If the game involves more than two players then the bilevel game has a more complex structure where the set of the players is split into two subfamilies, one (the followers) interacting between them in a non cooperative way thus defining a Nash game which is parameterized by the decisions of the

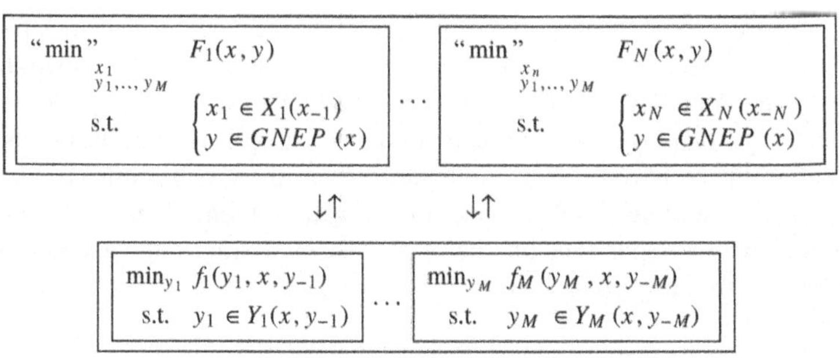

Fig. 3.1 MLMFG

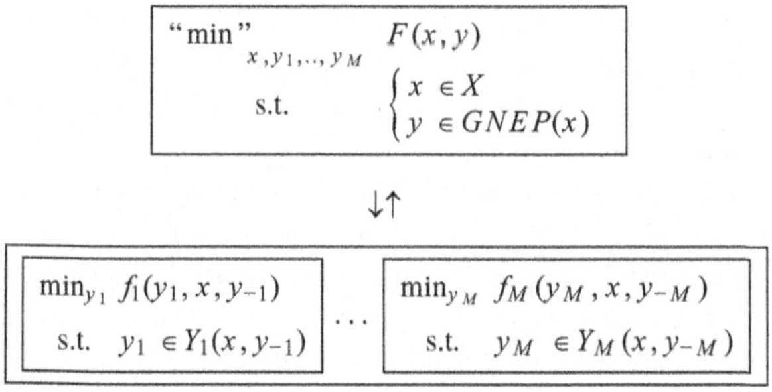

Fig. 3.2 SLFMG

complementary subset of players (the leaders) who also play a Nash game between them. Such structure is called a Multi-Leader-Multi-Follower game and is one of the most complex systems of interdependent optimization problems in the literature. For a game with N leaders and M followers, with their respective variables $x_1, \ldots x_N$ and $y_1, \ldots y_M$, the MLMFG can be expressed as:

But let us first concentrate on the important case where there is only one leader, that is, Single-Leader-Multi-Follower games SLMFG. This can be represented by the following diagram (Fig. 3.2):
where for any $j = 1, \ldots, M$, the set-valued map $Y_j(x, y_{-j})$ expresses the constraints, parameterized by x and y_{-j} that the decision variable of player j must satisfy and where $GNEP(x)$ stands for the set of (generalized) Nash equilibria of the non cooperative game between the followers. The notation "min" is used here to enlighten that this is simply a first rough definition that is not free of ambiguities. In particular, if the reaction of the followers is not uniquely determined, the leader cannot anticipate which (GNEP)-reaction will take place and thus the upper level problem becomes ill-posed/ambiguous. This kind of question will be addressed in Sect. 3.3.

As observed in [4], if none of the constraint maps Y_j nor the objectives ϕ_j depend on the decision variable of the other followers then the SLMFG admits an equivalent reformulation as a classical bilevel problem with only one follower. This can be seen by defining a (unique) follower's variable as $y := (y_1, \ldots, y_M)$, the objective $f(x, y) := \sum_{j=1}^{M} f_j(y_j, x)$ and the aggregated constraint map Y by $Y(x) := \{y \mid y_j \in Y_j(x), \forall j\}$. Thus, under this particular structure, several analyses on the single-leader-single-follower case can be directly extended to multiple followers, while in general having multiple followers does bring new difficulties.

One application for which the model SLMFG has proved its efficiency is the optimal design of *industrial eco-parks (IEP)*. This new way to design industrial parks is based on the sharing of fluids (water, vapor, ...) or of energy between companies. The aim of such an organization of an industrial park is twofold: first each participating company want to reduce his own production cost (in this sense "eco" stands for *economy*); second, and this is the novel part, the target is to reduce the ecological impact of the industrial park (and then "eco" stands for *ecological*).

Here the ecological impact of the park is measured by the total amount of wastes and/or of the incoming raw materials. For example and to simplify, let us consider the case of water sharing. Assume that the companies of the park use fresh water for some production processes and that the water is not "consumed" by the process but get out "polluted/contaminated" with a certain contamination rate. Thus in a stand-alone situation (no water sharing) each company would have to pay, at a quite high price, both for the fresh water and for the contaminated water disposal. In an IEP situation, the contaminated water of a company (possibly mixed with fresh water) is used as incoming water flow by another company. The contamination rate of the mixed incoming flow being of course compatible with the maximum tolerance of the industrial process which receives this flow. Note that regeneration units can also be included. The IEP structure (connection tubes, flows, installation of regeneration units) is then decided in order to minimize the benefit of each company and, at the same time the ecological impact.

This problem, already considered in the 60s, has been treated in the literature using the *multi-objective optimization approach*. However, this technique has shown its limits in particular because it requires a selection process among the obtained Pareto points. This selection is almost always based on a prioritization scheme between the companies (through weighted sum, goal programming, etc.) which seriously limits its applicability. Recently in [5], a Multi-Leader-Follower approach has been proposed with success. In such a model, the followers are the companies, interacting in a non cooperative way (GNEP), each of them aiming to reduce their production cost. The unique leader is the designer/manager of the industrial park whose target is to minimize the ecological impact. The designer will also ensure the clearance and confidentiality of the decision process. Thus for example in the case of the design of the water network developed in [5] the variable of the designer is the vector x of flows of clear water coming to each process of each company while the variable of each company j is the vector y_j of shared flows between the processes of the company j and the processes of the concurrent companies. The resulting SLMFG model is as follows

$$\min_{x,y} \sum_i x_i$$
$$s.t. \quad \begin{cases} x_l \geq 0, \quad \forall i \\ \forall j, \quad y_j \text{ solution of:} \quad \min_{z_j} \; cost_j(z_j, x, y_{-j}) \\ \qquad s.t. \quad \begin{cases} \text{water balance equation} \\ \text{contamination bounds} \\ \text{mass balance of contaminants} \\ \text{other technical assumptions} \end{cases} \end{cases}$$

Thanks to the use of this approach, an important reduction of the global water consumption has been obtained while ensuring the reduction of the production cost of all of the participating companies. Other recent developments of this approach can be found in [6–8]. Note that even if historically industrial eco-parks have been focusing on water exchanges, several other things can also be shared between the

$$
\begin{array}{|c|}
\hline
\begin{array}{|ll|}
\hline
\text{“}\min\text{”}_{x_1, y} & F_1(x,y) \\
\text{s.t.} & \begin{cases} x_1 \in X_1(x_{-1}) \\ y \in opt(x) \end{cases} \\
\hline
\end{array}
\;\cdots\;
\begin{array}{|ll|}
\hline
\text{“}\min\text{”}_{x_N, y} & F_N(x,y) \\
\text{s.t.} & \begin{cases} x_N \in X_N(x_{-N}) \\ y \in opt(x) \end{cases} \\
\hline
\end{array} \\
\hline
\end{array}
$$

$$\downarrow\uparrow$$

$$
\begin{array}{|ll|}
\hline
\min_y & f(y,x) \\
\text{s.t.} & y \in Y(x) \\
\hline
\end{array}
$$

Fig. 3.3 MLSFG

companies, see e.g. [9] for an optimal design of the energy and water exchange in an eco-park.

Symmetrically to the SLMFG, whenever the set of followers is reduced to only one player, then the “bilevel model” leads to the so-called Multi-Leader-Single-Follower game (Fig. 3.3):

where $opt(x)$ denotes the set of global optima of the (unique) follower's problem, which depends on the decision variable of the leaders.

Those difficult models cover a very large class of applications in different real-life fields and in particular in the management of energy. For example the MLSFG provides a perfect model for the description of so-called *day-ahead electricity markets*. In such electricity markets, that are implemented in many countries around the world, 1 day before delivery and in the morning, suppliers and retailers make respectively some sale offers and purchase offers. Then market bidding is closed and then the regulator of the market, called the Independent System Operator (ISO), computes an optimal dispatch by maximizing the total welfare of the market or equivalently minimizing the total cost of production if the total demand is assumed to be fixed (assumption of no elasticity on the market). He also ensures the clearance of the decision process. This computation is usually done in the middle of the day and then, during the afternoon and based on the acceptance/rejection of their sale (respectively purchase) offers, the suppliers (respectively retailers) decide about the production plan that they will use the day latter (day of delivery), this latter work being called *unit commitment*.

Of course each of the suppliers/retailers aims to maximize his benefit and they interact in a non cooperative way. A Nash model is thus well adapted. But all of the maximization problems of this Nash game are actually depending of the decision vector of the ISO, itself being determined by the optimization problem of the ISO which maximizes the total welfare while ensuring the demand/offer balance. This dependence is thus perfectly handled by a Multi-Leader-Single-Follower structure in which the leaders are the suppliers and retailers whose decision variables x_i are market offers (usually energy/price blocks or affine bid curves) while the unique and common follower is the ISO (see [10–18] and references therein). The regulator/follower variable is the vector y of decisions (acceptances/rejections) of the bids of the producers. As a by product of the resolution of the follower problem, the Lagrange multiplier associated to the balance constraint will be the unit marginal price of electricity on the market. The corresponding MLSFG, in a simplified form, is thus as follows:

$$\text{For any } i, \quad \min_{x_i, y}\ profit(x_i, y, x_{-i})$$
$$s.t. \begin{cases} x_i \text{ admissible bid} \\ y \text{ solution of: } \min_z\ Total\_welfare(z, x) \\ \qquad s.t. \begin{cases} \forall k, z_k \text{ decision concerning} \\ \quad \text{bids of producer } k \\ \text{demand/offer balance} \end{cases} \end{cases}$$

It can be clearly noticed here that if the regulator/follower's problem admits possibly more than one solution for a given leader strategy x, then the overall MLSFG problem is ill-posed, carrying some ambiguity; see beginning of Sect. 3.4. In electricity market modelling, the uniqueness of the solution of the regulator/follower's problem is guaranteed by some strict convexity of the "total_welfare" function with regard to variable z thanks to specific assumptions on the bid structure (strictly convex quadratic bid curves—see e.g. [10, 13, 17, 18]) or some *equity property* on the decision process (see [11, 12]).

3.3 Single-Leader-Multi-Follower Games

In this section we consider the case where there is a single leader and multiple followers, which we refer to as a SLMFG and we use the notations of the corresponding diagram of Sect. 3.2.

If, for any decision x of the leader, there exists (implicitly or explicitly) a unique equilibrium $y(x) = (y_1(x), \dots, y_M(x))$ then, the SLMFG can be treated as a classical mathematical programming problem

$$\min_x F(x, y(x)), \quad \text{with } x \in X$$

where of course some good properties (semi-continuity, differentiability, convexity...) of the response function $y(x)$ must be satisfied for the reformulation to be useful. But in general case the formulation of SLMFG carries some ambiguities. This ambiguity coming from the possible non-uniqueness of the lower level equilibrium problem, which is already present in the case of one leader and one follower, is in our setting of several followers an even more inevitable situation. Indeed, since the lower level is an equilibrium problem (GNEP), the uniqueness of an equilibrium can rarely be ensured, and it cannot be avoided simply by assuming strict convexity, see for instance the examples in [19]. Despite this argument for general problems, there are some cases where the lower level problem might have unique responses as in [20] and others.

The most common approach to tackle this ambiguity is the *optimistic* approach, which consists in considering the best equilibrium reaction of the followers with regards to the leader's objective. It can be argued as a kind of cooperation of the followers with the leader. In fact, it is often the case in applications that the leader is assumed to take his decision before the followers, and thus he can after having computed his optimal decision suggest the followers to take certain equilibrium reaction that is convenient to him. Each of the followers will then have no incentive to unilateral deviate from the proposed equilibrium strategy, because of the nature of equilibria.

Definition 3.3.1 We say that $(\bar{x}, \bar{y}) \in \mathbb{R}^n \times \mathbb{R}^m$ is an *optimistic equilibrium* of the SLMFG if it is a solution of the following optimization problem

$$\min_{x,y} \quad F(x, y)$$

$$\begin{cases} x \in X, \\ y \in \mathrm{GNEP}(x) \end{cases} \qquad \triangle$$

An opposite approach is the *pessimistic* one, which consists for the leader in minimising the worst possible equilibrium reaction with regard to the leader objective. Thus, it is based on a minimax problem.

Definition 3.3.2 We say that $(\bar{x}, \bar{y}) \in \mathbb{R}^n \times \mathbb{R}^m$ is an *pessimistic equilibrium* of the SLMFG if it is a solution of the following minimax problem

$$\min_{x} \max_{y} \quad F(x, y)$$

$$\begin{cases} x \in X, \\ y \in \mathrm{GNEP}(x) \end{cases} \qquad \triangle$$

Apart from these two approaches, there are other possibilities based on selections of the lower-level problem (see e.g. [13]) and on set-valued optimisation but we do not discuss them here. Note also that an alternative approach has been developed in [21] in a specific context.

3.3.1 Existence of Equilibria in SLMFG

Here we discuss conditions under which a SLMFG admits at least one equilibrium. We present a positive result for the case of optimistic equilibrium. Nevertheless, for the pessimistic case it has been shown an example of an apparently very well behaved problem (linear and compact) which admits no equilibria (see [19, Example 3.2]).

The mathematical tools that are often used in this analysis when the lower-level reaction is not unique (which is mostly the case in our setting, see previous paragraphs) are part the so-called set-valued maps theory. We recall here some basic definitions and results of this theory concerning semi-continuity properties and refer the reader to the book [22] for further details.

A set-valued map $T : \mathbb{R}^p \rightrightarrows \mathbb{R}^q$ is basically a function that associates to each point $x \in \mathbb{R}^q$ a subset $T(x)$ of $\mathbb{R}^q$. The *graph* of the set-valued map T is defined as $\text{gph}(T) := \{(x, y) \in \mathbb{R}^p \times \mathbb{R}^q : y \in T(x)\}$. Note that $\text{gph}(T)$ is a subset of $\mathbb{R}^n \times \mathbb{R}^m$.The domain of the map T is denoted by $\text{dom}\, T = \{x \in \mathbb{R} : T(x) \neq \emptyset\}$ and the map is said to be closed if $\text{gph}(T)$ is a closed set of $\mathbb{R}^p \times \mathbb{R}^q$.

Definition 3.3.3 The set-valued map $T : \mathbb{R}^p \rightrightarrows \mathbb{R}^q$ is said to be *Lower Semi-Continuous* (LSC, for short) at $\bar{x} \in \mathbb{R}^n$ if for each open set V in $\mathbb{R}^m$ satisfying $T(\bar{x}) \cap V \neq \emptyset$, there exists an open neighbourhood U of $\bar{x}$ such that

$$T(x) \cap V \neq \emptyset, \quad \forall x \in U.$$

We say that T is Lower Semi-Continuous, if it is so at every $\bar{x} \in X$. △

We present now a slight refinement of [19, Theorem 3.1] (see also [23, Corollary 4.4] for analysis in the case of strategy sets that are subsets of reflexive Banach spaces). It is based on continuity properties of both functions and set-valued maps defining the SLMFG. In particular, it assumes the lower semi-continuity of the set-valued maps that defines the feasible set of the followers, that is, for $j = 1, \ldots, M$ the set-valued map $Y_j : \mathbb{R}^n \times \mathbb{R}^{m-j} \rightrightarrows \mathbb{R}^{m_j}$.

Theorem 3.3.4 *Assume for the SLMFG that*

(1) F is lower semi-continuous and X is closed,
(2) for each $j = 1, \ldots, M$, f_j is continuous,
(3) for each $j = 1, \ldots, M$, Y_j is lower semi-continuous relative to its non-empty domain and has closed graph, and
(4) either F is coercive or, X is compact and at least for one j, the images of Y_j are uniformly bounded.

If the graph of the lower level GNEP *is non-empty, then the SLMFG admits an optimistic equilibrium.* △

Proof As in the proof of [19, Theorem 3.1], assumptions (2) and (3) ensure the closedness of the constraints of the leader's problem in both variables x and y. Thus, the classical Weierstrass theorem can be applied to prove the existence of a minimum of the leaders'optimization problem, which constitutes an optimistic equilibrium of the SLMFG. □

Usually the constraints of the followers are described as level sets of certain functions: $Y_j(x, y_{-j}) := \{y_j \in \mathbb{R}^{m_j} : g_j(x, y) \leq 0\}$, with $g_j : \mathbb{R}^n \times \mathbb{R}^m \to \mathbb{R}^{p_j}$. We recall what is the MFCQ for the parametric optimization problems of the followers (see, e.g. the appendix in [24]).

Definition 3.3.5 The MFCQ for the followers' problems is satisfied at $(\bar{x}, \bar{y})$ if for each j the family of gradients there exist $a_j \in \mathbb{R}^{m_j}$ such that

$$\nabla_{y_j} g_{jk}(\bar{x}, \bar{y}) a_j < 0 \quad \vee \quad g_{jk}(\bar{x}, \bar{y}) < 0, \quad \forall k = 1, \ldots, p_j. \qquad \triangle$$

We provide conditions on these data functions that ensure the existence of optimistic equilibrium of the SLMFG. As a particular case we recover Theorem 5.2 in [25]. The forthcoming corollary is just a consequence of our previous result, since the conditions on the data functions imply the continuity properties of the constraint set-valued maps (see [25, Theorem 4.3]).

Corollary 3.3.6 *Let us assume that*

(1) F is lower semi-continuous and X is closed,
(2) for each $j = 1, \ldots, M$, f_j is continuous,
(3) for each $j = 1, \ldots, M$, Dom Y_j is non-empty, g_j is continuous on $\mathbb{R}^n \times \mathbb{R}^m$ and satisfy MFCQ at each feasible point, and
(4) either F is coercive or, X is compact and at least for one j, the images of Y_j are uniformly bounded.

If the graph of the lower level GNEP *is nonempty, then the SLMFG admits an optimistic equilibrium.* $\triangle$

Remark 3.3.7 Condition (4) in Theorem 3.3.4 and Corollary 3.3.6 is assumed to obtain the compactness of the graph of the set-valued map GNEP which assigns to a leader strategy x the set of solutions of the lower-level GNEP(x). $\triangle$

3.3.2 Reformulations

Reformulating a SLMFG is a way of considering it within a framework where a well-developed theory exists for either finding an equilibrium or better understanding the properties of the problem.

We will restrict our discussion here to reformulations of the optimistic approach of the SLMFG, though for the pessimistic approach corresponding reformulations can also be considered.

Two reformulations of the SLMFG are the most classical and can be considered as particular cases of Mathematical Programs with Equilibrium Constraints (MPECs) in certain references. In both cases the reformulation is based on the replacement of the lower-level (generalized) Nash equilibrium problem by a related problem.

A first possibility is to replace the lower-level problem by the (quasi)Variational Inequality problem (VI) associated to the gradients of the objectives (assuming the objective functions are differentiable), or the normal operator (see [26–28] for quasiconvex objective functions), while keeping the constraints. The resulting problem is the so-called Optimization Problem with Variational Inequality Constraints (OPVIC).

3.3.2.1 OPQVIC Reformulation

An OPQVIC (see for instance [23, 29, 30]) is a problem of the form

$$\min_{x,y} \quad F(x, y)$$

$$\begin{cases} x \in X, \\ y \in \mathrm{QVI}(T(x, \cdot), K(x, \cdot)) \end{cases}$$

where $\mathrm{QVI}(T(x, \cdot), K(x, \cdot))$ stands for the solutions set of the following parametric Quasi Variational Inequality problem: find $y \in K(x, y)$ such that

$$\langle T(x, y), y - z\rangle \geq 0, \quad \forall z \in K(x, y). \tag{3.3.1}$$

The OPQVIC reformulation of an optimistic SLMFG consists in considering a parametric QVI defined by $T(x, y) := (\nabla_{y_j} f_j(x, y))_{j=1}^{M}$ and $K(x, y) := \prod_{j=1}^{M} Y_j(x, y_{-j})$.

In the case where the lower-level is a parametric (non-generalized) Nash equilibrium problem, the resulting reformulation reduces to an OPVIC (Variational Inequality Constraints). This specific problem has received much more attention since it is a more tractable case. See for instance [31, 32].

It is easy to see that the OPQVIC reformulation is equivalent to the (optimistic) SLMFG whenever the problems of the followers satisfy the following parametric convexity assumption:

for each follower j the objective function $f_j(x, \cdot, y_{-j})$ is pseudoconvex with respect to y_j and the constraint sets $Y_j(x, y_{-j})$ are convex.

Thus, under these convexity assumptions the existence of optimistic equilibria for the SLMFG could be deduced also from [33]. Note that Theorem 3.3.4 does not require any such convexity assumption.

In some cases it is possible to write the OPVIC as a nonlinear program (see [34]), but some usual constraint qualifications like the MFCQ, are in general not satisfied for that nonlinear program. Therefore, some well adapted constraint qualification for this class of optimization problems have been developed in the literature [35].

3.3.2.2 MPCC Reformulation

Another classical technique consists in replacing the lower-level GNEP by the concatenation of the associated parametric KKT conditions of each of the followers and obtaining a so-called Mathematical Program with Complementarity Constraints (MPCC).

In fact, to each follower's problem we can associate its KKT optimality conditions, that is (y, μ_j) satisfying

$$\begin{cases} \nabla_{y_j} f_j(x,y) + \sum_{k=1}^{p_j} \mu_{jk} \nabla_{y_j} g_{jk}(x,y) = 0 \\ 0 \leq \mu_j \perp -g_j(x,y) \geq 0 \end{cases}$$

We denote by $KKT(x)$ the set of solutions of the concatenation of KKT conditions of all the followers, that is, (y, μ) such that, for each $j = 1, \ldots, M$, (y_j, μ_j) solves the KKT system given the parameters (x, y_{-j}).

Thus the MPCC reformulation of the SLMFG consists of the following optimization problem

$$\min_{x,y,\mu} \quad F(x,y)$$

$$\begin{cases} x \in X, \\ (y,\mu) \in KKT(x) \end{cases}$$

Numerical methods for such a reformulation can be found for instance in [36]. See also [37, 38].

An important difference between the OPVIC and the MPCC reformulations is that in the latter a new variable, the Lagrange multipliers μ, appears as part of the definition of the optimization problem. Moreover, to consider the MPCC reformulation it is important to analyse constraint qualifications of the lower-level problem for the existence of Lagrange multipliers and their well-behaviour.

To be more precise, in order to have a notion of equivalence between the global solutions of the initial SLMFG and those of its MPCC reformulation an (in general) infinite number of constraint qualifications have to be verified. This fact was first noticed in [39].

We make the following basic hypotheses:

(H_1) (Follower's differentiability) For any $j \in J$ and any $(x, y_{-j}) \in X \times \mathbb{R}^{m-j}$, $f_j(x, \cdot, y_{-j})$ and $g_j(x, \cdot, y_{-j})$ are differentiable;

(H_2) (Follower's player convexity) For any $j \in J$ and any $(x, y_{-j}) \in X \times \mathbb{R}^{m-j}$, $f_j(x, \cdot, y_{-j})$ is convex and the components of $g_j(x, \cdot, y_{-j})$ are quasiconvex functions.

Since the lower-level equilibrium problem is player convex the concatenated KKT optimality conditions are sufficient. If we somehow knew that the KKT conditions were also necessary, then it is quite simple to deduce that global solutions of SLMFG yields solutions of the MPCC reformulation, and vice versa.

Theorem 3.3.8 *Assume (H_1) and (H_2). The relation between solutions of SLMFG and its MPCC reformulation are as follows.*

1. *If* $(\bar{x}, \bar{y}) \in$ *SLMFG and* $\bar{\mu} \in \Lambda(\bar{x}, \bar{y})$*, then* $(\bar{x}, \bar{y}, \bar{\mu}) \in$ *(MPCC).*
2. *Assume that for each leader's strategy* $x \in X$*, for each follower* $j \in J$*, and for each joint strategy* $y = (y_j, y_{-j})$ *which is feasible for all followers the Guignard's CQ holds for the constraint "*$g_j(x, \cdot, y_{-j}) \le 0$*" at the point* y_j*. If* $(\bar{x}, \bar{y}, \bar{\mu}) \in$ *(MPCC), then* $(\bar{x}, \bar{y}) \in$ *SLMFG.* △

Remark 3.3.9 Let us observe that the assumptions of Theorem 3.3.8 are not really tractable i.e. quite hard to verify. Indeed, the Guignard's constraint qualification should hold true for each joint strategy being feasible for all follower, that is such that $g_l(x, y) \le 0$, for every follower $l \in J$. On the other hand these assumptions (of Theorem 3.3.8) are in some sense minimal. Indeed, the weakest condition that makes SLMFG equivalent to its MPCC reformulation independently of the objective of the leader is that

$$\text{GNEP}(x) = KKT(x), \quad \forall x \in X. \tag{3.3.2}$$

Moreover, for a given $x \in X$, the weakest condition that makes $\text{GNEP}(x) = KKT(x)$, independently of the objectives of the followers, is in fact the huge set of Guignard's CQs described in the assumptions of Theorem 3.3.8. △

However, using the techniques developed in [40], we can reduce significantly the conditions to be verified in order to have the desired equivalence, as we explain now.

Assume that the followers' constraint functions g_{jk} are jointly convex with respect to the vector (x, y).

Definition 3.3.10 Let $j \in J$. An opponent strategy $(\hat{x}, \hat{y}_{-j}) \in \mathbb{R}^n \times \mathbb{R}^{m_j}$ is said to be

- an *admissible opponent strategy* (for player j) if $(\hat{x}, \hat{y}_{-j}) \in \mathcal{A}_j := \text{dom}\, Y_j$, that is, such that there exists $y_j \in X$ with $g_j(\hat{x}, y_j, \hat{y}_{-j}) \le 0$;
- an *interior opponent strategy* if it is in $int(\mathcal{A}_j)$;
- a *boundary opponent strategy* if it is in $bd(\mathcal{A}_j)$. △

Theorem 3.3.11 *Assume* (H_1)*,* (H_2) *and that for each* $j \in J$*, the three following properties hold:*

(1) (Joint Convexity) Each g_{jk} *is jointly convex with respect to* (x, y)*;*
(2) (Joint Slater's CQ) There exists a joint strategy $(\tilde{x}(j), \tilde{y}(j))$ *such that*

$$g_j(\tilde{x}(j), \tilde{y}(j)) < 0;$$

(3) (Guignard's CQs for boundary opponent strategies) For any boundary opponent strategy $(\hat{x}, \hat{y}_{-j}) \in bd(\mathcal{A}_j)$ *Guignard's CQ is satisfied at any feasible point* $y_j \in Y_j(\hat{x}, \hat{y}_{-j})$*.*

If $(\bar{x}, \bar{y}, \bar{\mu}) \in$ *(MPCC), then* $(\bar{x}, \bar{y}) \in$ *SLMFG.* △

Proof Let $j \in J$ and $x \in X$, and take a $y = (y_j, y_{-j})$ that is feasible for all followers, so that (x, y_{-j}) is an admissible parameter. Let us now verify that Guignard's CQ holds for the constraint "$g_j(x, \cdot, y_{-j}) \leq 0$" at the point y_j. If (x, y_{-j}) is a boundary opponent strategy we know from Assumption (3) that Guignard's CQ is satisfied at y_j. Otherwise, (x, y_{-j}) is an interior opponent strategy. Then by Proposition 2.1 in [40], Slater's CQ holds for this parameter, which itself imply Guignard's CQ at y_j. Thus the conclusion follows by applying Theorem 3.3.8.

Definition 3.3.12 We say that the lower-level of a SLMFG is *fully feasible* if for any follower $j \in J$, $\mathcal{A}_j = \mathbb{R}^n \times \mathbb{R}^{m_{-j}}$, that is, for any opponent strategy $(x, y_{-j}) \in \mathbb{R}^n \times \mathbb{R}^{m_{-j}}$, there exists $y_j \in Y_j(x, y_{-j})$. △

The above definition does not allow boundary opponent strategies to exist. Thus, Assumption (3.3.11) of Theorem 3.3.11 is trivially satisfied, leading to the following corollary.

Corollary 3.3.13 *Assume* (H_1), (H_2) *and that the lower level is fully feasible (in the sense of Definition 3.3.12). For each* $j \in J$ *we make the following assumptions:*

1. *(Joint Convexity) Each* g_j *is jointly convex with respect to* (x, y)*;*
2. *(Joint Slater's CQ) There exists a joint strategy* $(\tilde{x}(j), \tilde{y}(j))$ *such that*

$$g_j(\tilde{x}(j), \tilde{y}(j)) < 0.$$

If $(\bar{x}, \bar{y}, \bar{\mu}) \in$ *(MPCC), then* $(\bar{x}, \bar{y}) \in$ *SLMFG.* △

3.3.3 Algorithms

There exist actually very few algorithms tackling directly the SLMFG model. In the seminal paper [20] where the case of an oligopoly was studied, a first simple algorithm was proposed. The idea of the algorithm was first to divide the interval of strategies of the leader into finite subinterval, in each of them a linearisation of the lower-level reaction function is considered, and to minimize the leader's objective composed with the linearisation of the lower level problem in the subinterval. The new points are added to the grid. When a termination criterion is satisfied, the best point of the grid is the proposed approximate solution. This idea was then adapted to the case where there is an uncertainty in the problem of the leader [41].

Apart from this direct algorithm most of the papers first start with a reformulation and then use algorithms for solving the corresponding reformulation. In [42] an MPCC reformulation was considered and then the problem was solved using a smoothing approach of the complementarity constraints.

The MPCC reformulation is commonly preferred (see discussion in [19]) since it benefits from a more explicit expression. On the other hand, the OPVIC reformulation, being a more direct one, is preferred whenever the constraint qualifications of the lower-level problem cannot be established or are too difficult to be proven.

Numerical approaches for the OPQVIC reformulation have been considered in [29, 30] for the general case, while OPVIC have been considered in [35].

On the other hand algorithms developed for the resolution of the MPCC reformulation face the difficulty of the treatment of the complementarity constraints involving the Lagrange multipliers. The main numerical techniques are the smoothing, the decomposition, the penalization, and the relaxation approaches.

Simply to illustrate one of these approaches we give below the main steps of the application of the relaxation method to SLMFG. In the KKT system, the constraints of the form $0 \leq \mu_j \perp -g_j(x, y) \geq 0$ can be described by the nonlinear system

$$\begin{cases} -\mu_j g_j(x, y) \leq 0, \ \ \forall j = 1, \ldots, M \\ \mu_j \geq 0, \ -g_j(x, y) \geq 0, \ \ \forall j = 1, \ldots, M. \end{cases}$$

The source of main difficulties is the product constraint. One approach due to Scholtes is to enlarge the feasible set by imposing instead $-\mu_j g_j(x, y) \leq \varepsilon$. By doing this the relaxed problem might now satisfy some CQs and some usual methods (like interior point method used in [43]) for solving the new nonlinear problem can be applied.

The new family of problems would be

$$\min_{x,y,\mu} \ F(x, y)$$

$$\begin{cases} x \in X, \\ \nabla_{y_j} f_j(x, y) + \sum_{k=1}^{p_j} \mu_{jk} \nabla_{y_j} g_{jk}(x, y) = 0 \\ 0 \leq \mu_j, 0 \leq -g_j(x, y) \\ -\mu_j g_j(x, y) \leq \varepsilon \end{cases}$$

with $\varepsilon > 0$ tending to 0.

The limit of a solution of such problems as ε tends to 0 is a C-stationary solution of the usual MPCC, under suitable constraint qualifications, see e.g. [44].

3.4 Multi-Leader-Multi-Follower Games

Let us now focus on Multi-Leader-Follower games in which there are several leaders. Let us first make some comments on the "a priori" most simple model of this kind corresponding to Multi-Leader-Single-Follower games, that is the case where there is only one (common) follower and that the rest of the players play a GNEP game between them.

MLMFG models intrinsically carry several ambiguities in their formulation. First and as for Single-Leader-Multi-Follower games, for any decision vector $x = (x_1, \ldots, x_n)$ of the leaders, there can exist more than one optimal response of the follower. And thus any leader can consider either an optimistic point of view ($\min_{x_i} \min_y$) or a pessimistic positioning ($\min_{x_i} \max_y$). Therefore a precise formulation of MLMFG could generate either a "multi-optimistic version" or a "multi-pessimistic" one or even any other combination of optimistic and pessimistic positioning.

Moreover, and even if we fix a "multi-optimistic version" of MLMFG, we still have some ambiguities in order to define precisely what should be a solution of the problem. To see this, for any leader $i = 1, \ldots, n$, let us denote by $S_i(x_i, x_{-i})$ the set of solutions of the following problem

$$\begin{array}{l}\min_y \; F_i(x_i, x_{-i}, y)\\ \qquad y \in opt(x)\end{array}$$

Then if we set $\varphi_i(x)$ being the value function of the above optimization problem then the MLMFG can be reformulated as a generalized Nash game in which each of the player's problem is

$$\begin{array}{l}\min_{x_i} \; \varphi_i(x_i, x_{-i})\\ \qquad x_i \in X_i(x_{-i})\end{array}$$

But with a given equilibrium $\bar{x} = (\bar{x}_1, \ldots, \bar{x}_n)$ of this GNEP, one will usually face the unexpected situation that for $i \neq j$, $S_i(\bar{x}_i, \bar{x}_{-i})$ and $S_j(\bar{x}_j, \bar{x}_{-j})$ may differ, thus making difficult to find a common $y \in opt(x)$.

As a simple example one can consider the following MLMFG: let us define a game with two leaders and a follower for which respective variables and objective functions are x_1, x_2, y and $F_1(x_1, x_2, y) = (x_1 - 2)^2 - y$, $F_2(x_2, x_1, y) = (x_2 - 2)^2 + y$, $f(y, x) = x_1 x_2 - (y - 1)^2 + 1$. Let us assume that the only constraint on the variables is that the three of them are non negative. Then $\bar{x} = (2, 2)$ while $S_1(2, 2) = \{2\}$ and $S_2(2, 2) = \{0\}$. This difficulty/ambiguity which is fundamental and intrinsically associated to MLMFG with possibly several optimal responses for the follower's problem, is unfortunately often neglected in the literature, in particular in works dedicated to applications.

The same difficulties/ambiguities occur in the most general case of Multi-Leader-Multi-Follower games MLMFG. Indeed in this case $S_i(x_i, x_{-i})$ is the set of generalized Nash Equilibrium of the non cooperative game between the followers. And it is well known that these equilibria are rarely unique. Recently, Kulkarni-Shanbhag proposed in [45] some alternative formulations in order to avoid these difficulties in MLMFG models.

One way to avoid such ambiguities is to satisfy conditions ensuring the uniqueness of the optimal response (for MLSFG) or equilibrium response (for MLMFG) for the follower's problem for any leaders' decision x, thus leading to a response function $y : x \mapsto y(x)$. This choice has been done for example in [11, 12, 15–17].

3.4.1 Existence

In the literature, existence results for MLMFG games are scarce and most of them (if not all, see for instance [46, 47]) are based on a technique that we present now. The technique is basically to reduce the MLMFG to a Nash equilibrium problem by 'plugging' the unique lower-level response into the leaders' objectives, and then trying to prove some good properties of the resulting Nash equilibrium problem:

$$\boxed{\begin{array}{ll} \min_{x_1} & F_1(x, y(x)) \\ \text{s.t.} & x_1 \in X_1(x_{-1}) \end{array}} \;\cdots\; \boxed{\begin{array}{ll} \min_{x_N} & F_N(x, y(x)) \\ \text{s.t.} & x_N \in X_N(x_{-N}) \end{array}}$$

The general assumptions for this technique are:

(A1) for each leaders' profile of strategies x there exists a unique lower level response $y(x)$,

(A2) for any i, the leaders' objectives F_i and the best response function y are continuous,

(A3) for any i, there exist a nonempty, convex and compact set $K_i \in \mathbb{R}^{n_i}$ such that the set-valued map $X_i : K_i \rightrightarrows K_{-i}$ is both upper and lower semicontinuous with nonempty closed and convex values, where $K_{-i} := \prod_{k \neq i} K_k$,

(A4) for any i, the composition functions $\tilde{F}_i(x) := F_i(x, y(x))$ are quasiconvex with respect to x_i.

Proposition 3.4.1 *Assume the above conditions (A1)–(A4). Then the MLMFG admits a solution.* △

Proof According to [48] the Nash equilibrium problem defined by the objectives $\tilde{F}_i$, $i = 1, \ldots, N$ admits an equilibrium. Thus, $\bar{x}$ along with the corresponding reaction of the followers $\bar{y} := y(\bar{x})$ yield an equilibrium $(\bar{x}, \bar{y})$ of the MLMFG. □

The most intricate condition is (A4). In fact, since usually $y(x)$ is only described implicitly, verifying the quasiconvexity of that composition is very difficult in general, but in some cases it is though possible as has been shown by some researchers.

Sherali in [47, Theorem 2] provided, to the best of our knowledge, the first existence result for a particular class of MLFG, by somehow using this technique. In the context of an oligopolistic Stackelberg-Nash-Cournot competition, a group of firms (the leaders) have objectives $F_i(x, y) := x_i p(\sum_k x_k + \sum_j y_j) - c_i(x_i)$ while the rest of the firms (the followers) have objectives $f_j(x, y) := y_j p(\sum_i x_i +$

$\sum_l y_l) - c_j(y_j)$, where p is the inverse demand function and the c_i, c_j are cost functions.

It is proved in [47, Lemma 1, Theorem 3] that under some reasonable assumption on the inverse demand function p and on the cost functions, the $\tilde{F}_i$'s are convex in x_i. The corresponding existence result [47, Theorem 2] can be then expressed as follows:

Theorem 3.4.2 *Assume that p is strictly decreasing, twice differentiable and $p'(z) + zp''(z) \leq 0$ for each $z \geq 0$, and that c_i and c_j are non-negative, non-decreasing, convex and twice differentiable and there exists $z_u > 0$ such that $c_i'(z) \geq p(z)$ and $c_j'(z) \geq p(z)$ for all $z \in [0, z_u]$. If the map $x \mapsto \sum_j y_j(x)$ is convex (if for instance, p is linear), where $y(x)$ is the unique equilibrium response of the followers, then the MLMFG has at least one equilibrium.* △

Fukushima and Hu's existence results (Theorems 4.3 and 4.4 in [49]) are also obtained using the same technique but in a more general setting that considers uncertainty in both levels and a robust approach.

A different technique has been proposed in [50] which is based on the ideas of potential game theory, see [51]. A first possibility is again based on the uniqueness of the lower-level responses, that is, condition (A1). A MLMFG is *implicitly potential* if there exists a so-called *potential* function π for the game defined by the functions $\tilde{F}_i$, that is, for all i and for all $x = (x_i, x_{-i})$ and x_i' it holds

$$\tilde{F}_i(x_i, x_{-i}) - \tilde{F}_i(x_i', x_{-i}) = \pi(x_i, x_{-i}) - \pi(x_i', x_{-i}). \tag{3.4.1}$$

Let us notice that, as in the previous technique, the existence of the potential for the implicit description of the functions $\tilde{F}_i$ is also an intricate condition. A variant of this approach was proposed also in [50], where it is not assumed that the lower-level responses are unique. The game is said to be a quasi-potential game if there exist functions h and π such that the functions F_i have the following structure

$$F_i(x, y) := \phi_i(x) + h(x, y) \tag{3.4.2}$$

and the family of functions ϕ_i, $i = 1, \ldots, N$, admit π as a potential function, that is

$$\phi_i(x_i, x_{-i}) - \phi_i(x_i', x_{-i}) = \pi(x_i, x_{-i}) - \pi(x_i', x_{-i}). \tag{3.4.3}$$

The existence of equilibria for the MLMFG can be deduced, in the first case from the existence of a global minimizer of the potential function, as usual in potential games. In the second case, the existence of equilibria for MLFG can be deduced from the minimization of $\pi + h$, which is not strictly speaking a potential function for the F_i. In fact, $\pi + h$ is defined in the space $X_1 \times \ldots \times X_N \times Y$, while a potential function for the quasi-potential game should be defined on the product of the strategy spaces $(X_1 \times Y) \times \ldots \times (X_N \times Y)$, for instance as

$\psi(x_1, y_1, \ldots, x_N, y_N) := \pi(x) + \sum_i h(x, y_i)$. We can thus call the function $\pi + h$ a quasi-potential function for the game.

The following theorem [50] shows a way of computing an equilibrium in a pseudo potential MLMFG.

Theorem 3.4.3 *Assume that the MLMFG is a pseudo-potential game, and that the constraints set of player i is a constant set equal to a nonempty compact and convex K_i. Then any minimizer of the pseudo potential function $\pi + h$ corresponds to a solution of the MLMFG.* △

3.4.2 Reformulations

As explained previously in this section, the analysis of MLMFG in the literature is mostly focused on the case of a unique lower-level response. Under this assumption, the lower-level response can be plugged into the leaders objectives transforming the initial MLMFG "simply" into a Nash equilibrium problem, though with quite complicated objective functions, in general non-smooth and non-convex. Then, the usual techniques used for solving Nash equilibrium problems can be used for this formulation.

In particular, in [46, 49] the function of (unique) lower-level responses are linear with respect to the leaders variables and can be somehow plugged into the leaders objective because of the specific structure that is considered (there is a term in the leaders' objective that is also present in the followers' objective but with negative sign). The resulting Nash equilibrium problem is reformulated as a Variational Inequality and a forward-backward splitting method (see [52]) is applied to solve the variational inequality.

In the case of non-uniqueness of the lower-level problem, and considering the (possibly inconsistent) multi-optimistic MLMFG, we can extend the approach of the SLMFG case by replacing the lower-level equilibrium problem by the concatenation of KKT conditions of the followers. The resulting reformulation is a so-called Equilibrium Problem with Equilibrium Constraints (EPEC, for short).

A first question to address concerns the equivalence between the initial MLMFG and its EPCC[1] reformulation, facing similar arguments as for SLMFG, that is requiring an infinite number of CQs to be verified. The analysis made for the case of one leader in the previous section can be easily extended to the case of multiple leaders as done in [19, Theorem 4.9], by considering the joint convexity of the followers' constraint functions and the joint Slater's CQ.

[1]Equilibrium Problem with Complementarity Constraints

3.4.3 *Algorithms*

The above mentioned EPEC reformulation of the MLMFG is an equilibrium problem (among leaders) so that we are tempted to use the machinery for equilibrium problems to solve the MLFG. Nevertheless, considering that the specific type of constraints (equilibrium constraints) are very ill-behaved non-convex and non-smooth, solving the equilibrium problem is in fact extremely challenging.

If we are facing a pseudo-potential game (see previous subsection), then the problem can be solved by minimizing the pseudo-potential function, constrained by the equilibrium problem, thus going back to the SLMF case (see Theorem 3.4.3).

3.5 Conclusion and Future Challenges

The aim of the present chapter was to present the recent advances for different kinds of Multi-Leader-Follower games. The cases of a single leader game SLMFG and of a single follower MLSFG play a particular roles in applications and it is one of the reasons why the general case MLMFG has been actually less investigated in the literature. However we have seen that, for all the models, a special attention must be addressed to avoid an ill-posed problem and to fix possible ambiguities. Let us add that, as observed for MLMFG in [16], those ambiguities are even more tricky when one deals with reformulation involving Lagrange multipliers. Applications of MLFG are numerous and have been well explored (energy or water management, economics, pollution control, telecommunications, metro pricing [53], etc.) but from a theoretical point of view a lot of questions are still open concerning SLMFG, MLSFG and of course even more for MLMFG. For example, to our knowledge, very few papers (see e.g. [54]) consider sensitivity/stability analysis for MLFG. In the same vein, gap functions has not been studied for this class of problems.

We restricted ourself to deterministic versions of MLFG because considering stochastic models would have been beyond the scope of this chapter. But it is important to mention that some models and results in settings with uncertainties or random variables have been recently studied, see e.g. [21, 55].

Models with more than two levels were also not considered here. Some preliminary studies appeared (see e.g. [56, 57] but applications are calling for more analysis of such models.

Finally we would like to emphasize that one keystone to push further the analysis of MLFG could be to consider, at least as a first step, some specific structures/models like the concept of Multi-Leader-Disjoint-Follower problem presented in [58]. Indeed in those particular interactions between leaders and followers could intrinsically carry properties that allow to obtain more powerful results.

Acknowledgements This research benefited from the support of the FMJH Program Gaspard Monge in optimization and operation research, and from the support to this program from EDF. The second author was also benefited by a CONICYT grant CONICYT-PFCHA/Doctorado Nacional/2018 N21180645

References

1. H. von Stackelberg, *Marktform und Gleichgewicht* (Springer, Berlin, 1934)
2. H. von Stackelberg, D. Bazin, L. Urch, R. Hill, *Market Structure and Equilibrium* (Springer, Heidelberg, 2011)
3. F. Facchinei, J.-S. Pang, *Finite-Dimensional Variational Inequalities and Complementarity Problems*, vol. I and II (Springer, Berlin, 2003)
4. H.I. Calvete, C. Galé, Linear bilevel multi-follower programming with independent followers. J. Glob. Optim. **39**(3), 409–417 (2007)
5. M.A. Ramos, M. Boix, D. Aussel, L. Montastruc, S. Domenech, Water integration in eco-industrial parks using a multi-leader-follower approach. Comput. Chem. Eng. **87**, 190–207 (2016)
6. M. Ramos, M. Boix, D. Aussel, L. Montastruc, P. Vilamajo, S. Domenech, Water exchanges in eco-industrial parks through multiobjective optimization and game theory. Comput. Aided Chem. Eng. **37**, 1997–2002 (2015)
7. D. Salas, V.K. Cao, L. Montastruc, D. Aussel, Optimal design of exchange networks with blind inputs—part 1: theoretical analysis (2019). Preprint
8. V.K. Cao, D. Salas, L. Montastruc, D. Aussel, Optimal design of exchange networks with blind inputs—Part 2: applications to ECO-industrial parks (2019). Preprint
9. M.A. Ramos, M. Rocafull, M. Boix, D. Aussel, L. Montastruc, S. Domenech, Utility network optimization in eco-industrial parks by a multi-leader follower game methodology. Comput. Chem. Eng. **112**, 132–153 (2018)
10. X. Hu, D. Ralph, Using EPECs to model bilevel games in restructured electricity markets with locational prices. Oper. Res. **55**(5), 809–827 (2007)
11. D. Aussel, P. Bendotti, M. Pištěk, Nash equilibrium in pay-as-bid electricity market: Part 1—existence and characterisation. Optimization **66**(6), 1013–1025 (2017)
12. D. Aussel, P. Bendotti, M. Pištěk, Nash equilibrium in pay-as-bid electricity market: Part 2—best response of producer. Optimization **66**(6), 1027–1053 (2017)
13. J.F. Escobar, A. Jofré, Monopolistic competition in electricity networks with resistance losses. Econom. Theory **44**(1), 101–121 (2010)
14. J.F. Escobar, A. Jofré, Equilibrium analysis of electricity auctions (Department of Economics, Stanford University, Stanford, 2008)
15. D. Aussel, R. Correa, M. Marechal, Electricity spot market with transmission losses. J. Ind. Manag. Optim. **9**(2), 275–290 (2013)
16. D. Aussel, M. Červinka, M. Marechal, Deregulated electricity markets with thermal losses and production bounds: models and optimality conditions. RAIRO Oper. Res. **50**(1), 19–38 (2016)
17. E. Allevi, D. Aussel, R. Riccardi, On an equilibrium problem with complementarity constraints formulation of pay-as-clear electricity market with demand elasticity. J. Glob. Optim. **70**(2), 329–346 (2018)
18. R. Henrion, J. Outrata, T. Surowiec, Analysis of M-stationary points to an EPEC modeling oligopolistic competition in an electricity spot market. ESAIM Control Optim. Calc. Var. **18**(2), 295–317 (2012)
19. D. Aussel, A. Svensson, Some remarks about existence of equilibria, and the validity of the EPCC reformulation for multi-leader-follower games. J. Nonlinear Convex Anal. **19**(7), 1141–1162 (2018)

20. H.D. Sherali, A.L. Soyster, F.H. Murphy, Stackelberg-Nash-Cournot equilibria: characterizations and computations. Oper. Res. **31**(2), 253–276 (1983)
21. W. van Ackooij, J. De Boeck, B. Detienne, S. Pan, M. Poss, Optimizing power generation in the presence of micro-grids. Eur. J. Oper. Res. **271**(2), 450–461 (2018)
22. J.-P. Aubin, H. Frankowska, *Set-valued Analysis* (Springer, Berlin, 2009)
23. M.B. Lignola, J. Morgan, Existence of solutions to generalized bilevel programming problem, in *Multilevel Optimization: Algorithms and Applications* (Springer, Berlin, 1998), pp. 315–332
24. B. Bank, J. Guddat, D. Klatte, B. Kummer, K. Tammer, *Non-linear Parametric Optimization* (Akademie, Berlin, 1982)
25. S. Dempe, *Foundations of Bilevel Programming* (Springer, New York, 2002)
26. D. Aussel, N. Hadjisavvas, Adjusted sublevel sets, normal operator, and quasi-convex programming. SIAM J. Optim. **16**(2), 358–367 (2005)
27. D. Aussel, J, Dutta, Generalized Nash equilibrium, variational inequality and quasiconvexity. Oper. Res. Lett. **36**(4), 461–464 (2008). [Addendum in Oper. Res. Lett. **42** (2014)]
28. D. Aussel, New developments in quasiconvex optimization, in *Fixed Point Theory, Variational Analysis, and Optimization* (CRC Press, Boca Raton, 2014), pp. 171–205
29. J. Wu, L.W. Zhang, A smoothing Newton method for mathematical programs constrained by parameterized quasi-variational inequalities. Sci. China Math. **54**(6), 1269–1286 (2011)
30. J. Wu, L. Zhang, Y. Zhang, An inexact newton method for stationary points of mathematical programs constrained by parameterized quasi-variational inequalities. Numer. Algorithm. **69**(4), 713–735 (2015)
31. J. Outrata, J. Zowe, A numerical approach to optimization problems with variational inequality constraints. Math. Prog. **68**(1–3), 105–130 (1995)
32. J.J. Ye, X.Y. Ye, Necessary optimality conditions for optimization problems with variational inequality constraints. Math. Oper. Res. **22**(4), 977–997 (1997)
33. P.T. Harker, J.-S. Pang, Existence of optimal solutions to mathematical programs with equilibrium constraints. Oper. Res. Lett. **7**(2), 61–64 (1988)
34. J.J. Ye, D.L. Zhu, Q.J. Zhu, Exact penalization and necessary optimality conditions for generalized bilevel programming problems. SIAM J. Optim. **7**(2), 481–507 (1997)
35. J.J. Ye, Constraint qualifications and necessary optimality conditions for optimization problems with variational inequality constraints. SIAM J. Optim. **10**(4), 943–962 (2000)
36. L. Guo, G.-H. Lin, J. Ye Jane, Solving mathematical programs with equilibrium constraints. J. Optim. Theory Appl. **166**(1), 234–256 (2015)
37. Z.-Q. Luo, J.-S. Pang, D. Ralph, *Mathematical Programs with Equilibrium Constraints* (Cambridge University, Cambridge, 1996)
38. M. Fukushima, G.-H. Lin, Smoothing methods for mathematical programs with equilibrium constraints, in *International Conference on Informatics Research for Development of Knowledge Society Infrastructure, 2004 (ICKS 2004)* (IEEE, Silver Spring, 2004), pp. 206–213
39. A. Ehrenmann, *Equilibrium Problems with Equilibrium Constraints and their Application to Electricity Markets*. PhD thesis (Citeseer, Princeton, 2004)
40. D. Aussel, A. Svensson, Towards tractable constraint qualifications for parametric optimisation problems and applications to generalised Nash games. J. Optim. Theory Appl. **182**(1), 404–416 (2019)
41. D. De Wolf, Y. Smeers, A stochastic version of a Stackelberg-Nash-Cournot equilibrium model. Manag. Sci. **43**(2), 190–197 (1997)
42. H. Xu, An MPCC approach for stochastic Stackelberg–Nash–Cournot equilibrium. Optimization **54**(1), 27–57 (2005)
43. V. DeMiguel, M.P. Friedlander, F.J. Nogales, S. Scholtes, A two-sided relaxation scheme for mathematical programs with equilibrium constraints. SIAM J. Optim. **16**(2), 587–609 (2005)
44. S. Scholtes, Convergence properties of a regularization scheme for mathematical programs with complementarity constraints. SIAM J. Optim. **11**(4), 918–936 (2001)
45. A.A. Kulkarni, U.V. Shanbhag, A shared-constraint approach to multi-leader multi-follower games. Set-Valued Var. Anal. **22**(4), 691–720 (2014)

46. M. Hu, M. Fukushima, Existence, uniqueness, and computation of robust Nash equilibria in a class of multi-leader-follower games. SIAM J. Optim. **23**(2), 894–916 (2013)
47. H.D. Sherali, A multiple leader Stackelberg model and analysis. Oper. Res. **32**(2), 390–404 (1984)
48. T. Ichiishi, M. Quinzii, Decentralization for the core of a production economy with increasing return. Int. Econ. Rev., 397–412 (1983)
49. M. Hu, M. Fukushima, Multi-leader-follower games: models, methods and applications. J. Oper. Res. Soc. Japan **58**(1), 1–23 (2015)
50. A.A. Kulkarni, U.V. Shanbhag, An existence result for hierarchical Stackelberg v/s Stackelberg games. IEEE Trans. Autom. Control **60**(12), 3379–3384 (2015)
51. D. Monderer, L.S. Shapley, Potential games. Games Econom. Behav. **14**(1), 124–143 (1996)
52. F. Facchinei, J.-S. Pang, *Finite-Dimensional Variational Inequalities and Complementarity Problems* (Springer, New York, 2007)
53. A. Printezis, A. Burnetas, G. Mohan, Pricing and capacity allocation under asymmetric information using Paris metro pricing. Int. J. Oper. Res. **5**(3), 265–279 (2009)
54. W. Jia, S. Xiang, J. He, Y. Yang, Existence and stability of weakly Pareto-Nash equilibrium for generalized multiobjective multi-leader–follower games. J. Global Optim. **61**(2), 397–405 (2015)
55. L. Mallozzi, R. Messalli, Multi-leader multi-follower model with aggregative uncertainty. Games **8**(3), Paper No. 25, 14 (2017)
56. D. Aussel, L. Brotcorne, S. Lepaul, L. von Niederhäusern, A trilevel model for best response in energy demand-side management. Eur. J. Oper. Res. **281**(2), 299–315 (2020)
57. D. Aussel, S. Lepaul, L. von Niederhäusern, A multi-leader-follower game for energy demand-side management (2019). Preprint
58. D. Aussel, G. Bouza, S. Dempe, S. Lepaul, Genericity analysis of multi-leader-follower games (2019). Preprint

Chapter 4
Regularization and Approximation Methods in Stackelberg Games and Bilevel Optimization

Francesco Caruso, M. Beatrice Lignola, and Jacqueline Morgan

Abstract In a two-stage Stackelberg game, depending on the leader's information about the choice of the follower among his optimal responses, one can associate different types of mathematical problems. We present formulations and solution concepts for such problems, together with their possible roles in bilevel optimization, and we illustrate the crucial issues concerning these solution concepts. Then, we discuss which of these issues can be positively or negatively answered and how managing the latter ones by means of two widely used approaches: regularizing the set of optimal responses of the follower, via different types of approximate solutions, or regularizing the follower's payoff function, via the Tikhonov or the proximal regularizations. The first approach allows to define different kinds of regularized problems whose solutions exist and are stable under perturbations assuming sufficiently general conditions. Moreover, when the original problem has no solutions, we consider suitable regularizations of the second-stage problem, called *inner regularizations*, which enable to construct a surrogate solution, called *viscosity solution*, to the original problem. The second approach permits to overcome the non-uniqueness of the follower's optimal response, by constructing sequences of Stackelberg games with a unique second-stage solution which approximate in some sense the original game, and to select among the solutions by using appropriate constructive methods.

F. Caruso
Department of Economics and Statistics, University of Naples Federico II, Naples, Italy
e-mail: francesco.caruso@unina.it

M. Beatrice Lignola
Department of Mathematics and Applications R. Caccioppoli, University of Naples Federico II, Naples, Italy
e-mail: lignola@unina.it

J. Morgan (✉)
Department of Economics and Statistics & Center for Studies in Economics and Finance, University of Naples Federico II, Naples, Italy
e-mail: morgan@unina.it

S. Dempe, A. Zemkoho (eds.), *Bilevel Optimization*, Springer Optimization and Its Applications 161, https://doi.org/10.1007/978-3-030-52119-6_4

Keywords Stackelberg game · Bilevel optimization problem · Subgame perfect Nash equilibrium problem · Approximate solution · Inner regularization and viscosity solution · Tikhonov regularization · Proximal point method

4.1 Introduction

In this chapter we are interested in describing interactions that could occur between two agents (or players) which act sequentially in two stages. More precisely, we consider that

1. **in the first stage:** one player, henceforth called the *leader*, chooses an action x in his action set X;
2. **in the second stage:** one player, henceforth called the *follower*, observes the action x chosen by the leader in the first stage and then chooses an action y in his action set Y;
3. **after the two-stage interaction:** the leader receives $L(x, y)$ where $L: X \times Y \to \mathbb{R}$ is the leader's payoff function, and the follower receives $F(x, y)$ where $F: X \times Y \to \mathbb{R}$ is the follower's payoff function.

Such a situation defines a *Stackelberg game*, denoted by $\Gamma = (X, Y, L, F)$, which has been modelled for the first time by von Stackelberg in [109] in an economic framework.

Assume that players aim to minimize their payoff functions, as traditionally used in optimization literature where the payoff functions embed players' costs. However, since $\max f(\cdot) = -\min\{-f(\cdot)\}$, we could assume equivalently that players maximize their payoff functions (which usually appears in economics contexts, where payoff functions represent players' profits). At the moment, we set no assumptions on the structure of the actions sets X and Y: they could be of finite or infinite cardinalities, subsets of finite or infinite dimensional spaces, strict subsets or whole spaces, etc. Their nature will be specified each time in the sequel of the chapter.

The follower, once he has seen the action x chosen by the leader, faces the optimization problem

$$(P_x): \min_{y \in Y} F(x, y),$$

i.e. finding $y \in Y$ such that $F(x, y) = \inf_{z \in Y} F(x, z)$. Then, he reacts to the leader's choice by picking an optimal response to x, that is an action belonging to the set, called the *follower's optimal reaction set*

$$M(x) = \operatorname*{Arg\,min}_{y \in Y} F(x, y) = \left\{ y \in Y \text{ such that } F(x, y) = \inf_{z \in Y} F(x, z) \right\}.$$

The set-valued map $M: X \rightrightarrows Y$ that assigns to any action x of the leader the set $M(x)$ defined above is the *follower's best reply correspondence*. For sake of brevity, it will be also called *argmin map*.

The leader, if he can foresee the optimal response $y(x) \in \operatorname{Arg\,min}_{y \in Y} F(x, y)$ chosen for any action x by the follower, will pick an action in the set $\operatorname{Arg\,min}_{x \in X} L(x, y(x))$ according to such a forecast.

Given the informative structure of the game, predicting the follower's optimal responses (or, equivalently, the response function $y(\cdot)$ of the follower) is typically a hard task for the leader, unless some additional information is available. Therefore, depending on the additional information of the leader about the single-valuedness of M or, when M is not single-valued, about how the follower chooses an optimal response in the set $M(x)$ for any $x \in X$, one can associate to the Stackelberg game Γ different kinds of mathematical problems.

In Sect. 4.2 such mathematical problems and the related solution concepts are presented, together with illustrative examples and their possible roles in bilevel optimization.

In Sect. 4.3 we describe and discuss the main crucial issues arising in the just mentioned problems, focusing on the following ones: *existence* of solutions, *"variational" stability*, *well-posedness*, *approximation*, *numerical approximation* and *selection* of solutions. The crucial issues negatively answered are illustrated by counterexamples.

In order to overcome the drawbacks connected to the (negatively answered) crucial issues, two basic and widely used approaches are identified. They lead to two classes of regularization methods which allow to obtain key-properties for the regularized problems which, in general, are not satisfied in the original problem.

- Regularizing the follower's optimal reaction sets;
- Regularizing the follower's payoff function.

In Sect. 4.4 the first of the above approaches is presented and investigated for some of the problems defined in Sect. 4.2, and also for other types of two-stage games where the second-stage problem to solve is defined by a variational or a quasi-variational inequality, by a Nash equilibrium problem or by a quasi-equilibrium problem. Such an approach permits to define different kinds of regularized problems whose solutions exist and are stable under perturbations assuming sufficiently general conditions. Moreover, when the original problem has no solutions, we present suitable regularizations of the second-stage problem, called *inner regularizations*, which allow to construct a surrogate solution, called *viscosity solution*, to the original problem.

In Sect. 4.5 we investigate the second approach, that enables both to overcome the non-uniqueness of the follower's optimal response and to select among the solutions. To get these goals, we construct sequences of Stackelberg games with a unique second-stage solution which approximate in some sense the original game by exploiting first the *Tikhonov regularization* and then the *proximal regularization*, two well-known regularization techniques in convex optimization.

A conclusive section concerns various extensions of the Stackelberg game notion considered in this chapter.

4.2 Models and Solution Concepts

We will illustrate five problems and solution concepts connected to the Stackelberg game defined in the previous section. The first one appears if the follower's best reply correspondence M is assumed to be single-valued (and the leader knows it), whereas the others concern situations where M is not single-valued. Their connections will be emphasized and illustrated by examples. In this section, we only refer to seminal papers and to the first ones on regularization and approximation methods.

4.2.1 Stackelberg Problems

When the follower's optimal response to any choice of the leader is unique, the leader can fully anticipate the reaction of the follower taking this into account before choosing his action and the follower behaves answering to the leader in the optimal (expected) way. Therefore, assumed that M is single-valued and $M(x) = \{m(x)\}$ for any $x \in X$, the leader faces the so-called *Stackelberg problem*

$$(SP) \begin{cases} \min\limits_{x\in X} L(x, m(x)) \\ \text{where } m(x) \text{ is the solution to } (P_x). \end{cases}$$

The players acting in this way are often referred to be engaged in a *classical Stackelberg game*, terminology due to von Stackelberg who first investigated such an interaction in [109] where an economic model involving two firms that compete sequentially on quantities to produce is presented. Then, properties of Stackelberg solutions together with extensions of the classical Stackelberg games to a dynamic framework have been investigated in [24, 105].

Let us recall the solution concepts associated to (SP).

Definition 4.2.1

- The infimum value $\inf_{x\in X} L(x, m(x))$ is the *Stackelberg value* of (SP).
- A leader's action $\bar{x} \in X$ is a *Stackelberg solution* to (SP) if

$$\bar{x} \in \operatorname*{Arg\,min}_{x\in X} L(x, m(x)).$$

- An action profile $(\bar{x}, \bar{y}) \in X \times Y$ is a *Stackelberg equilibrium* of (SP) if $\bar{x}$ is a Stackelberg solution and $\bar{y} = m(\bar{x})$. △

Stackelberg Leadership Model
The Stackelberg leadership model described in [109] regards two firms in a market which sell homogeneous products and have to choose the quantities to produce. Firm 1, acting as the leader, chooses first a quantity $q_1 \geq 0$; then firm 2, acting as the follower, observes the quantity q_1 and chooses a quantity $q_2 \geq 0$. The firms aim to maximize their profit functions defined on $[0, +\infty[^2$ by

$$L(q_1, q_2) = q_1 P(Q) - C_1(q_1) \quad \text{and} \quad F(q_1, q_2) = q_2 P(Q) - C_2(q_2),$$

respectively, where $Q = q_1 + q_2$ is the total quantity in the market, $P(\cdot)$ is the inverse demand function and $C_i(\cdot)$ is the cost function of firm $i \in \{1, 2\}$. As often happens in economics contexts, we assume linear inverse demand function $P(Q) = \max\{0, a - bQ\}$ with $a, b \in]0, +\infty[$ and linear cost functions $C_i(q_i) = cq_i$ for any $i \in \{1, 2\}$ with $c \in]0, a[$. Hence, the optimal response of firm 2 to any quantity chosen by firm 1 is unique and given by

$$m(q_1) = \max\left\{0, \frac{a - bq_1 - c}{2b}\right\}, \quad \text{for any } q_1 \in [0, +\infty[$$

and, since $\operatorname{Arg\,max}_{q_1 \in [0, +\infty[} L(q_1, m(q_1)) = \{(a - c)/2b\}$, the quantities produced at equilibrium are

$$\bar{q}_1 = \frac{a - c}{2b} \quad \text{and} \quad \bar{q}_2 = \frac{a - c}{4b}.$$

In Stackelberg leadership models the equilibrium quantity of the firm acting as the leader is always greater than equilibrium quantity of the firm acting as the follower (analogously for the equilibrium profits).

We point out that in [67, 72, 103] the following more general Stackelberg problem has been treated:

$$(GSP) \begin{cases} \displaystyle\min_{x \in X,\, g(x, m(x)) \leq 0} L(x, m(x)) \\ \text{where } m(x) \text{ is the solution to } (P_x), \end{cases}$$

where $g \colon X \times Y \to \mathbb{R}$. In such a problem the leader's constraints depend also on the follower's best reply function m, which makes it harder to manage than (SP) from the mathematical point of view. In the sequel of the chapter we will consider only constraints for the leader which do not depend on the actions of the follower.

From now on, in the next subsections of this section we assume that M is not single-valued.

4.2.2 Pessimistic Leader: Weak Stackelberg Problem

Firstly, we consider an "extreme" situation where the pessimistic leader believes that the follower could choose the worst action for him in the set of optimal responses. The leader, who would prevent himself against the worst he can obtain when the follower plays optimally, will face the *weak Stackelberg problem*, also called *pessimistic bilevel optimization problem*

$$(PB) \begin{cases} \min_{x\in X} \sup_{y\in M(x)} L(x, y) \\ \text{where } M(x) = \operatorname{Arg\,min}_{y\in Y} F(x, y). \end{cases}$$

One of the first investigations of such a problem is due to Leitmann in [46], where it is denoted as *generalized Stackelberg* problem. The solution concepts associated to (PB) are reminded below.

Definition 4.2.2

- The infimum value $w = \inf_{x\in X} \sup_{y\in M(x)} L(x, y)$ is the *security (or pessimistic) value* of (PB).
- A leader's action $\bar{x} \in X$ is a *weak Stackelberg (or pessimistic) solution* to (PB) if

$$\bar{x} \in \operatorname{Arg\,min}_{x\in X} \sup_{y\in M(x)} L(x, y),$$

  that is, $\sup_{y\in M(\bar{x})} L(\bar{x}, y) = \inf_{x\in X} \sup_{y\in M(x)} L(x, y)$.
- An action profile $(\bar{x}, \bar{y}) \in X \times Y$ is a *weak Stackelberg (or pessimistic) equilibrium* of (PB) if $\bar{x}$ is a weak Stackelberg solution and $\bar{y} \in M(\bar{x})$. △

For first investigations about regularization and approximation of problem (PB) see [51, 69, 71, 89].

4.2.3 Optimistic Leader: Strong Stackelberg Problem

In a second "extreme" situation, the leader is optimistic and believes that the follower will choose the best action for the leader in the set of his optimal responses (or the leader is able to force the follower to choose, among all the optimal responses, the best one for the leader). The leader will deal with the *strong Stackelberg problem*, also called *optimistic bilevel optimization problem*

$$(OB) \begin{cases} \min\limits_{x\in X} \inf\limits_{y\in M(x)} L(x,y) \\ \text{where } M(x) = \operatorname*{Arg\,min}\limits_{y\in Y} F(x,y). \end{cases}$$

The solution concepts associated to (OB) are reminded below.

Definition 4.2.3

- The infimum value $s = \inf_{x\in X} \inf_{y\in M(x)} L(x,y)$ is the *optimistic value* of (OB).
- A leader's action $\bar{x} \in X$ is a *strong Stackelberg (or optimistic) solution* to (OB) if

$$\bar{x} \in \operatorname*{Arg\,min}_{x\in X} \inf_{y\in M(x)} L(x,y),$$

  that is, $\inf_{y\in M(\bar{x})} L(\bar{x},y) = \inf_{x\in X} \inf_{y\in M(x)} L(x,y)$.
- An action profile $(\bar{x},\bar{y}) \in X \times Y$ is a *strong Stackelberg (or optimistic) equilibrium* of (OB) if $\bar{x}$ is a strong Stackelberg solution and $\bar{y} \in \operatorname{Arg\,min}_{y\in M(\bar{x})} L(\bar{x},y)$. △

The use of terms "weak Stackelberg" and "strong Stackelberg" goes back to [20] where a concept of sequential Stackelberg equilibrium in a general dynamic games framework is introduced. For a first investigation on regularization and approximation of problem (OB) see [52].

We point out that $(\bar{x},\bar{y})$ is a strong Stackelberg equilibrium of (OB) if and only if it is a (global) solution of the following problem:

$$\min_{x\in X,\, y\in M(x)} L(x,y)$$

that is, $\bar{x} \in X$, $\bar{y} \in M(\bar{x})$ and $L(\bar{x},\bar{y}) = \inf_{x\in X,\, y\in M(x)} L(x,y)$. Moreover, the value $s' = \inf_{x\in X,\, y\in M(x)} L(x,y)$ coincides with s, the optimistic value of (OB), and the problem is frequently called "bilevel optimization problem" (for first results, see for example [14, 26, 85, 96, 108, 111]).

The following examples show that the pessimistic and the optimistic behaviours of the leader can design different solutions, values and equilibria.

A Stackelberg Game with Finite Action Sets

Consider the Stackelberg game in [93, Example 2.1] where the leader has two available actions T and B, that is $X = \{T, B\}$, the follower has two available actions Q and R, that is $Y = \{Q, R\}$, and the payoff functions are defined by

$$L(T, Q) = 2, \quad L(T, R) = 3, \quad L(B, Q) = 1, \quad L(B, R) = 4,$$
$$F(T, Q) = 0, \quad F(T, R) = 0, \quad F(B, Q) = 0, \quad F(B, R) = 0.$$

Such a game can be also described by using the following "extensive form representation":

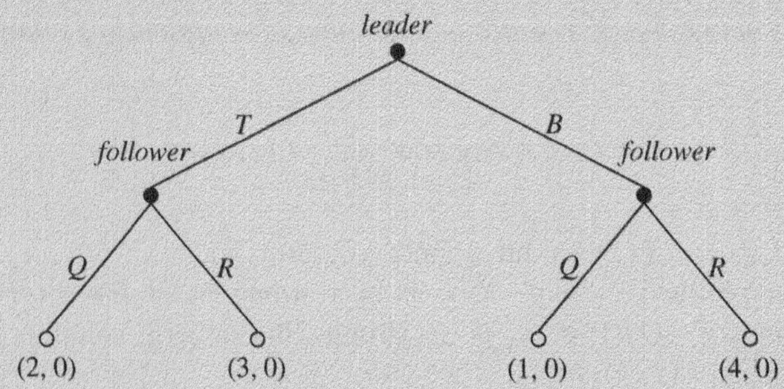

The follower's best reply correspondence is defined by $M(T) = \{Q, R\}$ and $M(B) = \{Q, R\}$. Since

$$x = T \Longrightarrow \sup_{y \in M(x)} L(x, y) = 3 \quad \text{and} \quad \inf_{y \in M(x)} L(x, y) = 2,$$

$$x = B \Longrightarrow \sup_{y \in M(x)} L(x, y) = 4 \quad \text{and} \quad \inf_{y \in M(x)} L(x, y) = 1,$$

then $\inf_{x \in \{T,B\}} \sup_{y \in M(x)} L(x, y) = 3$ and $\inf_{x \in \{T,B\}} \inf_{y \in M(x)} L(x, y) = 1$. Therefore, the weak Stackelberg solution is T, the pessimistic value is 3 and the weak Stackelberg equilibria are (T, Q) and (T, R). Whereas, the strong Stackelberg solution is B, the optimistic value is 1 and the strong Stackelberg equilibrium is (B, Q).

A Stackelberg Game with Infinite Action Sets

Consider the Stackelberg game in [93, Example 3.4] where $X = [-2, 2]$, $Y = [-1, 1]$,

$$L(x, y) = -x - y \quad \text{and} \quad F(x, y) = \begin{cases} (x + 7/4)y, & \text{if } x \in [-2, -7/4[\\ 0, & \text{if } x \in [-7/4, 7/4[\\ (x - 7/4)y, & \text{if } x \in [7/4, 2]. \end{cases}$$

The follower's best reply correspondence M is defined on $[-2, 2]$ by

$$M(x) = \begin{cases} \{1\}, & \text{if } x \in [-2, -7/4[\\ [-1, 1], & \text{if } x \in [-7/4, 7/4] \\ \{-1\}, & \text{if } x \in]7/4, 2]. \end{cases}$$

On the one hand, since

$$\sup_{y \in M(x)} L(x, y) = \begin{cases} -x - 1, & \text{if } x \in [-2, -7/4[\\ -x + 1, & \text{if } x \in [-7/4, 2], \end{cases}$$

then $\inf_{x \in X} \sup_{y \in M(x)} L(x, y) = -1$. Hence, the weak Stackelberg solution is 2, the pessimistic value is -1 and the weak Stackelberg equilibrium is $(2, -1)$.

On the other hand, as

$$\inf_{y \in M(x)} L(x, y) = \begin{cases} -x - 1, & \text{if } x \in [-2, 7/4] \\ -x + 1, & \text{if } x \in]7/4, 2], \end{cases}$$

then $\inf_{x \in X} \inf_{y \in M(x)} L(x, y) = -11/4$. Therefore, the strong Stackelberg solution is $7/4$, the optimistic value is $-11/4$ and the strong Stackelberg equilibrium is $(7/4, 1)$.

4.2.4 Intermediate Stackelberg Problem

The problems (PB) and (OB) reflect the two possible extreme behaviours of the leader regarding how the follower chooses his action in the second stage. However, an intermediate situation could occur: the leader has some information on the follower's choice in his set of optimal responses which allows to attribute a probability distribution reflecting his beliefs on $M(x)$ for any $x \in X$. Assume here, in order to simplify, that $M(x)$ has a finite number of elements for any $x \in X$, namely

$$M(x) = \{y_1(x), \dots, y_{k(x)}(x)\}.$$

Therefore, denoted with $p_i(x)$ the probability (enforced by the leader) that the follower chooses $y_i(x)$ in $M(x)$ for any $x \in X$ and defined $D = \{D(x) \mid x \in X\}$ where $D(x) = \{p_1(x), \dots, p_{k(x)}(x)\}$, the leader will face the so-called *intermediate Stackelberg problem* with respect to D

$$(IS_D) \quad \min_{x \in X} \sum_{i=1}^{k(x)} p_i(x) L(x, y_i(x)).$$

Problem (IS_D) and the related solution concepts were introduced by Mallozzi and Morgan in [82, 83] also in the general case where $M(x)$ is not a discrete set for any $x \in X$ (see also [84] for an application to oligopolistic markets). Note that the *strong-weak Stackelberg problem* proposed in [1] includes particular intermediate problems where the leader attributes probability 0 to any point in $M(x)$ except for the best one and the worst one for him. Afterwards, an approach similar to [83] has been examined in [21] where a "partial cooperative" attitude of the follower is modelled in Stackelberg games with linear payoff functions.

Let us recall the solution concepts related to (IS_D).

Definition 4.2.4

- The infimum value $v_D = \inf_{x \in X} \sum_{i=1}^{k(x)} p_i(x) L(x, y_i(x))$ is the *intermediate value* of (IS_D).
- A leader's action $\bar{x} \in X$ is an *intermediate Stackelberg solution* with respect to D if

$$\bar{x} \in \operatorname{Arg\,min}_{x \in X} \sum_{i=1}^{k(x)} p_i(x) L(x, y_i(x)).$$

- An action profile $(\bar{x}, \bar{y}) \in X \times Y$ is an *intermediate Stackelberg equilibrium* with respect to D if $\bar{x}$ is an intermediate Stackelberg solution with respect to D and $\bar{y} \in M(\bar{x})$. △

Depending on the probability distribution $D(x)$ on $M(x)$ for any $x \in X$, problem (IS_D) could become (PB) or (OB). Examples on the connections among weak, strong and intermediate Stackelberg solutions are illustrated in [83], one of which is presented below for the sake of completeness.

About Comparing Weak, Strong and Intermediate Stackelberg Solutions
Consider $X = Y = [-1, 1]$, $L(x, y) = x + y$ and $F(x, y) = |y^2 - x^4|$. The follower's best reply correspondence M is defined on [-1,1] by $M(x) = \{x^2, -x^2\}$ and let $D(x) = \{\alpha, 1 - \alpha\}$ with $\alpha \in [0, 1]$ be the probability distribution attributed by the leader on $M(x)$, for any $x \in X$. Hence, the intermediate Stackelberg solution is

$$\operatorname*{Arg\,min}_{x\in X} \left(\alpha L(x, x^2) + (1-\alpha)L(x, -x^2)\right) = \begin{cases} \{-1\}, & \text{if } \alpha \in [0, 3/4] \\ \{-1/2(2\alpha - 1)\}, & \text{if } \alpha \in]3/4, 1]. \end{cases}$$

We point out that the solution to (IS_D) coincides with the solution to (PB) if $\alpha = 1$ and with the solution to (OB) if $\alpha \leq 3/4$. Furthermore, regarding the values, we have

$$v_D = \inf_{x\in X} \left(\alpha L(x, x^2) + (1-\alpha)L(x, -x^2)\right) = \begin{cases} 2\alpha - 2, & \text{if } \alpha \in [0, 3/4] \\ -1/4(2\alpha - 1), & \text{if } \alpha \in]3/4, 1], \end{cases}$$

and, reminded that we denoted with w and s, respectively, the pessimistic value and the optimistic value, it follows

$$-2 = s \leq v_D \leq w = -1/4$$

for any $\alpha \in [0, 1]$. The inequality above relies on a general property: the intermediate value is always between the optimistic value and the pessimistic value (see [83, Remark 2.1]).

4.2.5 *Subgame Perfect Nash Equilibrium Problem*

The equilibrium concepts presented until now consist of action profiles, that is pairs composed by one action of the leader and one action of the follower and, moreover, they are focused primarily on the leader's perspectives. Such concepts are broadly used in an engineering setting or by optimization practitioners, see for example [10, 28, 41, 81]. However, as traditional in a game-theoretical framework, we are now interested in an equilibrium concept which takes more into account the strategic aspects of the game. The equilibrium concept naturally fitting such goal is the *subgame perfect Nash equilibrium* (henceforth SPNE), which represents the most widely employed solution concept in an economic setting. The SPNE solution concept was introduced by the Nobel laureate Selten in [101] who suggested that players should act according to the so-called *principle of sequential rationality*:

"the equilibria to select are those whereby the players behave optimally from any point of the game onwards". For the notions of strategy, subgame and SPNE in a general game-theoretical framework and for further discussion on the behavioural implications and procedures to find SPNEs see, for example, [39, Chapter 3] and [87, Chapter 7].

We first recall the notion of *player's strategy* only for a two-player Stackelberg game $\Gamma = (X, Y, L, F)$. In such a game, the set of leader's strategies coincides with the set of leader's actions X, whereas the set of follower's strategies is the set of all the functions from X to Y, i.e. $Y^X = \{\varphi : X \to Y\}$.

The set of strategy profiles is $X \times Y^X$ and the definition of subgame perfect Nash equilibrium is characterized in the following way.

Definition 4.2.5 A strategy profile $(\bar{x}, \bar{\varphi}) \in X \times Y^X$ is an SPNE of Γ if the following conditions are satisfied:

(SG1) for each choice x of the leader, the follower minimizes his payoff function,

$$\bar{\varphi}(x) \in M(x) \quad \text{for any } x \in X;$$

(SG2) the leader minimizes his payoff function taking into account his hierarchical advantage, i.e.

$$\bar{x} \in \operatorname*{Arg\,min}_{x \in X} L(x, \bar{\varphi}(x)).$$

△

Note that the denomination "subgame perfect Nash equilibrium" is due to the following key features: first the SPNE notion is a refinement of the Nash equilibrium solution concept (introduced by the Nobel laureate Nash in [95]), secondly the restriction of an SPNE to any subgame constitutes a Nash equilibrium.

We emphasize that an SPNE consists of a strategy profile, that is a pair composed by one strategy of the leader (or equivalently one action of the leader) and one strategy of the follower (which is a function from the actions set of the leader to the actions set of the follower), as illustrated in Definition 4.2.5 and differently from all the equilibrium concepts defined before where only action profiles are involved. However, we can connect SPNEs and the equilibrium notions described above in this section, as pointed out in the following examples.

SPNEs of the Classical Stackelberg Games

The set of SPNEs of a Stackelberg game $\Gamma = (X, Y, L, F)$ where the follower's optimal response to any choice of the leader is unique can be fully characterized in terms of the solutions to the associate problem (SP) defined in Sect. 4.2.1.

(continued)

In fact, assumed that M is single-valued and set $M(x) = \{m(x)\}$ for any $x \in X$, by exploiting the Definitions 4.2.1 and 4.2.5, the following equivalence holds:

$$(\bar{x}, \bar{\varphi}) \in X \times Y^X \text{ is an SPNE of } \Gamma \iff \begin{array}{l} \bar{x} \text{ is a solution to } (SP) \text{ and} \\ \bar{\varphi}(x) = m(x) \text{ for any } x \in X \end{array}$$

(in such a framework m is the unique function that can be chosen as follower's strategy in an SPNE).

Moreover

$$(\bar{x}, \bar{\varphi}) \in X \times Y^X \text{ is an SPNE of } \Gamma \Longrightarrow (\bar{x}, \bar{\varphi}(\bar{x})) \in X \times Y \text{ is a Stackelberg equilibrium.}$$

For instance, in the Stackelberg leadership model presented in the example on page 81, the unique SPNE is the strategy profile $(\bar{q}_1, \bar{m})$ where $\bar{q}_1 \in [0, +\infty[$ and $\bar{m} \colon [0, +\infty[\to [0, +\infty[$ are given by

$$\bar{q}_1 = \frac{a-c}{2b} \quad \text{and} \quad \bar{m}(q_1) = \max\left\{0, \frac{a - bq_1 - c}{2b}\right\}.$$

We remind that $\bar{q}_1$ is the unique Stackelberg solution and, moreover, $(\bar{q}_1, \bar{m}(\bar{q}_1))$ is the unique Stackelberg equilibrium.

Non Single-Valued Follower Best Reply: SPNEs of a Game with Finite Action Sets

In the case where the follower's best reply correspondence is not single-valued, the solutions to the associate problem (PB) as well as to the associate problem (OB) defined in Sects. 4.2.2 and 4.2.3, respectively, can be part of an SPNE.

For instance, in the game of the example on page 84 (similarly to what is pointed out in [93, Example 2.1]) there are two SPNEs whose first component is a weak Stackelberg solution, namely $(T, \bar{\varphi}_1)$ and $(T, \bar{\varphi}_2)$ where $\bar{\varphi}_1$ and $\bar{\varphi}_2$ are the functions defined by

$$\bar{\varphi}_1(x) = \begin{cases} R, & \text{if } x = T \\ R, & \text{if } x = B \end{cases} \quad \text{and} \quad \bar{\varphi}_2(x) = \begin{cases} Q, & \text{if } x = T \\ R, & \text{if } x = B, \end{cases}$$

respectively; there are two SPNEs whose first component is a strong Stackelberg solution, namely $(B, \bar{\varphi}_3)$ and $(B, \bar{\varphi}_4)$ where $\bar{\varphi}_3$ and $\bar{\varphi}_4$ are the functions

(continued)

defined by

$$\bar{\varphi}_3(x) = \begin{cases} Q, & \text{if } x = T \\ Q, & \text{if } x = B \end{cases} \quad \text{and} \quad \bar{\varphi}_4(x) = \begin{cases} R, & \text{if } x = T \\ Q, & \text{if } x = B, \end{cases}$$

respectively. Focusing on the SPNEs $(T, \bar{\varphi}_1)$ and $(B, \bar{\varphi}_3)$, since

$$\{\bar{\varphi}_1(T)\} = \operatorname*{Arg\,max}_{y \in \{Q,R\}} L(T, y) \quad \text{and} \quad \{\bar{\varphi}_1(B)\} = \operatorname*{Arg\,max}_{y \in \{Q,R\}} L(B, y),$$

$$\{\bar{\varphi}_3(T)\} = \operatorname*{Arg\,min}_{y \in \{Q,R\}} L(T, y) \quad \text{and} \quad \{\bar{\varphi}_3(B)\} = \operatorname*{Arg\,min}_{y \in \{Q,R\}} L(B, y),$$

such SPNEs naturally reflect the pessimistic and the optimistic behaviours of the leader.

Therefore, having also in mind to deal with the issue of how reducing the number of SPNEs, we will take into account these latter types of SPNEs "induced" by weak or strong Stackelberg solutions.

Definition 4.2.6

- A strategy profile $(\bar{x}, \bar{\varphi}) \in X \times Y^X$ is an SPNE of Γ *induced by a weak Stackelberg solution* if $(\bar{x}, \bar{\varphi})$ is an SPNE of Γ which satisfies

  $$\bar{x} \text{ is a solution to problem } (PB) \quad \text{and} \quad \bar{\varphi}(x) \in \operatorname*{Arg\,max}_{y \in M(x)} L(x, y) \text{ for any } x \in X.$$

- A strategy profile $(\bar{x}, \bar{\varphi}) \in X \times Y^X$ is an SPNE of Γ *induced by a strong Stackelberg solution* if $(\bar{x}, \bar{\varphi})$ is an SPNE of Γ which satisfies

  $$\bar{x} \text{ is a solution to problem } (OB) \quad \text{and} \quad \bar{\varphi}(x) \in \operatorname*{Arg\,min}_{y \in M(x)} L(x, y) \text{ for any } x \in X.$$

△

Coming back to the second example on page 89, $(T, \bar{\varphi}_1)$ is an SPNE induced by a weak Stackelberg solution, $(B, \bar{\varphi}_3)$ is an SPNE induced by a strong Stackelberg solution, $(T, \bar{\varphi}_2)$ and $(B, \bar{\varphi}_4)$ are SPNEs that are not induced either by a weak or by a strong Stackelberg solution and the game has no further SPNEs.

Let us provide below an example where the SPNEs induced by weak or strong Stackelberg solutions are derived in a continuous setting, differently from the above example.

SPNEs Induced in a Game with Infinite Action Sets

In the game of the example on page 85, there is one weak Stackelberg solution to (PB) and one strong Stackelberg solution to (OB) and each one induces one SPNE: the strategy profiles $(2, \bar{\varphi})$ and $(7/4, \bar{\psi})$ where $\bar{\varphi}\colon [-2, 2] \to [-1, 1]$ and $\bar{\psi}\colon [-2, 2] \to [-1, 1]$ are defined respectively by

$$\bar{\varphi}(x) = \begin{cases} 1, & \text{if } x \in [-2, -7/4[\\ -1, & \text{if } x \in [-7/4, 2] \end{cases} \quad \text{and} \quad \bar{\psi}(x) = \begin{cases} 1, & \text{if } x \in [-2, 7/4] \\ -1, & \text{if } x \in]7/4, 2], \end{cases}$$

are the SPNEs induced by the weak Stackelberg solution and the strong Stackelberg solution, respectively. As in the previous example, also in this case there are SPNEs not induced either by weak or by strong Stackelberg solutions, like the strategy profile $(2, \bar{\vartheta})$ where $\bar{\vartheta}\colon [-2, 2] \to [-1, 1]$ is defined by

$$\bar{\vartheta}(x) = \begin{cases} 1, & \text{if } x \in [-2, -7/4[\\ -4x/7, & \text{if } x \in [-7/4, 7/4[\\ -1, & \text{if } x \in [7/4, 2]. \end{cases}$$

4.3 Crucial Issues in Stackelberg Games

After having presented the mathematical problems associated to a Stackelberg game interaction, we describe the main issues arising when one deals with the different kinds of problems illustrated in the previous section. In particular, we will focus on the following topics:

- **Existence:** do the solutions of the problem exist under not too restrictive assumptions on the actions sets and on the payoff functions (possibly discontinuous, as not unusual in economic frameworks)?
- **"Variational" stability:** the data of the problem could be affected by uncertainty, hence given a perturbation of the original problem acting on the actions sets and/or on the payoff functions, does a sequence of solutions of the perturbed problems converge to a solution of the original problem? And what about the values? Such analysis, which helps to predict the "behaviour" of the game when one is not able to deal with exact equilibria, and similar analyses that compare outcomes before and after a change in exogenous parameters are known in economics as *comparative statics*.

- **Well-posedness:** does the problem have a unique solution? And does any method which constructs an "approximating sequence" automatically allow to approach a solution?
- **Approximation:** how constructing appropriate concepts of approximate solutions in order to both obviate the lack of solutions of a problem and to manage problems where infinite dimensional spaces are involved?
- **Approximation via numerical methods:** how transforming the original problem into a better-behaved equivalent problem? And how overcoming the numerical difficulties which derive from the possible non-single-valuedness of the follower's best reply correspondence?
- **Selection:** if the problem has more than one solution, is it possible to construct a procedure which allows to reduce the number of solutions or, better yet, to pick just one solution? Which are the motivations that would induce the players to act according to such a procedure in order to reach the designed solution?

Obviously the same issues appear also for the equilibria.

Now, we discuss which of the issues displayed above can be positively or negatively answered.

4.3.1 Stackelberg Problems

Problem (SP) is well-behaved regarding all the topics just summarized. In particular, by applying the maximum theorem in [6, 16], the existence of Stackelberg solutions and Stackelberg equilibria is guaranteed provided that the action sets X and Y are compact subsets of two Euclidean spaces and that the payoff functions L and F are continuous over $X \times Y$.

Remark 4.3.1 However, we point out that the functions involved in many real world situations are not always continuous and in infinite dimensional frameworks requiring the continuity is very restrictive. So, as proved in propositions 4.1 and 5.1 in [91], one can only assume that

- L is lower semicontinuous over $X \times Y$,
- F is lower semicontinuous over $X \times Y$,
- for any $(x, y) \in X \times Y$ and any sequence $(x_n)_n$ converging to x in X, there exists a sequence $(\tilde{y}_n)_n$ in Y such that

$$\limsup_{n \to +\infty} F(x_n, \tilde{y}_n) \leq F(x, y).$$

Furthermore, it is worth to highlight that the second and the third condition stated above do not imply the continuity of F (see, for example, [49, Example 3.1.1]). △

As regards variational stability of the Stackelberg solution, equilibrium and value of (SP) under perturbations of the payoff functions, consider the perturbed problems

$$(SP)_n \begin{cases} \min_{x \in X} L_n(x, m_n(x)) \\ \text{where } m_n(x) \text{ is the solution to } \min_{y \in Y} F_n(x, y), \end{cases}$$

with L_n and F_n real-valued functions defined on $X \times Y$, for any $n \in N$. Then, the following results hold.

Proposition 4.3.2 ([73, Prop. 3.1, 3.2 and 3.3]) *Assume that X and Y are compact subsets of two Euclidean spaces and that the sequences $(L_n)_n$ and $(F_n)_n$ continuously converge to L and F, respectively, i.e. for any $(x, y) \in X \times Y$ and any sequence $(x_n, y_n)_n$ converging to (x, y) in $X \times Y$, we have*

$$\lim_{n \to +\infty} L_n(x_n, y_n) = L(x, y) \quad \textit{and} \quad \lim_{n \to +\infty} F_n(x_n, y_n) = F(x, y).$$

Then, about the second-stage problem, for any sequence $(x_n)_n$ converging to x in X,

$$\lim_{n \to +\infty} m_n(x_n) = m(x) \quad \textit{and} \quad \lim_{n \to +\infty} F_n(x_n, m_n(x_n)) = F(x, m(x)).$$

Moreover, regarding the first-stage problem,

- *for any sequence $(\bar{x}_n)_n$ such that $\bar{x}_{n_k} \in \operatorname{Arg\,min}_{x \in X} L_{n_k}(x, m_{n_k}(x))$ and $(\bar{x}_{n_k})_k$ converges to $\bar{x}$ in X for a selection of integers $(n_k)_k$, we have*

$$\bar{x} \in \operatorname*{Arg\,min}_{x \in X} L(x, m(x)),$$

- $\limsup_{n \to +\infty} \left(\inf_{x \in X} L_n(x, m_n(x)) \right) \leq \inf_{x \in X} L(x, m(x)).$

Finally, if $(\bar{x}_n, \bar{y}_n) \in X \times Y$ is a Stackelberg equilibrium of $(SP)_n$ for any $n \in \mathbb{N}$, then any convergent subsequence of the sequence $(\bar{x}_n, \bar{y}_n)_n$ in $X \times Y$ has a limit $(\bar{x}, \bar{y})$ which is a Stackelberg equilibrium of (SP) and satisfies

$$\lim_{n \to +\infty} L_n(\bar{x}_n, \bar{y}_n) = L(\bar{x}, \bar{y}) \quad \textit{and} \quad \lim_{n \to +\infty} F_n(\bar{x}_n, \bar{y}_n) = F(\bar{x}, \bar{y}). \qquad \triangle$$

Remark 4.3.3 Stability results illustrated in Proposition 4.3.2 hold even by replacing the convergence assumptions with weaker convergence requirements, as proved in [73]. In particular

- the continuous convergence of $(L_n)_n$ to L can be substituted by the following two conditions:

(a) for any $(x, y) \in X \times Y$ and any sequence $(x_n, y_n)_n$ converging to (x, y) in $X \times Y$, we have

$$\liminf_{n\to+\infty} L_n(x_n, y_n) \geq L(x, y);$$

(b) for any $x \in X$ there exists a sequence $(\tilde{x}_n)_n$ converging to x in X such that, for any $y \in Y$ and any sequence $(y_n)_n$ converging to y in Y, we have

$$\limsup_{n\to+\infty} L_n(\tilde{x}_n, y_n) \leq L(x, y);$$

- whereas, the continuous convergence of $(F_n)_n$ to F can be substituted by the following two conditions:

(c) for any $(x, y) \in X \times Y$ and any sequence $(x_n, y_n)_n$ converging to (x, y) in $X \times Y$, we have

$$\liminf_{n\to+\infty} F_n(x_n, y_n) \geq F(x, y);$$

(d) for any $(x, y) \in X \times Y$ and any sequence $(x_n)_n$ converging to x in X, there exists a sequence $(\tilde{y}_n)_n$ in Y such that

$$\limsup_{n\to+\infty} F_n(x_n, \tilde{y}_n) \leq F(x, y).$$

Examples and counterexamples on how conditions (a)–(d) are connected with known convergence notions can be found in remarks 2.2 and 2.3 in [73] where perturbations on the constraints of the leader and the follower are also considered. △

Results on well-posedness of (SP) can be found in [68, 91], whereas in [28, Chapter 6] numerical approximation and algorithmic issues are widely investigated.

Concerning problem (GSP) defined in Sect. 4.2.1, we mention that in [103] a first approximation technique based on a barrier method has been proposed, then in [67] a general approximation scheme involving conditions of minimal character has been introduced together with applications to barrier methods, whereas in [72] also external penalty methods have been considered.

4.3.2 *Weak Stackelberg (or Pessimistic Bilevel Optimization) Problem*

Problem (PB) is the worst-behaved among the problems illustrated in Sect. 4.2. In fact, the compactness of the action sets and the continuity of the payoff functions do not guarantee, in general, the existence of weak Stackelberg solutions and weak

Stackelberg equilibria even in a finite dimensional setting, as proved in many folk examples in literature (see [15, 80] and [10, Remark 4.3]) and in the following one.

Weak Stackelberg Solutions May Not Exist
Consider $X = Y = [-1, 1]$, $L(x, y) = -x + y$ and $F(x, y) = -xy$. The follower's best reply correspondence M is defined on [-1,1] by

$$M(x) = \begin{cases} \{-1\}, & \text{if } x \in [-1, 0[\\ [-1, 1], & \text{if } x = 0 \\ \{1\}, & \text{if } x \in]0, 1]. \end{cases} \tag{4.3.1}$$

Since

$$\sup_{y \in M(x)} L(x, y) = \begin{cases} -x - 1, & \text{if } x \in [-1, 0[\\ -x + 1, & \text{if } x \in [0, 1], \end{cases}$$

then $\text{Arg}\min_{x\in X} \sup_{y\in M(x)} L(x, y) = \emptyset$. Hence such a (PB) has no solution and no weak Stackelberg equilibrium.

Nevertheless, existence results for very special classes of problem (PB) can be found in [2, 3, 80].

In a more general framework, to overcome the lack of solutions, for any $\epsilon > 0$ the concept of ϵ-approximate solution of (PB) has been investigated in a sequential setting by Loridan and Morgan in [71], where the existence of such solutions is analyzed under mild convexity assumptions on the follower's payoff function, and in [74] where quasi-convexity assumptions are considered.

Then, in [76] the notion of strict ϵ-approximate solution has been introduced and related existence results have been provided without requiring convexity assumptions. A crucial property of these two concepts is the convergence of the approximate pessimistic values towards the pessimistic value of (PB) as ϵ tends to zero. We point out that all the results in the papers just mentioned have been obtained in the case of non-parametric constraints.

Afterwards, Lignola and Morgan in [51] extensively investigated all the possible kinds of constraints (including those defined by parametric inequalities) in a topological setting. More recently, new notions of approximate solutions have been introduced in [63, 64] in an easier-to-manage sequential setting which allow to construct a surrogate for the solution of (PB) called *viscosity solution*. Such concepts will be presented in Sect. 4.4 together with their properties.

Regarding the variational stability issue of (PB), consider the following perturbed problems

$$(PB)_n \quad \begin{cases} \min\limits_{x \in X} \sup\limits_{y \in M_n(x)} L_n(x, y) \\ \text{where } M_n(x) = \operatorname{Arg\,min}\limits_{y \in Y} F_n(x, y), \end{cases}$$

with L_n and F_n real-valued functions defined on $X \times Y$, for any $n \in N$. Even requiring compactness and heavy convergence assumptions, variational stability of the weak Stackelberg solutions and the pessimistic values under perturbations of the payoff functions does not hold in general, as described in the example below.

Pessimistic Values May Not Converge

Consider $X = Y = [-1, 1]$, $L(x, y) = x + y$, $F(x, y) = 0$ and let $L_n(x, y) = x + y + 1/n$ and $F_n(x, y) = y/n$ for any $n \in \mathbb{N}$.

Although the sequence of functions $(L_n)_n$ continuously converges to L and the sequence $(F_n)_n$ continuously converges to F, the sequence of pessimistic values of $(PB)_n$ does not converge to the pessimistic value of (PB).

In fact, since the follower's best reply correspondences in $(PB)_n$ and in (PB) are defined respectively by $M_n(x) = \{-1\}$ and by $M(x) = [-1, 1]$ for any $x \in X$, then

$$\inf_{x \in X} \sup_{y \in M_n(x)} L_n(x, y) = -2 + 1/n \quad \text{and} \quad \inf_{x \in X} \sup_{y \in M(x)} L(x, y) = 0.$$

Hence

$$-2 = \lim_{n \to +\infty} \left(\inf_{x \in X} \sup_{y \in M_n(x)} L_n(x, y) \right) \neq \inf_{x \in X} \sup_{y \in M(x)} L(x, y) = 0.$$

However, convergence results of the exact solutions to $(PB)_n$ towards a solution to (PB) have been obtained in [69] for a special class of problems.

In a more general framework, in order to obviate the lack of stability, a first attempt to manage the stability issue (intended in the Hausdorff's sense) of (PB) can be found in [89] by means of approximate solutions. Afterwards, "variational" stability (in the sense of the above example) for such approximate solutions has been investigated under conditions of minimal character and assuming non-parametric constraints in [71] (under mild convexity assumptions on the follower's payoff function) and in [74] (under quasi-convexity assumptions). For applications to interior penalty methods see [70, Section 5] and to exterior penalty methods see [74, Section 5].

Then, a comprehensive analysis on all the types of follower's constraints has been provided in [51] for a general topological framework. Such results will be presented in Sect. 4.4.

Furthermore, (PB) is hard to manage even regarding well-posedeness properties, investigated in [91].

Concerning the numerical approximation issue of (PB), various regularization methods involving both the follower's optimal reaction set and the follower's payoff function have been proposed in order to approach problem (PB) via a sequence of Stackelberg problems; for first results, see [29, 75, 77, 78, 88]. These methods, their properties and the results achieved will be presented in Sect. 4.5.

4.3.3 Strong Stackelberg (or Optimistic Bilevel Optimization) Problem

Problem (OB) is ensured to have strong Stackelberg solutions under the same assumptions stated for the existence of solutions of (SP), i.e. by assuming that the action sets X and Y are compact subsets of two Euclidean spaces and that the payoff functions L and F are continuous over $X \times Y$.

Remark 4.3.4 The continuity assumptions can be weakened as in Remark 4.3.1, by applying Proposition 4.1.1 in [49] about the lower semicontinuity of the marginal function. △

We remind that results concerning the existence of strong Stackelberg solutions have been first obtained in [53].

As regards to the variational stability issue of (OB), consider the following perturbed problems

$$(OB)_n \quad \begin{cases} \min\limits_{x \in X} \inf\limits_{y \subset M_n(x)} L_n(x, y) \\ \text{where } M_n(x) = \operatorname*{Arg\,min}\limits_{y \in Y} F_n(x, y), \end{cases}$$

with L_n and F_n real-valued functions defined on $X \times Y$, for any $n \in N$.

We point out that problem (OB), as well as (PB), can exhibit under perturbation a lack of convergence of the strong Stackelberg solutions and of the optimistic values, as illustrated in Example 4.1 in [52], rewritten below for the sake of completeness.

Optimistic Values May Not Converge
Consider $X = Y = [0, 1]$, $L(x, y) = x - y + 1$, $F(x, y) = x$ and let $L_n(x, y) = L(x, y)$ and $F_n(x, y) = y/n + x$ for any $n \in \mathbb{N}$.

Though the sequences of functions $(L_n)_n$ and $(F_n)_n$ continuously converge to L and F, respectively, the sequence of optimistic values of $(OB)_n$ does not converge to the optimistic value of (OB).

Indeed, as the follower's best reply correspondences in $(OB)_n$ and in (OB) are defined respectively by $M_n(x) = \{0\}$ and by $M(x) = [0, 1]$ for any $x \in X$, then

$$\inf_{x \in X} \inf_{y \in M_n(x)} L_n(x, y) = 1 \quad \text{and} \quad \inf_{x \in X} \inf_{y \in M(x)} L(x, y) = 0.$$

Therefore,

$$\lim_{n \to +\infty} \left(\inf_{x \in X} \inf_{y \in M_n(x)} L_n(x, y) \right) \neq \inf_{x \in X} \inf_{y \in M(x)} L(x, y).$$

However, notions of approximate solutions have been introduced and investigated in [52] in order to face the lack of stability in (OB), whereas achievements on well-posedness can be derived by applying results obtained in [54, 55].

Obviously, as well as in (PB), the non-single-valuedness of the follower's best reply correspondence M gives rise to difficulties from the numerical point of view in (OB).

4.3.4 *Intermediate Stackelberg Problem*

Sufficient conditions for the existence of intermediate Stackelberg solutions of (IS_D) are investigated in [83, Section 3] in the general case where $M(x)$ is not a discrete set for at least one $x \in X$. However, we point out that problem (IS_D) inherits all the difficulties illustrated for (PB) about existence, stability, well-posedness and approximation issues.

4.3.5 *Subgame Perfect Nash Equilibrium Problem*

When the follower's best reply correspondence M is not a single-valued map, infinitely many SPNEs could come up in Stackelberg games, as shown in the following example.

SPNEs May Be Infinite

Consider $X = Y = [-1, 1]$, $L(x, y) = x + y$ and $F(x, y) = -xy$. The follower's best reply correspondence M is given in (4.3.1). Denoted with φ^σ the function defined on $[-1, 1]$ by

$$\varphi^\sigma(x) = \begin{cases} -1, & \text{if } x \in [-1, 0[\\ \sigma, & \text{if } x = 0 \\ 1, & \text{if } x \in]0, 1], \end{cases}$$

where $\sigma \in [-1, 1]$, it follows that $\varphi^\sigma(x) \in M(x)$ for any $x \in [-1, 1]$ and $\text{Arg}\min_{x \in X} L(x, \varphi^\sigma(x)) = \{-1\}$. Therefore, the strategy profile $(-1, \varphi^\sigma)$ is an SPNE of Γ for any $\sigma \in [-1, 1]$, so Γ has infinitely many SPNEs.

Hence, restricting the number of SPNEs becomes essential and the selection issue arises. For further discussion on the theory of equilibrium selection in games, see [40].

Remark 4.3.5 We showed via the second example on page 89 and the example on page 91 that starting from a solution to the associated (PB) or to the associated (OB) one can induce an SPNE motivated according to each of the two different behaviours of the leader. Finding such SPNEs induced by weak or strong Stackelberg solutions, as defined in Definition 4.2.6, provides two methods to select an SPNE in Stackelberg games. Nevertheless such methods can be exploited only in the two extreme situations illustrated before. Anyway they require to the leader of knowing the follower's best reply correspondence and, at the best of our knowledge, they do not exhibit a manageable constructive approach to achieve an SPNE. △

Furthermore, also the numerical approximation issue matters for the SPNEs in Stackelberg games since M can be not single-valued. Therefore, it would be desirable to design constructive methods in order to select an SPNE with the following features: relieving the leader of knowing M, allowing to overcome the difficulties deriving from the possible non-single-valuedness of M, providing some behavioural motivations of the players (different from the extreme situations described above). Results in this direction have been first obtained in [93], where an SPNE selection method based on Tikhonov regularization is proposed, and more recently in [22], where proximal regularization is employed.

In the rest of the chapter we will again present results under easy-to-manage continuity or convergence assumptions, but we point out that the papers quoted in the statements involve conditions of minimal character, as in Remarks 4.3.1 and 4.3.3.

4.4 Regularizing the Follower's Optimal Reaction Set in Stackelberg Games

The first attempts for approximating a Stackelberg game $\Gamma = (X, Y, L, F)$ by a general sequence of perturbed or regularized games have been carried out by Loridan and Morgan. They started in [67, 68, 73] with the case where the second-stage problem (P_x) has a unique solution for any x and there is no difference between the optimistic and the pessimistic bilevel optimization problem; then, they extensively investigated in [69–71, 74–77] pessimistic bilevel problems in which the follower's constraints do not depend on the leader's actions.

In this section, we illustrate regularization methods for the bilevel problems (PB) and (OB) and also for bilevel problems in which the follower's problem is modeled by an *equilibrium* or a *quasi-equilibrium* problem. Due to the relevance of inequality constraints investigation in optimization theory, we assume that the follower's constraints are described by a finite number of inequalities whereas, for the sake of brevity, the leader's constraints are assumed to be constant. Then, we set

$$K(x) = \bigcap_{i=1}^{k} K_i(x) = \bigcap_{i=1}^{k} \{y \in Y \text{ such that } g_i(x, y) \leq 0\}, \qquad (4.4.1)$$

where X and Y are nonempty subsets of two Euclidean spaces, g_i is a function from $X \times Y$ to $\mathbb{R}$ for $i = 1, \ldots, k$, and we denote by:

- $\mathcal{S}$ the set of solutions to the optimistic bilevel problem (OB), i.e. the set of strong Stackelberg solutions of (OB).
- $\mathcal{W}$ the set of solutions to the pessimistic bilevel problem (PB), i.e. the set of weak Stackelberg solutions of (PB).

4.4.1 *Regularization of Pessimistic Bilevel Optimization Problems*

First investigations on pessimistic bilevel optimization problems mainly consisted in trying to obviate the lack of solutions, which may arise, as shown in the example on page 95, even in very simple models. In fact, in that example the marginal function

$$e(x) = \sup_{y \in M(x)} L(x, y)$$

is not lower semicontinuous and this is a possible consequence of the fact that the map M does not enjoy an important property, namely *lower semicontinuity*.

Definition 4.4.1 ([6]) Let X and Y be nonempty subsets of two Euclidean spaces.

A set-valued map $T : x \in X \rightrightarrows T(x) \subseteq Y$ is lower semicontinuous over X if for any $x \in X$, for any sequence $(x_n)_n$ converging to x in X and any $y \in T(x)$ there exists a sequence $(y_n)_n$ converging to y in Y such that $y_n \in T(x_n)$ for n sufficiently large, that is

$$T(x) \subseteq \operatorname*{Lim\,inf}_n T(x_n).$$

A set-valued map T from X to Y is closed over X if the graph of T is closed in $X \times Y$, that is for any sequence $(x_n, y_n)_n$ converging to $(x, y) \in X \times Y$, such that $y_n \in T(x_n)$ for $n \in \mathbb{N}$, one has $y \in T(x)$, equivalent to

$$\operatorname*{Lim\,sup}_n T(x_n) \subseteq T(x).$$

Here, Lim inf and Lim sup denote, respectively, the Painlevé-Kuratowski lower and upper limit of a sequence of sets. △

Then, let us recall the well-known *maximum theorem* [6, 16].

Lemma 4.4.2 *If the following assumptions are satisfied:*

- *the function L is lower semicontinuous over $X \times Y$;*
- *the set-valued map $T : X \rightrightarrows Y$ is lower semicontinuous over X;*

then the marginal function defined by

$$\sup_{y \in T(x)} L(x, y)$$

is lower semicontinuous over X. △

4.4.1.1 Approximate and Strict Approximate Solutions

An attempt to present a complete discussion, which aimed to unify all previous results in a general scheme, was given in [51] in the setting of topological vector spaces and by analyzing the second-stage problem in the presence and in the absence of general or inequality constraints. Unfortunately, using Γ-limits and working in a topological setting have restricted the readability and the applicability of that paper, so, here we recall the main idea and the principal results in a simplified form but under more restrictive assumptions.

Let $\varepsilon > 0$. The ε-argmin map

$$M^{\varepsilon} : x \in X \rightrightarrows M^{\varepsilon}(x) = \left\{ y \in K(x) \text{ such that } F(x, y) \leq \inf_{z \in K(x)} F(x, z) + \varepsilon \right\},$$

can be a potential candidate to substitute the argmin map M but, unfortunately, also M^{ε} may fail to be lower semicontinuous even when the constraints do not depend on x. However, employing strict inequalities (and passing to large inequalities when possible) is the right way to proceed in order to get the lower semicontinuity property of the ε-minima. Indeed, denoted by $\widetilde{M}^{\varepsilon}$ the map

$$\widetilde{M}^{\varepsilon} : x \in X \rightrightarrows \widetilde{M}^{\varepsilon}(x) = \left\{ y \in K(x) \text{ such that } F(x,y) < \inf_{z \in K(x)} F(x,z) + \varepsilon \right\},$$

in the following example (Example 2.3 in [63]) $\widetilde{M}^{\varepsilon}$ is lower semicontinuous even if the maps M and M^{ε} are not.

About Lower Semicontinuity of the ε-Minima

Let $X = Y = [0,1]$, $K(x) = [0,1]$ for any $x \in X$. Consider $F(x,y) = -y^2 + (1+x)y - x$ for every $(x,y) \in [0,1]^2$ and any $\varepsilon \in [0, 1/4[$. Then, one gets:

- $M(0) = \{0,1\}$ and $M(x) = \{0\}$ for every $x \in]0,1]$;
- $M^{\varepsilon}(x) = [0, \alpha^{\varepsilon}(x)] \cup [\beta^{\varepsilon}(x), 1]$, if $x \in [0, \varepsilon]$;
- $M^{\varepsilon}(x) = [0, \alpha^{\varepsilon}(x)]$, if $x \in]\varepsilon, 1]$;

where $\alpha^{\varepsilon}(x) = \frac{1}{2}\left(x + 1 - \sqrt{(x+1)^2 - 4\varepsilon}\right)$,
$\beta^{\varepsilon}(x) = \frac{1}{2}\left(x + 1 + \sqrt{(x+1)^2 - 4\varepsilon}\right)$ solve the equation

$$y^2 - (x+1)y + \varepsilon = 0.$$

So, both the maps M and M^{ε} are not lower semicontinuous over X: M lacks the lower semicontinuity property at the point $x = 0$ whereas M^{ε} at the point $x = \varepsilon$. Moreover,

- $\widetilde{M}^{\varepsilon}(x) = [0, \alpha^{\varepsilon}(x)[\cup]\beta^{\varepsilon}(x), 1]$, if $x \in [0, \varepsilon[$
- $\widetilde{M}^{\varepsilon}(x) = [0, \alpha^{\varepsilon}(x)[$, if $x \in [\varepsilon, 1]$.

Then, it is easy to see that $\widetilde{M}^{\varepsilon}$ is lower semicontinuous on whole X.

In order to prove the lower semicontinuity of $\widetilde{M}^{\varepsilon}$, a crucial role is played both by the lower semicontinuity and closedness of the constraints map K and by the continuity of the follower's payoff function F, as shown by the next lemmas, which concern, respectively, the continuity and convexity properties of inequality constraints maps and the lower semicontinuity of $\widetilde{M}^{\varepsilon}$ and M^{ε}. The first lemma can be found in [13, theorems 3.1.1 and 3.1.6].

Lemma 4.4.3 *If the set X is closed and the following assumption is satisfied:*

(C_1) *for every $i = 1, \dots, k$, the function g_i is lower semicontinuous over $X \times Y$; then, the set-valued map K is closed on X.*

If the following assumptions are satisfied:

(C_2) *for every $i = 1, \dots, k$ and $y \in Y$, the function $g_i(\cdot, y)$ is upper semicontinuous over X*

(C_3) *for every $i = 1, \dots, k$ and $x \in X$, the function $g_i(x, \cdot)$ is strictly quasiconvex (see [13]) and upper semicontinuous on Y, which is assumed to be convex;*

(C_4) *for every $x \in X$ there exists $y \in Y$ such that $\max_{i=1,\dots,k} g_i(x, y) < 0$;*

then, the set-valued map K is lower semicontinuous and convex-valued over X. △

Remark 4.4.4 We emphasize that the strict quasiconvexity of the function $g_i(x, \cdot)$ guarantees that

$$K_i(x) \subseteq \text{cl}\,(\text{int}\, K_i(x))$$

where the map K_i is defined in (4.4.1). This condition is crucial in proving the lower semicontinuity of the map K_i, but K_i can be lower semicontinuous even if the function $g_i(x, \cdot)$ is not strictly quasiconvex as it occurs, for example, with $g_i(x, y) = x^2 - y^2$. △

Lemma 4.4.5 ([51, Th. 4.3]) *Under the assumptions $(C_1) - (C_4)$, if the set X is closed, the set Y is convex and compact and the function F is continuous over $X \times Y$, then the map $\widetilde{M}^\varepsilon$ is lower semicontinuous over X.* △

In order to get the lower semicontinuity of M^ε, a crucial role is also played by suitable *convexity* properties of the data which guarantee that $M^\varepsilon(x) \subseteq \text{cl}\left(\widetilde{M}^\varepsilon(x)\right)$ for any $x \in X$.

Lemma 4.4.6 ([51, Th. 5.4]) *Under the assumptions of Lemma 4.4.5, if also the following condition holds:*

(H_1) *the function $F(x, \cdot)$ is strictly quasiconvex on $K(x)$, for every $x \in X$*

then, the map M^ε is lower semicontinuous over X. △

The next step consists in achieving the lower semicontinuity of the marginal functions

$$e^\varepsilon : x \in X \to \sup_{y \in M^\varepsilon(x)} L(x, y) \qquad \widetilde{e}^\varepsilon : x \in X \to \sup_{y \in \widetilde{M}^\varepsilon(x)} L(x, y),$$

using Lemma 4.4.2, and then the existence of two types of approximate solutions for the problem (PB), $\mathcal{W}^\varepsilon$ and $\widetilde{\mathcal{W}}^\varepsilon$, defined by

$$x^\varepsilon \in \mathcal{W}^\varepsilon \Longleftrightarrow e^\varepsilon(x^\varepsilon) = \inf_{x \in X} e^\varepsilon(x) \Longleftrightarrow \sup_{y \in M^\varepsilon(x^\varepsilon)} L(x^\varepsilon, y) = \inf_{x \in X} \sup_{y \in M^\varepsilon(x)} L(x, y),$$

$$\widetilde{x}^\varepsilon \in \widetilde{\mathcal{W}}^\varepsilon \Longleftrightarrow \widetilde{e}^\varepsilon(\widetilde{x}^\varepsilon) = \inf_{x \in X} \widetilde{e}^\varepsilon(x) \Longleftrightarrow \sup_{y \in \widetilde{M}^\varepsilon(\widetilde{x}^\varepsilon)} L(\widetilde{x}^\varepsilon, y) = \inf_{x \in X} \sup_{y \in \widetilde{M}^\varepsilon(x)} L(x, y).$$

Proposition 4.4.7 *Assume that the set X is compact and the function L is lower semicontinuous over $X \times Y$.*

If assumptions of Lemma 4.4.5 hold, then the approximate solutions set $\widetilde{\mathcal{W}}^\varepsilon$ is nonempty for any $\varepsilon > 0$.

If assumptions of Lemma 4.4.6 hold, then the approximate solutions set $\mathcal{W}^\varepsilon$ is nonempty for any $\varepsilon > 0$. △

Now, two questions naturally arise:

- do the approximate security values w^ε and $\widetilde{w}^\varepsilon$, defined respectively by

$$w^\varepsilon = \inf_{x \in X} \sup_{y \in M^\varepsilon(x)} L(x, y) \qquad \widetilde{w}^\varepsilon = \inf_{x \in X} \sup_{y \in \widetilde{M}^\varepsilon(x)} L(x, y),$$

  converge towards the security value w of the original problem when ε tends to zero?
- are such approximate solutions *stable* with respect to perturbations of the data, for any fixed $\varepsilon > 0$?

The term *stable* should be intended in the following sense. Assume that the data of the problem (PB) are asymptotically approached by the data of the perturbed problems

$$(PB)_n \quad \text{find } x_n \in X \text{ such that} \quad \sup_{y \in M_n(x_n)} L_n(x_n, y) = \inf_{x \in X} \sup_{y \in M_n(x)} L_n(x, y),$$

where $(F_n)_n$ and $(L_n)_n$ are sequences of functions from $X \times Y$ to $\mathbb{R}$. Let, for any n, M_n^ε is the ε-argmin map of F_n with constraints K_n. More specifically, given k sequences of real valued functions $(g_{i,n})_n$ defined on $X \times Y$, we set

$$K_n(x) = \bigcap_{i=1}^{k} \{y \in Y \text{ such that } g_{i,n}(x, y) \leq 0\} \tag{4.4.2}$$

and we denote by M_n^ε the map

$$M_n^\varepsilon : x \in X \rightrightarrows M_n^\varepsilon(x) = \left\{ y \in K_n(x) : F_n(x, y) \leq \inf_{z \in K_n(x)} F_n(x, z) + \varepsilon \right\}$$

and by $\mathcal{W}_n^\varepsilon$ the set

$$\mathcal{W}_n^\varepsilon = \left\{ x \in X : \sup_{y \in M_n^\varepsilon(x)} L_n(x, y) = \inf_{u \in X} \sup_{y \in M_n^\varepsilon(u)} L_n(u, y) = w_n^\varepsilon \right\}.$$

Then, one asks under which assumptions any convergent sequence $(x_n^\varepsilon)_n$ of approximate solutions for $(PB)_n$ converges to $x^\varepsilon \in \mathcal{W}^\varepsilon$ and one has

$$\operatorname*{Limsup}_n \mathcal{W}_n^\varepsilon \subseteq \mathcal{W}^\varepsilon, \quad \forall\, \varepsilon > 0. \tag{4.4.3}$$

The following propositions answer to both questions on the previous page.

Proposition 4.4.8 ([63, Cor. 2.2]) *Under the assumptions of Lemma 4.4.5, if in addition the function L is upper semicontinuous over $X \times Y$, then*

$$\lim_{\varepsilon \to 0} \widetilde{w}^\varepsilon = w = \lim_{\varepsilon \to 0} w^\varepsilon.$$

△

We remark that:

- under the assumptions of Proposition 4.4.8, the map M^ε may not be lower semicontinuous over X, so this property is not necessary for the approximate security values convergence;
- no convexity assumption is made on g_i, F and L.

Proposition 4.4.9 ([51, Th. 7.6]) *Assume that X is compact and Y is convex and compact. If the sequences of functions $(F_n)_n$ and $(g_{i,n})_n$ continuously converge over $X \times Y$ to the functions F and g_i respectively, L is lower semicontinuous over $X \times Y$, assumptions $(C_1) - (C_4)$ and the following hold for any $n \in \mathbb{N}$:*

- $(C_{3,n})$ *for every $i = 1, \ldots, k$ and $x \in X$, the function $g_{i,n}(x, \cdot)$ is strictly quasi-convex on Y;*
- $(C_{4,n})$ *for every $x \in X$ there exists $y \in Y$ such that $\max_{i=1,\ldots,k} g_{i,n}(x, y) < 0$;*
- $(H_{2,n})$ *for any $(x, y) \in X \times Y$ and any sequence $(x_n, y_n)_n$ converging to (x, y) in $X \times Y$ one has*

$$\limsup_{n \to +\infty} L_n(x_n, y_n) \le L(x, y);$$

then, for any $\varepsilon > 0$, we get:

- *the ε-solutions are stable, i.e. inclusion in* (4.4.3) *holds,*
- *the ε-values are stable, i.e.* $\lim_{n \to +\infty} w_n^\varepsilon = w^\varepsilon$. △

We stress that, in general, one cannot prove a result analogous to (4.4.3) also for the strict approximate solutions $\widetilde{\mathcal{W}}^\varepsilon$, due to the *open* nature of them. A natural extension of the previous investigations was the research of possible candidates to substitute the lacking weak Stackelberg solutions to (PB), which will be presented in the next subsection.

We mention that the approximate solutions presented above have been afterwards employed in [1, 4] (about the *strong-weak Stackelberg problem*, which generalizes problems (PB) and (OB)) and in [45].

4.4.1.2 Inner Regularizations and Viscosity Solutions

The following procedure seems to be quite natural to obviate the lack of solutions for pessimistic bilevel optimization problems:

- defining regularizing problems which have solutions under not too restrictive conditions;
- utilizing sequences of solutions to the regularizing problems admitting convergent subsequences;
- considering a limit point of one of such subsequences as a *surrogate* solution to (PB) provided that the supremum values of the objective functions calculated in such approximating solutions converge to the security value w.

To accomplish the first step, we identify the set-valued map properties that are helpful in this regularization process.

Definition 4.4.10 ([63, Def. 2.1]) A family of set-valued maps $\mathfrak{T} = \{\mathcal{T}^{\varepsilon}, \varepsilon > 0\}$, where

$$\mathcal{T}^{\varepsilon} : x \in X \rightrightarrows \mathcal{T}^{\varepsilon}(x) \subseteq Y,$$

is an inner regularization for the family of minimum problems $\{(P_x), x \in X\}$ if the conditions below are satisfied:

(R_1) $M(x) \subseteq \mathcal{T}^{\varepsilon}(x) \subseteq \mathcal{T}^{\eta}(x)$ for every $x \in X$ and $0 < \varepsilon < \eta$;

(R_2) for any $x \in X$, any sequence $(x_n)_n$ converging to x in X and any sequence of positive numbers $(\varepsilon_n)_n$ decreasing to zero, one has

$$\operatorname*{Lim\,sup}_n \mathcal{T}^{\varepsilon_n}(x_n) \subseteq M(x);$$

(R_3) $\mathcal{T}^{\varepsilon}$ is a lower semicontinuous set-valued map on X, for every $\varepsilon > 0$. △

The above definition is related to the pessimistic nature of the problem (PB) and we are aware that different problems at the upper level would require different definitions of inner regularizations for the family $\{(P_x), x \in X\}$.

The strict and large ε-argmin maps, $\widetilde{\mathfrak{M}} = \left\{\widetilde{M}^{\varepsilon}, \varepsilon > 0\right\}$ and $\mathfrak{M} = \{M^{\varepsilon}, \varepsilon > 0\}$, constitute inner regularization classes under appropriate conditions.

Proposition 4.4.11 ([63, Cor. 2.1]) *The family $\widetilde{\mathfrak{M}}$ is an inner regularization whenever the assumptions of Lemma 4.4.5 are satisfied.*

The family $\mathfrak{M}$ is an inner regularization whenever the assumptions of Lemma 4.4.6 are satisfied. △

Other inner regularization classes can be found in [63] and also in [64], where the constraints of the second-stage problem are allowed to be violated. Here, we only mention the following ones:

$$M_d^\varepsilon \colon x \in X \rightrightarrows M_d^\varepsilon(x) = \left\{ y \in Y : d(y, K(x)) \leq \varepsilon,\ F(x, y) \leq \inf_{z \in K(x)} F(x, z) + \varepsilon \right\},$$

$$\widetilde{M}_d^\varepsilon \colon x \in X \rightrightarrows \widetilde{M}_d^\varepsilon(x) = \left\{ y \in Y : d(y, K(x)) < \varepsilon,\ F(x, y) < \inf_{z \in K(x)} F(x, z) + \varepsilon \right\}.$$

Proposition 4.4.12 *The family $\widetilde{\mathfrak{M}}_d = \left\{\widetilde{M}_d^\varepsilon, \varepsilon > 0\right\}$ is an inner regularization whenever the assumptions of Lemma 4.4.5 are satisfied.*

The family $\mathfrak{M}_d = \left\{M_d^\varepsilon, \varepsilon > 0\right\}$ is an inner regularization whenever the assumptions of Lemma 4.4.6 are satisfied. △

Proof The result follows by arguing as in [64] and using Lemma 4.4.3.

However, assumption (C_4), which is crucial for the lower semicontinuity of the constraints map K, could be not satisfied, so a good alternative to M_d^ε and $\widetilde{M}_d^\varepsilon$ could be the maps:

$$G^\varepsilon \colon x \in X \rightrightarrows G^\varepsilon(x) = \bigcap_{i=1}^{k} \left\{ y \in Y : g_i(x, y) \leq \varepsilon,\ F(x, y) \leq \inf_{z \in K(x)} F(x, z) + \varepsilon \right\},$$

$$\widetilde{G}^\varepsilon \colon x \in X \rightrightarrows \widetilde{G}^\varepsilon(x) = \bigcap_{i=1}^{k} \left\{ y \in Y : g_i(x, y) < \varepsilon,\ F(x, y) < \inf_{z \in K(x)} F(x, z) + \varepsilon \right\}.$$

Proposition 4.4.13 *Assume that Y is convex and compact and F is continuous over $X \times Y$. Then:*

- *the family $\widetilde{\mathfrak{G}} = \left\{\widetilde{G}^\varepsilon, \varepsilon > 0\right\}$ is an inner regularization whenever assumptions (C_1)–(C_3) and the following hold:*

  (C_5) *for every $\varepsilon > 0$ and $x \in X$ there exists $y \in Y$ such that $\max_{i=1,\ldots,k} g_i(x, y) < \varepsilon$;*

- *the family $\mathfrak{G} = \{G^\varepsilon, \varepsilon > 0\}$ is an inner regularization whenever the assumptions (C_1)–(C_3), (C_5) and (H_1) are satisfied.* △

Proof Assumptions (C_1), (C_2) and (C_5) guarantee that the map

$$x \in X \rightrightarrows \bigcap_{i=1}^{k} \{y \in Y : g_i(x, y) < \varepsilon\}$$

is lower semicontinuous over $X \times Y$, as well assumptions (C_1)–(C_3) and (C_5) guarantee that the map

$$x \in X \rightrightarrows \bigcap_{i=1}^{k} \{y \in Y : g_i(x, y) \leq \varepsilon\}$$

is lower semicontinuous over $X \times Y$. Then, the result can be proved arguing as in propositions 2.1, 2.2 and 2.4 in [64]. □

The next steps consist in introducing the concept of *viscosity solution* for pessimistic bilevel optimization problems related to an inner regularization class and in stating a related existence result.

Definition 4.4.14 ([63, Def. 2.2]) Let $\mathfrak{T}$ be an inner regularization for the family $\{(P_x), x \in X\}$. A point $\bar{x} \in X$ is a $\mathfrak{T}$-viscosity solution for (PB) if for every sequence $(\varepsilon_n)_n$ decreasing to zero there exists $(x^{\varepsilon_n})_n$, $x^{\varepsilon_n} \in X$ for any $n \in \mathbb{N}$, such that:

(V_1) a subsequence of $(x^{\varepsilon_n})_n$ converges to $\bar{x}$;

(V_2) $\sup\limits_{y \in \mathcal{T}^{\varepsilon_n}(x^{\varepsilon_n})} L(x^{\varepsilon_n}, y) = \inf\limits_{x \in X} \sup\limits_{y \in \mathcal{T}^{\varepsilon_n}(x)} L(x, y) \ \ \forall\, n \in \mathbb{N}$;

(V_3) $\lim\limits_{n \to +\infty} \sup\limits_{y \in \mathcal{T}^{\varepsilon_n}(x^{\varepsilon_n})} L(x^{\varepsilon_n}, y) = \inf\limits_{x \in X} \sup\limits_{y \in M(x)} L(x, y) = w.$ △

Theorem 4.4.15 ([63, Prop. 4.1]) *If $\mathfrak{T} = \{\mathcal{T}^{\varepsilon}, \varepsilon > 0\}$ is an inner regularization for the family $\{(P_x), x \in X\}$, the sets X and Y are compact and the function L is continuous over $X \times Y$, then there exists a $\mathfrak{T}$-viscosity solution for the pessimistic problem (PB).* △

Suitable existence results of viscosity solutions with respect to each of the families considered above can be derived from Theorem 4.4.15 and Propositions 4.4.11, 4.4.12, and 4.4.13.

The next example illustrates the above procedure in a simple case.

Computing Viscosity Solutions

Let $X = [0, 1]$, $Y = [-1, 1]$, $L(x, y) = x + y$ and $F(x, y) = xy$ for every $(x, y) \in X \times Y$. Consider $i = 1$ and $g_1(x, y) = x^2 - y^2$, so that $K(x) = [-1, -x] \cup [x, 1]$. If $\varepsilon \in]0, 2]$, it is easy to check that:

- the argmin map M of the minima to (P_x) is not lower semicontinuous at $x = 0$, since $M(0) = Y$ and $M(x) = \{-1\}$ for every $x \in]0, 1]$;
- the marginal function $e(\cdot)$ is not lower semicontinuous at $x = 0$, since $e(0) = 1$ and $e(x) = x - 1$ for every $x \in]0, 1]$;
- the problem (PB) does not have a solution since $w = \inf\limits_{x \in X} \sup\limits_{y \in M(x)} L(x, y) = -1$ but $e(x) > -1$ for every $x \in [0, 1]$;

(continued)

- the map $\widetilde{M}_d^\varepsilon$ is lower semicontinuous over X since $\widetilde{M}_d^\varepsilon(x) = [-1, 1]$ if $x \in [0, \varepsilon/2[$ and $\widetilde{M}_d^\varepsilon(x) = [-1, -1 + \varepsilon/x[$ if $x \in [\varepsilon/2, 1]$;
- the minimum point of the marginal function $\sup_{y \in \widetilde{M}_d^\varepsilon(\cdot)} L(\cdot, y)$ is $\widetilde{x}_d^\varepsilon = \sqrt{\varepsilon}$ and

$$\widetilde{w}_d^\varepsilon = \inf_{x \in X} \sup_{y \in \widetilde{M}_d^\varepsilon(x)} L(x, y) = -1 + 2\sqrt{\varepsilon};$$

- the family $\widetilde{\mathfrak{M}}_d$ is an inner regularization even if the function $g_1(x, \cdot)$ is not strictly quasiconvex (see Remark 4.4.4);
- the data satisfy the conditions in Theorem 4.4.15;
- the $\widetilde{\mathfrak{M}}_d$-viscosity solution for (PB) is $x = 0$.

4.4.2 Regularization of Optimistic Bilevel Optimization Problems

As we observed in Sect. 4.3.3, a strong Stackelberg equilibrium exists under not too restrictive conditions. However, when such a problem has to be approached by a sequence of perturbed problems (e.g. in discretization process of (OB) in an infinite dimensional setting), the perturbed solutions may not converge towards a solution of the original problem, that is such problems are lacking in stability in the sense specified in Sect. 4.4.1.1.

More precisely, assume that the data of the problem (OB) are asymptotically approached by the data of the perturbed problems

$$(OB)_n \quad \text{find } x_n \in X \text{ such that } \inf_{y \in M_n(x_n)} L_n(x_n, y) = \inf_{x \in X} \inf_{y \in M_n(x)} L_n(x, y),$$

where $(F_n)_n$ and $(L_n)_n$ are sequences of functions from $X \times Y$ to $\mathbb{R}$ and, for any n, K_n is defined as in (4.4.2) and M_n is the argmin map of F_n with constraints K_n. Then, denoted by $\mathcal{S}_n$ the set of solutions to $(OB)_n$, for any $n \in \mathbb{N}$, one asks if one has

$$\operatorname*{Lim\,sup}_n \mathcal{S}_n \subseteq \mathcal{S}$$

and the answer is negative, in general.

Strong Stackelberg Solutions May Not Be Stable

Let $X = Y = [0, 1]$, $K(x) = K_n(x) = Y$, $F_n(x, y) = y/n$ and $L_n(x, y) = -xy + 1/n$. It is easy to see that:

- $(F_n)_n$ and $(L_n)_n$ uniformly converge, and therefore also continuously converge, to the functions defined by $F(x, y) = 0$ and $L(x, y) = -xy$ respectively;
- $M(x) = [0, 1]$ and, for any $n \in \mathbb{N}$, $M_n(x) = \{0\}$ for every $x \in [0, 1]$;
- $\mathcal{S} = \{1\}$ and, for any $n \in \mathbb{N}$, $\mathcal{S}_n = [0, 1]$.

Then, $\operatorname{Lim\,sup}_n \mathcal{S}_n \not\subseteq \mathcal{S}$.

Nevertheless, also in the optimistic case, regularizing the follower's reaction set turns out to be useful to obtain satisfactory approximation results for the strong Stackelberg values and solutions.

In line with the pessimistic case, we define:

$$s^\varepsilon = \inf_{x\in X} \inf_{y\in M^\varepsilon(x)} L(x, y) \quad \text{and} \quad \widetilde{s}^\varepsilon = \inf_{x\in X} \inf_{y\in \widetilde{M}^\varepsilon(x)} L(x, y),$$

$$\mathcal{S}^\varepsilon = \left\{ x \in X : \inf_{y\in M^\varepsilon(x)} L(x, y) = \inf_{u\in X} \inf_{y\in M^\varepsilon(u)} L(u, y) \right\},$$

$$\widetilde{\mathcal{S}}^\varepsilon = \left\{ x \in X : \inf_{y\in \widetilde{M}^\varepsilon(x)} L(x, y) = \inf_{u\in X} \inf_{y\in \widetilde{M}^\varepsilon(u)} L(u, y) \right\},$$

$$\mathcal{S}_n^\varepsilon = \left\{ x \in X : \inf_{y\in M_n^\varepsilon(x)} L_n(x, y) = \inf_{u\in X} \inf_{y\in M_n^\varepsilon(u)} L_n(u, y) = s_n^\varepsilon \right\}.$$

Proposition 4.4.16 ([52, Th. 3.2]) *Under the assumptions of Lemma 4.4.5, if in addition: the function L is lower semicontinuous over $X \times Y$, then*

$$\lim_{\varepsilon\to 0} s^\varepsilon = s = \lim_{\varepsilon\to 0} \widetilde{s}^\varepsilon \quad \textit{and} \quad \operatorname{Lim\,sup}_{\varepsilon\to 0} \mathcal{S}^\varepsilon \subseteq \mathcal{S}.$$

△

Proposition 4.4.17 ([52, Th. 5.6]) *Assume that X is compact and Y is convex and compact. If the sequences of functions $(F_n)_n$ and $(g_{i,n})_n$ continuously converge over $X \times Y$ to the functions F and g_i respectively, assumptions $(C_1) - (C_4)$, $(C_{3,n})$, $(C_{4,n})$, (H_1), $(H_{2,n})$ and the following hold for any $n \in \mathbb{N}$:*

$(H_{3,n})$ *for any $(x, y) \in X \times Y$ and any sequence $(x_n, y_n)_n$ converging to (x, y) in $X \times Y$ one has*

$$L(x, y) \leq \liminf_{n\to+\infty} L_n(x_n, y_n);$$

then, for any $\varepsilon > 0$

$$\operatorname*{Lim\,sup}_n \mathcal{S}_n^\varepsilon \subseteq \mathcal{S}^\varepsilon \text{ and } \lim_{n\to+\infty} s_n^\varepsilon = s^\varepsilon.$$

Moreover, one has

$$\operatorname*{Lim\,sup}_{\varepsilon\to 0}\left(\operatorname*{Lim\,sup}_n \mathcal{S}_n^\varepsilon\right) \subseteq \mathcal{S}. \qquad \triangle$$

However, in this approximation process, the role of ε and n cannot be inverted and, in general, one cannot get a result of the type

$$\operatorname*{Lim\,sup}_n \mathcal{S}_n^{\varepsilon_n} \subseteq \mathcal{S}, \tag{4.4.4}$$

with $(\varepsilon_n)_n$ decreasing to zero, as shown by the next example.

Caution in Regularizing and Perturbing Simultaneously
Let $X = Y = [0, 1]$, $K(x) = K_n(x) = Y$, $F_n(x, y) = y/n$ and $L_n(x, y) = x(x - y) + 1/n$. It can be computed that:

- $(F_n)_n$ and $(L_n)_n$ continuously converge on $X \times Y$ to the functions defined by $F(x, y) = 0$ and $L(x, y) = x(x - y)$ respectively;
- $M(x) = M^\varepsilon(x) = [0, 1]$ for every $x \in X$;
- for any $n \in \mathbb{N}$, $M_n^\varepsilon(x) = [0, n\varepsilon]$ if $n\varepsilon \le 1$ and $M_n^\varepsilon(x) = [0, 1]$ if $n\varepsilon > 1$;
- $\mathcal{S} = \mathcal{S}^\varepsilon = \{1/2\}$ and, for any $n \in \mathbb{N}$, $\mathcal{S}_n^\varepsilon = \{n\varepsilon/2\}$ if $n\varepsilon \le 1$ and $\mathcal{S}_n^\varepsilon = \{1/2\}$ if $n\varepsilon > 1$.

Therefore,

$$\operatorname*{Lim\,sup}_{\varepsilon\to 0}\left(\operatorname*{Lim\,sup}_n \mathcal{S}_n^\varepsilon\right) = \mathcal{S} \quad \text{and} \quad \operatorname*{Lim\,sup}_n\left(\operatorname*{Lim\,sup}_{\varepsilon\to 0} \mathcal{S}_n^\varepsilon\right) = \{0\} \not\subseteq \mathcal{S},$$

and we remark that $\operatorname*{Lim\,sup}_n \mathcal{S}_n^{\varepsilon_n} \subseteq \mathcal{S}$ if the sequence $(\varepsilon_n)_n$ is infinitesimal of the first order.

It would be interesting defining suitable viscosity solutions also for optimistic bilevel optimization problems aimed to reach an inclusion analogous to (4.4.4).

This is a still unexplored topic, but given an optimistic bilevel problem approached by a sequence of approximating optimistic bilevel problems $(OB)_n$, it seems to be reasonable investigating the limit points of sequences of solutions to $(OB)_n$ that are not so far from the solution set to (OB).

We mention that algorithms to approach approximate solutions to (OB) have been developed in [66], where the reformulation of (OB) via the optimal value function is exploited, and in [44], where a reformulation of (OB) as a generalized Nash equilibrium problem is employed.

4.4.3 *Regularization of Bilevel Problems with Equilibrium Constraints*

Stackelberg games are models that can be naturally extended to the case where there are one leader and one, or more than one, follower who solve a second-stage problem described by a parametric variational inequality, Nash equilibrium, quasi-variational inequality or more generally quasi-equilibrium. The critical issues of pessimistic and optimistic approach are still present and regularizing the problem solved by the follower(s) is useful also in this case. First we present the case where one solves a variational inequality in the second stage. The set of solutions of such a parametric variational inequality can be considered as the follower's constraints in the first stage, so traditionally the associate bilevel problem is called *bilevel problem with variational inequality constraints*, and analogously for the other problems considered in the second stage.

4.4.3.1 Variational Inequality Constraints

Assume that A is a function from $X \times Y$ to Y and consider the family of variational inequalities $\{(V_x), x \in X\}$, where any problem (V_x) consists in finding, see [12],

$$y \in K(x) \ \text{ such that } \ \langle A(x, y), y - w \rangle \leq 0 \ \forall \, w \in K(x).$$

Then, the pessimistic and the optimistic bilevel problems with variational inequality in the second stage are respectively defined by:

$$(PBVI) \quad \text{find } x \in X \text{ such that } \sup_{y \in \mathcal{V}(x)} L(x, y) = \inf_{u \in X} \sup_{y \in \mathcal{V}(u)} L(u, y),$$

$$(OBVI) \quad \text{find } x \in X \text{ such that } \inf_{y \in \mathcal{V}(x)} L(x, y) = \inf_{u \in X} \inf_{y \in \mathcal{V}(u)} L(u, y),$$

where, for any x, $\mathcal{V}(x)$ is the set of solutions to the variational inequality (V_x).

In this case, given a positive number ε, fruitful regularizations of the follower's reaction set consist in finding

$$y \in K(x) \ \text{ such that } \ \langle A(x, y), y - w \rangle \leq \varepsilon \ \forall w \in K(x)$$

or

$$y \in K(x) \text{ such that } \langle A(x,w), y-w\rangle \le \varepsilon \;\forall w \in K(x),$$

together with their *strict* versions. This double possibility of regularizing the map $\mathcal{V}$ follows from using the Minty type variational inequality, which consists in finding

$$y \in K(x) \text{ such that } \langle A(x,w), y-w\rangle \le 0 \;\forall w \in K(x),$$

use which is essential in infinite dimensional spaces, as displayed in [12].

The above regularizations have been employed for investigating, in a theoretical setting, well-posedness in [54] and stability properties in [50] of $(OBVI)$ in Banach spaces; moreover, we mention that an exact penalization scheme has been proposed in [112] for deriving necessary optimality conditions of $(OBVI)$. Convergence properties of the approximate weak Stackelberg values of $(PBVI)$ can be found in [57] in finite dimensional spaces. They have been also considered, in applied framework as truss topology optimization in [36] or climate regulation policy in [34].

4.4.3.2 Nash Equilibrium Constraints

Consider $Y = \prod_{j=1}^{l} Y_j$ and l real-valued functions $F_1, \ldots, F_l$ defined on $X \times Y$. Then, for any $x \in X$, the Nash equilibrium problem (NE_x) consists in finding

$$\mathbf{y} \in Y \text{ such that } F_j(x,\mathbf{y}) \le \inf_{u_j \in Y_j} F_j(x, u_j, \mathbf{y}_{-j}) \;\forall j = 1, .., l,$$

where $\mathbf{y}_{-j} = (y_1, \ldots, y_{j-1}, y_{j+1}, \ldots, y_l)$, and the pessimistic and the optimistic bilevel problems with Nash equilibrium problem in the second stage are respectively defined by:

$$(PBNE) \quad \text{find } x \in X \text{ such that } \sup_{\mathbf{y}\in\mathcal{N}(x)} L(x,\mathbf{y}) = \inf_{u\in X}\sup_{\mathbf{y}\in\mathcal{N}(u)} L(u,\mathbf{y}),$$

$$(OBNE) \quad \text{find } x \in X \text{ such that } \inf_{\mathbf{y}\in\mathcal{N}(x)} L(x,\mathbf{y}) = \inf_{u\in X}\inf_{\mathbf{y}\in\mathcal{N}(u)} L(u,\mathbf{y}),$$

where, for any x, $\mathcal{N}(x)$ is the set of solutions to the Nash equilibrium problem (NE_x).

Regularizing the follower's reaction set consists, in this case, in considering:

$$\mathcal{N}^{\varepsilon}(x) = \left\{\mathbf{y} \in Y : F_j(x,\mathbf{y}) \le \inf_{u_j\in Y_j} F_j(x,u_j,\mathbf{y}_{-j}) + \varepsilon \;\forall j = 1, .., l\right\},$$

that is the classical set of approximate Nash equilibria (see, for example, [10]), or

$$\widehat{\mathcal{N}}^{\varepsilon}(x)=\left\{y\in Y:\sum_{j=1}^{l}F_j(x,y)\le\inf_{u_j\in Y_j}\sum_{j=1}^{l}F_j(x,u_j,y_{-j})+\varepsilon\right\},$$

introduced and investigated in [94]. These sets do not coincide in general, as shown in [94, Ex. 3.6], but both can be used to study parametric well-posedness properties of Nash equilibrium problems since, as proven in [55],

$$\lim_{\varepsilon\to 0}\operatorname{diam}\mathcal{N}^{\varepsilon}(x)=0\iff\lim_{\varepsilon\to 0}\operatorname{diam}\widehat{\mathcal{N}}^{\varepsilon}(x)=0.$$

Well-posedness of optimistic bilevel problems with Nash equilibrium problem in the second stage has been investigated in that paper, whereas viscosity solutions for pessimistic ones have been introduced in [60] as a prototype for the concepts of viscosity solutions for (PB) later developed.

4.4.3.3 Quasi-Variational Inequality Constraints

Assume that T is a set-valued map from $X\times Y$ to Y, A is a function from $X\times Y$ to Y, and consider the family of quasi-variational inequalities $\{(Q_x), x\in X\}$, where any (Q_x) consists in finding, see [12],

$$y\in Y\ \text{such that}\ y\in T(x,y)\ \text{and}\ \langle A(x,y),y-w\rangle\le 0\ \ \forall w\in T(x,y).$$

Then, the pessimistic and the optimistic bilevel problems with quasi-variational inequality in the second stage are defined respectively by:

$$(PBQI)\quad \text{find } x\in X \text{ such that }\sup_{y\in\mathcal{Q}(x)}L(x,y)=\inf_{u\in X}\sup_{y\in\mathcal{Q}(u)}L(u,y),$$

$$(OBQI)\quad \text{find } x\in X \text{ such that }\inf_{y\in\mathcal{Q}(x)}L(x,y)=\inf_{u\in X}\inf_{y\in\mathcal{Q}(u)}L(u,y),$$

where, for any x, $\mathcal{Q}(x)$ is the set of solutions to the quasi-variational inequality (Q_x).

In this case, given a positive number ε, fruitful regularizations of the follower's reaction set consist in finding

$$y\in Y\ \text{such that}\ d\,(y,T(x,y))\le\varepsilon\ \text{and}\ \langle A(x,y),y-w\rangle\le\varepsilon\ \ \forall w\in T(x,y)$$

or

$$y\in Y\ \text{such that}\ d(y,T(x,y))\le\varepsilon\ \text{and}\ \langle A(x,w),y-w\rangle\le\varepsilon\ \ \forall w\in T(x,y)$$

together with their *strict* versions.

Requiring that an approximate solution to (Q_x) may violate the constraints, provided that it remains in a neighborhood of them, is quite necessary in quasi-variational inequality setting where a fixed-point problem is involved, as shown in [48, Ex. 2.1]. The above regularizations have been used in [59] for establishing convergence results of the weak Stackelberg values of $(PBQI)$ in finite dimensional spaces, and in [61] for a complete investigation in infinite dimensional spaces of convergence properties of approximate solutions and values of $(OBQI)$ problems, there called *semi-quasivariational optimistic bilevel problems*.

4.4.3.4 Quasi-Equilibrium Constraints

Let h be a real-valued function defined in $X \times Y \times Y$ and let T be a set-valued map from $X \times Y$ to Y with nonempty values. For any $x \in X$, the quasi-equilibrium problem (QE_x) (also called *quasi-variational problem* in [58]) consists in finding $y \in Y$ such that

$$y \in T(x, y) \text{ and } h(x, y, w) \leq 0 \ \forall w \in T(x, y).$$

The class of such problems encompasses several problems arising in optimization, in game theory and in variational analysis, as illustrated in [58].

In [62], the concept of viscosity solutions relative to an inner regularization class for the family $\{(QE_x), x \in X\}$ was introduced and it was proved that the map $\widetilde{\mathcal{Q}}^{\varepsilon}$ defined by

$$\widetilde{\mathcal{Q}}^{\varepsilon}(x) = \{ y \in Y : d(y, T(x, y)) < \varepsilon \text{ and } h(x, y, w) < \varepsilon \ \forall w \in T(x, y)\}$$

generates an inner regularization class under suitable assumptions and that the *large* version of $\widetilde{\mathcal{Q}}^{\varepsilon}$ may fail to be an inner regularization under the same hypotheses. The results were established in infinite dimensional Banach spaces, so it had been necessary to balance the use of weak and strong convergence in the assumptions, but, in finite dimensional spaces, the statements can be simplified and made more readable.

4.5 Regularizing the Follower's Payoff Function in Stackelberg Games

The general non-single-valuedness of the follower's best reply correspondence M brings out the difficulties in the numerical approximation of problems (PB) and (OB), as mentioned in Sect. 4.3.3, as well as the need to define constructive methods for selecting an SPNE in Stackelberg games, as emphasized in Sect. 4.3.5. For these reasons and possibly also for behavioural motivations (see, for example,

[22]), regularization methods involving the follower's payoff function have been introduced. Such methods allow to construct sequences of perturbed problems where the solution of the second-stage problem is unique by exploiting well-known regularization techniques in convex optimization. For the sake of brevity, we will present only two approaches for regularizing the follower's payoff function: the first one is based on the Tikhonov regularization, the second one on the proximal regularization.

In this section we assume that the leader's and follower's action sets X and Y are subsets of the Euclidean spaces $\mathbb{X}$ and $\mathbb{Y}$, respectively, and we denote by $\|\cdot\|_{\mathbb{X}}$ and $\|\cdot\|_{\mathbb{Y}}$ the norm of $\mathbb{X}$ and $\mathbb{Y}$, respectively. Furthermore, for the sake of brevity, both leader's and follower's constraints are assumed to be constant.

4.5.1 *Regularizing the Follower's Payoff Function via Tikhonov Method*

Let us recall preliminarily the Tikhonov regularization method for the approximation of solutions to convex optimization problems. Then, we illustrate how it has been employed to regularize bilevel optimization problems (PB) or (OB) and to select SPNEs.

4.5.1.1 Tikhonov Regularization in Convex Optimization

Assume we deal with the minimization problem

$$(P): \min_{a \in A} J(a),$$

where J is a real-valued function defined on a subset A of a Euclidean space $\mathbb{A}$ with norm $\|\cdot\|_{\mathbb{A}}$. In [107] Tikhonov introduced, in an optimal control framework, the following perturbed minimization problems

$$(P)^{\mathcal{T}}_{\lambda_n}: \min_{a \in A} \left(J(a) + \frac{1}{2\lambda_n} \|a\|^2_{\mathbb{A}} \right),$$

where $n \in \mathbb{N}$ and $\lambda_n > 0$, and he proved the connections between the solutions to problem $(P)^{\mathcal{T}}_{\lambda_n}$ and the solutions to problem (P), which are recalled in the next well-known result.

Theorem 4.5.1 *Assume that the set A is compact and convex, the function J is lower semicontinuous and convex over A and* $\lim_{n \to +\infty} \lambda_n = +\infty$.

Then, problem $(P)^{\mathcal{T}}_{\lambda_n}$ *has a unique solution* $\bar{a}^{\mathcal{T}}_n \in A$, *for any* $n \in \mathbb{N}$, *and the sequence* $(\bar{a}^{\mathcal{T}}_n)_n$ *is convergent to the minimum norm element of the set of the minimizers of* J *over* A, *so*

$$\left\| \lim_{n\to+\infty} \bar{a}^{\mathcal{T}}_n \right\|_{\mathbb{A}} = \inf_{a\in V} \|a\|_{\mathbb{A}},$$

where $V = \operatorname{Arg\,min}_{z\in A} J(z)$. △

Therefore, the Tikhonov regularization method allows to approach a solution of a convex minimization problem by constructing a sequence of perturbed problems, in general better-behaved than the original problem, that have a unique solution. Furthermore, the limit of the sequence generated accordingly is uniquely characterized in the set of solutions to the original problem.

We mention that, afterwards, Tikhonov regularization has been broadly exploited for finding Nash equilibria in one-stage games where players move simultaneously: see [7, 47] for zero-sum games (with applications also to differential games) and [93, Section 4], [25, 43] for general N-players games.

4.5.1.2 Tikhonov Regularization of the Follower's Payoff Function in Stackelberg Games

Let $\Gamma = (X, Y, L, F)$ be a Stackelberg game. The first attempt to regularize the follower's payoff function by employing Tikhonov regularization is due to Loridan and Morgan in [77]. Having in mind to approximate problem (PB), for any $x \in X$ they considered the following Tikhonov-regularized second-stage problems

$$(P_x)^{\mathcal{T}}_{\lambda_n}: \min_{y\in Y} \left(F(x, y) + \frac{1}{2\lambda_n} \|y\|^2_{\mathbb{Y}} \right),$$

where $n \in \mathbb{N}$ and $\lambda_n > 0$. The next result states preliminarily properties about the connections between the perturbed problems $(P_x)^{\mathcal{T}}_{\lambda_n}$ and problem (P_x) as n goes to $+\infty$.

Proposition 4.5.2 ([77, Prop. 3.1]) *Assume that the set* Y *is compact and convex, the function* F *is continuous over* $X \times Y$ *and* $\lim_{n\to+\infty} \lambda_n = +\infty$.

Then, for any $x \in X$ *and any sequence* $(x_n)_n$ *converging to* x *in* X, *we have*

- $\displaystyle\lim_{n\to+\infty} \inf_{y\in Y} \left(F(x_n, y) + \tfrac{1}{2\lambda_n} \|y\|^2_{\mathbb{Y}} \right) = \inf_{y\in Y} F(x, y);$
- $\displaystyle\operatorname{Lim\,sup}_{n} \operatorname{Arg\,min}_{y\in Y} \left(F(x_n, y) + \tfrac{1}{2\lambda_n} \|y\|^2_{\mathbb{Y}} \right) \subseteq M(x).$ △

Remark 4.5.3 We highlight that the sequence of functions generated by Tikhonov-regularizing the follower's payoff function satisfies assumptions $(F1)$, $(F2)$ and $(\widetilde{F}2)$ in [71, Section 4], therefore additional results involving ε-solutions and strict ε-solutions hold by applying propositions 4.2, 4.4 and 6.1 in [71]. For instance, the inclusion stated in Proposition 4.5.2 above is still valid even when considering ε-argmin maps. △

More specific results concerning problems $(P_x)^{\mathcal{T}}_{\lambda_n}$ and the connections with (P_x) are achieved by adding a convexity assumption, usual in a Tikhonov regularization framework.

Proposition 4.5.4 ([77, Prop. 3.2 and 3.4]) *Assume that the function $F(x,\cdot)$ is convex over Y for any $x \in X$ and hypotheses of Proposition 4.5.2 are satisfied.*

Then, for any $x \in X$,

- *problem $(P_x)^{\mathcal{T}}_{\lambda_n}$ has a unique solution $\bar{\varphi}^{\mathcal{T}}_n(x) \in Y$ for any $n \in \mathbb{N}$, that is*

$$\{\bar{\varphi}^{\mathcal{T}}_n(x)\} = \operatorname*{Arg\,min}_{y \in Y} \left(F(x, y) + \frac{1}{2\lambda_n} \|y\|^2_{\mathbb{Y}} \right); \tag{4.5.1}$$

- $\lim_{n \to +\infty} \bar{\varphi}^{\mathcal{T}}_n(x) = \hat{\varphi}(x)$*, where $\hat{\varphi}(x)$ is the minimum norm element of the set $M(x)$, that is*

$$\{\hat{\varphi}(x)\} = \operatorname*{Arg\,min}_{y \in M(x)} \|y\|_Y. \tag{4.5.2}$$

Furthermore, for any $n \in \mathbb{N}$ and any sequence $(x_k)_k$ converging to x in X, we have

$$\lim_{k \to +\infty} \bar{\varphi}^{\mathcal{T}}_n(x_k) = \bar{\varphi}^{\mathcal{T}}_n(x) \quad \textit{and} \quad \lim_{k \to +\infty} F(x_k, \bar{\varphi}^{\mathcal{T}}_n(x_k)) = F(x, \bar{\varphi}^{\mathcal{T}}_n(x)).$$

△

Remark 4.5.5 Under the assumptions of Proposition 4.5.4 additional results regarding ε-solutions have been proved in [77, Proposition 3.3] for the Tikhonov-regularized problem $(P_x)^{\mathcal{T}}_{\lambda_n}$. △

We point out that, in general, the sequence $(\bar{\varphi}^{\mathcal{T}}_n)_n$ is not continuously convergent to $\hat{\varphi}$ (for the definition of continuous convergence see Proposition 4.3.2), as shown in [77, Remark 3.1] and in the example below.

The Sequence $(\bar{\varphi}_n^{\mathcal{T}})_n$ May Not Be Continuously Convergent
Consider $X = Y = [-1, 1]$ and $F(x, y) = -xy$. The follower's best reply correspondence M is given in (4.3.1). Choosing $\lambda_n = n$, for any $x \in X$ we obtain

$$\bar{\varphi}_n^{\mathcal{T}}(x) = \begin{cases} -1, & \text{if } x \in [-1, -1/n[\\ nx, & \text{if } x \in [-1/n, 1/n] \\ 1, & \text{if } x \in]1/n, 1] \end{cases} \quad \text{and} \quad \hat{\varphi}(x) = \begin{cases} -1, & \text{if } x \in [-1, 0[\\ 0, & \text{if } x = 0 \\ 1, & \text{if } x \in]0, 1]. \end{cases}$$

Therefore, the sequence of functions $(\bar{\varphi}_n^{\mathcal{T}})_n$ is not continuously convergent to $\hat{\varphi}$, as the function $\hat{\varphi}$ is not continuous at $x = 0$.

Now, let us consider the following Tikhonov-regularized bilevel problems:

$$(SP)_{\lambda_n}^{\mathcal{T}} \quad \begin{cases} \min\limits_{x \in X} L(x, \bar{\varphi}_n^{\mathcal{T}}(x)) \\ \text{where } \bar{\varphi}_n^{\mathcal{T}}(x) \text{ is the solution to } (P_x)_{\lambda_n}^{\mathcal{T}}. \end{cases}$$

Provided that assumptions of Proposition 4.5.4 are satisfied, $(SP)_{\lambda_n}^{\mathcal{T}}$ is a Stackelberg problem for any $n \in \mathbb{N}$ since the solution to the second-stage problem is unique. Two questions arise:

1. Are there solutions to $(SP)_{\lambda_n}^{\mathcal{T}}$?
2. What happens when $n \to +\infty$?

The next proposition provides the answer to the first question.

Proposition 4.5.6 ([77, Prop. 3.5]) *Assume that the set X is compact, Y is compact and convex, the functions L and F are continuous over $X \times Y$ and the function $F(x, \cdot)$ is convex over Y for any $x \in X$.*

Then, the Tikhonov-regularized Stackelberg problem $(SP)_{\lambda_n}^{\mathcal{T}}$ has at least one solution, for any $n \in \mathbb{N}$. △

As regards to investigation on the connections between $(SP)_{\lambda_n}^{\mathcal{T}}$ and (PB) as n goes to $+\infty$, the second question is addressed in the following result.

Proposition 4.5.7 ([77, Prop. 3.6]) *Assume that $\lim_{n \to +\infty} \lambda_n = +\infty$ and hypotheses of Proposition 4.5.6 are satisfied. Denote with $\bar{x}_n^{\mathcal{T}}$ a solution to $(SP)_{\lambda_n}^{\mathcal{T}}$ for any $n \in \mathbb{N}$.*

Then, any convergent subsequence of the sequence $(\bar{x}_n^{\mathcal{T}}, \bar{\varphi}_n^{\mathcal{T}}(\bar{x}_n^{\mathcal{T}}))_n$ in $X \times Y$ has a limit $(\bar{x}, \bar{y})$ which satisfies

$$L(\bar{x}, \bar{y}) \leq w \quad \text{and} \quad \bar{y} \in M(\bar{x}), \tag{4.5.3}$$

where w is the security value of (PB) defined in Definition 4.2.2. △

Definition 4.5.8 ([69]) An action profile $(\bar{x}, \bar{y}) \in X \times Y$ satisfying (4.5.3) is a *lower Stackelberg equilibrium pair* for problem (PB). △

The lower Stackelberg equilibrium pair solution concept was introduced in [69, Remark 5.3] in the context of approximation of (PB) via ε-solutions. Note that a result analogous to Proposition 4.5.7 has been obtained in [75] for the *least-norm regularization* of the second-stage problem of (PB). In fact such a regularization, which is an adaptation of the method introduced in [104] and which involves the regularization both of the follower's optimal reaction set and of the follower's payoff function, generates sequences of action profiles whose limit points are lower Stackelberg equilibrium pairs, as proved in propositions 2.6 and 3.5 in [75].

We observe that if $(\bar{x}, \bar{y})$ is a weak Stackelberg equilibrium of (PB), then $(\bar{x}, \bar{y})$ is a lower Stackelberg equilibrium pair. The converse is not true, in general, as illustrated in the following example.

Weak Stackelberg Equilibrium and Lower Stackelberg Equilibrium Pair May Not Coincide

Consider $X = [1/2, 2]$, $Y = [-1, 1]$,

$$L(x, y) = -x - y \quad \text{and} \quad F(x, y) = \begin{cases} 0, & \text{if } x \in [1/2, 1[\\ (x-1)y, & \text{if } x \in [1, 2]. \end{cases}$$

The pair $(1, 1)$ is a lower Stackelberg equilibrium, whereas the unique weak Stackelberg equilibrium is $(2, -1)$.

Remark 4.5.9 It is worth to emphasize that, in general, the sequence $(\bar{x}_n^{\mathcal{T}})_n$, where $\bar{x}_n^{\mathcal{T}}$ is a solution to $(SP)^{\mathcal{T}}_{\lambda_n}$, converges neither to a weak Stackelberg (or pessimistic) solution of (PB) nor to a strong Stackelberg (or optimistic) solution of (OB); analogously, the sequence $(\bar{x}_n^{\mathcal{T}}, \bar{\varphi}_n^{\mathcal{T}}(\bar{x}_n^{\mathcal{T}}))_n$ converges, in general, neither to a weak Stackelberg equilibrium nor to a strong Stackelberg equilibrium, as illustrated in the next example also used in [93]. △

The Sequence $(\bar{x}_n^{\mathcal{T}}, \bar{\varphi}_n^{\mathcal{T}}(\bar{x}_n^{\mathcal{T}}))_n$ May Converge Neither to a Weak Stackelberg Equilibrium Nor to a Strong Stackelberg Equilibrium

Consider $X = [-1/2, 1/2]$, $Y = [-1, 1]$,

$$L(x, y) = -x - y \quad \text{and} \quad F(x, y) = \begin{cases} (x + 1/4)y, & \text{if } x \in [-1/2, -1/4[\\ 0, & \text{if } x \in [-1/4, 1/4] \\ (x - 1/4)y, & \text{if } x \in]1/4, 1/2]. \end{cases}$$

(continued)

The sequences $(\bar{x}_n^{\mathcal{T}})_n$ and $(\bar{x}_n^{\mathcal{T}}, \bar{\varphi}_n^{\mathcal{T}}(\bar{x}_n^{\mathcal{T}}))_n$ converge to $-1/4$ and $(-1/4, 1)$, respectively. Instead, the strong Stackelberg solution is $1/4$ and the strong Stackelberg equilibrium is $(1/4, 1)$, whereas the weak Stackelberg solution and the weak Stackelberg equilibrium do not exist.

Afterwards, the Tikhonov regularization approach illustrated up to now has been employed in [27] where, by requiring stronger assumptions on the payoff functions, further regularity properties for the Tikhonov-regularized second-stage problem $(P_x)^{\mathcal{T}}_{\lambda_n}$ have been shown. Moreover, an algorithm have been designed for the solutions of the Tikhonov-regularized Stackelberg problem $(SP)^{\mathcal{T}}_{\lambda_n}$ as n goes to $+\infty$.

We mention that a regularization approach in the same spirit of $(P_x)^{\mathcal{T}}_{\lambda_n}$ consists in considering the following perturbed second-stage problems

$$\min_{y\in Y}\left(F(x,y)+\frac{1}{2\lambda_n}L(x,y)\right),$$

which have a unique solution for any $n \in \mathbb{N}$, provided that the function $L(x,\cdot)$ is strongly convex on Y for any $x \in X$. In [29] such a Tikhonov-like approach has been investigated for problem (OB), in order to circumvent the non-uniqueness of the solutions to the second stage problem, and an algorithm has been proposed. Further discussions on this kind of regularization can be found in [28, Subsection 7.3.2].

We highlight that the idea of using the leader's payoff function in the regularization of the second-stage problem goes back to Molodtsov in [88]: he introduced a "mixed" regularization method, by combining both the regularization of the follower's reaction set and the regularization of the follower's payoff function, which allows to approximate problem (PB) via a sequence of strong Stackelberg problems. Then, more general properties of the *Molodtsov regularization* have been obtained in [78], where perturbations of the data of problem (PB) have been also considered.

4.5.1.3 Selection of SPNEs via Tikhonov Regularization

A Stackelberg game where the follower's best reply correspondence M is not single-valued could have infinitely many SPNEs, as illustrated in the example on page 99. Therefore, let us focus on how restricting the number of SPNEs or, better yet, picking just one. The first attempt to select an SPNE via a constructive approach that allows to overcame the non-single-valuedness of M is due to Morgan and Patrone in [93], where the Tikhonov regularization approach presented in Sect. 4.5.1.2 was exploited.

Let $\Gamma = (X, Y, L, F)$ be a Stackelberg game and consider the following Tikhonov-regularized Stackelberg games

$$\Gamma_{\lambda_n}^{\mathcal{T}} = (X, Y, L, F_{\lambda_n}^{\mathcal{T}}),$$

where $F_{\lambda_n}^{\mathcal{T}} : X \times Y \to \mathbb{R}$ is defined on $X \times Y$ by

$$F_{\lambda_n}^{\mathcal{T}}(x, y) = F(x, y) + \frac{1}{2\lambda_n}\|y\|_{\mathbb{Y}}^2,$$

for any $n \in \mathbb{N}$, i.e. $\Gamma_{\lambda_n}^{\mathcal{T}}$ is the game obtained from Γ by replacing the follower's payoff function F with the objective function of the Tikhonov-regularized problem $(P_x)_{\lambda_n}^{\mathcal{T}}$. Propositions 4.5.4 and 4.5.6 imply the following properties for $\Gamma_{\lambda_n}^{\mathcal{T}}$.

Corollary 4.5.10 *Under the assumptions of Proposition 4.5.6 we have, for any* $n \in \mathbb{N}$,

- *the follower's best reply correspondence in* $\Gamma_{\lambda_n}^{\mathcal{T}}$ *is single-valued;*
- *the game* $\Gamma_{\lambda_n}^{\mathcal{T}}$ *has at least one SPNE: the strategy profile* $(\bar{x}_n^{\mathcal{T}}, \bar{\varphi}_n^{\mathcal{T}}(\cdot))$, *where* $\bar{\varphi}_n^{\mathcal{T}} : X \to Y$ *is defined in* (4.5.1) *and* $\bar{x}_n^{\mathcal{T}} \in X$ *is a solution to* $(SP)_{\lambda_n}^{\mathcal{T}}$, *is an SPNE of* $\Gamma_{\lambda_n}^{\mathcal{T}}$.

Moreover, the sequence of strategies $(\bar{\varphi}_n^{\mathcal{T}})_n$ *is pointwise convergent to the function* $\hat{\varphi}$ *defined in* (4.5.2). △

It is natural to ask if the limit of the sequence of SPNEs $(\bar{x}_n^{\mathcal{T}}, \bar{\varphi}_n^{\mathcal{T}}(\cdot))_n$, associated to the sequence of perturbed Stackelberg games $(\Gamma_{\lambda_n}^{\mathcal{T}})_n$, is an SPNE of the original game Γ. The answer is negative, in general, as displayed in the example below.

The Sequence of SPNEs $(\bar{x}_n^{\mathcal{T}}, \bar{\varphi}_n^{\mathcal{T}}(\cdot))_n$ May Not Converge to an SPNE of the Original Game

Consider X, Y, L and F defined as in the example on page 95. The follower's best reply correspondence M is given in (4.3.1). Choosing $\lambda_n = n$, the SPNE of $\Gamma_{\lambda_n}^{\mathcal{T}}$ is $(\bar{x}_n^{\mathcal{T}}, \bar{\varphi}_n^{\mathcal{T}}(\cdot))$ where $\bar{x}_n^{\mathcal{T}} = -1/n$ and the function $\bar{\varphi}_n^{\mathcal{T}}$ is derived in the example on page 119; moreover

$$\lim_{n\to+\infty} \bar{x}_n^{\mathcal{T}} = 0 \quad \text{and} \quad \lim_{n\to+\infty} \bar{\varphi}_n^{\mathcal{T}}(x) = \hat{\varphi}(x) \text{ for any } x \in X,$$

(continued)

where the function $\hat{\varphi}$ is derived in the example on page 119. Regarding the strategy profile $(0, \hat{\varphi}(\cdot))$, we have

- $\hat{\varphi}(x) \in M(x)$ for any $x \in X$, so condition (SG1) in Definition 4.2.5 is satisfied,
- $0 \notin \operatorname*{Arg\,min}_{x \in X} L(x, \hat{\varphi}(x)) = \emptyset$, so condition (SG2) in Definition 4.2.5 does not hold,

hence, $(0, \hat{\varphi}(\cdot))$ is not an SPNE of Γ.

In order to achieve an SPNE of Γ we need to take into account the limit of the sequence of action profiles $(\bar{x}_n^{\mathcal{T}}, \bar{\varphi}_n^{\mathcal{T}}(\bar{x}_n^{\mathcal{T}}))_n$.

Theorem 4.5.11 [93, Th. 3.1]) *Assume that* $\lim_{n\to+\infty} \lambda_n = +\infty$ *and hypotheses of Proposition 4.5.6 are satisfied. Denote with* $(\bar{x}_n^{\mathcal{T}}, \bar{\varphi}_n^{\mathcal{T}}(\cdot))$ *an SPNE of* $\Gamma_{\lambda_n}^{\mathcal{T}}$, *as defined in Corollary 4.5.10, for any* $n \in \mathbb{N}$.

If the sequence $(\bar{x}_n^{\mathcal{T}}, \bar{\varphi}_n^{\mathcal{T}}(\bar{x}_n^{\mathcal{T}}))_n$ *converges to* $(\bar{x}^{\mathcal{T}}, \bar{y}^{\mathcal{T}})$ *in* $X \times Y$, *then the strategy profile* $(\bar{x}^{\mathcal{T}}, \widetilde{\varphi}^{\mathcal{T}}(\cdot))$ *where*

$$\widetilde{\varphi}^{\mathcal{T}}(x) = \begin{cases} \bar{y}^{\mathcal{T}}, & \text{if } x = \bar{x}^{\mathcal{T}} \\ \hat{\varphi}(x), & \text{if } x \neq \bar{x}^{\mathcal{T}} \end{cases}$$

with $\hat{\varphi}(x)$ *defined in* (4.5.2), *is an SPNE of* Γ. △

Remark 4.5.12 Coming back to the above example, we have $(\bar{x}_n^{\mathcal{T}}, \bar{\varphi}_n^{\mathcal{T}}(\bar{x}_n^{\mathcal{T}})) = (-1/n, -1)$ for any $n \in \mathbb{N}$ and $(\bar{x}^{\mathcal{T}}, \bar{y}^{\mathcal{T}}) = (0, -1)$. Therefore, the SPNE selected according to Theorem 4.5.11 is $(0, \widetilde{\varphi}^{\mathcal{T}}(\cdot))$ where the function $\widetilde{\varphi}^{\mathcal{T}} : X \to Y$ is defined on X by

$$\widetilde{\varphi}^{\mathcal{T}}(x) = \begin{cases} -1, & \text{if } x \in [-1, 0] \\ 1, & \text{if } x \in]0, 1]. \end{cases}$$

In fact, $\widetilde{\varphi}^{\mathcal{T}}(x) \in M(x)$ for any $x \in X$ and $0 \in \operatorname{Arg\,min}_{x\in X} L(x, \widetilde{\varphi}^{\mathcal{T}}(x))$. △

The following example shows that the SPNE selected according to Theorem 4.5.11 does not coincide, in general, with the SPNEs induced by a weak or a strong Stackelberg solution, as defined in Definition 4.2.6.

The SPNE Selected via Tikhonov Regularization May Not Coincide with SPNEs Induced by Weak or by Strong Stackelberg Solutions
Consider X, Y, L and F defined as in the example on page 91. The sequence $(\bar{x}_n^{\mathcal{T}}, \bar{\varphi}_n^{\mathcal{T}}(\bar{x}_n^{\mathcal{T}}))_n$ converges to $(7/4, 0)$, so the SPNE selected according to Theorem 4.5.11 is the strategy profile $(7/4, \widetilde{\varphi}^{\mathcal{T}}(\cdot))$ where

$$\widetilde{\varphi}^{\mathcal{T}}(x) = \begin{cases} 1, & \text{if } x \in [-2, -7/4[\\ 0, & \text{if } x \in [-7/4, 7/4] \\ -1, & \text{if } x \in]7/4, 2]. \end{cases}$$

Such an SPNE is different from the SPNEs induced by weak and strong Stackelberg solutions, which are given in the example on page 91. Moreover, since Γ has multiple SPNEs and only one of which is obtained via the method displayed above, the selection method for SPNEs designed via Tikhonov regularization is effective.

The Tikhonov approach to select SPNEs presented in this subsection has been extended also to the class of *one-leader two-follower Stackelberg games* in [93, Section 4]. In this case the two followers, after having observed the action x chosen by the leader, face a parametric two-player Nash equilibrium problem (NE_x) (see Sect. 4.4.3.2). By Tikhonov-regularizing the followers' payoff functions in problem (NE_x), a sequence of perturbed one-leader two-follower Stackelberg games has been defined where the solution to the Nash equilibrium problem in the second-stage is unique. This allows to select an SPNE similarly to Theorem 4.5.11, as proved in theorems 4.1 and 4.2 in [93].

4.5.2 *Regularizing the Follower's Payoff Function via Proximal Method*

Let us recall firstly the proximal regularization method for approximating the solutions of a convex optimization problem. Then, we show how it can be employed to regularize bilevel optimization problems (PB) or (OB) and to select SPNEs involving behavioural motivations.

4.5.2.1 Proximal-Point Algorithm in Convex Optimization

Assume we deal with the minimization problem

$$(P): \min_{a \in A} J(a),$$

where J is a real-valued function defined on a subset A of a Euclidean space $\mathbb{A}$ with norm $\|\cdot\|_{\mathbb{A}}$. The algorithm defined below has been introduced by Martinet in [86] for regularizing variational inequalities and by Rockafellar in [99] for finding zeros of maximal monotone operators.

Proximal point algorithm (PA)

Fix an initial point $\bar{a}_0 \in A$ and define for any $n \in \mathbb{N}$

$\{\bar{a}_n^{\mathcal{P}}\} = \operatorname*{Arg\,min}_{a \in A} \left(J(a) + \frac{1}{2\gamma_n} \|a - \bar{a}_{n-1}^{\mathcal{P}}\|_{\mathbb{A}}^2 \right),$

where $\gamma_n > 0$ for any $n \in \mathbb{N}$.

The well-definedness of algorithm (*PA*) and its convergence properties are recalled in the following well-known result stated in [99].

Theorem 4.5.13 *Assume that the set A is compact and convex, the function J is lower semicontinuous and convex over A and* $\sum_{n=1}^{+\infty} \gamma_n = +\infty$.

Then, algorithm (PA) *is well-defined and the sequence* $(\bar{a}_n^{\mathcal{P}})_n$ *is convergent to a solution of problem* (P)*, that is*

$$\lim_{n \to +\infty} \bar{a}_n^{\mathcal{P}} \in \operatorname*{Arg\,min}_{a \in A} J(a).$$

△

Remark 4.5.14 We point out that algorithm (*PA*) and its denomination rely on the *Moreau–Yosida regularization* of J with parameter $\gamma > 0$, which is the function $J_\gamma^{\mathcal{P}} : A \to [-\infty, +\infty]$ defined on A by

$$J_\gamma^{\mathcal{P}}(a) = \inf_{x \in A} \left(J(x) + \frac{1}{2\gamma} \|x - a\|_{\mathbb{A}}^2 \right),$$

introduced previously by Moreau in [90], and also called *Moreau envelope* or *proximal approximation*. △

Therefore, the proximal point algorithm allows to approximate a solution of a convex minimization problem by constructing recursively a sequence of proximal-perturbed problems, namely

$$(P)_{\gamma_n}^{\mathcal{P}} : \min_{a \in A} \left(J(a) + \frac{1}{2\gamma_n} \|a - \bar{a}_{n-1}^{\mathcal{P}}\|_{\mathbb{A}}^2 \right),$$

in general better-behaved than the original problem and that have a unique solution. Looking at the differences with respect to the Tikhonov approach, we have

- the proximal-regularized problems $(P)_{\gamma_n}^{\mathcal{P}}$ are recursively defined, differently from the Tikhonov-regularized problems $(P)_{\lambda_n}^{\mathcal{T}}$ in Sect. 4.5.1.1;

- the limit of the sequence generated by (PA) is "just" a minimizer of J over A, while in the Tikhonov approach the limit point is characterized as the minimum norm element in the set of minimizers of J over A;
- as regards to the assumptions ensuring the convergence of algorithm (PA) and of Tikhonov method, the hypothesis on the sequence of parameters $(\gamma_n)_n$ in Theorem 4.5.13 is weaker than the corresponding hypothesis on $(\lambda_n)_n$ in Theorem 4.5.1.

In a one-stage games framework, we mention that algorithm (PA) has been used in [86, 99] also for approximating saddle-points in zero-sum games and in [38] in general N-player games. More recently, in [5] an alternating proximal minimization algorithm has been proposed for two-player weighted potential games and in [17] Moreau–Yosida regularization has been exploited to define a new equilibrium concept to model strategic uncertainty among players.

4.5.2.2 Proximal Regularization of the Follower's Payoff Function in Stackelberg Games

Let $\Gamma = (X, Y, L, F)$ be a Stackelberg game. We first display some preliminarily results on employing proximal regularization only in the second-stage problem similarly as presented in Sect. 4.4.1.2 for the Tikhonov regularization. We do this in order to make easier to illustrate the methodology for selecting an SPNE of Γ we will use in the next subsection which involves the regularization of both players' payoff functions. Consider the following approach: fixed an initial point $\bar{y}_0 \in Y$, define for any $n \in \mathbb{N}$

$$\{\bar{\varphi}_n^{\mathcal{P}}(x)\} = \operatorname*{Arg\,min}_{y \in Y} \left(F(x, y) + \frac{1}{2\gamma_n} \|y - \bar{\varphi}_{n-1}^{\mathcal{P}}(x)\|_{\mathbb{Y}}^2 \right) \quad \text{for any } x \in X, \tag{4.5.4}$$

where $\gamma_n > 0$ for any $n \in \mathbb{N}$ and $\bar{\varphi}_0^{\mathcal{P}}(x) = \bar{y}_0$ for any $x \in X$.

Remark 4.5.15 Assume that Y is compact and convex, the function F is continuous over $X \times Y$, the function $F(x, \cdot)$ is convex over Y for any $x \in X$ and $\sum_{n=1}^{+\infty} \gamma_n = +\infty$. From Theorem 4.5.13 it straightforwardly follows that, for any $x \in X$, the sequence $(\bar{\varphi}_n^{\mathcal{P}}(x))_n$ is well-defined and $\lim_{n \to +\infty} \bar{\varphi}_n^{\mathcal{P}}(x) \in M(x)$. △

Therefore, we can approach the follower's problem (P_x) via a sequence of proximal-regularized second-stage problems recursively defined by

$$(P_x)_{\gamma_n}^{\mathcal{P}}: \ \operatorname*{Arg\,min}_{y \in Y} \left(F(x, y) + \frac{1}{2\gamma_n} \|y - \bar{\varphi}_{n-1}^{\mathcal{P}}(x)\|_{\mathbb{Y}}^2 \right),$$

and consequently we can define the following proximal-regularized bilevel problems:

$$(SP)^{\mathcal{P}}_{\gamma_n} \begin{cases} \min_{x \in X} L(x, \bar{\varphi}^{\mathcal{P}}_n(x)) \\ \text{where } \bar{\varphi}^{\mathcal{P}}_n(x) \text{ is the solution to } (P_x)^{\mathcal{P}}_{\gamma_n} \end{cases}$$

which come out to be Stackelberg problems, since the solution to the second-stage problem of $(SP)^{\mathcal{P}}_{\gamma_n}$ is unique for any $n \in \mathbb{N}$. Analogously to Sect. 4.5.1.2, we are interested in two questions.

1. Are there solutions to $(SP)^{\mathcal{P}}_{\gamma_n}$?
2. What happens when $n \to +\infty$?

By means of the maximum theorem, the function $\bar{\varphi}^{\mathcal{P}}_n : X \to Y$ defined in (4.5.4) is continuous over X, so the next proposition provides the answer to the first question.

Proposition 4.5.16 *Assume that the set X is compact, Y is compact and convex, the functions L and F are continuous over $X \times Y$ and the function $F(x, \cdot)$ is convex over Y for any $x \in X$.*

Then, the proximal-regularized Stackelberg problem $(SP)^{\mathcal{P}}_{\gamma_n}$ has at least one solution, for any $n \in \mathbb{N}$. △

Concerning the second question, the limit of the sequence generated by the solutions to $(SP)^{\mathcal{P}}_{\gamma_n}$ for any $n \in \mathbb{N}$ has no connections, in general, either with weak Stackelberg (or pessimistic) solutions to (PB) or with strong Stackelberg (or optimistic) solutions to (OB), as illustrated in the following example also used in [22].

The Sequence of Solutions to $(SP)^{\mathcal{P}}_{\gamma_n}$ May Not Converge to Weak or to Strong Stackelberg Solutions

Consider $X = [1/2, 2]$, $Y = [-1, 1]$,

$$L(x, y) = x + y \quad \text{and} \quad F(x, y) = \begin{cases} 0, & \text{if } x \in [1/2, 1[\\ (x-1)y, & \text{if } x \in [1, 2]. \end{cases}$$

Choosing $\bar{y}_0 = 1$ and $\gamma_n = n$, we obtain

$$\bar{\varphi}^{\mathcal{P}}_n(x) = \begin{cases} 1, & \text{if } x \in [1/2, 1] \\ c_n - c_n x + 1, & \text{if } x \in]1, 1 + 2/c_n] \\ -1, & \text{if } x \in]1 + 2/c_n, 2], \end{cases}$$

(continued)

where the sequence $(c_n)_n$ is recursively defined by $c_1 = 1$ and $c_{n+1} = c_n + n + 1$ for any $n \geq 1$.

Hence, the solution to $(SP)^{\mathcal{P}}_{\gamma_n}$ is

$$\bar{u}_n^{\mathcal{P}} = \begin{cases} 1/2, & \text{if } n = 1 \\ 1 + 2/c_n, & \text{if } n \geq 2. \end{cases}$$

Since $\lim_{n\to+\infty} c_n = +\infty$, the sequence $(\bar{u}_n^{\mathcal{P}})_n$ converges to 1, which is neither a weak nor a strong Stackelberg solution. In fact, a solution to (PB) does not exist, whereas the solution to (OB) is $1/2$. Consequently, even the sequence $(\bar{u}_n^{\mathcal{P}}, \bar{\varphi}_n^{\mathcal{P}}(\bar{u}_n^{\mathcal{P}}))_n$ does not converge either to a weak or to a strong Stackelberg equilibrium.

Example above shows also that, in general, the sequence of functions $(\bar{\varphi}_n^{\mathcal{P}})_n$ is not continuously convergent. In fact, coming back to such an example, it is sufficient to note that the pointwise limit of $(\bar{\varphi}_n^{\mathcal{P}})_n$ defined by

$$\lim_{n\to+\infty} \bar{\varphi}_n^{\mathcal{P}}(x) = \begin{cases} 1, & \text{if } x \in [1/2, 1] \\ -1, & \text{if } x \in]1, 2], \end{cases}$$

is not a continuous function.

We mention that methods involving proximal regularization have been recently proposed in [100, 102] for solving the *simple bilevel optimization problem*

$$(SBP) \begin{cases} \min_{x\in E} l(x) \\ \text{where } E = \operatorname{Arg\,min}_{x\in\Delta} f(x), \end{cases}$$

with Δ convex subset of $\mathbb{R}^n$ and l and f real-valued convex functions defined on $\mathbb{R}^n$. Such a problem, firstly studied in [106] and labelled as "simple" in [30], involves a non-parametric second-stage problem differently from the general definitions of (PB) and (OB), and its solutions clearly coincide with the weak and strong Stackelberg solutions. Consequently, problem (SBP) does not entail the same kind of inherent difficulties that bilevel optimization problems (PB) and (OB) exhibit regarding the regularization issue, as illustrated in this subsection and in Sect. 4.5.1.2.

4.5.2.3 Selection of SPNEs via Proximal Regularization

Let us present now how proximal regularization has been employed in Stackelberg games in order to select SPNEs. Caruso, Ceparano and Morgan introduced in [22] a behavioural motivated constructive method which involves the proximal regularization both in the second-stage and in the first-stage problem (the additional regularization of the leader's payoff function has been done in order to strongly motivate the behavioural interpretation of the method). Anyway, if the proximal regularization is carried out only in the second-stage problem, by arguing analogously as in the proofs in [22], a selection result for SPNEs can be proved even in such a case.

Let $\Gamma = (X, Y, L, F)$ be a Stackelberg game and consider the following procedure.

Stackelberg game proximal procedure (SGP)

Fix an initial point $(\bar{x}_0, \bar{y}_0) \in X \times Y$ and define for any $n \in \mathbb{N}$

$$(\mathcal{A}_n) \quad \begin{cases} \{\bar{\varphi}_n^{\mathcal{P}}(x)\} = \operatorname*{Arg\,min}_{y \in Y} F_{\gamma_n}^{\mathcal{P}}(x, y), \text{ for any } x \in X \\ \bar{x}_n^{\mathcal{P}} \in \operatorname*{Arg\,min}_{x \in X} L_{\beta_n}^{\mathcal{P}}(x, \bar{\varphi}_n^{\mathcal{P}}(x)) \end{cases}$$

where for any $(x, y) \in X \times Y$

$F_{\gamma_n}^{\mathcal{P}}(x, y) = F(x, y) + \frac{1}{2\gamma_n} \|y - \bar{\varphi}_{n-1}^{\mathcal{P}}(x)\|_{\mathbb{Y}}^2$

$L_{\beta_n}^{\mathcal{P}}(x, y) = L(x, y) + \frac{1}{2\beta_n} \|x - \bar{x}_{n-1}^{\mathcal{P}}\|_{\mathbb{X}}^2,$

with $\gamma_n > 0$ for any $n \in \mathbb{N}$ and $\lim_{n\to+\infty} \gamma_n = +\infty$,

$\beta_n > 0$ for any $n \in \mathbb{N}$ and $\lim_{n\to+\infty} \beta_n = +\infty$,

and $\bar{\varphi}_0^{\mathcal{P}}(x) = \bar{y}_0$ for any $x \in X$.

Procedure (SGP) allows to construct recursively a sequence of strategies $(\bar{\varphi}_n^{\mathcal{P}}(\cdot))_n$, where $\bar{\varphi}_n^{\mathcal{P}}$ is a function from X to Y, and a sequence of leader's actions $(\bar{x}_n^{\mathcal{P}})_n$, where $\bar{x}_n^{\mathcal{P}} \in X$. We point out that looking only at the follower's stage in procedure (SGP), one can recognize the same approach illustrated in Sect. 4.5.2.2 (for this reason here we kept unchanged the notation $\bar{\varphi}_n^{\mathcal{P}}$ used previously).

In the next proposition, the well-definedness of procedure (SGP) and its regularity and convergence properties are stated.

Proposition 4.5.17 ([22, Prop. 1 and 2]) *Under the assumptions of Proposition 4.5.16 we have*

- *procedure (SGP) is well-defined;*
- *for any sequence $(x_k)_k$ converging to x in X, we have*

$$\lim_{k\to+\infty} \bar{\varphi}_n^{\mathcal{P}}(x_k) = \bar{\varphi}_n^{\mathcal{P}}(x) \quad \text{for any } n \in \mathbb{N};$$

- *for any $x \in X$ the sequence $(\bar{\varphi}_n^{\mathcal{P}}(x))_n$ is convergent in Y to a solution of problem (P_x).* $\triangle$

Thanks to the well-definedness of procedure (SGP), we can construct recursively a sequence of proximal-regularized Stackelberg games $(\Gamma^{\mathcal{P}}_{\upsilon_n})_n$ where

$$\Gamma^{\mathcal{P}}_{\upsilon_n} = (X, Y, L^{\mathcal{P}}_{\beta_n}, F^{\mathcal{P}}_{\gamma_n})$$

is the game obtained from Γ by replacing the players' payoff functions L and F with the proximal-regularized functions $L^{\mathcal{P}}_{\beta_n}$ and $F^{\mathcal{P}}_{\gamma_n}$ and $\upsilon_n = (\beta_n, \gamma_n)$. Then, Proposition 4.5.17 shows in particular that for any $n \in \mathbb{N}$ the follower's best reply correspondence in $\Gamma^{\mathcal{P}}_{\upsilon_n}$ is single-valued and that the strategy profile $(\bar{x}^{\mathcal{P}}_n, \bar{\varphi}^{\mathcal{P}}_n(\cdot)) \in X \times Y^X$ generated at step $(\mathcal{A}_n)$ of procedure (SGP) is an SPNE of $\Gamma^{\mathcal{P}}_{\upsilon_n}$ obtained as update of $(\bar{x}^{\mathcal{P}}_{n-1}, \bar{\varphi}^{\mathcal{P}}_{n-1}(\cdot))$, SPNE of $\Gamma^{\mathcal{P}}_{\upsilon_{n-1}}$.

Before stating the SPNE selection result, we highlight that procedure (SGP) has a behavioural interpretation linked to the costs that players face when deviating from their current actions, which is presented below.

Interpretation of the Procedure At the generic step $(\mathcal{A}_n)$ of procedure (SGP), the follower chooses his strategy $\bar{\varphi}^{\mathcal{P}}_n$ taking into account his previous strategy $\bar{\varphi}^{\mathcal{P}}_{n-1}$. In making such a choice, he finds an action that compromises between minimizing $F(x, \cdot)$ and being near to $\bar{\varphi}^{\mathcal{P}}_{n-1}(x)$, for any $x \in X$. The latter purpose is motivated according to an *anchoring effect*, explained in [5, Subsection 1.4], which is formulated by adding a quadratic slight cost to move that reflects the difficulty of changing the previous action. The coefficient γ_n is linked to the *per unit of distance cost to move* of the follower and it is related to the trade-off parameter between minimizing $F(x, \cdot)$ and minimizing the distance from $\bar{\varphi}^{\mathcal{P}}_{n-1}(x)$. Since the same arguments apply for the preceding steps until going up to step $(\mathcal{A}_1)$, it follows that $\bar{\varphi}^{\mathcal{P}}_n(x)$ as well as the limit of $\bar{\varphi}^{\mathcal{P}}_n(x)$ embed the willingness of being near to $\bar{y}_0$. Analogous observations hold also for the leader, who chooses an action having in mind to be near to his previous choices, and therefore even with the purpose of being near to $\bar{x}_0$.

By taking into account the limit of the sequence of action profiles $(\bar{x}^{\mathcal{P}}_n, \bar{\varphi}^{\mathcal{P}}_n(\bar{x}^{\mathcal{P}}_n))_n$, the following existence result of an SPNE's selection holds.

Theorem 4.5.18 ([22, Th. 1]) *Assume that hypotheses of Proposition 4.5.16 are satisfied and let $(\bar{x}^{\mathcal{P}}_n, \bar{\varphi}^{\mathcal{P}}_n(\cdot))_n$ be the sequence of strategy profiles generated by procedure (SGP).*

If the sequence $(\bar{x}^{\mathcal{P}}_n, \bar{\varphi}^{\mathcal{P}}_n(\bar{x}^{\mathcal{P}}_n))_n$ converges to $(\bar{x}^{\mathcal{P}}, \bar{y}^{\mathcal{P}})$ in $X \times Y$, then the strategy profile $(\bar{x}^{\mathcal{P}}, \widetilde{\varphi}^{\mathcal{P}}(\cdot))$ where

$$\widetilde{\varphi}^{\mathcal{P}}(x) = \begin{cases} \bar{y}^{\mathcal{P}}, & \text{if } x = \bar{x}^{\mathcal{P}} \\ \lim\limits_{n \to +\infty} \bar{\varphi}^{\mathcal{P}}_n(x), & \text{if } x \neq \bar{x}^{\mathcal{P}}, \end{cases}$$

is an SPNE of Γ. △

Theorem 4.5.18 points out that the sequence $(\bar{x}_n^{\mathcal{P}}, \bar{\varphi}_n^{\mathcal{P}}(\cdot))_n$ of SPNEs of the proximal-regularized games $(\Gamma_{\upsilon_n}^{\mathcal{P}})_n$ could not converge to an SPNE of Γ. This happens because the follower's strategy $\widetilde{\varphi}^{\mathcal{P}}$ in the SPNE defined according to Theorem 4.5.18 could differ from the pointwise limit, whose existence is ensured by Proposition 4.5.17, of sequence $(\bar{\varphi}_n^{\mathcal{P}})_n$ in one point. In fact, denoted with $\bar{\varphi}\colon X \to Y$ such a pointwise limit, if the two limits

$$\lim_{n\to+\infty} \bar{\varphi}_n^{\mathcal{P}}(\bar{x}_n^{\mathcal{P}}) \quad \text{and} \quad \lim_{n\to+\infty} \bar{\varphi}_n^{\mathcal{P}}(\bar{x}^{\mathcal{P}}), \tag{4.5.5}$$

where $\bar{x}^{\mathcal{P}} = \lim_{n\to+\infty} \bar{x}_n^{\mathcal{P}}$, coincide, then $\widetilde{\varphi}^{\mathcal{P}}(x) = \bar{\varphi}(x)$ for any $x \in X$ and the strategy profile $(\bar{x}^{\mathcal{P}}, \bar{\varphi})$ is an SPNE of Γ in light of Theorem 4.5.18. Instead, if the two limits in (4.5.5) do not coincide, then $\widetilde{\varphi}^{\mathcal{P}}(\bar{x}^{\mathcal{P}}) \neq \bar{\varphi}(\bar{x}^{\mathcal{P}})$ and the strategy profile $(\bar{x}^{\mathcal{P}}, \bar{\varphi})$ could be not an SPNE of Γ, hence we need the follower's strategy $\widetilde{\varphi}^{\mathcal{P}}$ as in statement of Theorem 4.5.18 in order to get an SPNE. Examples 2 and 3 in [22] illustrate the two cases described above: in the first one the two limits in (4.5.5) are equal, whereas, in the second one the two limits in (4.5.5) are different and the pair $(\bar{x}^{\mathcal{P}}, \bar{\varphi})$ comes out to be not an SPNE of the game. Such examples show also that, in general, the sequence $(\bar{\varphi}_n^{\mathcal{P}})_n$ is not continuously convergent to $\bar{\varphi}$.

As regards to the connections with other methods for selecting SPNEs, the following example (examples 3, 5 and 6 in [22]) shows that the selection method based on proximal regularization presented in this subsection, the selection method relying on Tikhonov regularization described in Sect. 4.5.1.3 and the way of selecting via weak and strong Stackelberg solutions illustrated in Remark 4.3.5 do not generate, in general, the same SPNE.

The Considered Selection Methods May Select All Different SPNEs

Consider X, Y, L and F defined as in the example on page 127. We have that

- the SPNE selected via proximal regularization is $(1, \widetilde{\varphi}^{\mathcal{P}})$, where

$$\widetilde{\varphi}^{\mathcal{P}}(x) = \begin{cases} 1, & \text{if } x \in [1/2, 1[\\ -1, & \text{if } x \in [1, 2]; \end{cases}$$

- the SPNE selected by using the Tikhonov regularization is $(1, \widetilde{\varphi}^{\mathcal{T}})$, where

$$\widetilde{\varphi}^{\mathcal{T}}(x) = \begin{cases} 0, & \text{if } x \in [1/2, 1[\\ -1, & \text{if } x \in [1, 2]; \end{cases}$$

(continued)

- there are no SPNEs induced by a weak Stackelberg solution, as problem (PB) has no solutions;
- there exists one SPNE induced by the (unique) strong Stackelberg solution: the pair $(1/2, \bar{\psi})$ where $\bar{\psi}(x) = -1$ for any $x \in [1/2, 2]$.

Finally, we mention that convergence results involving "costs of change" and proximal methods in Stackelberg games have been recently investigated in [37, Section 5].

4.6 Conclusion

In this chapter we considered two-stage Stackelberg games, the natural environment for different kinds of mathematical problems that can arise depending on the leader's information about the optimal responses of the follower. We discussed crucial issues related to such problems and we provided two approaches for managing them. The first one consists in regularizing the follower's optimal reaction set via the introduction of appropriate solution concepts which allowed to obviate both the lack of existence and stability of the weak Stackelberg (or pessimistic bilevel) problem and the lack of stability of the strong Stackelberg (or optimistic bilevel) problem. The second one consists in regularizing the follower's payoff function by employing the Tikhonov and the proximal regularizations which enabled both to overcome the non-single-valuedness of the follower's best reply correspondence and to select subgame perfect Nash equilibria.

We note that the issues and the approaches (for facing such issues) we focused in this chapter are not the unique ones. For instance, the question of *sensitivity analysis* for problems (PB) and (OB), has been recently investigated in [31, 32]. Moreover, beyond the two regularizing approaches examined in this chapter, one can construct classes of *mixed* regularization methods which exploit simultaneously both the idea of regularizing the follower's optimal reaction set and of regularizing the follower's payoff function; first approximation schemes of this type have been defined in [75, 78, 88].

For the sake of brevity, we dealt with Stackelberg games only in a static framework (that is any player has actions which do not evolve over time) where players have a single objective and just one leader acts in the first stage. However, such a Stackelberg game model has been extended in many directions:

- **Dynamic:** players' actions are functions depending on time. First results on *dynamic Stackelberg games* can be found in [10, 20, 24, 68, 105, 110], while applications to economic models are described in [11, 33].
- **Multiobjective:** the follower has a vector-valued payoff function or solves vector (quasi-)variational inequalities. Methods for solving (optimistic) *multiobjective*

bilevel optimization problems have been investigated in [35] (where scalarization techniques are used) and in [113] (where a model based on set-valued optimization is firstly proposed to manage strong Stackelberg problems and then extended to a multiobjective setting). Concerning results on approximate solutions which could be useful in the analysis of Stackelberg games with second stage described by vector problems, we remind [56, 65, 79, 92]. Furthermore, a penalty scheme has been introduced to approach (optimistic) *semivectorial bilevel optimization problems* in [18] and, in a dynamic framework, existence results have been first obtained for the (optimistic and pessimistic) *semivectorial bilevel optimal control problems* in [19].

- **Multileader:** more than one leader acts in the first stage. For general *multi-leader-follower games* see, for example, [42, 98]. The situation where each follower can observe the action of one leader has been investigated in [97] for an economic model and in [23] for deriving a selection result. An increasing literature is also devoted to the class of *multi-leader-common-follower games* and to its applications to electricity markets, see for example [8, 9].

We emphasize that for the situations illustrated above it would be interesting to provide new concepts and results which derive from the analogous exploitation of the two types of regularization approaches presented in the chapter.

References

1. A. Aboussoror, P. Loridan, Strong-weak Stackelberg problems in finite dimensional spaces. Serdica Math. J. **21**, 151–170 (1995). http://www.math.bas.bg/serdica/n2_95.html
2. A. Aboussoror, P. Loridan, Existence of solutions to two-level optimization problems with nonunique lower-level solutions. J. Math. Anal. Appl. **254**, 348–357 (2001). https://doi.org/10.1006/jmaa.2000.7001
3. A. Aboussoror, A. Mansouri, Weak linear bilevel programming problems: existence of solutions via a penalty method. J. Math. Anal. Appl. **304**, 399–408 (2005). https://doi.org/10.1016/j.jmaa.2004.09.033
4. A. Aboussoror, S. Adly, F.E. Saissi, Strong-weak nonlinear bilevel problems: existence of solutions in a sequential setting. Set-Valued Var. Anal. **25**, 113–132 (2017). https://doi.org/10.1007/s11228-016-0369-4
5. H. Attouch, P. Redont, A. Soubeyran, A new class of alternating proximal minimization algorithms with costs-to-move. SIAM J. Optim. **18**, 1061–1081 (2007). https://doi.org/10.1137/060657248
6. J.-P. Aubin, H. Frankowska, *Set-Valued Analysis* (Birhäuser, Boston, 1990)
7. A. Auslender, Recherche des points de selle d'une fonction, in *Symposium on Optimization*, ed. by A.V. Balakrishnan, M. Contensou, B.F. de Veubeke, P. Krée, J.L. Lions, N.N. Moiseev (Springer, Berlin, 1970), pp. 37–52. https://doi.org/10.1007/BFb0066671
8. D. Aussel, P. Bendotti, M. Pištěk, Nash equilibrium in a pay-as-bid electricity market: part 1—existence and characterization. Optimization **66**, 1013–1025 (2017). https://doi.org/10.1080/02331934.2016.1227981
9. D. Aussel, P. Bendotti, M. Pištěk, Nash equilibrium in a pay-as-bid electricity market: part 2—best response of a producer. Optimization **66**, 1027–1053 (2017). https://doi.org/10.1080/02331934.2016.1227982

10. T. Başar, G. Olsder, *Dynamic Noncooperative Game Theory* (Society for Industrial and Applied Mathematics, Philadelphia, 1998)
11. A. Bagchi, *Stackelberg Differential Games in Economic Models*. Lecture Notes in Control and Information Sciences, vol. 64 (Springer, Berlin, 1984)
12. C. Baiocchi, A. Capelo, *Variational and Quasi-Variational Inequalities* (John Wiley and Sons, New York, 1984)
13. B. Bank, J. Guddat, D. Klatte, B. Kummer, K. Tammer, *Non-Linear Parametric Optimization* (Birkhäuser, Basel, 1983)
14. J.F. Bard, An algorithm for solving the general bilevel programming problem. Math. Oper. Res. **8**, 260–272 (1983). https://doi.org/10.1287/moor.8.2.260
15. J.F. Bard, J.E. Falk, An explicit solution to the multi-level programming problem. Comput. Oper. Res. **9**, 77–100 (1982). https://doi.org/10.1016/0305-0548(82)90007-7
16. C. Berge, *Espaces topologiques et fonctions multivoques* (Dunod, 1959) [Translation: E.M. Patterson. Topological Spaces) (Oliver & Boyd, 1963)]
17. P. Bich, Strategic uncertainty and equilibrium selection in discontinuous games. J. Econ. Theory **183**, 786–822 (2019). https://doi.org/10.1016/j.jet.2019.08.001
18. H. Bonnel, J. Morgan, Semivectorial bilevel optimization problem: penalty approach. J. Optim. Theory Appl. **131**, 365–382 (2006). https://doi.org/10.1007/s10957-006-9150-4
19. H. Bonnel, J. Morgan, Semivectorial bilevel convex optimal control problems: existence results. SIAM J. Control Optim. **50**, 3224–3241 (2012). https://doi.org/10.1137/100795450
20. M. Breton, A. Alj, A. Haurie, Sequential Stackelberg equilibria in two-person games. J. Optim. Theory Appl. **59**, 71–97 (1988). https://doi.org/10.1007/BF00939867
21. D. Cao, L.C. Leung, A partial cooperation model for non-unique linear two-level decision problems. Eur. J. Oper. Res. **140**, 134–141 (2002). https://doi.org/10.1016/S0377-2217(01)00225-9
22. F. Caruso, M.C. Ceparano, J. Morgan, Subgame perfect Nash equilibrium: a learning approach via costs to move. Dyn. Games Appl. **9**, 416–432 (2019). https://doi.org/10.1007/s13235-018-0277-3
23. M.C. Ceparano, J. Morgan, Equilibrium selection in multi-leader-follower games with vertical information. TOP **25**, 526–543 (2017). https://doi.org/10.1007/s11750-017-0444-5
24. C. Chen, J.B. Cruz, Stackelberg solution for two-person games with biased information patterns. IEEE Trans. Autom. Control **17**, 791–798 (1972). https://doi.org/10.1109/TAC.1972.1100179
25. J.B. Clempner, A.S. Poznyak, Computing the strong Nash equilibrium for Markov chains games. Appl. Math. Comput. **265**, 911–927 (2015). https://doi.org/10.1016/j.amc.2015.06.005
26. S. Dempe, A simple algorithm for the linear bilevel programming problem. Optimization **18**, 373–385 (1987). https://doi.org/10.1080/02331938708843247
27. S. Dempe, A bundle algorithm applied to bilevel programming problems with non-unique lower level solutions. Comput. Optim. Appl. **15**, 145–166 (2000). https://doi.org/10.1023/A:1008735010803
28. S. Dempe, *Foundations of Bilevel Programming* (Springer, New York, 2002)
29. S. Dempe, H. Schmidt, On an algorithm solving two-level programming problems with nonunique lower level solutions. Comput. Optim. Appl. **6**, 227–249 (1996). https://doi.org/10.1007/BF00247793
30. S. Dempe, N. Dinh, J. Dutta, Optimality conditions for a simple convex bilevel programming problem, in: *Variational Analysis and Generalized Differentiation in Optimization and Control: In Honor of Boris S. Mordukhovich*, ed. by R.S. Burachik, J.-C. Yao (Springer, New York, 2010), pp. 149–161. https://doi.org/10.1007/978-1-4419-0437-9_7
31. S. Dempe, B. Mordukhovich, A.B. Zemkoho, Sensitivity analysis for two-level value functions with applications to bilevel programming. SIAM J. Optim. **22**, 1309–1343 (2012). https://doi.org/10.1137/110845197
32. S. Dempe, B.S. Mordukhovich, A.B. Zemkoho, Two-level value function approach to nonsmooth optimistic and pessimistic bilevel programs. Optimization **68**, 433–455 (2019). https://doi.org/10.1080/02331934.2018.1543294

33. E.J. Dockner, S. Jorgensen, N.V. Long, G. Sorger, *Differential Games in Economics and Management Science* (Cambridge University Press, Cambridge, 2000)
34. L. Drouet, A. Haurie, F. Moresino, J.-P. Vial, M. Vielle, L. Viguier, An oracle based method to compute a coupled equilibrium in a model of international climate policy. Comput. Manag. Sci. **5**, 119–140 (2008). https://doi.org/10.1007/s10287-007-0043-2
35. G. Eichfelder, Multiobjective bilevel optimization. Math. Program. **123**, 419–449 (2010). https://doi.org/10.1007/s10107-008-0259-0
36. A. Evgrafov, M. Patriksson, On the convergence of stationary sequences in topology optimization, Int. J. Numer. Methods Eng. **64**, 17–44 (2005). https://doi.org/10.1002/nme.1359
37. S.D. Flåm, On games and cost of change. Ann. Oper. Res. (2020). https://doi.org/10.1007/s10479-020-03585-w
38. S.D. Flåm, G.H. Greco, Non-cooperative games; methods of subgradient projection and proximal point, in *Advances in Optimization*, ed. by W. Oettli, D. Pallaschke (Springer, Berlin, 1992), pp. 406–419. https://doi.org/10.1007/978-3-642-51682-5_27
39. D. Fudenberg, J. Tirole, *Game Theory* (MIT Press, Cambridge, 1991)
40. J.C. Harsanyi, R. Selten, *A General Theory of Equilibrium Selection in Games* (The MIT Press, Cambridge, 1988)
41. A. Haurie, J.B. Krawczyk, G. Zaccour, *Games and Dynamic Games* (World Scientific Publishing Company, Singapore, 2012)
42. M. Hu, M. Fukushima, Existence, uniqueness, and computation of robust Nash equilibria in a class of multi-leader-follower games. SIAM J. Optim. **23**, 894–916 (2013). https://doi.org/10.1137/120863873
43. A. Kannan, U. Shanbhag, Distributed computation of equilibria in monotone Nash games via iterative regularization techniques. SIAM J. Optim. **22** (2012), 1177–1205. https://doi.org/10.1137/110825352
44. L. Lampariello, S. Sagratella, Numerically tractable optimistic bilevel problems. Comput. Optim. Appl. **76**, 277–303 (2020). https://doi.org/10.1007/s10589-020-00178-y
45. L. Lampariello, S. Sagratella, O. Stein, The standard pessimistic bilevel problem. SIAM J. Optim. **29**, 1634–1656 (2019). https://doi.org/10.1137/18M119759X
46. G. Leitmann, On generalized Stackelberg strategies. J. Optim. Theory Appl. **26**, 637–643 (1978). https://doi.org/10.1007/BF00933155
47. B. Lemaire, Jeux dans les equations aux derivees partielles, in, *Symposium on Optimization*, ed. by A.V. Balakrishnan, M. Contensou, B.F. de Veubeke, P. Krée, J.L. Lions, N.N. Moiseev (Springer, Berlin, 1970), pp. 181–195. https://doi.org/10.1007/BFb0066682
48. M.B. Lignola, Well-posedness and l-well-posedness for quasivariational inequalities. J. Optim. Theory Appl. **128**, 119–138 (2006). https://doi.org/10.1007/s10957-005-7561-2
49. M.B. Lignola, J. Morgan, Semi-continuities of marginal functions in a sequential setting. Optimization **24**, 241–252 (1992). https://doi.org/10.1080/02331939208843793
50. M.B. Lignola, J. Morgan, Approximate solutions to variational inequalities and applications. Le Matematiche **49**, 281–293 (1995). https://lematematiche.dmi.unict.it/index.php/lematematiche/article/view/528
51. M.B. Lignola, J. Morgan, Topological existence and stability for Stackelberg problems. J. Optim. Theory Appl. **84**, 145–169 (1995). https://doi.org/10.1007/BF02191740
52. M.B. Lignola, J. Morgan, Stability of regularized bilevel programming problems. J. Optim. Theory Appl. **93**, 575–596 (1997). https://doi.org/10.1023/A:1022695113803
53. M.B. Lignola, J. Morgan, Existence of solutions to generalized bilevel programming problem, in *Multilevel Optimization: Algorithms and Applications*, ed. by A. Migdalas, P.M. Pardalos, P. Värbrand (Springer, Boston, 1998), pp. 315–332. https://doi.org/10.1007/978-1-4613-0307-7_14
54. M.B. Lignola, J. Morgan, Well-posedness for optimization problems with constraints defined by variational inequalities having a unique solution. J. Glob. Optim. **16**, 57–67 (2000). https://doi.org/10.1023/A:1008370910807

55. M.B. Lignola, J. Morgan, α-well-posedness for Nash equilibria and for optimization problems with Nash equilibrium constraints. J. Glob. Optim. **36**, 439–459 (2006). https://doi.org/10.1007/s10898-006-9020-5
56. M.B. Lignola, J. Morgan, Vector quasi-variational inequalities: approximate solutions and well-posedness. J. Convex Anal. **13**, 373–384 (2006). http://www.heldermann.de/JCA/JCA13/JCA132/jca13031.htm
57. M.B. Lignola, J. Morgan, Approximate values for mathematical programs with variational inequality constraints. Comput. Optim. Appl. **53**(2), 485–503 (2012). https://doi.org/10.1007/s10589-012-9470-2
58. M.B. Lignola, J. Morgan, Stability in regularized quasi-variational settings. J. Convex Anal. **19**, 1091–1107 (2012). http://www.heldermann.de/JCA/JCA19/JCA194/jca19058.htm
59. M.B. Lignola, J. Morgan, Approximating values of minsup problems with quasi-variational inequality constraints. Pac. J. Optim. **10**, 749–765 (2014). http://www.ybook.co.jp/online2/pjov10.html
60. M.B. Lignola, J. Morgan, Viscosity solutions for bilevel problems with Nash equilibrium constraints. Far Est J. Appl. Math. **88**, 15–34 (2014). http://www.pphmj.com/abstract/8683.htm
61. M.B. Lignola, J. Morgan, Asymptotic behavior of semi-quasivariational optimistic bilevel problems in Banach spaces. J. Math. Anal. Appl. **424**, 1–20 (2015). https://doi.org/10.1016/j.jmaa.2014.10.059
62. M.B. Lignola, J. Morgan, A method to bypass the lack of solutions in minsup problems under quasi-equilibrium constraints. Optim. Lett. **10**, 833–846 (2016). https://doi.org/10.1007/s11590-015-0956-6
63. M.B. Lignola, J. Morgan, Inner regularizations and viscosity solutions for pessimistic bilevel optimization problems. J. Optim. Theory Appl. **173**, 183–202 (2017). https://doi.org/10.1007/s10957-017-1085-4
64. M.B. Lignola, J. Morgan, Further on inner regularizations in bilevel optimization. J. Optim. Theory Appl. **180**, 1087–1097 (2019). https://doi.org/10.1007/s10957-018-1438-7
65. M.B. Lignola, J Morgan, V. Scalzo, Lower convergence of approximate solutions to vector quasi-variational problems. Optimization **59**, 821–832 (2010). https://doi.org/10.1080/02331930902863707
66. G.-H. Lin, M. Xu, J.J. Ye, On solving simple bilevel programs with a nonconvex lower level program. Math. Program. **144**, 277–305 (2014). https://doi.org/10.1007/s10107-013-0633-4
67. P. Loridan, J. Morgan, Approximation results for two-level optimization problem and application to penalty methods, Preprint n.52-1985. Dipartimento di Matematica e Applicazioni R. Caccioppoli, Università di Napoli, December 1984. https://www.researchgate.net/publication/335619596_Approximation_results_for_a_two-level_optimization_problem_and_application_to_penalty_methods
68. P. Loridan, J. Morgan, Approximation of the Stackelberg problem and applications in control theory. Annu. Rev. Autom. Program. **13**, 121–124 (1985). https://doi.org/10.1016/0066-4138(85)90472-0
69. P. Loridan, J. Morgan, Approximate solutions for two-level optimization problems, in *Trends in Mathematical Optimization: 4th French-German Conference on Optimization* 1986, ed. by K.-H. Hoffmann, J. Zowe, J.-B. Hiriart-Urruty, C. Lemarechal (Birkhäuser, Basel, 1988), pp. 181–196. https://doi.org/10.1007/978-3-0348-9297-1_13
70. P. Loridan, J. Morgan, ϵ-regularized two-level optimization problems: approximation and existence results, in *Optimization: Proceedings of the Fifth French–German Conference Held in Castel-Novel (Varetz), October 3–8, 1988*, ed. by S. Dolecki (Springer, Berlin, 1989), pp. 99–113. https://doi.org/10.1007/BFb0083589
71. P. Loridan, J. Morgan, New results on approximate solution in two-level optimization. Optimization **20**, 819–836 (1989). https://doi.org/10.1080/02331938908843503
72. P. Loridan, J. Morgan, A sequential stability result for constrained Stackelberg problems. Ric. Mat. **38**, 19–32 (1989). https://www.researchgate.net/publication/266591292_A_sequential_stability_result_for_constrained_Stackelberg_problems

73. P. Loridan, J. Morgan, A theoretical approximation scheme for Stackelberg problems. J. Optim. Theory Appl. **61**, 95–110 (1989). https://doi.org/10.1007/BF00940846
74. P. Loridan, J. Morgan, Quasi convex lower level problem and applications in two level optimization, in *Proceedings of the International Workshop on "Generalized Concavity, Fractional Programming and Economic Applications" Held at the University of Pisa, Italy, May 30–June 1, 1988*, ed. by A. Cambini, E. Castagnoli, L. Martein, P. Mazzoleni, S. Schaible (Springer, Berlin, 1990), pp. 325–341. https://doi.org/10.1007/978-3-642-46709-7_23
75. P. Loridan, J. Morgan, Least-norm regularization for weak two-level optimization problems, in *Optimization, Optimal Control and Partial Differential Equations*, ed. by V. Barbu, D. Tiba, J.F. Bonnans (Birkhäuser, Basel, 1992), pp. 307–318. https://doi.org/10.1007/978-3-0348-8625-3_28
76. P. Loridan, J. Morgan, On strict-solution for a two level optimization problem, in *Proceedings of The international Conference On Operation Research*, ed. by G. Feichtinger, W. Bühler, F. Radermacher, P. Stähly (Springer Verlag, Vienna, 1992), pp. 165–172. https://doi.org/10.1007/978-3-642-77254-2_19
77. P. Loridan, J. Morgan, Regularizations for two-level optimization problems, in *Advances in Optimization*, ed. by W. Oettli, D. Pallaschke (Springer, Berlin, 1992), pp. 239–255. https://doi.org/10.1007/978-3-642-51682-5_16
78. P. Loridan, J. Morgan, Weak via strong Stackelberg problem: new results. J. Glob. Optim. 263–287 (1996). https://doi.org/10.1007/BF00121269
79. P. Loridan, J. Morgan, Convergence of approximate solutions and values in parametric vector optimization, in *Vector Variational Inequalities and Vector Equilibria: Mathematical Theories*, ed. by F. Giannessi (Springer, Boston, 2000), pp. 335–350. https://doi.org/10.1007/978-1-4613-0299-5_19
80. R. Lucchetti, F. Mignanego, G. Pieri, Existence theorems of equilibrium points in Stackelberg games with constraints. Optimization **18**, 857–866 (1987). https://doi.org/10.1080/02331938708843300
81. Z.-Q. Luo, J.-S. Pang, D. Ralph, *Mathematical Programs with Equilibrium Constraints* (Cambridge University Press, Cambridge, 1996)
82. L. Mallozzi, J. Morgan, *Problema di Stackelberg con risposte pesate*. Atti del XIX Convegno AMASES (1995), pp. 416–425
83. L. Mallozzi, J. Morgan, Hierarchical systems with weighted reaction set, in *Nonlinear Optimization and Applications*, ed. by G. Di Pillo, F. Giannessi (Springer, Boston, 1996), pp. 271–282. https://doi.org/10.1007/978-1-4899-0289-4_19
84. L. Mallozzi, J. Morgan, Oligopolistic markets with leadership and demand functions possibly discontinuous. J. Optim. Theory Appl. **125**, 393–407 (2005). https://doi.org/10.1007/s10957-004-1856-6
85. P. Marcotte, G Savard, A note on the Pareto optimality of solutions to the linear bilevel programming problem, Comput. & Oper. Res. **18**, 355–359 (1991). https://doi.org/10.1016/0305-0548(91)90096-A
86. B. Martinet, Brève communication. régularisation d'inéquations variationnelles par approximations successives. Rev. Fr. Inform. Rech. Opér. Sér. Rouge **4**, 154–158 (1970). http://www.numdam.org/item/M2AN_1970__4_3_154_0/
87. M. Maschler, E. Solan, S. Zamir, *Game Theory* (Cambridge University Press, Cambridge, 2013)
88. D.A. Molodtsov, The solution of a class of non-antagonistic games. USSR Comput. Math. Math. Phys. **16**, 67–72 (1976). https://doi.org/10.1016/0041-5553(76)90042-2
89. D.A. Molodtsov, V.V. Fedorov, Approximation of two-person games with information exchange. USSR Comput. Math. Math. Phys. **13**, 123–142 (1973). https://doi.org/10.1016/0041-5553(73)90010-4
90. J.J. Moreau, Proximité et dualité dans un espace hilbertien. Bull. Soc. Math. Fr. **93**, 273–299 (1965). https://doi.org/10.24033/bsmf.1625
91. J. Morgan, Constrained well-posed two-level optimization problems, in *Nonsmooth Optimization and Related Topics*, ed. by F.H. Clarke, V.F. Dem'yanov, F. Giannessi (Springer, Boston, 1989), pp. 307–325. https://doi.org/10.1007/978-1-4757-6019-4_18

92. J. Morgan, Approximations and well-posedness in multicriteria games. Ann. Oper. Res. **137**, 257–268 (2005). https://doi.org/10.1007/s10479-005-2260-9
93. J. Morgan, F. Patrone, Stackelberg problems: subgame perfect equilibria via Tikhonov regularization, in *Advances in Dynamic Games: Applications to Economics, Management Science, Engineering, and Environmental Management*, ed. by A. Haurie, S. Muto, L.A. Petrosjan, T.E.S. Raghavan (Birkhäuser, Boston, 2006), pp. 209–221. https://doi.org/10.1007/0-8176-4501-2_12
94. J. Morgan, R. Raucci, Lower semicontinuity for approximate social Nash equilibria. Int. J. Game Theory **31**, 499–509 (2003). https://doi.org/10.1007/s001820300134
95. J.F. Nash, Equilibrium points in n-person games. Proc. Natl. Acad. Sci. **36**, 48–49 (1950). https://doi.org/10.1073/pnas.36.1.48
96. J.V. Outrata, On the numerical solution of a class of Stackelberg problems. Z. Oper. Res. **34**, 255–277 (1990). https://doi.org/10.1007/BF01416737
97. M. Pagnozzi, S. Piccolo, Vertical separation with private contracts. Econ. J. **122**, 173–207 (2012). https://doi.org/10.1111/j.1468-0297.2011.02471.x
98. J.-S. Pang, M. Fukushima, Quasi-variational inequalities, generalized Nash equilibria, and multi-leader-follower games. Comput. Manag. Sci. **2**, 21–56 (2005). https://doi.org/10.1007/s10287-004-0010-0
99. R.T. Rockafellar, Monotone operators and the proximal point algorithm. SIAM J. Control Optim. **14**, 877–898 (1976). https://doi.org/10.1137/0314056
100. S. Sabach, S. Shtern, A first order method for solving convex bilevel optimization problems. SIAM J. Optim. **27**, 640–660 (2017). https://doi.org/10.1137/16M105592X
101. R. Selten, Spieltheoretische Behandlung eines Oligopolmodells mit Nachfrageträgheit, Z. Gesamte Staatswiss. **121**, 301–324 (1965). https://www.jstor.org/stable/40748884
102. Y. Shehu, P.T. Vuong, A. Zemkoho, An inertial extrapolation method for convex simple bilevel optimization. Optim Methods Softw. 1–19 (2019). https://doi.org/10.1080/10556788.2019.1619729
103. K. Shimizu, E. Aiyoshi, A new computational method for Stackelberg and min–max problems by use of a penalty method. IEEE Trans. Autom. Control **26**, 460–466 (1981). https://doi.org/10.1109/TAC.1981.1102607
104. V.F. Sholohovich, Unstable extremal problems and geometric properties of Banach spaces. Dokl. Akad. Nauk **195**, 289–291 (1970). http://www.mathnet.ru/php/archive.phtml?wshow=paper&jrnid=dan&paperid=35782&option_lang=eng
105. M. Simaan, J.B. Cruz, On the Stackelberg strategy in nonzero-sum games. J. Optim. Theory Appl. **11**, 533–555 (1973). https://doi.org/10.1007/BF00935665
106. M. Solodov, An explicit descent method for bilevel convex optimization. J. Convex Anal. **14**, 227–237 (2007). http://www.heldermann.de/JCA/JCA14/JCA142/jca14016.htm
107. A.N. Tikhonov, Methods for the regularization of optimal control problems. Sov. Math. Dokl. **6**, 761–763 (1965)
108. L. Vicente, G. Savard, J. Júdice, Descent approaches for quadratic bilevel programming, J. Optim Theory Appl. **81**, 379–399 (1994). https://doi.org/10.1007/BF02191670
109. H. von Stackelberg, *Marktform und Gleichgewicht* (Verlag von Julius Springer, Berlin, 1934) [Translation: D. Bazin, R. Hill, L. Urch. *Market Structure and Equilibrium* (Springer, Berlin, 2011)]
110. J.J. Ye, Necessary conditions for bilevel dynamic optimization problems. SIAM J. Control Optim. **33**, 1208–1223 (1995). https://doi.org/10.1137/S0363012993249717
111. J.J. Ye, D.L. Zhu, Optimality conditions for bilevel programming problems. Optimization **33**, 9–27 (1995). https://doi.org/10.1080/02331939508844060
112. J.J. Ye, D.L. Zhu, and Q.J. Zhu, Exact penalization and necessary optimality conditions for generalized bilevel programming problems. SIAM J. Optim. **7**, 481–507 (1997). https://doi.org/10.1137/S1052623493257344
113. A.B. Zemkoho, Solving ill-posed bilevel programs. Set-Valued Var. Anal. **24**, 423–448 (2016). https://doi.org/10.1007/s11228-016-0371-x

Chapter 5
Applications of Bilevel Optimization in Energy and Electricity Markets

Sonja Wogrin, Salvador Pineda, and Diego A. Tejada-Arango

Abstract Ever since the beginning of the liberalization process of the energy sector and the arrival of electricity markets, decision making has gone away from centralized planners and has become the responsibility of many different entities such as market operators, private generation companies, transmission system operators and many more. The interaction and sequence in which these entities make decisions in liberalized market frameworks have led to a renaissance of Stackelberg-type games referred to as bilevel problems, which capture the natural hierarchy in many decision-making processes in power systems. This chapter aims at demonstrating the crucial insights that such models can provide, and at providing a broad overview of the plethora of the applications of bilevel programming in energy and electricity markets. Finally, a numerical example is included for illustrative purposes.

Keywords Transmission expansion planning · Generation expansion planning · Energy storage systems · Strategic bidding · Electricity markets · Energy applications

5.1 General Overview of Bilevel Programming in Energy

The liberalization of the electricity sector and the introduction of electricity markets first emerged in the 1980s in countries like Chile, the United Kingdom, and New Zealand. Nowadays, the vast majority of all developed countries have undergone such a liberalization process, which has complicated the organization of the electricity and energy sector greatly. In regulated electricity systems, many tasks—such as expansion planning, for example—are usually carried out by a centralized

S. Wogrin (✉) · D. A. Tejada-Arango
Comillas Pontifical University, Madrid, Spain
e-mail: Sonja.Wogrin@comillas.edu

S. Pineda
University of Malaga, Malaga, Spain
e-mail: spineda@uma.es

S. Dempe, A. Zemkoho (eds.), *Bilevel Optimization*, Springer Optimization and Its Applications 161, https://doi.org/10.1007/978-3-030-52119-6_5

planner who minimizes total cost while meeting a future demand forecast, reliability constraints, and environmental requirements identified by the government. Planning the investment and operation of such a regulated system can be regarded as stable, relatively predictable, and essentially risk-free for all entities involved.

However, under a liberalized framework, many decisions are no longer regulated but to a large extent, up to the responsibility of individual entities or private companies that act strategically and might have opposing objectives. Within the realm of strategic decision making, the hierarchy in which decisions are taken becomes extremely important. A strategic storage investor maximizing profits is not going to invest the same amount as a centralized planner that maximizes social welfare, nor will the storage facility be operated in the same way. In order to characterize this strategic behavior, bilevel models inspired by Stackelberg games become necessary and important decision-support tools in electricity and energy markets.

Bilevel models—which have first been used in the electricity sector to formulate electricity markets for example by Cardell et al. [10], Berry et al. [5], Weber and Overbye [70], Hobbs et al. [26] or Ramos et al. [56] just to name a few—allow us to represent a sequential decision-making process as opposed to single-level models where all decisions are considered to be taken simultaneously, which can be a gross simplification of reality and distort model outcomes.

In energy and electricity markets, there are numerous different applications of interesting bilevel optimization and bilevel equilibrium problems. The purpose of this chapter is to provide a general overview of the areas in energy and electricity markets where bilevel programming is being used in a meaningful way. Hence, the remainder of this chapter is organized as follows. Section 5.1 contains a detailed literature review of bilevel applications in energy. Sections 5.2 and 5.3 give a brief outlook on existing solution techniques and pending challenges in bilevel models in the power sector. In Sect. 5.4 , we formulate a bilevel model that takes strategic investment decisions and compare it to a traditional model in a numerical case study. Finally, Sect. 5.5 concludes the chapter.

Section 5.1 provides a detailed but not exhaustive literature review on many different applications of bilevel programming in energy and electricity markets. We first identify several important areas to which bilevel optimization is applied: Transmission Expansion Planning, Generation Expansion Planning (conventional and renewable), Energy Storage Systems, Strategic Bidding, Natural Gas Markets, and Others. In the remainder of this section, we analyze each of those areas separately and point out what type of bilevel games can be encountered in the literature. Please note that these areas are not necessarily mutually exclusive. Therefore, relevant references might appear more than once if applicable.

5.1.1 Transmission Expansion Planning

At the core of every multi-level optimization or equilibrium problem, there lies a sequential decision-making process. When it comes to transmission expansion planning (TEP) in liberalized electricity markets, the main factor of decision hierarchy stems from the generation expansion planning (GEP). Does the transmission planner take its decisions after the generation has been decided and sited, or do generation companies plan their investments after transmission assets have been decided? What comes first, the chicken or the egg? The sequence of TEP versus GEP defines two different philosophies: proactive TEP; and, reactive TEP.

Under proactive TEP approaches, the transmission company (TRANSCO) is the Stackelberg leader, whereas generation companies (GENCOs) are the Stackelberg followers. This means that TRANSCO has the first-mover advantage and decides the best possible TEP taking into account the feedback from generation companies. In other words, the network planner can influence generation investment and, furthermore, the spot market behavior. Note that while the market operation is an essential part of TEP and GEP problems, and corresponding modeling details have to be discussed in order to establish possible gaps in the literature, in terms of decision-making sequence, market decisions happen after both GEP and TEP. Hence, the market is not the main focus in the bilevel TEP/GEP discussion.

On the other hand, under a reactive TEP approach, the network planner assumes that generation capacities are given (GENCOs are Stackelberg leaders), and then the network planner optimizes transmission expansion based only on the subsequent market operation. Reactive planning is thus represented by a model with multi-leader GENCOs and one or several TRANSCOs as followers.

Additionally, there exist some alternative TEP approaches, such as [30], where the upper level represents both GEP and TEP simultaneously, while the lower level represents the market operation (MO). In [3], the authors decide optimal wind and transmission expansion in the upper level, subject to a market-clearing. Such an approach is more suitable for a centralized power system where the network planner and the generation companies belong to the same decision-making agent. This is usually not the case in liberalized electricity markets. The work of [57] explores the interactions between an ISO with the security of supply objectives with private profit-maximizing entities deciding GEP and TEP investment simultaneously. The entire problem, however, is never explicitly formulated as a bilevel problem. Instead, an iterative solution procedure is proposed. Since our main focus lies in the sequence between TEP and GEP, we continue our detailed literature review on bilevel TEP problems distinguishing between proactive and reactive approaches.

In practice [65], most of the TRANSCOs in the world follow a reactive TEP approach, but most of the TEP-GEP literature is on proactive planning. However, Pozo et al. [54] mention other approaches that are close to proactive planning. For example, a regulation was approved in the US that includes the concept of anticipative (proactive) transmission planning to obtain a higher social welfare [20]. Additionally, in the current European context, ENTSOE plays the role of a centralized agent that proposes future planning pathways, in which regional coordination takes place, and then nationally, generation companies can react to its decisions. Thus, under this regulatory context, a proactive planning approach would make more sense.

5.1.1.1 Proactive Transmission Expansion Planning

Sauma and Oren [61] present the first work on proactive TEP we want to discuss. In [61], the authors extend their work in [60] and explore optimal transmission expansion given different objectives. They also consider a spot market where the distinctive ownership structures are reflected, as proposed in [78]. The authors consider a three-level structure: first TEP, then GEP, and finally, the market. The market equilibrium and strategic GEP already yield an equilibrium problem with equilibrium constraints (EPEC). An additional layer of TEP is set on top of that.

Pozo et al. [52] extend the groundwork of [60, 61], and propose a first complete model formulation of the three-level TEP problem. The three levels are still: first TEP, then GEP, and finally, the market-clearing. In the second level (the GEP stage), all GENCOs' investment strategies are enumerated, expressing Nash equilibria as a set of inequalities. While this work represents a significant step forward in terms of actually formulating a three-level problem in closed form, the proposed model does not seem to be computationally efficient.

Pozo et al. [53] extend their work of [52] by including uncertainty in the demand, and apply the model to a realistic power system in Chile. The same authors overcome the computational shortcoming of their previously proposed models in [54] by proposing a novel column-and-row decomposition technique for solving the arising EPECs, which ultimately yields the global optimum. Furthermore, the authors propose a pessimistic and optimistic network planner to describe all possible outcomes of the EPEC. The authors conclude that in practice, if multiple generation expansion exists in the equilibrium, proactive planning does not always yield the best welfare results, and it can even reduce social welfare.

Jin and Ryan [31] also propose a three-level TEP problem, but they extend previous approaches by considering Cournot's strategic decisions in the market. They propose a hybrid approach to solve the three-level problem, which proves computationally efficient; however, it does not guarantee to find the global optimum.

Motamedi et al. [42] present another multi-level approach to TEP. The authors characterize their problem as a four-level problem; however, since two of those levels refer to problems that are considered simultaneously, mathematically speaking, the framework boils down to a three-level problem: TEP, GEP, and market (pairs of price and quantity bids). The authors propose an iterative algorithm using search-based techniques and agent-based modeling to solve the arising problem.

In Taheri et al. [67], the authors also tackle a three-level TEP problem: TEP, GEP, and market clearing through price and offer quantities. The temporal representation of this work resorts to a load duration curve, which does not allow for modeling intertemporal constraints such as ramping or commitment constraints.

Apart from the three-level proactive approaches mentioned above, there are two-level approaches where transmission investment decisions are taken first, and then generation investment and operation decisions are taken simultaneously. On the one hand, the work in [29] models the perfectly competitive market clearing in the lower level, which includes GEP investment variables as well. Additionally, they consider a network fee so the TRANSCO can recover investments in the case of a flow-based fee regulation typically used in the US. The authors linearize the arising MPEC using [22] and solve it as a mixed-integer linear program (MILP)—a method frequently used in bilevel energy applications.

Pisciella et al. [51] also present a proactive TEP approach where the TSO takes investment decisions subject to a lower level market equilibrium of GENCOs and a market operator that exchange price and quantity bids and a market-clearing.

Later, Weibelzahl and Märtz [71] extend the work of Jenabi and Fatemi Ghomi [29] and choose a pessimistic TRANSCO. The authors prove some subsequent uniqueness properties. Since their model uses chronological time steps, they additionally consider battery expansion in their framework. In [69], the authors present various types of TEP-GEP models, one of which represents a proactive approach with a pessimistic TEP in the upper level, and strategic GENCOs in the lower level. On the other hand, Maurovich et al. [39] consider a stochastic bilevel model with a merchant investor of transmission in the upper level, and GEP including wind expansion and a Cournot market in the lower level.

Finally, Gonzalez-Romero et al. [23] apply the same structure, but they consider storage expansion and Cournot competition in the lower level. Both works [23, 39] find counterintuitive results when considering Cournot competition in the lower level compared to a perfect competition case.

5.1.1.2 Reactive Transmission Expansion Planning

One of the first reactive TEP approaches was proposed in [61], showing some theoretical results and some practical results for fixed transmission plans. Unfortunately, the subsequent research on this regard is limited. In general, under this approach, several GENCOs are considered as the leaders and a single TRANSCO as a follower. However, it could also be the case that only one GENCO is the leader, and the remaining GENCOs and TRANSCO(s) are followers, yielding a *one leader multiple followers* structure which is easier to solve numerically speaking.

In [69], Tohidi and Hesamzadeh propose a comparison between the proactive and the reactive approach. In contrast to [53], authors in [69] do not consider anticipation of market outcomes by GENCOs and propose the elimination of the multiple Nash equilibria by considering a pessimistic or optimistic TRANSCO.

Dvorkin et al. [17] present a reactive TEP with three levels: upper level representing a merchant storage owner; middle level that carries out centralized TEP; and, a lower level simulating the market-clearing, to our best knowledge mathematically speaking this is not a three-level structure as the middle (TEP), and the lower (market) level are solved simultaneously. Mathematically speaking, this belongs to the bilevel structures when merchant investors are in the upper level, and then TEP and market decisions are taken simultaneously in the lower level. The arising problem is solved numerically efficiently using a column generation algorithm, which is applied to a real-size case study. The authors conclude that the co-planning of storage and transmission leads to more significant cost savings than independent planning.

5.1.2 Generation Expansion Planning

Generation expansion planning (GEP), in general, refers to the expansion of thermal, renewable, or storage generation capacity in a power system. As we have discussed in the previous Sect. 5.1.1, many bilevel TEP problems include GEP in some way or another. In this section, however, we focus on expansion planning problems that do not include network expansion. Moreover, we also omit the discussion of the literature regarding storage investment, as this is discussed in detail in Sect. 5.1.3.

The emphasis on [11, 73] is on a single strategic generation company (GENCO) that maximizes total profits deciding generation capacity investment in conventional thermal technologies, subject to a strategic market equilibrium using conjectural variations. The use of conjectural variations allows the authors to emulate different degrees of strategic market behavior, ranging from perfect competition to the Cournot oligopoly. The market equilibrium is obtained as the system of KKT conditions of all market players, i.e., also the GENCOs, that simultaneously decide production decisions while maximizing market profits. The hierarchy of decision making in these problems involves separating the type of decision (investment versus operation) and not the type of players. The upper-level player is also a lower level player.

Wogrin et al. [74–76] extend their work of [11, 73] going from MPECs to EPECs by including multiple strategic GENCOs that take investment decisions. The arising EPECs are solved using digitalization, or for some small problem instances, they are solved exactly in [76]. These models capture the fact that GENCOs that maximize their profit have a clear incentive to under-invest, thereby manipulating prices and revenues in the market, even if that market is perfectly competitive. Such behavior cannot be captured with a single-level model. Moreover, other interesting counterintuitive results are obtained: there exist cases where a perfectly competitive market even yields lower social welfare than a Cournot-type market. This is because, under perfect competition, prices are low, which creates an unattractive investment environment for GENCOs as it yields minor profits.

In the work of Baringo and Conejo [4], the authors consider a strategic investor in wind power that maximizes its total profits in the upper level, deciding capacity investment and offers for the day-ahead markets, subject to many different scenarios of the day-ahead market clearing in the lower level.

Pineda and Morales [47] explore several different multi-level (bilevel and three-level) optimization problems emphasizing stochastic power generators, for example, wind power producers. They analyze how imbalances impact the capacity expansion of renewable producers due to power forecast errors considering day-ahead and balancing markets. As an example, a strategic investor maximizes total profits deciding generation capacity subject to the impact on lower-level dispatch decisions.

In Pineda et al. [49], the authors study generation expansion in renewables; however, the main focus of this work lies in policy analysis.

5.1.3 Energy Storage Systems

When it comes to Energy Storage Systems (ESS), there exist many topics on bilevel applications. Some articles [17, 71] explore the interactions between ESS and the transmission network. Others [19, 24, 44, 45] analyze the impact of hierarchical decision making in problems of optimal bidding and scheduling involving ESS.

Nasrolahpour et al. [44] decide the optimal price and quantity bids for a strategic profit-maximizing storage owner, subject to different scenarios of the day ahead and balancing markets. Fang et al. [19] propose a bilevel model whose primary objective is to maximize a load-serving entity's profit by optimally scheduling the ESS charging/discharging profile via an economic dispatch. Pandžić and Kuzle [45] study the impact of a profit-maximizing storage owner in the day-ahead market.

Another important family of energy-related problems that are modeled using bilevel programming, and that we discuss in more detail in this section, consist of determining optimal investment decisions for generation or storage units by anticipating the impact of such decisions on market outcomes. The discussion on how to determine the sequence in which the involved players make their decisions makes no sense in the problems described here, as it is evident that investments are necessarily made before the market is cleared and electricity prices are revealed.

Most planning models to determine optimal investment decisions in liberalized electricity markets are formulated as bilevel optimization problems. The upper level usually maximizes the foreseen profits of potential generation of the storage units to be built, while the lower level represents the variability of the market conditions throughout the lifetime of such devices. We have, therefore, two types of players that make decisions and interact with each other as follows. The GENCO makes investment decisions first and then represents the leader in this Stackelberg game. Afterward, the market operator (MO) clears the market to maximize the social welfare for a given electricity demand level and subject to the available generation capacity in the system. The market-clearing outcome would depend on the investment decisions made by the GENCO. For instance, if GENCO's investment decisions result in a capacity inadequate power system, part of the demand cannot be satisfied. Similarly, the revenue of the investments by the GENCO is also profoundly affected by the market-clearing electricity prices.

More recent references [16, 25, 43, 46] propose similar bilevel optimization problems to decide the location, size, and technology of energy storage devices that are to be operated within a decentralized market environment. Next, we explain in more detail the similarities and differences among these research works focusing on the storage applications.

The authors of [16] propose a bilevel optimization problem that determines the optimal location and size of storage devices to maximize social welfare while ensuring that storage devices collect sufficient profits to recover their investment costs. Nasrolahpour et al. [43] extend this model to incorporate uncertainty related to competitive offering strategies and future demand levels, then yielding a stochastic bilevel optimization problem that is solved using Benders' decomposition techniques. In the same line, Hassan and Dvorkin [25] present a bilevel program to decide the optimal site and size of storage units in the distribution system. Because of the AC power flow equation, the solution procedure requires the use of second-order cone programming. Finally, Pandžić and Bobanac [46] propose a model to optimize investments in energy storage units that compete in the joint energy and reserve market.

5.1.4 Strategic Operation, Bidding or Scheduling

The problems presented in this section correspond to the optimal operation of strategic players. In particular, Ruiz and Conejo [58] formulate an MPEC problem of a strategic producer that maximizes profits deciding the optimal price and quantity bids subject to the market-clearing maximizing social welfare with a DC optimal power flow (OPF). Since the lower level is a linear problem, it is replaced by its KKT conditions and linearized by Fortuny-Amat and McCarl [22]. The bilinear term, representing revenues, in the upper-level objective function, is elegantly replaced by a linear equivalent employing the strong duality theorem. Ruiz, Conejo, and Smeers [59] extend this work to an EPEC framework where there are multiple strategic generators.

A classic paper by Hu and Ralph [27] analyzes bilevel games, pricing and dispatch taking to account an optimal power flow. Theoretical results, as well as diagonalization procedures are presented. Escobar and Jofre [18] and Aussel et al. [2] assess bilevel electricity market structures including information of transmission losses. Allevi et al. [1] model a pay-as-clear electricity market that yields an equilibrium problem with complementarity constraints where GENCOs are the leaders and the ISO is the follower.

Hartwig and Kockar [24] consider a strategic bidder maximizing profits deciding price and quantity bids. However, the bidder is operating a storage facility instead of a conventional thermal plant. The lower level represents the market-clearing and DC power-flow constraints. The final MPEC is also solved as a MIP.

Karkados et al. [34] addresses the optimal bidding strategy problem of a commercial virtual power plant, which comprises of distributed energy resources, battery storage systems, and electricity consumers, and participates in the day-ahead electricity market. A virtual power plant maximizes profits in the upper level and carries out the market-clearing in the lower level.

Finally, Moiseeva et al. [40] formulate a bilevel equilibrium game (EPEC) where strategic producers decide ramping levels in the upper level maximizing their profits, and then choose production levels in the lower level. For simple instances, a closed-form solution to this problem can be found. For large-scale problems, diagonalization is employed to find equilibrium points.

5.1.5 Natural Gas Market

One of the first applications of bilevel programming to the natural gas market can be found in De Wolf and Smeers' work [12]. The bilevel problem is a stochastic Stackelberg game, where a leader decides its output first, and then the followers compete a la Cournot.

There exist several articles related to the cash-out problem in natural gas markets. The cash-out problem reflects the shipper's difficulties in delivering the agreed-upon amount of gas at several points. If an imbalance (between what is actually delivered and what should have been delivered) occurs, the shipper is penalized by the pipeline. Hence, a shipper has to solve the problem of operation while minimizing the incurred penalties. In the work of Dempe et al. [15], a shipper maximizes its revenues in the upper level, subject to the pipeline's decisions at the lower level. The authors apply a penalty function method [72] to solve the arising MPEC because the problem has discrete variables in the lower level, that the authors move into the upper level. In [33], Kalashnikov et al. extend their previous work to consider a stochastic cash-out problem in natural gas. Moreover, [32] proposes a linearization approach to this problem, which is compared to the penalty method.

Siddiqui and Gabriel [64] propose a new SOS1-based approach to solving MPECs and apply it to the US natural gas market. It comprises a shale gas leader company that decides its output strategically as a Stackelberg leader maximizing its profits. The lower level represents a Cournot equilibrium among the follower firms.

Li et al. [38] propose an interesting security-constrained bilevel economic dispatch model for integrated natural gas and electricity systems considering wind power and power-to-gas process. The upper-level objective function minimizes total production cost in the economic dispatch of the electricity system, while the lower level optimally allocates natural gas. Its KKT conditions replace the lower level and transformed into an MIP.

In [13] del Valle et al. propose a model whose objective is to represent a realistic decision-making process for analyzing the optimal infrastructure investments in natural gas pipelines and regasification terminals within the EU framework under a market perspective. In the upper level, the network planner chooses investments in new pipelines and regasification capacity considering multiple regulatory objective functions. The lower level is defined as a generalized Nash-Cournot equilibrium modeling successive natural gas trade, taking into account both the up- and the downstream, and operation of the network infrastructure. To our knowledge, this is the only bilevel model in natural gas that is also multi-objective. The final MPEC is also solved using mixed-integer programming.

5.1.6 *Other Bilevel Applications in Energy*

In energy and electricity markets, there exist many other applications of bilevel programming; however, the majority of these articles do not make up an entire area of research in bilevel optimization. Therefore, we have not dedicated separate sections to each of these applications. Instead, this section comprises an overview of what else is out there.

Momber et al. [41] study the impact of different price signals that an aggregator provides to customers with plug-in electric vehicles. The arising MPEC is solved as an MIP. Another example of the bilevel programming of a retailer can be found in Zugno et al. [81], where retailers send price signals to customers in the lower level under price uncertainty. The final MPEC is also transformed into a MIP. In Pineda et al. [49], the authors analyze several policy options for capacity expansion in renewables. The hierarchy of the bilevel model arises due to the different support schemes. Le Cadre et al. [36] analyze different coordination schemes of TSO-DSO interaction.

5.2 Overview of Solution Methods for Bilevel Problems in Energy

As shown in the previous section, bilevel programming allows us to model a wide variety of situations that involve two decision-makers with their objectives and constraints. However, the computational complexity of bilevel programming problems is exceptionally high. The authors of [8] theoretically show that the bilevel knapsack problem is contained in the complexity class Σ_2^p. Simulation results supporting the high computational burden of this bilevel problem are reported in [9]. Bilevel problems are hard to solve even under linearity assumptions, and hence, most research efforts are currently focused on solving linear bilevel problems (LBP) [14].

Current methods to solve LBP highly depend on whether upper-level variables are discrete or continuous. If upper-level variables are discrete and lower-level variables continuous, then an equivalent single-level mixed-integer reformulation of the original bilevel can be obtained using the strong-duality theorem [80]. Although such equivalent problems belong to the NP-complete class, current commercial optimization software can be used to compute globally optimal solutions. This procedure is the one used in Sect. 5.4.3 provided that investment in generation and storage units are considered as discrete variables.

If both upper- and lower-level variables are continuous, the solution procedures proposed in the technical literature are not as straightforward. From a practical point of view, methods to solve LBP can be divided into two main categories. The first category includes those methods that make use of dedicated solution algorithms to solve bilevel problems [7, 37, 63]. While these methods are usually efficient and ensure global optimality, they involve substantial additional and ad-hoc coding work to be implemented in commercially available off-the-shelf optimization software. The second category includes the methods that can be implemented in or in combination with general-purpose optimization software without any further ado. Most of the existing applications of bilevel programming to energy-related problems rely on solution methods that belong to this second category, and therefore, we explain the most commonly used ones in more detail next.

All the solution methods described next are based on reformulating the original LBP as an equivalent single-level problem that can be solved using off-the-shelf optimization software. Therefore, the lower-level optimization problem is replaced by its necessary and sufficient Karush-Kuhn-Tucker (KKT) optimality conditions. Unfortunately, the KKT equations include nonlinear complementarity conditions, thus yielding a nonconvex single-level problem. Additionally, such a problem violates the Mangasarian-Fromovitz constraint qualification at every feasible point, and therefore, the computation of globally optimal solutions becomes extraordinarily difficult. Existing methods to solve these nonconvex nonregular optimization problems differ regarding their strategy to deal with the nonlinear complementary conditions.

In energy-related problems, the most commonly used method first proposed in [22] consists of reformulating the complementary constraints by an equivalent set of linear inequalities to obtain a single-level mixed-integer problem. This strategy has, however, two main drawbacks. Firstly, it requires a large number of extra binary variables that increase the computational burden of the problem. Secondly, the equivalence between the single-level reformulation and the original BLP is only guaranteed if valid bounds on lower-level dual variables can be found. It is worth mentioning that most research works in the technical literature do not pay much attention to this aspect, which may lead to highly suboptimal solutions, as proven in [48]. If investment decisions are assumed to be continuous variables, this is the procedure used at the end of Sect. 5.4.5 to obtain the single-level reformulation of the problem.

A related method, also based on the combinatorial nature of complementarity conditions, has been proposed in [64]. Such a method uses particular order sets variables, and its computational advantages are problem-dependent. Alternatively, a regularization approach to solving the nonconvex single-level reformulation of LBP was first introduced in [62] and further investigated in [55]. This method solves a set of regular nonconvex optimization problems in which the complementarity conditions are iteratively imposed easily.

Methods based on the combinatorial aspect of the complementarity conditions as those proposed in [22, 64] are computationally intensive but provide optimal solutions for valid bounds on the lower-level dual variables. On the other hand, the nonlinear-based methods investigated in [55, 62] are fast but only guarantee locally optimal solutions. Finally, the authors of [50] propose a method that combines the advantages of the two approaches mentioned above. Numerical simulations show that the method significantly improves the computational tractability of LBP.

All methods previously described aim at solving bilevel programming problems with two decision-makers. Since some of the works mentioned in Sect. 5.1 include hierarchical programming problems that include more levels or decision-makers, we concisely summarize next to the main strategies to solve such problems in the energy field. For instance, the authors of [52] propose a methodology to solve a three-level optimization problem to determine the optimal proactive transmission expansion of a power system. They do so by assuming that generation investment decisions are discrete and finite, which allows them to enumerate all combinations of investment strategies. Despite its performance, this solution methodology becomes computationally intractable for real-size problems. The related work [31] also proposes a methodology to solve a three-level transmission expansion problem. In this paper, the authors propose a hybrid approach to solve the three-level problem, which takes advantage of two different ways of solving the generation expansion equilibrium formulated in the second-level: the diagonalization method and formulating the equilibrium as a complementarity problem. While the proposed algorithm proves computationally efficient, it does not guarantee to find the global optimum. Finally, reference [67] describes an alternative method to solve a three-level optimization problem related to transmission expansion. In this case, the single-level reformulation is obtained by applying the KKT optimality conditions twice. The second set of KKT conditions are not sufficient for optimality, and therefore, the authors have to carry out an ex-post validation technique to check if the obtained point is an equilibrium, which questions the applicability of this approach.

5.3 Challenges of Bilevel Programming in Energy

The challenges for bilevel programming in energy markets are, in general, problem-dependent. However, there are specific topics that constitute gaps in the literature common to many of the works discussed in this chapter. We analyze each of these topics in the following paragraphs.

The first topic for discussion is the full optimal power flow in AC or AC-OPF. Many of the discussed articles, [3, 23, 39], to name a few, involve the representation of the transmission of the distribution network. The way that power flows in an electricity network is governed by Kirchhoff's Laws, which can be expressed mathematically as a set on nonlinear and nonconvex equations. This problem is often referred to as AC power flow, or if a specific objective is considered then AC-OPF.

In terms of mathematical programming, and especially when considering bilevel programming, taking into account the full AC-OPF constitutes a severe issue. For example, the AC-OPF cannot be considered in the lower level of a bilevel problem as it cannot be replaced by its KKT conditions, for example (since the problem is nonconvex). As a solution, in the literature, the AC-OPF is replaced by its linear and thereby convex approximation, the DC-OPF. The assumption upon which this simplification is based can fail in low-voltage networks (like distribution systems), or when the network is not well meshed with a high transfer of power between areas (for example, considering the transmission network of Spain and France which is not well-interconnected). When working with the DC-OPF formulation, important variables such as reactive power are not captured. In a future power system with increasing penetration of renewables and decreasing the amount of conventional synchronous machines that are important for system stability, not taking into account reactive power in planning models, for example, might lead to sub-optimal solutions. As one possible solution, the DC-OPF could be replaced by a convex quadratic problem called second-order cone program [36], which replicates the AC-OPF exactly under specific hypotheses; however, these hypotheses need to be evaluated very carefully for each application.

Another shortcoming of existing bilevel models, and in particular, GEP-TEP applications, is the fact that many of them disregard storage technologies. If the focus of the bilevel model lies on TEP, then usually the GEP corresponding to storage is omitted, for example, in [73]. Taking into account that storage technologies will very likely play an essential role in power systems of the future, they should be included in these models in order to correctly capture power system operation. Moreover, different types of storage technologies should be considered to cover the different range of services they can supply: For example, long-term storage technologies such as pumped hydro facilities that can carry out energy arbitrage over long time periods; and, short-term technologies such as batteries that have several charge and discharge cycles within 1 day. One of the practical reasons for disregarding storage is based on the fact that many planning models either use load periods [3, 73], representative days [17, 46] or just few sequential hours [31, 39, 54] in order to reduce the temporal dimension which brings us to another challenge which is appropriate representation of time in medium- and long-term models. All previously mentioned methods have difficulties in representing storage constraints. Traditional load period models cannot formulate short-term storage constraints unless considering improvements such as system states [77]. Models that use representative days can formulate short-term storage well; however, they fail to adequately represent long-term storage whose cycles go beyond the representative day. Tejada-Arango et al. [68] fix this shortcoming by introducing enhanced representative periods that are linked among each other.

An important issue that is also related to the previous one is the proper characterization of power system operations via unit commitment (UC)-type constraints, e.g., start-up, shut-down, ramping constraints. Such constraints either involve binary variables (e.g., start-up or shut-down) or require sequential time steps (e.g., ramping). The fact that UC constraints involve binary variables makes them very undesirable for lower-level problems as there are no meaningful KKT conditions. However, it is highly questionable to omit important constraints in bilevel problems just because it makes solving them even more difficult. Apart from involving binary variables, many UC constraints require sequential time steps, which makes them challenging for load-period models, for example. Some attempts have been made in the literature to tackle the issue of binary variables in equilibrium problems or bilevel problems [28, 79]; however, this is still an open field of research with much room for improvement.

Most current existing methods to efficiently solve bilevel problems assume that the upper-level objective function may include both upper-level and lower-level primal variables. However, some energy applications such as GEP are formulated as bilevel problems in which the upper-level objective function also includes lower-level dual variables. This happens because, following economic theory, the electricity price must be computed as the dual variable of the power balance equation at each network node. How to extend current solution methods to take this particularity into account is currently an open research question.

Finally, there is a general need to develop more powerful computational methods to solve bilevel problems more efficiently. Currently, and due to the lack of adequate computational methods, the vast majority of bilevel applications in energy do not classify as large-scale problems. This becomes even more important when considering stochasticity. Many existing works in the literature, e.g., [46, 52, 67, 71], disregard the stochastic nature of many investments and operating problems. Therefore, introducing stochasticity in many bilevel applications, and being able to solve them efficiently still constitutes a significant challenge in bilevel optimization.

5.4 Strategic Investment in Generation and Storage Units

This section contains a numerical example of a bilevel investment model that will demonstrate the importance of taking into account hierarchy in decision making. A classical application of bilevel optimization in energy consists of determining strategic investment decisions for generation and storage units participating in a deregulated electricity market. In such a case, the upper-level maximizes the foreseen profits obtained by the built units. Each profit is computed anticipating the market outcomes through a lower-level problem that maximizes social welfare while complying with the generation, storage, and network technical constraints.

In Sect. 5.1, we have reviewed a wide variety of research works that propose to determine strategic investment models through bilevel optimization problems. Here, in Sect. 5.4, we present a generic formulation that covers as many of these existing varieties of models as possible. Notwithstanding this, we also aim to keep the proposed model simple enough to be easily understood by the initiated reader. For these reasons, we made the following simplifying assumptions:

- *Single-bus approach.* It is assumed that investment decisions are to be made for a power system in which the network hardly never gets congested. Therefore, all network constraints are disregarded, and a single-bus representation of the system is considered.
- *Planning horizon.* The planning horizon considered spans one target year. The power system status throughout that year varies significantly depending on the electricity demand and the renewable generation, for example. We assume that the planning horizon is divided into a set of time periods t, whose duration is usually 1 h.
- *Generating units.* Each power generating unit g is characterized by a linear production cost C_g^G, a capacity P_g^G and a capacity factor ρ_{gt}. Electricity production of thermal-based units involves a cost ($C_g^G > 0$) and are fully dispatchable ($\rho_{gt} = 1$). Conversely, the production of renewable-based units is free ($C_g^G = 0$) but its capacity depends on varying weather conditions ($0 \leq \rho_{gt} \leq 1$). The subset of existing and candidate generating units are denoted by $\mathcal{G}^E$ and $\mathcal{G}^B$, respectively.
- *Storage units.* Each storage unit s is characterized by a capacity (P_s^S) and an energy capacity η_s. Similarly to generating units, the subset of existing and candidate storage units is denoted by $\mathcal{S}^E$ and $\mathcal{S}^B$, respectively.
- *Investment decisions.* In principle, investment decisions are modeled through binary variables u_g and v_s for generating and storage units, respectively. Then, available investment projects cannot be accepted partially. Annualized investment costs for generating and storage projects are denoted by I_g^G and I_s^S, respectively.
- *Electricity demand.* For each time period t the level of inflexible electricity demand d_t is assumed known. Demand can also be shed at a very high cost C^S.

In this section, we first introduce a centralized planning approach in Sect. 5.4.1 that will serve as a benchmark for the strategic bilevel investment model of Sect. 5.4.2. Since bilevel models are, in general, nonconvex models, we also show one way to linearize such a model in Sect. 5.4.3 in order to guarantee global optimality of solutions. An illustrative case study is presented in Sect. 5.4.4. An alternative formulation of the strategic MPEC is briefly mentioned in Sect. 5.4.5.

Nomenclature

Indices and Sets

g	Generating unit index
s	Storage unit index
t	Time period index
$\mathcal{G}^E$	Subset of existing generating units
$\mathcal{S}^E$	Subset of existing storage units

$\mathcal{G}^B$ Subset of generating units to be built
$\mathcal{S}^B$ Subset of storage units to be built

Parameters

η_s Energy capacity of storage unit s (h)
ρ_{gt} Capacity factor of generating unit g and time t (p.u.)
C_g^G Linear cost parameter of generating unit g (€/MWh)
D_t Demand level at time period t (MW)
I_g^G Annualized investment cost of generating unit g (€)
I_s^S Annualized investment cost of storage unit s (€)
P_g^G Capacity of generating unit g (MW)
P_s^S Capacity of storage unit s (MW)
C^S Load shedding cost (€/MWh)

Variables

p_{gt}^G Output of generating unit g in time t (MW)
p_{st}^S Output of storage unit s and time t. Discharge/charge if positive/negative (MW)
d_t Satisfied demand in time t (MW)
e_{st} Energy level of storage unit s and time t (MWh)
u_g Binary variable equal to 1 if generating unit g already exists or is built in the current planning period, and 0 otherwise
v_s Binary variable equal to 1 if storage unit s already exists or is built in the current planning period, and 0 otherwise
λ_t Electricity price at time t (€/MWh)

5.4.1 Centralized Approach

Before formulating the investment problem using bilevel optimization, let us first discuss how the problem looks if decisions are made to centrally minimize the total operating and investment costs for the power system. For these simplifying assumptions, model (5.4.1) formulates the centralized version of this investment problem as a single-level mixed-integer programming problem. The complete notation of this problem is provided in the Appendix of this chapter.

$$\min_{u_g, v_s, p^G_{gt}, p^S_{st}, e_{st}, d_t} \quad \sum_{t, g \in \mathcal{G}} C^G_g p^G_{gt} + \sum_{g \in \mathcal{G}^B} I^G_g u_g + \sum_{s \in \mathcal{S}^B} I^S_s v_s + \sum_t C^S (D_t - d_t) \tag{5.4.1a}$$

$$\text{s.t.} \quad u_g \in \{0, 1\}, \quad \forall g \tag{5.4.1b}$$

$$v_s \in \{0, 1\}, \quad \forall s \tag{5.4.1c}$$

$$u_g = 1, \quad \forall g \in \mathcal{G}^E \tag{5.4.1d}$$

$$v_s = 1, \quad \forall s \in \mathcal{S}^E \tag{5.4.1e}$$

$$\sum_g p^G_{gt} + \sum_s p^S_{st} = d_t, \quad \forall t \tag{5.4.1f}$$

$$0 \le d_t \le D_t, \quad \forall t \tag{5.4.1g}$$

$$0 \le p^G_{gt} \le u_g \rho_{gt} P^G_g, \quad \forall g, t \tag{5.4.1h}$$

$$-v_s P^S_s \le p^S_{st} \le v_s P^S_s, \quad \forall s, t \tag{5.4.1i}$$

$$e_{st} = e_{st-1} - p^S_{st}, \quad \forall s, t \tag{5.4.1j}$$

$$0 \le e_{st} \le \eta_s v_s P^S_s, \quad \forall s, t \tag{5.4.1k}$$

Decision variables include investment decisions in new generation and storage units (u_g, v_s), output of generating and storage units (p^G_{gt}, p^S_{st}), energy level of storage units (e_{st}) and satisfied demand (d_t). Objective function (5.4.1a) minimizes total operating and investment costs. Total costs include production cost (first term), investment costs in generating and storage units (second and third term) and load shedding cost (fourth term). Variables u_g and v_s are declared as binary in (5.4.1b) and (5.4.1c), respectively. For the sake of simplicity, constraints (5.4.1d) and (5.4.1e) prevent the decommission of any existing generating and storage units, respectively. The power balance at each time period t is ensured by (5.4.1f). Limits on satisfied demand, generating levels and storage output are imposed in (5.4.1g), (5.4.1h) and (5.4.1i), respectively. Constraint (5.4.1j) keeps track of the energy level of each storage unit, which is in turn restricted through (5.4.1k). The single-level mixed-integer programming problem (5.4.1) can be efficiently solved to global optimality using commercially available optimization software such as CPLEX.

5.4.2 *Strategic Bilevel Approach*

Since the electricity sector was liberalized, investment decisions in generating and storage units are not centrally made to minimize costs, and therefore, single-level optimization problem (5.4.1) is no longer valid. Instead, merchant investors (MI) can strategically invest in these assets to maximize the obtained profits.

This new paradigm for investment decision making can be appropriately modeled through bilevel optimization. For the list of simplifying assumptions presented at the beginning of this section, the strategic investment problem of one MI can be generically formulated using the bilevel optimization problem (5.4.2).

$$\max_{u_g, v_s} \quad \sum_{t, g \in \mathcal{G}^B} \left(\lambda_t - C_g^G\right) p_{gt}^G + \sum_{t, s \in \mathcal{S}^B} \lambda_t p_{st}^S - \sum_{g \in \mathcal{G}^B} I_g^G u_g - \sum_{s \in \mathcal{S}^B} I_s^S v_s \tag{5.4.2a}$$

$$\text{s.t.} \quad u_g \in \{0, 1\}, \quad \forall g \tag{5.4.2b}$$

$$v_s \in \{0, 1\}, \quad \forall s \tag{5.4.2c}$$

$$u_g = 1, \quad \forall g \in \mathcal{G}^E \tag{5.4.2d}$$

$$v_s = 1, \quad \forall s \in \mathcal{S}^E \tag{5.4.2e}$$

$$\min_{p_{gt}^G, p_{st}^S, e_{st}, d_t} \quad \sum_{gt} C_g^G p_{gt}^G + \sum_t C^S (D_t - d_t) \tag{5.4.2f}$$

$$\text{s.t.} \quad \sum_g p_{gt}^G + \sum_s p_{st}^S = d_t : \lambda_t, \quad \forall t \tag{5.4.2g}$$

$$0 \le d_t \le D_t : \underline{\alpha_t}, \overline{\alpha_t}, \quad \forall t \tag{5.4.2h}$$

$$0 \le p_{gt}^G \le u_g \rho_{gt} P_g^G : \underline{\beta}_{gt}, \overline{\beta}_{gt}, \quad \forall g, t \tag{5.4.2i}$$

$$-v_s P_s^S \le p_{st}^S \le v_s P_s^S : \underline{\gamma}_{st}, \overline{\gamma}_{st}, \quad \forall s, t \tag{5.4.2j}$$

$$e_{st} = e_{st-1} - p_{st}^S : \kappa_{st}, \quad \forall s, t \tag{5.4.2k}$$

$$0 \le e_{st} \le \eta_s v_s P_s^S : \underline{\mu}_{st}, \overline{\mu}_{st}, \quad \forall s, t \tag{5.4.2l}$$

Upper-level decision variables only include now the investment decisions in new generation and storage units, u_g and v_s, respectively. Besides, objective function (5.4.2a) maximizes now the obtained revenue from such units throughout the target year. If the electricity price for each time period t is denoted by λ_t, the terms of (5.4.2a) represent the revenue of new generating and storage units (first and second term) and the investment cost of new generating and storage units (third and fourth term). Constraints on investment decisions u_g and v_s are equal to those made in (5.4.1). The operation of the power system for each time period t is characterized by the lower-level problem (5.4.2f)–(5.4.2l), which represents a market-clearing algorithm. The lower-level objective function minimizes the cost of production costs plus load shedding costs subject to the same technical constraints of generating and storage constraints employed in (5.4.1). Notice that dual variables for each lower-level constraint are added to the formulation after a colon. As customary, the dual variable of the power balance constraint represents the marginal electricity price [66].

Although formulation (5.4.2) may not seem very different from (5.4.1) since they share all constraints, the procedure to obtain the optimal global solution for each case are indeed quite different. While (5.4.1) can be solved using available mixed-integer optimization software without any further ado, model (5.4.2) needs to be first reformulated as an equivalent single-level optimization problem as explained next.

For fixed values of upper-level variables u_g, v_s, the lower-level optimization problem is convex and satisfies the Slater condition and therefore, the KKT conditions are necessary and sufficient for optimality [6]. This means that the lower-level optimization problem (5.4.2f)–(5.4.2l) can be replaced by its KKT optimality conditions (5.4.3).

$$\sum_g p^G_{gt} + \sum_s p^S_{st} = d_t, \quad \forall t \tag{5.4.3a}$$

$$0 \le d_t \le D_t, \quad \forall t \tag{5.4.3b}$$

$$0 \le p^G_{gt} \le u_g \rho_{gt} P^G_g, \quad \forall g, t \tag{5.4.3c}$$

$$-v_s P^S_s \le p^S_{st} \le v_s P^S_s, \quad \forall s, t \tag{5.4.3d}$$

$$e_{st} = e_{st-1} - p^S_{st}, \quad \forall s, t \tag{5.4.3e}$$

$$0 \le e_{st} \le \eta_s v_s P^S_s, \quad \forall s, t \tag{5.4.3f}$$

$$-\overline{C}^S_g + \lambda_t - \underline{\alpha}_t + \overline{\alpha}_t = 0, \quad \forall t \tag{5.4.3g}$$

$$C^G_g - \lambda_t - \underline{\beta}_{gt} + \overline{\beta}_{gt} = 0, \quad \forall g, t \tag{5.4.3h}$$

$$-\lambda_t - \underline{\gamma}_{st} + \overline{\gamma}_{st} + \kappa_{st} = 0, \quad \forall s, t \tag{5.4.3i}$$

$$\kappa_{st} - \kappa_{st+1} - \underline{\mu}_{st} + \overline{\mu}_{st} = 0, \quad \forall s, t < T \tag{5.4.3j}$$

$$\kappa_{sT} - \underline{\mu}_{sT} + \overline{\mu}_{sT} = 0, \quad \forall s \tag{5.4.3k}$$

$$\underline{\alpha}_t, \overline{\alpha}_t, \underline{\beta}_{gt}, \overline{\beta}_{gt}, \underline{\gamma}_{st}, \overline{\gamma}_{st}, \underline{\mu}_{st}, \overline{\mu}_{st} \ge 0, \quad \forall g, s, t \tag{5.4.3l}$$

$$d_t \underline{\alpha}_t = 0, \quad \forall t \tag{5.4.3m}$$

$$(d_t - D_t)\overline{\alpha}_t = 0, \quad \forall t \tag{5.4.3n}$$

$$p^G_{gt} \underline{\beta}_{gt} = 0, \quad \forall g, t \tag{5.4.3o}$$

$$\left(p^G_{gt} - u_g \rho_{gt} P^G_g\right) \overline{\beta}_{gt} = 0, \quad \forall g, t \tag{5.4.3p}$$

$$\left(p^S_{st} + v_s P^S_s\right) \underline{\gamma}_{st} = 0, \quad \forall s, t \tag{5.4.3q}$$

$$\left(p^S_{st} - v_s P^S_s\right) \overline{\gamma}_{st} = 0, \quad \forall s, t \tag{5.4.3r}$$

$$e_{st}\underline{\mu}_{st} = 0, \quad \forall s, t \tag{5.4.3s}$$

$$\left(e_{st} - \eta_s v_s P_s^S\right)\overline{\mu}_{st} = 0, \quad \forall s, t \tag{5.4.3t}$$

Equations (5.4.3a)–(5.4.3f) ensure primal feasibility. The stationary conditions for each primal variable is ensured through (5.4.3g)–(5.4.3k). Dual feasibility is imposed by constraints (5.4.3l). Finally, constraints (5.4.3m)–(5.4.3t) represent the complementarity conditions. It can be observed that the complementarity conditions include the product of two continuous variables. Therefore, if the lower-level problem (5.4.2f)–(5.4.2l) is replaced by (5.4.3), the rendered optimization problem would be highly nonconvex and global optimality would not be guaranteed. This is a common problem in bilevel programming.

In order to overcome this difficulty, complementarity conditions can be equivalently replaced by the strong duality equation (5.4.4a) [80]. This implies that the optimality conditions of the lower-level problem (5.4.3a)–(5.4.3t) can be replaced by (5.4.4).

$$\sum_{gt} C_g^G p_{gt}^G + \sum_t C^S (D_t - d_t) = -\sum_t \left(\overline{\alpha}_t - C^S\right) D_t - \sum_{gt} \overline{\beta}_{gt} u_g \rho_{gt} P_g^G -$$
$$\sum_{st} \underline{\gamma}_{st} v_s P_s^S - \sum_{st} \overline{\gamma}_{st} v_s P_s^S - \sum_{st} \overline{\mu}_{st} v_s \eta_s P_s^S \tag{5.4.4a}$$

$$(5.4.3a) - (5.4.3l) \tag{5.4.4b}$$

5.4.3 *Linearization of MPEC*

Notice that strong duality equality constraint (5.4.4a) includes the product of continuous and binary variables which can be linearized as follows assuming valid upper and lower bounds of the continuous variables [21]:

$$\sum_{gt} C_g^G p_{gt}^G + \sum_t C^S (D_t - d_t) = -\sum_t \left(\overline{\alpha}_t - C^S\right) D_t - \sum_{gt} \left(\overline{\beta}_{gt} - \overline{\beta}_{gt}^{\text{aux}}\right) \rho_{gt} P_g^G -$$
$$\sum_{st} \left(\underline{\gamma}_{st} - \underline{\gamma}_{st}^{\text{aux}}\right) P_s^S - \sum_{st} \left(\overline{\gamma}_{st} - \overline{\gamma}_{st}^{\text{aux}}\right) P_s^S - \sum_{st} \left(\overline{\mu}_{st} - \overline{\mu}_{st}^{\text{aux}}\right) \eta_s P_s^S$$
$$\tag{5.4.5a}$$

$$u_g \overline{\beta}_{gt}^{\min} \leq \overline{\beta}_{gt} - \overline{\beta}_{gt}^{\text{aux}} \leq u_g \overline{\beta}_{gt}^{\max}, \quad \forall g, t \tag{5.4.5b}$$

$$(1 - u_g)\overline{\beta}_{gt}^{\min} \leq \overline{\beta}_{gt}^{\text{aux}} \leq (1 - u_g)\overline{\beta}_{gt}^{\max}, \quad \forall g, t \tag{5.4.5c}$$

$$v_s \underline{\gamma}_{st}^{\min} \leq \underline{\gamma}_{st} - \underline{\gamma}_{st}^{\text{aux}} \leq v_s \underline{\gamma}_{st}^{\max}, \quad \forall s, t \tag{5.4.5d}$$

$$(1 - v_s)\underline{\gamma}_{st}^{\min} \leq \underline{\gamma}_{st}^{\text{aux}} \leq (1 - v_s)\underline{\gamma}_{st}^{\max}, \quad \forall s, t \tag{5.4.5e}$$

$$v_s \overline{\gamma}_{st}^{\min} \le \overline{\gamma}_{st} - \overline{\gamma}_{st}^{\text{aux}} \le v_s \overline{\gamma}_{st}^{\max}, \quad \forall s, t \tag{5.4.5f}$$

$$(1 - v_s)\overline{\gamma}_{st}^{\min} \le \overline{\gamma}_{st}^{\text{aux}} \le (1 - v_s)\overline{\gamma}_{st}^{\max}, \quad \forall s, t \tag{5.4.5g}$$

$$v_s \overline{\mu}_{st}^{\min} \le \overline{\mu}_{st} - \overline{\mu}_{st}^{\text{aux}} \le v_s \overline{\mu}_{st}^{\max}, \quad \forall s, t \tag{5.4.5h}$$

$$(1 - v_s)\overline{\mu}_{st}^{\min} \le \overline{\mu}_{st}^{\text{aux}} \le (1 - v_s)\overline{\mu}_{st}^{\max}, \quad \forall s, t \tag{5.4.5i}$$

$$(5.4.3a) - (5.4.3l) \tag{5.4.5j}$$

If we now replace the lower-level problem (5.4.3a)–(5.4.3t) by its optimality conditions (5.4.5), then the MPEC given by (5.4.2) can we written as the equivalent single-level optimization problem (5.4.6). The mathematical format of (5.4.6) would be a mixed integer nonlinear program (MINLP).

$$\max_{u_g, v_s} \quad \sum_{t, g \in \mathcal{G}^B} \left(\lambda_t - C_g^G\right) p_{gt}^G + \sum_{t, s \in \mathcal{S}^B} \lambda_t p_{st}^S - \sum_{g \in \mathcal{G}^B} I_g^G u_g - \sum_{s \in \mathcal{S}^B} I_s^S v_s \tag{5.4.6a}$$

$$\text{s.t.} \quad u_g \in \{0, 1\}, \quad \forall g \tag{5.4.6b}$$

$$v_s \in \{0, 1\}, \quad \forall s \tag{5.4.6c}$$

$$u_g = 1, \quad \forall g \in \mathcal{G}^E \tag{5.4.6d}$$

$$v_s = 1, \quad \forall s \in \mathcal{S}^E \tag{5.4.6e}$$

$$(5.4.3a) - (5.4.3l) \tag{5.4.6f}$$

$$(5.4.5a) - (5.4.5i) \tag{5.4.6g}$$

Although all constraints of (5.4.6) are now linear, the objective function includes nonlinear terms $\lambda_t p_{gt}^G$ and $\lambda_t p_{st}^S$. In order to solve this problem computationally efficiently, it is desirable to convert the current MINLP (5.4.6) into a MILP. To that purpose, we must linearize the nonlinear terms in the objective function.

Let us first linearize the profit obtained by each generating unit $\sum_t \left(\lambda_t - C_g^G\right) p_{gt}^G$ using some of the optimality conditions of the lower-level problem as follows [58]:

$$\sum_t \left(\lambda_t - C_g^G\right) p_{gt}^G \overset{(5.4.3i)}{=} \sum_t \left(-\underline{\beta}_{gt} + \overline{\beta}_{gt}\right) p_{gt}^G = \sum_t -\underline{\beta}_{gt} p_{gt}^G + \overline{\beta}_{gt} p_{gt}^G \overset{((5.4.3o),\,(5.4.3p))}{=}$$

$$\sum_t 0 + u_g \overline{\beta}_{gt} \rho_{gt} P_g^G = \sum_t \left(\overline{\beta}_{gt} - \overline{\beta}_{gt}^{\text{aux}}\right) \rho_{gt} P_g^G \tag{5.4.7}$$

Similarly, we use next other set of optimality conditions of the lower-level problem to linearize the revenue obtained by each storage unit s as follows:

$$\sum_t \lambda_t p_{st}^S \overset{(5.4.3i)}{=} \sum_t \left(-\underline{\gamma}_{st} + \overline{\gamma}_{st} + \kappa_{st}\right) p_{st}^S \overset{((5.4.3q),(5.4.3r))}{=} \sum_t \underline{\gamma}_{st} v_s P_s^S + \sum_t \overline{\gamma}_{st} v_s P_s^S +$$

$$\sum_t \kappa_{st} p_{st}^S = \sum_t \left(\underline{\gamma}_{st} - \underline{\gamma}_{st}^{\text{aux}}\right) P_s^S + \sum_t \left(\overline{\gamma}_{st} - \overline{\gamma}_{st}^{\text{aux}}\right) P_s^S + \sum_t \kappa_{st} p_{st}^S \tag{5.4.8}$$

According to (5.4.8), the profit obtained by a storage unit s can be computed as the sum of three terms. While the two first terms do not include any product of variables, the third one does. We continue using optimality conditions to linearize the last term of (5.4.8) as follows:

$$\sum_t \kappa_t p_{st}^S = \kappa_{s1} p_{s1}^S + \sum_{t=2}^{T-1} \kappa_{st} p_{st}^S + \kappa_{sT} p_{sT}^S \overset{(5.4.3e)}{=} -\kappa_{s1} e_{s1} + \sum_{t=2}^{T-1} \kappa_{st}(e_{st-1} - e_{st}) +$$

$$\kappa_{sT}(e_{sT-1} - e_{sT}) = \sum_{t=1}^{T-1} e_{st}(\kappa_{st+1} - \kappa_{st}) - \kappa_{sT} e_{sT} \overset{((5.4.3j),(5.4.3k))}{=} \sum_{t=1}^{T-1} e_{st}\left(-\underline{\mu}_{st} + \overline{\mu}_{st}\right) +$$

$$e_{sT}\left(-\underline{\mu}_{sT} + \overline{\mu}_{sT}\right) = \sum_t e_{st}\left(-\underline{\mu}_{st} + \overline{\mu}_{st}\right) \overset{((5.4.3s),(5.4.3t))}{=} 0 + \sum_t \overline{\mu}_{st} v_s \eta_s P_s^S =$$

$$\sum_t \left(\overline{\mu}_{st} - \overline{\mu}_{st}^{\text{aux}}\right) \eta_s P_s^S \tag{5.4.9}$$

Using (5.4.7), (5.4.8) and (5.4.9) in objective function (5.4.6a), we obtain the following equivalent single-level MIP reformulation of the original bilevel problem (5.4.2):

$$\max_{u_g, v_s} \sum_{t,g \in \mathcal{G}^B} \left(\overline{\beta}_{gt} - \overline{\beta}_{gt}^{\text{aux}}\right) \rho_{gt} P_g^G + \sum_{t,s \in \mathcal{S}^B} \left(\underline{\gamma}_{st} - \underline{\gamma}_{st}^{\text{aux}}\right) P_s^S - \sum_{s \in \mathcal{S}^B} I_s^S v_s +$$

$$\sum_{t,s \in \mathcal{S}^B} \left(\overline{\gamma}_{st} - \overline{\gamma}_{st}^{\text{aux}}\right) P_s^S + \sum_{t,s \in \mathcal{S}^B} \left(\overline{\mu}_{st} - \overline{\mu}_{st}^{\text{aux}}\right) \eta_s P_s^S - \sum_{g \in \mathcal{G}^R} I_g^G u_g \tag{5.4.10a}$$

$$\text{s.t. } u_g \in \{0, 1\}, \quad \forall g \tag{5.4.10b}$$

$$v_s \in \{0, 1\}, \quad \forall s \tag{5.4.10c}$$

$$u_g = 1, \quad \forall g \in \mathcal{G}^E \tag{5.4.10d}$$

$$v_s = 1, \quad \forall s \in \mathcal{S}^E \tag{5.4.10e}$$

$$(5.4.3a) – (5.4.3l) \tag{5.4.10f}$$

$$(5.4.5a) – (5.4.5i) \tag{5.4.10g}$$

Table 5.1 Demand and solar capacity factor profiles for the representative day

	t01	*t02*	*t03*	*t04*	*t05*	*t06*	*t07*	*t08*	*t09*	*t10*	*t11*	*t12*
Demand (p.u.)	0.65	0.60	0.50	0.28	0.31	0.46	0.65	0.74	0.79	0.86	0.88	0.82
Solar (p.u.)	0.00	0.00	0.00	0.00	0.03	0.35	0.51	0.59	0.58	0.51	0.23	0.54

	t13	*t14*	*t15*	*t16*	*t17*	*t18*	*t19*	*t20*	*t21*	*t22*	*t23*	*t24*
Demand (p.u.)	0.69	0.59	0.56	0.66	0.79	0.94	1.00	0.98	0.88	0.75	0.69	0.65
Solar (p.u.)	0.28	0.34	0.45	0.69	0.70	0.61	0.32	0.02	0.00	0.00	0.00	0.00

Given valid bounds on some dual variables of the lower-level problem, optimization problem (5.4.10) is an equivalent single-level mixed-integer reformulation of the original bilevel problem (5.4.2) and can thus be solved using commercial software as CPLEX.[1] Importantly, this reformulation does not require additional binary variables, and then its computational complexity is comparable to that of its centralized version formulated in (5.4.1). However, it is important to bring to light that finding such valid bounds on dual variables is proven to be NP-hard even for linear bilevel problems [35]. Although some heuristic methods have been proposed to determine bounds of dual variables in a reasonable time, caution must always be exercised when claiming that the optimal global solution of (5.4.10) is also an optimal global solution of (5.4.2). In other words, if the selection of these bounds artificially limits the dual variables, the equivalence between the original bilevel problem and the single-level reformulation is no longer valid.

5.4.4 *Illustrative Case Study*

Next, we present an illustrative example to show the different outcomes of the centralized investment model (5.4.1) and the strategic investment model (5.4.2). For the sake of simplicity, the planning period spans a single target year, and the variability of the operating conditions is captured by a single representative day. Table 5.1 provides the electricity demand and the solar capacity factor for such a representative day. Finally, peak demand and load shedding costs are equal to 1000 MW and 300 €/MWh.

For the sake of illustration, we assume a greenfield approach, i.e., no initial generating capacity is considered. We consider three different technologies:

- *Thermal power generation* consisting of combined cycle gas turbine (CCGT) units with a capacity of 100 MW, a linear production cost of 60 €/MWh, and an annualized investment cost of 42,000 €/MW. Failures of thermal units are not considered and therefore $\rho_{gt} = 1$.

[1]The GAMS code of the linearized MPEC can be downloaded at https://github.com/datejada/SIGASUS.

- *Solar power generating* consisting of solar farms with a capacity of 100 MW, a variable production cost equal to 0 €/MWh, and an annualized investment cost of 85,000 €/MW. The capacity factor of these units is provided in Table 5.1.
- *Energy storage* consisting of Lithium-Ion batteries with a power capacity of 100 MW, the discharge time of 4 h, and an annualized investment cost of 4000 €/MW, which is associated with a low-cost projection of this technology in the following years.

Table 5.2 contains the capacities of each technology determined by the centralized and strategic investing models (5.4.1) and (5.4.2) together with the investment, operating and total costs, the average price and the profit obtained by the power producer. The numerical results collated in Table 5.2 are obtained by solving the corresponding models with a mixed-integer optimization solver such as CPLEX. These results allow us to draw the following conclusions:

- The MI withholds thermal and solar capacity to create scarcity in the system, which causes the total investment costs to be higher in the centralized approach.
- The lack of capacity investment of the MI leads to some demand shedding, which, in turn, increases the operating costs if compared with the centralized approach.
- If the investor behaves strategically, the total cost obtained is significantly higher due to the exercise of market power.
- In the strategic approach, the electricity price is always equal to C^S due to the load shedding actions caused by the limited investments in generation.
- The power producer profit is much higher for the strategic approach because of the price increase caused by withholding generating capacity.
- A centralized planner would have never captured the fact that a MI strategically withholds capacity to drive up market prices (and even cause load shedding) in order to increase profits.
- Bilevel models provide invaluable insight when exploring the strategic behavior of agents in electricity markets.

Table 5.2 Result comparison for strategic and centralized approach

	Strategic	Centralized
Thermal capacity (MW)	600	700
Solar capacity (MW)	300	1000
Storage capacity (MW)	400	300
Load shedding (%)	1.8	0
Investment costs (M€)	52	116
Operating costs (M€)	348	218
Total costs (M€)	400	334
Average price (€/MWh)	300	60
Power producer profit (M€)	1431	32

5.4.5 Alternative Formulation

For the sake of simplicity, we considered discrete investment decisions in this section. This makes sense for large thermal-based generating units that take advantage of the economy of scale. For example, most nuclear power plants have a capacity of 1000 MW approximately. On the contrary, investment in renewable-based power plants such as wind or solar farms allows for a broader range of economically profitable capacities. Hence, one may argue that investment decisions for some projects should be considered as continuous variables. Although this may look like a small variation, it completely changes both the procedure to obtain the equivalent single-level optimization problem and the computational burden of the solution strategy.

Let us revisit the strong duality condition (5.4.4a). If u_g and v_s are continuous variables, the products $\overline{\beta}_{gt}u_g$, $\underline{\gamma}_{st}v_s$, $\overline{\gamma}_{st}v_s$ and $\overline{\mu}_{st}v_s$ cannot be linearized as explained in [21]. Therefore, replacing the complementarity conditions by the strong duality condition does not imply any modeling advantage. Alternatively, we can use the method proposed in [22] to reformulate the KKT conditions (5.4.3a)–(5.4.3t) as follows:

$$(5.4.3a) - (5.4.3l) \tag{5.4.11a}$$

$$\underline{\alpha}_t \leq v_t^1 \underline{\alpha}_t^{\max}, \quad \forall t \tag{5.4.11b}$$

$$d_t \leq (1 - v_t^1)M^1, \quad \forall t \tag{5.4.11c}$$

$$\overline{\alpha}_t \leq v_t^2 \overline{\alpha}_t^{\max}, \quad \forall t \tag{5.4.11d}$$

$$D_t - d_t \leq (1 - v_t^2)M^2, \quad \forall t \tag{5.4.11e}$$

$$\underline{\beta}_{gt} \leq v_{gt}^3 \underline{\beta}_{gt}^{\max}, \quad \forall g, t \tag{5.4.11f}$$

$$p_{gt}^G \leq (1 - v_{gt}^3)M^3, \quad \forall g, t \tag{5.4.11g}$$

$$\overline{\beta}_{gt} \leq v_{gt}^4 \overline{\beta}_{gt}^{\max}, \quad \forall g, t \tag{5.4.11h}$$

$$u_g \rho_{gt} P_g^G - p_{gt}^G \leq (1 - v_{gt}^4)M^4, \quad \forall g, t \tag{5.4.11i}$$

$$\underline{\gamma}_{st} \leq v_{st}^5 \underline{\gamma}_{st}^{\max}, \quad \forall s, t \tag{5.4.11j}$$

$$p_{st}^S + v_s P_s^S \leq (1 - v_{st}^5)M^5, \quad \forall s, t \tag{5.4.11k}$$

$$\overline{\gamma}_{st} \leq v_{st}^6 \overline{\gamma}_{st}^{\max}, \quad \forall s, t \tag{5.4.11l}$$

$$v_s P_s^S - p_{st}^S \leq (1 - v_{st}^6)M^6, \quad \forall s, t \tag{5.4.11m}$$

$$\underline{\mu}_{st} \leq v_{st}^7 \underline{\mu}_{st}^{\max}, \quad \forall s, t \tag{5.4.11n}$$

$$e_{st} \leq (1 - v_{st}^7)M^7, \quad \forall s, t \tag{5.4.11o}$$

$$\overline{\mu}_{st} \leq v^8_{st}\overline{\mu}^{\max}_{st}, \quad \forall s, t \tag{5.4.11p}$$

$$\eta_s v_s P^S_s - e_{st} \leq (1 - v^8_{st})M^8, \quad \forall s, t \tag{5.4.11q}$$

$$v^1_t, v^2_{gt}, v^3_{gt}, v^4_{gt}, v^5_{st}, v^6_{st}, v^7_{st}, v^8_{st} \in \{0, 1\} \tag{5.4.11r}$$

Additionally, the procedure to linearize the profit (5.4.2a) using optimality conditions is not valid either. Therefore, replacing the lower-level (5.4.2f)–(5.4.2l) by (5.4.11) leads to the following single-level optimization problem:

$$\max_{u_g, v_s} \sum_{t, g \in \mathcal{G}^B} \left(\lambda_t - C^G_g\right) p^G_{gt} + \sum_{t, s \in \mathcal{S}^B} \lambda_t p^S_{st} - \sum_{g \in \mathcal{G}^B} I^G_g u_g - \sum_{s \in \mathcal{S}^B} I^S_s v_s \tag{5.4.12a}$$

$$\text{s.t.} \quad u_g \in \{0, 1\}, \quad \forall g \tag{5.4.12b}$$

$$v_s \in \{0, 1\}, \quad \forall s \tag{5.4.12c}$$

$$u_g = 1, \quad \forall g \in \mathcal{G}^E \tag{5.4.12d}$$

$$v_s = 1, \quad \forall s \in \mathcal{S}^E \tag{5.4.12e}$$

$$(5.4.11) \tag{5.4.12f}$$

Optimization problem (5.4.12) is significantly more complicated than (5.4.10) because: it includes a nonconvex objective function and therefore, globally optimal solutions can hardly ever be guaranteed; the nonconvex term in the objective function could be linearized by discretizing price, for example, however, this yields another problem of loss of feasible (and potentially optimal) solutions; it requires a large number of additional binary variables v^1_t, v^2_{gt}, v^3_{gt}, v^4_{gt}, v^5_{st}, v^6_{st}, v^7_{st}, v^8_{st}; it requires additional valid bounds on dual variables $\underline{\alpha}_t$, $\overline{\alpha}_t$, $\underline{\beta}_{gt}$, $\underline{\mu}_{st}$; it requires additional large enough constants M^1, M^2, M^3, M^4, M^5, M^6, M^7, M^8.

Hence, investigating methods to efficiently solve single-level reformulations of bilevel problems such as (5.4.12) is one of the current challenges discussed in Sect. 5.3.

5.5 Conclusions

This chapter provides a general overview of the many applications of bilevel optimization in energy and electricity markets, ranging from transmission and generation expansion planning, to strategic scheduling and bidding, to numerous applications involving energy storage, to natural gas markets, to TSO-DSO interactions or even to electric vehicle aggregators. After a brief discussion on available solution methods for bilevel programs, we point out the specific challenges of bilevel optimization in energy, topics such as: handling the nonconvexity of the full AC

OPF formulation; incorporating storage technologies and inter-temporal constraints, which are closely linked to the representation of time in and the computational complexity of power system models; incorporating UC constraints involving binary variables in lower levels of bilevel problems; developing computational methods to solve bilevel problems more efficiently; and, extending them to include stochasticity.

In order to demonstrate the important insights that can be gained from bilevel problems, we present an application of a strategic agent that can invest in thermal, renewable and storage units while considering the feedback from a perfectly competitive market. We formulate the arising MPEC of such a merchant investor and show how to linearize it efficiently. In an illustrative example we compare the strategic investment decisions to the ones of a centralized planner and identify that a strategic agent has incentives to withhold capacity in order to raise market prices. Such behavior could not have been captured by a centralized planning model. This simple example perfectly illustrates that bilevel models are extremely useful to capture trends and decisions made by strategic agents in energy and electricity markets.

Acknowledgements This work has been supported by Project Grants ENE2016-79517-R and ENE2017-83775-P, awarded by the Spanish Ministerio de Economia y Competitividad.

References

1. E. Allevi, D. Aussel, R. Riccardi, On an equilibrium problem with complementarity constraints formulation of pay-as-clear electricity market with demand elasticity. J. Glob. Optim. **70**(2), 329–346 (2018)
2. D. Aussel, M. Červinka, M. Marechal, Deregulated electricity markets with thermal losses and production bounds: models and optimality conditions. RAIRO-Oper. Res. **50**(1), 19–38 (2016)
3. L. Baringo, A.J. Conejo, Transmission and wind power investment. IEEE Trans. Power Syst. **27**(2), 885–893 (2012)
4. L. Baringo, A.J. Conejo, Strategic wind power investment. IEEE Trans. Power Syst. **29**(3), 1250–1260 (2014)
5. C.A. Berry, B.F. Hobbs, W.A. Meroney, R.P. O'Neill, W.R. Stewart, Understanding how market power can arise in network competition: a game theoretic approach. Util. Policy **8**(3), 139–158 (1999)
6. S.P. Boyd, L. Vandenberghe, *Convex Optimization* (Cambridge University Press, Cambridge, 2004)
7. H.I. Calvete, C. Galé, P.M. Mateo, A new approach for solving linear bilevel problems using genetic algorithms. Eur. J. Oper. Res. **188**(1), 14–28 (2008)
8. A. Caprara, M. Carvalho, A. Lodi, G.J. Woeginger, A complexity and approximability study of the bilevel knapsack problem, in *Integer Programming and Combinatorial Optimization. 16th International Conference, IPCO 2013*. Lecture Notes in Computer Science (including subseries Lecture Notes in Artificial Intelligence and Lecture Notes in Bioinformatics), vol. 7801 (Springer, Berlin, 2013), pp. 98–109
9. A. Caprara, M. Carvalho, A. Lodi, G.J. Woeginger, Bilevel Knapsack with interdiction constraints. INFORMS J. Comput. **28**(2), 319–333 (2016)

10. J.B. Cardell, C.C. Hitt, W.W. Hogan, Market power and strategic interaction in electricity networks. Resour. Energy Econ. **19**, 109–137 (1997)
11. E. Centeno, S. Wogrin, López-Pena, M. Vázquez, Analysis of investments in generation capacity: a bilevel approach. IET Gener. Transm. Distrib. **5**(8), 842 (2011)
12. D. De Wolf, Y. Smeers, A stochastic version of a Stackelberg-Nash-Cournot equilibrium model. Manag. Sci. **43**(2), 190–197 (1997)
13. A. del Valle, S. Wogrin, J. Reneses, Multi-objective bi-level optimization problem for the investment in new gas infrastructures. Technical report, Comillas Pontifical University (2018)
14. S. Dempe, Annotated bibliography on bilevel programming and mathematical programs with equilibrium constraints. Optimization **52**(3), 333–359 (2003)
15. S. Dempe, V. Kalashnikov, R.Z. Rıos-Mercado, Discrete bilevel programming: application to a natural gas cash-out problem. Eur. J. Oper. Res. **166**(2), 469–488 (2005)
16. Y. Dvorkin, R. Fernández-Blanco, D.S. Kirschen, H. Pandžić, J.P. Watson, C.A. Silva-Monroy, Ensuring profitability of energy storage. IEEE Trans. Power Syst. **32**(1), 611–623 (2017)
17. Y. Dvorkin, R. Fernández-Blanco, Y. Wang, B. Xu, D.S. Kirschen, H. Pandžić, J.P. Watson, C.A. Silva-Monroy, Co-planning of investments in transmission and merchant energy storage. IEEE Trans. Power Syst. **33**(1), 245–256 (2018)
18. J.F. Escobar, A. Jofré, Monopolistic competition in electricity networks with resistance losses. Econ. Theory **44**(1), 101–121 (2010)
19. X. Fang, F. Li, Y. Wei, H. Cui, Strategic scheduling of energy storage for load serving entities in locational marginal pricing market. IET Gener. Transm. Distrib. **10**(5), 1258–1267 (2016)
20. Federal Energy Regulatory Commission et al. Order no. 1000 - transmission planning and cost allocation (2012)
21. C.A. Floudas, *Nonlinear and Mixed-Integer Optimization: Fundamentals and Applications* (Oxford University Press, New York, 1995)
22. J. Fortuny-Amat, B. McCarl, A representation and economic interpretation of a two-level programming problem. J. Oper. Res. Soc. **32**(9), 783–792 (1981)
23. I.C. Gonzalez-Romero, S. Wogrin, T. Gomez, Proactive transmission expansion planning with storage considerations. Energy Strategy Rev. **24**, 154–165 (2019)
24. K. Hartwig, I. Kockar, Impact of strategic behavior and ownership of energy storage on provision of flexibility. IEEE Trans. Sustain. Energy **7**(2), 744–754 (2016)
25. A. Hassan, Y. Dvorkin, Energy storage siting and sizing in coordinated distribution and transmission systems. IEEE Trans. Sustain. Energy **9**(4), 1692–1701 (2018)
26. B.F. Hobbs, C.B. Metzler, J.-S. Pang, Calculating equilibria in imperfectly competitive power markets: an MPEC approach. IEEE Trans. Power Syst. **15**, 638–645 (2000)
27. X. Hu, D. Ralph, Using EPECs to model bilevel games in restructured electricity markets with locational prices. Oper. Res. **55**(5), 809–827 (2007)
28. D. Huppmann, S. Siddiqui, An exact solution method for binary equilibrium problems with compensation and the power market uplift problem. Eur. J. Oper. Res. **266**(2), 622–638 (2018)
29. M. Jenabi, S.M.T. Fatemi Ghomi, Y. Smeers, Bi-level game approaches for coordination of generation and transmission expansion planning within a market environment. IEEE Trans. Power Syst. **28**(3), 2639–2650 (2013)
30. S. Jin, S.M. Ryan, Capacity expansion in the integrated supply network for an electricity market. IEEE Trans. Power Syst. **26**(4), 2275–2284 (2011)
31. S. Jin, S.M. Ryan, A tri-level model of centralized transmission and decentralized generation expansion planning for an electricity market - part I. IEEE Trans. Power Syst. **29**(1), 132–141 (2014)
32. V.V. Kalashnikov, G.A. Pérez-Valdés, N.I. Kalashnykova, A linearization approach to solve the natural gas cash-out bilevel problem. Ann. Oper. Res. **181**(1), 423–442 (2010)
33. V.V. Kalashnikov, G.A. Pérez-Valdés, A. Tomasgard, N.I. Kalashnykova, Natural gas cash-out problem: bilevel stochastic optimization approach. Eur. J. Oper. Res. **206**(1), 18–33 (2010)
34. E.G. Kardakos, C.K. Simoglou, A.G. Bakirtzis, Optimal offering strategy of a virtual power plant: a stochastic bi-level approach. IEEE Trans. Smart Grid **7**(2), 794–806 (2016)

35. T. Kleinert, M. Labbé, F. Plein, M. Schmidt, There's no free lunch: on the hardness of choosing a correct Big-M in bilevel optimization (2019), pp. 11–13
36. H. Le Cadre, I. Mezghani, A. Papavasiliou, A game-theoretic analysis of transmission-distribution system operator coordination. Eur. J. Oper. Res. **274**(1), 317–339 (2019)
37. H. Li, L. Fang, An evolutionary algorithm for solving bilevel programming problems using duality conditions. Math. Probl. Eng. **2012**, 14 pp. (2012)
38. G. Li, R. Zhang, T. Jiang, H. Chen, L. Bai, X. Li, Security-constrained bi-level economic dispatch model for integrated natural gas and electricity systems considering wind power and power-to-gas process. Appl. Energy **194**, 696–704 (2017)
39. L. Maurovich-Horvat, T.K. Boomsma, A.S. Siddiqui, Transmission and wind investment in a deregulated electricity industry. IEEE Trans. Power Syst. **30**(3), 1633–1643 (2015)
40. E. Moiseeva, S. Wogrin, M.R. Hesamzadeh, Generation flexibility in ramp rates: strategic behavior and lessons for electricity market design. Eur. J. Oper. Res. **261**(2), 755–771 (2017)
41. I. Momber, S. Wogrin, T. Gomez, Retail pricing: a bilevel program for PEV aggregator decisions using indirect load control. IEEE Trans. Power Syst. **31**(1), 464–473 (2016)
42. A. Motamedi, H. Zareipour, M.O. Buygi, W.D. Rosehart, A transmission planning framework considering future generation expansions in electricity markets. IEEE Trans. Power Syst. **25**(4), 1987–1995 (2010)
43. E. Nasrolahpour, S.J. Kazempour, H. Zareipour, W.D. Rosehart, Strategic sizing of energy storage facilities in electricity markets. IEEE Trans. Sustain. Energy **7**(4), 1462–1472 (2016)
44. E. Nasrolahpour, J. Kazempour, H. Zareipour, W.D. Rosehart, A bilevel model for participation of a storage system in energy and reserve markets. IEEE Trans. Sustain. Energy **9**(2), 582–598 (2018)
45. H. Pandžić, I. Kuzle, Energy storage operation in the day-ahead electricity market, in *International Conference on the European Energy Market, EEM* (2015), pp. 1–6
46. H. Pandžić, Y. Dvorkin, M. Carrión, Investments in merchant energy storage: trading-off between energy and reserve markets. Appl. Energy **230**, 277–286 (2018)
47. S. Pineda, J.M. Morales, Capacity expansion of stochastic power generation under two-stage electricity markets. Comput. Oper. Res. **70**, 101–114 (2016)
48. S. Pineda, J.M. Morales, Solving linear bilevel problems using big-Ms: not all that glitters is gold. IEEE Trans. Power Syst. **34**(3), 2469–2471 (2019)
49. S. Pineda, T.K. Boomsma, S. Wogrin, Renewable generation expansion under different support schemes: a stochastic equilibrium approach. Eur. J. Oper. Res. **266**(3), 1086–1099 (2018)
50. S. Pineda, H. Bylling, J.M. Morales, Efficiently solving linear bilevel programming problems using off-the-shelf optimization software. Optim. Eng. **19**(1), 187–211 (2018)
51. P. Pisciella, M. Bertocchi, M.T. Vespucci, A leader-followers model of power transmission capacity expansion in a market driven environment. Comput. Manag. Sci. **13**, 87–118 (2016)
52. D. Pozo, J. Contreras, E. Sauma, If you build it, he will come: anticipative power transmission planning. Energy Econ. **36**, 135–146 (2013)
53. D. Pozo, E. Sauma, J. Contreras, A three-level static MILP model for generation and transmission expansion planning. IEEE Trans. Power Syst. **28**(1), 202–210 (2013)
54. D. Pozo, E. Sauma, J. Contreras, When doing nothing may be the best investment action: pessimistic anticipative power transmission planning. Appl. Energy **200**, 383–398 (2017)
55. D. Ralph, S.J. Wright, Some properties of regularization and penalization schemes for MPECs. Optim. Methods Softw. **19**(5), 527–556 (2004)
56. A. Ramos, M. Ventosa, M. Rivier, Modeling competition in electric energy markets by equilibrium constraints. Util. Policy **7**, 233–242 (1999)
57. J.H. Roh, M. Shahidehpour, Y. Fu, Market-based coordination of transmission and generation capacity planning. IEEE Trans. Power Syst. **22**(4), 1406–1419 (2007)
58. C. Ruiz, A.J. Conejo, Pool strategy of a producer with endogenous formation of locational marginal prices. IEEE Trans. Power Syst. **24**(4), 1855–1866 (2009)
59. C. Ruiz, A.J. Conejo, Y. Smeers, Equilibria in an oligopolistic electricity pool with stepwise offer curves. IEEE Trans. Power Syst. **27**(2), 752–761 (2012)

60. E.E. Sauma, S.S. Oren, Proactive planning and valuation of transmission investments in restructured electricity markets. J. Regul. Econ. **30**(3), 358–387 (2006)
61. E.E. Sauma, S.S. Oren, Economic criteria for planning transmission investment in restructured electricity markets. IEEE Trans. Power Syst. **22**(4), 1394–1405 (2007)
62. S. Scholtes, Convergence properties of a regularization scheme for mathematical programs with complementarity constraints. SIAM J. Optim. **11**(4), 918–936 (2001)
63. C. Shi, J. Lu, G. Zhang, H. Zhou, An extended branch and bound algorithm for linear bilevel programming. Appl. Math. Comput. **180**(2), 529–537 (2006)
64. S. Siddiqui, S.A. Gabriel, An SOS1-based approach for solving MPECs with a natural gas market application. Netw. Spat. Econ. **13**(2), 205–227 (2012)
65. E. Spyrou, J.L. Ho, B.F. Hobbs, R.M. Johnson, J.D. McCalley, What are the benefits of co-optimizing transmission and generation investment? Eastern interconnection case study. IEEE Trans. Power Syst. **32**(6), 4265–4277 (2017)
66. S. Stoft, *Power System Economics: Designing Markets for Electricity* (IEEE Press-Wiley, New York, 2002)
67. S.S. Taheri, J. Kazempour, S. Seyedshenava, Transmission expansion in an oligopoly considering generation investment equilibrium. Energy Econ. **64**, 55–62 (2017)
68. D.A. Tejada-Arango, M. Domeshek, S. Wogrin, E. Centeno, Enhanced representative days and system states modeling for energy storage investment analysis. IEEE Trans. Power Syst. **33**(6), 6534–6544 (2018)
69. Y. Tohidi, M.R. Hesamzadeh, F. Regairaz, Sequential coordination of transmission expansion planning with strategic generation investments. IEEE Trans. Power Syst. **32**(4), 2521–2534 (2017)
70. J.D. Weber, T.J. Overbye, A two-level optimization problem for analysis of market bidding strategies, in *IEEE Power Engineering Society Summer Meeting*, Edmonton (1999), pp. 682–687
71. M. Weibelzahl, A. Märtz, Optimal storage and transmission investments in a bilevel electricity market model. Ann. Oper. Res. **287**, 911–940 (2020)
72. D.J. White, G. Anandalingam, A penalty function approach for solving bi-level linear programs. J. Glob. Optim. **3**(4), 397–419 (1993)
73. S. Wogrin, E. Centeno, J. Barquín, Generation capacity expansion in liberalized electricity markets: a stochastic MPEC approach. IEEE Trans. Power Syst. **26**(4), 2526–2532 (2011)
74. S. Wogrin, J. Barquin, E. Centeno, Capacity expansion equilibria in liberalized electricity markets: an EPEC approach. IEEE Trans. Power Syst. **28**(2), 1531–1539 (2013)
75. S. Wogrin, E. Centeno, J. Barquin, Generation capacity expansion analysis: open loop approximation of closed loop equilibria. IEEE Trans. Power Syst. **28**(3), 3362–3371 (2013)
76. S. Wogrin, B.F. Hobbs, D. Ralph, E. Centeno, J. Barquín, Open versus closed loop capacity equilibria in electricity markets under perfect and oligopolistic competition. Math. Program. **140**(2), 295–322 (2013)
77. S. Wogrin, D. Galbally, J. Reneses, Optimizing storage operations in medium- and long-term power system models. IEEE Trans. Power Syst. **31**(4), 3129–3138 (2016)
78. J. Yao, I. Adler, S.S. Oren, Modeling and computing two-settlement oligopolistic equilibrium in a congested electricity network. Oper. Res. **56**(1), 34–47 (2008)
79. Y. Ye, D. Papadaskalopoulos, J. Kazempour, G. Strbac, Incorporating non-convex operating characteristics into bi-level optimization electricity market models. IEEE Trans. Power Syst. **35**(1), 163–176 (2020)
80. M.H. Zare, J.S. Borrero, B. Zeng, O.A. Prokopyev, A note on linearized reformulations for a class of bilevel linear integer problems. Ann. Oper. Res. **272**(4), 99–117 (2019)
81. M. Zugno, J.M. Morales, P. Pinson, H. Madsen, A bilevel model for electricity retailers' participation in a demand response market environment. Energy Econ. **36**, 182–197 (2013)

Chapter 6
Bilevel Optimization of Regularization Hyperparameters in Machine Learning

Takayuki Okuno and Akiko Takeda

Abstract Most of the main machine learning (ML) models are equipped with parameters that need to be prefixed. Such parameters are often called hyperparameters. Needless to say, prediction performance of ML models significantly relies on the choice of hyperparameters. Hence, establishing methodology for properly tuning hyperparameters has been recognized as one of the most crucial matters in ML. In this chapter, we introduce the role of bilevel optimization in the context of selecting hyperparameters in regression and classification problems.

Keywords Machine learning · Hyperparameter optimization · Nonsmooth bilevel optimization · Sparse regularizer · ℓ_q-regularizer

6.1 Introduction

Bilevel optimization is to minimize a given real-valued function subject to constraints related to an optimal solution set of another optimization problem. Many problems arising from various fields can be formulated as bilevel optimization problems. In this chapter, we introduce the role of bilevel optimization in the context of Machine Learning (ML) for, in particular, selecting regularization *hyperparameters* of ML problems (or models).

One of the main tasks of ML is, from given data, to design a model, which, thanks to the generalization mechanism, can predict the future. For this purpose, ML

T. Okuno (✉)
RIKEN AIP, Tokyo, Japan
e-mail: takayuki.okuno.ks@riken.jp

A. Takeda
Department of Creative Informatics, Graduate School of Information Science and Technology, The University of Tokyo, Tokyo, Japan
e-mail: takeda@mist.i.u-tokyo.ac.jp

S. Dempe, A. Zemkoho (eds.), *Bilevel Optimization*, Springer Optimization and Its Applications 161, https://doi.org/10.1007/978-3-030-52119-6_6

researchers have developed many efficient learning algorithms that are able to deal with huge amounts of data and, moreover, show robust performance even when the data contains some noise or lacks full information. Such learning algorithms often admit parameters that are tuned by the users manually. These parameters are called *hyperparameters*. They play an important role in allowing for a high prediction performance. If the hyperparameters are tuned properly, the predictive performance of the learning algorithms will be increased.

We can find various kinds of hyperparameters depending on ML problems. For example, the hyperparameters of a deep learning problem can be architectures of the deep neural networks, forms of the activation functions, and step size rules (see e.g., [1, 2], and the references therein). In decision tree approaches, the hyperparameters are the total number of leaves in the tree, the height of the tree, the minimum leaf split size, the number of leaves in each level of the tree. In this chapter, we will pay particular attention to so-called regularization hyperparameters that appear in regression and classification problems.

6.1.1 Regression and Classification Problems and Regularization Hyperparameters

Regression and classification are central tasks in supervised learning aimed at learning a mapping function from the input to the output data based on the data already labeled. We shall start by introducing the problem setting and the notation used throughout this chapter. Let $\mathcal{X} \subset \mathbb{R}^n$ be the input domain and suppose that a set of binary labels $\{+1, -1\}$ and the real space $\mathbb{R}$ are the output domains for classification and regression, respectively. The observed training samples are denoted by $(\boldsymbol{x}_i, y_i) \in \mathcal{X} \times \{+1, -1\}$ or $(\boldsymbol{x}_i, y_i) \in \mathcal{X} \times \mathbb{R}$, for $i = 1, \ldots, m_{tr}$ with $m_{tr} \in \mathbb{N}$.

A decision function $f_{\boldsymbol{w}} : \mathcal{X} \to \mathbb{R}$, which is parameterized by a vector $\boldsymbol{w} = (w_1, w_2, \ldots, w_n)^\top \in \mathbb{R}^n$, is estimated from the training samples. For instance, $f_{\boldsymbol{w}}$ is often set to be $\boldsymbol{x}^\top \boldsymbol{w}$ for the input vector $\boldsymbol{x} \in \mathbb{R}^n$. The classification label of the input $\boldsymbol{x}$ is predicted by $\mathrm{sign}(f_{\boldsymbol{w}}(\boldsymbol{x}))$, where sign is defined by $\mathrm{sign}(\xi) := 1$ if $\xi \geq 0$ and -1 otherwise, while the regression output is predicted by $f_{\boldsymbol{w}}(\boldsymbol{x})$. Then, the goal of the classification/regression problem is to find $\boldsymbol{w}$, i.e., $f_{\boldsymbol{w}}$, such that the total prediction error for unseen test samples is minimized. Vapnik's risk minimization theory provides us with a deep insight into the estimation of $f_{\boldsymbol{w}}$ from statistical learning theory. For example, see [3].

For finding such $\boldsymbol{w}$, we often solve optimization problems of the following form:

$$\min_{\boldsymbol{w} \in C} \left\{ g(\boldsymbol{w}) + \sum_{i=1}^{r} \lambda_i R_i(\boldsymbol{w}) \right\}, \tag{MLopt}$$

where $C \subseteq \mathbb{R}^n$ is a closed set and $g : \mathbb{R}^n \to \mathbb{R}$ and $R_i : \mathbb{R}^n \to \mathbb{R}$ $(i = 1, 2, \ldots, r)$ are functions called *loss* function and *regularization* functions, respectively.

Moreover, $\lambda_i > 0$ $(i = 1, 2, \ldots, r)$ are the regularization hyperparameters of our interest in this chapter. The loss function g is utilized to represent how well the function $f_{\boldsymbol{w}}$ fits the training samples $\{(\boldsymbol{x}_i, y_i)\}_{i=1}^{m_{tr}}$ and it is usually constructed from those training samples. Solely minimizing g leads to $f_{\boldsymbol{w}}$ fitting the training samples quite well, but possibly not other samples, thus suffering from poor generalization capabilities. This phenomenon is often called *overfitting* and such $f_{\boldsymbol{w}}$ is said to be too complex. The regularization functions R_i $(i = 1, 2, \ldots, r)$ are given to avoid this overfitting-effect by penalizing too complex functions $f_{\boldsymbol{w}}$.

The problem (MLopt) has two objectives. We want to fit the decision function $f_{\boldsymbol{w}}$ to the training samples well, but, on the other hand, we want to avoid too complex $f_{\boldsymbol{w}}$ by keeping the regularization small. The hyperparameters $(\lambda_1, \lambda_2, \ldots, \lambda_r)^\top =: \boldsymbol{\lambda}$ define a trade-off between these two objectives. We illustrate how the trade-off occurs in Fig. 6.1, where we plot the training samples (depicted by the filled circles) together with $f_{\boldsymbol{w}}$ (depicted by the curves) obtained by solving (MLopt) with three values of $\boldsymbol{\lambda}$. For the case of $\boldsymbol{\lambda} = \boldsymbol{0}$ (the top-left figure), we can see that $f_{\boldsymbol{w}}$ takes a complex form so that it fits with zero error as many training samples as possible

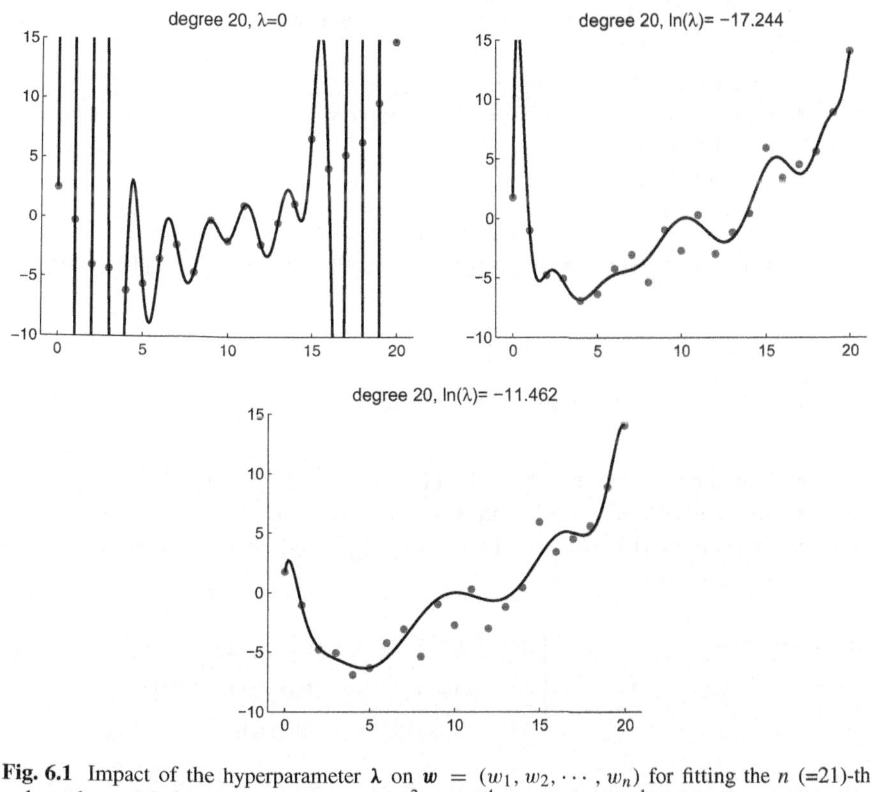

Fig. 6.1 Impact of the hyperparameter $\boldsymbol{\lambda}$ on $\boldsymbol{w} = (w_1, w_2, \cdots, w_n)$ for fitting the n (=21)-th order polynomial $f_{\boldsymbol{w}}(x) = w_1 + w_2 x + w_3 x^2 + w_4 x^4 + \ldots + w_n x^{n-1}$ via ridge regression, that is, by solving (MLopt) with $C = \mathbb{R}^n$, $r = 1$, g being the ℓ_2-loss function, and R_1 being the ℓ_2 regularizer

(overfitting). On the other hand, if $\boldsymbol{\lambda} > \mathbf{0}$ (the bottom and top-right figures), the number of sample points which f_w fits decreases, but the form of f_w becomes simpler. So, to avoid overfitting the given data samples and to enhance the prediction performance of (MLopt), a proper choice of the hyperparameters $\boldsymbol{\lambda}$ is needed.

There are various kinds of loss functions. Among them, typical examples of the function g are as follows:

Examples of the Loss Function g

(A) Loss functions for regression with training samples $(\boldsymbol{x}_i, y_i) \in \mathbb{R}^n \times \mathbb{R}, i = 1, \ldots, m_{tr}$:

- ℓ_2-loss: $g(\boldsymbol{w}) := \sum_{i=1}^{m_{tr}} (y_i - f_{\boldsymbol{w}}(\boldsymbol{x}_i))^2$
- Huber-loss [4]: $g(\boldsymbol{w}) := \sum_{i=1}^{m_{tr}} g_i^{\delta}(\boldsymbol{w})$, where $\delta > 0$ is prefixed and

$$g_i^{\delta}(\boldsymbol{w}) := \begin{cases} (y_i - f_{\boldsymbol{w}}(\boldsymbol{x}_i))^2 & \text{if } f_{\boldsymbol{w}}(\boldsymbol{x}_i) \in [y_i - \delta, y_i + \delta] \\ 2\delta(|y_i - f_{\boldsymbol{w}}(\boldsymbol{x}_i)| - \frac{\delta}{2}) & \text{otherwise} \end{cases}$$

  for each $i = 1, 2, \ldots, m_{tr}$.
- ε-insensitive loss: $g(\boldsymbol{w}) := \sum_{i=1}^{m_{tr}} \max(|y_i - f_{\boldsymbol{w}}(\boldsymbol{x}_i)| - \varepsilon, 0)$, where $\varepsilon > 0$ is prefixed.
- τ-quantile loss:

$$g(\boldsymbol{w}) := \sum_{i=1}^{m_{tr}} ((1 - \tau) \max(f_{\boldsymbol{w}}(\boldsymbol{x}_i) - y_i, 0) + \tau \max(y_i - f_{\boldsymbol{w}}(\boldsymbol{x}_i), 0)),$$

  where $\tau \in (0, 1)$ is prefixed.

(B) Loss functions for classification with training samples $(\boldsymbol{x}_i, y_i) \in \mathbb{R}^n \times \{+1, -1\}$, $i = 1, \ldots, m_{tr}$:

- Logistic loss: $g(\boldsymbol{w}) = \sum_{i=1}^{m_{tr}} \log(1 + \exp(-y_i f_{\boldsymbol{w}}(\boldsymbol{x}_i)))$
- Hinge loss: $g(\boldsymbol{w}) = \sum_{i=1}^{m_{tr}} \max(1 - y_i f_{\boldsymbol{w}}(\boldsymbol{x}_i), 0)$.
- Smoothing Hinge loss: $g(\boldsymbol{w}) = \sum_{i=1}^{m_{tr}} g_i(\boldsymbol{w})$, where for each $i = 1, 2, \ldots, m_{tr}$,

$$g_i(\boldsymbol{w}) := \begin{cases} 0 & \text{if } y_i f_{\boldsymbol{w}}(\boldsymbol{x}_i) \geq 1 \\ \frac{1}{2} - y_i f_{\boldsymbol{w}}(\boldsymbol{x}_i) & \text{if } y_i f_{\boldsymbol{w}}(\boldsymbol{x}_i) < 0 \\ \frac{1}{2}(1 - y_i f_{\boldsymbol{w}}(\boldsymbol{x}_i))^2 & \text{otherwise} \end{cases}$$

Let us list typical examples of regularizers below. See Figs. 6.2 and 6.3 for illustration.

Examples of $\sum_{i=1}^{r} \lambda_i R_i(\boldsymbol{w})$

(A) Regularizers including the ℓ_p function $\|\boldsymbol{w}\|_p^p := \sum_{i=1}^{n} |w_i|^p$ for $p\ (> 0)$ as below:

- ℓ_1 regularizer: $\lambda_1 \|\boldsymbol{w}\|_1$
- ℓ_2 regularizer: $\lambda_1 \|\boldsymbol{w}\|_2^2$
- ℓ_q regularizer: $\lambda_1 \|\boldsymbol{w}\|_q^q$ with $0 < q < 1$.
- Elastic net regularizer [5]: $\lambda_1 \|\boldsymbol{w}\|_1 + \lambda_2 \|\boldsymbol{w}\|_2^2$.

(B) The following regularizers are also popular:

- SCAD (smoothly clipped absolute deviation) regularizer [6]: $\lambda_1 \sum_{j=1}^{n} r_j^{\eta,a}(w_j)$, where $\eta > 0,\ a > 1$, and

$$r_j^{\eta,a}(w_j) := \begin{cases} \eta|w_j| & \text{if } |w_j| \le \eta \\ -\frac{|w_j|^2 - 2a\eta|w_j| + \eta^2}{2(a-1)} & \text{if } \eta \le |w_j| \le a\eta \\ \frac{(a+1)\eta^2}{2} & \text{otherwise.} \end{cases}$$

- MCP (minimax concave penalty) regularizer [7]: $\lambda_1 \sum_{j=1}^{n} p_j^{\eta,a}(w_j)$, where $\eta > 0,\ a > 1$, and

$$p_j^{\eta,a}(w_j) := \begin{cases} \eta|w_j| - \frac{w_j^2}{2a} & \text{if } |w_j| \le a\eta \\ \frac{a\eta^2}{2} & \text{otherwise.} \end{cases}$$

When $f_{\boldsymbol{w}}(\boldsymbol{x}) = \boldsymbol{x}^\top \boldsymbol{w}$, g is the ℓ_2-loss function and R_1 is the ℓ_2 regularizer, the problem (MLopt) is called ridge regression model. Since the ℓ_2 regularizer has the effect of avoiding overfitting (see Fig. 6.1 again), the ridge model is able to achieve a high prediction accuracy as a result. But, on the other hand, this model suffers from the drawback that it is hard to discern which feature has a significant meaning in the generation mechanism of the data. Lasso (least absolute shrinkage and selection operator) model [8], which is obtained by replacing the ℓ_2-regularizer in the ridge model with the ℓ_1 regularizer, was proposed to address the issue. Indeed, Lasso shrinks some of the coefficients , $\boldsymbol{w}$, of the less important features to zero thus removing some features altogether, while the ℓ_2 regularizer does not remove most of the features.

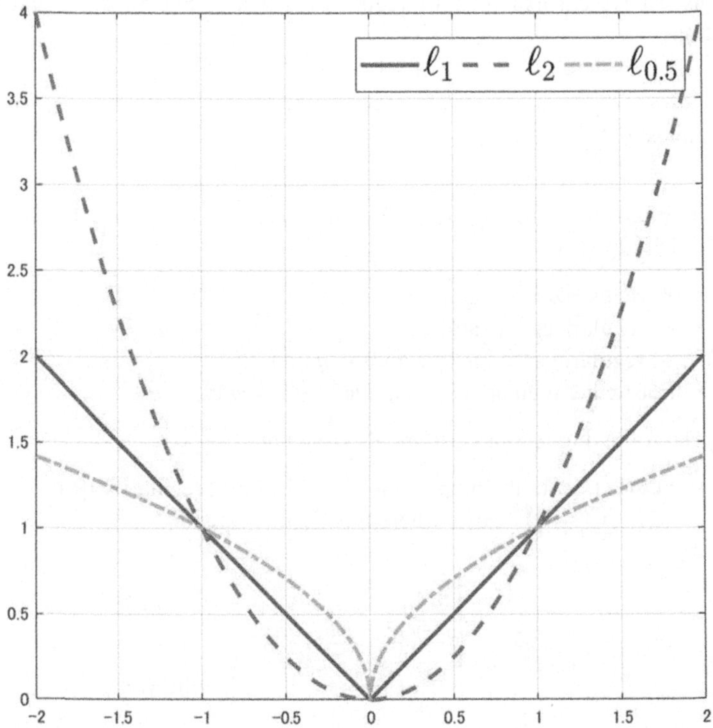

Fig. 6.2 $\ell_1, \ell_2, \ell_{0.5}$ with $n = 1$ and $\lambda_1 = 1$

Besides Lasso, many regularizers have been developed so far. The elastic net regularizer was proposed in [5] to mitigate some limitations of Lasso, e.g., Lasso is often outperformed by ridge regression if there exists some correlation among the features. Moreover, in the last decade, nonconvex regularizers such as ℓ_q, SCAD and MCP have attracted much attention among the researchers in ML, signal processing, and image processing. This is because, in comparison with convex regularizers including the ℓ_1, ℓ_2 and elastic net regularizers, these nonconvex regularizers show superior performance in gaining a sparse solution, namely, $\boldsymbol{w}$ with many zero components. In particular, (MLopt) equipped with such nonconvex regularizers is often called a sparse model. For a comprehensive survey of nonconvex regularizers, see the survey article [9].

6.1.2 Hyperparameter Optimization

In (MLopt), the solution $\boldsymbol{w}$ would differ as hyperparameter values of $\boldsymbol{\lambda}$ vary. Accordingly, in predicting unseen test samples, the performance of $f_{\boldsymbol{w}}$ changes.

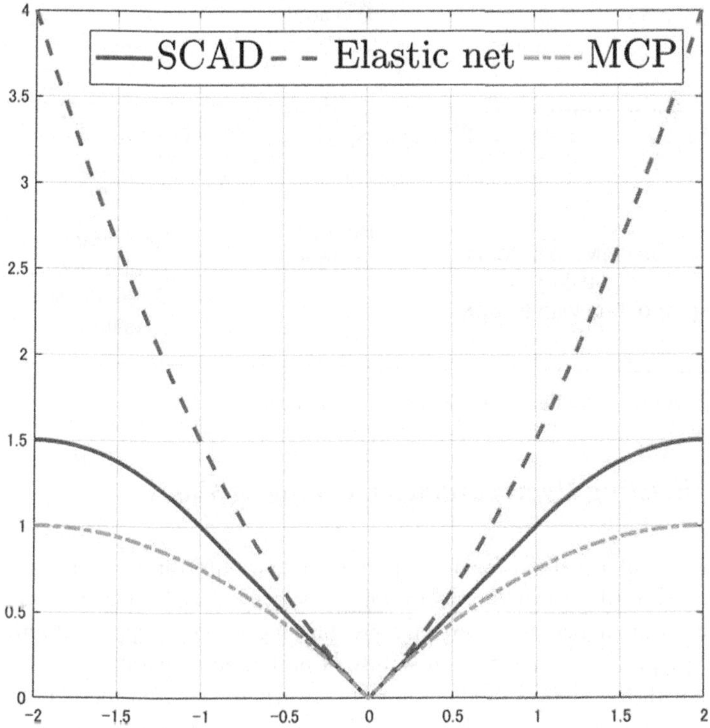

Fig. 6.3 SCAD, Elastic net, MCP with $n = 1$ and $(\lambda_1, \lambda_2) = (1, 0.5)$

Hence, finding good hyperparameters, in the sense that $f_{\boldsymbol{w}}$ has high prediction performance, is a crucial issue in ML. The task of finding as good as possible values of the hyperparameters is often referred to as *hyperparameter optimization* [10].

In supervised learning, for the hyperparameter optimization, there are a few methods as described in the next subsection. A common technique underlying those methods is as follows (Also, see Fig. 6.4.):

1. Given a data set consisting of observed sample data, divide it into three sets: train, validation, and test sets.
2. Consider (MLopt) with a loss function g_{tr}, called *training error* function, defined on the train set and solve it for various values of $\boldsymbol{\lambda}$. For each tried $\boldsymbol{\lambda}$, denote an obtained solution of (MLopt) by $\boldsymbol{w}_{\boldsymbol{\lambda}}$.
3. Consider a loss function g_{val} defined on the validation set and select the best $(\boldsymbol{\lambda}^*, \boldsymbol{w}_{\boldsymbol{\lambda}^*})$ such that $g_{val}(\boldsymbol{w}_{\boldsymbol{\lambda}})$, called *validation error* function, is minimized.

For measuring the quality of the obtained $(\boldsymbol{\lambda}^*, \boldsymbol{w}_{\boldsymbol{\lambda}^*})$, we use a loss function $g_{te}(\boldsymbol{w})$, called *test error* function, which is defined on the test set.

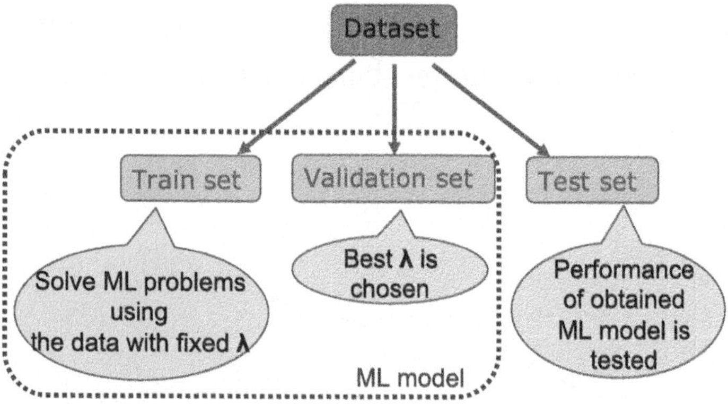

Fig. 6.4 Train set and validation set are inputs of our model

6.1.3 Existing Hyperparameter Tuning Methods

The most popular method for the hyperparameter optimization is grid search. For example, see [10]. The method is to train a learning model using training data for all values of the hyperparameters on a prechosen grid (i.e., a specified finite subset of the hyperparameter space) sequentially or preferably in parallel, and choose the best one with the highest prediction accuracy when tested on the validation data with, e.g., cross validation [10]. Cross-Validation is performed by splitting learning data (train set + validation set) into M subsets with almost the same size and using one subset for validation and the other $M - 1$ subsets as train set. By changing the train and validation set among M subsets, machine learning problems such as (MLopt) can be trained M times and the best parameter values $\boldsymbol{\lambda}$ in terms of the average validation error can be chosen.

There is another technique for hyperparameter optimization, random search [11], that evaluates learning models for randomly sampled hyperparameter values or more sophisticated method called Bayesian optimization [12, 13]. To find $f_{\boldsymbol{w}_\lambda}$ with good prediction performance, it is reasonable to minimize the validation error $\bar{f}(\boldsymbol{\lambda}) := g_{val}(\boldsymbol{w}_\lambda)$ in hyperparameters $\boldsymbol{\lambda}$. Here, $\boldsymbol{w}_\lambda$ is regarded as a function of $\boldsymbol{\lambda}$. However, in general, we do not know an explicit expression of $\bar{f}$ in terms of $\boldsymbol{\lambda}$, although we can obtain its value for given $\boldsymbol{\lambda}$ by computing a corresponding $\boldsymbol{w}_\lambda$. Therefore, it is difficult to minimize $\bar{f}$ in the $\boldsymbol{\lambda}$-space using an algorithm exploiting functional information such as the gradient of $\bar{f}$ w.r.t $\boldsymbol{\lambda}$. For such a black-box (meaning unknown) objective function $\bar{f}$, Bayesian optimization algorithms try to minimize $\bar{f}$ with exploration (i.e., the acquisition of new knowledge) and exploitation (i.e., the use of existing knowledge to drive improvement). They use previous observations $\bar{f}(\boldsymbol{\lambda})$ of the function at some hyperparameter values $\boldsymbol{\lambda}$ to determine the next point $\boldsymbol{\lambda}^+$ to evaluate based on the assumption that the objective function can be described by a Gaussian process as a prior. For more detail on the Bayesian optimization technique used for minimizing $\bar{f}$, see, e.g., [13, 14].

6.1.4 Formulation as a Bilevel Optimization Problem

The procedure 1–3 of hyperparameter optimization explained in Sect. 6.1.2 can be naturally formulated as the following bilevel optimization problem comprising the training error function g_{tr} and validation error function g_{val} that are defined in procedure 2–3:

$$\begin{aligned} \min_{\boldsymbol{w}_{\boldsymbol{\lambda}},\boldsymbol{\lambda}} \quad & g_{val}(\boldsymbol{w}_{\boldsymbol{\lambda}}) \\ \text{s.t.} \quad & \boldsymbol{w}_{\boldsymbol{\lambda}} \in \arg\min_{\boldsymbol{w}\in C} \left\{ g_{tr}(\boldsymbol{w}) + \sum_{i=1}^{r} \lambda_i R_i(\boldsymbol{w}) \right\}, \\ & \boldsymbol{\lambda} \geq \boldsymbol{0}. \end{aligned} \tag{6.1.1}$$

We note that, when considering hyperparameter selection with a measure of the validation error g_{val}, the above bilevel problem is the optimization problem that we aim at solving although implicitly. Crucially, many existing methods for hyperparameter selection can be viewed as heuristic methods tackling this bilevel problem implicitly.

The article [15] pioneered the idea of the bilevel approach for hyperparameter optimization. Since then, many researchers have advanced in this direction for optimizing hyperparameters in various kinds of ML models. For example, refer to articles [16–28].

An advantage of the bilevel optimization approach is that it is able to seek good hyperparameters *continuously* by means of continuous optimization methods such as a sequential quadratic programming method [29]. On the other hand, grid search is simple and fast, but it merely searches a finite discretized set given by the users a priori. As to Bayesian optimization, it looks for the best hyperparameters continuously as well as the bilevel approach. However, the number of hyperparameters Bayesian optimization is able to handle is not so large in practice and the method tends to be inefficient (see, e.g., [1] and Section 2.10 in [30] in addition to the numerical examples shown in Sect. 6.3.5) because it tries to find a global optimal solution.

6.1.5 Organization of this Chapter

The remaining part of this chapter is organized as follows. In Sect. 6.2, we overview relevant works on bilevel approaches to the hyperparameter optimization. In Sect. 6.3, we introduce the authors' recent work about a bilevel approach for ℓ_p-hyperparameter optimization with $0 < p \leq 1$ in detail. Particularly, we give scaled bilevel Karush-Kuhn-Tucker (KKT) conditions that are optimality conditions for the obtained bilevel optimization problem. In addition, we introduce a smoothing-type method generating a sequence converging to a point satisfying the scaled bilevel KKT conditions. In Sect. 6.4, we conclude this chapter with some remarks.

6.1.6 Notations

In this chapter, we often denote the vector $z \in \mathbb{R}^d$ by $z = (z_1, z_2, \ldots, z_d)^\top$. For a set of vectors $\{v_i\}_{i\in I} \subseteq \mathbb{R}^m$ with $I := \{i_1, i_2, \ldots, i_p\}$, we define $(v_i)_{i\in I} := (v_{i_1}, v_{i_2}, \ldots, v_{i_p}) \in \mathbb{R}^{m\times p}$.

For a differentiable function $h : \mathbb{R}^n \to \mathbb{R}$, we denote the gradient function from $\mathbb{R}^n$ to $\mathbb{R}^n$ by ∇h, i.e., $\nabla h(x) := (\frac{\partial h(x)}{\partial x_1}, \ldots, \frac{\partial h(x)}{\partial x_n})^\top \in \mathbb{R}^n$ for $x \in \mathbb{R}^n$, where $\frac{\partial h(x)}{\partial x_i}$ stands for the partial differential of h with respect to x_i for $i = 1, 2, \ldots, n$. To express the gradient of h with respect to a sub-vector $\tilde{x} := (x_i)_{i\in I}^\top$ of x with $I := \{i_1, i_2, \ldots, i_p\} \subseteq \{1, 2, \ldots, n\}$, we write $\nabla_{\tilde{x}} h(x) := \left(\frac{\partial h(x)}{\partial x_{i_1}}, \frac{\partial h(x)}{\partial x_{i_2}}, \ldots, \frac{\partial h(x)}{\partial x_{i_p}}\right)^\top \in \mathbb{R}^{|I|}$. We often write $\nabla g(x)|_{x=\bar{x}}$ ($\nabla_{\tilde{x}} g(x)|_{x=\bar{x}}$) or $\nabla h(\bar{x})$ ($\nabla_{\tilde{x}} h(\bar{x})$) to represent the (partial) gradient value of g at $x = \bar{x}$. Moreover, when h is twice differentiable, we denote the Hessian of h by $\nabla^2 h : \mathbb{R}^n \to \mathbb{R}^{n\times n}$, i.e., $\nabla^2 h(x) := \left(\frac{\partial^2 h(x)}{\partial x_i \partial x_j}\right)_{1\le i,j\le n} \in \mathbb{R}^{n\times n}$.

6.2 Existing Works on Bilevel Approaches to Hyperparameter Optimization

In this section, we briefly summarize the bilevel approaches for hyperparameter optimization that were studied in [15, 21].

6.2.1 MPEC Approach to SVR

Given the m_{tr} training samples $\{(x_i, y_i)\}_{i=1}^{m_{tr}} \subseteq \mathbb{R}^n \times \mathbb{R}$, consider the following support-vector-regression (SVR) problem:

$$\min_{\underline{w}\le w\le \overline{w}} \left\{\sum_{i=1}^{m_{tr}} \max(|y_i - x_i^\top w| - \varepsilon, 0) + \frac{\lambda}{2}\|w\|_2^2\right\},$$

where $\underline{w}, \overline{w} \in \mathbb{R}^n$ and $\varepsilon > 0$ are given. The parameter ε is called the tube parameter. The second-term is the ε-insensitive loss introduced in Sect. 6.1, and $\lambda > 0$ is a regularization hyperparameter.

For hyperparameter selection, we use m_{val} validation samples $\{(\tilde{\boldsymbol{x}}_i, \tilde{y}_i)\}_{i=1}^{m_{tr}} \subseteq \mathbb{R}^n \times \mathbb{R}$ and set a validation error function $g_{val}(\boldsymbol{w}) := \sum_{i=1}^{m_{val}} |\tilde{\boldsymbol{x}}_i^\top \boldsymbol{w} - \tilde{y}_i|$ in (6.1.1). Then, we obtain the following bilevel optimization problem:

$$\begin{aligned}
\min_{\boldsymbol{w}_\lambda, \lambda} \quad & \sum_{i=1}^{m_{val}} |\tilde{\boldsymbol{x}}_i^\top \boldsymbol{w}_\lambda - \tilde{y}_i| \\
\text{s.t.} \quad & \lambda \geq 0 \\
& \boldsymbol{w}_\lambda \in \operatorname*{arg\,min}_{\underline{\boldsymbol{w}} \leq \boldsymbol{w}_\lambda \leq \overline{\boldsymbol{w}}} \left\{ \sum_{i=1}^{m_{tr}} \max(|y_i - \boldsymbol{x}_i^\top \boldsymbol{w}_\lambda| - \varepsilon, 0) + \frac{\lambda}{2} \|\boldsymbol{w}_\lambda\|_2^2 \right\}.
\end{aligned} \tag{6.2.1}$$

Note that the lower-level problem is convex, but includes nonsmooth functions. In particular, when $\lambda > 0$, it is strongly convex due to $\lambda \|\boldsymbol{w}_\lambda\|_2^2/2$ and thus $\boldsymbol{w}_\lambda$ is uniquely determined. By use of slack variables and further replacing the lower-level problem with its Karush-Kuhn-Tucker conditions, the problem can be finally reformulated as a mathematical programming problem with equilibrium constraints (MPEC) [31] such that the included functions are all smooth.

In [15], Bennett et al. further integrated the cross validation technique into the problem (6.2.1) to obtain a more sophisticated formula. Also, the hyperparameters they actually handled are not restricted to the class of regularization hyperparameters, i.e., λ. They also include a wider class of hyperparameters such as upper and lower bounds $\overline{\boldsymbol{w}}, \underline{\boldsymbol{w}}$ of the interval constraint and the tube parameter ε. Such flexibility is a notable advantage of the bilevel formulation. They solved the obtained MPEC via a relaxation technique transforming the problem into a nonlinear optimization problem without complementarity constraints.

6.2.2 Newton-type Methods for ℓ_p-hyperparameter Optimization with $p = 1, 2$

In the context of image processing, in [21], Kunisch and Pock considered the ℓ_p-regularized squared regression problem $\min_{\boldsymbol{w}} \left\{ \frac{1}{2} \|\boldsymbol{w} - \boldsymbol{f}\|_2^2 + \frac{1}{r} \sum_{i=1}^{r} \lambda_i \|K_i \boldsymbol{w}\|_p^p \right\}$, which was referred to as a variational model therein. Here, $K_i \in \mathbb{R}^{m \times n}$ for $i = 1, 2, \ldots, r$ and $\boldsymbol{\lambda} := (\lambda_i)_{i=1}^r \in \mathbb{R}^r$. The term $\|K_i \boldsymbol{w}\|_p^p$ can enforce a certain structural sparsity on the coefficients in the solution $\boldsymbol{w}$. For example, with an appropriate K_i, $\|K_i \boldsymbol{w}\|_p^p$ can express $\sum_{i=2}^n |w_i - w_{i-1}|^p$, which penalizes the absolute differences in adjacent coordinates of $\boldsymbol{w}$. This specific $\|K_i \boldsymbol{w}\|_1$ leads to the so-called fused Lasso [32]. To determine an optimal $\boldsymbol{\lambda}$, they derived the following bilevel problem:

$$\begin{aligned} \min_{w_\lambda, \lambda} \quad & \|w_\lambda - g\|_2^2 \\ \text{s.t.} \quad & w_\lambda \in \arg\min_{w} \left\{ \frac{1}{2}\|w - f\|_2^2 + \frac{1}{r}\sum_{i=1}^{r} \lambda_i \|K_i w\|_p^p \right\} \\ & \lambda \geq 0. \end{aligned} \tag{6.2.2}$$

The authors considered the above problem with $p = 1, 2$ and proposed algorithms for each case. When $p = 2$, the lower level problem is smooth and strongly convex. Since the lower level problem has the unique minimizer

$$w_\lambda = (\frac{2}{r}\sum_{i=1} \lambda_i K_i^\top K_i + I)^{-1} f$$

with $I \in \mathbb{R}^{n\times n}$ being an identity matrix, the problem (6.2.2) is equivalent to

$$\min_{\lambda \geq 0} \left\{ \hat{g}_{val}(\lambda) := \left\| \left(\frac{2}{r}\sum_{i=1} \lambda_i K_i^\top K_i + I \right)^{-1} f - g \right\|_2^2 \right\},$$

which reduces to solving the following nonsmooth equations obtained via the KKT conditions:

$$\nabla \hat{g}_{val}(\lambda) - \eta = 0, \; \eta - \max(\eta - \lambda, 0) = 0.$$

For solving this equation, the authors in [21] proposed a semismooth Newton method and analyzed its superlinear convergence property in detail.

For the case of $p = 1$, the lower-level problem is strongly convex, but nonsmooth. As a remedy, the authors employed smoothing approximation techniques (called regularization therein) to the absolute value function $| \cdot |$. Specifically, they used the smoothing function $n_\varepsilon : \mathbb{R} \to \mathbb{R}$ defined as

$$n_\varepsilon(t) := \begin{cases} -\frac{1}{8\varepsilon^3}t^4 + \frac{3}{\varepsilon}t^2 + \frac{3\varepsilon}{8} & \text{if } |t| < \varepsilon, \\ |t| & \text{else}, \end{cases}$$

where $\varepsilon > 0$ is a parameter. The function n_ε is twice continuously differentiable and satisfies $n_\varepsilon''(t) \in [0, \frac{3}{2\varepsilon}]$ and $n_\varepsilon \geq 1$ for all $t \in \mathbb{R}$. Then, they considered the problem (6.2.2) with the lower level problem replaced by

$$\min_{w} \left\{ \frac{1}{2}\|w - f\|_2^2 + \frac{1}{r}\sum_{i=1}^{r}\sum_{j=1}^{n} \lambda_i n_\varepsilon((K_i w)_j) \right\}.$$

The authors proposed a semismooth Newton method and a reduced Newton method for solving certain nonsmooth equations equivalent to the first optimality conditions for the obtained smoothed problem. Furthermore, they studied sufficient conditions exploiting the properties of n_ε for the superlinear convergence. A detailed analysis as $\varepsilon \to 0$ was deferred to a future work.

The proposed algorithmic framework for $p = 1$ can be extended to the case of $p < 1$ straightforwardly. The authors exhibited the numerical results for $p = 1/2$.

6.3 Bilevel Optimization of ℓ_p-hyperparameter with $0 < p \leq 1$

In this section, we focus on ℓ_p-hyperparameter optimization with $0 < p \leq 1$ and introduce the authors' recent work [26].

6.3.1 (MLopt) and the Bilevel Problem to be Considered

Throughout this section, we consider (MLopt) with ℓ_p-regularizer R_1, with $0 < p \leq 1$, namely,

$$R_1(\boldsymbol{w}) := \|\boldsymbol{w}\|_p^p.$$

Moreover, the other regularizers $R_2, \cdots, R_r$, and the training error function g_{tr} are assumed to be twice continuously differentiable functions. Furthermore, in (6.1.1), we assume that $C = \mathbb{R}^n$ and the validation function g_{val} is smooth.

As mentioned in Sect. 6.1.1, the ℓ_1 regularizer plays a meaningful role in Lasso model. The ℓ_p regularizer with $0 < p < 1$ also performs well particularly in constructing a sparse model. We believe that this regularizer was first considered in [33], as a generalization of the ℓ_2 regularizer. Its actual efficiency is evidenced both theoretically and empirically in fields such as low-rank matrix completion and signal reconstruction. For example, see [34, 35].

The ℓ_2 and logistic-loss functions are typical examples for the above loss functions g_{tr} and g_{val}. For example, in this case, g_{tr} can be defined as follows:

- ℓ_2-loss function for regression: $g_{tr}(\boldsymbol{w}) = \sum_{i=1}^{m_{tr}} (y_i - \boldsymbol{x}_i^\top \boldsymbol{w})^2$ for training samples $(y_i, \boldsymbol{x}_i) \in \mathbb{R} \times \mathbb{R}^n, i = 1, \cdots, m_{tr}$
- Logistic-loss function for classification: $g_{tr}(\boldsymbol{w}) = \sum_{i=1}^{m_{tr}} \log(1 + \exp(-y_i \boldsymbol{x}_i^\top \boldsymbol{w}))$ for training samples $(y_i, \boldsymbol{x}_i) \in \{+1, -1\} \times \mathbb{R}^n, i = 1, \cdots, m_{tr}$.

The following regularizers satisfy all the assumptions on the regularizer term $\sum_{i=1}^{r} \lambda_i R_i(\boldsymbol{w})$:

- ℓ_p-regularizers ($0 < p \leq 1$): $\sum_{i=1}^{r} \lambda_i R_i(\boldsymbol{w}) = \lambda_1 \|\boldsymbol{w}\|_p^p$ with $r = 1$
- Elastic net regularizer: $\sum_{i=1}^{r} \lambda_i R_i(\boldsymbol{w}) = \lambda_1 \|\boldsymbol{w}\|_1 + \lambda_2 \|\boldsymbol{w}\|_2^2$ with $r = 2$.

To clarify the smooth and non-smooth parts in (MLopt) under the current assumptions, we define the function

$$G(\boldsymbol{w}, \bar{\boldsymbol{\lambda}}) := g_{tr}(\boldsymbol{w}) + \sum_{i=2}^{r} \lambda_i R_i(\boldsymbol{w})$$

with $\bar{\boldsymbol{\lambda}} := (\lambda_2, \ldots, \lambda_r)^\top \in \mathbb{R}^{r-1}$. In terms of this function, (MLopt) is expressed as

$$\min_{\boldsymbol{w} \in \mathbb{R}^n} \left\{ G(\boldsymbol{w}, \bar{\boldsymbol{\lambda}}) + \lambda_1 R_1(\boldsymbol{w}) \right\}. \tag{6.3.1}$$

After all, we will consider the following bilevel program:

$$\min \left\{ g_{val}(\boldsymbol{w}_{\boldsymbol{\lambda}}) \text{ s.t. } \boldsymbol{\lambda} \geq \boldsymbol{0},\ \boldsymbol{w}_{\boldsymbol{\lambda}} \in \mathop{\arg\min}_{\boldsymbol{w} \in \mathbb{R}^n} \left\{ G(\boldsymbol{w}, \bar{\boldsymbol{\lambda}}) + \lambda_1 R_1(\boldsymbol{w}) \right\} \right\}. \tag{6.3.2}$$

6.3.2 Replacement of the Lower-Level Problem with Its First-order Optimality Conditions

For tackling the problem (6.3.2), one reasonable approach is to replace the lower level problem (6.3.1) constraint with its first-order optimality condition [36, Theorem 10.1] using a general subdifferential as follows:

> Alternative formulation of (6.3.2) using a subdifferential

$$\min_{\boldsymbol{w}, \boldsymbol{\lambda}} \left\{ g_{val}(\boldsymbol{w}) \text{ s.t. } \boldsymbol{0} \in \partial_{\boldsymbol{w}} (G(\boldsymbol{w}, \bar{\boldsymbol{\lambda}}) + \lambda_1 R_1(\boldsymbol{w})),\ \boldsymbol{\lambda} \geq \boldsymbol{0} \right\}. \tag{6.3.3}$$

In (6.3.3), $\partial_{\boldsymbol{w}}(G(\boldsymbol{w}, \bar{\boldsymbol{\lambda}}) + \lambda_1 R_1(\boldsymbol{w}))$ stands for the subdifferential of $G(\boldsymbol{w}, \bar{\boldsymbol{\lambda}}) + \lambda_1 R_1(\boldsymbol{w})$ with respect to $\boldsymbol{w}$. For the definition of subdifferential, see [36, Chapter 8]. Since $G(\boldsymbol{w}, \bar{\boldsymbol{\lambda}}) + \lambda_1 R_1(\boldsymbol{w})$ is not necessarily convex with respect to $\boldsymbol{w}$, the feasible region of (6.3.3) is possibly wider than that of the original problem (6.3.2). Indeed, not only the global optimal solutions of the lower-level problem (6.3.1) but also its local optimal solutions are feasible to (6.3.3). So, the problem (6.3.3) differs from the original one (6.3.2). But, note that solving (6.3.3) means searching for the best hyperparameters $\boldsymbol{\lambda}$ in the wider space than (6.3.2). This may, as a result, lead to better prediction performance for unseen test samples. In other words, we may obtain $\boldsymbol{\lambda}$ with a better value of the test error function g_{te} (see Sect. 6.1.2 for the definition).

In [26], the authors considered the following transformation of (6.3.2) using the *scaled first-order necessary condition* of the lower-level problem.

Alternative formulation of (6.3.2) using the scaled first-order condition

$$\min_{\boldsymbol{w},\boldsymbol{\lambda}} \left\{ g_{val}(\boldsymbol{w}) \text{ s.t. } \boldsymbol{W}\nabla_{\boldsymbol{w}} G(\boldsymbol{w}, \bar{\boldsymbol{\lambda}}) + p\lambda_1 |\boldsymbol{w}|^p = \boldsymbol{0},\ \boldsymbol{\lambda} \geq \boldsymbol{0} \right\}, \tag{6.3.4}$$

where $\boldsymbol{W} := \mathrm{diag}(\boldsymbol{w})$ and $|\boldsymbol{w}|^p := (|w_1|^p, |w_2|^p, \ldots, |w_n|^p)^\top$.

The scaled first-order optimality condition was originally presented by [34, 37, 38] for certain optimization problems admitting non-Lipschitz functions. The equation $\boldsymbol{W}\nabla_{\boldsymbol{w}} G(\boldsymbol{w}, \bar{\boldsymbol{\lambda}}) + p\lambda_1 |\boldsymbol{w}|^p = \boldsymbol{0}$ is nothing but the scaled first-order optimality condition for the problem (6.3.1) and any local optimum of (6.3.1) satisfies this condition. As well as (6.3.3), the feasible region of (6.3.4) includes not only the global optimal solution of the lower-level problem in the original problem (6.3.2) but also its local solutions. Notice that the problem (6.3.4) is still nonsmooth due to the existence of $|\boldsymbol{w}|^q$.

According to [26], the following proposition highlights the relationship between the feasible regions of the two formulations (6.3.3) and (6.3.4).

Proposition 6.3.1 *Let* $p \leq 1$. *For* $\boldsymbol{w} \in \mathbb{R}^n$ *and* $\boldsymbol{\lambda} \geq \boldsymbol{0}$, *if* $\boldsymbol{0} \in \partial_{\boldsymbol{w}}(G(\boldsymbol{w}, \bar{\boldsymbol{\lambda}}) + \lambda_1 R_1(\boldsymbol{w}))$, *then* $\boldsymbol{W}\nabla_{\boldsymbol{w}} G(\boldsymbol{w}, \bar{\boldsymbol{\lambda}}) + p\lambda_1 |\boldsymbol{w}|^p = \boldsymbol{0}$. *In particular, when* $p < 1$, *the converse is also true.* △

Below, we summarize the inclusion relation among the feasible regions of (6.3.2), (6.3.3), and (6.3.4).

Relationships among the feasible regions of (6.3.2), (6.3.3), and (6.3.4)

$$\text{Feasible region of (6.3.2)} \subseteq \text{Feasible region of (6.3.3)}$$

$$\left\{ \begin{matrix} \subseteq_{p=1} \\ =_{p<1} \end{matrix} \right\} \text{Feasible region of (6.3.4)}$$

Thus, if $p < 1$, (6.3.3) and (6.3.4) are exactly equivalent problems. Unfortunately, it is a hard business to solve both the problems (6.3.3) and (6.3.4) as they are. Actually, solving such problems that include subdifferentials or non-Lipschitz equations in constraints seems to be beyond the capabilities of the existing algorithms.

In [26], the authors presented new optimality conditions, named *scaled bilevel Karush-Kuhn-Tucker (SB-KKT)* conditions. The conditions were proved to be necessary conditions which any local optimum of (6.3.4) should satisfy under certain constraint qualification-like assumptions. Furthermore, they proposed a smoothing-type method for solving (6.3.3) and showed that the sequence it generates converges to a point satisfying the SB-KKT conditions. In the subsequent sections, we will overview those results.

6.3.3 Scaled Bilevel KKT Conditions

In this section, we introduce the *scaled bilevel Karush-Kuhn-Tucker (SB-KKT)* conditions proposed in [26]. The conditions are formally defined as follows:

Scaled Bilevel-KKT (SB-KKT) Conditions

We say that the scaled bilevel Karush-Kuhn-Tucker (SB-KKT) conditions hold at $(\boldsymbol{w}^*, \boldsymbol{\lambda}^*) \in \mathbb{R}^n \times \mathbb{R}^r$ for the problem (6.3.2) when there exists a pair of vectors $(\boldsymbol{\zeta}^*, \boldsymbol{\eta}^*) \in \mathbb{R}^n \times \mathbb{R}^r$ such that

$$W_*^2 \nabla g_{val}(\boldsymbol{w}^*) + H(\boldsymbol{w}^*, \boldsymbol{\lambda}^*)\boldsymbol{\zeta}^* = \mathbf{0}, \tag{6.3.5}$$

$$W_* \nabla_{\boldsymbol{w}} G(\boldsymbol{w}^*, \bar{\boldsymbol{\lambda}}^*) + p\lambda_1^* |\boldsymbol{w}^*|^p = \mathbf{0}, \tag{6.3.6}$$

$$p \textstyle\sum_{i \notin I(\boldsymbol{w}^*)} \operatorname{sgn}(w_i^*)|w_i^*|^{p-1}\zeta_i^* = \eta_1^*, \tag{6.3.7}$$

$$\zeta_i^* = 0 \ (i \in I(\boldsymbol{w}^*)), \tag{6.3.8}$$

$$\nabla R_i(\boldsymbol{w}^*)^\top \boldsymbol{\zeta}^* = \eta_i^* \ \ (i = 2, 3, \ldots, r), \tag{6.3.9}$$

$$\mathbf{0} \leq \boldsymbol{\lambda}^*, \ \mathbf{0} \leq \boldsymbol{\eta}^*, \ (\boldsymbol{\lambda}^*)^\top \boldsymbol{\eta}^* = 0, \tag{6.3.10}$$

where $W_* := \operatorname{diag}(\boldsymbol{w}^*)$, and we write $\operatorname{sgn}(\xi) := 0 \ (\xi = 0), 1 \ (\xi > 0), -1 \ (\xi < 0)$ for $\xi \in \mathbb{R}$ and

$$H(\boldsymbol{w}, \boldsymbol{\lambda}) := W^2 \nabla_{\boldsymbol{w}}^2 G(\boldsymbol{w}, \bar{\boldsymbol{\lambda}}) + \lambda_1 p(p-1)\operatorname{diag}(|\boldsymbol{w}|^p)$$

with $W := \operatorname{diag}(\boldsymbol{w})$ for $\boldsymbol{w} \in \mathbb{R}^n$ and $\boldsymbol{\lambda} \in \mathbb{R}^r$. In addition, we define

$$I(\boldsymbol{w}) := \{i \in \{1, 2, \ldots, n\} \mid w_i = 0\}, \ |\boldsymbol{w}|^q := \left(|w_1|^q, |w_2|^q, \ldots, |w_n|^q\right)^\top$$

for $\boldsymbol{w} \in \mathbb{R}^n$. In particular, we call a point $(\boldsymbol{w}^*, \boldsymbol{\lambda}^*) \in \mathbb{R}^n \times \mathbb{R}^r$ satisfying the above conditions (6.3.5)–(6.3.10) an SB-KKT point for the problem (6.3.2).

The following theorem states that the SB-KKT conditions are nothing but necessary optimality conditions for the problem (6.3.4), i.e., by Proposition 6.3.1, those for the problem (6.3.3) when $p < 1$.

Theorem 6.3.2 (SB-KKT Conditions as Necessary Optimality Conditions) *Let point* $(\boldsymbol{w}^*, \boldsymbol{\lambda}^*) \in \mathbb{R}^n \times \mathbb{R}^r$ *be a local optimum of* (6.3.4). *Then,* $(\boldsymbol{w}^*, \boldsymbol{\lambda}^*)$ *together with some vectors* $\boldsymbol{\zeta}^* \in \mathbb{R}^n$ *and* $\boldsymbol{\eta}^* \in \mathbb{R}^r$ *satisfies the SB-KKT conditions* (6.3.5)–(6.3.10) *under an appropriate constraint qualification concerning the constraints* $\frac{\partial G(\boldsymbol{w},\bar{\boldsymbol{\lambda}})}{\partial w_i} + p\,\mathrm{sgn}(w_i)\lambda_1|w_i|^{p-1} = 0\ (i \notin I(\boldsymbol{w}^*)),\ w_i = 0\ (i \in I(\boldsymbol{w}^*))$, *and* $\boldsymbol{\lambda} \geq \boldsymbol{0}$.
△

Sketch of the Proof Notice that $(\boldsymbol{w}^*, \boldsymbol{\lambda}^*)$ is a local optimum of the following problem:

$$
\begin{aligned}
\min_{\boldsymbol{w},\boldsymbol{\lambda}} \quad & g_{val}(\boldsymbol{w}) \\
\text{s.t.} \quad & \frac{\partial G(\boldsymbol{w}, \bar{\boldsymbol{\lambda}})}{\partial w_i} + p\,\mathrm{sgn}(w_i)\lambda_1|w_i|^{p-1} = 0\ (i \notin I(\boldsymbol{w}^*)) \\
& w_i = 0\ (i \in I(\boldsymbol{w}^*)) \\
& \boldsymbol{\lambda} \geq \boldsymbol{0}.
\end{aligned}
\tag{6.3.11}
$$

This is because $(\boldsymbol{w}^*, \boldsymbol{\lambda}^*)$ is also feasible to (6.3.11) and the feasible region of (6.3.4) is larger than that of (6.3.11). Hence, the KKT conditions hold for (6.3.11) under constraint qualifications. Finally, these KKT conditions can be equivalently transformed into the desired SB-KKT conditions. □

6.3.4 *Algorithm for Bilevel Problem with ℓ_p Regularizers*

In this section, we introduce the smoothing-type algorithm proposed in [26] for solving (6.3.2).

6.3.4.1 Smoothing Method for the Bilevel Problem (6.3.2)

The smoothing method is one of the most powerful methodologies developed for solving nonsmooth equations, nonsmooth optimization problems, and so on. Fundamentally, the smoothing method solves smoothed optimization problems or equations sequentially to produce a sequence converging to a point that satisfies some optimality conditions of the original nonsmooth problem. The smoothed problems solved therein are obtained by replacing the nonsmooth functions with so-called smoothing functions defined as follows. Let $\varphi_0 : \mathbb{R}^n \to \mathbb{R}$ be a nonsmooth function. Then, we say that $\varphi : \mathbb{R}^n \times \mathbb{R}_+ \to \mathbb{R}$ is a smoothing function of φ_0 when (i) $\varphi(\cdot, \cdot)$ is continuous and $\varphi(\cdot, \mu)$ is continuously differentiable for any $\mu > 0$; (ii) $\lim_{\tilde{\boldsymbol{w}} \to \boldsymbol{w}, \mu \to 0+} \varphi(\tilde{\boldsymbol{w}}, \mu) = \varphi_0(\boldsymbol{w})$ for any $\boldsymbol{w} \in \mathbb{R}^n$.

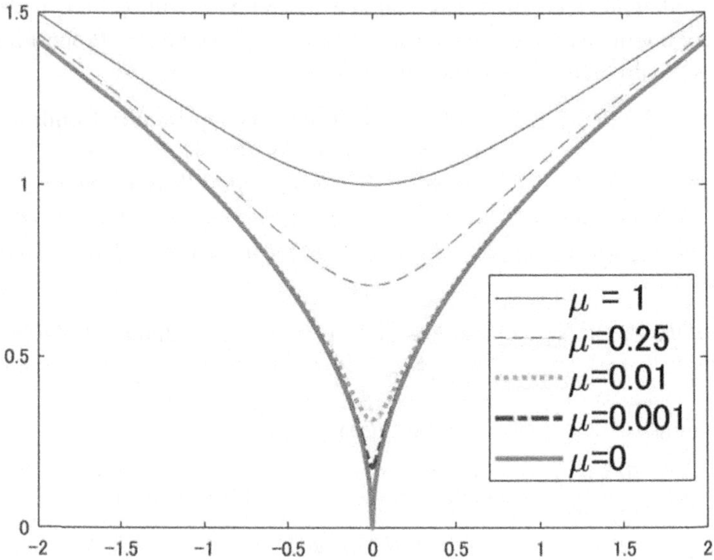

Fig. 6.5 φ_μ with $p = 0.5$ and $\mu = 1, 0.25, 0.01, 0.001, 0$

In particular, we call $\mu \geq 0$ a smoothing parameter. For more details on the smoothing method, see the comprehensive survey article [39] and the relevant articles [40, 41]. In the smoothing method [26], we replace the ℓ_p function $R_1(\boldsymbol{w}) = \|\boldsymbol{w}\|_p^p$ in (6.3.2) by the following smoothing function (Fig. 6.5):

$$\varphi_\mu(\boldsymbol{w}) := \sum_{i=1}^{n} (w_i^2 + \mu^2)^{\frac{p}{2}}.$$

We then have the following bilevel problem approximating the original one (6.3.2):

$$\min \left\{ g_{val}(\boldsymbol{w}_{\boldsymbol{\lambda}}) \text{ s.t. } \boldsymbol{w}_{\boldsymbol{\lambda}} \in \operatorname*{arg\,min}_{\boldsymbol{w} \in \mathbb{R}^n} \left\{ G(\boldsymbol{w}, \bar{\boldsymbol{\lambda}}) + \lambda_1 \varphi_\mu(\boldsymbol{w}) \right\},\ \boldsymbol{\lambda} \geq \boldsymbol{0} \right\}$$

which naturally leads to the following one-level problem:

$$\begin{aligned} \min_{\boldsymbol{w}, \boldsymbol{\lambda}} \quad & g_{val}(\boldsymbol{w}) \\ \text{s.t.} \quad & \nabla_{\boldsymbol{w}} G(\boldsymbol{w}, \bar{\boldsymbol{\lambda}}) + \lambda_1 \nabla \varphi_\mu(\boldsymbol{w}) = \boldsymbol{0} \\ & \boldsymbol{\lambda} \geq \boldsymbol{0}. \end{aligned} \tag{6.3.12}$$

Note that the problem (6.3.12) is smooth since the function φ_μ is twice continuously differentiable when $\mu \neq 0$.[1] Since the problem (6.3.12) can be regarded as an approximation to the problem (6.3.3), as $\mu \to 0$, a KKT point of (6.3.12) is expected to approach a point related to (6.3.3). In fact, as shown in Sect. 6.3.4.2, an accumulation point of the KKT points satisfies the SB-KKT conditions, namely, necessary optimality conditions for (6.3.3) when $p < 1$. See Proposition 6.3.1 and Theorem 6.3.2.

Let us explain the smoothing method [26] precisely. To this end, for a parameter $\hat{\varepsilon} > 0$, we define an $\hat{\varepsilon}$-approximate KKT point for the problem (6.3.12). We say that $(\boldsymbol{w}, \boldsymbol{\lambda}, \boldsymbol{\zeta}, \boldsymbol{\eta}) \in \mathbb{R}^n \times \mathbb{R}^r \times \mathbb{R}^n \times \mathbb{R}^r$ is an $\hat{\varepsilon}$-approximate KKT point for (6.3.12) if there exists a vector $(\boldsymbol{\varepsilon}_1, \varepsilon_2, \boldsymbol{\varepsilon}_3, \boldsymbol{\varepsilon}_4, \varepsilon_5) \in \mathbb{R}^n \times \mathbb{R} \times \mathbb{R}^{r-1} \times \mathbb{R}^n \times \mathbb{R}$ such that

$$\nabla g_{val}(\boldsymbol{w}) + \left(\nabla^2_{\boldsymbol{ww}} G(\boldsymbol{w}, \bar{\boldsymbol{\lambda}}) + \lambda_1 \nabla^2 \varphi_\mu(\boldsymbol{w})\right) \boldsymbol{\zeta} = \boldsymbol{\varepsilon}_1, \tag{6.3.13}$$

$$\nabla \varphi_\mu(\boldsymbol{w})^\top \boldsymbol{\zeta} - \eta_1 = \varepsilon_2, \tag{6.3.14}$$

$$\nabla R_i(\boldsymbol{w})^\top \boldsymbol{\zeta} - \eta_i = (\boldsymbol{\varepsilon}_3)_i \;\; (i = 2, 3, \ldots, r), \tag{6.3.15}$$

$$\nabla_{\boldsymbol{w}} G(\boldsymbol{w}, \bar{\boldsymbol{\lambda}}) + \lambda_1 \nabla \varphi_\mu(\boldsymbol{w}) = \boldsymbol{\varepsilon}_4, \tag{6.3.16}$$

$$\boldsymbol{0} \le \boldsymbol{\lambda}, \; \boldsymbol{0} \le \boldsymbol{\eta}, \; \boldsymbol{\lambda}^\top \boldsymbol{\eta} = \varepsilon_5, \tag{6.3.17}$$

and

$$\|(\boldsymbol{\varepsilon}_1, \varepsilon_2, \boldsymbol{\varepsilon}_3, \boldsymbol{\varepsilon}_4, \varepsilon_5)\| \le \hat{\varepsilon},$$

where $\nabla^2_{\boldsymbol{ww}} G(\boldsymbol{w}, \bar{\boldsymbol{\lambda}})$ is the Hessian of G with respect to $\boldsymbol{w}$. Notice that an $\hat{\varepsilon}$-approximate KKT point is nothing but a KKT point for the problem (6.3.12) if $\hat{\varepsilon} = 0$. Hence, $\boldsymbol{\zeta} \in \mathbb{R}^n$ and $\boldsymbol{\eta} \in \mathbb{R}^r$ are regarded as approximate Lagrange multiplier vectors corresponding to the equality constraint $\nabla_{\boldsymbol{w}} G(\boldsymbol{w}, \bar{\boldsymbol{\lambda}}) + \lambda_1 \nabla \varphi_\mu(\boldsymbol{w}) = \boldsymbol{0}$ and the inequality constraints $\boldsymbol{\lambda} \ge \boldsymbol{0}$, respectively. The proposed algorithm produces a sequence of $\hat{\varepsilon}$-approximate KKT points for the problem (6.3.12) while decreasing $\hat{\varepsilon}$ and μ to 0. It is formally described as in Algorithm 1.

Algorithm 1 Smoothing Method for Nonsmooth Bilevel Program

Require: Choose $\mu_0 \neq 0$, $\beta_1, \beta_2 \in (0, 1)$ and $\hat{\varepsilon}_0 \ge 0$. Set $k \leftarrow 0$.
repeat
  Find an $\hat{\varepsilon}_k$-approximate KKT point $(\boldsymbol{w}^{k+1}, \boldsymbol{\lambda}^{k+1}, \boldsymbol{\zeta}^{k+1}, \boldsymbol{\eta}^{k+1})$ for the problem (6.3.12) with $\mu = \mu_k$ by means of e.g. the sequential quadratic programming (SQP) method [29].
  Update the smoothing and error parameters by $\mu_{k+1} \leftarrow \beta_1 \mu_k$ and $\hat{\varepsilon}_{k+1} \leftarrow \beta_2 \hat{\varepsilon}_k$.
  $k \leftarrow k + 1$.
until convergence of $(\boldsymbol{w}^k, \boldsymbol{\lambda}^k, \boldsymbol{\zeta}^k, \boldsymbol{\eta}^k)$.

[1]Huber's function [41] is a popular smoothing function of $R_1(\cdot)$, but is not twice continuously differentiable.

In the numerical experiment, we used the MATLAB `fmincon` solver implementing the SQP to compute an $\hat{\varepsilon}$-approximate KKT point for (6.3.12). But, regarding how to compute such a point, there is a lot of room to explore other approaches.

As for practical stopping criteria of Algorithm 1, we make use of the scaled bilevel (SB-)KKT conditions.

Needless to say, we do not require any smoothing techniques if the nonsmooth regularizer ℓ_q is absent. In such situation, existing SQP methods for smooth constrained optimization are directly applicable to the problem (6.3.3).

6.3.4.2 Convergence of Algorithm 1 to an SB-KKT Point

In this section, we exhibit some key propositions proved in [26]. For a precise proof of each proposition or theorem, we refer to [26].

Hereafter, we suppose that the algorithm is well-defined in the sense that an $\hat{\varepsilon}_k$-approximate KKT point of (6.3.12) can be found in Step 2 at every iteration, and it generates an infinite number of iteration points. In addition, we make the following assumptions:

Assumption A Let $\{(\boldsymbol{w}^k, \boldsymbol{\lambda}^k, \boldsymbol{\zeta}^k, \boldsymbol{\eta}^k)\} \subseteq \mathbb{R}^n \times \mathbb{R}^r \times \mathbb{R}^n \times \mathbb{R}^r$ be a sequence produced by the proposed algorithm. Then, the following properties hold:

A1: $\liminf_{k\to\infty} \lambda_1^k > 0$.
A2: The sequence $\{(\boldsymbol{w}^k, \boldsymbol{\lambda}^k, \boldsymbol{\zeta}^k, \boldsymbol{\eta}^k)\}$ is bounded.
A3: Let $p = 1$ and $(\boldsymbol{w}^*, \boldsymbol{\lambda}^*)$ be an arbitrary accumulation point of the sequence $\{(\boldsymbol{w}^k, \boldsymbol{\lambda}^k)\}$. It then holds that $\lambda_1^* \neq \left|\frac{\partial G(\boldsymbol{w}^*, \bar{\boldsymbol{\lambda}}^*)}{\partial w_i}\right|$ for any $i \in I(\boldsymbol{w}^*)$.

Assumption A1 means that the ℓ_p-regularization term, i.e., the function R_1, works effectively. In the presence of a certain assumption similar to the linear independence constraint qualification, Assumption A2 can be weakened to the boundedness of only $\{(\boldsymbol{w}^k, \boldsymbol{\lambda}^k)\}$. Assumption A3 is a technical assumption for the case of $p = 1$. It means that, at $(\boldsymbol{w}^*, \boldsymbol{\lambda}^*)$, for all $i \in I(\boldsymbol{w}^*)$, zero is not situated on the boundary of the subdifferential of $G(\boldsymbol{w}, \boldsymbol{\lambda}) + \lambda\|\boldsymbol{w}\|_1$ w.r.t. w_i. Interestingly, for the case of $p < 1$, we can establish the convergence property in the absence of A3.

The next proposition plays a key role in examining the limiting behavior of a generated sequence.

Proposition 6.3.3 *Let Assumptions A1–A3 hold. Let $(\boldsymbol{w}^*, \boldsymbol{\lambda}^*)$ be an arbitrary accumulation point of $\{(\boldsymbol{w}^k, \boldsymbol{\lambda}^k)\}$ and $\{(\boldsymbol{w}^k, \boldsymbol{\lambda}^k)\}_{k\in K} (\subseteq \{(\boldsymbol{w}^k, \boldsymbol{\lambda}^k)\})$ be an arbitrary subsequence converging to $(\boldsymbol{w}^*, \boldsymbol{\lambda}^*)$. Then, there exists some $\gamma > 0$ such that*

$$\mu_{k-1} \geq \gamma |w_i^k|^{\frac{1}{2-p}} \quad (i \in I(\boldsymbol{w}^*)) \tag{6.3.18}$$

for all $k \in K$ sufficiently large. △

This proposition means that the smoothing parameter μ_{k-1} gradually approaches 0 with a speed not faster than $\max_{i\in I(\boldsymbol{w}^*)} |w_i^k|^{\frac{1}{2-p}}$.

Thanks to these results, the following theorem can be obtained:

Theorem 6.3.4 (Global Convergence to a SB-KKT Point) *Let Assumptions A1–A3 hold. Any accumulation point of* $\{(\boldsymbol{w}^k, \boldsymbol{\lambda}^k, \boldsymbol{\zeta}^k, \boldsymbol{\eta}^k)\}$ *satisfies the SB-KKT conditions* (6.3.5)–(6.3.10) *for the problem* (6.3.2). △

6.3.5 Numerical Examples

To demonstrate the performance of Algorithm 1, we show partial results of the numerical experiments conducted in [26]. Algorithm 1 requires that an $\hat{\varepsilon}$-approximate KKT point of (6.3.12) is found in Step 2 at every iteration. We used `fmincon` with "MaxIterations= 10^7" in MATLAB for solving the problem by the SQP method. For the sake of the hot-start effect, we set the previous point as a starting point of the SQP after the second iteration. The parameter setting $(\mu_0, \beta_1) = (1, 0.95)$ was used for Algorithm 1. The other parameters $\hat{\varepsilon}_0$ and β_2 were just ignored, since we solved (6.3.12) by `fmincon` with the default optimality tolerance every iteration. We used $(\lambda_1^0, \boldsymbol{w}^0) = (10, \boldsymbol{0})$ as an initial solution for experiments using the ℓ_p regularizer, and $(\lambda_1^0, \lambda_2^0, \boldsymbol{w}^0) = (10, 10, \boldsymbol{0})$ for the elastic net regularizer.

The termination criteria of our method are that, at a resulting solution $\boldsymbol{w}^*$, SB-KKT conditions (6.3.5), (6.3.6) and (6.3.7) divided by $\max_{i=1,\ldots,n} |w_i^*|$ are within the error of $\epsilon = 10^{-5}$ or $\max\left(\|\boldsymbol{\lambda}^{k+1} - \boldsymbol{\lambda}^k\|, \mu_{k+1}\right) \leq \epsilon$. We also checked if the other SB-KKT conditions (6.3.8)–(6.3.10) are satisfied.

We used `bayesopt` in MATLAB with "MaxObjectiveEvaluations=30" for Bayesian optimization. At each iteration of `bayesopt`, we need to solve the lower-level problem of (6.3.2) with a given $\boldsymbol{\lambda}$. We used `fmincon` again for solving the problem. We executed all numerical experiments on a personal computer with Intel Core i7-4790 CPU 3.60 GHz and 8.00 GB memory. We implemented our proposed algorithms as well as the one we use for comparisons with MATLAB R2017a.

6.3.6 Problem Setting and Results for UCI Datasets

In the experiment, the following bilevel problem concerning squared linear regression was solved

$$\begin{aligned} \min_{\boldsymbol{w},\boldsymbol{\lambda}} \quad & \|A_{val}\boldsymbol{w} - \boldsymbol{b}_{val}\|_2^2 \\ \text{s.t.} \quad & \boldsymbol{w} \in \operatorname*{argmin}_{\hat{\boldsymbol{w}}} \left\{ \|A_{tr}\hat{\boldsymbol{w}} - \boldsymbol{b}_{tr}\|_2^2 + \lambda_1 \|\hat{\boldsymbol{w}}\|_p^p \left(+ \lambda_2 \|\hat{\boldsymbol{w}}\|_2^2 \right) \right\} \\ & \boldsymbol{\lambda} \geq 0, \end{aligned} \tag{6.3.19}$$

where $A_{val} \in \mathbb{R}^{m_{val} \times n}$, $\boldsymbol{b}_{val} \in \mathbb{R}^{m_{val}}$, $A_{tr} \in \mathbb{R}^{m_{tr} \times n}$, $\boldsymbol{b}_{tr} \in \mathbb{R}^{m_{tr}}$. Also, as a test error function (see Sect. 6.1.2), the following function was utilized:

$$\|A_{te}\boldsymbol{w} - \boldsymbol{b}_{te}\|_2^2,$$

where $A_{te} \in \mathbb{R}^{m_{te} \times n}$ and $\boldsymbol{b}_{te} \in \mathbb{R}^{m_{te}}$. The data matrices $A_{\{val,tr,te\}}$ and vectors $\boldsymbol{b}_{\{val,tr,te\}}$ were taken from the UCI machine learning repository [42]. See Table 6.1 for the data sets we utilized, in which the parts in bold are used to refer to each data set. For each data set, Algorithm 1 and `bayesopt` were applied to (6.3.19), respectively. The obtained results are shown in Tables 6.2, 6.3, 6.4, 6.5, and 6.6, adopting the following notations ($\boldsymbol{w}^*$: output solution):

- $\mathrm{Err}_{te} := \|A_{te}\boldsymbol{w}^* - \boldsymbol{b}_{te}\|_2^2$
- $\mathrm{Err}_{val} := \|A_{val}\boldsymbol{w}^* - \boldsymbol{b}_{te}\|_2^2$
- $p = 1, 0.8, 0, 5$: $\ell_1, \ell_{0.8}, \ell_{0.5}$-regularizers
- EN: elastic net regularizer defined by $\lambda_1\|\boldsymbol{w}\|_1 + \lambda_2\|\boldsymbol{w}\|_2^2$
- time(s): running time in seconds

The bold entries in Tables 6.2, 6.3, 6.4, 6.5, and 6.6 indicate the best values of time(s) and Err_{te} between the proposed method and `bayesopt`. Moreover, the hyphen "–" indicates that an algorithm could not terminate within 27000 s.

Table 6.1 Name of data, the number of data samples($= m_{tr} + m_{te} + m_{val}$), the number of features($= n$)

Data	♯ sample	♯ feature (variable size n)
BlogFeedback	60,021	280
Facebook comment volume	40,949	53
Insurance company benchmark	9000	85
Student performance (math)	395	272
Communities and crime	1994	1955

Table 6.2 Facebook ($n = 280$, $m_{tr} \approx 20000$)

	Bilevel algorithm			bayesopt		
p	Err_{te}	Err_{val}	Time(s)	Err_{te}	Err_{val}	Time(s)
1	9.815	4.624	**34.10**	**9.813**	4.624	217.53
0.8	**9.979**	4.697	**37.32**	10.246	4.706	93.78
0.5	**10.247**	4.811	53.89	22.353	6.108	**51.03**
EN	**9.806**	4.634	**56.78**	9.812	4.624	216.17

Table 6.3 BlogFeedback ($n = 58$, $m_{tr} \approx 13500$)

	Bilevel algorithm			bayesopt		
p	Err_{te}	Err_{val}	Time(s)	Err_{te}	Err_{val}	Time(s)
1	**6.038**	5.618	**1432.69**	6.048	5.617	17014.32
0.8	6.028	5.626	**2279.44**	**5.998**	5.618	4886.72
0.5	**6.019**	5.639	**2212.54**	6.047	5.647	4507.78
EN	**6.093**	5.616	**2539.26**	6.188	5.62	16131.29

Table 6.4 Insurance ($n = 85$, $m_{tr} \approx 3000$)

	Bilevel algorithm			bayesopt		
p	Err_{te}	Err_{val}	Time(s)	Err_{te}	Err_{val}	Time(s)
1	108.192	76.737	**10.52**	**107.876**	76.576	89.32
0.8	108.368	76.921	**11.94**	**108.071**	76.436	61.69
0.5	**108.206**	76.936	**21.23**	116.313	78.305	26.43
EN	108.196	76.734	**11.57**	**107.874**	76.564	93.91

Table 6.5 Student ($n = 272$, $m_{tr} \approx 130$)

	Bilevel algorithm			bayesopt		
p	Err_{te}	Err_{val}	Time(s)	Err_{te}	Err_{val}	Time(s)
1	**1.125**	0.778	**66.74**	1.158	0.794	221.44
0.8	**1.082**	0.724	**61.92**	1.106	0.734	223.28
0.5	**1.082**	0.724	49.01	2.554	0.968	**27.48**
EN	**1.125**	0.777	**32.39**	1.153	0.788	180.00

Table 6.6 Communities ($n = 1955$, $m_{tr} \approx 665$)

	Bilevel algorithm			bayesopt		
p	Err_{te}	Err_{val}	Time(s)	Err_{te}	Err_{val}	Time(s)
1	**5.523**	5.733	**11821.06**	–	–	–
0.8	**5.515**	5.832	**22387.79**	–	–	–
0.5	**5.577**	5.830	25806.36	7.374	6.966	**1043.97**
EN	**5.523**	5.733	**11172.45**	–	–	–

Table 6.2 shows that our bilevel algorithm runs faster than the other method while our prediction performance, i.e., the value of Err_{te} seems slightly better. Especially for ill-posed problems with $n > m_{tr}$ such as Student (Table 6.5) and Communities (Table 6.6), the prediction performance improved by our algorithm. As p decreases, non-smoothness becomes stronger and, as a result, our method tends to be superior to `bayesopt` particularly in time(s).

6.4 Concluding Remarks

In this chapter, we have briefly reviewed bilevel approaches to optimize regularization hyperparameters in machine learning (ML). The problem of finding the best hyperparameters is naturally formulated as a bilevel optimization problem. Interestingly, when considering nonsmooth regularizers such as ℓ_p with $0 < p \leq 1$, a bilevel problem whose lower level problem is nonsmooth and possibly non-Lipschitz naturally arises. This is not the most standard case in the literature as far as we know. To cope with such hard bilevel problem, existing algorithms and theory seem still weak and need to be reinforced.

One of the advantages of the bilevel approach is that it allows us to apply state-of-the art methods for bilevel optimization to hyperparameter optimization.

In comparison with existing ML methods including grid search and Bayesian optimization, such sophisticated bilevel algorithms may bring us a high chance of finding better hyperparameters more accurately and faster. On the other hand, the bilevel approach suffers from the drawback that, as the data-size increases, the resulting bilevel problems tend to be more intractable. However, those issues can be expected to be settled somehow. For example, it is a promising way to solve the unconstrained problems obtained by applying penalty functions to the constraints of the bilevel problems.

Acknowledgements We express our gratitude to the anonymous referee for his/her valuable comments and suggestions.

References

1. S. Albelwi, A. Mahmood, A framework for designing the architectures of deep convolutional neural networks. Entropy **19**(6), 242 (2017)
2. I. Hovden, *Optimizing Artificial Neural Network Hyperparameters and Architecture* (University of Oslo, Oslo, 2019)
3. V. Vapnik, *The Nature of Statistical Learning Theory* (Springer, New York, 2013)
4. P.J. Huber, *Robust Estimation of a location Parameter* (Springer, New York, 1992), pp. 492–518
5. H. Zou, T. Hastie, Regularization and variable selection via the elastic net. J. R. Stat. Soc. Series B (Stat. Methodol.) **67**(2), 301–320 (2005)
6. J. Fan, R. Li, Variable selection via nonconcave penalized likelihood and its oracle properties. J. Am. Stat. Assoc. **96**(456), 1348–1360 (2001)
7. C.-H. Zhang et al. Nearly unbiased variable selection under minimax concave penalty. Annal. Stat. **38**(2), 894–942 (2010)
8. R. Tibshirani, Regression shrinkage and selection via the lasso. J. R. Stat. Soc. Series B (Methodolog.) **58**(1), 267–288 (1996)
9. F. Wen, L. Chu, P. Liu, R.C. Qiu, A survey on nonconvex regularization-based sparse and low-rank recovery in signal processing, statistics, and machine learning. IEEE Access **6**, 69883–69906 (2018)
10. M. Feurer, F. Hutter, Hyperparameter optimization, in *Automated Machine Learning* (Springer, Berlin, 2019), pp. 3–33
11. J. Bergstra, Y. Bengio, Random search for hyper-parameter optimization. J. Mach. Learn. Res. **13**, 281–305 (2012)
12. J. Mockus, V. Tiesis, A. Zilinskas, The application of bayesian methods for seeking the extremum. Towards Global Optim. **2**, 117–129 (1978)
13. P.I. Frazier, A tutorial on bayesian optimization (2018). arXiv preprint:1807.02811
14. C.E. Rasmussen, Gaussian processes in machine learning, in *Summer School on Machine Learning* (Springer, Berlin, 2003), pp. 63–71
15. K.P. Bennett, J. Hu, X. Ji, G. Kunapuli, J. Pang, Model selection via bilevel optimization, in *The 2006 IEEE International Joint Conference on Neural Network Proceedings*, pp. 1922–1929 (2006)
16. K.P. Bennett, G. Kunapuli, J. Hu, J. Pang, Bilevel optimization and machine learning, in *Computational Intelligence: Research Frontiers (WCCI 2008)*. Lecture Notes in Computer Science, vol. 5050 (Springer, Berlin, 2008)

17. G.M. Moore, C. Bergeron, K.P. Bennett, Nonsmooth bilevel programming for hyperparameter selection, in *Proceedings of the 2009 IEEE International Conference on Data Mining Workshops* (2009), pp. 374–381
18. G.M. Moore, *Bilevel Programming Algorithms for Machine Learning Model Selection* (Rensselaer Polytechnic Institute, New York, 2010)
19. G.M. Moore, C. Bergeron, K.P. Bennett, Model selection for primal SVM. Mach. Learn. **85**(1), 175–208 (2011)
20. S. Rosset, Bi-level path following for cross validated solution of kernel quantile regression. J. Mach. Learn. Res. **10**, 2473–2505 (2009)
21. K. Kunisch, T. Pock, A bilevel optimization approach for parameter learning in variational models. SIAM J. Imag. Sci. **6**(2), 938–983 (2013)
22. P. Ochs, R. Ranftl, T. Brox, T. Pock, Bilevel optimization with nonsmooth lower level problems, in *Proceedings of the International Conference on Scale Space and Variational Methods in Computer Vision* (Springer, Berlin, 2015), pp. 654–665
23. N. Couellan, W. Wang, On the convergence of stochastic bi-level gradient methods. Optimization. http://www.optimization-online.org/
24. F. Pedregosa, Hyperparameter optimization with approximate gradient, in *Proceedings of the 33rd International Conference on Machine Learning*, vol. 48, ed. by M.F. Balcan, K.Q. Weinberger. Proceedings of Machine Learning Research (PMLR, New York, 2016), pp. 737–746
25. J. Frecon, S. Salzo, M. Pontil, Bilevel learning of the group lasso structure, in *Advances in Neural Information Processing Systems*, vol. 31, ed. by S. Bengio, H. Wallach, H. Larochelle, K. Grauman, N. Cesa-Bianchi, R. Garnett (Curran Associates Inc., Red Hook, 2018), pp. 8301–8311
26. T. Okuno, A. Takeda, A. Kawana, Hyperparameter learning via bilevel nonsmooth optimization (2018). arXiv preprint:1806.01520
27. G. Kunapuli, K. Bennett, J. Hu, J.-S. Pang, Classification model selection via bilevel programming. Optim. Methods Softw. **23**(4), 475–489 (2008)
28. L. Franceschi, P. Frasconi, S. Salzo, R. Grazzi, M. Pontil, Bilevel programming for hyperparameter optimization and meta-learning, in *Proceedings of the International Conference on Machine Learning* (2018), pp. 1563–1572
29. J. Nocedal, S. Wright, *Numerical Optimization* (Springer, New York, 2006)
30. M. Gelbart, *Constrained Bayesian Optimization and Applications*. Ph.D. Thesis (Harvard University, Cambridge, 2015)
31. Z.-Q. Luo, J.-S. Pang, D. Ralph, *Mathematical Programs with Equilibrium Constraints* (Cambridge University Press, Cambridge, 1996)
32. R. Tibshirani, M. Saunders, S. Rosset, J. Zhu, K. Knight, Sparsity and smoothness via the fused lasso. J. R. Stat. Soc. Series B (Stat. Methodol.) **67**(1), 91–108 (2005)
33. L.E. Frank, J.H. Friedman, A statistical view of some chemometrics regression tools. Technometrics **35**(2), 109–135 (1993)
34. X. Chen, F. Xu, Y. Ye, Lower bound theory of nonzero entries in solutions of ℓ_2-ℓ_p minimization. SIAM J. Sci. Comput. **32**(5), 2832–2852 (2010)
35. G. Marjanovic, V. Solo, On ℓ_q optimization and matrix completion. IEEE Trans. Signal Process. **60**(11), 5714–5724 (2012)
36. R.T. Rockafellar, R.J.B. Wets, *Variational Analysis*, vol. 317 (Springer, New York, 2009)
37. X. Chen, L. Niu, Y. Yuan, Optimality conditions and a smoothing trust region Newton method for nonLipschitz optimization. SIAM J. Optim. **23**(3), 1528–1552 (2013)
38. W. Bian, X. Chen, Optimality and complexity for constrained optimization problems with nonconvex regularization. Math. Oper. Res. **42**(4), 1063–1084 (2017)
39. X. Chen, Smoothing methods for nonsmooth, nonconvex minimization. Math. Program. **134**(1), 71–99 (2012)

40. Y. Nesterov, Smooth minimization of non-smooth functions. Math. Program. **103**(1), 127–152 (2005)
41. A. Beck, M. Teboulle, Smoothing and first order methods: a unified framework. SIAM J. Optim. **22**(2), 557–580 (2012)
42. M. Lichman, *UCI Machine Learning Repository*, University of California, Irvine, School of Information and Computer Sciences (2013). http://archive.ics.uci.edu/ml

Part II
Theory and Methods for Linear and Nonlinear Bilevel Optimization

Chapter 7
Bilevel Optimization and Variational Analysis

Boris S. Mordukhovich

Abstract This chapter presents a self-contained approach of variational analysis and generalized differentiation to deriving necessary optimality conditions in bilevel optimization with Lipschitzian data. We mainly concentrate on optimistic models, although the developed machinery also applies to pessimistic versions. Some open problems are posed and discussed.

Keywords Bilevel optimization · Variational analysis · Nondifferentiable programming · Generalized differentiation · Lipschitzian functions and mappings

7.1 Introduction

Bilevel optimization has been well recognized as a theoretically very challenging and practically important area of applied mathematics. We refer the reader to the monographs [2, 6, 14, 30, 41], the extensive bibliographies and commentaries therein, as well as to the advanced material included in this book for various approaches, theoretical and numerical results, and a variety of practical applications of bilevel optimization and related topics.

One of the characteristic features of bilevel optimization problems is their intrinsic *nonsmoothness*, even if their initial data are described by linear functions. This makes natural to develop an approach of modern *variational analysis and generalized differentiation* to the study and applications of major models in bilevel optimization. It has been done in numerous publications, which are presented and analyzed in the author's recent book [30].

The main goal we pursue here is to overview this approach together with the corresponding machinery of variational analysis and to apply it to deriving necessary optimality conditions in optimistic bilevel models with *Lipschitzian* data

B. S. Mordukhovich (✉)
Department of Mathematics, Wayne State University, Detroit, MI, USA
e-mail: boris@math.wayne.edu

S. Dempe, A. Zemkoho (eds.), *Bilevel Optimization*, Springer Optimization and Its Applications 161, https://doi.org/10.1007/978-3-030-52119-6_7

while also commenting on other versions in bilevel optimization with posting open questions. To make this chapter largely *self-contained* and more accessible for the reader, we present here the basic background from variational analysis and generalized differentiation, which is needed for applications to bilevel optimization. For brevity and simplicity we confine ourselves to problems in finite-dimensional spaces.

The rest of this work is organized as follows. In Sect. 7.2 we recall those constructions of *generalized differentiation* in variational analysis, which are broadly used in the subsequent text. Section 7.3 presents the fundamental *extremal principle* that is behind generalized differential calculus and applications to optimization in the geometric approach to variational analysis developed in [29, 30]. Section 7.4 is devoted to deriving—via the extremal principle—the two basic *calculus rules*, which are particularly useful for applications to optimality conditions. In Sect. 7.5 we establish *subdifferential evaluations* and efficient conditions that ensure the local *Lipschitz continuity* of *optimal value functions* in general problems of parametric optimization. These results are crucial for variational applications to bilevel programming.

To proceed with such applications, we first consider in Sect. 7.6 problems of *nondifferentiable programming* with Lipschitzian data. Subdifferential necessary optimality conditions for Lipschitzian programs are derived there by using the extremal principle and calculus rules. Section 7.7 contains the formulation of the bilevel optimization problems under consideration and the description of the *variational approach* to their study. Based on this approach and subdifferentiation of the optimal value functions for lower-level problems, we establish in Sect. 7.8 necessary optimality conditions for *Lipschitzian bilevel programs*. The other developments in this direction for bilevel optimization problems with Lipschitzian data is presented in Sect. 7.9 by using the *subdifferential difference rule* based on a certain variational technique. The concluding Sect. 7.10 discusses *further perspectives* of employing concepts and techniques of variational analysis to bilevel optimization with formulations of some *open questions*.

Throughout this chapter we use the standard notation and terminology of variational analysis and generalized differentiation; see, e.g., [29, 30, 40].

7.2 Basic Constructions of Generalized Differentiation

Here we present the basic definitions of generalized normals to sets, coderivatives of set-valued mappings, and subgradients of extended-real-valued functions initiated by the author [26] that are predominantly used in what follows. The reader is referred to the books [29, 30, 40] for more details. Developing a *geometric approach* to generalized differentiation, we start with normals to sets, then continue with coderivatives of (set-valued and single-valued) mappings, and finally pass to subgradients of extended-real-valued functions.

Given a nonempty set $\Omega \subset \mathbb{R}^n$, we always suppose without loss of generality that it is locally closed around the reference point $\bar{x} \in \Omega$. For each $x \in \mathbb{R}^n$ close to $\bar{x}$ consider its (nonempty) Euclidean projector to Ω defined by

$$\Pi(x;\Omega) := \big\{ w \in \Omega \big|\ \|x - w\| = \min_{u \in \Omega} \|x - u\| \big\}.$$

Then the (basic, limiting, Mordukhovich) *normal cone* to Ω at $\bar{x}$ is

$$N(\bar{x};\Omega) := \Big\{ v \in \mathbb{R}^n \Big|\ \exists\, x_k \to \bar{x},\ \exists\, w_k \in \Pi(x_k;\Omega),\ \exists\, \alpha_k \geq 0 \text{ such that } \alpha_k(x_k - w_k) \to v \text{ as } k \to \infty \Big\}. \tag{7.2.1}$$

The normal cone (7.2.1) is always closed while may be nonconvex in standard situations; e.g., when Ω is the graph of the simplest nonsmooth convex function $|x|$ at $\bar{x} = (0,0) \in \mathbb{R}^2$. Nevertheless, this normal cone and the associated coderivatives of mappings and subdifferentials of functions enjoy comprehensive calculus rules due to variational/extremal principles of variational analysis. Note that $N(\bar{x};\Omega) \neq \{0\}$ if and only if $\bar{x}$ is a boundary point of Ω.

There is a useful representation of the normal cone (7.2.1) in terms of convex collections of (pre)normal vectors to Ω at points nearby $\bar{x}$. Given $x \in \Omega$ close to $\bar{x}$, the *prenormal cone* to Ω at x (known also as the regular or Fréchet normal cone) is defined by

$$\hat{N}(x;\Omega) := \Big\{ v \in \mathbb{R}^n \Big|\ \limsup_{u \xrightarrow{\Omega} x} \frac{\langle v, u - x \rangle}{\|u - x\|} \leq 0 \Big\}, \tag{7.2.2}$$

where the symbol $u \xrightarrow{\Omega} x$ means that $u \to x$ with $u \in \Omega$. Then the prenormal cone (7.2.2) is always closed and convex while may collapse to $\{0\}$ at boundary points of closed sets, which in fact contradicts the very meaning of generalized normals. If Ω is convex, then both normal and prenormal cones reduce to the normal cone of convex analysis. In general we have

$$N(\bar{x};\Omega) = \big\{ v \in \mathbb{R}^n \big|\ \exists\, x_k \xrightarrow{\Omega} \bar{x},\ v_k \in \hat{N}(x_k;\Omega) \text{ with } v_k \to v \text{ as } k \to \infty \big\}. \tag{7.2.3}$$

bNote that the limiting representation (7.2.3) keeps holding if the prenormal cone (7.2.2) therein is expanded to its ε_k*-enlargements* $\hat{N}_{\varepsilon_k}$ as $\varepsilon_k \downarrow 0$, where the latter expansions are defined by replacing 0 with ε_k on the right-hand side of (7.2.2).

Let $F\colon \mathbb{R}^n \rightrightarrows \mathbb{R}^m$ be a set-valued mapping/multifunction with values $F(x) \subset \mathbb{R}^m$ and with its domain and graph defined, respectively, by

$$\operatorname{dom} F := \big\{ x \in \mathbb{R}^n \big|\ F(x) \neq \emptyset \big\} \text{ and } \operatorname{gph} F := \big\{ (x,y) \in \mathbb{R}^n \times \mathbb{R}^m \big|\ y \in F(x) \big\}.$$

When F is single-valued, we use the standard notation $F\colon \mathbb{R}^n \to \mathbb{R}^m$. Assuming that the graph of F is locally closed around $(\bar{x},\bar{y}) \in \operatorname{gph} F$, we define the *coderivative* of F at this point via the normal cone (7.2.1) to the graph of F by

$$D^*F(\bar{x},\bar{y})(w) := \big\{v \in \mathbb{R}^n \big|\ (v,-w) \in N\big((\bar{x},\bar{y}); \operatorname{gph} F\big)\big\}, \quad w \in \mathbb{R}^m. \tag{7.2.4}$$

Thus $D^*F(\bar{x},\bar{y})\colon \mathbb{R}^m \rightrightarrows \mathbb{R}^n$ is a set-valued positively homogeneous mapping, which reduces to the adjoint/transposed Jacobian for all single-valued mappings $F\colon \mathbb{R}^n \to \mathbb{R}^m$ that are smooth around $\bar{x}$, where $\bar{y} = F(\bar{x})$ is dropped in this case in the coderivative notation:

$$D^*F(\bar{x})(w) = \big\{\nabla F(\bar{x})^* w\big\} \text{ for all } w \in \mathbb{R}^m.$$

Besides a full calculus available for the coderivative (7.2.4), this construction plays an important role in variational analysis and its applications since it provides complete characterizations of fundamental *well-posedness* properties of multifunctions concerning Lipschitzian stability, metric regularity, and linear openness/covering. In this work we deal with the *Lipschitz-like* (Aubin, pseudo-Lipschitz) property of $F\colon \mathbb{R}^n \rightrightarrows \mathbb{R}^m$ around $(\bar{x},\bar{y}) \in \operatorname{gph} F$ defined as follows: there exist neighborhoods U of $\bar{x}$ and V of $\bar{y}$ and a constant $\ell \geq 0$ such that

$$F(x) \cap V \subset F(u) + \ell\|x-u\|\mathbb{B} \text{ for all } x,u \in U, \tag{7.2.5}$$

where $\mathbb{B}$ stands for the closed unit ball of the space in question. If $V = \mathbb{R}^m$ in (7.2.5), then it reduces to the classical local Lipschitzian property of F around $\bar{x}$. The *coderivative characterization* of (7.2.5), which is called in [40] the Mordukhovich criterion, tells us that F is Lipschitz-like around $(\bar{x},\bar{y})$ if and only if we have

$$D^*F(\bar{x},\bar{y})(0) = \{0\}. \tag{7.2.6}$$

Furthermore, the exact bound (infimum) of all the Lipschitz constant $\{\ell\}$ in (7.2.5) is calculated as the norm $\|D^*F(\bar{x},\bar{y})\|$ of the positively homogeneous coderivative mapping $w \mapsto D^*F(\bar{x},\bar{y})(w)$. The reader can find in [28–30, 40] different proofs of this result with numerous applications.

Consider finally an extended-real-valued function $\varphi\colon \mathbb{R}^n \to \overline{\mathbb{R}} := (-\infty,\infty]$ finite at $\bar{x}$ and lower semicontinuous (l.s.c.) around this point. Denote by

$$\operatorname{dom}\varphi := \big\{x \in \mathbb{R}^n \big|\ \varphi(x) < \infty\big\} \text{ and } \operatorname{epi}\varphi := \big\{(x,\mu) \in \mathbb{R}^n \times \mathbb{R} \big|\ \mu \geq \varphi(x)\big\}$$

the domain and epigraph of φ, respectively. Given $\bar{x} \in \operatorname{dom}\varphi$ and using the normal cone (7.2.1) to the epigraph of φ at $(\bar{x},\varphi(\bar{x}))$, we define the two types of the subdifferentials of φ at $\bar{x}$: the *basic subdifferential* and the *singular subdifferential* by, respectively,

$$\partial\varphi(\bar{x}) := \big\{v \in \mathbb{R}^n \big|\ (v, -1) \in N\big((\bar{x}, \varphi(\bar{x})); \operatorname{epi}\varphi\big)\big\}, \tag{7.2.7}$$

$$\partial^\infty\varphi(\bar{x}) := \big\{v \in \mathbb{R}^n \big|\ (v, 0) \in N\big((\bar{x}, \varphi(\bar{x})); \operatorname{epi}\varphi\big)\big\}. \tag{7.2.8}$$

The basic subdifferential (7.2.7) reduces to the gradient $\{\nabla\varphi(\bar{x})\}$ for smooth functions and to the subdifferential of convex analysis if φ is convex. Observe that $\partial\varphi(\bar{x}) = D^*E_\varphi(\bar{x}, \varphi(\bar{x}))(1)$ and $\partial^\infty\varphi(\bar{x}) = D^*E_\varphi(\bar{x}, \varphi(\bar{x}))(0)$ via the coderivative (7.2.4) of the epigraphical multifunction $E_\varphi\colon \mathbb{R}^n \rightrightarrows \mathbb{R}$ defined by $E_\varphi(x) := \{\mu \in \mathbb{R} |\ \mu \ge \varphi(x)\}$. Thus the coderivative characterization (7.2.6) of the Lipschitz-like property of multifunctions implies that a lower semicontinuous function φ is *locally Lipschitzian* around $\bar{x}$ if and only if

$$\partial^\infty\varphi(\bar{x}) = \{0\}. \tag{7.2.9}$$

Note also that, given any (closed) set $\Omega \subset \mathbb{R}^n$ with its *indicator function* $\delta(x; \Omega) = \delta_\Omega(x)$ equal 0 for $x \in \Omega$ and ∞ otherwise, we have that

$$\partial\delta(\bar{x}; \Omega) = \partial^\infty\delta(\bar{x}; \Omega) = N(\bar{x}; \Omega) \ \text{ whenever } \ \bar{x} \in \Omega. \tag{7.2.10}$$

Both subdifferentials (7.2.7) and (7.2.8) admit limiting representations in terms of the *presubdifferential*, or *regular subdifferential*

$$\hat{\partial}\varphi(x) := \Big\{v \in \mathbb{R}^n \Big|\ \liminf_{u \to x} \frac{\varphi(u) - \varphi(x) - \langle v, u - x\rangle}{\|u - x\|} \ge 0\Big\} \tag{7.2.11}$$

of φ at points x close to $\bar{x}$. Namely, we have

$$\partial\varphi(\bar{x}) = \big\{v \in \mathbb{R}^n \big|\ \exists x_k \xrightarrow{\varphi} \bar{x},\ \exists v_k \to v \ \text{ with } \ v_k \in \hat{\partial}\varphi(x_k) \ \text{ as } \ k \to \infty\big\}, \tag{7.2.12}$$

$$\partial^\infty\varphi(\bar{x}) = \big\{v \in \mathbb{R}^n \big|\ \ \exists x_k \xrightarrow{\varphi} \bar{x},\ \exists \lambda_k \downarrow 0,\ \exists v_k \to v \ \text{ with } \\ v_k \in \lambda_k\hat{\partial}\varphi(x_k) \ \text{ as } \ k \to \infty\big\}, \tag{7.2.13}$$

where the symbol $x \xrightarrow{\varphi} \bar{x}$ indicates that $x \to \bar{x}$ with $\varphi(x) \to \varphi(\bar{x})$. Note that the presubdifferential (7.2.11) is related to the prenormal cone (7.2.2) as in (7.2.7) and is also used in variational analysis under the names of the Fréchet subdifferential and the viscosity subdifferential. Similarly to the case of basic normals in (7.2.3) it is not hard to observe that we still have the subdifferential representations in (7.2.12) and (7.2.13) if the presubdifferential (7.2.11) therein is expanded by its ε_k-*enlargements* $\hat{\partial}_{\varepsilon_k}\varphi$ defined with replacing 0 on the right-hand side of (7.2.11) by $-\varepsilon_k$.

7.3 Extremal Principle in Variational Analysis

In this section we recall, following [22], the notion of locally extremal points for systems of finitely many sets and then derive the fundamental *extremal principle*, which gives us necessary conditions for extremality of closed set systems in $\mathbb{R}^n$.

Definition 7.3.1 Let $\Omega_1, \ldots, \Omega_s$ as $s \geq 2$ be nonempty subsets of $\mathbb{R}^n$, which are assumed to be locally closed around their common point $\bar{x}$. We say that $\bar{x}$ is a Locally Extremal Point of the set system $\{\Omega_1, \ldots, \Omega_s\}$ if there exist a neighborhood U of $\bar{x}$ and sequences of vectors $a_{ik} \in \mathbb{R}^n$, $i = 1, \ldots, s$, such that $a_{ik} \to 0$ as $k \to \infty$ for all $i \in \{1, \ldots, s\}$ and

$$\bigcap_{i=1}^{s} \big(\Omega_i - a_{ik}\big) \cap U = \emptyset \text{ whenever } k = 1, 2, \ldots. \tag{7.3.1}$$

△

Geometrically Definition 7.3.1 says that $\bar{x}$ is a locally extremal point of the set system $\{\Omega_1, \ldots, \Omega_s\}$ if these sets can be pushed apart from each other locally around $\bar{x}$. Observe also that for the case of two sets Ω_1, Ω_2 containing $\bar{x}$ the above definition can be equivalently reformulated as follows: there is a neighborhood U of $\bar{x}$ such that for any $\varepsilon > 0$ there exists a vector $a \in \mathbb{R}^n$ with $\|a\| \leq \varepsilon$ and $(\Omega_1 - a) \cap \Omega_2 \cap U = \emptyset$.

It is easy to see that for a closed set Ω and any of its boundary points $\bar{x}$, we have that $\bar{x}$ is a locally extremal point of the set system $\{\Omega, \{\bar{x}\}\}$. Furthermore, the introduced notion of set extremality covers various notions of optimality and equilibria in problems of scalar and vector optimization. In particular, a local minimizer $\bar{x}$ of the general constrained optimization problem

$$\text{minimize } \varphi(x) \text{ subject to } x \in \Omega \subset \mathbb{R}^n,$$

where φ is l.s.c. and Ω is closed around $\bar{x}$, corresponds to the locally extremal point $(\bar{x}, \varphi(\bar{x}))$ of the sets $\Omega_1 := \operatorname{epi}\varphi$ and $\Omega_2 := \Omega \times \{\varphi(\bar{x})\}$. As we see below, extremal systems naturally arise in deriving calculus rules of generalized differentiation.

Now we are ready to formulate and prove the basic extremal principle of variational analysis for systems of finitely many closed sets in $\mathbb{R}^n$ by using the normal cone construction (7.2.1).

Theorem 7.3.2 *Let $\bar{x}$ be a locally extremal point of the system $\{\Omega_1, \ldots, \Omega_s\}$ of nonempty subsets of $\mathbb{R}^n$, which are locally closed around $\bar{x}$. Then there exist generalized normals $v_i \in N(\bar{x}; \Omega_i)$ for $i = 1, \ldots, s$, not equal to zero simultaneously, such that we have the generalized Euler equation*

$$v_1 + \ldots + v_s = 0. \tag{7.3.2}$$

△

Proof Using Definition 7.3.1, suppose without loss of generality that $U = \mathbb{R}^n$. Taking the sequences $\{a_{ik}\}$ therein, for each $k = 1, 2, \ldots$ consider the unconstrained optimization problem:

$$\text{minimize } \varphi_k(x) := \Big[\sum_{i=1}^{s} d^2(x + a_{ik}; \Omega_i)\Big]^{1/2} + \|x - \bar{x}\|^2, \quad x \in \mathbb{R}^n, \tag{7.3.3}$$

where $d(x; \Omega)$ indicates the Euclidean distance between x and Ω. Since φ_k is continuous and the level sets of it are bounded, we deduce from the classical Weierstrass theorem that there exists an optimal solution x_k to each problem (7.3.3) as $k = 1, 2, \ldots$. It follows from the crucial extremality requirement (7.3.1) in Definition 7.3.1 that

$$\gamma_k := \Big[\sum_{i=1}^{s} d^2(x_k + a_{ik}; \Omega_i)\Big]^{1/2} > 0. \tag{7.3.4}$$

The optimality of x_k in (7.3.3) tells us that

$$0 < \gamma_k + \|x_k - \bar{x}\|^2 = \varphi_k(x_k) \le \varphi_k(\bar{x}) = \Big[\sum_{i=1}^{s} \|a_{ik}\|^2\Big]^{1/2} \downarrow 0,$$

and so $\gamma_k \downarrow 0$ and $x_k \to \bar{x}$ as $k \to \infty$. By the closedness of the sets Ω_i, $i = 1, \ldots, s$, around $\bar{x}$, we pick $w_{ik} \in \Pi(x_k + a_{ik}; \Omega_i)$ and for each k form another unconstrained optimization problem:

$$\text{minimize } \psi_k(x) := \Big[\sum_{i=1}^{s} \|x + a_{ik} - w_{ik}\|^2\Big]^{1/2} + \|x - \bar{x}\|^2, \quad x \in \mathbb{R}^n \tag{7.3.5}$$

which obviously has the same optimal solution x_k. In contrast to φ_k in (7.3.3), the function ψ_k in (7.3.5) is *differentiable* at x_k due to (7.3.4). Thus applying the Fermat rule in (7.3.5) tells us that

$$\nabla \psi_k(x_k) = \sum_{i=1}^{s} v_{ik} + 2(x_k - \bar{x}) = 0 \tag{7.3.6}$$

with $v_{ik} := (x_k + a_{ik} - w_{ik})/\gamma_k$, $i = 1, \ldots, s$, satisfying

$$\|v_{1k}\|^2 + \ldots + \|v_{sk}\|^2 = 1 \text{ for all } k = 1, 2, \ldots. \tag{7.3.7}$$

Remembering the compactness of the unit sphere in $\mathbb{R}^n$, we get by passing to the limit as $k \to \infty$ in (7.3.6) and (7.3.7) that there exist $v_1, \ldots, v_s$, not equal to zero simultaneously, for which (7.3.2) holds. Finally, it follows directly from the above constructions and the normal cone definition (7.2.1) that $v_i \in N(\bar{x}; \Omega_i)$ for all $i = 1, \ldots, s$. This completes the proof of the theorem. □

Since for convex sets Ω the normal cone (7.2.1) reduces to the normal cone of convex analysis

$$N(\bar{x}; \Omega) := \big\{v \in \mathbb{R}^n \big|\ \langle v, x - \bar{x}\rangle \le 0 \ \text{ whenever } \ x \in \Omega\big\},$$

the extremal principle of Theorem 7.3.2 can be treated as a *variational extension* of the classical separation theorem to the case of finitely many *nonconvex* sets in $\mathbb{R}^n$.

7.4 Fundamental Calculus Rules

Employing the extremal principle, we derive here two fundamental rules of generalized differential calculus, which are broadly used in this chapter and from which many other calculus rules follow; see [29, 30]. The first result is the *intersection rule* for basic normals (7.2.1).

Theorem 7.4.1 *Let $\Omega_1, \ldots, \Omega_s$ be nonempty subsets of $\mathbb{R}^n$, which are locally closed around their common point $\bar{x}$. Assume the validity of the following qualification condition:*

$$\big[x_i \in N(\bar{x}; \Omega_i),\ x_1 + \ldots + x_s = 0\big] \Longrightarrow x_i = 0 \ \textit{for all} \ i = 1, \ldots, s. \tag{7.4.1}$$

Then we have the normal cone intersection rule

$$N\Big(\bar{x}; \bigcap_{i=1}^{s} \Omega_i\Big) \subset N(\bar{x}; \Omega_1) + \ldots + N(\bar{x}; \Omega_s). \tag{7.4.2}$$

△

Proof Arguing by induction, we first verify the result for $s = 2$. Pick any $v \in N(\bar{x}; \Omega_1 \cap \Omega_2)$ and use the normal cone representation (7.2.3). It gives us sequences $x_k \to \bar{x}$ with $x_k \in \Omega_1 \cap \Omega_2$ and $v_k \to v$ with $v_k \in \ddot{N}(\bar{x}, \Omega_1 \cap \Omega_2)$ as $k \to \infty$. Take any sequence $\varepsilon_k \downarrow 0$ and construct the sets

$$\Theta_1 := \Omega_1 \times \mathbb{R}_+,$$
$$\Theta_{2k} := \big\{(x, \alpha) \big|\ x \in \Omega_2,\ \langle v_k, x - x_k\rangle - \varepsilon_k \|x - x_k\| \ge \alpha\big\} \ \text{ for any } \ k = 1, 2, \ldots.$$

These sets are obviously closed around $(x_k, 0) \in \Theta_1 \cap \Theta_{2k}$ for all k sufficiently large. Furthermore, it follows from the prenormal cone definition (7.2.2) that there exists a neighborhood U of x_k with

$$\Theta_1 \cap \big(\Theta_{2k} - (0, \gamma)\big) \cap (U \times \mathbb{R}) = \emptyset$$

for small numbers $\gamma > 0$. Thus the pair $(x_k, 0)$ is a locally extremal point of the set system $\{\Theta_1, \Theta_{2k}\}$ for such k. Applying to this system the extremal principle from Theorem 7.3.2 gives us pairs (u_k, λ_k) from the unit sphere in $\mathbb{R}^{n+1}$ for which

$$(u_k, \lambda_k) \in N\big((x_k, 0); \Theta_1\big) \text{ and } (-u_k, -\lambda_k) \in N\big((x_k, 0); \Theta_{2k}\big). \tag{7.4.3}$$

Considering a subsequence of the unit vectors (u_k, λ_k) if needed, we get $(u_k, \lambda_k) \to (u, \lambda)$ as $k \to \infty$ for some $(u, \lambda) \in \mathbb{R}^{n+1}$ with $\|(u, \lambda)\| = 1$. Passing to the limit in the first inclusion of (7.4.3) gives us $(u, \lambda) \in N((\bar{x}, 0); \Theta_1)$, which immediately implies—by the easily verifiable product rule equality for normals to Cartesian products (see, e.g., [29, Proposition 1.2])—that $u \in N(\bar{x}; \Omega_1)$ and $\lambda \le 0$. On the other hand, the limiting procedure in the second inclusion of (7.4.3) leads us by the structure of Θ_{2k} to

$$(-\lambda v - u, \lambda) \in N\big((\bar{x}, 0); \Omega_2 \times \mathbb{R}_+\big). \tag{7.4.4}$$

To verify (7.4.4), for each $k \in \mathbb{N}$ rewrite the set Θ_{2k} as

$$\Theta_{2k} = \big\{\big(x, \langle v_k, x - x_k\rangle - \varepsilon_k\|x - x_k\| - \alpha\big) \,\big|\, x \in \Omega_2,\ \alpha \in \mathbb{R}_+\big\}.$$

Using representation (7.2.3) of basic normals via limits of regular ones, for each fixed $k \in \mathbb{N}$ we find sequences $x_{km} \to x_k$, $\alpha_{km} \downarrow 0$, $u_{km} \to u_k$ and $\lambda_{km} \to \lambda_k$ as $m \to \infty$ such that

$$(-u_{km}, -\lambda_{km}) \in \hat{N}\big((x_{km}, \langle v_k, x_{km} - x_k\rangle - \varepsilon_k\|x_{km} - x_k\| - \alpha_{km}); \Theta_{2k}\big)$$

whenever $m \in \mathbb{N}$. The latter means by definition (7.2.2) of regular normals that

$$\limsup_{\substack{x \xrightarrow{\Omega_2} x_{km} \\ \alpha \xrightarrow{\mathbb{R}_+} \alpha_{km}}} \left\{ \frac{\langle -u_{km}, x - x_{km}\rangle}{\|x - x_{km}\|(1 + \|v_k\| + \varepsilon_k) + |\alpha - \alpha_{km}|} - \frac{\lambda_{km}\big[\langle v_k, x - x_{km}\rangle - (\alpha - \alpha_{km}) - \varepsilon_k(\|x - x_k\| - \|x_{km} - x_k\|)\big]}{\|x - x_{km}\|(1 + \|v_k\| + \varepsilon_k) + |\alpha - \alpha_{km}|} \right\} \le 0.$$

Thus for all $k \in \mathbb{N}$ and $m \in \mathbb{N}$ we arrive at the limiting condition

$$\limsup_{\substack{x \xrightarrow{\Omega_2} x_{km} \\ \alpha \xrightarrow{\mathbb{R}_+} \alpha_{km}}} \left\{ \frac{\langle -u_{km} - \lambda_{km} v_k, x - x_{km}\rangle + \lambda_{km}(\alpha - \alpha_{km})}{(1 + \|v_k\| + \varepsilon_k)(\|x - x_{km}\| + |\alpha - \alpha_{km}|)} - \frac{|\lambda_{km}|\varepsilon_k(\|x - x_{km}\| + |\alpha - \alpha_{km}|)}{(1 + \|v_k\| + \varepsilon_k)(\|x - x_{km}\| + |\alpha - \alpha_{km}|)} \right\} \le 0,$$

which can be rewritten in terms of ε-enlargements of the prenormal cone as follows:

$$(-u_{km} - \lambda_{km} v_k, \lambda_{km}) \in \hat{N}_{\varepsilon_{km}}\big((x_{km}, \alpha_{km}); \Omega_2 \times \mathbb{R}_+\big) \text{ with } \varepsilon_{km} := \frac{|\lambda_{km}|\varepsilon_k}{1 + \|v_k\| + \varepsilon_k}.$$

Employing the standard diagonal process and remembering the representation of basic normals via limits of ε-enlargements of regular ones bring us to the claimed inclusion (7.4.4).

Observe that having $\lambda = 0$ in (7.4.4) contradicts the qualification condition (7.4.1) for $s = 2$. Thus $\lambda < 0$, which readily implies that $v \in N(\bar{x}; \Omega_1) + N(\bar{x}; \Omega_2)$.

To proceed finally by induction for $s > 2$, we observe that the induction assumption for (7.4.2) in the previous step yields the validity of the qualification condition (7.4.1) needed for the current step of induction. This completes the proof of the theorem. □

Next we derive the *subdifferential sum rules* concerning both basic subdifferential (7.2.7) and singular subdifferential (7.2.8). For our subsequent applications to bilevel optimization, it is sufficient to consider the case where *all but one* of the functions involved in summation are locally *Lipschitzian* around the reference point. This case allows us to obtain the subdifferential sum rules *without any* qualification conditions.

Theorem 7.4.2 *Let $\varphi_1 : \mathbb{R}^n \to \overline{\mathbb{R}}$ be l.s.c. around $\bar{x} \in \operatorname{dom} \varphi_1$, and let $\varphi_i : \mathbb{R}^n \to \overline{\mathbb{R}}$ for $i = 2, \ldots, s$ and $s \geq 2$ be locally Lipschitzian around $\bar{x}$. Then we have the sum rules*

$$\partial\Big(\sum_{i=1}^{s} \varphi_i\Big)(\bar{x}) \subset \sum_{i=1}^{s} \partial\varphi_i(\bar{x}), \tag{7.4.5}$$

$$\partial^\infty\Big(\sum_{i=1}^{s} \varphi_i\Big)(\bar{x}) = \partial^\infty \varphi_1(\bar{x}). \tag{7.4.6}$$

△

Proof We consider the case where only two functions are under summation since the general case of finitely many functions obviously follows by induction. Let us start with the basic subdifferential sum rule (7.4.5) for $s = 2$ therein.

Pick any $v \in \partial(\varphi_1 + \varphi_2)(\bar{x})$ and get by definition (7.2.7) that

$$(v, -1) \in N\big((\bar{x}, (\varphi_1 + \varphi_2)(\bar{x})); \operatorname{epi}(\varphi_1 + \varphi_2)\big).$$

Then construct the sets

$$\Omega_i := \big\{(x, \mu_1, \mu_2) \in \mathbb{R}^n \times \mathbb{R} \times \mathbb{R} \big| \; \mu_i \geq \varphi_i(x)\big\} \text{ for } i = 1, 2.$$

Denoting $\bar{\mu}_i := \varphi_i(\bar{x})$, $i = 1, 2$, we obviously have that the sets Ω_1 and Ω_2 are locally closed around the triple $(\bar{x}, \bar{\mu}_1, \bar{\mu}_2) \in \Omega_1 \cap \Omega_2$. It is easy to check that $(v, -1, -1) \in N((\bar{x}, \bar{\mu}_1, \bar{\mu}_2); \Omega_1 \cap \Omega_2)$. Applying now to this set intersection the normal cone intersection rule from Theorem 7.4.1, we observe that the qualification condition (7.4.1) is automatically satisfied in this case due to the singular subdifferential characterization (7.2.9) of the local Lipschitz continuity. Hence we get pairs $(v_i, -\lambda_i) \in N((\bar{x}, \bar{\mu}_i); \operatorname{epi}\varphi_i)$ for $i = 1, 2$ satisfying the condition

$$(v, -1, -1) = (v_1, -\lambda_1, 0) + (v_2, 0, -\lambda_2),$$

which implies that $v = v_1 + v_2$ and $\lambda_1 = \lambda_2 = -1$. Therefore it shows that $v_i \in \partial\varphi_i(\bar{x})$ for $i = 1, 2$, and thus the sum rule (7.4.5) is verified.

Next we proceed with the proof of (7.4.6) for $s = 2$ starting with verifying the inclusion "$\subset$" therein. Pick $v \in \partial^\infty(\varphi_1 + \varphi_2)(\bar{x})$ and find by definition sequences $\gamma_k \downarrow 0$, $(x_k, \mu_k) \overset{\operatorname{epi}(\varphi_1+\varphi_2)}{\longrightarrow} (\bar{x}, (\varphi_1 + \varphi_2)(\bar{x}))$, $v_k \to v$, $\nu_k \to 0$, and $\eta_k \downarrow 0$ such that

$$\langle v_k, x - x_k\rangle + \nu_k(\mu - \mu_k) \le \gamma_k(\|x - x_k\| + |\mu - \mu_k|)$$

whenever $(x, \mu) \in \operatorname{epi}(\varphi_1 + \varphi_2)$ with $x \in x_k + \eta_k\mathbb{B}$ and $|\mu - \mu_k| \le \eta_k$ as $k = 1, 2, \ldots$. Taking a Lipschitz constant $\ell > 0$ of φ_2 around $\bar{x}$, denote $\tilde{\eta}_k := \eta_k/2(\ell+1)$ and $\widetilde{\mu}_k := \mu_k - \varphi_2(x_k)$. Then $(x_k, \widetilde{\mu}_k) \overset{\operatorname{epi}\varphi_1}{\longrightarrow} (\bar{x}, \varphi_1(\bar{x}))$ and

$$(x, \mu + \varphi_2(x)) \in \operatorname{epi}(\varphi_1 + \varphi_2), \quad |(\mu + \varphi_2(x)) - \mu_k| \le \eta_k$$

for all $(x, \mu) \in \operatorname{epi}\varphi_1$, $x \in x_k + \widetilde{\eta}_k\mathbb{B}$, and $|\mu - \tilde{\mu}_k| \le \widetilde{\eta}_k$. Therefore

$$\langle v_k, x - x_k\rangle + \nu_k(\mu - \tilde{\mu}_k) \le \varepsilon_k(\|x - x_k\| + |\mu - \tilde{\mu}_k|) \text{ with } \varepsilon_k := \gamma_k(1 + \ell) + |\nu_k|\ell$$

if $(x, \mu) \in \operatorname{epi}\varphi_1$ with $x \in x_k + \tilde{\eta}_k\mathbb{B}$ and $|\mu - \tilde{\mu}_k| \le \tilde{\eta}_k$. It yields $(v_k, \nu_k) \in \widehat{N}_{\varepsilon_k}((x_k, \tilde{\mu}_k); \operatorname{epi}\varphi_1)$ for all $k = 1, 2, \ldots$, and so $(v, 0) \in N((\bar{x}, \varphi(\bar{x})); \operatorname{epi}\varphi_1)$ since $\varepsilon_k \downarrow 0$ as $k \to \infty$. This verifies the inclusion "$\subset$" in (7.4.6). Applying it to the sum $\varphi_1 = (\varphi_1 + \varphi_2) + (-\varphi_2)$ yields $\partial^\infty\varphi_1(\bar{x}) \subset \partial^\infty(\varphi_2 + \varphi_1)(\bar{x})$, which justifies the equality in (7.4.6) and thus completes the proof. □

7.5 Subdifferentials and Lipschitz Continuity of Value Functions

In this section we consider the class of extended-real-valued functions $\vartheta : \mathbb{R}^n \to \overline{\mathbb{R}}$ defined by

$$\vartheta(x) := \inf\{\varphi(x, y) \big| \; y \in F(x)\}, \quad x \in \mathbb{R}^n, \tag{7.5.1}$$

where $\varphi\colon \mathbb{R}^n \times \mathbb{R}^m \to \overline{\mathbb{R}}$ is an l.s.c. function, and where $F\colon \mathbb{R}^n \rightrightarrows \mathbb{R}^m$ is a set-valued mapping of closed graph. We can view (7.5.1) as the *optimal value function* in the problem of parametric optimization described as follows:

$$\text{minimize } \varphi(x, y) \text{ subject to } y \in F(x)$$

with the cost function φ and the constraint mapping F, where y and x are the decision and parameter variables, respectively. Functions of this type are also known in variational analysis under the name of "marginal functions." A characteristic feature of such functions is their nonsmoothness regardless of the smoothness of the cost function φ and the simplicity of the constraint mapping F that may nicely behave on the parameter x.

As seen below, functions of type (7.5.1) play a crucial role in applications to bilevel optimization while revealing *intrinsic nonsmoothness* of the latter class of optimization problems. This section presents evaluations of both basic and singular subdifferentials of (7.5.1), which are equally important for the aforementioned applications. *Singular subdifferential* evaluations are used for establishing the *Lipschitz continuity* of $\vartheta(x)$ with respect to the parameter x that allows us to reduce the bilevel model under consideration to a single-level problem of Lipschitzian programming. On the other hand, *basic subdifferential* evaluations open the gate to derive in this way *necessary optimality conditions* for Lipschitzian bilevel programs.

To proceed, let us consider the *argminimum mapping* $M\colon \mathbb{R}^n \rightrightarrows \mathbb{R}^m$ associated with (7.5.1) by

$$M(x) := \big\{y \in F(x)\big|\ \varphi(x, y) = \vartheta(x)\big\}, \quad x \in \mathbb{R}^n, \tag{7.5.2}$$

and recall that this mapping is *inner semicontinuous* at $(\bar{x}, \bar{y}) \in \operatorname{gph} M$ if for every sequence $x_k \overset{\operatorname{dom} M}{\longrightarrow} \bar{x}$ there exists a sequence $y_k \in M(x_k)$ that converges to $\bar{y}$ as $k \to \infty$. Observe that the inner semicontinuity of M at $(\bar{x}, \bar{y})$ is implied by its Lipschitz-like property at this point.

The following theorem gives us efficient upper estimates of both basic and singular subdifferentials of the optimal value function ϑ needed for subsequent applications. We confine ourselves to the case of local Lipschitz continuity of the cost function φ in (7.5.1) that is sufficient to derive necessary optimality conditions for bilevel programs in Sects. 7.8 and 7.9.

Theorem 7.5.1 *Let the argminimum mapping* (7.5.2) *be inner semicontinuous at the point* $(\bar{x}, \bar{y}) \in \operatorname{gph} M$, *and let the cost function* φ *be locally Lipschitzian around this point. Then we have*

$$\partial\vartheta(\bar{x}) \subset \bigcup_{(v,w)\in\partial\varphi(\bar{x},\bar{y})} \Big[v + D^*F(\bar{x}, \bar{y})(w)\Big], \tag{7.5.3}$$

$$\partial^\infty\vartheta(\bar{x}) \subset D^*F(\bar{x}, \bar{y})(0). \tag{7.5.4}$$

△

Proof To start with the verification of (7.5.3), consider the extended-real-valued function $\psi\colon \mathbb{R}^n \times \mathbb{R}^m \to \overline{\mathbb{R}}$ defined via the indicator function of the set gph F by

$$\psi(x, y) := \varphi(x, y) + \delta\big((x, y); \text{gph}\, F\big) \quad \text{for all} \quad (x, y) \in \mathbb{R}^n \times \mathbb{R}^m \tag{7.5.5}$$

and prove first the fulfillment of the estimate

$$\partial\vartheta(\bar{x}) \subset \big\{v \in \mathbb{R}^n \big|\ (v, 0) \in \partial\psi(\bar{x}, \bar{y})\big\}. \tag{7.5.6}$$

Indeed, pick any subgradient $v \in \partial\vartheta(\bar{x})$ and get from its representation in (7.2.12) sequences $x_k \xrightarrow{\vartheta} \bar{x}$ and $v_k \to v$ with $v_k \in \hat{\partial}\vartheta(x_k)$ as $k \to \infty$. Based on definition (7.2.11), for any sequence $\varepsilon_k \downarrow 0$ there exists $\eta_k \downarrow 0$ as $k \to \infty$ such that

$$\langle v_k, x - x_k\rangle \le \vartheta(x) - \vartheta(x_k) + \varepsilon_k\|x - x_k\| \quad \text{whenever} \quad x \in x_k + \eta_k\mathbb{B}, \quad k = 1, 2, \ldots.$$

This ensures by using the constructions above that

$$\langle (v_k, 0), (x, y) - (x_k, y_k)\rangle \le \psi(x, y) - \psi(x_k, y_k) + \varepsilon_k\big(\|x - x_k\| + \|y - y_k\|\big)$$

for all $y_k \in M(x_k)$ and $(x, y) \in (x_k, y_k) + \eta_k\mathbb{B}$. This tells us that $(v_k, 0) \subset \hat{\partial}_{\varepsilon_k}\psi(x_k, y_k)$ for all $k = 1, 2, \ldots$. Employing further the inner semicontinuity of the argminimum mapping M at $(\bar{x}, \bar{y})$, we find a sequence of $y_k \in M(x_k)$ converging to $\bar{y}$ as $k \to \infty$. It follows from imposed convergence $\vartheta(x_k) \to \vartheta(\bar{x})$ that $\psi(x_k, y_k) \to \psi(\bar{x}, \bar{y})$. Hence we arrive at $(v, 0) \in \partial\psi(\bar{x}, \bar{y})$ by passing to the limit as $k \to \infty$, which verifies therefore the validity of the upper estimate (7.5.6). To derive from (7.5.6) the one in (7.5.3) claimed in the theorem, it remains to use in (7.5.6) the basic subdifferential sum rule (7.4.5) from Theorem 7.4.2 combining it with subdifferentiation of the indicator function in (7.2.10) and the coderivative definition in (7.2.4).

Next we verify the singular subdifferential estimate

$$\partial^\infty\vartheta(\bar{x}) \subset \big\{v \in \mathbb{R}^n \big|\ (v, 0) \in \partial^\infty\psi(\bar{x}, \bar{y})\big\} \tag{7.5.7}$$

for optimal value function (7.5.1) in terms of the auxiliary function (7.5.5) under the assumptions made. Picking $v \in \partial^\infty\vartheta(\bar{x})$ and taking any sequence $\varepsilon_k \downarrow 0$, find by (7.2.13) sequences $x_k \xrightarrow{\vartheta} \bar{x}$, $(v_k, \nu_k) \to (v, 0)$, and $\eta_k \downarrow 0$ as $k \to \infty$ satisfying

$$\langle v_k, x - x_k\rangle + \nu_k(\mu - \mu_k) \le \varepsilon_k\big(\|x - x_k\| + |\mu - \mu_k|\big)$$

for all $(x, \mu) \in \text{epi}\,\vartheta$, $x \in x_k + \eta_k\mathbb{B}$, and $|\mu - \mu_k| \le \eta_k$. The assumed inner semicontinuity of (7.5.2) ensures the existence of sequences $y_k \xrightarrow{M(x_k)} \bar{y}$ and $\mu_k \downarrow \psi(\bar{x})$ such that

$$(v_k, 0, \nu_k) \in \hat{N}_{\varepsilon_k}\big((x_k, y_k, \mu_k); \text{epi}\,\psi\big) \quad \text{for all} \quad k = 1, 2, \ldots,$$

via the ε_k-enlargements $\hat{N}_{\varepsilon_k}$ of the prenormal cone to the epigraph of ψ. This gives us (7.5.7) by passing to the limit as $k \to \infty$. Applying finally to $\partial^\infty \psi$ in (7.5.7) the singular subdifferential relation (7.4.6) from Theorem 7.4.2 with taking into account the singular subdifferential calculation in (7.2.10) together with the coderivative definition (7.2.4), we arrive at the claimed upper estimate (7.5.4) and thus complete the proof of the theorem. □

As mentioned above, in our applications to bilevel programming we need to have verifiable conditions that ensure the local Lipschitz continuity of the optimal value function (7.5.1). This is provided by the following corollary, which is a direct consequence of Theorem 7.5.1 and the coderivative criterion (7.2.6) for the Lipschitz-like property.

Corollary 7.5.2 *In addition to the assumptions of Theorem 7.5.1, suppose that the constraint mapping F is Lipschitz-like around $(\bar{x}, \bar{y}) \in \text{gph}\, M$ in* (7.5.2)*. Then the optimal value function* (7.5.1) *is locally Lipschitzian around $(\bar{x}, \bar{y})$.* △

Proof We know from the coderivative criterion (7.2.6) that F is Lipschitz-like around $(\bar{x}, \bar{y})$ if and only if $D^*F(\bar{x}, \bar{y})(0) = \{0\}$. Applying it to (7.5.4) tells us that the assumed Lipschitz-like property of the constraint mapping F in (7.5.1) ensures that $\partial^\infty \vartheta(\bar{x}) = \{0\}$. Furthermore, it easily follows from the assumptions made that the optimal value function is l.s.c. around $\bar{x}$. Thus ϑ is locally Lipschitzian around $\bar{x}$ by the characterization of this property given in (7.2.9). □

7.6 Problems of Lipschitzian Programming

Before deriving necessary optimality conditions in Lipschitzian problems of bilevel optimization in the subsequent sections, we devote this section to problems of single-level Lipschitzian programming. The results obtained here are based on the extremal principle and subdifferential characterization of local Lipschitzian functions while being instrumental for applications to bilevel programs given in Sect. 7.8.

The mathematical program under consideration here is as follows:

$$\begin{aligned} &\text{minimize } \varphi_0(x) \text{ subject to} \\ &\varphi_i(x) \le 0 \text{ for all } i = 1, \ldots, m, \end{aligned} \tag{7.6.1}$$

where the functions $\varphi_i : \mathbb{R}^n \to \mathbb{R}$, $i = 0, \ldots, m$, are locally Lipschitzian around the reference point $\bar{x}$. The next theorem provides necessary optimality conditions in problem (7.6.1) of Lipschitzian programming that are expressed in terms of the basic subdifferential (7.2.7).

Theorem 7.6.1 *Let $\bar{x}$ be a feasible solution to problem* (7.6.1) *that gives a local minimum to the cost function φ_0 therein. Then there exist multipliers $\lambda_0, \ldots, \lambda_m$*

satisfying the sign conditions

$$\lambda_i \geq 0 \text{ for all } i = 0, \ldots, m, \tag{7.6.2}$$

the nontriviality conditions

$$\lambda_0 + \ldots + \lambda_m \neq 0, \tag{7.6.3}$$

the complementary slackness conditions

$$\lambda_i \varphi_i(\bar{x}) = 0 \text{ whenever } i = 1, \ldots, m, \tag{7.6.4}$$

and the subdifferential Lagrangian inclusion

$$0 \in \sum_{i=0}^{m} \lambda_i \partial \varphi_i(\bar{x}). \tag{7.6.5}$$

Assume in addition that

$$\Big[\sum_{i \in I(\bar{x})} \lambda_i v_i = 0,\ \lambda_i \geq 0\Big] \Longrightarrow \Big[\lambda_i = 0 \text{ for all } i \in I(\bar{x})\Big] \tag{7.6.6}$$

whenever $v_i \in \partial \varphi_i(\bar{x})$ *with* $I(\bar{x}) := \big\{i \in \{1, \ldots, m\} \big| \ \varphi_i(\bar{x}) = 0\big\}$. *Then the necessary optimality conditions formulated above hold with* $\lambda_0 = 1$. △

Proof Supposing without loss of generality that $\varphi_0(\bar{x}) = 0$, consider the point $(\bar{x}, 0) \in \mathbb{R}^n \times \mathbb{R}^{m+1}$ and form the following system of $m + 2$ sets in the space $\mathbb{R}^n \times \mathbb{R}^{m+1}$:

$$\Omega_i := \big\{(x, \mu_0, \ldots, \mu_m) \in \mathbb{R}^n \times \mathbb{R}^{m+1} \big| (x, \mu_i) \in \operatorname{epi} \varphi_i\big\}, i = 0, \ldots, m, \tag{7.6.7a}$$

$$\Omega_{m+1} := \mathbb{R}^n \times \{0\}. \tag{7.6.7b}$$

It is obvious that $(\bar{x}, 0) \in \Omega_0 \cap \ldots \cap \Omega_{m+1}$ and that all the sets Ω_i, $i = 0, \ldots, m+1$, are locally closed around $(\bar{x}, 0)$. Furthermore, there exists a neighborhood U of the local minimizer $\bar{x}$ such that for any $\varepsilon > 0$ we find $\nu \in (0, \varepsilon)$ ensuring that

$$\big(\Omega_0 + a\big) \bigcap_{i=1}^{m+1} \Omega_i \cap \big(U \times \mathbb{R}^{m+1}\big) = \emptyset, \tag{7.6.8}$$

where $a := (0, \nu, 0, \ldots, 0) \in \mathbb{R}^n \times \mathbb{R}^{m+1}$ with $\nu \in \mathbb{R}$ standing at the first position after $0 \in \mathbb{R}^n$. Indeed, the negation of (7.6.8) contradicts the local minimality of $\bar{x}$ in (7.6.1). Having (7.6.8) gives us (7.3.1) for the set system in (7.6.7a), (7.6.7b) and thus verifies that $(\bar{x}, 0)$ is a locally extremal point of these sets. Applying now the extremal principle from Theorem 7.3.2 to $\{\Omega_0, \ldots, \Omega_{m+1}\}$ at $(\bar{x}, 0)$ with taking into account the structures of Ω_i, we get $v_0, \ldots, v_m \in \mathbb{R}^n$, $\lambda_0, \ldots, \lambda_m \in \mathbb{R}$, and $\xi = (\xi_1, \ldots, \xi_{m+1}) \in \mathbb{R}^{m+1}$ satisfying the inclusions

$$(v_i, -\lambda_i) \in N\big((\bar{x}, 0); \operatorname{epi} \varphi_i\big) \ \text{ for all } \ i = 0, \ldots, m \tag{7.6.9}$$

together with the relationships

$$\sum_{i=0}^{m} \|v_i\| + \sum_{i=0}^{m} |\lambda_i| + \|\xi\| \neq 0, \tag{7.6.10}$$

$$\sum_{i=0}^{m} v_i = 0, \ \text{ and } \lambda_i = \xi_{i+1} \ \text{ as } \ i = 0, \ldots, m. \tag{7.6.11}$$

It easily follows from (7.6.9) and the structure of the epigraphical sets in (7.6.9) that the sign conditions (7.6.2) are satisfied. If we suppose while arguing by contradiction to (7.6.3) that $\lambda_0 = \ldots = \lambda_m = 0$, then $\xi = 0$ by (7.6.11). Furthermore, the singular subdifferential criterion (7.2.9) for the local Lipschitz continuity of the functions φ_i as $i = 0, \ldots, m$ being combined with the sign conditions (7.6.2) and the definitions (7.2.7) and (7.2.8) of the basic and singular subdifferentials, respectively, tells us that the inclusions in (7.6.9) are equivalent to

$$v_i \in \lambda_i \partial \varphi_i(\bar{x}) \ \text{ for all } \ i = 0, \ldots, m, \tag{7.6.12}$$

and hence we get $v_0 = \ldots = v_m = 0$. This clearly violates (7.6.10) and thus verifies (7.6.3). The generalized Euler equation in (7.6.11) clearly reduces to the Lagrangian inclusion (7.6.5).

To check further the complementary slackness conditions in (7.6.4), fix $i \in \{1, \ldots, m\}$ and suppose that $\varphi_i(\bar{x}) < 0$. Then the continuity of φ_i at $\bar{x}$ ensures that the pair $(\bar{x}, 0)$ is an interior point of the epigraphical set $\operatorname{epi} \varphi_i$. It readily implies that $N((\bar{x}, 0); \operatorname{epi} \varphi_i) = (0, 0)$, and hence $\lambda_i = 0$. This yields $\lambda_i \varphi_i(\bar{x}) = 0$, which justifies (7.6.4).

To complete the proof of the theorem, it remains to check that the validity of (7.6.6) ensures that $\lambda_0 = 1$ in (7.6.5). We easily arrive at this assertion while arguing by contradiction. □

If the constraint functions φ_i, $i = 1, \ldots, m$, are smooth around the reference point $\bar{x}$, condition (7.6.6) clearly reduces to the classical Mangasarian-Fromovitz constraint qualification. It suggests us to label this condition (7.6.6) as the *generalized Mangasarian-Fromovitz constraint qualification*, or the *generalized MFCQ*.

7.7 Variational Approach to Bilevel Optimization

This section is devoted to describing some models of *bilevel programming* and a *variational approach* to them that involves *nondifferentiable* optimal value functions in single-level problems of parametric optimization.

Let us first consider the following problem of *parametric optimization* with respect to the decision variable $y \in \mathbb{R}^m$ under each fixed parameter $x \in \mathbb{R}^n$:

$$\text{minimize } \varphi(x, y) \text{ subject to } y \in F(x) \text{ with fixed } x \in \mathbb{R}^n, \tag{7.7.1}$$

where $\varphi\colon \mathbb{R}^n \times \mathbb{R}^m \to \mathbb{R}$ is the cost function and $F\colon \mathbb{R}^n \to \mathbb{R}^m$ is the constraint mapping in (7.7.1), which is called the *lower-level problem* of parametric optimization. Denoting by

$$S(x) := \text{argmin}\big\{\varphi(x, y)\big|\ y \in F(x)\big\} \tag{7.7.2}$$

the parameterized *solution set* for (7.7.1) for each $x \in \mathbb{R}^n$ and given yet another cost function $\psi\colon \mathbb{R}^n \times \mathbb{R}^m \to \mathbb{R}$, we consider the *upper-level* parametric optimization problem of minimizing $\psi(x, y)$ over the lower-level solution map $S\colon \mathbb{R}^n \rightrightarrows \mathbb{R}^m$ from (7.7.2) written as:

$$\text{minimize } \psi(x, y) \text{ subject to } y \in S(x) \text{ for each } x \in \Omega. \tag{7.7.3}$$

The *optimistic bilevel programming* model is defined by

$$\text{minimize } \mu(x) \text{ subject to } x \in \Omega, \text{ where } \mu(x) := \inf\big\{\psi(x, y)\big|\ y \in S(x)\big\}, \tag{7.7.4}$$

and where $\Omega \subset \mathbb{R}^n$ be a given constraint set. On the other hand, the *pessimistic bilevel programming* model is defined as follows:

$$\text{minimize } \eta(x) \text{ subject to } x \in \Omega, \text{ where } \eta(x) := \sup\big\{\psi(x, y)\big|\ y \in S(x)\big\}. \tag{7.7.5}$$

We refer the reader to [5–8, 11–16, 19, 30, 45, 47, 48], and the bibliographies therein for more details on both optimistic and pessimistic versions in bilevel programming, their local and global solutions as well as reformulations, modifications and relationships with other classes of optimization problems, theoretical and numerical developments, and various applications in finite-dimensional spaces. Investigations and applications of bilevel optimization problems in infinite dimensions can be found, e.g., in [1, 3, 8, 15, 17, 18, 20, 21, 23, 24, 34, 36, 37, 42, 43].

The main attention in this and subsequent sections is paid to the application of the machinery and results of variational analysis and generalized differentiation to problems of bilevel optimization by implementing the *value function approach.*

This approach is based on reducing bilevel programs to single-level problems of mathematical programming by using the nonsmooth *optimal value function* $\vartheta(x)$ of the lower-level problem defined in (7.5.1). Such a device was initiated by Outrata [35] for a particular class of bilevel optimization problems and was used by him for developing a numerical algorithm to solve bilevel programs. Then this approach was strongly developed by Ye and Zhu [44] who employed it to derive necessary optimality conditions for optimistic bilevel programs by using Clarke's generalized gradients of optimal value functions. More advanced necessary optimality conditions for optimistic bilevel programs in terms of the author's generalized differentiation reviewed in Sect. 7.2 were developed in [11, 12, 30, 34, 46, 47]. Optimality and stability conditions for pessimistic bilevel models were derived in [13, 16].

We restrict ourselves in what follows to implementing variational analysis and the aforementioned machinery of generalized differentiation within the value function approach to optimistic bilevel models with Lipschitzian data in finite-dimensional spaces. This allows us to most clearly communicate the basic variational ideas behind this approach, without additional technical complications. The variational results presented in the previous sections make our presentation self-contained and complete.

For simplicity we consider the optimistic bilevel model (7.7.4) with only *inequality constraints* on the lower and upper levels described by

$$F(x) := \{y \in \mathbb{R}^m \mid f_i(x, y) \le 0 \text{ for } i = 1, \ldots, r\}, \tag{7.7.6}$$

$$\Omega := \{x \in \mathbb{R}^n \mid g_j(x) \le 0 \text{ for } j = 1, \ldots, s\}. \tag{7.7.7}$$

The reduction of the bilevel program (7.7.4) with the constraints (7.7.6) and (7.7.7) to a single-level problem of nondifferentiable programming and deriving in this way necessary optimality conditions for it are given in the next section.

7.8 Optimality Conditions for Lipschitzian Bilevel Programs

The optimal value function (7.5.1) for the lower-level program (7.7.1) with the inequality constraints specified in (7.7.6) reads as

$$\vartheta(x) = \inf\{\varphi(x, y) \mid f_i(x, y) \le 0,\ i = 1, \ldots, r\}, \quad x \in \mathbb{R}^n. \tag{7.8.1}$$

With the upper-level cost function ψ given in (7.7.3) and the upper-level constraints taken from (7.7.7), consider the following single-level mathematical program with inequality constraints:

$$\begin{aligned} &\text{minimize } \psi(x, y) \text{ subject to } g_j(x) \le 0,\ j = 1, \ldots, s, \\ &f_i(x, y) \le 0,\ i = 1, \ldots, r, \text{ and } \varphi(x, y) \le \vartheta(x). \end{aligned} \tag{7.8.2}$$

We can easily observe that global optimal solutions to (7.8.2) agree with those to problem (7.7.4), (7.7.6), and (7.7.7). Although this is not always the case for local minimizers, it holds under the inner semicontinuity assumption on the solution map associated with the optimistic bilevel program (7.7.4); see [12, Proposition 6.9] for the precise statement.

Looking at (7.8.2), observe that this problem is of type (7.6.1) for which necessary optimality conditions are given in Theorem 7.6.1, provided that all the functions involved are locally Lipschitzian. However, the direct application of Theorem 7.6.1 to problem (7.8.2) is not efficient due to the structure of the last constraint therein defined via the lower-level optimal value function (7.8.1). Indeed, it has been realized in bilevel programming that this constraint prevents the fulfillment of conventional constraint qualifications; in particular, the generalized MFCQ (7.6.6). To avoid this obstacle, Ye and Zhu [44] introduced the following property postulating an appropriate behavior of the cost function in (7.8.2) with respect to linear perturbations of the constraint $\varphi(x, y) \leq \vartheta(x)$. Consider the problem:

$$\begin{aligned} &\text{minimize } \psi(x, y) \text{ subject to } g_j(x) \leq 0, \ j = 1, \ldots, s, \\ &f_i(x, y) \leq 0, \ i = 1, \ldots, r, \ \text{ and } \ \varphi(x, y) - \vartheta(x) + \nu = 0 \ \text{ as } \ \nu \in \mathbb{R}. \end{aligned} \tag{7.8.3}$$

Definition 7.8.1 Problem (7.8.2) is Partially Calm at its local optimal solution $(\bar{x}, \bar{y})$ if there exist a constant $\kappa > 0$ and a neighborhood U of $(\bar{x}, \bar{y}, 0) \in \mathbb{R}^n \times \mathbb{R}^m \times \mathbb{R}$ such that

$$\psi(x, y) - \psi(\bar{x}, \bar{y}) + \kappa|\nu| \geq 0 \tag{7.8.4}$$

for all the triples $(x, y, \nu) \in U$ feasible to (7.8.3). △

There are various efficient conditions, which ensure the fulfillment of the partial calmness property for (7.8.2). They include the uniform sharp minimum condition [44], linearity of the lower-level problem with respect to the decision variable [9], the kernel condition [30], etc. On the other hand, partial calmness may fail in rather common situations; see [19, 30] for more results and discussions on partial calmness and related properties.

The main impact of partial calmness to deriving necessary optimality conditions for (7.8.2) is its equivalence to the possibility of transferring the troublesome constraint $\varphi(x, y) \leq \vartheta(x)$ into the *penalized* cost function as in the following proposition.

Proposition 7.8.2 *Let problem (7.8.2) be partially calm at its local optimal solution $(\bar{x}, \bar{y})$ with ψ being continuous at this point. Then $(\bar{x}, \bar{y})$ is a local optimal solution to the penalized problem*

$$\begin{aligned} &\text{minimize } \psi(x, y) + \kappa\big(\varphi(x, y) - \vartheta(x)\big) \text{ subject to} \\ &g_j(x) \le 0, \; j = 1, \ldots, s, \text{ and } f_i(x, y) \le 0, \; i = 1, \ldots, r, \end{aligned} \tag{7.8.5}$$

where the number $\kappa > 0$ is taken from (7.8.4). △

Proof Taking κ and U from Definition 7.8.1 and using the continuity of ψ at $(\bar{x}, \bar{y})$, we find $\gamma > 0$ and $\eta > 0$ with $\tilde{U} := [(\bar{x}, \bar{y}) + \eta\mathbb{B}] \times (-\gamma, \gamma) \subset U$ and

$$|\psi(x, y) - \psi(\bar{x}, \bar{y})| \le \kappa\gamma \text{ for all } (x, y) - (\bar{x}, \bar{y}) \in \eta\mathbb{B}.$$

Let us employ it to verify that

$$\psi(x, y) - \psi(\bar{x}, \bar{y}) + \kappa\big(\varphi(x, y) - \vartheta(x)\big) \ge 0 \text{ if } g_j(x) \le 0, \; j = 1, \ldots, s, \tag{7.8.6}$$

and $(x, y) \in [(\bar{x}, \bar{y}) + \eta\mathbb{B}] \cap \operatorname{gph} F$ with F taken from (7.7.6). If $(x, y, \vartheta(x) - \varphi(x, y)) \in \tilde{U}$, then (7.8.6) follows from (7.8.4). In the remaining case where $(x, y, \vartheta(x) - \varphi(x, y)) \notin \tilde{U}$, we get that

$$\varphi(x, y) - \vartheta(x) \ge \gamma, \text{ and hence } \kappa\big(\varphi(x, y) - \vartheta(x)\big) \ge \kappa\gamma,$$

which also yields (7.8.6) by $\psi(x, y) - \psi(\bar{x}, \bar{y}) \ge -\kappa\gamma$. The feasibility of $(\bar{x}, \bar{y})$ to (7.8.2) tells us that $\varphi(\bar{x}, \bar{y}) - \vartheta(\bar{x}) = 0$, which thus verifies the claimed statement. □

Thus the imposed partial calmness allows us to deduce the original problem of optimistic bilevel optimization to the single-level mathematical program (7.8.5) with conventional inequality constraints, where the troublesome term $\varphi(x, y) - \vartheta(x)$ enters the penalized cost function. To derive necessary optimality conditions for (7.8.5), let us reformulate the generalized MFCQ (7.6.6) in conventional bilevel terms as in the case of bilevel programs with smooth data [6].

We say that $(\bar{x}, \bar{y}) \in \mathbb{R}^n \times \mathbb{R}^m$ is *lower-level regular* if it satisfies the generalized MFCQ in the lower-level problem (7.7.1). This means due to the structure of (7.7.1) that

$$\Big[\sum_{i \in I(\bar{x}, \bar{y})} \lambda_i v_i = 0, \; \lambda_i \ge 0\Big] \Longrightarrow \Big[\lambda_i = 0 \text{ for all } i \in I(\bar{x}, \bar{y})\Big] \tag{7.8.7}$$

whenever $(u_i, v_i) \in \partial f_i(\bar{x}, \bar{y})$ with some $u \in \mathbb{R}^n$ and

$$I(\bar{x}, \bar{y}) := \big\{i \in \{1, \ldots, r\} \big| \; f_i(\bar{x}, \bar{y}) = 0\big\}.$$

Similarly, a point $\bar{x} \in \mathbb{R}^n$ satisfying the upper-level constraints in (7.7.7) is *upper-level regular* if

$$\Big[0 \in \sum_{j \in J(\bar{x})} \mu_j \partial g_j(\bar{x}),\ \mu_j \geq 0\Big] \Longrightarrow \Big[\mu_j = 0 \text{ whenever } j \in J(\bar{x})\Big] \tag{7.8.8}$$

with the active constraint indexes $J(\bar{x}) := \big\{j \in \{1,\ldots,s\} \big|\ g_j(\bar{x}) = 0\big\}$.

Now we are ready for the application of Theorem 7.6.1 to the optimistic bilevel program in the equivalent form (7.8.5). To proceed, we have to verify first that the optimal value function $\vartheta(x)$ in the cost function of (7.8.5) is locally Lipschitzian around the reference point and then to be able to upper estimate the basic subdifferential (7.2.7) of the function $-\vartheta(\cdot)$. Since the basic subdifferential $\partial\vartheta(\bar{x})$ does not possess the plus-minus symmetry while its convex hull $\text{co}\,\partial\vartheta(\bar{x})$ does, we need to convexify in the proof the set on the right-hand side of the upper estimate in (7.5.3). In this way we arrive at the following major result.

Theorem 7.8.3 *Let $(\bar{x}, \bar{y})$ be a local optimal solution to the optimistic bilevel program in the equivalent form* (7.8.2) *with the lower-level optimal value function $\vartheta(x)$ defined in* (7.8.1)*. Assume that all the functions φ, ψ, f_i, g_j are locally Lipschitzian around the reference point, that the lower-level solution map S from* (7.7.2) *is inner semicontinuous at $(\bar{x}, \bar{y})$, that the lower-level regularity* (7.8.7) *and upper-level regularity* (7.8.8) *conditions hold, and that problem* (7.8.2) *is partially calm at $(\bar{x}, \bar{y})$ with constant $\kappa > 0$. Then there exist multipliers $\lambda_1, \ldots, \lambda_r$, $\mu_1, \ldots, \mu_s$, and $\nu_1, \ldots, \nu_r$ satisfying the sign and complementary slackness conditions*

$$\lambda_i \geq 0,\ \ \lambda_i f_i(\bar{x}, \bar{y}) = 0 \ \textit{for}\ i = 1, \ldots, r, \tag{7.8.9}$$

$$\mu_j \geq 0,\ \ \mu_j g_j(\bar{x}) = 0 \ \textit{for}\ j = 1, \ldots, s, \tag{7.8.10}$$

$$\nu_i \geq 0,\ \ \nu_i f_i(\bar{x}, \bar{y}) = 0 \ \textit{for}\ i = 1, \ldots, r, \tag{7.8.11}$$

together with the following relationships, which involve some vector $u \in \text{co}\,\partial\vartheta(\bar{x})$:

$$(u, 0) \in \text{co}\,\partial\varphi(\bar{x}, \bar{y}) + \sum_{i=1}^{r} \nu_i \,\text{co}\,\partial f_i(\bar{x}, \bar{y}), \tag{7.8.12}$$

$$(u, 0) \in \partial\varphi(\bar{x}, \bar{y}) + \kappa^{-1}\partial\psi(\bar{x}, \bar{y}) + \sum_{i=1}^{r} \lambda_i \partial f_i(\bar{x}, \bar{y}) + \sum_{j=1}^{s} \mu_j\Big(\partial g_j(\bar{x}), 0\Big). \tag{7.8.13}$$

△

Proof It follows from Proposition 7.8.2 that $(\bar{x}, \bar{y})$ is a local minimizer of the mathematical program (7.8.5) with inequality constraints. To show that it belongs to problems of Lipschitzian programming considered in Sect. 7.6, we need to check that the optimal value function (7.8.1) is locally Lipschitzian around $\bar{x}$ under the assumptions made. This function is clearly l.s.c. around $\bar{x}$, and thus its Lipschitz continuity around this point is equivalent to the condition $\partial^\infty\vartheta(\bar{x}) = \{0\}$. It follows from the upper estimate (7.5.4) of Theorem 7.5.1 and the assumed inner semicontinuity of the solution map (7.7.2) at $(\bar{x}, \bar{y})$ that

$$\partial^\infty\vartheta(\bar{x}) \subset D^*F(\bar{x})(0) \text{ with } F(x) = \big\{y \in \mathbb{R}^m \big|\ f_i(x, y) \le 0,\ i = 1, \ldots, r\big\}.$$

Corollary 7.5.2 tells us that the local Lipschitz continuity of ϑ around $\bar{x}$ follows from the Lipschitz-like property of the mapping $F\colon \mathbb{R}^n \rightrightarrows \mathbb{R}^m$. Since

$$\operatorname{gph} F = \big\{(x, y) \in \mathbb{R}^n \times \mathbb{R}^m \big|\ f_i(x, y) \le 0,\ i = 1, \ldots, r\big\},$$

we deduce that $D^*F(\bar{x}, \bar{y})(0) = \{0\}$ from the assumed lower-level regularity due to the coderivative definition and the normal come intersection rule in Theorem 7.4.1. Thus F is Lipschitz-like around $(\bar{x}, \bar{y})$ by the coderivative criterion (7.2.6), and the Lipschitz continuity of $\vartheta(\cdot)$ is verified.

Now we can apply to problem (7.8.5) the necessary optimality conditions for Lipschitzian programs obtained in Theorem 7.6.1. The assumed lower-level regularity and upper-level regularity clearly imply that the generalized MFCQ condition (7.6.6) holds. Thus there are multipliers $\lambda_1, \ldots, \lambda_r$ and $\mu_1, \cdots, \mu_s$ satisfying the sign and complementary slackness conditions in (7.8.9) and (7.8.10) for which we have

$$\begin{aligned} 0 \in\ & \partial\psi(\bar{x}, \bar{y}) + \kappa\partial\varphi(\bar{x}, \bar{y}) + \Big(\kappa\partial(-\vartheta)(\bar{x}), 0\Big) \\ & + \sum_{i=1}^{r} \lambda_i \partial f_i(\bar{x}, \bar{y}) + \sum_{j=1}^{s} \mu_j \Big(\partial g_j(\bar{x}), 0\Big). \end{aligned} \tag{7.8.14}$$

To estimate $\partial(-\vartheta)(\bar{x})$ in (7.8.14), recall that

$$\partial(-\vartheta)(\bar{x}) \subset \overline{\partial}(-\vartheta)(\bar{x}) = -\overline{\partial}\vartheta(\bar{x}) = -\operatorname{co}\partial\vartheta(\bar{x}),$$

where $\overline{\partial}$ stands for Clarke's generalized gradient of locally Lipschitzian functions that possesses the plus-minus symmetry property [4]. Using it in (7.8.14), we get $u \in \operatorname{co}\partial\vartheta(\bar{x})$ such that

$$\kappa(u, 0) \in \partial\psi(\bar{x}, \bar{y}) + \partial\varphi(\bar{x}, \bar{y}) + \sum_{i=1}^{r} \lambda_i \partial f_i(\bar{x}, \bar{y}) + \sum_{j=1}^{s} \Big(\mu_j \partial g_j(\bar{x}), 0\Big). \tag{7.8.15}$$

Applying the convexified subdifferential estimates (7.5.3) from Theorem 7.5.1 to the optimal value function (7.8.1) allows us to find multipliers $\nu_1, \ldots, \nu_r$ satisfying the sign and complementary slackness conditions in (7.8.11) that ensure the validity of (7.8.12). To verify finally (7.8.13), we divide (7.8.15) by $\kappa > 0$ with keeping the same notation for the scaled multipliers λ_i and μ_j. $\square$

The next section presents an independent set of necessary optimality conditions for optimistic bilevel programs with Lipschitzian data that are obtained without using any convexification while employing instead yet another variational device and subdifferential calculus rule.

7.9 Bilevel Optimization via Subdifferential Difference Rule

Considering the single-level problem (7.8.5), which we are finally dealing with while deriving necessary optimality conditions for optimistic bilevel programs, note that the objective therein contains the *difference* of two nonsmooth functions. The basic subdifferential (7.2.7) does not possesses any special rule for difference of nonsmooth functions, but the regular subdifferential (7.2.11) does, as was first observed in [33] by using a smooth variational description of regular subgradients. Here we employ this approach to establish necessary optimality conditions for Lipschitzian bilevel programs that are different from those in Theorem 7.8.3.

The derivation of these necessary optimality conditions is based on the following two results, which are certainly of their independent interest. The first one provides a *smooth variational description* of regular subgradients of arbitrary functions $\varphi\colon \mathbb{R}^n \to \overline{\mathbb{R}}$.

Lemma 7.9.1 *Let $\varphi\colon \mathbb{R}^n \to \overline{\mathbb{R}}$ be finite at $\bar{x}$, and let $v \in \hat{\partial}\varphi(\bar{x})$. Then there exists a neighborhood U of $\bar{x}$ and a function $\psi\colon U \to \mathbb{R}$ such that $\psi(\bar{x}) = \varphi(\bar{x})$, that ψ is Fréchet differentiable at $\bar{x}$ with $\nabla\psi(\bar{x}) = v$, and that the difference $\psi - \varphi$ achieves at $\bar{x}$ its local maximum on U.* $\triangle$

Proof We proceed geometrically due to the relationship

$$\hat{\partial}\varphi(\bar{x}) = \big\{v \in \mathbb{R}^n \big|\ (v, -1) \in \hat{N}\big((\bar{x}, \varphi(\bar{x})); \operatorname{epi}\varphi\big)\big\},$$

which reduces the claimed assertion to the following: $w \in \hat{N}(\bar{z}; \Omega)$ if and only if there exists a neighborhood U of $\bar{z}$ and a function $\psi\colon \mathbb{R}^n \to \mathbb{R}$ such that ψ is Fréchet differentiable at $\bar{z}$ with $\nabla\psi(\bar{z}) = w$ while achieving at $\bar{z}$ its local maximum relative to Ω.

To verify the latter, observe that for any $\psi\colon U \to \mathbb{R}$ satisfying the listed properties we get

$$\psi(z) = \psi(\bar{z}) + \langle w, z - \bar{z}\rangle + o(\|z - \bar{z}\|) \le \psi(\bar{z}) \quad \text{whenever } z \in U.$$

It shows that $\langle w, z - \bar{z}\rangle + o(\|z - \bar{z}\|) \le 0$, and thus $w \in \hat{N}(\bar{z}; \Omega)$ by definition (7.2.2). Conversely, pick $w \in \hat{N}(\bar{z}; \Omega)$ and define the function

$$\psi(z) := \begin{cases} \min\{0, \langle w, z - \bar{z}\rangle\} & \text{if } z \in \Omega, \\ \langle w, z - \bar{z}\rangle & \text{otherwise.} \end{cases}$$

It is easy to check that this function enjoys all the properties listed above. □

The second lemma gives us the aforementioned *difference rule* for regular subgradients.

Lemma 7.9.2 *Consider two arbitrary functions $\varphi_1, \varphi_2 \colon \mathbb{R}^n \to \overline{\mathbb{R}}$ that are finite at $\bar{x}$ and assume that $\hat{\partial}\varphi_2(\bar{x}) \neq \emptyset$. Then we have the inclusions*

$$\hat{\partial}(\varphi_1 - \varphi_2)(\bar{x}) \subset \bigcap_{v \in \hat{\partial}\varphi_2(\bar{x})} \Big[\hat{\partial}\varphi_1(\bar{x}) - v\Big] \subset \hat{\partial}\varphi_1(\bar{x}) - \hat{\partial}\varphi_2(\bar{x}). \tag{7.9.1}$$

It implies, in particular, that any local minimizer $\bar{x}$ of the difference function $\varphi_1 - \varphi_2$ satisfies the necessary optimality condition

$$\hat{\partial}\varphi_2(\bar{x}) \subset \hat{\partial}\varphi_1(\bar{x}). \tag{7.9.2}$$

△

Proof Starting with (7.9.1), pick $v \in \hat{\partial}\varphi_2(\bar{x})$. Then the smooth variational description of v from Lemma 7.9.1 gives us $\psi \colon U \to \mathbb{R}$ on a neighborhood U of $\bar{x}$ that is differentiable at $\bar{x}$ with

$$\psi(\bar{x}) = \varphi_2(\bar{x}), \quad \nabla\psi(\bar{x}) = v, \quad \text{and } \psi(x) \le \varphi_2(x) \text{ whenever } x \in U.$$

Fix further an arbitrary vector $w \in \hat{\partial}(\varphi_1 - \varphi_2)(\bar{x})$ and for any $\varepsilon > 0$ find $\gamma > 0$ such that

$$\begin{aligned} \langle w, x - \bar{x}\rangle &\le \varphi_1(x) - \varphi_2(x) - \big(\varphi_1(\bar{x}) - \varphi_2(\bar{x})\big) + \varepsilon\|x - \bar{x}\| \\ &\le \varphi_1(x) - \psi(x) - \big(\varphi_1(x) - \psi(x)\big) + \varepsilon\|x - \bar{x}\| \end{aligned}$$

if $\|x - \bar{x}\| \le \gamma$. Due to the differentiability of ψ at $\bar{x}$ we use the elementary sum rule for the regular subgradients (see, e.g., [29, Proposition 1.107(i)]) and immediately get the relationships

$$w \in \hat{\partial}(\varphi_1 - \psi)(\bar{x}) = \hat{\partial}\varphi_1(\bar{x}) - \nabla\psi(\bar{x}) = \hat{\partial}\varphi_1(\bar{x}) - v,$$

which clearly verify both inclusions in (7.9.1). Picking further any $v \in \hat{\partial}\varphi_2(\bar{x})$, we deduce from (7.9.1) and the obvious Fermat stationary rule via regular subgradients that

$$0 \in \hat{\partial}(\varphi_1 - \varphi_2)(\bar{x}) \subset \hat{\partial}\varphi_1(\bar{x}) - v.$$

It shows that $v \in \hat{\partial}\varphi_1(\bar{x})$ and thus verifies the fulfillment of (7.9.2). □

Now we are in a position to derive refined necessary optimality conditions for optimistic bilevel programs with Lipschitzian data.

Theorem 7.9.3 *Let $(\bar{x}, \bar{y})$ be a local optimal solution to the optimistic bilevel program in the equivalent form (7.8.2) without upper level constraints. Suppose that all the functions φ, ψ, f_i are locally Lipschitzian around the reference point, that the lower-level solution map S in (7.7.2) is inner semicontinuous at $(\bar{x}, \bar{y})$, that the lower-level regularity (7.8.7) condition holds, and that problem (7.8.2) is partially calm at $(\bar{x}, \bar{y})$ with constant $\kappa > 0$. Assume in addition that $\hat{\partial}\vartheta(\bar{x}) \neq \emptyset$ for the optimal value function (7.8.1). Then there exists a vector $u \in \hat{\partial}\vartheta(\bar{x})$ together with multipliers $\lambda_1, \ldots, \lambda_r$ and $\nu_1, \ldots, \nu_r$ for $i = 1, \ldots, r$ satisfying the sign and complementary slackness conditions in (7.8.11) and (7.8.9), respectively, such that*

$$(u, 0) \in \partial\varphi(\bar{x}, \bar{y}) + \sum_{i=1}^{r} \nu_i \partial f_i(\bar{x}, \bar{y}), \tag{7.9.3}$$

$$(u, 0) \in \partial\varphi(\bar{x}, \bar{y}) + \kappa^{-1}\partial\psi(\bar{x}, \bar{y}) + \sum_{i=1}^{r} \lambda_i \partial f_i(\bar{x}, \bar{y}). \tag{7.9.4}$$

△

Proof We get from the partial calmness penalization in Proposition 7.8.2 employed together with the infinite penalization of the lower-level constraints that $(\bar{x}, \bar{y})$ a local minimizer for the unconstrained optimization problem

$$\text{minimize } \psi(x, y) + \kappa\big(\varphi(x, y) - \vartheta(x)\big) + \delta\big((x, y); \text{gph } F\big) \tag{7.9.5}$$

with the mapping $F\colon \mathbb{R}^n \rightrightarrows \mathbb{R}^m$ defined in (7.7.6). Then applying to (7.9.5) the difference rule (7.9.2) from Proposition 7.9.2 gives us the inclusion

$$\big(\kappa\hat{\partial}\vartheta(\bar{x}), 0\big) \subset \hat{\partial}\big(\psi(\cdot) + \kappa\varphi(\cdot) + \delta(\cdot; \text{gph } F)\big)(\bar{x}, \bar{y}). \tag{7.9.6}$$

At the same time it follows from the proof of Theorem 7.5.1 that

$$\big(\hat{\partial}\vartheta(\bar{x}), 0\big) \subset \hat{\partial}\big(\varphi(\cdot) + \delta(\cdot; \text{gph } F)\big)(\bar{x}, \bar{y}). \tag{7.9.7}$$

Replacing $\hat{\partial}$ by the larger ∂ on the right-hand sides of (7.9.6) and (7.9.7) and then using the basic subdifferential sum rule from Theorem 7.4.2 yield the inclusions

$$\begin{aligned} &\Big(\kappa\hat{\partial}\vartheta(\bar{x}),0\Big) \subset \partial\psi(\bar{x},\bar{y}) + \kappa\partial\varphi(\bar{x},\bar{y}) + N\big((\bar{x},\bar{y});\operatorname{gph} F\big),\\ &\Big(\hat{\partial}\vartheta(\bar{x}),0\Big) \subset \partial\varphi(\bar{x},\bar{y}) + N\big((\bar{x},\bar{y});\operatorname{gph} F\big). \end{aligned} \tag{7.9.8}$$

Employing in these inclusions Theorem 7.4.1 under the lower-level regularity of $(\bar{x},\bar{y})$ with the usage of the singular subdifferential characterization of Lipschitzian functions in (7.2.9), we get

$$N\big((\bar{x},\bar{y});\operatorname{gph} F\big) \subset \bigcup\Big\{\sum_{i=1}^{r}\lambda_i\partial f_i(\bar{x},\bar{y})\Big|\ \lambda_i \geq 0,\ \lambda_i f_i(\bar{x},\bar{y}) = 0\ \text{ as }\ i = 1,\ldots,r\Big\}.$$

It allows us to deduce from (7.9.8) the existence of $u \in \hat{\partial}\vartheta(\bar{x})$ ensuring the validity of (7.9.3) and

$$\kappa(u,0) \in \partial\psi(\bar{x},\bar{y}) + \kappa\partial\varphi(\bar{x},\bar{y}) + \sum_{i=1}^{r}\lambda_i\partial f_i(\bar{x},\bar{y}).$$

Dividing the latter by $\kappa > 0$, we arrive at (7.9.4) and thus complete the proof of the theorem. □

Observe that we always have $\hat{\partial}\vartheta(\bar{x}) \neq \emptyset$ if the optimal value function (7.8.1) is *convex*, which is surely the case when all the functions φ and f_i therein are convex. If in addition the upper-level data are also convex, more general results were derived in [17] for problems of *semi-infinite programming* with arbitrary number of inequality constraints in locally convex topological vector spaces by reducing them to problems of *DC programming* with objectives represented as *differences of convex* functions. Note further that the necessary optimality conditions obtained in Theorems 7.8.3 and 7.9.3 are independent of each other even in the case of bilevel programs with smooth data. In particular, we refer the reader to [30, Example 6.24] for illustrating this statement and for using the obtained results to solve smooth bilevel programs. Finally, we mention the possibility to replace the inner semicontinuity assumption on the solution map $S(x)$ imposed in both Theorems 7.8.3 and 7.9.3 by the *uniform boundedness* of this map in finite-dimensions, or by its *inner semicompactness* counterpart in infinite-dimensional spaces; cf. [11, 16, 30, 34] for similar transitions in various bilevel settings.

7.10 Concluding Remarks and Open Questions

In this self-contained chapter of the book we described a variational approach to bilevel optimization with its implementation to deriving advanced necessary optimality conditions for optimistic bilevel programs in finite-dimensional spaces. The entire machinery of variational analysis and generalized differentiation (including the fundamental extremal principle, major calculus rules, and subdifferentiation of optimal value functions), which is needed for this device, is presented here with the proofs. The given variational approach definitely has strong perspectives for further developments. Let us briefly discuss some open questions in this direction.

- The major difference between the optimistic model (7.7.4) and pessimistic model (7.7.5) in bilevel programming is that the latter invokes the *supremum* marginal/optimal value function instead of the infimum type in (7.7.4). Subdifferentiation of the supremum marginal functions is more involved in comparison with that of the infimum type. Some results in this vein for problems with Lipschitzian data can be distilled from the recent papers [31, 32, 38], while their implementation in the framework of pessimistic bilevel programs is a challenging issue.
- The given proof of the necessary optimality conditions for bilevel programs in Theorem 7.8.3 requires an upper estimate of $\partial(-\vartheta)(\bar{x})$, which cannot be directly derived from that for $\partial\vartheta(\bar{x})$ since $\partial(-\vartheta)(\bar{x}) \neq -\partial\vartheta(\bar{x})$. To obtain such an estimate, we used the subdifferential convexification and the fact that the convexified/Clarke subdifferential of Lipschitz continuous functions possesses the plus-minus symmetry. However, there is a nonconvex subgradient set that is much smaller than Clarke's one while having this symmetry. It is the *symmetric subdifferential* $\partial^0\vartheta(\bar{x}) := \partial\vartheta(\bar{x}) \cup (-\partial(-\vartheta)(\bar{x}))$, which enjoys full calculus induced by the basic one. Efficient evaluations of $\partial^0\vartheta(\bar{x})$ for infimum and supremum marginal functions would lead us to refined optimality conditions for both optimistic and pessimistic models in bilevel optimization.
- The partial calmness property used in both Theorems 7.8.3 and 7.9.3 seems to be rather restrictive when the lower-level problem is nonlinear with respect to the decision variable. It is a challenging research topic to relax this assumption and to investigate more the uniform weak sharp minimum property and its modifications that yield partial calmness.
- One of the possible ways to avoid partial calmness in bilevel programming is as follows. Having the solution map $S\colon \mathbb{R}^n \rightrightarrows \mathbb{R}^m$ to the lower-level problem, consider the constrained upper-level problem given by

$$\text{minimize } \Psi(x) := \psi\big(x, S(x)\big) \text{ subject to } x \in \Omega, \tag{7.10.1}$$

where $\psi : \mathbb{R}^n \times \mathbb{R}^m \to \overline{\mathbb{R}}$ is the cost function on the upper level with the upper-level constraint set $\Omega \subset \mathbb{R}^n$, and where the minimization of $\Psi\colon \mathbb{R}^n \rightrightarrows \mathbb{R}$ is understood with respect to the standard order on $\mathbb{R}$. Then (7.10.1) is a problem of *set-valued optimization* for which various necessary optimality conditions of the coderivative and subdifferential types can be found in [30] and the references therein. Evaluating the coderivatives and subdifferentials of the composition $\psi(x, S(x))$ in terms of the given lower-level and upper-level data of bilevel programs would lead us to necessary optimality conditions in both optimistic and pessimistic models. There are many open questions arising in efficient realizations of this approach for particular classes of problems in bilevel optimization even with smooth initial data. We refer the reader to the paper by Zemkoho [47] for some recent results and implementations in this direction.

Note that a somewhat related approach to bilevel optimization was developed in [1], where a lower-level problem was replaced by the corresponding KKT system described by a certain *generalized equation* of the Robinson type [39]. Applying to the latter necessary optimality conditions for upper-level problems with such constraints allowed us to establish verifiable results for the original nonsmooth problem of bilevel programming.

- Henrion and Surowiec suggested in [19] a novel approach to derive necessary optimality conditions for optimistic bilevel programs with C^2-smooth data and convex lower-level problems. Their approach used a reduction to *mathematical programs with equilibrium constraints* (MPECs) and allowed them to significantly relax the partial calmness assumption. Furthermore, in this way they obtained new necessary optimality conditions for the bilevel programs under consideration, which are described via the Hessian matrices of the program data.

  A challenging direction of the future research is to develop the approach and results from [19] to nonconvex bilevel programs with nonsmooth data. It would be natural to replace in this way the classical Hessian in necessary optimality conditions by the generalized one (known as the *second-order subdifferential*) introduced by the author in [27] and then broadly employed in variational analysis and its applications; see, e.g., [30] and the references therein.
- As it has been long time realized, problems of bilevel optimization are generally *ill-posed*, which creates serious computational difficulties for their numerical solving; see, e.g., [5, 6, 47] for more details and discussions. Furthermore, various regularization methods and approximation procedures devised in order to avoid ill-posedness have their serious drawbacks and often end up with approximate solutions, which may be far enough from optimal ones. Thus it seems appealing to deal with ill-posed bilevel programs how they are and to develop *numerical algorithms* based on the obtained necessary optimality conditions. Some results in this vein are presented in [47] with involving necessary optimality conditions of the type discussed here, while much more work is required to be done in this very important direction with practical applications.

Acknowledgements This research was partly supported by the USA National Science Foundation under grants DMS-1512846 and DMS-1808978, by the USA Air Force Office of Scientific Research under grant 15RT04, and by Australian Research Council Discovery Project DP-190100555.

The author is indebted to an anonymous referee and the editors for their very careful reading of the original presentation and making many useful suggestions and remarks, which resulted in essential improvements of the paper.

References

1. T.Q. Bao, P. Gupta, B.S. Mordukhovich, Necessary conditions for multiobjective optimization with equilibrium constraints. J. Optim. Theory Appl. **135**, 179–203 (2007)
2. J.F. Bard, *Practical Bilevel Optimization: Algorithms and Applications* (Kluwer, Dordrecht, 1998)
3. F. Benita, S. Dempe, P. Mehlitz, Bilevel optimal control problems with pure state constraints and finite-dimensional lower level. SIAM J. Optim. **26**, 564–588 (2016)
4. F.H. Clarke, *Optimization and Nonsmooth Analysis* (Wiley-Interscience, New York, 1983)
5. B. Colson, P. Marcotte, G. Savard, An overview of bilevel optimization. Ann. Oper. Res. **153**, 235–256 (2007)
6. S. Dempe, *Foundations of Bilevel Programming* (Kluwer, Dordrecht, 2002)
7. S. Dempe, J. Dutta, Is bilevel programming a special case of mathematical programming with complementarity constraints? Math. Program. **131**, 37–48 (2012)
8. S. Dempe, P. Mehlitz, Semivectorial bilevel programming versus scalar bilevel programming. Optimization (2019). https://doi.org/10.1080/02331934.2019.1625900
9. S. Dempe, A.B. Zemkoho, The bilevel programming problem: reformulations, constraint qualifications and optimality conditions. Math. Program. **138**, 447–473 (2013)
10. S. Dempe, A.B. Zemkoho, KKT reformulation and necessary conditions for optimality in nonsmooth bilevel optimization. SIAM J. Optim. **24**, 1639–1669 (2014)
11. S. Dempe, J. Dutta, B.S. Mordukhovich, New necessary optimality conditions in optimistic bilevel programming. Optimization **56**, 577–604 (2007)
12. S. Dempe, B.S. Mordukhovich, A.B. Zemkoho, Sensitivity analysis for two-level value functions with applications to bilevel programming. SIAM J. Optim. **22**, 1309–1343 (2012)
13. S. Dempe, B.S. Mordukhovich, A.B. Zemkoho, Necessary optimality conditions in pessimistic bilevel programming. Optimization **63**, 505–533 (2014)
14. S. Dempe, V. Kalashnikov, G.A. Pérez-Valdés, N. Kalashnikova, *Bilevel Programming Problems* (Springer, New York, 2015)
15. S. Dempe, F. Harder, P. Mehlitz, G. Wachsmuth, Solving inverse optimal control problems via value functions to global optimality. J. Global Optim. **74**, 297–325 (2019)
16. S. Dempe, B.S. Mordukhovich, A.B. Zemkoho, Two-level value function approach to optimistic and pessimistic bilevel programs. Optimization **68**, 433–455 (2019)
17. N. Dinh, B.S. Mordukhovich, T.T.A. Nghia, Subdifferentials of value functions and optimality conditions for some classes of DC and bilevel infinite and semi-infinite programs. Math. Program. **123**, 101–138 (2010)
18. F. Harder, G. Wachsmuth, Optimality conditions for a class of inverse optimal control problems with partial differential equations. Optimization **68**, 615–643 (2019)
19. R. Henrion, T. Surowiec, On calmness conditions in convex bilevel programming. Appl. Anal. **90**, 951–970 (2011)
20. G. Holler, K. Kunisch, R.C. Barnard, A bilevel approach for parameter learning in inverse problems. Inverse Probl. **34**, 1–28 (2018)
21. K. Knauer, C. Büskens, Hybrid solution methods for bilevel optimal control problems with time dependent coupling, in *Recent Advances in Optimization and Its Applications in Engineering*, ed. by M. Diehl et al. (Springer, Berlin, 2010), pp. 237–246
22. A.Y. Kruger, B.S. Mordukhovich, Extremal points and the Euler equation in nonsmooth optimization. Dokl. Akad. Nauk BSSR **24**, 684–687 (1980)

23. M.B. Lignola, J. Morgan, Inner regularizations and viscosity solutions for pessimistic bilevel optimization problems. J. Optim. Theory Appl. **173**, 183–202 (2017)
24. P. Mehlitz, G. Wachsmuth, Weak and strong stationarity in generalized bilevel programming and bilevel optimal control. Optimization **65**, 907–935 (2017)
25. K. Mombaur, A. Truong, J.-P. Laumond, From human to humanoid locomotion–an inverse optimal control approach. Autom. Robots **28**, 369–383 (2010)
26. B.S. Mordukhovich, Maximum principle in problems of time optimal control with nonsmooth constraints. J. Appl. Math. Mech. **40**, 960–969 (1976)
27. B.S. Mordukhovich, Sensitivity analysis in nonsmooth optimization, in *Theoretical Aspects of Industrial Design*, ed. by D.A. Field, V. Komkov. SIAM Proc. Appl. Math., vol. 58 (Philadelphia, Pennsylvania, 1992), pp. 32–46
28. B.S. Mordukhovich, Complete characterization of openness, metric regularity, and Lipschitzian properties of multifunctions. Trans. Am. Math. Soc. **340**, 1–35 (1993)
29. B.S. Mordukhovich, *Variational Analysis and Generalized Differentiation, I: Basic Theory, II: Applications* (Springer, Berlin, 2006)
30. B.S. Mordukhovich, *Variational Analysis and Applications* (Springer, Cham, 2018)
31. B.S. Mordukhovich, T.T.A. Nghia, Subdifferentials of nonconvex supremum functions and their applications to semi-infinite and infinite programs with Lipschitzian data. SIAM J. Optim. **23**, 406–431 (2013)
32. B.S. Mordukhovich, T.T.A. Nghia, Nonsmooth cone-constrained optimization with applications to semi-infinite programming. Math. Oper. Res. **39**, 301–337 (2014)
33. B.S. Mordukhovich, N.M. Nam, N.D. Yen, Fréchet subdifferential calculus and optimality conditions in nondifferentiable programming. Optimization **55**, 685–396 (2006)
34. B.S. Mordukhovich, N.M. Nam, H.M. Phan, Variational analysis of marginal functions with applications to bilevel programming. J. Optim. Theory Appl. **152**, 557–586 (2011)
35. J.V. Outrata, On the numerical solution of a class of Stackelberg problems. ZOR–Methods Models Oper. Res. **34**, 255–277 (1990)
36. K.D. Palagachev, M. Gerdts, Exploitation of the value function in a bilevel optimal control problem, in *System Modeling and Optimization*, ed. by L. Bociu et al. (Springer, Cham, 2016), pp. 410–419
37. K.D. Palagachev, M. Gerdts, Numerical approaches towards bilevel optimal control problems with scheduling tasks, in *Math for the Digital Factory*, ed. by L. Ghezzio et al. (Springer, Cham, 2017), pp. 205–228
38. P. Pérez-Aros, Subdifferential formulae for the supremum of an arbitrary family of functions. SIAM J. Optim. **29**, 1714–1743 (2019)
39. S.M. Robinson, Generalized equations and their solutions, I: basic theory. Math. Program. Study **10**, 128–141 (1979)
40. R.T. Rockafellar, R.J-B. Wets, *Variational Analysis* (Springer, Berlin, 1998)
41. K. Shimizu, Y. Ishizuka, J.F. Bard, *Nondifferentiable and Two-Level Mathematical Programming* (Kluwer, Dordrecht, 1997)
42. J.J. Ye, Necessary conditions for bilevel dynamic optimization problems. SIAM J. Control Optim. **33**, 1208–1223 (1995)
43. J.J. Ye, Optimal strategies for bilevel dynamic problems. SIAM J. Control Optim. **35**, 512–531 (1997)
44. J.J. Ye, D.L. Zhu, Optimality conditions for bilevel programming problems. Optimization **33**, 9–27 (1995)
45. J.J. Ye, D.L. Zhu, New necessary optimality conditions for bilevel programs by combining MPEC and the value function approach. SIAM J. Optim. **20**, 1885–1905 (2010)
46. A.J. Zaslavski, Necessary optimality conditions for bilevel minimization problems. Nonlinear Anal. **75**, 1655–1678 (2012)
47. A. Zemkoho, Solving ill-posed bilevel programs. Set-Valued Var. Anal. **24**, 423–448 (2016)
48. R. Zhang, Multistage bilevel programming problems. Optimization **52**, 605–616 (2003)

Chapter 8
Constraint Qualifications and Optimality Conditions in Bilevel Optimization

Jane J. Ye

Abstract In this paper we study constraint qualifications and optimality conditions for bilevel programming problems. We strive to derive checkable constraint qualifications in terms of problem data and applicable optimality conditions. For the bilevel program with convex lower level program we discuss drawbacks of reformulating a bilevel programming problem by the mathematical program with complementarity constraints and present a new sharp necessary optimality condition for the reformulation by the mathematical program with a generalized equation constraint. For the bilevel program with a nonconvex lower level program we propose a relaxed constant positive linear dependence (RCPLD) condition for the combined program.

Keywords Bilevel programs · Necessary optimality conditions · Constraint qualifications

8.1 Introduction

In this paper we consider the following bilevel program:

$$\begin{aligned} (\mathrm{BP}) \qquad & \min\ F(x,y) \\ & \text{s.t.}\ \ y \in S(x),\ G(x,y) \le 0,\ H(x,y) = 0, \end{aligned}$$

where $S(x)$ denotes the solution set of the lower level program

$$(\mathrm{P}_x) \qquad \min_{y\in\Gamma(x)}\ f(x,y),$$

J. J. Ye (✉)
Department of Mathematics and Statistics, University of Victoria, Victoria, BC, Canada
e-mail: janeye@uvic.ca

S. Dempe, A. Zemkoho (eds.), *Bilevel Optimization*, Springer Optimization and Its Applications 161, https://doi.org/10.1007/978-3-030-52119-6_8

where $\Gamma(x) := \{y \in \mathbb{R}^m : g(x, y) \le 0, h(x, y) = 0\}$ is the feasible region of the lower level program, $F : \mathbb{R}^n \times \mathbb{R}^m \to \mathbb{R}$, $G : \mathbb{R}^n \times \mathbb{R}^m \to \mathbb{R}^p$ and $H : \mathbb{R}^n \times \mathbb{R}^m \to \mathbb{R}^q$ $f : \mathbb{R}^n \times \mathbb{R}^n \to \mathbb{R}$, $g : \mathbb{R}^n \times \mathbb{R}^m \to \mathbb{R}^r, h : \mathbb{R}^n \times \mathbb{R}^m \to \mathbb{R}^s$. Throughout the paper, for simplicity we assume that $S(x) \ne \emptyset$ for all x.

In economics literature, a bilevel program is sometimes referred to as a Stackelberg game due to the introduction of the concept by Stackelberg [28]. Although it can be used to model a game between the leader and the follower of a two level hierarchical system, the bilevel program has been used to model much wider range of applications; see e.g. [4, 5]. Recently, it has been applied to hyper-parameters selection in machine learning; see e.g. [17, 18].

The classical approach or the first order approach to study optimality conditions for bilevel programs is to replace the lower level problem by its Karush-Kuhn-Tucker (KKT) conditions and minimize over the original variables as well as the multipliers. The resulting problem is a so-called mathematical program with complementarity constraints or mathematical program with equilibrium constraints. The class of mathematical program with complementarity/equilibrium constraints has been studied intensively in the last three decades; see e.g. [19, 24] and the reference within.

There are two issues involved in using the first order approach. Firstly, since the KKT condition is only a sufficient but not necessary condition for optimality, the first order approach can only be used when the lower level problem is a convex program. Secondly, even when the lower level is a convex program if the lower level problem has more than one multiplier, the resulting problem is not equivalent to the original bilevel program if local optimality is considered. In this paper we discuss these issues and present some strategies to deal with this problem. These strategies including using the value function approach, the combined approach and the generalized equation approach.

For a stationary condition to hold at a local optimal solution, usually certain constraint qualifications are required to hold. There are some weak constraint qualifications which are not checkable since they are defined implicitly; e.g. Abadie constraint qualification. In this paper we concentrate on only those checkable constraint qualifications.

The following notation will be used throughout the paper. We denote by $B(\bar{x}; \delta)$ the closed ball centered at $\bar{x}$ with radius δ and by B the closed unit ball centered at 0. We denote by $\mathbb{B}_\delta(\bar{x})$ the open ball centered at $\bar{x}$ with radius δ. For a matrix A, we denote by A^T its transpose. The inner product of two vectors x, y is denoted by $x^T y$ or $\langle x, y\rangle$ and by $x \perp y$ we mean $\langle x, y\rangle = 0$. The polar cone of a set Ω is $\Omega^\circ = \{x | x^T v \le 0 \, \forall v \in \Omega\}$. For a set Ω, we denote by coΩ the convex hull of Ω. For a differentiable mapping $P : \mathbb{R}^d \to \mathbb{R}^s$, we denote by $\nabla P(z)$ the Jacobian matrix of P at z if $s > 1$ and the gradient vector if $s = 1$. For a function $f : \mathbb{R}^d \to \mathbb{R}$, we denote by $\nabla^2 f(\bar{z})$ the Hessian matrix of f at $\bar{z}$. Let $M : \mathbb{R}^d \rightrightarrows \mathbb{R}^s$ be an arbitrary set-valued mapping. We denote its graph by gph$M := \{(z, w) | w \in M(z)\}$. $o : \mathbb{R}_+ \to \mathbb{R}$ denotes a function with the property that $o(\lambda)/\lambda \to 0$ when $\lambda \downarrow 0$. By $z_k \xrightarrow{\Omega} z$ we mean that $z_k \in \Omega$ and $z_k \to z$.

8.2 Preliminaries on Variational Analysis

In this section, we gather some preliminaries in variational analysis and optimization theories that will be needed in the paper. The reader may find more details in the monographs [3, 21, 27] and in the papers we refer to.

Definition 8.2.1 (Tangent Cone and Normal Cone) Given a set $\Omega \subseteq \mathbb{R}^d$ and a point $\bar{z} \in \Omega$, the (Bouligand-Severi) *tangent/contingent cone* to Ω at $\bar{z}$ is a closed cone defined by

$$T_\Omega(\bar{z}) := \limsup_{t \downarrow 0} \frac{\Omega - \bar{z}}{t} = \left\{ u \in \mathbb{R}^d \,\middle|\, \exists t_k \downarrow 0,\ u_k \to u \text{ with } \bar{z} + t_k u_k \in \Omega\ \forall k \right\}.$$

The (Fréchet) *regular normal cone* and the (Mordukhovich) *limiting/basic normal cone* to Ω at $\bar{z} \in \Omega$ are closed cones defined by

$$\widehat{N}_\Omega(\bar{z}) := (T_\Omega(\bar{z}))^\circ$$

$$\text{and} \quad N_\Omega(\bar{z}) := \left\{ z^* \in \mathbb{R}^d \mid \exists z_k \stackrel{\Omega}{\to} \bar{z} \text{ and } z_k^* \to z^* \text{ such that } z_k^* \in \widehat{N}_\Omega(z_k)\ \forall k \right\},$$

respectively. △

When the set Ω is convex, the regular and the limiting normal cones are equal and reduce to the classical normal cone of convex analysis, i.e.,

$$N_\Omega(\bar{z}) := \{z^* | \langle z^*, z - \bar{z} \rangle \leq 0 \quad \forall z \in \Omega\}.$$

We now give definitions for subdifferentials.

Definition 8.2.2 (Subdifferentials) Let $f : \mathbb{R}^d \to \bar{\mathbb{R}}$ be an extended value function, $\bar{x} \in \mathbb{R}^d$ and $f(\bar{x})$ is finite. The regular subdifferential of f at $\bar{x}$ is the set defined by

$$\widehat{\partial} f(\bar{x}) := \{v \in \mathbb{R}^d | f(x) \geq f(\bar{x}) + \langle v, x - \bar{x} \rangle + o(\|x - \bar{x}\|)\}.$$

The limiting subdifferential of f at $\bar{x}$ is the set defined by

$$\partial f(\bar{x}) := \{v \in \mathbb{R}^d | v = \lim_k v_k, v_k \in \widehat{\partial} f(x_k), x_k \to \bar{x}, f(x_k) \to f(\bar{x})\}.$$

Suppose that f is Lipschitz continuous at $\bar{x}$. Then the Clarke subdifferential of f at $\bar{x}$ is the set defined by

$$\partial^c f(\bar{x}) = conv \partial f(\bar{x}).$$

△

When the function f is convex, all the subdifferentials defined above are equal and reduce to the classical subgradient of convex analysis, i.e.,

$$\partial f(\bar{x}) := \{v \in \mathbb{R}^d | f(x) \geq f(\bar{x}) + \langle v, x - \bar{x}\rangle\}.$$

Definition 8.2.3 (Coderivatives) For a set-valued map $\Phi : \mathbb{R}^d \rightrightarrows \mathbb{R}^s$ and a point $(\bar{x}, \bar{y}) \in \text{gph}\Phi$, the Fréchet coderivative of Φ at $(\bar{x}, \bar{y})$ is a multifunction $\widehat{D}^*\Phi(\bar{x}, \bar{y}) : \mathbb{R}^s \rightrightarrows \mathbb{R}^d$ defined as

$$\widehat{D}^*\Phi(\bar{x}, \bar{y})(w) := \left\{\xi \in \mathbb{R}^d | (\xi, -w) \in \widehat{N}_{\text{gph}\Phi}(\bar{x}, \bar{y})\right\}.$$

And the limiting (Mordukhovich) coderivative of Φ at $(\bar{x}, \bar{y})$ is a multifunction $D^*\Phi(\bar{x}, \bar{y}) : \mathbb{R}^s \rightrightarrows \mathbb{R}^d$ defined as

$$D^*\Phi(\bar{x}, \bar{y})(w) := \left\{\xi \in \mathbb{R}^d | (\xi, -w) \in N_{\text{gph}\Phi}(\bar{x}, \bar{y})\right\}. \qquad \triangle$$

We now review some concepts of stability of a set-valued map.

Definition 8.2.4 (Aubin [2]) Let $\Sigma : \mathbb{R}^n \rightrightarrows \mathbb{R}^d$ be a set-valued map and $(\bar{\alpha}, \bar{x}) \in \text{gph}\Sigma$. We say that Σ is pseudo-Lipschitz continuous at $(\bar{\alpha}, \bar{x})$ if there exist a neighborhood $\mathbb{V}$ of $\bar{\alpha}$, a neighborhood $\mathbb{U}$ of $\bar{x}$ and $\kappa \geq 0$ such that

$$\Sigma(\alpha) \cap \mathbb{U} \subseteq \Sigma(\alpha') + \kappa \|\alpha' - \alpha\| B, \ \ \forall \alpha', \alpha \in \mathbb{V}. \qquad \triangle$$

Definition 8.2.5 (Robinson [25]) Let $\Sigma : \mathbb{R}^n \rightrightarrows \mathbb{R}^d$ be a set-valued map and $\bar{\alpha} \in \mathbb{R}^n$. We say that Σ is upper-Lipschitz continuous at $\bar{\alpha}$ if there exist a neighborhood $\mathbb{V}$ of $\bar{\alpha}$ and $\kappa \geq 0$ such that

$$\Sigma(\alpha) \subseteq \Sigma(\bar{\alpha}) + \kappa \|\alpha - \bar{\alpha}\| B, \ \ \forall \alpha \in \mathbb{V}. \qquad \triangle$$

Definition 8.2.6 (Ye and Ye [35]) Let $\Sigma : \mathbb{R}^n \rightrightarrows \mathbb{R}^d$ be a set-valued map and $(\bar{\alpha}, \bar{x}) \in \text{gph}\Sigma$. We say that Σ is calm (or pseudo upper-Lipschitz continuous) at $(\bar{\alpha}, \bar{x})$ if there exist a neighborhood $\mathbb{V}$ of $\bar{\alpha}$, a neighborhood $\mathbb{U}$ of $\bar{x}$ and $\kappa \geq 0$ such that

$$\Sigma(\alpha) \cap \mathbb{U} \subseteq \Sigma(\bar{\alpha}) + \kappa \|\alpha - \bar{\alpha}\| B, \ \ \forall \alpha \in \mathbb{V}. \qquad \triangle$$

Note that the terminology of calmness was suggested by Rockafellar and Wets in [27].

It is clear that both the pseudo-Lipschitz continuity and the upper-Lipschitz continuity are stronger than the pseudo upper-Lipschitz continuity. It is obvious that if $\Sigma : \mathbb{R}^n \to \mathbb{R}^d$ is a continuous single-valued map, then the pseudo-Lipschitz continuity at $(\bar{\alpha}, \bar{x})$ reduces to the Lipschitz continuity at $\bar{\alpha}$, while the calmness/pseudo upper-Lipschitz continuity reduces to the calmness at $\bar{\alpha}$, i.e., there exist a neighborhood $\mathbb{V}$ of $\bar{\alpha}$ and a constant $\kappa \geq 0$ such that

$$\|\Sigma(\alpha) - \Sigma(\bar{\alpha})\| \leq \kappa \|\alpha - \bar{\alpha}\| \quad \forall \alpha \in \mathbb{V}.$$

Hence it is easy to see that the calmness/pseudo upper-Lipschitz continuity is a much weaker stability condition than the pseudo-Lipschitz continuity condition.

Many optimization problems can be written in the following form:

$$\min_x f(x) \quad \text{s.t. } 0 \in \Phi(x), \tag{8.2.1}$$

where $f : \mathbb{R}^d \to \mathbb{R}$ is Lipschitz continuous around the point of interest and $\Phi : \mathbb{R}^d \rightrightarrows \mathbb{R}^n$ is a set-valued map with a closed graph.

Let $\bar{x}$ be a feasible solution for the above optimization problem. We say that Mordukhovich (M-) stationary condition holds at $\bar{x}$ if there exists η such that

$$0 \in \partial f(\bar{x}) + D^*\Phi(\bar{x}, 0)(\eta), \tag{8.2.2}$$

respectively.

We now discuss the constraint qualifications under which a local optimal solution $\bar{x}$ satisfies the M-stationary condition. For this purpose we consider the perturbed feasible solution mapping

$$\Sigma(\alpha) := \Phi^{-1}(\alpha) = \{x \in \mathbb{R}^d | \alpha \in \Phi(x)\}. \tag{8.2.3}$$

The property of the calmness of set-valued map $\Sigma(\cdot)$ at $(0, \bar{x}) \in \text{gph}\, \Sigma$ is equivalent with the property of the metric subregularity of its inverse map $\Sigma^{-1}(x) = \Phi(x)$ at $(\bar{x}, 0)$, cf. [9]. This justifies the terminology defined below.

Definition 8.2.7 Let $0 \in \Phi(\bar{x})$. We say that the metric subregularity constraint qualification (MSCQ) holds at $\bar{x}$ if the perturbed feasible solution mapping defined by (8.2.3) is calm at $(0, \bar{x})$. △

Theorem 8.2.8 (Ye and Ye [35, Theorem 3.1]) *Let $\bar{x}$ be a local optimal solution of problem (8.2.1). Suppose that MSCQ holds at $\bar{x}$. Then the M-stationary condition (8.2.2) holds at $\bar{x}$.* △

In the case where $\Phi(x) := (h(x), g(x) + \mathbb{R}^r_+)$ with $g : \mathbb{R}^n \to \mathbb{R}^r, h : \mathbb{R}^n \to \mathbb{R}^s$ being smooth, problem (8.2.1) is the nonlinear program with equality and inequality constraints and (8.2.2) reduces to the KKT condition for the nonlinear program. We say that a feasible solution $\bar{x}$ of problem (8.2.1) satisfies the linear independence constraint qualification (LICQ) if the gradients

$$\{\nabla g_i(\bar{x})\}_{i \in I_g} \cup \{\nabla h_i(\bar{x})\}_{i=1}^s, \quad I_g = I_g(\bar{x}) := \{i | g_i(\bar{x}) = 0\}$$

are linearly independent. We say that the positive linear independence constraint qualification (PLICQ) or no nonzero abnormal constraint qualification (NNAMCQ) holds at $\bar{x}$ if there is no nonzero vector (η^h, η^g) such that

$$0 = \sum_{i \in I_g} \nabla g_i(\bar{x})\eta_i^g + \sum_{i=1}^{s} \nabla h_i(\bar{x})\eta_i^h, \quad \eta^g \geq 0.$$

By an alternative theorem, it is well-known that PLICQ/NNAMCQ is equivalent to the Mangasarian Fromovitz constraint qualification (MFCQ): the gradients $\{\nabla h_i(\bar{x})\}_{i=1}^s$ is linearly independent and $\exists v \in \mathbb{R}^n$ such that

$$\nabla g_i(\bar{x})^T v < 0 \, \forall i \in I_g, \quad \nabla h_i(\bar{x})^T v = 0 \, \forall i = 1, \ldots, s.$$

LICQ is stronger than MFCQ which is equivalent to saying that the perturbed feasible solution map $\Sigma(\cdot)$ is pseudo-Lipschitz continuous at $(0, \bar{x})$ and hence stronger than the MSCQ/calmness condition.

We will also need the following definition.

Definition 8.2.9 (Generalized Linearity Space) Given an arbitrary set $C \subseteq \mathbb{R}^d$, we call a subspace L the generalized linearity space of C and denote it by $\mathcal{L}(C)$ provided that it is the largest subspace $L \subseteq \mathbb{R}^d$ such that $C + L \subseteq C$. △

In the case where C is a convex cone, the linearity space of C is the largest subspace contained in C and can be calculated as $\mathcal{L}(C) = (-C) \cap C$.

8.3 Bilevel Program with Convex Lower Level Program

In this section we consider the case where given x the lower level problem (P_x) is a convex program. We first discuss the challenges for such a problem and follow by considering two special cases where the first one is a problem where the lower level problem is completely linear and the second one is a problem where the lower level constraint is independent of the upper level variable.

To concentrate the main idea, for simplicity in this section we omit the upper level constraints and lower level equality constraints and consider

$$(\mathrm{BP})_1 \quad \min_{x,y} \; F(x, y)$$
$$\text{s.t.} \;\; y \in \arg\min_{y'}\{f(x, y')|g(x, y') \leq 0\},$$

where $g(x, y)$ is either affine in y or convex in y and the Slater condition holds, i.e., for each x there is $y(x)$ such that $g(x, y(x)) < 0$. We assume that $F : \mathbb{R}^n \times \mathbb{R}^m \to \mathbb{R}$, is continuously differentiable, $f : \mathbb{R}^n \times \mathbb{R}^m \to \mathbb{R}$, $g : \mathbb{R}^n \times \mathbb{R}^m \to \mathbb{R}^r$ are twice continuously differentiable in variable y.

Under the assumptions we made, the KKT condition is necessary and sufficient for optimality. So

$$y \in S(x) \Longleftrightarrow \exists \lambda \quad \text{s.t. } 0 = \nabla_y f(x, y) + \nabla_y g(x, y)^T \lambda, \quad 0 \leq -g(x, y) \perp \lambda \geq 0.$$

A common approach in the bilevel program literature is to replace "$\exists \lambda$" by "$\forall \lambda$" in the above and hence consider solving the following mathematical program with complementarity constraints (MPCC) instead.

$$\begin{aligned} \text{(MPCC)} \qquad \min_{x,y,\lambda} \;& F(x, y) \\ \text{s.t. } \;& 0 = \nabla_y f(x, y) + \nabla_y g(x, y)^T \lambda, \\ & 0 \leq -g(x, y) \perp \lambda \geq 0. \end{aligned}$$

Problem (MPCC) looks like a standard mathematical program. However if one treats it as a mathematical program with equality and inequality constraints, then the usual constraint qualification such as MFCQ fails at each feasible solution (see Ye and Zhu [36, Proposition 3.2]). This observation leads to the introduction of weaker stationary conditions such as Weak (W-), Strong (S-), Mordukhovich (M-) and Clarke (C-) stationary conditions for (MPCC); see e.g. Ye [31] for more discussions. We recall the definitions of various stationary conditions there.

Definition 8.3.1 (Stationary Conditions for MPCC) Let $(\bar{x}, \bar{y}, \bar{\lambda})$ be a feasible solution for problem (MPCC). We say that $(\bar{x}, \bar{y}, \bar{\lambda})$ is a weak stationary point of (MPCC) if there exist $w \in \mathbb{R}^m$, $\xi \in \mathbb{R}^r$ such that

$$0 = \nabla_x F(\bar{x}, \bar{y}) - \nabla^2_{yx} f(\bar{x}, \bar{y}) w - \nabla^2_{yx} (\bar{\lambda}^T g)(\bar{x}, \bar{y}) w + \nabla_x g(\bar{x}, \bar{y})^T \xi,$$

$$0 = \nabla_y F(\bar{x}, \bar{y}) - \nabla^2_{yy} f(\bar{x}, \bar{y}) w - \nabla^2_{yy} (\bar{\lambda}^T g)(\bar{x}, \bar{y}) w + \nabla_y g(\bar{x}, \bar{y})^T \xi,$$

$$\xi_i = 0 \;\text{ if } g_i(\bar{x}, \bar{y}) < 0, \bar{\lambda}_i = 0, \tag{8.3.1}$$

$$\nabla_y g_i(\bar{x}, \bar{y})^T w = 0 \;\text{ if } g_i(\bar{x}, \bar{y}) = 0, \bar{\lambda}_i > 0. \tag{8.3.2}$$

We say that $(\bar{x}, \bar{y}, \lambda)$ is a S-, M-, C- stationary point of (MPCC) if there exist $w \in \mathbb{R}^m$, $\xi \in \mathbb{R}^r$ such that the above conditions and the following condition holds

$$\xi_i \geq 0, \nabla_y g_i(\bar{x}, \bar{y})^T w \leq 0 \;\text{ if } g_i(\bar{x}, \bar{y}) = \bar{\lambda}_i = 0,$$

$$\text{either } \xi_i > 0, \nabla_y g_i(\bar{x}, \bar{y})^T w < 0 \text{ or } \xi_i \nabla_y g_i(\bar{x}, \bar{y})^T w = 0 \;\text{ if } g_i(\bar{x}, \bar{y}) = \bar{\lambda}_i = 0, \tag{8.3.3}$$

$$\xi_i \nabla_y g_i(\bar{x}, \bar{y})^T w \leq 0 \;\text{ if } g_i(\bar{x}, \bar{y}) = \bar{\lambda}_i = 0,$$

respectively. △

For a mathematical program, it is well-known that under certain constraint qualification, a local optimal solution must be a stationary point and hence a stationary point is a candidate for a local optimal solution. Unfortunately as pointed out by Dempe and Dutta in [6], this is not true for bilevel programs even when the lower level is convex. Precisely, it is possible that $(\bar{x}, \bar{y}, \bar{\lambda})$ is a local optimal solution of (MPCC) but $(\bar{x}, \bar{y})$ is not a local optimal solution of $(BP)_1$. Note that since (MPCC) is a nonconvex program, one usually only hope to find a local optimal solution and hence this is very bad news. This observation indicates that extreme care should be taken when using MPCC reformulation in the case where the lower level problem has non-unique multipliers.

8.3.1 The Bilevel Program Where the Lower Level Program Is Completely Linear

We now discuss the special case of $(BP)_1$ where the lower level program is completely linear. That is, $f(x, y) = a^T x + b^T y$ and $g(x, y) = Cx + Dy - q$ with $a \in \mathbb{R}^n$, $b \in \mathbb{R}^m$, $C \in \mathbb{R}^{r\times n}$, $D \in \mathbb{R}^{r\times m}$, $q \in \mathbb{R}^r$. It is easy to see that $(BP)_1$ can be equivalently written as the following problem

$$
\begin{aligned}
(\mathrm{VP})_1 \qquad & \min \ F(x, y) \\
& \text{s.t. } a^T x + b^T y - V(x) \le 0, \\
& \qquad Cx + Dy - q \le 0,
\end{aligned}
$$

where $V(x) := \inf_{y'}\{a^T x + b^T y' \mid Cx + Dy' - q \le 0\}$ is the value function of the lower level problem. Then by convex analysis, the value function V(x) is a polyhedral convex function and we have an explicit expression for its subgradient.

Proposition 8.3.2 (See e.g., [34, Proposition 4.1]) *Let $\bar{y} \in S(\bar{x})$ and suppose that $f(x, y) = a^T x + b^T y$ and $g(x, y) = Cx + Dy - q$. Then $V(x)$ is convex with $\partial V(\bar{x}) \neq \emptyset$ and*

$$
\partial V(\bar{x}) = \left\{ a + C^T v | 0 = b + D^T v, 0 \le v \perp -(C\bar{x} + D\bar{y} - q) \ge 0 \right\}. \qquad \triangle
$$

Since the function $a^T x + b^T y - V(x)$ is a concave function, it was shown in [30] that the nonsmooth weak reverse constraint qualification holds for problem $(\mathrm{VP})_1$ and hence by using the nonsmooth multiplier rule and the expression for the subgradient of the value function the following optimality condition holds.

Theorem 8.3.3 (Ye [30, Corollary 4.1]) *Let* $(\bar{x}, \bar{y})$ *be a local optimal solution of* $(VP)_1$*. Then there exists* $\delta \geq 0$, $\bar{\nu} \in \mathbb{R}^r$ *and* $\alpha \in \mathbb{R}^r$ *such that*

$$\begin{aligned}
0 &= \nabla_x F(\bar{x}, \bar{y}) + C^T(\alpha - \delta\bar{\nu}),\\
0 &= \nabla_y F(\bar{x}, \bar{y}) + D^T(\alpha - \delta\bar{\nu}),\\
0 &\leq \alpha \perp -(C\bar{x} + D\bar{y} - q) \geq 0,\\
0 &\leq \bar{\nu} \perp -(C\bar{x} + D\bar{y} - q) \geq 0.
\end{aligned}$$

△

Let $\xi = \alpha - \delta\bar{\nu}$ where $\alpha, \delta, \bar{\nu}$ are those found in Theorem 8.3.3 and $w = 0$ in Definition 8.3.1. It is easy to verify that $(\bar{x}, \bar{y}, \bar{\nu})$ is an S-stationary point of the corresponding (MPCC). Unlike using (MPCC) reformulation by which usually a constraint qualification such as the MPCC LICQ is needed to ensure that a local optimal solution is an S-stationary point (see e.g. [31]), $\bar{\nu}$ is a multiplier for the lower level problem (P_x) selected automatically from the subdifferential of the value function. For this particular multiplier $\bar{\nu}$, the S-stationary condition holds under *no* constraint qualification.

For bilevel programs where the lower level problem is not completely linear but is convex with a convex value function, the reader is referred to [30, 32] for more detailed discussions and results.

8.3.2 *The Case Where the Lower Level Constraint Is Independent of the Upper Level*

As we see in the previous discussion, the difficulty of using the first order approach occurs when the lower level has non-unique multipliers. In this subsection we consider a special case where the lower level constraint $\Gamma(x) = \Gamma$ is independent of x. Then $y \in S(x)$ if and only if the generalized equation $0 \in \nabla_y f(x, y) + N_\Gamma(y)$ holds. So we next consider the mathematical program with equilibrium constraints (MPEC) which is an equivalent reformulation of $(BP)_1$ when g is independent of x:

$$\text{(MPEC)} \quad \min_{x,y} \; F(x, y) \quad \text{s.t.} \;\; 0 \in \nabla_y f(x, y) + N_\Gamma(y),$$

where $\Gamma := \{y : g(y) \leq 0\}$ and $g : \mathbb{R}^m \to \mathbb{R}^r$ is either affine or convex with a Slater point.

For problem (MPEC), Ye and Ye [35] showed that the pseudo upper-Lipschitz continuity/calmness/MSCQ guarantees M-stationarity of solutions.

Theorem 8.3.4 (Ye and Ye [35, Theorem 3.2]) *Let $(\bar{x}, \bar{y})$ be a local optimal solution of (MPEC). Suppose that the perturbed feasible solution map*

$$\Sigma(\alpha) := \{(x, y) \mid \alpha \in \nabla_y f(x, y) + N_\Gamma(y)\}$$

is calm/pseudo upper-Lipschitz continuous at $(0, \bar{x}, \bar{y})$ (i.e., MSCQ holds at $(\bar{x}, \bar{y})$). Then $(\bar{x}, \bar{y})$ is an M-stationary point of problem (MPEC), i.e., there exist $w \in \mathbb{R}^m$ such that

$$0 = \nabla_x F(\bar{x}, \bar{y}) + \nabla^2_{yx} f(\bar{x}, \bar{y})w,$$

$$0 = \nabla_y F(\bar{x}, \bar{y}) + \nabla^2_{yy} f(\bar{x}, \bar{y})w + D^* N_\Gamma(\bar{y}, -\nabla_y f(\bar{x}, \bar{y}))(w). \qquad \triangle$$

In the case where $\nabla_y f(x, y)$ is affine and Γ is a convex polyhedral set, the set-valued map $\Sigma(\cdot)$ is a polyhedral multifunction which means that its graph is the union of finitely many polyhedral convex sets. According to Robinson [26], $\Sigma(\cdot)$ is upper Lipschitz continuous which implies that MSCQ holds automatically at each feasible solution. How to check MSCQ for the general case? We now describe a sufficient condition for MSCQ derived by Gfrerer and Ye in [12]. First we start with some notation. Let

$$\bar{\Lambda} := \{\lambda \mid 0 = \nabla_y f(\bar{x}, \bar{y}) + \nabla g(\bar{y})^T \lambda, 0 \le -g(\bar{y}) \perp \lambda \ge 0\}$$

be the multiplier set for the lower level problem $(P_{\bar{x}})$ at $\bar{y}$. Define the critical cone for Γ at $\bar{y}$ as

$$\bar{K}_\Gamma := \{v \mid \nabla g(\bar{y})v \in T_{\mathbb{R}^q_-}(g(\bar{y})), \nabla_y f(\bar{x}, \bar{y})^T v = 0\}.$$

For every $v \in \bar{K}_\Gamma$, define the directional multiplier set at direction v as

$$\bar{\Lambda}(v) := \arg\max \left\{v^T \nabla^2(\lambda^T g)(\bar{y})v \mid \lambda \in \bar{\Lambda}\right\} = \{\lambda | v^T \nabla^2 g(\bar{y})v \in N_{\bar{\lambda}}(\lambda)\}.$$

Let $\bar{I} := \{i \mid g_i(\bar{y}) = 0\}$ be the index of constraints active at $\bar{y}$. For every $\lambda \in \bar{\Lambda}$, we define the index set of strongly active constraints

$$\bar{J}^+(\lambda) := \{i | g_i(\bar{y}) = 0, \lambda_i > 0\}.$$

Under our assumption of this section, the multiplier set $\bar{\Lambda}$ is nonempty and hence the critical cone for Γ at $\bar{y}$ can be represented as

$$\bar{K}_\Gamma = \left\{v \mid \nabla g_i(\bar{y})^T v \begin{cases} = 0 & i \in \bar{J}^+(\bar{\Lambda}) \\ \le 0 & i \in \bar{I} \setminus \bar{J}^+(\bar{\Lambda}) \end{cases}\right\},$$

where $\bar{J}^+(\bar{\Lambda}) := \cup_{\lambda \in \bar{\Lambda}} \bar{J}^+(\lambda)$. For every $\bar{v} \in \bar{K}_\Gamma$, we denote the index set of active constraints for the critical cone at $\bar{v}$ as

$$\bar{I}(\bar{v}) := \{i \in \bar{I} | \nabla g_i(\bar{y})^T \bar{v} = 0\}.$$

Denote by $\bar{\mathcal{E}}$ the collection of all the extreme points of the closed and convex set of multipliers $\bar{\Lambda}$ and recall that $\lambda \in \bar{\Lambda}$ belongs to $\bar{\mathcal{E}}$ if and only the family of gradients $\{\nabla g_i(\bar{y}) | i \in \bar{J}^+(\lambda)\}$ is linearly independent. Specializing the result from [12] we have the following checkable constraint qualification for problem (MPEC).

Theorem 8.3.5 ([12, Theorems 4 and 5]) *Let $(\bar{x}, \bar{y})$ be a feasible solution of problem (MPEC). Assume that there do not exist $(u, v) \neq 0$, $\lambda \in \bar{\Lambda}(v) \cap \bar{\mathcal{E}}$ and $w \neq 0$ satisfying*

$$-\nabla^2_{yx} f(\bar{x}, \bar{y})u - \nabla^2_{yy} f(\bar{x}, \bar{y})v - \nabla^2(\lambda^T g)(\bar{y})v \in N_{\bar{K}_\Gamma}(v),$$

$$\nabla^2_{xy} f(\bar{x}, \bar{y})w = 0,$$

$$\nabla g_i(\bar{y})^T w = 0, i \in \bar{J}^+(\lambda), \ w^T \left(\nabla^2_{yy} f(\bar{x}, \bar{y}) + \nabla^2(\lambda^T g)(\bar{y})\right) w \leq 0.$$

Then MSCQ for problem (MPEC) holds at $(\bar{x}, \bar{y})$. △

We would like to comment that recently [1] has compared the calmness condition for the two problems (MPEC) and (MPCC). They have shown that in general the calmness condition for (MPEC) is weaker than the one for the corresponding (MPCC).

Now we consider the M-stationary condition in Theorem 8.3.4. The expression of the M-stationary condition involves the coderivative of the normal cone mapping $N_\Gamma(\cdot)$. Precise formulae for this coderivative in terms of the problem data can be found in [15, Proposition 3.2] if Γ is polyhedral, in [16, Theorem 3.1] if LICQ holds at $\bar{y}$ for the lower level problem, in [10, Theorem 3] under a relaxed MFCQ combined with the so-called 2-regularity.

Recently Gfrerer and Ye [13] have derived a necessary optimality condition that is sharper than the M-stationary condition under the following 2-nondegeneracy condition.

Definition 8.3.6 Let $v \in \bar{K}_\Gamma$. We say that g is *2-nondegenerate in direction v at $\bar{y}$* if

$$\nabla^2(\mu^T g)(\bar{y})v \in N_{\bar{K}_\Gamma}(v) - N_{\bar{K}_\Gamma}(v), \ \mu \in \text{span}\,(\bar{\Lambda}(v) - \bar{\Lambda}(v)) \implies \mu = 0.$$ △

In the case where the directional multiplier set $\bar{\Lambda}(v)$ is a singleton, span $(\bar{\Lambda}(v) - \bar{\Lambda}(v)) = \{0\}$ and hence g is 2-nondegenerate in this direction v.

Theorem 8.3.7 ([13, Theorem 6]) *Assume that $(\bar{x}, \bar{y})$ is a local minimizer for problem (MPEC) fulfilling MSCQ at $(\bar{x}, \bar{y})$. Further assume that g is 2-*

nondegenerate in every nonzero critical direction $0 \neq v \in \bar{K}_\Gamma$. *Then there are a critical direction* $\bar{v} \in \bar{K}_\Gamma$, *a directional multiplier* $\bar{\lambda} \in \bar{\Lambda}(\bar{v})$, *index sets* $\mathcal{J}^+$, $\mathcal{J}$, $\mathcal{I}^+$, *and* $\mathcal{I}$ *with* $\bar{J}^+(\bar{\lambda}) \subseteq \mathcal{J}^+ \subseteq \mathcal{J} \subseteq \bar{J}^+(\bar{\Lambda}(\bar{v})) \subseteq \bar{J}^+(\bar{\Lambda}) \subseteq \mathcal{I}^+ \subseteq \mathcal{I} \subseteq \bar{I}(\bar{v})$ *and elements* $w \in \mathbb{R}^m$, $\eta, \xi \in \mathbb{R}^q$ *such that*

$$
\begin{aligned}
&0 = \nabla_x F(\bar{x}, \bar{y}) - \nabla^2_{xy} f(\bar{x}, \bar{y})w,\\
&0 = \nabla_y F(\bar{x}, \bar{y}) - \nabla^2_{yy} f(\bar{x}, \bar{y})w - \nabla^2(\bar{\lambda}^T g)(\bar{y})w + \nabla g(\bar{y})^T \xi + 2\nabla^2(\eta^T g)(\bar{y})\bar{v},\\
&\xi_i = 0 \text{ if } i \notin \mathcal{I},\\
&\xi_i \geq 0, \nabla g_i(\bar{y})^T w \leq 0 \text{ if } i \in \mathcal{I} \setminus \mathcal{I}^+,\\
&\nabla g_i(\bar{y})^T w = 0 \text{ if } i \in \mathcal{I}^+,\\
&\nabla g(\bar{y})^T \eta = 0,\ \eta_i = 0, i \notin \mathcal{J},\ \eta_i \geq 0, i \in \mathcal{J} \setminus \mathcal{J}^+.
\end{aligned}
$$

△

In the case where the multiplier set $\bar{\Lambda} = \{\bar{\lambda}\}$ is a singleton, the 2-nondegeneracy condition holds automatically and the η in the optimality condition becomes zero. In this case we have the following result.

Corollary 8.3.8 *([13, Corollary 1]) Assume that* $(\bar{x}, \bar{y})$ *is a local minimizer for problem (MPEC) fulfilling MSCQ at* $(\bar{x}, \bar{y})$ *and the lower level multiplier is unique, i.e.,* $\bar{\Lambda} = \{\bar{\lambda}\}$. *Then there are a critical direction* $\bar{v} \in \bar{K}_\Gamma$, *index sets* $\mathcal{I}^+$ *with* $\bar{J}^+(\bar{\lambda}) \subseteq \mathcal{I}^+ \subseteq \bar{I}(\bar{v})$ *and elements* $w \in \mathbb{R}^m$, $\xi \in \mathbb{R}^q$ *such that*

$$
\begin{aligned}
&0 = \nabla_x F(\bar{x}, \bar{y}) - \nabla^2_{xy} f(\bar{x}, \bar{y})w,\\
&0 = \nabla_y F(\bar{x}, \bar{y}) - \nabla^2_{yy} f(\bar{x}, \bar{y})w - \nabla^2(\bar{\lambda}^T g)(\bar{y})w + \nabla g(\bar{y})^T \xi,\\
&\xi_i = 0 \text{ if } i \notin \bar{I}(\bar{v}), && (8.3.4)\\
&\xi_i \geq 0, \nabla g_i(\bar{y})^T w \leq 0 \text{ if } i \in \bar{I}(\bar{v}) \setminus \mathcal{I}^+,\\
&\nabla g_i(\bar{y})^T w = 0 \text{ if } i \in \mathcal{I}^+. && (8.3.5)
\end{aligned}
$$

△

Actually we can show that the stationary condition in Theorem 8.2.8 is stronger than the M-stationary condition for (MPCC). Suppose that $(\bar{x}, \bar{y}, \bar{\lambda})$ satisfies the stationary condition in Theorem 8.2.8 and let $w \in \mathbb{R}^m$, $\xi \in \mathbb{R}^q$ be those found in Theorem 8.2.8. Then since

$$
i \notin \bar{I}(\bar{v}) \Longleftrightarrow \begin{array}{l} \text{either } g_i(\bar{y}) = 0, \nabla g_i(\bar{y})^T \bar{v} < 0, \bar{\lambda}_i = 0 \\ \text{or } g_i(\bar{y}) < 0, \bar{\lambda}_i = 0 \end{array},
$$

(8.3.4) and (8.3.5) imply that $\xi_i = 0$ if $\bar{\lambda}_i = 0$ and $\nabla g_i(\bar{y})^T w = 0$ if $\bar{\lambda}_i > 0$. It follows that (8.3.1), (8.3.2), (8.3.3) hold. Therefore $(\bar{x}, \bar{y}, \bar{\lambda})$ must satisfy the M-stationary condition for (MPCC) as well. Hence in the case where the lower level multiplier is unique, the above stationary condition is in general stronger than the M-stationary condition of (MPCC).

Finally in the rest of this section, we will discuss the S-stationary condition for (MPEC). Let

$$\bar{\mathcal{N}} := \{v \in \mathbb{R}^m \mid \nabla g_i(\bar{y})^T v = 0 \,\forall i \in \bar{I}\}$$

be the nullspace of gradients of constraints active at $\bar{y}$. Define for each $v \in \bar{\mathcal{N}}$, the sets

$$\bar{\mathcal{W}}(v) := \left\{w \in \bar{K}_\Gamma \mid w^T \nabla^2((\lambda^1 - \lambda^2)^T g)(\bar{y})v = 0, \forall \lambda^1, \lambda^2 \in \bar{\Lambda}(v)\right\},$$

$$\tilde{\Lambda}(v) := \begin{cases} \bar{\Lambda}(v) \cap \bar{\mathcal{E}} & \text{if } v \neq 0 \\ conv(\cup_{0 \neq u \in \bar{K}_\Gamma} \bar{\Lambda}(u) \cap \bar{\mathcal{E}}) & \text{if } v = 0, \bar{K}_\Gamma \neq \{0\}, \end{cases}$$

and for each $w \in \bar{K}_\Gamma$,

$$\bar{L}(v; w) := \begin{cases} \{-\nabla^2(\lambda^T g)(\bar{y})w \mid \lambda \in \tilde{\Lambda}(v)\} + \bar{K}_\Gamma^\circ & \text{if } \bar{K}_\Gamma \neq \{0\} \\ \mathbb{R}^m & \text{if } \bar{K}_\Gamma = \{0\} \end{cases}.$$

The following theorem is a slight improvement of [11, Theorem 8] in that the assumption is weaker.

Theorem 8.3.9 *Assume that $(\bar{x}, \bar{y})$ is a local minimizer for problem (MPEC) fulfilling MSCQ at $(\bar{x}, \bar{y})$ and the generalized linear independence constraint qualification holds:*

$$\nabla P(\bar{x}, \bar{y})\mathbb{R}^{n+m} + \mathcal{L}\left(T_{gph\, N_\Gamma}(P(\bar{x}, \bar{y}))\right) = \mathbb{R}^{2m},$$

where $P(x, y) := (y, -\nabla_y f(x, y))$ and $\mathcal{L}(C)$ is the generalized linearity space of set C as defined in Definition 8.2.9. Then $(\bar{x}, \bar{y})$ is an S-stationary point for (MPEC), i.e., there exists elements w such that

$$0 = \nabla_x F(\bar{x}, \bar{y}) + \nabla^2_{xy} f(\bar{x}, \bar{y})w,$$

$$0 \in \nabla_y F(\bar{x}, \bar{y}) + \nabla^2_{yy} f(\bar{x}, \bar{y})w + \widehat{D}N_\Gamma(\bar{y}, -\nabla_y f(\bar{x}, \bar{y}))(w). \quad (8.3.6)$$

In particular, we have $w \in -\bigcap_{v \in \bar{\mathcal{N}}} \bar{\mathcal{W}}(v)$ and

$$0 = \nabla_x F(\bar{x}, \bar{y}) + \nabla^2_{xy} f(\bar{x}, \bar{y})w,$$

$$0 \in \nabla_y F(\bar{x}, \bar{y}) + \nabla^2_{yy} f(\bar{x}, \bar{y})w + \bigcap_{v \in \bar{I}(v)} \bar{L}(v; -w).$$

△

Proof Since $(\bar{x}, \bar{y})$ is a local minimizer for problem (MPEC) which can be rewritten as

$$\text{(MPEC)} \quad \min_{x,y} \; F(x, y)$$
$$\text{s.t.} \quad P(x, y) := (y, -\nabla_y f(x, y)) \in D := \operatorname{gph} N_\Gamma,$$

by the basic optimality condition

$$0 \in \nabla F(\bar{x}, \bar{y}) + \widehat{N}_{\mathcal{F}}(\bar{x}, \bar{y}),$$

where $\mathcal{F} := \{(x, y) : P(x, y) \in D\}$. By [11, Theorem 4], under MSCQ and the generalized LICQ,

$$\widehat{N}_{\mathcal{F}}(\bar{x}, \bar{y}) = \nabla P(\bar{x}, \bar{y})^T \widehat{N}_D(P(\bar{x}, \bar{y}))$$

holds. It follows that the S-stationary condition

$$0 \in \nabla F(\bar{x}, \bar{y}) + \nabla P(\bar{x}, \bar{y})^T \widehat{N}_D(P(\bar{x}, \bar{y}))$$

holds. Since

$$\nabla P(\bar{x}, \bar{y})^T \widehat{N}_D(P(\bar{x}, \bar{y})) = \left\{ \begin{pmatrix} \nabla^2_{xy} f(\bar{x}, \bar{y}) w \\ \nabla^2_{yy} f(\bar{x}, \bar{y}) w + w^* \end{pmatrix} \middle| \; (w^*, -w) \in \widehat{N}_{\operatorname{gph} N_\Gamma}(\bar{y}, -\nabla_y f(\bar{x}, \bar{y})) \right\},$$

and by definition of the co-derivative,

$$(w^*, -w) \in \widehat{N}_{\operatorname{gph} N_\Gamma}(\bar{y}, -\nabla_y f(\bar{x}, \bar{y})) \Longleftrightarrow w^* \in \widehat{D} N_\Gamma(\bar{y}, -\nabla_y f(\bar{x}, \bar{y}))(w),$$

(8.3.6) follows. By Gfrerer and Outrata [11, Proposition 5], we have

$$\widehat{N}_D(P(\bar{x}, y)) = T_D(P(\bar{x}, \bar{y}))^\circ$$
$$\subseteq \left\{ (w^*, w) \,|\, w \in -\bigcap_{v \in \bar{\mathcal{N}}} \bar{\mathcal{W}}(v), \, w^* \in \bigcap_{v \in \bar{\mathcal{N}}} \bar{L}(v; -w) \right\}. \qquad \square$$

8.4 Bilevel Program with Nonconvex Lower Level Program

In this section we consider the general bilevel program (BP) as stated in the introduction and assume that $F : \mathbb{R}^n \times \mathbb{R}^m \to \mathbb{R}$, $G : \mathbb{R}^n \times \mathbb{R}^m \to \mathbb{R}^p$ and

$H : \mathbb{R}^n \times \mathbb{R}^m \to \mathbb{R}^q$ are continuously differentiable, $f : \mathbb{R}^n \times \mathbb{R}^m \to \mathbb{R}$, $g : \mathbb{R}^n \times \mathbb{R}^m \to \mathbb{R}^r, h : \mathbb{R}^n \times \mathbb{R}^m \to \mathbb{R}^s$ are twice continuously differentiable in variable y.

In the bilevel programming literature, in particular in early years, the first order approach has been popularly used even when the lower level is nonconvex. But if the lower level program (P_x) is not convex, the optimality condition

$$0 \in \nabla_y f(x, y) + \widehat{N}_{\Gamma(x)}(y)$$

is only necessary but not sufficient for $y \in S(x)$. That is, the inclusion

$$S(x) \subseteq \left\{y | 0 \in \nabla_y f(x, y) + \widehat{N}_{\Gamma(x)}(y)\right\}$$

may be strict. However, It was pointed out by Mirrlees [20] that an optimal solution of the bilevel program may not even be a stationary point of the reformulation by the first order approach.

Ye and Zhu [36] proposed to investigate the optimality condition based on the value function reformulation first proposed by Outrata [23]. By the value function approach, one would replace the original bilevel program (BP) by the following equivalent problem:

$$\begin{aligned} \text{(VP)} \qquad \min \quad & F(x, y) \\ \text{s.t.} \quad & f(x, y) - V(x) < 0, \\ & g(x, y) \leq 0, h(x, y) = 0, \\ & G(x, y) \leq 0, \ H(x, y) = 0, \end{aligned}$$

where

$$V(x) := \inf_{y'}\{f(x, y') \mid g(x, y') \leq 0, h(x, y') = 0\}$$

is the value function of the lower level program.

There are two issues involved in using the value function approach. First, problem (VP) is a nonsmooth optimization problem since the value function $V(x)$ is in general nonsmooth. Moreover it is an implicit function of problem data. To ensure the lower semi-continuity of the value function we need the following assumption.

Definition 8.4.1 (See [3, Hypothesis 6.5.1] or [14, Definition 3.8]) We say that the restricted inf-compactness holds around $\bar{x}$ if $V(\bar{x})$ is finite and there exist a compact Ω and a positive number ϵ_0 such that, for all $x \in \mathbb{B}_{\epsilon_0}(\bar{x})$ for which $V(x) < V(\bar{x}) + \epsilon_0$, the problem (P_x) has a solution in Ω. △

The restricted inf-compactness condition is very weak. It does not even require the existence of solutions of problem (P_x) for all x near $\bar{x}$. A sufficient condition for the restricted inf-compactness to hold around $\bar{x}$ is the inf-compactness condition: there exist $\alpha > 0, \delta > 0$ and a bounded set C such that $\alpha > V(\bar{x})$ and

$$\{y | g(x, y) \leq 0, h(x, y) = 0, f(x, y) \leq \alpha, x \in \mathbb{B}_\delta(\bar{x})\} \subseteq C.$$

To ensure the Lipschitz continuity of the value function, we also need the following regularity condition.

Definition 8.4.2 For $\bar{y} \in S(\bar{x})$, we say that $(\bar{x}, \bar{y})$ is quasi-normal if there is no nonzero vector (λ^g, λ^h) such that

$$0 = \nabla g(\bar{x}, \bar{y})^T \lambda^g + \nabla h(\bar{x}, \bar{y})^T \lambda^h, \quad \lambda^g \geq 0$$

and there exists $(x^k, y^k) \to (\bar{x}, \bar{y})$ such that

$$\lambda_i^g > 0 \Rightarrow \lambda_i^g g_i(x^k, y^k) > 0,$$
$$\lambda_i^h \neq 0 \Rightarrow \lambda_i^h h_i(x^k, y^k) > 0.$$

△

It is easy to see that the quasinormality is weaker than MFCQ: there is no nonzero vector (λ^g, λ^h) such that

$$0 = \nabla g(\bar{x}, \bar{y})^T \lambda^g + \nabla h(\bar{x}, \bar{y})^T \lambda^h,$$
$$0 \leq -g(\bar{x}, \bar{y}) \perp \lambda^g \geq 0.$$

Now we can state a sufficient condition which ensures the Lipschitz continuity of the value function and an upper estimate for the limiting subdifferential of the value function. Since MFCQ is stronger than the quasi-normality and the set of quasi-normal multipliers is smaller than the classical multipliers, the following estimate is sharper and holds under weaker conditions than the classical counterpart in [3, Corollary 1 of Theorem 6.5.2].

Proposition 8.4.3 ([14, Corollary 4.8]) *Assume that the restricted inf-compactness holds around $\bar{x}$ and for each $\bar{y} \in S(\bar{x})$, $(\bar{x}, \bar{y})$ is quasi-normal. Then the value function $V(x)$ is Lipschitz continuous around $\bar{x}$ with*

$$\partial V(\bar{x}) \subseteq \widetilde{W}(\bar{x}), \tag{8.4.1}$$

where

$$\widetilde{W}(\bar{x}) := \bigcup_{\bar{y} \in S(\bar{x})} \left\{ \nabla_x f(\bar{x}, \bar{y}) + \nabla_x g(\bar{x}, \bar{y})^T \lambda^g + \nabla_x h(\bar{x}, \bar{y})^T \lambda^h : (\lambda^g, \lambda^h) \in \mathcal{M}(\bar{x}, \bar{y}) \right\}, \tag{8.4.2}$$

where $S(\bar{x})$ denotes the solution set of the lower level program $(P_{\bar{x}})$ and $\mathcal{M}(\bar{x}, \bar{y})$ is the set of quasi-normal multipliers, i.e.,

$$\mathcal{M}(\bar{x}, \bar{y}) := \left\{ (\lambda^g, \lambda^h) \middle| \begin{array}{l} 0 = \nabla_y f(\bar{x}, \bar{y}) + \nabla_y g(\bar{x}, \bar{y})^T \lambda^g + \nabla_y h(\bar{x}, \bar{y})^T \lambda^h, \ \lambda^g \geq 0 \\ \text{there exists } (x^k, y^k) \to (\bar{x}, \bar{y}) \text{ such that} \\ \lambda_i^g > 0 \Rightarrow \lambda_i^g g_i(x^k, y^k) > 0, \\ \lambda_i^h \neq 0 \Rightarrow \lambda_i^h h_i(x^k, y^k) > 0 \end{array} \right\}.$$

In addition to the above assumptions, if $\widetilde{W}(\bar{x}) = \{\zeta\}$, then $V(x)$ is strictly differentiable at $\bar{x}$ and $\nabla V(\bar{x}) = \{\zeta\}$. △

Note that moreover if the solution map of the lower level program $S(x)$ is semi-continuous at $(\bar{x}, \bar{y})$ for some $\bar{y} \in S(\bar{x})$, then the union $\bigcup_{\bar{y} \in S(\bar{x})}$ sign can be omitted in (8.4.2); see [21, Corollary 1.109].

Secondly, is a local optimal solution of problem (BP) a stationary point of problem (VP)? For problem (VP), suppose that $(\bar{x}, \bar{y})$ is a local optimal solution and if the value function $V(x)$ is Lipschitz continuous at $\bar{x}$, then the Fritz John type necessary optimality condition in terms of limiting subdifferential holds. That is, there exist multipliers $\alpha \geq 0$, $\mu \geq 0$, $\lambda^g, \lambda^h, \lambda^G, \lambda^H$ not all equal to zero such that

$$\begin{aligned} &0 \in \alpha \nabla_x F(\bar{x}, \bar{y}) + \mu \partial_x (f - V)(\bar{x}, \bar{y}) \\ &\quad + \nabla_x g(\bar{x}, \bar{y})^T \lambda^g + \nabla_x h(\bar{y}, \bar{y})^T \lambda^h + \nabla_x G(\bar{y}, \bar{y})^T \lambda^G + \nabla_x H(\bar{y}, \bar{y})^T \lambda^H, \\ &0 \in \alpha \nabla_y F(\bar{x}, \bar{y}) + \mu \nabla_y f(\bar{x}, \bar{y}) \\ &\quad + \nabla_y g(\bar{y}, \bar{y})^T \lambda^g + \nabla_y h(\bar{y}, \bar{y})^T \lambda^h + \nabla_y G(\bar{y}, \bar{y})^T \lambda^G + \nabla_y H(\bar{y}, \bar{y})^T \lambda^H, \\ &0 \leq -g(\bar{x}, \bar{y}) \perp \lambda^g \geq 0, \quad 0 \leq -G(\bar{x}, \bar{y}) \perp \lambda^G \geq 0. \end{aligned}$$

However it is easy to see that every feasible solution (x, y) of (VP) must be an optimal solution to the optimization problem

$$\begin{aligned} \min_{x', y'} \quad & f(x', y') - V(x') \\ \text{s.t.} \quad & g(x', y') \leq 0, h(x', y') = 0. \end{aligned}$$

By Fritz-John type optimality condition, there exists $\tilde{\mu} \geq 0, \tilde{\lambda}^g, \tilde{\lambda}^h$ not all equal to zero such that

$$\begin{aligned} &0 \in \tilde{\mu} \partial (f - V)(x, y) + \nabla g(x, y)^T \tilde{\lambda}^g + \nabla h(x, y)^T \tilde{\lambda}^h \\ &0 \leq -g(x, y) \perp \tilde{\lambda}^g \geq 0. \end{aligned}$$

This means that there always exists a nonzero abnormal multiplier $(0, \tilde{\mu}, \tilde{\lambda}^g, \tilde{\lambda}^h, 0, 0)$ for the problem (VP) at each feasible solution, i.e., the no nonzero abnormal multiplier constraint qualification (NNAMCQ) fails at each feasible point of the problem (VP). Therefore unlike the standard nonlinear programs, we can not derive the KKT condition (i.e., the Fritz John condition when $\alpha = 1$) from lack of nonzero abnormal multipliers. As we can see that the reason why NNAMCQ fails is the existence of the value function constraint $f(x, y) - V(x) \leq 0$. To address this issue, Ye and Zhu [36] proposed the following partial calmness condition.

Definition 8.4.4 Let $(\bar{x}, \bar{y})$ be a local optimal solution of problem (VP). We say that (VP) is partially calm at $(\bar{x}, \bar{y})$ provided that there exist $\delta > 0$, $\mu > 0$ such that for all $\alpha \in \mathbb{B}_\delta$ and all $(x, y) \in \mathbb{B}_\delta(\bar{x}, \bar{y})$ which are feasible for the partially perturbed problem

$$
\begin{aligned}
(\mathrm{VP}_\alpha) \qquad & \min \ F(x, y) \\
& \text{s.t. } f(x, y) - V(x) + \alpha = 0, \\
& \qquad g(x, y) \leq 0, h(x, y) = 0, \\
& \qquad G(x, y) \leq 0, \ H(x, y) = 0,
\end{aligned}
$$

there holds $F(x, y) - F(\bar{x}, \bar{y}) + \mu\|\alpha\| \geq 0$. △

It is obvious that the partial calmness is equivalent to the exact penalization, i.e., (VP) is partially calm at $(\bar{x}, \bar{y})$ if and only if for some $\mu > 0$, $(\bar{x}, \bar{y})$ is a local solution of the penalized problem

$$
\begin{aligned}
(\widetilde{\mathrm{VP}}) \qquad & \min \ F(x, y) + \mu(f(x, y) - V(x)) \\
& \text{s.t. } g(x, y) \leq 0, h(x, y) = 0, \\
& \qquad G(x, y) \leq 0, \ H(x, y) = 0.
\end{aligned}
$$

Since the difficult constraint $f(x, y) - V(x) \leq 0$ is replaced by a penalty in the objective function, the usual constraint qualification such as MFCQ or equivalently NNAMCQ can be satisfied for problem $(\widetilde{\mathrm{VP}})$. Consequently using a nonsmooth multiplier rule for problem $(\widetilde{\mathrm{VP}})$, one can derive a KKT type optimality condition for problem (VP). Such an approach has been used to derive necessary optimality condition of (BP) by Ye and Ye in [36] and later in other papers such as [7, 8, 22]. For this approach to work, however, one needs to ensure the partial calmness condition. In [36], it was shown that for the minmax problem and the bilevel program where the lower level is completely linear, the partial calmness condition holds automatically. In [7, Theorem 4.2], the last result was improved to conclude that the partial calmness condition holds automatically for any bilevel program where for each x, the lower level problem is a linear program. In [36, Proposition 5.1], the uniform weak sharp minimum is proposed as a sufficient condition for partial calmness and under certain conditions, the bilevel program with a quadratic program as the lower level program is shown to satisfy the partial calmness condition in [36, Proposition 5.2] (with correction in [37]).

Apart from the issue of constraint qualification, we may ask a question on how likely an optimal solution of (VP) is a stationary point of (VP). In the case where there are no upper and lower level constraints, the stationary condition of (VP) at $(\bar{x}, \bar{y})$ means the existence of $\mu \geq 0$ such that

$$0 \in \nabla_x F(\bar{x}, \bar{y}) + \mu \partial_x (f - V)(\bar{x}, \bar{y}),$$
$$0 = \nabla_y F(\bar{x}, \bar{y}) + \mu \nabla_y f(\bar{x}, \bar{y}).$$

But this condition is very strong. It will not hold unless $0 = \nabla_y F(\bar{x}, \bar{y})$.

As suggested by Ye and Zhu in [38], we may consider the combined program

$$\begin{aligned}
\text{(CP)} \qquad \min_{x,y,u,v} \quad & F(x, y) \\
\text{s.t.} \quad & f(x, y) - V(x) \leq 0, \\
& 0 = \nabla_y L(x, y, u, v) := \nabla_y f(x, y) + \nabla_y g(x, y)^T u + \nabla_y h(x, y)^T v, \\
& h(x, y) = 0, \quad 0 \leq -g(x, y) \perp u \geq 0, \\
& G(x, y) \leq 0, \; H(x, y) = 0.
\end{aligned}$$

The motivation is clear since if the KKT conditions hold at each optimal solution of the lower level problem, then the KKT condition is a redundant condition. By adding the KKT condition we have not changed the feasible region of (BP). Note that this reformulation requires the existence of the KKT condition at the optimal solution of the lower level program; see [6] for examples where the KKT condition does not hold at a lower level optimal solution.

Similarly as in the case of using MPCC to reformulate a bilevel program, when the lower level multipliers are not unique, it is possible that $(\bar{x}, \bar{y}, \bar{u}, \bar{v})$ is a local solution of (CP) but $(\bar{x}, \bar{y})$ is not a local optimal solution of (BP).

Due to the existence of the value function constraint $f(x, y) - V(x) \leq 0$, similarly to the analysis with problem (VP), NNAMCQ will never hold at a feasible solution of (CP) and hence in [38] the following partial calmness condition for problem (CP) is suggested as a condition to deal with the problem.

Definition 8.4.5 Let $(\bar{x}, \bar{y}, \bar{u}, \bar{v})$ be a local optimal solution of problem (CP). We say that (CP) is partially calm at $(\bar{x}, \bar{y}, \bar{u}, \bar{v})$ provided that there exist $\mu > 0$ such that $(\bar{x}, \bar{y}, \bar{u}, \bar{v})$ is a local solution of the partially perturbed problem:

$$\begin{aligned}
(\text{CP}_\mu) \qquad \min \quad & F(x, y) + \mu(f(x, y) - V(x)) \\
\text{s.t.} \quad & 0 = \nabla_y L(x, y, u, v), \\
& h(x, y) = 0, \quad 0 \leq -g(x, y) \perp u \geq 0, \\
& G(x, y) \leq 0, \; H(x, y) = 0.
\end{aligned}$$

△

Since there are more constraints in (CP) than in (VP), the partial calmness for (CP) is a weaker condition than the one for (VP).

Given a feasible vector $(\bar{x}, \bar{y}, \bar{u}, \bar{v})$ of the problem (CP), define the following index sets:

$$I_G = I_G(\bar{x}, \bar{y}) := \{i : G_i(\bar{x}, \bar{y}) = 0\},$$
$$I_g = I_g(\bar{x}, \bar{y}, \bar{u}) := \{i : g_i(\bar{x}, \bar{y}) = 0, \bar{u}_i > 0\},$$
$$I_0 = I_0(\bar{x}, \bar{y}, \bar{u}) := \{i : g_i(\bar{x}, \bar{y}) = 0, \bar{u}_i = 0\},$$
$$I_u = I_u(\bar{x}, \bar{y}, \bar{u}) := \{i : g_i(\bar{x}, \bar{y}) < 0, \bar{u}_i = 0\}.$$

Definition 8.4.6 (M-Stationary Condition for (CP) Based on the Value Function) A feasible point $(\bar{x}, \bar{y}, \bar{u}, \bar{v})$ of problem (CP) is called an M-stationary point based on the value function if there exist $\mu \geq 0$, $\beta \in \mathbb{R}^s$, $\lambda^G \in \mathbb{R}^p$, $\lambda^H \in \mathbb{R}^q$, $\lambda^g \in \mathbb{R}^m$, $\lambda^h \in \mathbb{R}^n$ such that the following conditions hold:

$$0 \in \partial F(\bar{x}, \bar{y}) + \mu \partial (f - V)(\bar{x}, \bar{y}) + \nabla G(\bar{x}, \bar{y})^T \lambda^G + \nabla H(\bar{x}, \bar{y})^T \lambda^H$$
$$+ \nabla_{x,y}(\nabla_y L)(\bar{x}, \bar{y})^T \beta + \nabla g(\bar{x}, \bar{y})^T \lambda^g + \nabla h(\bar{x}, \bar{y})^T \lambda^h,$$
$$\lambda_i^G \geq 0 \ i \in I_G, \ \lambda_i^G = 0 \ i \notin I_G,$$
$$\lambda_i^g = 0 \ i \in I_u, \ (\nabla_y g(\bar{x}, \bar{y})\beta)_i = 0 \ i \in I_g,$$
$$either \quad \lambda_i^g > 0, (\nabla_y g(\bar{x}, \bar{y})\beta)_i > 0, \quad \text{or} \quad \lambda_i^g (\nabla_y g(\bar{x}, \bar{y})\beta)_i = 0 \quad i \in I_0. \quad \triangle$$

In [38, Theorem 4.1], it was shown that under the partial calmness condition and certain constraint qualifications, a local optimal solution of (CP) must be an M-stationary point based on the value function provided the value function is Lipschitz continuous.

Recently Xu and Ye [29] introduced a nonsmooth version of the relaxed constant positive linear dependence (RCPLD) condition and apply it to (CP). We now describe the RCPLD condition.

In the following definition, we rewrite all equality constraints of problem (CP) by the equality constraint below:

$$0 = \Phi(x, y, u, v) := \begin{pmatrix} \nabla_y L(x, y, u, v) \\ h(x, y) \\ H(x, y) \end{pmatrix}.$$

We denote by $\{0\}^n$ the zero vector in $\mathbb{R}^n$ and by e_i the unit vector with the ith component equal to 1.

Definition 8.4.7 Suppose that the value function $V(x)$ is Lipschitz continuous at $\bar{x}$. Let $(\bar{x}, \bar{y}, \bar{u}, \bar{v})$ be a feasible solution of (CP). We say that RCPLD holds at $(\bar{x}, \bar{y}, \bar{u}, \bar{v})$ if the following conditions hold.

(I) The vectors

$$\{\nabla\Phi_i(x,y,u,v)\}_{i=1}^{m+s+q} \cup \{\nabla g_i(x,y)\times\{0^{r+s}\}\}_{i\in I_g} \cup \{(0^{n+m}, e_i, 0^s)\}_{i\in I_u}$$

have the same rank for all (x,y,u,v) in a neighbourhood of $(\bar{x},\bar{y},\bar{u},\bar{v})$.

(II) Let $\mathcal{I}_1 \subseteq \{1,\cdots,m+s+q\}$, $\mathcal{I}_2 \subseteq I_g$, $\mathcal{I}_3 \subseteq I_u$ be such that the set of vectors $\{\nabla\Phi_i(\bar{x},\bar{y},\bar{u},\bar{v})\}_{i\in\mathcal{I}_1} \cup \{\nabla g_i(\bar{x},\bar{y})\times\{0\}\}_{i\in\mathcal{I}_2} \cup \{(0,e_i,0)\}_{i\in\mathcal{I}_3}$ is a basis for

$$\operatorname{span}\left\{\nabla\Phi_i(x,y,u,v)\}_{i=1}^{m+s+q}\cup\{\nabla g_i(\bar{x},\bar{y})\times\{0^{r+s}\}\}_{i\in I_g}\cup\{(0^{n+m},e_i,0^s)\}_{i\in I_u}\right\}.$$

For any index sets $\mathcal{I}_4 \subseteq I_G, \mathcal{I}_5, \mathcal{I}_6 \subseteq I_0$, the following conditions hold.

(i) If there exists a nonzero vector $(\lambda^V,\lambda^\Phi,\lambda^G,\lambda^g,\lambda^u) \in \mathbb{R}\times\mathbb{R}^{m+s+q}\times\mathbb{R}^p\times\mathbb{R}^r\times\mathbb{R}^r$ satisfying $\lambda^V \geq 0,\ \lambda^G \geq 0$ and either $\lambda_i^g > 0, \lambda_i^u > 0$ or $\lambda_i^g\lambda_i^u = 0, \forall i \in I_0,\ \xi^* \in \partial(f-V)(\bar{x},\bar{y})$ such that

$$0 = \lambda^V\xi^* + \sum_{i\in\mathcal{I}_1}\lambda_i^\Phi\nabla\Phi_i(\bar{x},\bar{y},\bar{u},\bar{v}) + \sum_{i\in\mathcal{I}_4}\lambda_i^G\nabla G_i(\bar{x},\bar{y})\times\{0^{r+s}\}$$
$$+ \sum_{i\in\mathcal{I}_2\cup\mathcal{I}_5}\lambda_i^g\nabla g_i(\bar{x},\bar{y})\times\{0^{r+s}\} - \sum_{i\in\mathcal{I}_3\cup\mathcal{I}_6}\lambda_i^u(0^{n+m},e_i,0^s)$$

and $(x^k,y^k,u^k,v^k,\xi^k) \to (\bar{x},\bar{y},\bar{u},\bar{v},\xi^*)$ as $k\to\infty$, $\xi^k \in \partial(f-V)(x^k,y^k)$ then the set of vectors

$$\{\xi^k\}\cup\{\nabla\Phi_i(x^k,y^k,u^k,v^k)\}_{i\in\mathcal{I}_1}\cup\{\nabla G_i(x^k,y^k)\times\{0^{r+s}\}\}_{i\in\mathcal{I}_4}$$
$$\cup\{\nabla g_i(x^k,y^k)\times\{0^{r+s}\}\}_{i\in\mathcal{I}_2\cup\mathcal{I}_5}\cup\{(0^{n+m},e_i,0^s)\}_{i\in\mathcal{I}_3\cup\mathcal{I}_6},$$

where k is sufficiently large and $(x^k,y^k,u^k,v^k) \neq (\bar{x},\bar{y},\bar{u},\bar{v})$, is linearly dependent.

(ii) If there exists a nonzero vector $(\lambda^\Phi,\lambda^G,\lambda^g,\lambda^u) \in \mathbb{R}^{m+s+q}\times\mathbb{R}^p\times\mathbb{R}^r\times\mathbb{R}^r$ satisfying $\lambda^G \geq 0$ and either $\lambda_i^g > 0, \lambda_i^u > 0$ or $\lambda_i^g\lambda_i^u = 0, \forall i \in I_0$ such that

$$0 = \sum_{i\in\mathcal{I}_1}\lambda_i^\Phi\nabla\Phi_i(\bar{x},\bar{y},\bar{u},\bar{v}) + \sum_{i\in\mathcal{I}_4}\lambda_i^G\nabla G_i(\bar{x},\bar{y})\times\{0^{r+s}\}$$
$$+ \sum_{i\in\mathcal{I}_2\cup\mathcal{I}_5}\lambda_i^g\nabla g_i(\bar{x},\bar{y})\times\{0^{r+s}\} - \sum_{i\in\mathcal{I}_3\cup\mathcal{I}_6}\lambda_i^u(0^{n+m},e_i,0^s),$$

and $(x^k,y^k,u^k,v^k) \to (\bar{x},\bar{y},\bar{u},\bar{v})$, as $k\to\infty$. Then the set of vectors

$$\{\nabla\Phi_i(x^k,y^k,u^k,v^k)\}_{i\in\mathcal{I}_1}\cup\{\nabla G_i(x^k,y^k)\times\{0^{r+s}\}\}_{i\in\mathcal{I}_4}$$
$$\cup\{\nabla g_i(x^k,y^k)\times\{0^{r+s}\}\}_{i\in\mathcal{I}_2\cup\mathcal{I}_5}\cup\{(0^{n+m},e_i,0^s)\}_{i\in\mathcal{I}_3\cup\mathcal{I}_6},$$

where k is sufficiently large and $(x^k, y^k, u^k, v^k) \neq (\bar{x}, \bar{y}, \bar{u}, \bar{v})$, is linearly dependent. △

Since

$$\begin{aligned}\partial(f - V)(\bar{x}, \bar{y}) &\subseteq \partial^c(f - V)(\bar{x}, \bar{y}) \\ &= \nabla f(\bar{x}, \bar{y}) - \partial^c V(\bar{x}) \times \{0\} \subseteq \nabla f(\bar{x}, \bar{y}) - conv\widetilde{W}(\bar{x}) \times \{0\},\end{aligned}$$

where $\widetilde{W}(\bar{x})$ is the upper estimate of the limiting subdifferential of the value function at $\bar{x}$ defined as in (8.4.2), we can replace the set $\partial(f-V)(\bar{x}, \bar{y})$ by its upper estimate $\nabla f(\bar{x}, \bar{y}) - conv\widetilde{W}(\bar{x}) \times \{0\}$ in RCPLD and obtain a sufficient condition for RCPLD. Moreover if the solution map of the lower level program $S(x)$ is semi-continuous at $(\bar{x}, \bar{y})$, then the set $\partial(f-V)(\bar{x}, \bar{y})$ can be replaced by its upper estimate $\nabla f(\bar{x}, \bar{y}) - \widetilde{W}(\bar{x}) \times \{0\}$.

Theorem 8.4.8 ([29]) *Let $(\bar{x}, \bar{y}, \bar{u}, \bar{v})$ be a local solution of (CP) and suppose that the value function $V(x)$ is Lipschitz continuous at $\bar{x}$. If RCPLD holds at $(\bar{x}, \bar{y}, \bar{u}, \bar{v})$, then $(\bar{x}, \bar{y}, \bar{u}, \bar{v})$ is an M-stationary point of problem (CP) based on the value function.* △

In the following result, the value function constraint $f(x, y) - V(x) \leq 0$ is not needed in the verification.

Theorem 8.4.9 ([29]) *Let $(\bar{x}, \bar{y}, \bar{u}, \bar{v})$ be a local solution of (CP) and suppose that the value function $V(x)$ is Lipschitz continuous at $\bar{x}$. If the rank of the matrix*

$$J^* = \begin{bmatrix} \nabla(\nabla_y f + \nabla_y g^T \bar{u} + \nabla_y h^T \bar{v})(\bar{x}, \bar{y}) & \nabla_y h(\bar{x}, \bar{y})^T & \nabla_y g_{I_g \cup I_0}(\bar{x}, \bar{y})^T \\ \nabla h(\bar{x}, \bar{y}) & 0 & 0 \\ \nabla H(\bar{x}, \bar{y}) & 0 & 0 \\ \nabla g_{I_g}(\bar{x}, \bar{y}) & 0 & 0 \end{bmatrix}$$

is equal to $m+n+r+s-|I_u|$. Then RCPLD holds and $(\bar{x}, \bar{y}, \bar{u}, \bar{v})$ is an M-stationary point of problem (CP) based on the value function. △

In the last part of this section we briefly summarize some necessary optimality conditions obtained in [33] using the combined approach. For any given $\bar{x}$, define the set

$$W(\bar{x}) := \bigcup_{\bar{y} \in S(\bar{x})} \left\{ \begin{array}{l} \nabla_x f(\bar{x}, \bar{y}) + \nabla_x g(\bar{x}, \bar{y})^T \lambda^g + \nabla_x h(\bar{x}, \bar{y})^T \lambda^h : \\ 0 = \nabla_y g(\bar{x}, \bar{y})^T \lambda^g + \nabla_y h(\bar{x}, \bar{y})^T \lambda^h \\ 0 \leq -g(x, y) \perp \lambda^g \geq 0 \end{array} \right\}.$$

It is easy to see that $\widetilde{W}(\bar{x}) \subseteq W(\bar{x})$ and under the assumption made in Proposition 8.4.3, it is an upper estimate of the limiting subdifferential of the value function.

Definition 8.4.10 Let $(\bar{x}, \bar{y}, \bar{u}, \bar{v})$ be a feasible solution to (CP). We say that (CP) is weakly calm at $(\bar{x}, \bar{y}, \bar{u}, \bar{v})$ with modulus $\mu > 0$ if

$$[\nabla F(\bar{x}, \bar{y}) + \mu \nabla f(\bar{x}, \bar{y})]^T (d_x, d_y) - \mu \min_{\xi \in W(\bar{x})} \xi d_x \geq 0 \quad \forall d \in \mathcal{L}^{MPEC}((\bar{x}, \bar{y}, \bar{u}, \bar{v}); \widetilde{F}),$$

where $\widetilde{F}$ is the feasible region of problem (CP_μ) and $\mathcal{L}^{MPEC}((\bar{x}, \bar{y}, \bar{u}, \bar{v}); \widetilde{F})$ is the MPEC linearized cone of $\widetilde{F}$ defined by

$$\mathcal{L}^{MPEC}((\bar{x}, \bar{y}, \bar{u}, \bar{v}); \widetilde{F})$$

$$:= \left\{ \begin{array}{ll} (d_x, d_y, d_u, d_v) | & \\ \nabla_{x,y}(\nabla_y L)(\bar{x}, \bar{y}, \bar{u}, \bar{v})(d_x, d_y) + \nabla_y g(\bar{x}, \bar{y})^T d_u + \nabla_y h(\bar{x}, \bar{y})^T d_v = 0 & \\ \nabla G_i(\bar{x}, \bar{y})^T (d_x, d_y) \leq 0, & i \in I_G \\ \nabla H_i(\bar{x}, \bar{y})^T (d_x, d_y) = 0, & \\ \nabla g_i(\bar{x}, \bar{y})^T (d_x, d_y) = 0, & i \in I_g \\ (d_u)_i = 0, & i \in I_u \\ \nabla g_i(\bar{x}, \bar{y})^T (d_x, d_y) \cdot (d_u)_i = 0, \nabla g_i(\bar{x}, \bar{y})^T (d_x, d_y) \leq 0, (d_u)_i \geq 0 & i \in I_0. \end{array} \right\}.$$

△

Definition 8.4.11 (M-Stationary Condition for (CP) Based on the Upper Estimate) A feasible point $(\bar{x}, \bar{y}, \bar{u}, \bar{v})$ of problem (CP) is called an M-stationary point based on an upper estimate if there exist $\mu \geq 0$, $\beta \in \mathbb{R}^s$, $\lambda^G \in \mathbb{R}^p$, $\lambda^H \in \mathbb{R}^q$, $\lambda^g \in \mathbb{R}^m$, $\lambda^h \in \mathbb{R}^n$ such that the following conditions hold:

$$\begin{aligned} & 0 \in \partial F(\bar{x}, \bar{y}) + \mu[\nabla f(\bar{x}, \bar{y}) - conv W(\bar{x}) \times \{0\}] + \nabla G(\bar{x}, \bar{y})^T \lambda^G + \nabla H(\bar{x}, \bar{y})^T \lambda^H \\ & \quad + \nabla(\nabla_y f + \nabla_y g^T \bar{u} + \nabla_y h^T \bar{v})(\bar{x}, \bar{y})^T \beta + \nabla g(\bar{x}, \bar{y})^T \lambda^g + \nabla h(\bar{x}, \bar{y})^T \lambda^h, \\ & \lambda_i^G \geq 0 \ i \in I_G, \ \ \lambda_i^G = 0 \ i \notin I_G, \\ & \lambda_i^g = 0 \ i \in I_u, \ (\nabla_y g(\bar{x}, \bar{y})\beta)_i = 0 \ i \in I_g, \\ & either \quad \lambda_i^g > 0, (\nabla_y g(\bar{x}, \bar{y})\beta)_i > 0, \quad \text{or} \quad \lambda_i^g (\nabla_y g(\bar{x}, \bar{y})\beta)_i = 0 \quad i \in I_0. \end{aligned}$$

△

Let $(\bar{x}, \bar{y}, \bar{u}, \bar{v})$ be a feasible solution of problem (CP). By [33, Theorem 4.3], if the set $W(\bar{x})$ is nonempty and compact and (CP) is MPEC-weakly calm at $(\bar{x}, \bar{y}, \bar{u}, \bar{v})$, then $(\bar{x}, \bar{y}, \bar{u}, \bar{v})$ is an M-stationary point of problem (CP) based on an upper estimate. Note that it is obvious that the M-stationary condition based on an upper estimate is weaker than the corresponding M-stationary condition based on the value function.

Acknowledgements The research of this author was partially supported by NSERC. The author would like to thank an anonymous referee for the helpful suggestions and comments that have helped to improve the presentation of the paper.

References

1. L. Adam, R. Henrion, J. Outrata, On M-stationarity conditions in MPECs and the associated qualification conditions. Math. Program. **168**, 229–259 (2018)
2. J. Aubin, Lipschitz behavior of solutions to convex minimization problems. Math. Opera. Res. **9**, 87–111 (1984)
3. F.H. Clarke, *Optimization and Nonsmooth Analysis* (Wiley-Interscience, New York, 1983)
4. S. Dempe, *Foundations of Bilevel Programming* (Kluwer Academic Publishers, New York, 2002)
5. S. Dempe, Annotated bibliography on bilevel programming and mathematical programs with equilibrium constraints. Optimization **52**, 333–359 (2003)
6. S. Dempe, J. Dutta, Is bilevel programming a special case of a mathematical program with complementarity constraints?. Math. Program. **131**, 37–48 (2012)
7. S. Dempe, A.B. Zemkoho, The bilevel programming problems: reformulations, constraint qualifications and optimality conditions. Math. Program. **138**, 447–473 (2013)
8. S. Dempe, J. Dutta, B.S. Mordukhovich, New necessary optimality conditions in optimistic bilevel programming. Optimization **56**, 577–604 (2007)
9. A.L. Dontchev, R.T. Rockafellar, Regularity and conditioning of solution mappings in variational analysis. Set-Valued Anal. **12**, 79–109 (2004)
10. H. Gfrerer, J.V. Outrata, On computation of limiting coderivatives of the normal-cone mapping to inequality systems and their applications. Optimization **65**, 671–700 (2016)
11. H. Gfrerer, J.V. Outrata, On computation of generalized derivatives of the normal-cone mapping and their applications. Math. Oper. Res. **41**, 1535–1556 (2016)
12. H. Gfrerer, J.J. Ye, New constraint qualifications for mathematical programs with equilibrium constraints via variational analysis. SIAM J. Optim. **27**, 842–865 (2017)
13. H. Gfrerer, J.J. Ye, New sharp necessary optimality conditions for mathematical programs with equilibrium constraints. Set-Valued Var. Anal. **28**, 395–426 (2020)
14. L. Guo, G-H. Lin, J.J. Ye, J. Zhang, Sensitivity analysis of the value functions for parametric mathematical programs with equilibrium constraints. SIAM J. Optim. **24**, 1206–1237 (2014)
15. R. Henrion, W. Römisch, On M-stationary points for a stochastic equilibrium problem under equilibrium constraints in electricity spot market modeling. Appl. Math. **52**, 473–494 (2007)
16. R. Henrion, J. Outrata, T. Surowiec, On the co-derivative of normal cone mappings to inequality systems. Nonlinear Anal. Theory Methods Appl. **71**, 1213–1226 (2009)
17. G. Kunapuli, K.P. Bennett, J. Hu, J-S. Pang, Classification model selection via bilevel programming. Optim. Meth. Softw. **23**, 475–489 (2008)
18. Y.-C. Lee, J.-S. Pang, J.E. Mitchell, Global resolution of the support vector machine regression parameters selection problem with LPCC. EURO J. Comput. Optim. **3**, 197–261 (2015)
19. Z.-Q. Luo, J.-S. Pang, *Mathematical Programs with Equilibrium Constraints* (Cambridge University Press, Cambridge, 1996)
20. J. Mirrlees, The theory of moral hazard and unobservable behaviour– part I. Rev. Econ. Stud. **66**, 3–22 (1999)
21. B.S. Mordukhovich, *Variational Analysis and Generalized Differentiation, Vol. 1: Basic Theory, Vol. 2: Applications* (Springer, Berlin, 2006)
22. B.S. Mordukhovich, N.M. Nam, H.M. Phan, Variational analysis of marginal functions with applications to bilevel programming. J. Optim. Theory Appl. **152**, 557–586 (2012)
23. J.V. Outrata, On the numerical solution of a class of Stackelberg problems. ZOR-Math. Methods Oper. Res. **34**, 255–277 (1990)
24. J.V. Outrata, M. Kočvara, J. Zowe, *Nonsmooth Approach to Optimization Problems with Equilibrium Constraints: Theory, Applications and Numerical Results* (Kluwer Academic Publishers, Dordrecht, 1998)
25. S.M. Robinson, Stability theory for systems of inequalities. Part I: Linear systems. SIAM J. Numer. Anal. **12**, 754–769 (1975)

26. S.M. Robinson, Some continuity properties of polyhedral multifunctions. Math. Program. Stud. **14**, 206–214 (1981)
27. R.T. Rockafellar, R.J.-B. Wets, *Variational Analysis* (Springer, Berlin, 1998)
28. H. von Stackelberg, *Marktform and Gleichgewicht* (Springer, Berlin, 1934). Engl. transl.: *The Theory of the Market Economy* (Oxford University Press, Oxford, 1954)
29. M. Xu, J.J. Ye, Relaxed constant positive linear dependence constraint qualification and its application to bilevel programs. J. Glob. Optim. **78**, 181–205 (2020)
30. J.J. Ye, Nondifferentiable multiplier rules for optimization and bilevel optimization problems. SIAM J. Optim. **15**, 252–274 (2004)
31. J.J. Ye, Necessary and sufficient optimality conditions for optimization programs with equilibrium constraints. J. Math. Anal. Appl. **307**, 350–369 (2005)
32. J.J. Ye, Constraint qualifications and KKT conditions for bilevel programming problems. Math. Oper. Res. **31**, 811–824 (2006)
33. J.J. Ye, Necessary optimality conditions for multiobjective bilevel programs. Math. Oper. Res. **36**, 165–184 (2011)
34. J.J. Ye, S.Y. Wu, First order optimality conditions for generalized semi-infinite programming problems. J. Optim. Theory Appl. **137**, 419–434 (2008)
35. J.J. Ye, X.Y. Ye, Necessary optimality conditions for optimization problems with variational inequality constraints. Math. Oper. Res. **22**, 977–997 (1997)
36. J.J. Ye, D.L. Zhu, Optimality conditions for bilevel programming problems. Optimization **33**, 9–27 (1995)
37. J.J. Ye, D.L. Zhu, A note on optimality conditions for bilevel programming problems. Optimization **39**, 361–366 (1997)
38. J.J. Ye, D.L. Zhu, New necessary optimality conditions for bilevel programs by combining the MPEC and value function approaches. SIAM J. Optim. **20**, 1885–1905 (2010)

Chapter 9
Algorithms for Simple Bilevel Programming

Joydeep Dutta and Tanushree Pandit

Abstract In this article we focus on algorithms for solving simple bilevel programming problems. Simple bilevel problems consist of minimizing a convex function over the solution set of another convex optimization problem. Though the problem is convex the bilevel structure prevents the direct application of the standard methods of convex optimization. Hence several algorithms have been developed in the literature to tackle this problem. In this article we discuss several such algorithms including recent ones.

Keywords Bilevel programming problems · Convex optimization · Projected gradient method · Gradient Lipschitz condition · Proximal point method · Convex functions · Strongly convex functions

9.1 Introduction

In this article we focus on the following convex optimization problem (SBP)

$$\begin{aligned}
&\min f(x)\\
&\text{subject to } x \in S,\\
&\text{where } S = \arg\min\{g(x) : x \in C\}.
\end{aligned}$$

Here $f, g : \mathbb{R}^n \to \mathbb{R}$ are convex functions and C a closed convex set in $\mathbb{R}^n$. However we could consider $f, g : U \to \mathbb{R}$, where U is an open convex set containing C. Further we can also assume that $f : \mathbb{R}^n \to \bar{\mathbb{R}}$ and $g : \mathbb{R}^n \to \bar{\mathbb{R}}$, where $\bar{\mathbb{R}} = \mathbb{R} \cup \{-\infty, +\infty\}$ are proper, lower semi-continuous, convex functions. The problem stated above is called (SBP) to denote the phrase Simple Bilevel Program.

J. Dutta (✉) · T. Pandit
Indian Institute of Technology Kanpur, Kanpur, India
e-mail: jdutta@iitk.ac.in; tpandit@iitk.ac.in

S. Dempe, A. Zemkoho (eds.), *Bilevel Optimization*, Springer Optimization and Its Applications 161, https://doi.org/10.1007/978-3-030-52119-6_9

The fact that (SBP) has a bilevel structure is evident, while we call it simple since it is represented by only one decision variable in contrast to the two-variable representation in standard bilevel problem. For a detailed discussion on bilevel programming see Dempe [3]. Those readers with an experience in optimization may immediately make the following observation. In general a convex optimization problem may have more than one solution and hence uncountably infinite number of solutions. The decision maker usually would prefer only one among these many solutions. A best way to pick a single solution is to devise a strongly convex objective which is minimized over the solution set of the convex optimization. That gives rise to a simple bilevel problem. The above discussion is explained through the following example.

Example
Consider the following linear programming problem (LP)

$$\begin{aligned}&\min \langle c, x\rangle\\&\text{subject to } Ax = b,\\&x \geq 0.\end{aligned}$$

In many situations one needs to find a minimum norm solution of the problem (LP). Thus we may formulate the following simple bilevel problem

$$\begin{aligned}&\min \|x\|_2^2\\&\text{such that } x \in \arg\min\{\langle c, z\rangle : Az = b, z \geq 0\}\end{aligned}$$

On the other hand if we are seeking a sparse solution then we can formulate the problem as

$$\begin{aligned}&\min \|x\|_1\\&x \in \arg\min\{\langle c, z\rangle : Az = b, z \geq 0\}.\end{aligned}$$

Solodov [22] shows that any standard convex optimization problem can be expressed as a simple bilevel optimization problem. Consider the convex optimization problem (CP)

$$\begin{aligned}&\min f(x)\\&\text{subject to } Ax = b,\end{aligned}$$

where $f : \mathbb{R}^n \to \mathbb{R}$ is convex, A is an $m \times n$ matrix, $b \in \mathbb{R}^m$ and each $h_i : \mathbb{R}^n \to \mathbb{R}$, $i = 1, 2, \cdots, k$ is convex. Consider the following simple bilevel problem

$$\min f(x)$$
$$\text{subject to } x \in \underset{\mathbb{R}^n}{\arg\min}\, g,$$

where $g(x) = \|Ax - b\|^2 + \|\max\{h(x), 0\}\|^2$, $h(x) = (h_1(x), \cdots, h_k(x))$ and the maximum is taken component-wise i.e.

$$\max\{h(x), 0\} = (\max\{h_1(x), 0\}, \cdots, \max\{h_k(x), 0\})$$

Let x^* be a feasible point of (CP). Then $g(x^*) = 0$ and $x^* \in \arg\min_{\mathbb{R}^n} g$, since $g(x) \geq 0$ for all $x \in \mathbb{R}^n$. This shows that if F is the feasible set of (CP), then

$$F \subseteq \underset{\mathbb{R}^n}{\arg\min}\, g.$$

Further if $g(x^*) = 0$, then $x^* \in F$; showing that $F = \arg\min_{\mathbb{R}^n} g$. This implies that if $\bar{x}$ is a solution of (CP), then $f(\bar{x}) = \min\{f(x) : x \in \arg\min_{\mathbb{R}^n} g\}$.

There is another class of problems where the (SBP) formulation may play role in estimating the distance between a feasible point and the solution set of a convex programming problem. This was discussed in Dempe et al. [7]. However for completeness we will also discuss it here through the following example.

Example
Consider a convex programming problem (CP1)

$$\min f(x)$$
$$x \in C,$$

where $C = \{x \in \mathbb{R}^n : g_i(x) \leq 0; i = 1, \cdots, m\}$. For this problem the solution set is given as

$$\underset{C}{\arg\min}\, f = \{x : f(x) \leq \bar{\alpha}, g_1(x) \leq 0, \cdots, g_m(x) \leq 0\}, \qquad (9.1.1)$$

where $\bar{\alpha} = \inf_{x \in C} f$. Given any $\hat{x} \in \mathbb{R}^n$, suppose we want to estimate $d(\hat{x}, \arg\min_C f)$. In general this may not be so easily done. If we consider $\arg\min_C f$ as a system of convex inequalities as shown above in (9.1.1), we can not use the error bound theory in the literature (see for example Lewis

(continued)

and Pang [17] and the references therein) to establish an upper bound to $d(\hat{x}, \arg\min_C f)$, as the Slater's condition fails for this system of convex inequalities. One way out of this is to use the simple bilevel programming to estimate $d(\hat{x}, \arg\min_C f)$ numerically. In fact we can consider the following problem for a given $\hat{x} \in \mathbb{R}^n$,

$$\min_y \frac{1}{2}\|y - \hat{x}\|^2$$

$$y \in \arg\min_C f.$$

Since $\arg\min_C f$ is closed and convex if f and g_i's are finite-valued and $\frac{1}{2}\|y - \hat{x}\|^2$ is coercive in y and thus a minimizer exists and is unique since $\frac{1}{2}\|y - \hat{x}\|^2$ is strongly convex in y. Let y^* be that unique solution, then

$$d(\hat{x}, \arg\min_C f) = \|y^* - \hat{x}\|.$$

Thus if we want to devise an effective stopping criteria for solving the problem (CP1), one can estimate the distance from the solution set by using the approach of simple bilevel programming. The key idea in developing the algorithms for the simple bilevel programming problem rests on the idea that we shall be able to solve the problem without separately solving the lower level problem.

Note that if in (CP1), f is strongly convex, then an upper bound to $d(\hat{x}, \arg\min_C f)$ can be computed using the techniques of gap function. See for example Fukushima [24]. If f is convex but not strongly convex then the gap function approach does not seem to work and it appears that at least at present the simple bilevel programming approach is a nice way to estimate $d(\hat{x}, \arg\min_C f)$.

However one may argue that if we consider a linear programming problem, then we can simply use the celebrated Hoffman error bound to devise an upper bound to the solution set. Consider the following linear programming problem

$$\min c^T x,$$

$$\text{subject to } \langle a_j, x\rangle \geq b_j, \quad j = 1, \cdots, m$$

$$x_i \geq 0, \quad i = 1, \cdots, n;$$

where $c \in \mathbb{R}^n$, $a_j \in \mathbb{R}^n$, $b_j \in \mathbb{R}$, x_i is the i-th component of the vector $x \in \mathbb{R}^n$.

(continued)

Note that we have the solution set given as

$$\mathrm{sol}(LP) = \{x \in \mathbb{R}^n : c^T x \leq \hat{\alpha}, \langle a_j, x\rangle - b_j \geq 0, j = 1, \cdots, m; x_i \geq 0, i = 1, \cdots, n\},$$

where $\hat{\alpha} = \inf_{x \in F} c^T x$ and F is the feasible set of (LP). Though we can use the Hoffmann error bound to get an upper bound to $d(\hat{x}, \arg\min_C f)$, $\hat{x} \in \mathbb{R}^n$; that upper bound will depend on $\hat{\alpha}$ which is never known before hand. Thus even for the case of (LP) we want to argue that the approach through the simple bilevel formulation is better to estimate $d(\hat{x}, \arg\min_C f)$. Before we end our discussion on this issue we would like to mention that even by using the simple bilevel formulation we are indeed providing an upper bound to $d(\hat{x}, \arg\min_C f)$. Note that while computing the minimizer in the simple bilevel programming formulation, we usually choose an iterate say $y^k \in C$ as an approximate solution to the bilevel formulation. Hence

$$d(\hat{x}, \arg\min_C f) \leq \|y^k - \hat{x}\|.$$

We hope we have been able to convince the reader about the usefulness of the simple bilevel problem. Let us now list down the topics that we shall discuss in rest of the article.

- Section 9.2: Optimality and links with MPCC
- Section 9.3: Algorithms for (SBP) with smooth data
- Section 9.4: Algorithms for (SBP) with non-smooth data.

Let us remind the reader in the beginning that we will discuss the algorithms for (SBP) from the point of view of convergence analysis. We shall not study any complexity related issues or present any numerical experimentation.

9.2 Optimality and Links with MPCC

Optimality conditions lie at the heart of any analysis of optimization problems. In standard bilevel programming, where there are two decision vectors i.e. the leader's and the follower's decision vectors, it has been a challenging task to write optimality conditions, since standard constraint qualifications like Mangasarian-Fromovitz constraint qualification (MFCQ) fails at each feasible point of the usual bilevel programming problem. See for example Dempe [3], Dempe et al. [5] and Dutta [10]. The question that arises here is whether the simple bilevel programming problem (SBP) also faces similar hurdles. Let us now look into this issue. When we are concerned with optimality conditions, one of the key step is to reformulate it as an equivalent single-level problem. Assume that the lower-level problem in (SBP) has a finite lower bound. Let us set $\alpha = \inf_{x \in C} g(x)$. Then (SBP) can be equivalently reformulated as

$$\begin{aligned}&\min f(x)\\&\text{subject to } g(x) \leq \alpha,\\&x \in C.\end{aligned}$$

Those who are well-versed with the standard bilevel programming problem would immediately realize that above reformulation is the analogue of the value-function reformulation in the standard case. It is simple to see that the Slater's condition fails here. If $\hat{x} \in C$ such that $g(\hat{x}) < \alpha$, then it contradicts the fact that $\alpha = \inf_{x \in C} g(x)$. Since Slater's condition fails one may doubt whether the KKT conditions would hold. One might also feel that KKT conditions might hold under more weaker conditions. The KKT condition for the (SBP), is defined to be the KKT condition for the reformulated problem which we will term as (r-SBP). The KKT conditions at any x feasible to (r-SBP) postulates the existence of $\lambda \geq 0$, such that

1. $0 \in \partial f(x) + \lambda \partial g(x) + N_C(x)$
2. $\lambda(g(x) - \alpha) = 0$

If x^* is the solution of (SBP), then $g(x^*) = \alpha$ and hence (2) automatically holds. Thus if x^* solves (SBP), we have

$$0 \in \partial f(x^*) + \lambda \partial g(x) + N_C(x), \tag{9.2.1}$$

as the required KKT condition. This has to be both necessary and sufficient. Proving sufficiency is not very difficult while proving necessity needs quite involved qualification conditions. See for example Dempe et al. [6], Dhara and Dutta [9] and Dempe et al.[7].

Since (SBP) is a convex programming problem we expect it to have a necessary and sufficient optimality condition which ever way we approach it. However it was shown in Dempe et al. [6], that this is not the case. Consider g to be differentiable convex. In such a case one can equivalently write (SBP) as follows

$$\begin{aligned}&\min f(x)\\&\text{subject to } 0 \in \nabla g(x) + N_C(x).\end{aligned}$$

It has been shown in [6] that necessary optimality condition obtained is not sufficient.

Let us now focus our attention to the case where the lower level problem is a linear programming problem, i.e. consider the (SBP) given as

$$\begin{aligned}&\min f(x)\\&\text{subject to } x \in \arg\min\{\langle c, x\rangle : Ax = b\},\end{aligned}$$

where as before $c \in \mathbb{R}^n$, A is a $m \times n$ matrix and $b \in \mathbb{R}^m$. For simplicity let us assume that f is differentiable. Assume that

$$\alpha = \inf\{\langle c, x\rangle : Ax = b\}.$$

Then we can write down the associated (r-SBP) as

$$\begin{aligned}&\min f(x)\\&\text{subject to } \langle c, x\rangle \leq \alpha,\\&Ax = b.\end{aligned}$$

Since (r-SBP) has linear constraints it is well-known that there is no requirement of a constraint qualification for KKT condition to hold. They hold automatically (see Guler [12]).

Let x^* be a solution of (SBP) with the lower level problem as a linear problem as given above. Then there exists $\lambda \geq 0$ and $\mu \in \mathbb{R}^m$ such that

$$0 = \nabla f(x^*) + \lambda c - A^T \mu$$

i.e.

$$\nabla f(x^*) = A^T \mu - \lambda c. \tag{9.2.2}$$

Let x^* be a feasible point of (SBP) and assume that there exists $\mu \in \mathbb{R}^m$ and $\lambda \geq 0$, such that (9.2.2) holds. Then is x^* a solution of (SBP)? To answer this question it is important to note that x^* is also feasible for (r-SBP) as $\langle c, x\rangle = \alpha$. Now we have the following simple steps which we provide for completeness. For any feasible x for (SBP) we have

$$\begin{aligned}f(x) - f(x^*) &\geq \langle \nabla f(x^*), x - x^*\rangle,\\ \langle c, x\rangle - \langle c, x^*\rangle &= \langle c, x - x^*\rangle.\end{aligned}$$

These two together imply that

$$f(x) - f(x^*) + \lambda\langle c, x\rangle - \lambda\langle c, x^*\rangle \geq \langle \nabla f(x^*) + \lambda c, x - x^*\rangle.$$

Now as $\langle c, x\rangle = \langle c, x^*\rangle = \alpha$, using (9.2.2) we have,

$$f(x) - f(x^*) \geq \langle A^T\mu, x - x^*\rangle.$$

This implies that

$$f(x) - f(x^*) \geq \langle \mu, Ax - Ax^*\rangle.$$

Again as $Ax = Ax^* = b$, we get $f(x) - f(x^*) \geq 0$. Therefore

$$f(x) \geq f(x^*),$$

which implies that x^* is a solution of the (SBP) problem.

In particular, let $f(x) = \frac{1}{2}\|x\|^2$ and $\alpha = \inf\{\langle c, x\rangle : Ax = b\}$. Then x^* is a solution of the minimum-norm solution problem of the linear programming problem

$$\min\langle c, x\rangle$$
$$\text{subject to } Ax = b,$$

if and only if there exists $\mu \in \mathbb{R}^n$ and $\lambda \geq 0$ such that

$$x^* = A^T\mu - \lambda c$$
$$Ax^* = b$$
$$\langle c, x^*\rangle = \alpha.$$

So the key idea is to first solve the linear programming problem and find α, and then solve the above system of equations to find the minimum norm solution.

9.2.1 *Simple Bilevel Programming Problem and MPCC*

This subsection focuses in the relation between simple bilevel programming problem (9.2.3) and its corresponding MPCC problem. Here we present some results showing that (SBP) problem exactly follows the standard optimistic bilevel programming problem in this context (see [4] for details). We will show that if the lower level problem of the (SBP) satisfies the Slater condition then it is equivalent to it's corresponding MPCC problem. Lack of Slater condition leads to counter-examples where (SBP) and related MPCC are not connected in solution sense. The results of this subsection were first presented in the thesis of Pandit [18].

Let us consider the following structure of Simple Bilevel Programming (SBP) problem

$$\begin{aligned} &\text{minimize } f(x) \\ &\text{subject to} \\ &x \in \arg\min\{h(x) : g_i(x) \leq 0;\ i = 1, \cdots, m\}, \end{aligned} \tag{9.2.3}$$

where $f, h : \mathbb{R}^n \to \mathbb{R}$ are convex functions and for each $i \in \{1, \cdots, m\}$ the functions $g_i : \mathbb{R}^n \to \mathbb{R}$ are also convex functions. The corresponding MPCC is given by

$$\begin{aligned} &\text{minimize } f(x) \\ &\text{subject to} \\ &\nabla h(x) + \sum_{i=1}^{m} \lambda_i \nabla g_i(x) = 0 \\ &g_i(x) \leq 0,\ \lambda_i \geq 0 \\ &\lambda_i g_i(x) = 0 \text{ for all } i \in \{1, \cdots, m\}. \end{aligned} \tag{9.2.4}$$

(SBP) problem not only follows the optimistic bilevel programming problem, in fact we don't get any better results for (SBP) problem in spite of the extra convexity properties which lacked in bilevel programming problem. Here we present some results showing that Slater condition is sufficient to establish equivalence between (SBP) and simple MPCC problem.

For any $x \in \mathbb{R}^n$ such that $g(x) \leq 0$, let us define

$$\Lambda(x) := \{\lambda \in \mathbb{R}^m : \lambda_i \geq 0, \lambda_i g_i(x) = 0 \text{ for all } i \in \{1, \cdots, m\}; \\ \nabla h(x) + \sum_{i=1}^{m} \lambda_i \nabla g_i(x) = 0\}.$$

Then (x, λ) for $\lambda \in \Lambda(x)$ is a feasible point of the problem (9.2.4).

Theorem 9.2.1 *Let $\bar{x}$ be a global minimizer of the simple bilevel programming problem (9.2.3) and assume that the lower level problem satisfies the Slater condition. Then for any $\lambda \in \Lambda(\bar{x})$, the point $(\bar{x}, \lambda)$ is a global minimizer of the corresponding MPCC Problem (9.2.4).* △

Proof Since $\bar{x}$ is the minimizer of the (SBP) problem (9.2.3), thus it is a solution of the lower level problem. Since the Slater condition holds for the lower level problem in (9.2.3), we conclude that

$$\Lambda(\bar{x}) \neq \emptyset.$$

This immediately shows that for any $\lambda \in \Lambda(\bar{x})$, we have $(\bar{x}, \lambda)$ to be a feasible point of the (MPCC) problem (9.2.4). Consider any (x, λ') to be feasible point of the (MPCC) problem (9.2.4), then it is simple to observe by noting the convexity of h and g_i's we conclude that x is feasible for the (SBP) problem (9.2.3). Hence $f(x) \geq f(\bar{x})$ showing us that $(\bar{x}, \lambda)$, for any $\lambda \in \Lambda(\bar{x})$ is a solution of the (MPCC). □

Theorem 9.2.2 *Assume that the Slater condition hold true for the lower level problem of the (SBP) problem (9.2.3). Let $(\bar{x}, \bar{\lambda})$ is a local solution of the MPCC problem (9.2.4). Then $\bar{x}$ is a global solution of the SBP problem.* △

Proof Let $(\bar{x}, \bar{\lambda})$ is a local minimizer of the problem (9.2.4). Then by the definition of a local minimizer, there exists $\delta > 0$, such that

$$f(\bar{x}) \leq f(x), \text{ for all } (x, \lambda), \text{ with } \lambda \in \Lambda(x) \text{ and } \|(x, \lambda) - (\bar{x}, \bar{\lambda})\| < \delta. \tag{9.2.5}$$

Since the Slater condition holds for the lower level problem, it is well-known fact that if x and $\bar{x}$ are two different global minimizers of the lower level problem of (9.2.3), then

$$\Lambda(x) = \Lambda(\bar{x}). \tag{9.2.6}$$

For a proof of (9.2.6) see for example in Dhara and Dutta [9]. Since $(\bar{x}, \bar{\lambda})$ is feasible for MPCC it is clear that $\bar{x}$ is a solution of the lower level problem of the (SBP) problem (9.2.3). We will now demonstrate that $\bar{x}$ is a local solution of the (SBP) problem and hence it would be a global one as (SBP) is a convex programming problem.

Let x be feasible for the (SBP) problem (9.2.3) such that

$$\|x - \bar{x}\| < \delta.$$

Slater condition guarantees that $\Lambda(x) \neq \emptyset$. Hence by (9.2.6), $\bar{\lambda} \in \Lambda(x)$. Now consider the same δ as above. Then

$$\|(x, \lambda) - (\bar{x}, \lambda)\| < \delta.$$

Thus from our discussion above we have $f(x) \geq f(\bar{x})$. Hence $\bar{x}$ is a local minimizer of the (SBP) problem and thus a global one. □

Corollary 9.2.3 *Let the Slater's condition holds for the lower level problem of the SBP (9.2.3). Then $(\bar{x}, \bar{\lambda})$ is a local solution of the corresponding simple MPCC problem (9.2.4), which implies that $(\bar{x}, \bar{\lambda})$ is a global solution of the problem (9.2.4).* △

We omit the simple proof which is an application of Theorem 9.2.2 followed by Theorem 9.2.1. Corollary 9.2.3 tells us that the (MPCC) problem (9.2.4) is an example of a non-convex problem whose local minimizers are global minimizers also if the Slater condition holds for the lower level problem of the (SBP) problem (9.2.3).

In the above results one must have observed the key role played by the Slater condition. However if the Slater condition fails, the solution of one of the problems (9.2.3) or (9.2.4) may not lead to a solution of the other.

Example
(Slater's condition holds and the solution of SBP and simple MPCC are same).
Let

$$f(x) = (x - \frac{1}{2})^2$$
$$h(x) = x^2$$
$$g_1(x) = -x$$
$$g_2(x) = x - 3$$

Then Slater's condition holds as $g_1(2) < 0$ and $g_2(2) < 0$.
Here the feasible set for the simple MPCC problem is

$$\{(x, \lambda_1, \lambda_2) : x = 0, \lambda_1 \geq 0, \lambda_2 = 0\}$$

Hence the global optimal solution for the simple MPCC problem is $x = 0$ with optimal value $f(0) = \frac{1}{4}$.
The feasible set of the SBP problem is

$$\arg\min\{h(x) : 0 \leq x \leq 3\} = \{x = 0\}$$

Therefore the global solution of the SBP problem is same as the simple MPCC i.e. $x = 0$.

Example
SBP has unique solution but corresponding simple MPCC is not feasible (Slater's condition is not satisfied).
Let

$$f(x_1, x_2) = x_1 + x_2$$
$$h(x_1, x_2) = x_1$$

(continued)

$$g_1(x_1, x_2) = x_1^2 - x_2$$
$$g_2(x_1, x_2) = x_2$$

Clearly, $g_1(x_1, x_2) \leq 0$ and $g_2(x_1, x_2) \leq 0$ together imply that $x_1 = 0 = x_2$. Which implies that Slater's condition fails for the lower level problem of the SBP.

Now, the feasible set for the SBP problem is

$$\arg\min\{h(x_1, x_2) : x_1 = 0 = x_2\} = \{(0, 0)\}$$

Therefore, (0, 0) is the solution of the SBP problem.

But for $x_1 = 0 = x_2$, there does not exists $\lambda_1 \geq 0$ and $\lambda_2 \geq 0$ such that

$$\nabla h(x_1, x_2) + \lambda_1 \nabla g_1(x_1, x_2) + \lambda_2 \nabla g_2(x_1, x_2) = 0$$

Therefore the simple MPCC problem is not feasible even when the SBP has unique solution in case of the failure of Slater's condition.

9.3 Algorithms for Smooth Data

To the best of our knowledge the first algorithmic analysis was carried out by Solodov [22] in 2007, though the more simpler problem of finding the least norm solution was tackled by Beck and Sabach [1] in 2014 and then a strongly convex upper level objective was considered by Sabach and Shtern [20] in 2017. While writing an exposition it is always a difficult task as to how to present the subject. Shall one concentrate on the simpler problem first or shall one present in a more historical way. After some deliberations we choose to maintain the chronological order of the development so the significance of the simplifications would be more clear.

Let us recall the problem (SBP) given as

$$\min f(x)$$
$$\text{subject to } x \in \arg\min\{g(z) : z \in C\},$$

where $f : \mathbb{R}^n \to \mathbb{R}$ and $g : \mathbb{R}^n \to \mathbb{R}$ are smooth convex functions and C is a closed, convex set.

We shall present here the key ideas from Solodov [22] in a manner that highlights the difficulties in developing a convergent scheme for (SBP). The key idea in Solodov's approach is that of penalization. He constructs the function

$$\varphi_{\varepsilon}(x) = g(x) + \varepsilon f(x), \quad \varepsilon > 0;$$

where ε varies along iteration. Hence we have sequence of positive numbers $\{\varepsilon_k\}$, and

$$\lim_{k\to\infty} \varepsilon_k = 0 \quad \text{and} \quad \sum_{k=0}^{\infty} \varepsilon_k = +\infty. \tag{9.3.1}$$

In effect we are going minimize at each step functions of the form

$$\varphi_{\varepsilon_k}(x) = g(x) + \varepsilon_k f(x), \quad \varepsilon_k > 0$$

with ε_k satisfying (9.3.1) over the set C.

Solodov then applies the projected gradient method and develops a scheme for (SBP). Let us try to understand why such a penalization scheme is interesting for (SBP). We believe the key lies in the reformulated version of (SBP), i.e. (r-SBP), which we write as

$$\begin{aligned} &\min f(x) \\ &g(x) \le \alpha \\ &x \in C. \end{aligned}$$

The usual penalization schemes may not work as α is involved but a direct penalization can be done and one can write down the following penalty problem

$$\min_{x\in C} f(x) + \frac{1}{\varepsilon}(g(x) - \alpha).$$

Now for each value of ε, $\frac{\alpha}{\varepsilon}$ is a fixed quantity, thus it is enough to minimize $f(x)+\frac{1}{\varepsilon}g(x)$. The minimizer of this function will also minimize $\varepsilon f(x) + g(x)$ and thus we can use this format to design algorithms for (SBP). Let us understand that this approach will remain very fundamental to design algorithms for (SBP).

Solodov's Scheme (Algo-SBP-1 [22])

- Step 0: Choose parameters $\bar{\beta} > 0, \vartheta \in (0, 1)$ and $\eta \in (0, 1)$. Choose an initial point $x^0 \in C,\ \varepsilon > 0,\ k := 0$.
- Step 1: Given x^k, let us compute $x^{k+1} = z^k(\beta_k)$; where

$$z^k(\beta) = \text{Proj}_C(x^k - \beta \nabla \varphi_{\varepsilon_k}(x^k)) \tag{9.3.2}$$

and $\beta_k = \eta^{m_k}\bar{\beta}$, where m_k is the smallest non-negative integer for which the following Armijo type criteria holds, i.e.

$$\varphi_{\varepsilon_k}(z^k(\eta^{m_k}\bar{\beta})) \leq \varphi_{\varepsilon_k}(x^k) + \vartheta \langle \nabla \varphi_{\varepsilon_k}(x^k), z^k(\eta^{m_k}\bar{\beta}) - x^k \rangle. \tag{9.3.3}$$

- Step 2: Set $0 < \varepsilon_{k+1} < \varepsilon_k$; set $k := k + 1$ and repeat.

The important issue is how to stop the algorithm. Of course if we have for some k,

$$x^k = x^{k+1} = z^k(\beta_k) = \text{Proj}_C(x^k - \beta_k \nabla \varphi_{\varepsilon_k}(x^k)),$$

where $\beta_k = \eta^{m_k}\bar{\beta}$, where m_k is as given in the scheme then it is clear that x^k is the solution. However this is not the usual scenario and hence one needs to have a stopping criteria for the above algorithm while implementing the algorithm in practice. A good criteria may be to set a threshold $\varepsilon_0 > 0$ and stop when for a given k,

$$\|x^{k+1} - x^k\| < \varepsilon_0.$$

One needs to be cautioned that this algorithm is useful for those cases of (SBP) where the projection on the set C can be computed easily.

Let us now discuss the convergence analysis. One of the key assumptions is that the gradients of f and g are Lipschitz continuous over bounded sets. However for simplicity in our exposition, we shall consider ∇f and ∇g are Lipschitz continuous on $\mathbb{R}^n$ with Lipschitz constant $L > 0$. So our assumption (B) will be as follows.

Assumption B ∇f and ∇g are Lipschitz continuous on $\mathbb{R}^n$ with Lipschitz constant $L > 0$.

For examples of functions for which the gradient is Lipschitz continuous over $\mathbb{R}^n$, see Beck and Sabach [1]).

Further the fact that ∇f and ∇g are Lipschitz with Lipschitz constant $L > 0$ is equivalent to the fact that

$$f(y) - f(x) - \langle \nabla f(x), y - x \rangle \leq \frac{L}{2}\|y - x\|^2,$$

$$g(y) - g(x) - \langle \nabla g(x), y - x \rangle \leq \frac{L}{2}\|y - x\|^2,$$

for any $x, y \in \mathbb{R}^n$.

For the convergence analysis Solodov considers the following assumptions which we mark here as Assumption C; which is as follows.

Assumption C $\bar{f} > -\infty$, where $\bar{f} = \inf\{f(x) : x \in C\}$ and $\bar{g} > -\infty$, where $\bar{g} = \inf\{g(x) : x \in C\}$. Also we assume that C is non-empty.

A key feature to observe is that we will generate the same iterates if instead of φ_{ε_k}, $k \in \mathbb{N}$, we successively minimize

$$\tilde{\varphi}_{\varepsilon_k}(x) = (g(x) - \bar{g}) + \varepsilon(f(x) - \bar{f})$$

over the set C.

Of course we will begin by assuming that the (SBP) is solvable and hence the second assumption in Assumption C automatically holds.

Solodov [22] begins his analysis by proving a key lemma which says that finding the step size using the Armijo type rule indeed terminates, thus showing that algorithm described above is well-defined. The following lemma is presented in Solodov [22].

Lemma 9.3.1 *Let C be a closed convex set and the Assumption B holds. Then there is finite non-negative integer m_k, such that*

$$\beta_k = \eta^{m_k}\bar{\beta} \geq \min\{\bar{\beta}, \frac{2(1-\vartheta)}{(1+\varepsilon_k)L}\} > 0$$

and hence the step-size selection procedure at the k-th iteration terminates. △

Proof We shall outline the main ideas of the proof for completeness. Using (9.3.2) from the algorithm we have

$$\langle x^k - \beta\nabla\varphi_{\varepsilon_k}(x^k) - z^k(\beta), x^k - z^k(\beta)\rangle \leq 0,$$

implying that

$$\|z^k(\beta) - x^k\|^2 \leq \beta\langle \nabla\varphi_{\varepsilon_k}(x^k), x^k - z^k(\beta)\rangle. \tag{9.3.4}$$

By Assumption B, $\nabla\varphi_{\varepsilon_k}$ is Lipschitz on $\mathbb{R}^n$ with Lipschitz constant $(1+\varepsilon_k)L$. This shows that

$$\varphi_{\varepsilon_k}(z^k(\beta)) \leq \varphi_{\varepsilon_k}(x^k) + (1 - \frac{L(1+\varepsilon_k)\beta}{2})\langle \nabla\varphi_{\varepsilon_k}(x^k), z^k(\beta) - x^k\rangle.$$

If $1 - \frac{L(1+\varepsilon_k)\beta}{2} \leq \vartheta$, then $\beta \geq \frac{2(1-\vartheta)}{(1+\varepsilon_k)L}$. Now $\beta_k \leq \bar{\beta}$ by construction. Hence

$$\beta_k = \eta^{m_k}\bar{\beta} \geq \min\{\bar{\beta}, \frac{2(1-\vartheta)}{(1+\varepsilon_k)L}\}.$$

□

It is important to have a closer look at the last expression of the proof. If $\bar{\beta} < \frac{2(1-\vartheta)}{(1+\varepsilon_k)L}$, then $\beta_k \geq \bar{\beta}$. But as $\beta_k \leq \bar{\beta}$, in such a case $\beta_k = \bar{\beta}$. This case may not be very likely in real implementation.

We shall now look into the issue of convergence analysis. We shall present the key convergence result and then outline the key steps of proof and highlighting the key points which are hallmark of the (SBP) problem.

Theorem 9.3.2 ([22]) *Let us consider the problem (SBP), where f and g are convex differentiable functions. Let the Assumptions B and C hold. Let the solution set of (SBP), denoted as sol(SBP) be closed and bounded. Then for any sequence x^k generated by the algorithm (Algo-SBP-1), with the sequence $\{\varepsilon_k\}$, $\varepsilon_k \downarrow 0$, satisfying (9.3.2), we have*

$$d_{sol(SBP)}(x^k) \to 0 \quad as\ k \to \infty.$$

Here $d_C(x)$ represents the distance of the set C from the point x. △

Proof *(Outline of the Main Ideas)* The key idea of the proof lies in replacing φ_{ε_k} with $\widetilde{\varphi_{\varepsilon_k}}$, for each k and rewriting (Algo-SBP-1) in terms of $\widetilde{\varphi_{\varepsilon_k}}$. In such a case the Armijo type criteria shows that

$$\vartheta\langle \nabla\tilde{\varphi}_{\varepsilon_k}(x^k), x^k - x^{k+1}\rangle \leq \tilde{\varphi}_{\varepsilon_k}(x^k) - \tilde{\varphi}_{\varepsilon_k}(x^{k+1}).$$

This shows that

$$\begin{aligned}&\vartheta\langle \nabla\tilde{\varphi}_{\varepsilon_k}(x^k), x^k - x^{k+1}\rangle \\ &\leq \varepsilon_k(f(x^k) - \bar{f}) + (g(x^k) - \bar{g}) - \varepsilon_k(f(x^{k+1}) - \bar{f}) - (g(x^{k+1}) - \bar{g}),\end{aligned}$$

i.e.

$$\begin{aligned}&\vartheta\langle \nabla\tilde{\varphi}_{\varepsilon_k}(x^k), x^k - x^{k+1}\rangle \\ &\leq \varepsilon_k(f(x^k) - \bar{f}) - \varepsilon_k(f(x^{k+1}) - \bar{f}) + (g(x^k) - \bar{g}) - (g(x^{k+1}) - \bar{g}).\end{aligned} \tag{9.3.5}$$

Now by the definition of $\bar{f}$, $\bar{f} \leq f(x^k)$ for all $k \in \mathbb{N}$ as $x^k \in C$. Also note that $\varepsilon_{k+1} \leq \varepsilon^k$. Therefore for any $k \in \mathbb{N}$

$$0 \leq \varepsilon_{k+1}(f(x^{k+1}) - \bar{f}) \leq \varepsilon_k(f(x^{k+1}) - \bar{f})$$

Then from (9.3.5), we get

$$\begin{aligned} &\vartheta\langle \nabla\tilde{\varphi}_{\varepsilon_k}(x^k), x^k - x^{k+1}\rangle \\ &\leq \varepsilon_k(f(x^k) - \bar{f}) - \varepsilon_{k+1}(f(x^{k+1}) - \bar{f}) + (g(x^k) - \bar{g}) - (g(x^{k+1}) - \bar{g}). \end{aligned}$$

Since this holds for all $k \in \mathbb{N}$, by summing over from $k = 0$ to $k = \hat{k}$, we have

$$\begin{aligned} &\vartheta \sum_{k=0}^{\hat{k}} \langle \nabla\tilde{\varphi}_{\varepsilon_k}(x^k), x^k - x^{k+1}\rangle \\ &\leq \varepsilon_0(f(x^0) - \bar{f}) + (g(x^0) - \bar{g}) - \varepsilon_{\hat{k}+1}(f(x^{\hat{k}+1}) - \bar{f}) - (g(x^{\hat{k}+1}) - \bar{g}) \\ &\leq \varepsilon_0(f(x^0) - \bar{f}) + (g(x^0) - \bar{g}). \end{aligned}$$

We urge the reader to check the above inequality themselves. Now as $\hat{k} \to \infty$, we conclude that

$$\sum_{k=0}^{\infty} \langle \nabla\tilde{\varphi}_{\varepsilon_k}(x^k), x^k - x^{k+1}\rangle \leq \frac{1}{\vartheta}[\varepsilon_0(f(x^0) - \bar{f}) + (g(x^0) - \bar{g})] < +\infty.$$

Thus the series $\sum_{k=0}^{\infty} \langle \nabla\tilde{\varphi}_{\varepsilon_k}(x^k), x^k - x^{k+1}\rangle$ is summable and hence

$$\langle \nabla\tilde{\varphi}_{\varepsilon_k}(x^k), x^k - x^{k+1}\rangle \to 0 \quad \text{as } k \to \infty. \tag{9.3.6}$$

From the point of view of convergence analysis it is important that the sequence generated by the algorithm is bounded. In that case an effort should be made to show that any limit point of that sequence is in the solution set of the problem under consideration. In fact in the setting of bilevel programming it is slightly tricky to show that $\{x^k\}$ is bounded. Rather it is much easier to show that if $\{x^k\}$ is bounded, then every limit point will be a solution of the lower level problem. This is the approach taken by Solodov [22] and we will outline this approach here. We shall then show that $\{x^k\}$ is bounded and the accumulation points indeed belong to sol(SBP). Our first step would be to show that for any bounded $\{x^k\}$ generated by (Algo-SBP-1) a limit point is always a solution of the lower level problem. Now from the algorithm we know that $0 < \varepsilon_{k+1} < \varepsilon_k < \varepsilon_0$. Thus

$$\beta_k \geq \min\{\bar{\beta}, \frac{2(1-\vartheta)}{(1+\varepsilon_0)L}\} = \hat{\beta} \text{ (say)}. \tag{9.3.7}$$

Using (9.3.4) and (9.3.6) we conclude that $\|x^k - x^{k+1}\| \to 0$ and hence

$$x^k - x^{k+1} \to 0.$$

Thus by definition, as $k \to 0$

$$x^k - \text{proj}_C(x^k - \beta_k \nabla \tilde{\varphi}_{\varepsilon_k}(x^k)) \to 0.$$

Hence let $\tilde{x}$ be an accumulation point or limit point of $\{x^k\}$ then as $k \to 0$

$$\tilde{x} - \text{proj}_C(\tilde{x} - \tilde{\beta} \nabla \tilde{\varphi}_{\varepsilon_k}(\tilde{x})) = 0,$$

where $\beta_k \to \tilde{\beta} > 0$ (Noting that it is simple to show that β_k is bounded). Hence $\tilde{x} = \text{proj}_C(\tilde{x} - \tilde{\beta} \nabla \tilde{\varphi}_{\varepsilon_k}(\tilde{x}))$ showing that $\tilde{x}$ is the solution of the lower level problem. But before we establish that $\tilde{x}$ also lies in sol(SBP), we shall show that $\{x^k\}$ is bounded. Using the fact that the sol(SBP) is non empty, closed and bounded; Solodov starts the analysis of this part by considering an $\bar{x} \in$ sol(SBP). Now convexity of φ_{ε_k} will show us that

$$\langle \nabla \tilde{\varphi}_{\varepsilon_k}, \bar{x} - x^k \rangle \leq \varepsilon_k (f(\bar{x}) - f(x^k)). \tag{9.3.8}$$

Solodov then employs the following identity

$$\|x^{k+1} - \bar{x}\|^2 - \|x^{k+1} - x^k\|^2 = \|x^k - \bar{x}\|^2 + 2\langle x^{k+1} - x^k, x^{k+1} - \bar{x} \rangle. \tag{9.3.9}$$

Note that,

$$\begin{aligned}
&\langle x^{k+1} - x^k, x^{k+1} - \bar{x} \rangle \\
&= \langle x^{k+1} - x^k + \beta_k \nabla \tilde{\varphi}_{\varepsilon_k}(x^k) - \beta_k \nabla \tilde{\varphi}_{\varepsilon_k}(x^k), x^{k+1} - \bar{x} \rangle \\
&= \langle x^{k+1} - x^k + \beta_k \nabla \tilde{\varphi}_{\varepsilon_k}(x^k), x^{k+1} - \bar{x} \rangle - \beta_k \langle \nabla \tilde{\varphi}_{\varepsilon_k}(x^k), x^{k+1} - \bar{x} \rangle.
\end{aligned}$$

Since by the algorithm

$$x^{k+1} = \text{proj}_C(x^k - \beta_k \nabla \tilde{\varphi}_{\varepsilon_k}(x^k)), \tag{9.3.10}$$

by the well known property of projection (see e.g. Hirriart Urruty and Lemarechal [14]) we have

$$\begin{aligned}
&\langle x^k - \beta_k \nabla \tilde{\varphi}_{\varepsilon_k}(x^k) - x^{k+1}, \bar{x} - x^{k+1} \rangle \leq 0 \\
&i.e. \ \langle x^k - \beta_k \nabla \tilde{\varphi}_{\varepsilon_k}(x^k) - x^{k+1}, x^{k+1} - \bar{x} \rangle \geq 0
\end{aligned}$$

Hence from this we can deduce by noting that $\beta_k \leq \bar{\beta}$ and (9.3.8),

$$\langle x^{k+1} - x^k, x^{k+1} - \bar{x}\rangle \leq \bar{\beta}\langle \nabla\tilde{\varphi}_{\varepsilon_k}(x^k)), x^k - x^{k+1}\rangle + \beta_k\varepsilon_k(f(\bar{x}) - \bar{f}).$$

Now from (9.3.9) and using the above relations, we have

$$\begin{aligned} &\|x^{k+1} - \bar{x}\|^2 \\ &\leq \|x^k - \bar{x}\|^2 + 2\bar{\beta}\langle \nabla\tilde{\varphi}_{\varepsilon_k}(x^k)), x^k - x^{k+1}\rangle + 2\beta_k\varepsilon_k(f(\bar{x}) - f(x^k)). \end{aligned} \tag{9.3.11}$$

We need to focus on the quantity $(f(\bar{x}) - f(x^k))$, an information about the second term on the right hand side is given by (9.3.6). Now one needs to look into two separate cases formally.

Case I : There exists k_0, such that one has $f(\bar{x}) \leq f(x^k)$, for all $k \geq k_0$.
Case II : For any $k \in \mathbb{N}$, there exists $k_1 \in \mathbb{N}$ and $k_1 \geq k$ such that $f(\bar{x}) > f(x^{k_1})$.

We first deal with Case I. It is shown in Solodov [22] that using case I, (9.3.11) gives

$$\|x^{k+1} - \bar{x}\|^2 \leq \|x^k - \bar{x}\|^2 + 2\bar{\beta}\langle \nabla\tilde{\varphi}_{\varepsilon_k}(x^k)), x^k - x^{k+1}\rangle \tag{9.3.12}$$

This shows that $\{\|x^k - \bar{x}\|^2\}$ converges since if $\{a_k\}$ and $\{b_k\}$ are sequences of non-negative real numbers such that $a_{k+1} \leq a_k + b_k$ and $\sum_{k=1}^{\infty} b_k < +\infty$, then $\{a_k\}$ converges. Hence $\{x^k\}$ is bounded. Solodov [22] then proves that

$$\liminf_{k\to\infty} f(x^k) = f(\bar{x}) \tag{9.3.13}$$

Hence we can find a convergent subsequence of $\{x^{k_j}\}$ of $\{x^k\}$ such that

$$\liminf_{j\to\infty} f(x^{k_j}) = f(\bar{x}).$$

Let $x^{k_j} \to \tilde{x}$ as $j \to \infty$. Hence $f(\tilde{x}) = f(\bar{x})$. We have seen before that $\tilde{x} \in \arg\min_C g$ and thus $\tilde{x} \in \text{sol(SBP)}$.

Solodov [22] then shows that if we set $\tilde{x} = \bar{x}$, then we can demonstrate that $x^k \to \tilde{x}$ as $k \to \infty$. Now looking at the proof above we can show that $\{\|x^k - \tilde{x}\|^2\}$ converges. Since $x^{k_j} \to \tilde{x}$; there is a subsequence of $\{\|x^k - \tilde{x}\|^2\}$ that converges to 0. Hence the sequence $\{\|x^k - \tilde{x}\|^2\}$ converges to 0, which implies that $x^k \to \tilde{x} \in \text{sol(SBP)}$. This argument is a very crucial one as one observes. For details see Solodov [22]. Handling the second case is more complicated since we have to deal essentially with subsequences. One of the main tricks of this approach is to define the following quantity i_k for each $k \in \mathbb{N}$ as

$$i_k = \max\{i \leq k : f(\bar{x}) > f(x^i)\}$$

An important thing to observe is that for a given k one may have $\{i \leq k : f(\bar{x}) > f(x^i)\} = \emptyset$ and in that case i_k can not be defined. Thus a better way to write is that let $\hat{k} \in \mathbb{N}$ be first integer for which

$$\{i \leq \hat{k} : f(\bar{x}) > f(x^i)\}$$

is non-empty. Then for all $k \geq \hat{k}$ and $k \in \mathbb{N}$ we can define

$$i_k = \max\{i \leq k : f(\bar{x}) > f(x^i)\}.$$

Thus we shall only consider now $k \geq \hat{k}$. Also from the description of Case II, we can conclude that $i_k \to \infty$ as $k \to \infty$.

The key step now is to show that $\{x^{i_k}\}_{k=\hat{k}}^{\infty}$ is bounded. Let us observe that

$$\begin{aligned} \text{sol(SBP)} &= \{x \in \arg\min_C g : f(x) \leq f(\bar{x})\} \\ &= \{x \in C : \max\{f(x) - f(\bar{x}), g(x) - \bar{g}\} \leq 0\}. \end{aligned}$$

The function $\psi(x) = \max\{f(x) - f(\bar{x}), g(x) - \bar{g}\}$ is a convex function. This shows that

$$\text{sol(SBP)} = \{x \in C : \psi(x) \leq 0\}.$$

Now sol(SBP) is assumed to be non empty and bounded. The level set $\{x \in C : \psi(x) \leq 0\} = L_\psi(0)$ of the convex function ψ is non empty and bounded. Further it is well known that in such a case for any $q \in \mathbb{R}$,

$$L_\psi(q) := \{x \in C : \psi(x) \leq q\}$$

is also bounded. Solodov [22] then shows that

$$0 \leq \tilde{\varphi}_{\varepsilon_{k+1}}(x^{k+1}) \leq \tilde{\varphi}_{\varepsilon_k}(x^{k+1}) \leq \tilde{\varphi}_{\varepsilon_k}(x^k).$$

This is a key step since it shows that the sequence $\{\tilde{\varphi}_{\varepsilon_k}(x^k)\}$ is non-increasing and bounded below which establishes the fact that $\{\tilde{\varphi}_{\varepsilon_k}(x^k)\}$ is a convergent sequence and hence bounded. This shows that the sequence $\{g(x^k) - \bar{g}\}$ is bounded. In the very next step Solodov [22] uses this boundedness of $\{g(x^k) - \bar{g}\}$ in a profitable way. Let $q \geq 0$ be such that $g(x^k) - \bar{g} \leq q$ for all $k \in \mathbb{N}$.

Now from the definition of i_k we have for all $k \geq \hat{k}$

$$\varepsilon_k(f(x^{i_k}) - f(\bar{x})) < 0$$

and thus

$$\varepsilon_k(f(x^{i_k}) - f(\bar{x})) + (g(x^{i_k}) - g(\bar{x})) \leq q$$

implying that

$$\tilde{\varphi}_{\varepsilon_{i_k}}(x^{i_k}) \leq q \quad \forall \quad k \geq \hat{k}.$$

Hence $x^{i_k} \in L_{\tilde{\varphi}_{\varepsilon_{i_k}}}(q)$ for all $k \geq \hat{k}$. Thus $\{x^{i_k}\}$ is bounded. Now for any $k \geq \hat{k}$, the definition i_k tells us that

$$f(\bar{x}) \leq f(x^i)$$

for $i = i_k + 1, \cdots, k$; if $k > i_k$. Thus from (9.3.11), we get for $i = i_k + 1, \cdots, k$

$$\|x^{i+1} - \bar{x}\|^2 \leq \|x^i - \bar{x}\|^2 + 2\bar{\beta}\langle \nabla\tilde{\varphi}_{\varepsilon_i}(x^i), x^i - x^{i+1}\rangle.$$

This holds true for all cases where $k > i_k$. Thus we have

$$\|x^k - \bar{x}\|^2 \leq \|x^{i_k} - \bar{x}\|^2 + 2\bar{\beta} \sum_{i=i_k+1}^{k-1} \langle \nabla\tilde{\varphi}_{\varepsilon_i}(x^i), x^i - x^{i+1}\rangle,$$

for all k for which $i_k < k$.

Let us now consider the case where $i_k = k$. In that case we will proceed in the following way. For any $k \in \mathbb{N}$ we have

$$x^{k+1} = \text{proj}_C(x^k - \beta_k \nabla\tilde{\varphi}_{\varepsilon_k}(x^k)).$$

From the definition of projection we can conclude that

$$\|x^{k+1} - x^k\|^2 < \beta_k\langle \nabla\tilde{\varphi}_{\varepsilon_k}(x^k), x^k - x^{k+1}\rangle.$$

Thus we have for all $k \in \mathbb{N}$

$$\langle \nabla\tilde{\varphi}_{\varepsilon_k}(x^k), x^k - x^{k+1}\rangle \geq 0.$$

Observe however that this shows that

$$0 \leq 2\bar{\beta} \sum_{j=i_k+1}^{k-1} \langle \nabla\tilde{\varphi}_{\varepsilon_j}(x^j), x^j - x^{j+1}\rangle. \tag{9.3.14}$$

Thus for k with $i_k = k$ we have $\|x^k - \bar{x}\|^2 = \|x^{i_k} - \bar{x}\|^2$. Hence from (9.3.14)

$$\|x^k - \bar{x}\|^2 \leq \|x^{i_k} - \bar{x}\|^2 + 2\bar{\beta} \sum_{j=i_k+1}^{k-1} \langle \nabla \tilde{\varphi}_{\varepsilon_j}(x^j), x^j - x^{j+1} \rangle.$$

So we conclude that for all k we have

$$\|x^k - \bar{x}\|^2 \leq \|x^{i_k} - \bar{x}\|^2 + 2\bar{\beta} \sum_{j=i_k+1}^{\infty} \langle \nabla \tilde{\varphi}_{\varepsilon_j}(x^j), x^j - x^{j+1} \rangle. \quad (9.3.15)$$

Now $i_k \to \infty$ as $k \to \infty$ we see that

$$\sum_{j=i_k+1}^{\infty} \langle \nabla \tilde{\varphi}_{\varepsilon_j}(x^j), x^j - x^{j+1} \rangle \to 0 \quad \text{as } k \to \infty.$$

Thus using the boundedness of $\{x^{i_k}\}$ we conclude that $\{x^k\}$ is also bounded. We can immediately conclude that all accumulation points of $\{x^k\}$ lie in $\arg\min_C g$. Further for any accumulation point x^* of $\{x^{i_k}\}$ we have $f(\bar{x}) \geq f(x^*)$. But this shows that $x^* \in \text{sol(SBP)}$. In fact all accumulation points of $\{x^{i_k}\}$ must be in sol(SBP). If x^* is an accumulation point of $\{x^{i_k}\}$, thus there exists a convergent subsequence $\{x^{i_{k_j}}\}$ of $\{x^{i_k}\}$ such that $x^{i_{k_j}} \to x^*$ as $j \to \infty$. Further we have

$$d(x^{i_{k_j}}, \text{sol(SBP)}) \leq \|x^{i_{k_j}} - x^*\|.$$

Thus as $j \to \infty$ we have

$$d(x^{i_{k_j}}, \text{sol(SBP)}) \to 0. \quad (9.3.16)$$

We can also see that for any convergent subsequence of $\{x^{i_k}\}$ (9.3.16) holds, this shows that

$$d(x^{i_k}, \text{sol(SBP)}) \to 0 \quad \text{as } k \to \infty.$$

Now for each k, Solodov [22] defines

$$\bar{x}^k = \text{proj}_{\text{sol(SBP)}}(x^{i_k}).$$

Now using (9.3.15) with $\bar{x} = \bar{x}^k$ we have

$$\begin{aligned} d^2(x^k, \text{sol(SBP)}) &\leq \|x^k - \bar{x}^k\|^2 \\ &\leq d^2(x^{i_k}, \text{sol(SBP)}) + 2\bar{\beta} \sum_{j=i_k+1}^{\infty} \langle \nabla \tilde{\varphi}_{\varepsilon_i}(x^i), x^i - x^{i+1} \rangle. \end{aligned}$$

Now as $k \to \infty$ we have $d^2(x^k, \text{sol(SBP)}) \to 0$, hence $d(x^k, \text{sol(SBP)}) \to 0$. □

We shall now discuss a work by Beck and Sabach [1], which focuses on what we believe as the most important class of simple bilevel problems. In this case the upper-level objective is always strongly convex. We shall denote such problems as (SBP-1). Thus an (SBP-1) problem is given as

$$\begin{aligned} &\min \phi(x) \\ &\text{Subject to } x \in S \\ &S = \arg\min\{g(y) : y \in C\}, \end{aligned}$$

where $\phi : \mathbb{R}^n \to \mathbb{R}$ is strongly convex with modulus $\rho > 0$ and differentiable, while $g : \mathbb{R}^n \to \mathbb{R}$ is a convex and differentiable function with ∇g a Lipschitz function on $\mathbb{R}^n$ with Lipschitz rank (or Lipschitz constant) $L > 0$. When $\phi(x) = \frac{1}{2}\|x\|^2$, we are talking about the minimum norm problem. Thus (SBP-1) actually encompasses a large class of important problems.

Beck and Sabach [1] develops the Minimal norm gradient algorithm, which they present in two different versions. For the first version the Lipschitz rank of ∇g is known while for the second version it is assumed that the Lipschitz rank is not known and thus an Armijo type criteria is used. In Solodov [22] in comparison an Armijo type criteria is used even though the Lipschitz rank was known. Also note that in Beck and Sabach [1], the Lipschitz gradient assumption is only on g and not on ϕ. In fact Beck and Sabach [1] mention that a problem like (SBP-1) can be reformulated as follows

$$\begin{aligned} &\min \phi(x) \\ &\text{Subject to } g(x) \leq g^* \\ &x \in C, \end{aligned}$$

where $g^* = \inf_{x \in C} g$. We are already familiar with this reformulation and further they mention that as the Slater condition fails and g^* may not be known, the above problem may not be suitable to develop numerical schemes to solve (SBP-1). They mention that even if one could compute g^* exactly, the fact that Slater's condition fails would lead to numerical problems. As a result they take a completely different path by basing their approach on the idea of cutting planes. However we would like to mention that very recently in the thesis of Pandit [18], the above reformulation has been slightly modified to generate a sequence of problems whose solutions converge to that of (SBP) or (SBP-1), depending on the choice of the problem. The algorithm thus presented in the thesis of Pandit [18] has been a collaborative effort with Dutta and Rao and has also been presented in the preprint [19]. The approach in Pandit et al. [19] has been to develop an algorithm where no additional assumption on the upper level objective is assumed other than the fact that it is convex, where the lower level objective is differentiable but need not have a Lipschitz gradient. The algorithm in Pandit [18] or [19] is very simple and just relies on an Armijo type decrease criteria. Convergence analysis has been carried out and numerical experiments have given encouraging results. Further in Pandit et al. [19], the upper level objective need not be strongly convex.

Let us outline the work of Beck and Sabach [1]. Let us begin by mentioning some key tools in [1]. Let

$$x^* = \arg\min_{\mathbb{R}^n} \phi$$

and we call x^* the prox-center of ϕ. Without loss of generality we may assume that $\phi(x^*) = 0$. Then strong convexity of ϕ tells us that

$$\phi(x) \geq \frac{\rho}{2}\|x - x^*\|^2 \quad \forall x \in \mathbb{R}^n.$$

A key tool in the study of Beck and Sabach [1] is the Bregman distance. This is a non-Euclidean distance which makes the analysis a bit more involved.

Let $h : \mathbb{R}^n \to \mathbb{R}$ be a strictly convex function. Then the Bregman distance generated by h is given as

$$D_h(x, y) = h(x) - h(y) - \langle \nabla h(y), x - y\rangle.$$

It is clear that $D_h(x, y) \geq 0,\ \forall x, y \in \mathbb{R}^n$ and $D_h(x, y) = 0$ if and only if $x = y$. Further if h is strongly convex,

$$D_h(x, y) \geq \frac{\rho}{2}\|x - y\|^2.$$

A key inequality associated with a Bregman distance is the well-known three point inequality. If h is strongly convex and $x, y, z \in \mathbb{R}^n$, then

$$D_h(x, y) + D_h(y, z) - D_h(x, z) = \langle \nabla h(z) - \nabla h(y), x - y\rangle.$$

The following two vector-valued mappings play a central role in [1]. First comes the projected gradient mapping or proj-grad mapping. For $\mu > 0$, the proj-grad mapping is given as

$$P_M(x) = \text{proj}_C(x - \frac{1}{M}\nabla g(x)), \quad \forall x \in \mathbb{R}^n.$$

The gradient map is defined as

$$G_M(x) = M(x - P_M(x)) = M\big(x - \text{proj}_C(x - \frac{1}{M}\nabla g(x))\big).$$

If $C = \mathbb{R}^n$, then $G_M(x) = \nabla g(x)$, which is thus the source of its name. Further $G_M(x^*) = 0$ implies that x^* solves the lower level problem in (SBP-1). Further it was shown in [1] that the function

$$g(M) = \|G_M(x)\|, \quad M > 0;$$

is monotonically non-decreasing over $(0, \infty)$. The idea of their algorithm is based on the idea of cutting planes in convex optimization. If $x^* \in \mathbb{R}^n$ is a minimizer of a convex function $\phi : \mathbb{R}^n \to \mathbb{R}$, then for any $x \in \mathbb{R}^n$ we have

$$\langle \nabla\phi(x^*), x - x^*\rangle \geq 0.$$

By monotonicity, we have

$$\langle \nabla\phi(x), x - x^*\rangle \geq 0.$$

Thus if $\text{sol}(\phi, \mathbb{R}^n)$ denotes the solution set of the problem of minimizing ϕ over $\mathbb{R}^n$, then

$$\text{sol}(\phi, \mathbb{R}^n) \subseteq \{x^* \in \mathbb{R}^n : \langle \nabla\phi(x), x - x^*\rangle \geq 0, \forall x \in \mathbb{R}^n\}.$$

Therefore, given $x \in \mathbb{R}^n$, we can immediately eliminate the strict half space

$$\{x^* \in \mathbb{R}^n : \langle \nabla\phi(x), x - x^*\rangle < 0\}.$$

So one needs to check, given any $x \in \mathbb{R}^n$, the cut at x is given by the hyperplane

$$H_x^{\phi} = \{z \in \mathbb{R}^n : \langle \nabla\phi(x), x - z\rangle = 0\}.$$

This is a very important approach in convex optimization, see for example the fundamental paper by Kelley [15]. Following the idea of the cutting plane method,

Beck and Sabach introduces the following set whose motivation will be clear as we go along.

$$Q_{M,\alpha,x} = \{z \in \mathbb{R}^n : \langle G_M(x), x - z\rangle \geq \frac{1}{\alpha M}\|G_M(x)\|^2\}.$$

If $C = \mathbb{R}^n$, then

$$Q_{M,\alpha,x} = \{z \in \mathbb{R}^n : \langle \nabla g(x), x - z\rangle \geq \frac{1}{\alpha M}\|\nabla g(x)\|^2\}.$$

This is what they refer to as deep cuts and play a fundamental role in their algorithm which we describe below. The first algorithm considers the fact that the Lipschitz constant of ∇g is known. As before let us call the Lipschitz constant as $L > 0$. Here is the algorithm.

The Minimal Norm Gradient Algorithm (Lipschitz Constant Known)

- Step 0: Let $L > 0$ be the Lipschitz constant of ∇g. Let x^0= prox-center of $\phi = \arg\min_{\mathbb{R}^n} \phi$.
- Step 1: For $k = 1, 2, \cdots$

$$x^k = \mathrm{sol}(\phi, Q^k \cap W^k),$$

where

$$Q^k = Q_{L,\beta,x^{k-1}}$$

and

$$W^k = \{z \in \mathbb{R}^n : \langle \nabla w(x^{k-1}), z - x^{k-1}\rangle \geq 0\}.$$

Set $\beta = \frac{4}{3}$ if $C \neq \mathbb{R}^n$ and $\beta = 1$ if $C = \mathbb{R}^n$.

The reason for setting $\beta = \frac{4}{3}$ when $C \neq \mathbb{R}^n$ i.e. $C \subset \mathbb{R}^n$ is that $\mathrm{sol}(g, C) = \arg\min_C g \subset Q_{L,\frac{4}{3},x}$ and if $C = \mathbb{R}^n$, $\beta = 1$ as $\mathrm{sol}(g, \mathbb{R}^n) \subset Q_{L,1,x}$. These facts are proved in Lemma 2.3 and Lemma 2.4 in Beck and Sabach [1]. In Lemma 2.3 in [1], it is shown that

$$\langle G_L(x) - G_L(y), x - y\rangle \geq \frac{3}{4L}\|G_L(x) - G_L(y)\|^2.$$

If we set $y = x^* \in \arg\min_C g = \text{sol}(g, C)$, then $G_L(x^*) = 0$ and hence

$$\langle G_L(x), x - x^* \rangle \geq \frac{3}{4L} \|G_L(x)\|^2. \tag{9.3.17}$$

Thus $x^* \in Q_{L,\frac{4}{3},x}$. Since ∇g is Lipschitz continuous with constant L, then from [1] we see that

$$\langle \nabla g(x) - \nabla g(y), x - y \rangle \geq \frac{1}{L} \|\nabla g(x) - \nabla g(y)\|^2.$$

Putting $y = x^*$, where $x^* \in \arg\min_{\mathbb{R}^n} g = \text{sol}(g, \mathbb{R}^n)$, we have

$$\langle \nabla g(x), x - x^* \rangle \geq \frac{1}{L} \|\nabla g(x)\|.$$

This shows that

$$\arg\min_{\mathbb{R}^n} g = \text{sol}(g, \mathbb{R}^n) \subset Q_{L,1,x}.$$

If we do not have any knowledge of the Lipschitz constant of ∇g, then we use an Armijo type criteria in the Minimal norm gradient algorithm that we state here.

The Minimal Norm Gradient Algorithm (Lipschitz Constant is Not Known)

- Step 0: Initialization: x^0= prox-center of ϕ = $\arg\min_{\mathbb{R}^n} \phi$. Take $L_0 > 0$, $\eta > 1$.
- Step 1: For $k = 1, 2, \cdots$
  (a) Find the smallest non-negative integer i_k such that $\bar{L} = \eta^{i_k} L_{k-1}$ and the following inequality

  $$f(P_{\bar{L}}(x^{k-1})) \leq f(x^{k-1}) + \langle \nabla f(x^{k-1}), P_{\bar{L}}(x^{k-1}) - x^{k-1} \rangle + \frac{\bar{L}}{2} \|P_{\bar{L}}(x^{k-1}) - x^{k-1}\|^2$$

  is satisfied. Then set $L_k := \bar{L}$

  (b) Set $x^k = \text{sol}(\phi, Q^k \cap W^k)$, where

  $$Q^k = Q_{L_k,2,x^{k-1}} \quad \text{and} \quad W^k = \{z \in \mathbb{R}^n : \langle \nabla \phi(x^{k-1}), z - x^{k-1} \rangle \geq 0\}.$$

We shall now show that sol(g,C) $\subseteq Q_{\bar{L},2,x}$. It has been demonstrated in Lemma 2.5 in Beck and Sabach [1] that if

$$g(P_{\bar{L}}(x)) \leq g(x) + \langle \nabla g(x), P_{\bar{L}}(x) - x \rangle + \frac{\bar{L}}{2} \| P_{\bar{L}}(x) - x \|^2,$$

then for any $x^* \in$ sol(g,C), one has

$$\langle G_M(x), x - x^* \rangle \geq \frac{1}{2\bar{L}} \| G_M(x) \|^2$$

and thus sol(g,C) $\subseteq Q_{\bar{L},2,x}$.

Beck and Sabach [1] also shows that whichever form of the minimal norm gradient algorithm is used one has

$$\text{sol}(g, C) = \arg\min_C g \subseteq Q^k \cap W^k, \quad \forall k \in \mathbb{N}.$$

We will now state below the result of Beck and Sabach [1], describing the convergence of the minimum norm gradient algorithm.

Theorem 9.3.3 *Consider the (SBP-1) problem and let $\{x^k\}$ be a sequence generated by any of the two forms of the minimum norm gradient algorithm. Then*

(a) *$\{x^k\}$ is bounded.*
(b) *for any $k \in \mathbb{N}$,*

$$D_\phi(x^k, x^{k-1}) + D_\phi(x^{k-1}, x^0) \leq D_\phi(x^k, x^0).$$

(c) $x^k \to x^* = \arg\min\{\phi(y) : y \in sol(g, C)\}$ *i.e. x^* is a solution of (SBP-1).* △

Keeping in view our chronological approach to the development of algorithms for simple bilevel programming, we shall end this section by discussing the work of Sabach and Shtern [20], which appeared in 2017 and is in some sense a generalization of Beck and Sabach [1]. The problem they consider is the following

$$\begin{aligned} &\min \phi(x) \\ &\text{subject to } x \in S, \end{aligned}$$

where $\phi : \mathbb{R}^n \to \mathbb{R}$ is a strongly convex function and

$$S = \arg\min_{\mathbb{R}^n} \{f + g\}$$

where f is continuously differentiable and g is an extended-valued proper lower semi-continuous convex function. We shall call this problem (SBP-2). In fact strictly speaking we should consider this problem under the category of non-smooth simple bilevel problems since g can be non-smooth. However we are discussing this problem in this section since we view (SBP-2) as a generalization of (SBP-1). Observe that if we choose $g = \delta_C$, where δ_C is the indicator function of a closed convex set C, then (SBP-2) reduces to (SBP-1). Those readers who are aware of the algorithms of convex optimization might guess that a proximal gradient type approach needs to be used for solving (SBP-2). Indeed Sabach and Shtern [20], does develop an algorithm based on the prox-mapping. It is important to realize that since ϕ is strongly convex, (SBP-2) also has a unique minimizer which we may mark as x^*. In fact the method developed in [20], called BiG-SAM, is simpler and cheaper compared than the Minimal Norm Gradient (MNG) method of Beck and Sabach [1]. In the Big-SAM method one needs to compute $\nabla\phi$ only, while in the (MNG) method one needs to compute $\nabla\phi$ and also to minimize ϕ to compute the prox-center.

In Sabach and Shtern [20], the following assumptions are made

(1) f has Lipschitz-gradient, i.e. ∇f is Lipschitz on $\mathbb{R}^n$ with Lipschitz constant L_f.
(2) $S \neq \emptyset$.

One of the key approaches to solve the lower-level problem in (SBP-2) is the proximal gradient approach which depends on the prox-mapping. Given $h : \mathbb{R}^n \to \mathbb{R} \cup \{+\infty\}$, a proper lower semi-continuous prox-mapping is given as

$$\text{prox}_h(x) = \arg\min\{h(u) + \frac{1}{2}\|u - x\|^2 : u \in \mathbb{R}^n\}.$$

The proximal point algorithm generates the iterates for the lower level problem using the following scheme

$$x^{k+1} = \text{prox}_{tg}(x^k - t\nabla f(x^k)),$$

where $t > 0$ is the step-size. If we set

$$P_t(x) = \text{prox}_{tg}(x - t\nabla f(x)),$$

then x is a solution of the lower level problem if and only if $x = P_t(x)$. Further $P_t(x)$ is also non-expansive (see for example Lemma 1 in [22]). In [22], it is also assumed that the strongly convex upper level objective ϕ has a modulus $\rho > 0$ and also a Lipschitz gradient on $\mathbb{R}^n$, whose Lipschitz constant is given as L_ϕ. We are now in a position to state the Big-SAM method.

BiG-SAM Method [20]

Step 0: Initialization: $t \in (0, \frac{1}{L_f}]$, $\beta \in (0, \frac{2}{L_\phi + \rho})$ and $\{\alpha_k\}$ is a sequence satisfying $\alpha_k \in (0, 1]$, for all $k \in \mathbb{N}$; $\lim_{k\to\infty} \alpha_k = 0$ and $\sum_{k=1}^{\infty} \alpha_k = \infty$ and $\lim_{k\to\infty} \frac{\alpha_k+1}{\alpha_k} = 1$.

Set $x^0 \in \mathbb{R}^n$ as the initial guess solution.

Step 1: For $k = 1, 2, \cdots$, do

(i) $y^{k+1} = \text{prox}_{tg}(x^k - t\nabla f(x^k))$
(ii) $z^{k+1} = x^k - \beta\nabla\phi(x^k)$
(iii) $x^{k+1} = \alpha_k z^k + (1 - \alpha_k) y^k$.

The key result regarding the Big-SAM method, is given in Proposition 5 in [20] which says the iterates $\{x^k\}$ generated by this method converge to a point $\tilde{x} \in S$ such that

$$\langle \nabla\phi(\tilde{x}), x - \tilde{x} \rangle \geq 0, \quad \forall x \in C$$

showing that $\tilde{x}$ solves (SBP-2) and hence $\tilde{x} = x^*$. Here BiG-SAM means Bilevel gradient sequential averaging method, where the sequential averaging is reflected in (iii) of Step 1 in the above algorithm.

Very recently a modification of the BiG-SAM method was presented in [21].

9.4 Algorithms for Non-smooth Data

There are not many algorithmic analysis of the (SBP) problem in the literature and of course for the non-smooth (SBP) problem the situation is not better. The bundle method for non-smooth (SBP) problem presented by Solodov [23] is one of the initial work in this direction. He considered an (SBP) problem with both the upper and lower level objective function to be non-smooth and the lower level problem is unconstrained. The (SBP) problem considered by Solodov in [23] is given as

$$\begin{aligned} &\min f(x) \\ &\text{subject to } x \in \arg\min_{\mathbb{R}^n} g, \end{aligned} \tag{9.4.1}$$

where $f, g : \mathbb{R}^n \to \mathbb{R}$ are non-smooth convex functions. The main idea behind the algorithm presented in [23] is to minimize the penalized function

$$\varphi_\varepsilon(x) = g(x) + \varepsilon f(x)$$

over $\mathbb{R}^n$. As $\varepsilon \downarrow 0$, the solutions of the problem

$$\min \varphi_\varepsilon(x) \quad \text{subject to } x \in \mathbb{R}^n \tag{9.4.2}$$

tend to the solution of the (SBP) problem (9.4.1). But it is almost impossible to calculate the solution of (9.4.2) exactly. To avoid that, Solodov's idea was to make one step in the descent direction of the function φ_{ε_k} from the k-th iteration x^k and then update ε_k to ε_{k+1} and repeat the step. To realize the descent step Solodov considered the bundle method for unconstrained non-smooth optimization problem (see [16]). Also we will need ε-subdifferential to discuss the algorithm which we present here.

The *ε-subdifferential* of f, $\partial_\varepsilon f : \mathbb{R}^n \rightrightarrows \mathbb{R}^n$, is defined as

$$\partial_\varepsilon f(x) = \{v \in \mathbb{R}^n \ : \ f(y) \geq f(x) + \langle v, y - x\rangle - \varepsilon, \ \forall\, y \in \mathbb{R}^n\}.$$

The idea of a bundle method is to store information of the past in a *bundle*. At any k-th step let x^k be the iteration point and y^i $(i = 1, 2, \cdots, l-1)$ are the points generated by the algorithm so far. Note that all y^i's may not be accepted as the iteration point in the process. Which means there may be some y^i's which are not included in the sequence $\{x^k\}$. The information about all the y^i $(i = 1, 2, \cdots, l-1)$, such as the subgradient and some quantities required in this algorithm are stored in the *bundle*. Note that $\{x^n\}_{n=1}^k$ is actually a finite subsequence of the finite sequence $\{y^i\}_{i=1}^{l-1}$. Let there be l iterations before we choose to modify x^k and ε_k, then $k(l)$ is the index of the last iteration and $k = k(l)$. Let $\xi^i \in \partial f(y^i)$ and $\zeta^i \in \partial g(y^i)$. Then $\varepsilon_k \xi^i + \zeta^i \in \partial \phi_{\varepsilon_k}(y^i)$. In the bundle method, a linearisation of the function φ_{ε_k} is used (see for example [16]). Then $\varphi_{\varepsilon_k}(x)$ can be approximated by the following function

$$\psi(x) := \max\{\varphi_{\varepsilon_k}(x^k) + \langle w^k, x - x^k\rangle : w^k \in \partial\varphi_{\varepsilon_k}(x^k)\}.$$

But this includes the calculation of all the subgradients of φ_{ε_k} at x^k, which is again a tough job. So the following cutting-planes approximation ψ_l is used by Solodov of the function φ_{ε_k}.

$$\psi_l(x) := \max_{i<l}\{\varepsilon_k f(y^i) + g(y^i) + \langle \varepsilon_k \xi^i + \zeta^i, x - y^i\rangle\}. \tag{9.4.3}$$

This is a linearisation centred at y^i. To reduce the storage requirements the following expression is constructed, which is centred at x^k.

$$\begin{aligned}\psi_l(x) :=& \max_{i<l}\{\varepsilon_k f(x^k) + g(x^k) + \langle \varepsilon_k \xi^i + \zeta^i, x - x^k\rangle \\ & - \varepsilon_k[f(x^k) - f(y^i) - \langle \xi^i, x^k - y^i\rangle] - [g(x^k) - g(y^i) - \langle \zeta^i, x^k - y^i\rangle]\} \\ =& \varepsilon_k f(x^k) + g(x^k) + \max_{i<l}\{-\varepsilon_k e_f^{k,i} - e_g^{k,i} + \langle \varepsilon_k \xi^i + \zeta^i, x - x^k\rangle\}\end{aligned}$$

where

$$e_f^{k,i} = f(x^k) - f(y^i) - \langle \xi^i, x^k - y^i\rangle \tag{9.4.4}$$

and

$$e_g^{k,i} = g(x^k) - g(y^i) - \langle \zeta^i, x^k - y^i\rangle. \tag{9.4.5}$$

Clearly, by convexity of f and g; $e_f^{k,i}, e_g^{k,i} \geq 0$. Now as $\xi^i \in \partial f(y^i)$, for any $x \in \mathbb{R}^n$ we have

$$\begin{aligned} f(x) &\geq f(y^i) + \langle \xi^i, x - y^i\rangle \\ &= f(x^k) + \langle \xi^i, x - x^k\rangle - e_f^{k,i}. \end{aligned}$$

This implies that $\xi^i \in \partial_{e_f^{k,i}} f(x^k)$. Similarly, $\zeta^i \in \partial_{e_g^{k,i}} g(x^k)$. Then

$$\varepsilon_k \xi^i + \zeta^i \in \partial_{(\varepsilon_k e_f^{k,i} + e_g^{k,i})} \varphi_{\varepsilon_k}(x^k).$$

Choose $\mu_l > 0$, then the minimizer of the following convex optimization problem

$$\min_{x \in \mathbb{R}^n} \left\{\psi_l(x) + \frac{\mu_l}{2}\|x - x^k\|^2\right\} \tag{9.4.6}$$

can be taken as the next candidate point y^l. Note that (9.4.6) has unique solution since the objective function is strongly convex. If $\varphi_{\varepsilon_k}(y^l)$ is sufficiently small compared with $\varphi_{\varepsilon_k}(x^k)$, then y^l is considered to be a serious (or acceptable) step and $x^{k+1} := y^l$. In fact y^l is considered to be a serious step if the following decrease criteria holds, i.e.

$$\varphi_{\varepsilon_k}(y^l) \leq \varphi_{\varepsilon_k}(x^k) - \gamma \delta_l. \tag{9.4.7}$$

For $\gamma \in (0, 1)$, let

$$\delta_l = \hat{\varepsilon}^{k,l} + \frac{1}{2\mu_l}\|\hat{g}^l\|^2,$$

with,

$$\hat{\varepsilon}^{k,l} = \varphi_{\varepsilon_k}(x^k) - \varphi_{\varepsilon_k}(y^l) - \frac{1}{\mu_l}\|\hat{g}^l\|^2,$$

$$\hat{g}^l = \mu_l(x^k - y^l).$$

Therefore,

$$\delta_l = \varphi_{\varepsilon_k}(x^k) - \varphi_{\varepsilon_k}(y^l) - \frac{\mu_l}{2}\|x^k - y^l\|^2.$$

We can see that δ_l deals with the difference in the functional values of φ_{ε_k} at x^k and y^l and also keeps the distance between x^k and y^l into account. Note that $\delta_l \geq 0$ as $\hat{\varepsilon}^{k,l} \geq 0$, which is shown in Lemma 2.1 [23]. If y^l does not satisfy the condition (9.4.7), then y^l is considered as a null step and is not included in the sequence of iterates. But this null step then contributes to construct a better approximation ψ_{l+1} of the function φ_{ε_k} using (9.4.3). Then y^{l+1} is generated using (9.4.6) with $\mu_{l+1} > 0$, which can again be a serious step or a null step. The process is continued until x^{k+1} is generated.

Let us now discuss a little about *bundles*, where the information about the candidate points y^l (no matter serious step or null step) is stored. In [23], Solodov describes an efficient way to store all the information needed for the bundle method algorithm and also ensures that the number of elements in the *bundles* is under some pre-decided bound. Managing the bundle this way keeps the algorithm computationally more suitable and efficient. Solodov [23] has divided the information at any iteration l into two parts, one is called oracle bundle denoted by B_l^{oracle} and another one is aggregate bundle, B_l^{agg}. Both the bundles contain approximated subgradients of the objective functions f and g at x^k and are defined in the following way.

$$B_l^{oracle} \subset \bigcup_{i<l} \left\{ e_f^{k,i}, e_g^{k,i} \in \mathbb{R}_+, \xi^i \in \partial_{e_f^{k,i}} f(x^k), \zeta^i \in \partial_{e_g^{k,i}} g(x^k) \right\}$$

and

$$B_l^{agg} \subset \bigcup_{i<l} \left\{ \hat{\varepsilon}_f^{k,i}, \hat{\varepsilon}_g^{k,i} \in \mathbb{R}_+, \hat{\xi}^i \in \partial_{\hat{\varepsilon}_f^{k,i}} f(x^k), \hat{\zeta}^i \in \partial_{\hat{\varepsilon}_g^{k,i}} g(x^k) \right\},$$

where $e_f^{k,i}$, $e_g^{k,i}$ are given by (9.4.4) and (9.4.5) and let us set

$$\hat{\varepsilon}_f^{k,l} = \sum_{i \in B_l^{oracle}} \lambda_i^l e_f^{k,i} + \sum_{i \in B_l^{agg}} \hat{\lambda}_i^l \hat{\varepsilon}_f^{k,i}, \tag{9.4.8}$$

$$\hat{\varepsilon}_g^{k,l} = \sum_{i \in B_l^{oracle}} \lambda_i^l e_g^{k,i} + \sum_{i \in B_l^{agg}} \hat{\lambda}_i^l \hat{\varepsilon}_g^{k,i}, \tag{9.4.9}$$

with $\lambda_i^l, \hat{\lambda}_i^l \geq 0$ and $\sum_{i \in B_l^{oracle}} \lambda_i^l + \sum_{i \in B_l^{agg}} \hat{\lambda}_i^l = 1$.

Note that $\xi^i \in \partial f(y^i)$ and $\zeta^i \in \partial g(y^i)$, so we can see that the bundle B_l^{oracle} stores the subgradient information of the objective functions at some candidate points y^i. But there may not be any candidate point y^j, $j < l$ such that $\hat{\xi}^j \in \partial f(y^j)$ and $\hat{\zeta}^j \in \partial g(y^j)$. Also note that the construction of $\hat{\varepsilon}_f^{k,l}$ and $\hat{\varepsilon}_g^{k,l}$ justifies the term aggregate bundle. Even though the bundles B_l^{oracle} and B_l^{agg} do not restore all the subgradients at all the candidate points, still a cutting-plane approximation of φ_{ε_k} can be done using these bundle information which is given by

$$\begin{aligned}\psi_l(x) :=& \varepsilon_k f(x^k) + g(x^k) \\ &+ \max\left\{ \max_{i \in B_l^{oracle}} \left\{ -(\varepsilon_k e_f^{k,i} + e_g^{k,i}) + \langle \varepsilon_k \xi^i + \zeta^i, x - x^k \rangle \right\}, \right. \\ &\left. \max_{i \in B_l^{agg}} \left\{ -(\varepsilon_k \hat{\varepsilon}_f^{k,i} + \hat{\varepsilon}_g^{k,i}) + \langle \varepsilon_k \hat{\xi}^i + \hat{\zeta}^i, y - x^k \rangle \right\} \right\} \end{aligned} \tag{9.4.10}$$

As we can see the two approximations (9.4.3) and (9.4.10) of the penalized function φ_{ε_k} are different. The connection between these two approximations are well established in terms of the solution of the problem given in (9.4.6) and the function (9.4.10) in the following lemma (Lemma 2.1, [23]).

Lemma 9.4.1 *Let y^l be the solution of (9.4.6) with ψ_l, as given in (9.4.10). Then*

- $y^l = x^k - \frac{1}{\mu^l}(\varepsilon_k \hat{\xi}^l + \hat{\zeta}^l)$.
- $\hat{\xi}^l = \sum_{i \in B_l^{oracle}} \lambda_i^l \xi^i + \sum_{i \in B_l^{agg}} \hat{\lambda}_i^l \hat{\xi}^i$
  $\hat{\zeta}^l = \sum_{i \in B_l^{oracle}} \lambda_i^l \zeta^i + \sum_{i \in B_l^{agg}} \hat{\lambda}_i^l \hat{\zeta}^i$,
  where $\lambda_i^l, \hat{\lambda}_i^l \geq 0$ *and* $\sum_{i \in B_l^{oracle}} \lambda_i^l + \sum_{i \in B_l^{agg}} \hat{\lambda}_i^l = 1$.
- $\varepsilon_k \hat{\xi}^l + \hat{\zeta}^l \in \partial \psi_l(y^l)$.
- $\hat{\xi}^l \in \partial_{\hat{\varepsilon}_f^{k,l}} f(x^k)$, $\hat{\zeta}^l \in \partial_{\hat{\varepsilon}_g^{k,l}} g(x^k)$, *where* $\hat{\varepsilon}_f^{k,l}$ *and* $\hat{\varepsilon}_g^{k,l}$ *are from (9.4.8) and (9.4.9).*
- $\varepsilon_k \hat{\xi}^l + \hat{\zeta}^l = \hat{g}^l \in \partial_{\hat{\varepsilon}^{k,l}} \varphi_{\varepsilon_k}(x^k)$, *where* $\hat{\varepsilon}^{k,l} = \varepsilon_k \hat{\varepsilon}_f^{k,l} + \hat{\varepsilon}_g^{k,l}$.
- $\hat{\varepsilon}^{k,l} = \varphi_{\varepsilon_k}(x^k) - \varphi_{\varepsilon_k}(y^l) - \frac{1}{\mu_l}\|\hat{g}^l\|^2 \geq 0$. △

Here note that λ^l and $\hat{\lambda}^l$ in Lemma 9.4.1 are obtained as a part of subgradient calculation of ψ_l while solving the strongly convex optimization problem (9.4.6). We have already mentioned that Solodov [23] has developed two different bundles to maintain the bundle size under a fixed limit. Here we discuss how that is done. Let the maximum size that the bundle can attain together be $|B|_{max}$. So at any l-th step, if

$$|B_l^{oracle} \cup B_l^{agg}| = |B|_{max},$$

then at least two elements are removed from $B_{l+1}^{oracle} \cup B_{l+1}^{agg}$ and the aggregate information $(\hat{\varepsilon}_f^{k,l}, \hat{\varepsilon}_g^{k,l}, \hat{\xi}^l, \hat{\zeta}^l)$ is included in B_l^{agg} and renamed as B_{l+1}^{agg}. Also include $(e_f^{k,l}, e_g^{k,l}, \xi^l, \zeta^l)$ in B_l^{oracle} and called B_{l+1}^{oracle}. Now using the bundle information, Solodov constructed the following relations which satisfies all the results mentioned in Lemma 9.4.1.

$$\begin{aligned}
\text{For } i \in B_{l+1}^{oracle},\quad e_f^{k+1,i} &= e_f^{k,i} + f(x^{k+1}) - f(x^k) + \langle \xi^i, x^k - x^{k+1}\rangle,\\
e_g^{k+1,i} &= e_g^{k,i} + g(x^{k+1}) - g(x^k) + \langle \zeta^i, x^k - x^{k+1}\rangle\\
\text{and for } i \in B_{l+1}^{oracle}\quad \hat{\varepsilon}_f^{k+1,i} &= \hat{\varepsilon}_f^{k,i} + f(x^{k+1}) - f(x^k) + \langle \hat{\xi}^i, x^k - x^{k+1}\rangle,\\
\hat{\varepsilon}_g^{k+1,i} &= \hat{\varepsilon}_g^{k,i} + g(x^{k+1}) - g(x^k) + \langle \hat{\zeta}^i, x^k - x^{k+1}\rangle \quad . \qquad (9.4.11)
\end{aligned}$$

Now we are ready to present the algorithm for (SBP) problem.

Bundle Method Algorithm: [23]

Step 0: Choose $m \in (0,1)$, an integer $|B|_{max} \geq 2$, $x^0 \in \mathbb{R}^n$, $\varepsilon_0 > 0$ and $\beta_0 > 0$.
Set $y^0 := x^0$ and compute $f(y^0)$, $g(y^0)$, $\xi^0 \in \partial f(y^0)$, $\zeta^0 \in \partial g(y^0)$.
Set $k = 0, l = 1, e_f^{0,0} = e_g^{0,0} := 0$.
Define $B_l^{oracle} := \{(e_f^{0,0}, e_g^{0,0}, \xi^0, \zeta^0)\}$ and $B_l^{agg} := \emptyset$

Step 1: At any l-th step, choose $\mu_l > 0$ and compute y^l as the solution of (9.4.6) with ψ_l, as given in (9.4.10).
Compute $f(y^l)$, $g(y^l)$, $\xi^l \in \partial f(y^l)$, $\zeta^l \in \partial g(y^l)$ and $e_f^{k,l}, e_g^{k,l}$ using (9.4.4) and (9.4.5).

Step 2: If $\varphi_{\varepsilon_k}(y^l) \leq \varphi_{\varepsilon_k}(x^k) - m\delta_l$, then y^l is a serious step, otherwise null step.

Step 3: Check if $|B_l^{oracle} \cup B_l^{agg}| < |B|_{max}$, otherwise manage the bundle as mentioned earlier.

Step 4: If y^l is a serious step, then set $x^{k+1} := y^l$.
Choose $0 < \varepsilon_{k+1} < \varepsilon_k$ and $0 < \beta_{k+1} < \beta_k$. Update $e_f^{k+1,i}, e_g^{k+1,i}, \hat{\varepsilon}_f^{k+1,i}, \hat{\varepsilon}_g^{k+1,i}$ using (9.4.11).
Set $k = k + 1$ and go to Step 5.

If $\max \hat{\varepsilon}^{k,l}, \|\hat{g}^l\| \leq \beta_k \varepsilon_k$, choose $0 < \varepsilon_{k+1} < \varepsilon_k$ and $0 < \beta_{k+1} < \beta_k$. Set $x^{k+1} := x^k$, $k = k + 1$ and go to step 5.

Step 5: Set $l = l + 1$ and go to Step 1.

Note that $\max \hat{\varepsilon}^{k,l}, \|\hat{g}^l\| \leq \beta_k \varepsilon_k$ works as a stopping criteria to minimize the function ψ_k. The convergence result of this algorithm is presented through the following theorem from [23].

Theorem 9.4.2 *Let $f, g : \mathbb{R}^n \to \mathbb{R}$ are convex functions such that f is bounded below on $\mathbb{R}^n$ and the solution set of the (SBP) problem S is non-empty and bounded. Let us choose μ_l, ε_k, β_k such that the following conditions are satisfied at any $l, k \in \mathbb{N}$.*

1. *$\mu_{l+1} \leq \mu_l$ and there exists $\hat{\mu}, \bar{\mu} > 0$ such that*

$$0 < \hat{\mu} \leq \mu_l \leq \bar{\mu}.$$

2. *$\varepsilon_k \to 0$ as $k \to 0$ and $\sum_{k=0}^{\infty} \varepsilon_k = +\infty$.*
3. *$\beta_k \to 0$ as $k \to 0$.*

Then dist$(x^k, S) \to 0$ as $k \to 0$. This also implies that any accumulation point of the sequence $\{x^k\}$ is a solution of the (SBP) problem. △

Another algorithm is proposed by Dempe et al. [8] for the non-smooth simple bilevel programming problem. In [8], we present an algorithm for the (SBP) problem in more general scenario than that we have discussed so far. Note that in [23], the considered (SBP) problem is unconstrained in the lower level, whereas the (SBP) problem we consider, which given below, has constrained lower level problem.

$$\min f(x)$$

$$x \in \arg\min_{C} g,$$

where $f, g : \mathbb{R}^n \to \mathbb{R}$ are continuous, convex functions and $C \subseteq \mathbb{R}^n$ is closed, convex set. Our algorithm is inspired from the penalisation approach by Solodov [22]. But our algorithm is different from Solodov's approach in two aspects, one is that we don't assume any differentiability assumption on the upper and lower level objective functions which naturally helps us to get rid of the Lipschitz continuity assumption on the gradient of the objective functions that is made by Solodov [22], another one is the fact that our algorithm is based on proximal point method whereas Solodov's algorithm is based on projected gradient method. Our Algorithm is also inspired by Cabot [2] and Facchinei et al. [11] but again different in the sense that our algorithm deals with more general form of (SBP) compared with those. The only assumption we make is that the solution set of the (SBP) problem needs to be non-empty and bounded. This is done to ensure the boundedness of the iteration points which will finally lead to the existence of an accumulation point and hence a solution of the (SBP) problem.

In the proximal point algorithm that we present, we consider an approximate solution rather than the exact solution. For this purpose we need to use ε-subdifferential and ε-normal cone for $\varepsilon > 0$. We have already presented the ε-subdifferential while discussing the last algorithm. Here we present the ε-normal set.

The ε-normal set of $C \subseteq \mathbb{R}^n$ is defined as

$$N_C^{\varepsilon}(x) = \{v \in \mathbb{R}^n : \langle v, y - x \rangle \leq \varepsilon, \quad \forall y \in C\}.$$

Consider the following sequence of penalized functions

$$\varphi = g + \varepsilon_k f.$$

We aim to minimize this function approximately over the constraint set C to obtain the iteration point at any $(k+1)$-st step. Let $\eta_k > 0$ and $\lambda_k > 0$, then the $(k+1)$-st iterate comes from the following relation.

$$x^{k+1} \in \eta_k\text{-argmin}_C\{\varphi_k + \frac{1}{2\lambda_k}\|. - x^k\|^2\}$$

Considering the optimality condition for this approximate minimization problem we get our key step to design the algorithm (for details see [8]), which is given as

$$-\frac{1}{\lambda_k}(x^{k+1} - x^k) \in \partial_{\eta_k^1}\varphi_k(x^{k+1}) + N_C^{\eta_k^2}(x^{k+1}),$$

where $\eta_k^1 + \eta_k^2 \leq \eta_k$. Now we present the algorithm formally.

Algorithm for the (SBP) Problem

Step 0. Choose $x^0 \in C$, $\epsilon_0 \in (0, \infty)$ and let $k := 0$.
Step 1. Given x^k, $\lambda_k > 0$, $\varepsilon_k > 0$ and $\eta_k > 0$, choose $x^{k+1} \in C$ such that

$$-(\frac{x^{k+1} - x^k}{\lambda_k}) \in \partial_{\eta_k^1}(\varphi_k)(x^{k+1}) + N_C^{\eta_k^2}(x^{k+1})$$

where $\eta_k^1, \eta_k^2 \geq 0$ and $\eta_k^1 + \eta_k^2 \leq \eta_k$.

For the convergence analysis of this algorithm we make the following assumptions on the sequences $\{\varepsilon_k\}$, $\{\lambda_k\}$ and $\{\eta_k\}$. Then the convergence results of this algorithm is presented in the next theorem.

- $\{\varepsilon_k\}$ is a decreasing sequence such that $\lim_{n\to\infty} \varepsilon_n = 0$ and $\sum_{n=0}^{\infty} \varepsilon_n = +\infty$.
- There exists $\underline{\lambda} > 0$ and $\overline{\lambda} > 0$ such that $\underline{\lambda} \leq \lambda_n \leq \overline{\lambda}$ for all $n \in \mathbb{N}$.
- $\sum_{n=0}^{\infty} \eta_n < +\infty$.

Theorem 9.4.3 ([8]) *Let $C \subset \mathbb{R}^n$ be a non-empty closed convex set, $g : \mathbb{R}^n \to \mathbb{R}$ and $f : \mathbb{R}^n \to \mathbb{R}$ be convex functions. Assume that the functions g and f are bounded below, that the set $S_0 := \arg\min_C g$ is nonempty, and that the set $S_1 := \arg\min_{S_0} f$ is nonempty and bounded. Let $\{\varepsilon_n\}$, $\{\lambda_n\}$, $\{\eta_n\}$ be nonnegative sequences which satisfy the assumptions mentioned above. Then any sequence $\{x^n\}$ generated by the algorithm satisfies*

$$\lim_{n\to\infty} d(x^n, S_1) = 0.$$

This also implies that any accumulation point of the sequence $\{x^n\}$, is a solution of the (SBP) problem. △

The reference [13] in the list of reference was first brought to our notice by M. V. Solodov. In [13], the ε-subgradient is used to develop an inexact algorithmic scheme for the (SBP) problem. however the scheme and convergence analysis developed in [13] is very different from the one in [8]. It is important to note that in contrast to the convergence analysis in [8], the scheme in [13] needs more stringent condition. Also an additional assumption is made on the subgradient of the lower level objective (see Proposition 2, [13]). They also considered the case when the upper level objective is smooth and also has Lipschitz gradient.

References

1. A. Beck, S. Sabach, A first order method for finding minimal norm-like solutions of convex optimization problems. Math. Program. **147**(1–2), Ser. A, 25–46 (2014)
2. A. Cabot, Proximal point algorithm controlled by a slowly vanishing term: applications to hierarchical minimization. SIAM J. Optim. 15(2), 555–572 (2004/2005)
3. S. Dempe, *Foundations of Bilevel Programming*. (Kluwer Academic Publishers, Dordrecht, 2002)
4. S. Dempe, J. Dutta, Is bilevel programming a special case of a mathematical program with complementarity constraints? Math. Program. **131**(1–2), Ser. A, 37–48 (2012)
5. S. Dempe, J. Dutta, B.S. Mordukhovich, New necessary optimality conditions in optimistic bilevel programming. Optimization **56**(5–6), 577–604 (2007)
6. S. Dempe, N. Dinh, J. Dutta, Optimality conditions for a simple convex bilevel programming problem, in *Variational Analysis and generalized Differentiation in Optimization and Control*, ed. by R.S. Burachik, J.-C. Yao (Springer, Berlin, 2010), pp. 149–161
7. S. Dempe, N. Dinh, J. Dutta, T. Pandit, Simple bilevel programming and extensions. Mathematical Programming (2020). https://doi.org/10.1007/s10107-020-01509-x
8. S. Dempe, N. Dinh, J. Dutta, T. Pandit, Simple bilevel programming and extensions; Part-II: Algorithms (Preprint 2019). http://arxiv.org/abs/1912.06380
9. A. Dhara, J. Dutta, *Optimality Conditions in Convex Optimization: A Finite-Dimensional View*, With a foreword by S. Dempe (CRC Press, Boca Raton, 2012)
10. J. Dutta, Optimality conditions for bilevel programming: An approach through variational analysis, in *Generalized Nash Equilibrium Problems, Bilevel Programming and MPEC*. Forumfor Interdisciplinary Mathematics (Springer, Singapore, 2017), pp. 43–64

11. F. Facchinei, J.-S. Pang, G. Scutari, L. Lampariello, VI constrained hemivariational inequalities: Distributed algorithms and power control in ad-hoc networks. Math. Program. **145**, 59–96 (2014)
12. O. Güler, *Foundations of Optimization*. Graduate Texts in Mathematics, vol. 258 (Springer, New York, 2010)
13. E.S. Helou, L.E.A. Simões, ε-subgradient algorithms for bilevel convex optimization. Inverse Probl. **33**(5), 055020, 33 pp. (2017)
14. J.B. Hirriart Urruty, C. Lemarechal, *Convex Analysis and Minimization Algorithms. I. Fundamentals*. Grundlehren der Mathematischen Wissenschaften [Fundamental Principles of Mathematical Sciences], vol. 305 (Springer, Berlin, 1993)
15. J. Kelley, The cutting-plane method for solving convex programs, J. Soc. Indust. Appl. Math. **8**, 703–712 (1960)
16. K.C. Kiwiel, *Methods of Descent for Nondifferentiable Optimization*. Lecture Notes in Mathematics, vol. 1133 (Springer, Berlin, 1985)
17. A.S. Lewis, J.-S. Pang, Error bounds for convex inequality systems, in *Generalized Convexity, Generalized Monotonicity: Recent Results (Luminy, 1996)*. Nonconvex Optimization and Its Applications, vol. 27 (Kluwer Academic Publishers, Dordrecht, 1998), pp. 75–110
18. T. Pandit, A study of simple bilevel programming and related issues, Ph.D. Thesis, IIT Kanpur, 2019. http://172.28.64.70:8080/jspui/handle/123456789/18699
19. T. Pandit, J. Dutta, K.S.M. Rao, A new approach to algorithms for Simple Bilevel Programming (SBP), Problem (Preprint 2019)
20. S. Sabach, S. Shtern, A first order method for solving convex bilevel optimization problems. SIAM J. Optim. **27**(2), 640–660 (2017)
21. Y. Shehu, P.T. Vuong, A. Zemkoho, An inertial extrapolation method for convex simple bilevel optimization. Optim. Method Softw. (2019). https://doi.org/10.1080/10556788.2019.1619729
22. M.V. Solodov, An explicit descent method for bilevel convex optimization. J. Convex Anal. **14**(2), 227–237 (2007)
23. M.V. Solodov, A bundle method for a class of bilevel nonsmooth convex minimization problems. SIAM J. Optim. **18**(1), 242–259 (2007)
24. N. Yamashita, M. Fukushima, Equivalent unconstrained minimization and global error bounds for variational inequality problems. SIAM J. Control Optim. **35**(1), 273–284 (1997)

Chapter 10
Algorithms for Linear Bilevel Optimization

Herminia I. Calvete and Carmen Galé

Abstract This chapter addresses the linear bilevel optimization problem in which all the functions involved are linear. First, some remarks are made about the formulation of the problem, the difficulties which can arise when defining the concept of feasible solution or when proving the existence of optimal solution, and about its computational complexity. Then, the chapter focuses on the main algorithms proposed in the literature for solving the linear bilevel optimization problem. Most of them are exact algorithms, with only a few applying metaheuristic techniques. In this chapter, both kind of algorithms are reviewed according to the underlying idea that justifies them.

Keywords Linear bilevel programming · Enumerative algorithms · Karush-Kuhn-Tucker conditions · Metaheuristic algorithms

10.1 Linear Bilevel Program Formulation

Bilevel optimization programs involve two decision makers in a hierarchical framework. Each decision maker controls part of the variables and aims to optimize his/her objective function subject to constraints. The lower level (LL) decision maker optimizes his/her objective function knowing the values of the variables controlled by the upper level (UL) decision maker. This one, in return, having complete information on the reaction of the LL decision maker, selects the value of his/her variables so as to optimize his/her own objective function. As a consequence, bilevel programs are formulated as optimization problems which involve another optimization problem in the constraint set. The name bilevel or multilevel programming is credited to Candler and Norton [20] who present a linear

H. I. Calvete · C. Galé (✉)
Statistical Methods Department, IUMA, University of Zaragoza, Zaragoza, Spain
e-mail: herminia@unizar.es; cgale@unizar.es

S. Dempe, A. Zemkoho (eds.), *Bilevel Optimization*, Springer Optimization and Its Applications 161, https://doi.org/10.1007/978-3-030-52119-6_10

bilevel (LB) optimization problem in an economic context. Ben-Ayed [13] and Wen and Hsu [56] provide early reviews on LB problems.

Let $x \in \mathbb{R}^{n_1}$ and $y \in \mathbb{R}^{n_2}$ denote the variables controlled by the UL and the LL decision makers, respectively. In a LB optimization problem all the functions involved are linear. Thus, it can be formulated as:

$$\min_{x,y} \quad c_1 x + d_1 y \tag{10.1.1a}$$

s.t.

$$A_1 x + B_1 y \leqslant b_1 \tag{10.1.1b}$$

$$x \geqslant 0 \tag{10.1.1c}$$

where y solves:

$$\min_{y} \quad c_2 x + d_2 y \tag{10.1.1d}$$

s.t.

$$A_2 x + B_2 y \leqslant b_2 \tag{10.1.1e}$$

$$y \geqslant 0 \tag{10.1.1f}$$

where, for $i = 1, 2$, c_i is a row vector of dimension n_1, d_i is a row vector of dimension n_2, $b_i \in \mathbb{R}^{m_i}$, A_i is an $m_i \times n_1$ matrix, and B_i is an $m_i \times n_2$ matrix. Since once the UL decision maker selects the value of the variables x the first term in the LL objective function becomes a constant, it can be removed from the formulation of the problem.

Problem (10.1.1) is the most general formulation since constraints (10.1.1b) involve both levels' variables x and y, and so they are called coupling or joint constraints. However, it is worth pointing out that most papers in the literature consider the particular case in which those constraints only involve UL variables or do not appear in the formulation. In fact, this is so for the pioneering paper by Candler and Norton [20]. Constraints (10.1.1b) appeared for the first time in a paper by Aiyoshi and Shimizu [2] dealing with nonlinear bilevel problems. It should be noted that bilevel problems are very sensitive to the existence of coupling constraints. Audet et al. [5] and Mersha and Dempe [45] have investigated the consequences of shifting coupling constraints to the LL problem for LB problems.

Let R be the polyhedron defined by the UL constraints (10.1.1b)–(10.1.1c), S the polyhedron defined by the LL constraints (10.1.1e)–(10.1.1f), and $T = R \cap S$.

For a given x, the LL decision maker solves the problem

$$\min_{y} \quad d_2 y \tag{10.1.2a}$$

s.t.

$$B_2 y \leqslant b_2 - A_2 x \tag{10.1.2b}$$

$$y \geqslant 0 \tag{10.1.2c}$$

Let $S(x) = \{y \in \mathbb{R}^{n_2} : (x, y) \in S\}$ denote its feasible region and $M(x)$ be the set of optimal solutions, also called the LL rational reaction set. The feasible region of the bilevel problem, called inducible (or induced) region IR can be written as:

$$\mathrm{IR} = \{(x, y) : (x, y) \in T, y \in M(x)\} \tag{10.1.3}$$

Any point of IR is a bilevel feasible solution. A point $x \in \mathbb{R}^{n_1}$ is called permissible if a $y \in \mathbb{R}^{n_2}$ exists so that $(x, y) \in \mathrm{IR}$.

Unlike general mathematical programs, bilevel optimization problems may not possess a solution even when all the functions involved are continuous over compact sets. In particular, difficulties may arise when the rational reaction set $M(x)$ is not single-valued. Multiple optima do not affect the value of the LL objective function but can provide very different values of the UL objective function. Different approaches have been proposed in the literature to make sure that the bilevel problem is well posed. In the earlier papers dealing with bilevel optimization, a common approach was to assume that, for each value of the UL variables x, there is a unique solution to the LL problem (10.1.2), i.e., the set $M(x)$ is a singleton for all permissible x. Other approaches have focused on the way of selecting $y \in M(x)$, in order to evaluate the UL function, when $M(x)$ is not a singleton. The most popular is the optimistic rule, which assumes that the UL decision maker is able to influence the LL decision maker so that the latter always selects the variables y to provide the best value of the UL objective function. In fact, problem formulation (10.1.1) assumes the optimistic approach when minimizing over x and y. The pessimistic rule assumes that the LL decision maker always selects the optimal decision which gives the worst value of the UL objective function. This approach is much harder to deal with and thus less attention has been devoted to it in the literature. These and other approaches can be found in the book by Dempe [24].

Since the initial papers on LB problems by Candler and Townsley [21], Bialas and Karwan [15], and Bard and Falk [11], the difficulty of solving problem (10.1.1) has been recognized due to its nonconvexity and nondifferentiability. LB problems have been proved to be NP-hard by Jeroslow [37], Ben-Ayed and Blair [14] and Bard [10]. Hansen et al. [32] have strengthened this result proving that they are strongly NP-hard. Therefore, these problems are very demanding from the computational point of view.

In the next sections we will review the main algorithms developed for specifically solving the LB problem. Among the different taxonomies that can be chosen to classify them, we have selected to distinguish between exact and metaheuristic, which is quite a natural classification. Within each category, several subcategories have been established that address the main characteristic of the algorithm, although there are obviously algorithms that could be classified into several subcategories.

Thus, regarding exact methods, those algorithms that take advantage of the specific properties of the LB problem have been classified as enumerative. Another subcategory has been established for those algorithms that transform the LB problem into a single level problem using the Karush-Kuhn-Tucker (KKT) conditions. Finally, the third subcategory includes the algorithms that apply classical optimization theory. Concerning metaheuristic algorithms, most are evolutionary algorithms, so only this subcategory has been explicitly considered. The remaining algorithms have been grouped. It is worth mentioning that bilevel programming is receiving increasing attention in the recent years. Thus, a plethora of algorithms have been developed. Dempe [25], in his review until 2019, collected over 1500 papers on this topic. In this chapter, attention has been paid to algorithms specifically designed for solving the LB problem that have set out key ideas for handling the problem which have been applied in many subsequent works.

10.2 Enumerative Algorithms

LB problems have important properties which have allowed the development of ad-hoc algorithms for solving them. Next we review some of these properties which are relevant for the algorithms developed in the literature.

From the geometrical point of view, Savard [49] proved that, if there exists a finite optimal solution of the LB problem, then

$$\mathrm{IR} = R \cap \Gamma_M$$

where Γ_M is the graph of M, i.e.

$$\Gamma_M = \{(x, y) : (x, y) \in S, y \in M(x)\}$$

Since Γ_M is the union of a finite number of polyhedra, IR is not necessarily convex and, if there are coupling constraints, IR is not necessarily connected. Moreover, taking into account that IR is the union of a finite number of polyhedra and the UL objective function is linear, there exists an extreme point of IR that solves the LB problem. Since the extreme points of IR are extreme points of T, there exists an extreme point of T that solves problem (10.1.1).

Xu [58] achieved similar results under weaker assumptions via a penalty function approach. Also, papers [9, 15, 21] prove the extreme point optimal solution property assuming that no coupling constraints exist (therefore $T = S$), the polyhedron S is compact and $M(x)$ is a singleton for all x such that $(x, y) \in S$.

The relaxed problem formulated in (10.2.1) plays an important role when solving problem (10.1.1):

$$\min_{x,y} \quad c_1x + d_1y \tag{10.2.1a}$$

s.t.

$$A_1x + B_1y \leqslant b_1 \tag{10.2.1b}$$

$$A_2x + B_2y \leqslant b_2 \tag{10.2.1c}$$

$$x \geqslant 0, \quad y \geqslant 0 \tag{10.2.1d}$$

If the optimal solution of problem (10.2.1) is a bilevel feasible solution, then it is an optimal solution of problem (10.1.1). Otherwise, problem (10.2.1) provides a lower bound of the UL optimum objective function value.

Liu and Hart [42] consider LB problems without the constraints (10.1.1b) and prove that if there exists an extreme point of S not in IR which is an optimal solution of the relaxed problem, then there exists a boundary feasible extreme point that solves the LB problem, where a point $(x, y) \in$ IR is a boundary feasible extreme point if there exists an edge E of S such that (x, y) is an extreme point of E, and the other extreme point of E is not an element of IR.

The following algorithms to find an optimal solution of the LB problem are based on an enumerative scheme and/or previous geometrical properties.

10.2.1 Enumeration of Extreme Points

The Kth best algorithm is one of the most well-known algorithms for solving LB problems. Bearing in mind that there is an extreme point of T which solves the LB problem, an examination of all extreme points of the polyhedron T provides a procedure to find the optimal solution of the LB problem in a finite number of steps. This is unsatisfactory, however, since the number of extreme points of T is, in general, very large.

Bialas and Karwan [15] deal with the LB problem without coupling constraints assuming that S is compact and propose the Kth best algorithm to solve it. The idea is to select the best extreme point of S with respect to the UL objective function which is a bilevel feasible solution.

Hence, an optimal solution to the relaxed problem (10.2.1) (x_r^*, y_r^*) is first considered. If this is a point of IR, then it is an optimal solution of the LB problem. If this is not so, set $(x_{[1]}, y_{[1]}) = (x_r^*, y_r^*)$ and compute the set of its adjacent extreme points $W_{[1]}$.

Then, the extreme point in $W = W_{[1]}$ which provides the best value of the UL objective function is selected to test if it is a point of IR. Let this extreme point be $(x_{[2]}, y_{[2]})$. If it is in IR, the algorithm finishes. If this is not so, the point is eliminated from W and the set of its adjacent extreme points with a worse value of the UL objective function , $W_{[2]}$, is added to W.

The algorithm continues by selecting the best extreme point in W with respect to the UL objective function and repeating the process. Note that the algorithm follows an edge path in S from the optimal solution of the relaxed problem (x_r^*, y_r^*) to an extreme point of S which is an optimal solution of the LB problem and a boundary feasible extreme point.

The efficiency of the Kth best algorithm depends strongly on the 'closeness' of the optimal solution of the relaxed problem to the optimal solution of the LB problem.

Based also on the fact that the LB problem has an extreme point optimal solution, Dempe [23] develops an algorithm which uses the theory of subgradients to fully describe the feasible set in a neighborhood of a feasible point. The algorithm combines extreme point enumeration and a descent method.

10.2.2 Enumeration of Optimal Bases of the LL Problem

Candler and Townsley [21] under the same assumptions of the Kth algorithm, propose a more restrictive examination of extreme points, which only considers the bases which are submatrices of B_2.

Note that, if the LB problem has an optimal solution, then there exists an optimal solution (x^*, y^*) for which y^* is an optimal extreme point of the polyhedron $S(x^*)$. Hence, an optimal basis B^* of B_2 is associated with that extreme point y^*. The idea of the algorithm is to successively identify optimal bases of the LL problem which are of interest in the sense that they provide points in IR with better values of the UL objective function. It is worth pointing out that the values of the UL variables affect the feasibility of a basis B of B_2, but are not involved in the optimality condition. Therefore, each optimal basis B of B_2 has associated a subset of points $x \in S_1$, where S_1 is the projection of S onto $\mathbb{R}^{n_1}$, which makes it feasible for the corresponding LL problem. The best point of the IR associated with the optimal basis B can be computed by solving a linear program in x. Additional conditions on the optimal basis B are established leading to possibly better solutions and avoiding a return to any formerly analyzed basis.

Bard [8] provides the results of a computational experiment on this algorithm using LB instances for which the number of variables x and y ranges between 3 and 30, and the number of constraints is 16 or 25. From these results, the author concludes that the implicit search proposed by Candler and Townsley is generally unsatisfactory and its behavior is worse than that of the Kth best algorithm in terms of computational time.

10.2.3 Parametric Approach

Notice that the LL problem can be seen as an optimization problem parameterized in the UL variables. Following this idea, Faísca et al. [26] develop a global optimization approach for solving the LB problem based on parametric programming theory. Variables x are considered as a parameter vector of the LL problem (10.1.2). This problem is solved by a multiparametric programming algorithm in order to compute a function $y(x)$ as follows:

$$y(x) = \begin{cases} \alpha^1 + \beta^1 x, & \text{if } H^1 x \leqslant h^1, \\ \dots \\ \alpha^K + \beta^K x, & \text{if } H^K x \leqslant h^K, \end{cases}$$

where, α^k and h^k are real vectors and β^k and H^k are real matrices. The set of permissible x is divided into a set of K regions, each one with a set of rational sets. Including the expression $y(x)$ in the UL problem, the following K linear programming problems are solved:

$$\begin{aligned} &\min_x \; c_1 x + d_1(\alpha^k + \beta^k x_1) \\ &\text{s.t.} \\ &\qquad H^k x \leqslant h^k \end{aligned}$$

Then, the optimal solution which provides the minimum UL objective function value among the K solutions is an optimal solution of the LB problem. As indicated by the authors, the computational efficiency of this procedure depends on the performance of the underlying multiparametric programming algorithm.

10.3 Algorithms Based on the Karush-Kuhn-Tucker Conditions

The most common reformulation of the LB problem as a single level problem consists of substituting the LL problem (10.1.2) by its KKT conditions. This transformation takes into account that, for linear optimization, KKT conditions, also known as complementary slackness conditions, are necessary and sufficient for optimality. Denoting by u the m_2 row vector of dual variables associated with constraints (10.1.1e), the dual problem of the LL problem (10.1.2) is:

$$\max_u \quad u(A_2 x - b_2) \tag{10.3.1a}$$

s.t.

$$u B_2 \geqslant -d_2 \tag{10.3.1b}$$

$$u \geqslant 0 \tag{10.3.1c}$$

Let U denote the feasible region of the dual problem defined by constraints (10.3.1b)–(10.3.1c). Note that it does not depend on x. Therefore, the dual problems associated to all permissible x have the same feasible region.

Applying the KKT conditions, the LL problem (10.1.2) can be substituted by:

$$\begin{aligned}
&B_2 y \leqslant b_2 - A_2 x\\
&u B_2 \geqslant -d_2\\
&(d_2 + u B_2) y = 0\\
&u(b_2 - A_2 x - B_2 y) = 0\\
&y \geqslant 0, u \geqslant 0
\end{aligned}$$

Hence, the LB problem can be rewritten as:

$$\min_{x,y} \quad c_1 x + d_1 y \tag{10.3.2a}$$

s.t.

$$A_1 x + B_1 y \leqslant b_1 \tag{10.3.2b}$$

$$A_2 x + B_2 y \leqslant b_2 \tag{10.3.2c}$$

$$u B_2 \geqslant -d_2 \tag{10.3.2d}$$

$$(d_2 + u B_2) y = 0 \tag{10.3.2e}$$

$$u(b_2 - A_2 x - B_2 y) = 0 \tag{10.3.2f}$$

$$x \geqslant 0, y \geqslant 0 \tag{10.3.2g}$$

$$u \geqslant 0 \tag{10.3.2h}$$

A variety of exact algorithmic approaches have been developed for solving problem (10.3.2), which is nonlinear due to the complementary slackness conditions (10.3.2e) and (10.3.2f). In the following, we describe the main ways that have been considered in the literature to address this nonlinearity.

10.3.1 Mixed-Integer Linear Programming

One of the most common approaches to solve problem (10.3.2) takes advantage of the disjunctive nature of the complementary slackness conditions to obtain an equivalent mixed integer linear problem which can be solved by using well known techniques or optimization software. This reformulation was introduced by Fortuny-Amat and McCarl [27] for solving the bilevel quadratic programming problem, but it can also be applied to LB programs.

Constraints (10.3.2e) and (10.3.2f) consist of the sum of products of two terms which are nonnegative. Therefore, each product must equal zero, and they can be substituted by

$$
\begin{aligned}
&(d_2 + uB_2)_i \leqslant M\zeta_i, \quad i = 1, \ldots, n_2 \\
&y_i \leqslant M(1 - \zeta_i), \quad i = 1, \ldots, n_2 \\
&u_j \leqslant M\eta_j, \quad j = 1, \ldots, m_2 \\
&(b_2 - A_2x - B_2y)_j \leqslant M(1 - \eta_j), \quad j = 1, \ldots, m_2 \\
&\zeta_i, \eta_j \in \{0, 1\}, \quad i = 1, \ldots, n_2, j = 1, \ldots, m_2
\end{aligned}
\tag{10.3.3}
$$

where $(d_2 + uB_2)_i$, y_i stand for the ith component of vectors $d_2 + uB_2$, y, respectively; u_j, $(b_2 - A_2x - B_2y)_j$ refer to the jth component of vectors u, $b_2 - A_2x - B_2y$, respectively; and M is a large enough positive constant. Then, problem (10.3.2) can be rewritten as:

$$\min_{x,y} \quad c_1x + d_1y \tag{10.3.4a}$$

s.t.

$$A_1x + B_1y \leqslant b_1 \tag{10.3.4b}$$

$$A_2x + B_2y \leqslant b_2 \tag{10.3.4c}$$

$$uB_2 \geqslant -d_2 \tag{10.3.4d}$$

$$(d_2 + uB_2)_i \leqslant M\zeta_i, \quad i = 1, \ldots, n_2 \tag{10.3.4e}$$

$$y_i \leqslant M(1 - \zeta_i), \quad i = 1, \ldots, n_2 \tag{10.3.4f}$$

$$u_j \leqslant M\eta_j, \quad j = 1, \ldots, m_2 \tag{10.3.4g}$$

$$(b_2 - A_2x - B_2y)_j \leqslant M(1 - \eta_j), \quad j = 1, \ldots, m_2 \tag{10.3.4h}$$

$$x \geqslant 0, \; y \geqslant 0 \tag{10.3.4i}$$

$$u \geqslant 0 \tag{10.3.4j}$$

$$\zeta_i, \eta_j \in \{0, 1\}, \quad i = 1, \ldots, n_2, j = 1, \ldots, m_2 \tag{10.3.4k}$$

which is a mixed 0-1 integer programming problem.

It is worth mentioning that to determine a valid value of M is a nontrivial matter. Pineda and Morales [47] provide a counterexample to show that tuning by trial and error, which is a common heuristic to compute a big-M, may lead to solutions which are not optimal. The choice of M as a valid bound in order to preserve all bilevel optimal points has been studied by Kleinert et al. [40]. Their preliminary results encourage the determination of problem-specific bounds on the LL dual variables if this algorithm approach is used for solving a LB problem.

Instead of using the big-M linearization of the complementary constraints, Fortuny and McCarl [27] propose to solve problem (10.3.2) by using the special ordered sets (SOS) tool of some mathematical programming packages. Moreover, this approach does not increase the number of constraints.

A different approach to overcome the difficulty of finding the appropriate value of the big-M is addressed by Pineda et al. [48]. Note that problem (10.3.2) is a mathematical program with complementary conditions to which a regularization approach can be applied. Hence, they apply it for efficiently determining a local optimal solution of problem (10.3.2). Then, this local optimal solution is used to tune the big-M and reduce the computational effort of solving the mixed-integer reformulation which uses constraints (10.3.3).

Audet et al. [7] develop an exact and finite branch-and-cut algorithm for the LB problem which solves the equivalent mixed 0-1 integer programming problem (10.3.4). The authors introduce three new classes of valid Gomory like cuts and apply the branch-and-bound algorithm proposed by Hansen et al. [32]. A similar approach, but using disjunctive cuts, is proposed by Audet et al. [6].

10.3.2 Branch-and-Bound Algorithms

Bard and Moore [12] develop a branch and bound algorithm, in which branching is done on complementary slackness conditions (10.3.2e) and (10.3.2f). These authors address the bilevel linear quadratic problem without coupling constraints, but their algorithm is considered in the literature as one of the first branch-and-bound algorithms developed for solving LB problems.

In the initialization step, the algorithm solves the linear program which results from removing the nonlinear constraints from problem (10.3.2).

At each iteration, constraints (10.3.2e) and (10.3.2f) are checked in the current solution. If they are held, the current solution is a bilevel feasible solution, the branching process is stopped and the node is closed. Otherwise, the tree is branched in one of the violated complementary slackness constraints and two nodes are added to the tree. At each node, an additional constraint is included in the linear program solved at the predecessor node which enforces that one of the terms of the product equals 0. The procedure ends when all nodes of the tree have been closed. The branch and bound scheme is used to implicitly examine all possible products involved in the complementary slackness constraints.

10.3.3 Penalty Approaches

Bearing in mind that the LL problem is linear, it can be replaced either by its KKT conditions, as done in problem (10.3.2), or by the primal and dual feasibility constraints together with the strong-duality equation, i.e. the duality gap enforced to equal zero. Both sets of conditions are equivalent.

White and Anandalingam [57] propose a global optimal solution method for the LB problem without coupling constraints which applies an exact penalty function that uses a duality gap-penalty function. The complementary slackness terms are penalized to obtain the following single level problem:

$$\min_{x,y} \quad (c_1x + d_1y) + \mu[(d_2 + uB_2)y + u(b_2 - A_2x - B_2y)] \tag{10.3.5a}$$

s.t.

$$A_2x + B_2y \leqslant b_2 \tag{10.3.5b}$$

$$uB_2 \geqslant -d_2 \tag{10.3.5c}$$

$$x \geqslant 0,\ y \geqslant 0 \tag{10.3.5d}$$

$$u \geqslant 0 \tag{10.3.5e}$$

where $\mu \geqslant 0$ is a constant, and the coupling constraints have been removed. Taking into account that

$$(d_2 + uB_2)y + u(b_2 - A_2x - B_2y) = d_2y - u(A_2x - b_2)$$

the penalty function (10.3.5a) can be rewritten as:

$$(c_1x + d_1y) + \mu[d_2y - u(A_2x - b_2)] \tag{10.3.6}$$

Therefore, problem (10.3.5) minimizes the UL objective function plus a term which involves the duality gap of the LL problem. This objective function is nonlinear, but can be properly decomposed to provide a set of linear programs.

Next, we briefly describe the main results that justify the validity of the penalty method to solve the LB problem. Let S_e and U_e be the set of extreme points of the polyhedra S and U, feasible regions of the LL problem and its dual, respectively. Then, for a fixed μ, there is an optimal solution of problem (10.3.5) in $S_e \times U_e$. Moreover, there exists a finite value $\mu^* \geqslant 0$ for which an optimal solution of problem (10.3.5) with $\mu \geqslant \mu^*$ provides an optimal solution of the LB problem.

As a consequence, the penalty algorithm developed by White and Anandalingam [57] consists of successively solving problem (10.3.5) by monotonically increasing μ. This parameter is increased until the corresponding duality gap is zero, i.e. the optimality is reached. Nevertheless, the question of how to handle the bilinear problem that appears when the parameter μ is fixed in problem (10.3.5) still remains. In [57] a cone splitting algorithm is used to generate cones at local optimal solutions.

Campelo et al. [18] identify some difficulties with the penalty approach described above, since the assumption of boundedness of polyhedra S and U cannot be verified simultaneously. These authors obtain the same results as White and Anandalingam [57] under a weaker assumption: the polyhedron U is nonempty and the relaxed problem (10.2.1) has an optimal solution. However, in order to guarantee the convergence of the penalty algorithm it is also necessary to assume that U is compact. These authors also detect some mistakes in the set of cuts defined by White and Anandalingam to discard local optima, and thus redefine the cut set and the test to provide an accurate search for the remaining better solutions. Hence, the penalty approach proposed by White and Anandalingam [57] for the first time is justified and well-defined by Campelo et al. [18].

Amouzegar and Moshirvaziri [3] consider the LB problem (10.1.1) without coupling constraints and propose a penalty method that uses the exact penalty function (10.3.6). Then, they apply a global optimization approach developed for solving linear problems with an additional reverse convex constraint. The global optimization approach introduces appropriate dominating cuts (hyperplanes) at local optimal solutions, which allow the authors to reduce the complexity of the problems solved as the penalty method proceeds.

All the methods mentioned above assume that the optimistic rule is applied in case the LL problem has multiple optima. It is well known that both to study the existence of optimal solutions and to develop solution methods are difficult tasks when the pessimistic rule is adopted for dealing with bilevel optimization. Aboussoror and Mansouri [1], under the pessimistic rule assumption, apply the strong dual theorem of linear programming and a penalty method to prove some results about the existence of an optimal solution and how to find it.

10.3.4 Parametric Complementary Algorithms

One of the first algorithms proposed to solve the LB problem is due to Bialas and Karwan [16]. These authors develop the Sequential Linear Complementarity Problems (SLCP) algorithm based on the similarities existing between problem (10.3.2) and the linear complementary problem (LCP). Júdice and Faustino [38] prove that the SLCP is not convergent in general and so develop a modified convergent version of this algorithm which finds a global minimum for the LB problem without coupling constraints. It is assumed that S is bounded and the LB problem has an optimal solution. In this method, a parameter λ is introduced and the UL objective function is replaced by the constraint $c_1x + d_1y \leqslant \lambda$ to obtain the following parametric LCP(λ) problem:

$$\begin{aligned}
&\theta = b_2 - A_2x - B_2y \\
&\vartheta = d_2 + uB_2 \\
&\gamma = \lambda - c_1x - d_1y \\
&u\theta = 0 \\
&y\vartheta = 0 \\
&x \geqslant 0, y \geqslant 0, \vartheta \geqslant 0, u \geqslant 0, \theta \geqslant 0, \gamma \geqslant 0
\end{aligned} \tag{10.3.7}$$

where θ and ϑ are the vectors of slack variables associated with the LL problem and its dual, respectively. The global minimum of the LB problem considered is the solution of the LCP(λ^*) problem, where λ^* is the smallest value of λ for which the LCP(λ) problem has a solution.

After a solution of problem (10.3.7) is computed, the value of λ is parametrically decreased until the current complementary basis which solves the problem is no longer feasible. Then, the SLCP algorithm finds another complementary basis. The SLCP algorithm can also be seen as an implicit enumeration of the LL optimal basis.

The efficiency of these methods depends on the procedure used for solving the LCP problem (10.3.7). Júdice and Faustino [39] propose a hybrid enumerative method for solving it and compare the performance of the SLCP algorithm with applying a branch-and-bound method to deal with the complementary slackness conditions.

10.4 General Algorithmic Approaches

In addition to reformulation (10.3.2) based on the KKT conditions, other reformulation of the LB problem has been proposed in the literature that uses the optimal value function of the LL problem. Problem (10.1.1) can be written as:

$$
\begin{aligned}
\min_{x,y} \quad & c_1 x + d_1 y \\
\text{s.t.} \quad & \\
& A_1 x + B_1 y \leqslant b_1 \\
& A_2 x + B_2 y \leqslant b_2 \\
& \varphi(x) \geqslant d_2 y \\
& x \geqslant 0,\ y \geqslant 0
\end{aligned}
\tag{10.4.1}
$$

where $\varphi(x)$ provides the optimal value of the LL problem given the UL decision variable x. Problem (10.4.1) is a linear program with a facially reverse convex constraint, since function $\varphi(x)$ is convex.

Both reformulation (10.3.2) and reformulation (10.4.1) lead to a single level nonconvex problem. For handling this problem, some of the algorithms included in Sect. 10.3, such as the penalty approaches proposed by White and Anandalingam [57] and Campelo et al. [18], use local optimization methods of nonlinear programming to approach the optimal solution smoothly. In this section, we include other algorithms which apply general global optimization techniques and take full advantage of the specific structure of the LB problem.

Hansen et al. [32] propose a branch-and-bound algorithm which exploits necessary optimal conditions expressed in terms of the tightness of some LL constraints, similar to some conditions used in global optimization. They also analyze branching rules based on the logical relationships which involve some binary variables associated with the tightness of the constraints. The authors compare this algorithm with the algorithm in [12] and solve problems with up to 150 constraints, 250 UL decision variables and 150 LL decision variables.

Based on reformulation (10.4.1), D.C. optimization methods have also been applied to solve the LB problem. Tuy et al. [52] rewrite the LB problem as the reverse convex constrained problem (10.4.1) and prove that the function $\phi(x)$ is a convex polyhedral function. They then develop an algorithm which uses a polyhedral annexation method and exploits the bilevel structure to reduce the dimension of the global optimization problem to be solved in a space of much smaller dimension than $n_1 + n_2$. In a posterior paper [53], these authors propose a branch and bound algorithm, also in a global optimization framework, based on the reduction of problem (10.4.1) to a quasiconcave minimization problem in a space of dimension $1 + rank(A_2)$. Tuy and Ghannadan [51] develop a new branch and bound method based on reformulation (10.4.1) but, unlike the previous algorithms, the method works in the primal space. In order to numerically solve large-scale problems without coupling constraints, Gruzdeva and Petrova [31] also apply D.C. optimization to develop a local search method.

From the earliest days, attempts have been made in the literature to establish a relationship between bilevel programming and biobjective programming in the linear case. The algorithms developed by Bard [8] and Ünlü [54] and the sufficient condition proposed by Wen and Hsu [55] were proved to be incorrect [14, 19, 22, 33, 43, 49]. In fact, the optimal solution of a LB problem is not necessarily an efficient solution of the biobjective problem whose objective functions are the UL and the LL objective functions. Fülöp [28] was the first to realize that more than just two criteria are needed to establish the link between LB programming and multiobjective programming. Based on these results, Glackin et al. [30] develop an algorithm which uses simplex pivots on an expanded tableau. Their computational results show that this algorithm performs better than a standard branch-and-bound algorithm when the number of UL variables is small.

10.5 Metaheuristic Algorithms

In general, the exact algorithms described above are computationally very time consuming and can only tackle relatively small problems. Therefore, metaheuristic algorithms have been developed aiming to be robust in finding good solutions while being able to solve large problems in reasonable computational time. Next, we present some metaheuristic algorithms which have been proposed for solving the LB problem. As with exact algorithms, some of them are based on the geometric properties of the LB problem while others reformulate it as a single level problem.

10.5.1 Evolutionary Algorithms

Anandalingam et al. [4] develop GABBA, a Genetic Algorithm Based Bi-level programming Algorithm which explores the IR. GABBA (which is explained in more detail in [44]) encodes its structures as $n_1 + n_2$ strings of base-10 digits, which represent bilevel feasible solutions, not necessarily extreme points. The UL variables are generated and the LL variables are obtained by solving the corresponding LL problem. The initial population is generated within feasible lower and upper bounds, only mutation is applied and the selection strategy is elitist, i.e. the best members of the previous generation plus offspring plus randomly generated new chromosomes are selected for the next generation, always maintaining the population size. The algorithm was tested against Bard's grid search technique [8], and GABBA was preferred on the basis of providing a better near optimal solution.

Nishizaki et al. [46] transform the LB problem without coupling constraints into the single level problem (10.3.2) and use the linearization of the complementary slackness conditions given by (10.3.3). This results in a single level mixed 0-1 programming problem. In the evolutionary algorithm proposed, the chromosomes are encoded as $m_2 + n_2$ binary strings, where 1 is placed into n_2 bits and 0 is placed into the remaining bits, that represent the value of the 0-1 variables ζ_i, η_j. In order to compute the fitness, problem (10.3.4) is solved performing linear scaling and taking its inverse number. Crossover and mutation are applied and heuristic rules are given to improve the performance of the procedure. The performance of the algorithm is compared with the branch-and-bound algorithm proposed in [32] in problems ranging from 10 to 60 variables.

Hejazi et al. [35] also replace the second level problem by its KKT conditions. They then propose a genetic algorithm in which each chromosome is a string representing a bilevel feasible extreme point, which consists of $m_2 + n_2$ binary components. Taking into account constraints (10.3.2e) and (10.3.2f), checking the feasibility of a chromosome is equivalent to solving two linear problems. Moreover, bearing in mind the feasible regions of these problems, they propose rules to discard chromosomes which are not feasible. Mutation and crossover are applied. The crossover operator randomly selects a location in the parents and the offspring is obtained by maintaining the left part and interchanging the right part of the parents. The mutation operator changes one component of the randomly selected chromosome either from one to zero or from zero to one. The elitist strategy is applied to select the members of the next population. They compare the performance of the algorithm with that proposed in [29]. The experimental results show the efficiency of the algorithm both from the computational point of view and the quality of the solutions.

Calvete et al. [17] consider the LB problem without coupling constraints and develop an evolutionary algorithm that takes into account that an optimal solution of the LB problem can be found at an extreme point of S. Therefore, the chromosomes represent basic feasible solutions of the polyhedron S, encoded as m strings of integers whose components are the indices of the basic variables. The fitness of a chromosome evaluates its quality, penalizing the fact that the associated extreme point is not in IR. For the purpose of checking this, the relationship between the LL problem and its dual are applied. Two kinds of crossover operators are proposed. After randomly selecting a location in the parents, in the variable-to variable crossover, the right-hand indices are interchanged. The main drawback of this crossover operator is the possibility of getting chromosomes associated with solutions not in S, which must be rejected. The basis-to-basis crossover avoids this possibility by entering the variables associated with the right-hand indices one by one using the minimum ratio test rule of the simplex algorithm to select the variable leaving the basis. The mutation operator randomly selects a variable which is not in the incumbent chromosome. Then, this variable enters the basis associated with the chromosome and the minimum ratio test rule of the simplex algorithm is applied to determine the variable leaving the basis. Some improvements are also applied in the process of checking if a chromosome is feasible, which exploit the structure of the problem. For selecting the next population, the elitist strategy is applied. The computational results presented show the performance of the method in terms of the quality of the solution (comparing it with the Kth best algorithm and the algorithm proposed in [35]) as well as in terms of the computational time invested. Test instances involve up to 100 variables (excluding the slack variables) and up to 80 constraints.

10.5.2 Other Metaheuristic Approaches

Apart from the evolutionary algorithms presented in the previous section, other metaheuristic techniques (tabu search, simulated annealing, particle swarm optimization and artificial neural networks) have been applied in a few algorithms.

Gendreau et al. [29] develop a hybrid tabu-ascent algorithm which consists of three blocks: a startup phase, a local ascent phase and a tabu phase. The startup phase generates an initial solution using the penalized problem (10.3.5). They estimate the value of u and solve the penalized problem. If the optimal solution is not a bilevel feasible solution, the parameter μ is increased and the procedure repeated. As the authors indicate, this procedure performs exceptionally well providing solutions which the search phase is not able to improve. In the tabu phase, starting at a bilevel feasible solution, the aim is to obtain another bilevel feasible solution with the same UL objective function value. Three types of tabu tags are maintained which refer to entering variables, exiting variables and pivots. If the new bilevel feasible solution is obtained, the reverse of the heuristic algorithm used in the startup phase is applied aiming to improve it. Both, the startup phase and the tabu search are completed by a local search step aiming to improve the UL objective function while maintaining y in $M(x)$. They present the results of a computational experiment dealing with difficult test problems involving up to 200 variables and 200 constraints.

In the previously mentioned paper by Anandalingam et al. [4], the authors also develop SABBA, a Simulated Annealing Based Bi-level programming Algorithm.

At each iteration, a bilevel feasible solution is generated. Depending on its quality, the solution is maintained or changed with a certain probability. During the iterations, the temperature, which affects the above mentioned probability, is progressively decreased. Like GABBA, SABBA was also tested against Bard's Grid Search technique. Comparing the three algorithms, the authors concluded that GABBA was preferred.

Kuo and Huang [41] propose a generic particle swarm optimization algorithm to solve the LB problem in which the particles are associated with points in IR. For this purpose, the values of the x variables are randomly generated and then the values of the variables y are obtained by solving the LL problem. Four small problems taken from the literature, with either two or four variables are solved as a computational experiment.

Shih et al. [50] propose artificial neural networks as a promising technique for solving multilevel programming problems. In this paper, as well as in the papers by Hu et al. [36] and He et al. [34], a neural network is applied to solve problem (10.3.2). In all cases, only illustrative small instances are solved.

10.6 Conclusions

This chapter addresses the LB optimization problem by focusing on the essentials of the main algorithms proposed in the literature to solve them. The intention has been to pay attention to those algorithms that have really contributed either to better understand the problem or to initiate the application of some technique in this field, by giving an overview of the underlying ideas that justify them. The algorithms developed to solve LB problems have followed two lines of thinking. On the one hand, those methods which exploit the theoretical properties of the LB problem. On the other hand, those procedures that, after reformulating the problem as a single level one, apply some of the techniques of the wide and rich single level classical optimization theory.

From the initial works on LB optimization, the literature in this field has experienced a significant increase in the number of published works. This can be explained by the ability of bilevel problems to model hierarchical frameworks which appear very often in real systems. Thus, models involving nonlinear objective functions and/or constraints, integer and binary variables, uncertainty in the coefficients, multiple followers, etc. are nowadays promising lines of research.

Acknowledgements This research work has been funded by the Spanish Ministry of Economy and Competitiveness under grant ECO2016-76567-C4-3-R and by the Gobierno de Aragón under grant E41-17R (FEDER 2014–2020 "Construyendo Europa desde Aragón").

References

1. A. Aboussoror, A. Mansouri, Weak linear bilevel programming problems: existence of solutions via a penalty method. J. Math. Anal. Appl. **304**, 399–408 (2005)
2. E. Aiyoshi, K. Shimizu, Hierarchical decentralized systems and its new solution by a barrier method. IEEE Trans. Syst. Man Cybern. **11**, 444–449 (1981)
3. M. Amouzegar, K. Moshirvaziri, A penalty method for linear bilevel programming problems, in *Multilevel Optimization: Algorithms and Applications*, ed. by A. Migdalas, P.M. Pardalos, P. Varbrand (Kluwer Academic Publishers, Dordrecht, 1998), pp. 251–271
4. G. Anandalingam, R. Mathieu, C.L. Pittard, N. Sinha, Artificial intelligence based approaches for solving hierarchical optimization problems, in *Impacts of Recent Computer Advances on Operations Research*, ed. by R. Sharda, B.L. Golden, E. Wasil, O. Balci, W. Stewart (North-Holland, New York, 1989), pp. 289–301
5. C. Audet, J. Haddad, G. Savard, A note on the definition of a linear bilevel programming solution. Appl. Math. Comput. **181**, 351–355 (2006)
6. C. Audet, J. Haddad, G. Savard, Disjunctive cuts for continuous linear bilevel programming. Optim. Lett. **1**, 259–267 (2007)
7. C. Audet, G. Savard, W. Zghal, New branch-and-cut algorithm for bilevel linear programming. J. Optim. Theory Appl. **134**, 353–370 (2007)
8. J.F. Bard, An efficient point algorithm for a linear two-stage optimization problem. Oper. Res. **31**, 670–684 (1983)
9. J.F. Bard, Optimality conditions for the bilevel programming problem. Naval Res. Logist. Q. **31**, 13–26 (1984)
10. J.F. Bard, Some properties of the bilevel programming problem. J. Optim. Theory Appl. **68**(2), 371–378 (1991)
11. J.F. Bard, J.E. Falk, An explicit solution to the multilevel programming problem. Comput. Oper. Res. **9**(1), 77–100 (1982)
12. J.F. Bard, J.T. Moore, A branch and bound algorithm for the bilevel programming problem. SIAM J. Sci. Stat. Comput. **11**(2), 281–292 (1990)
13. O. Ben-Ayed, Bilevel linear programming. Comput. Oper. Res. **20**(5), 485–501 (1993)
14. O. Ben-Ayed, C.E. Blair, Computational difficulties of bilevel linear programming. Oper. Res. **38**, 556–560 (1990)
15. W.F. Bialas, M.H. Karwan, On two-level optimization. IEEE Trans. Autom. Control **27**, 211–214 (1982)
16. W.F. Bialas, M.H. Karwan, Two-level linear programming. Manag. Sci. **30**, 1004–1024 (1984)
17. H.I. Calvete, C. Galé, P.M. Mateo, A new approach for solving linear bilevel problems using genetic algorithms. Eur. J. Oper. Res. **188**, 14–28 (2008)
18. M. Campelo, S. Dantas, S. Scheimberg, A note on a penalty function approach for solving bilevel linear programs. J. Glob. Optim. **16**, 245–255 (2000)
19. W. Candler, A linear bilevel programming algorithm: a comment. Comput. Oper. Res. **15**(3), 297–298 (1988)
20. W. Candler, R. Norton, Multilevel programming. Technical Report 20, World Bank Development Research Center, Washington, DC (1977)
21. W. Candler, R. Townsley, A linear two-level programming problem. Comput. Oper. Res. **9**(1), 59–76 (1982)
22. P. Clarke, A. Westerberg, A note on the optimality conditions for the bilevel programming problem. Naval Res. Logist. **35**, 413–418 (1988)
23. S. Dempe, A simple algorithm for the linear bilevel programming problem. Optimization **18**, 373–385 (1987)
24. S. Dempe, *Foundations of Bilevel Programming* (Kluwer Academic Publishers, Dordrecht, 2002)
25. S. Dempe, Bilevel optimization: bibliography, in *Bilevel Optimization - Advances and Next Challenges*, ed. by S. Dempe, A.B. Zemkoho (Springer, Berlin, 2020)

26. N.P. Faísca, V. Dua, B. Rustem, P. Saraiva, E.N. Pistikopoulos, Parametric global optimisation for bilevel programming. J. Glob. Optim. **38**, 609–623 (2007)
27. J. Fortuny-Amat, B. McCarl, A representation and economic interpretation of a two-level programming problem. J. Oper. Res. Soc. **32**(9), 783–792 (1981)
28. J. Fülöp, On the equivalence between a linear bilevel programming problem linear optimization over the efficient set. Technical Report WP93-1 Laboratory of Operations Research and DecisionSystems, Computer and Automation Institute, Hungarian Academy of Sciences (1993)
29. M. Gendreau, P. Marcotte, G. Savard, A hybrid tabu-ascent algorithm for the linear bilevel programming problem. J. Glob. Optim. **9**, 1–14 (1996)
30. J. Glackin, J.G. Ecker, M. Kupferschmid, Solving bilevel linear programs using multiple objective linear programming. J. Optim. Theory Appl. **140**, 197–212 (2009)
31. T.V. Gruzdeva, E.G. Petrova, Numerical solution of a linear bilevel problem. Comput. Math. Math. Phys. **50**(10), 1631–1641 (2010)
32. P. Hansen, B. Jaumard, G. Savard, New branch-and-bound rules for linear bilevel programming. SIAM J. Sci. Stat. Comput. **13**, 1194–1217 (1992)
33. A. Haurie, G. Savard, D.J. White, A note on: an efficient point algorithm for a linear two-stage optimization problem. Oper. Res. **38**, 553–555 (1990)
34. X. He, C. Li, T. Huang, C. Li, J. Huang, A recurrent neural network for solving bilevel linear programming problem. IEEE Trans. Neural Netw. Learn. Syst. **25**(4), 824–830 (2014)
35. S.R. Hejazi, A. Memariani, G. Jahanshahloo, M.M. Sepehri, Linear bilevel programming solution by genetic algorithm. Comput. Oper. Res. **29**, 1913–1925 (2002)
36. T. Hu, X. Guo, X. Fu, Y. Lv, A neural network approach for solving linear bilevel programming problem. Knowl. Based Syst. **23**, 239–242 (2010)
37. R.G. Jeroslow, The polynomial hierarchy and a simple model for competitive analysis. Math. Program. **32**(2), 146–164 (1985)
38. J.J. Júdice, A.M. Faustino, The solution of the linear bilevel programming problem by using the linear complementary problem. Investigação Oper. **8**(1), 77–95 (1988)
39. J.J. Júdice, A.M. Faustino, A sequential LCP method for bilevel linear programming. Ann. Oper. Res. **34**, 89–106 (1992)
40. T. Kleinert, M. Labbé, F. Plein, M. Schmidt, There's no free lunch: on the hardness of choosing a correct big-M in bilevel optimization. Working paper hal-02106642, version 2 (2019). https://hal.inria.fr/hal-02106642
41. R.J. Kuo, C.C. Huang, Application of particle swarm optimization algorithm for solving bi-level linear programming problem. Comput. Math. Appl. **58**(4), 678–685 (2009)
42. Y.H. Liu, S.M. Hart, Characterizing an optimal solution to the linear bilevel programming problem. Eur. J. Oper. Res. **73**(1), 164–166 (1994)
43. P. Marcotte, G. Savard, A note on the Pareto optimality of solutions to the linear bilevel programming problem. Comput. Oper. Res. **18**, 355–359 (1991)
44. R. Mathieu, L. Pittard, G. Anandalingam, Genetic algorithms based approach to bi-level linear programming. RAIRO-Oper. Res. **28**(1), 1–22 (1994)
45. A.G. Mersha, S. Dempe, Linear bilevel programming with upper level constraints depending on the lower level solution. Appl. Math. Comput. **180**, 247–254 (2006)
46. I. Nishizaki, M. Sakawa, K. Niwa, Y. Kitaguchi, A computational method using genetic algorithms for obtaining Stackelberg solutions to two-level linear programming problems. Electron. Commun. Jpn. Pt 3 **85**(6), 55–62 (2002)
47. S. Pineda, J.M. Morales, Solving linear bilevel problems using big-Ms: not all that glitters is gold. IEEE Trans. Power Syst. **34**(3), 2469–2471 (2019)
48. S. Pineda, H. Bylling, J.M. Morales, Efficiently solving linear bilevel programming problems using off-the-shelf optimization software. Optim. Eng. **19**, 187–211 (2018)
49. G. Savard, Contribution à la programmation mathématique à deux niveaux. PhD thesis, Ecole Polytechnique de Montréal, Université de Montréal, Montréal, 1989
50. H.S. Shih, U.P. Wen, E.S. Lee, K.M. Lan, H.C. Hsiao, A neural network approach to multiobjective and multilevel programming problems. Comput. Math. Appl. **48**, 95–108 (2004)

51. H. Tuy, S. Ghannadan, A new branch and bound method for bilevel linear programs, in *Multilevel Optimization: Algorithms and Applications*, ed. by A. Migdalas, P.M. Pardalos, P. Varbrand (Kluwer Academic Publishers, Dordrecht, 1998), pp. 231–249
52. H. Tuy, A. Migdalas, P. Värbrand, A global optimization approach for the linear two-level program. J. Glob. Optim. **3**, 1–23 (1993)
53. H. Tuy, A. Migdalas, P. Värbrand, A quasiconcave minimization method for solving the linear two-level program. J. Glob. Optim. **4**, 243–263 (1994)
54. G. Ünlü, A linear bilevel programming algorithm based on bicriteria programming. Comput. Oper. Res. **14**(2), 173–179 (1987)
55. U.P. Wen, S.T. Hsu, A note on a linear bilevel programming algorithm based on bicriteria programming. Comput. Oper. Res. **16**(1), 79–83 (1989)
56. U.P. Wen, S.T. Hsu, Linear bi-level programming problems - a review. J. Oper. Res. Soc. **42**(2), 125–133 (1991)
57. D.J. White, G. Anandalingam, A penalty function approach for solving bi-level linear programs. J. Glob. Optim. **3**, 397–419 (1993)
58. Z.K. Xu, Deriving the properties of linear bilevel programming via a penalty function approach. J. Optim. Theory Appl. **103**, 441–456 (1999)

Chapter 11
Global Search for Bilevel Optimization with Quadratic Data

Alexander S. Strekalovsky and Andrei V. Orlov

Abstract This chapter addresses a new methodology for finding optimistic solutions in bilevel optimization problems (BOPs). In Introduction, we present our view of the classification for corresponding numerical methods available in the literature. Then we focus on the quadratic case and describe the reduction of BOPs with quadratic objective functions to one-level nonconvex problems and develop methods of local and global searches for the reduced problems. These methods are based on the new mathematical tools of global search in nonconvex problems: the Global Search Theory (GST). A special attention is paid to a demonstration of the efficiency of the developed methodology for numerical solution of test BOPs.

Keywords Quadratic bilevel optimization · Optimistic solution · KKT-approach · Penalty approach · Global search theory · Computational experiment

11.1 Introduction

Consider the problem of bilevel optimization in its general formulation [1, 2]:

$$(\mathcal{BP})\qquad \begin{array}{ll} \min\limits_{x,y} & F(x,y) \\ \text{s.t.} & (x,y) \in Z, \quad y \in \operatorname{Arg}\min\limits_{y}\{G(x,y) \mid y \in Y(x)\}. \end{array} \tag{11.1.1}$$

It has the decision makers at the upper level (the leader) and at the lower level (the follower). There are two most popular formulations of bilevel problems: optimistic (cooperative) and guaranteed (pessimistic), which lead to the corresponding definitions of solutions in BOPs [1]. In this paper, we will consider the optimistic case: problem $(\mathcal{BP})$.

A. S. Strekalovsky · A. V. Orlov (✉)
Matrosov Institute for System Dynamics & Control Theory SB RAS, Irkutsk, Russia
e-mail: strekal@icc.ru; anor@icc.ru

S. Dempe, A. Zemkoho (eds.), *Bilevel Optimization*, Springer Optimization and Its Applications 161, https://doi.org/10.1007/978-3-030-52119-6_11

It is known that a bilevel problem has an implicit structural nonconvexity even in its optimistic formulation. Moreover, this is true for the case of convexity (or linearity) of the functions $F(\cdot)$, $G(\cdot)$, and when the feasible sets at both levels are convex. Therefore, the study of various classes of BOPs from the theoretical point of view is a useful and important issue of modern Operations Research. At present, there are numerous publications in this area (see, for example, the surveys [3, 4]). Also, bilevel problems have a lot of applications in various areas of human life (see, for example, [1, 2, 5]). Thus, the study of BOPs from the numerical point of view, as well as development of new approaches and methods for finding solutions, remains a relevant and important problem.

Here we focus on works where new approaches and methods for solving BOPs are developed and corresponding numerical results are given. Below we present our view of the classification of approaches and methods for finding optimistic solutions in continuous bilevel problems, based on the available literature. It is clear that it is impossible to observe all of the works, and we mention only several papers from each class as examples.

First, we focus on several approaches of reducing BOPs to a single-level optimization problem ; then we look at the optimization methods that are used for solving the reduced problems.

(I) The most popular approach of the transformation of BOPs is carried out by replacing the convex lower level problem with its necessary and sufficient conditions such as Karush-Kuhn-Tucker conditions [1, 6, 7]. In this case, it yields an optimization problem with a complementarity constraint [8], which reveals the main complexity of the optimization problem in question.

(II) The second approach implies the additional reduction of the obtained single-level problem with the complementarity constraint to a penalized problem [2, 9–13]. Most often the penalized objective function is the sum of the upper-level objective function and the difference of objective functions of the original and dual function of the lower level problem with the penalty parameter (see also [14–16]). Here the principal nonconvexity of the problem is moved into the objective function.

(III) Next approach reduces the bilevel problem to a single-level mixed integer optimization problem [17].

(IV) Also, we can transform BOPs to single-level problems with the help of the optimal value function of the lower level problem [18–20]

(V) Finally, we can try to attack BOPs without any transformation of the original formulation [2, 21–23].

The methods which are applied to the transformed problems for finding optimistic solutions in BOPs can be divided into the following groups.

(1) Different local optimization algorithms which provide only a stationary point to the problem (at best, a local solution). For example, methods of descent directions for the upper level problem with gradient information obtained from the lower level problem [24, 25]. Trust region methods [26–28], where the

bilevel problem under consideration is replaced by a simpler model problem and the latter one is solved in some given (trust) domain. Then, the resulting solution is evaluated from the viewpoint of the original problem, and the current trust region is changed.

(2) Local optimization methods which, in addition, use some technique of "globalization" in order to find at least an approximate global solution to the reduced nonconvex problem. The author of [29, 30] use, for example, a special variant of sequential quadratic programming method with enumeration and brunch-and-bound techniques.
(3) Standard global optimization methods, such as the branch-and-bound method, cuts methods etc. [31–35], which can be applied both to the problem with the complementarity constraint and to the penalized problem. Even they can be applied to bilevel problems directly [2, 21–23].
(4) Methods from the field of mixed integer programming which are applied to the reduced single-level mixed integer optimization problem [17].
(5) Methods for solving bilevel problems with linear constraints at both levels by enumerating the vertices of a polyhedron that defines the feasible set of the bilevel problem [2, 13, 36]. In such problems, at least one solution is reached at the extreme point of the feasible set.
(6) Methods based on ideas from the multicriteria optimization [37, 38].
(7) Methods of "parametric programming" [39], when the feasible set of upper level problems is presented explicitly.
(8) Heuristic methods, which employ some elements of the neural networks theory and/or evolutionary algorithms applied to the reduced one-level problem with a complementarity constraint [40, 41].

In spite of the sufficiently wide set of developed methods, we should note that only a few results published so far deal with numerical solution of high-dimension bilevel problems (for example, up to 100 variables at each level for linear bilevel problems [17, 35]).

In most of the cases, the authors consider just illustrative examples with the dimension up to 10 (see, e.g., [11, 19, 23, 39]), which is insufficient for solving real-life applied problems. And only the works [2, 27, 29, 30] present some results on solving nonlinear bilevel problems of dimension up to 50 at each level. It can be explained, for example, by the facts that (1) convex optimization methods applied to essentially nonconvex bilevel problems directly, turn out to be inoperative, in general, when it comes to finding a global solution; (2) when we apply heuristics methods, we cannot guarantee that we will find a global solution to problems in question; (3) widespread methods of global optimization, in addition, suffer the so-called "curse of dimension", when the volume of computations grows exponentially along with the growth of the problem's dimension, and these methods deny the achievements of modern convex optimization.

To sum up, the development of new efficient numerical methods for finding optimistic solutions, even for rather simple classes of nonlinear bilevel problems, is now at the front edge of the recent advances in the modern mathematical optimization. In particular, according to J.-S. Pang [42], a distinguished expert in optimization, the development of methods for solving various problems with hierarchical structure is one of the three challenges faced by optimization theory and methods in the twenty-first century together with equilibrium problems and the dynamical optimization.

The principal objective of this paper is to present a new approach for solving continuous bilevel optimization problems. It can be classified as an approach of Group (II) from the above classification. Hence, we reduce the bilevel problem to the single-level optimization problem with an explicit nonconvexity in the feasible set.

Then, applying the exact penalty approach, we move the principal nonconvexity to the cost function and obtain a nonconvex problem with the difference of convex (d.c.) objective function and the convex feasible set. As for solution of the latter problem, we apply the new Global Search Theory (GST) based on the original Global Optimality Conditions (GOCs) developed by A.S. Strekalovsky [43–46].

This theory does not reject the achievements and methods of modern convex optimization, but, on the contrary, actively employs them (for example, "inside" local search). GST proposes procedures that allows us not only to escape the local pitfalls, but also to achieve a global solution in nonconvex problems. It is worth noting that the efficiency of global search procedures directly depends on the efficiency of the convex optimization methods used inside these procedures.

According to the GST, an algorithm for solving a nonconvex problem should consist of two principal stages:

(1) a special Local Search Method (LSM), which takes into account the structure of the problem in question [43];
(2) the procedure based on the Global Optimality Conditions (GOCs) [43–46], which allows us to improve the point provided by the LSM [43–45].

In our group we have a large and successful experience in the field of numerical solution of numerous problems with a nonconvex structure by the GST: bimatrix and hexamatrix games [47, 48], bilinear optimization problems [49] and many others [43–45]. In particular, our group have an experience in solving linear bilevel problems with up to 500 variables at each level [50] and quadratic-linear bilevel problems of dimension up to (150×150) [51].

In this paper, we present the process of developing local and global search methods for finding optimistic solutions to the quadratic bilevel problems according to the GST, and some promising numerical results for quadratic-linear and quadratic bilevel problems (see also [52]).

The paper is organized in the following way. Section 11.2 deals with the reduction of the original bilevel problem to the single-level nonconvex one. In Sect. 11.3 a special Local Search Method for the reduced problem is described. Section 11.4 is devoted to the corresponding global search procedure based on the GOCs. Section 11.5 provides numerical results obtained for specially generated test bilevel problems using the methods developed. Section 11.6 presents concluding remarks.

11.2 Quadratic Bilevel Problem and Its Reduction to the Nonconvex Problem

Consider the following quadratic bilevel optimization problem in its optimistic statement. In this case, according to the theory [1], the minimization at the upper level is carried out with respect to the both variables x and y:

$$(\mathcal{QBP})\quad \begin{array}{ll} \min\limits_{x,y} & F(x,y) := \frac{1}{2}\langle x, Cx\rangle + \langle c, x\rangle + \frac{1}{2}\langle y, Dy\rangle + \langle d, y\rangle \\ \text{s.t.} & (x,y) \in Z := \{(x,y) \in \mathbb{R}^m \times \mathbb{R}^n \mid Ax + By \leq b\}, \\ & y \in Y_*(x) := \operatorname{Arg}\min\limits_y \{\frac{1}{2}\langle y, D_1 y\rangle + \langle d_1, y\rangle + \langle x, Qy\rangle \mid \\ & \mid y \in Y(x)\}, \quad Y(x) \triangleq \{y \in \mathbb{R}^n \mid A_1 x + B_1 y \leq b_1\}, \end{array} \tag{11.2.1}$$

where $A \in \mathbb{R}^{p\times m}$, $B \in \mathbb{R}^{p\times n}$, $A_1 \in \mathbb{R}^{q\times m}$, $B_1 \in \mathbb{R}^{q\times n}$, $C \in \mathbb{R}^{m\times m}$, $D, D_1 \in \mathbb{R}^{n\times n}$, $Q \in \mathbb{R}^{m\times n}$, $c \in \mathbb{R}^m$, $d, d_1 \in \mathbb{R}^n$, $b \in \mathbb{R}^p$, $b_1 \in \mathbb{R}^q$. Besides, matrices C, D, and D_1 are symmetric and positive semidefinite. Note that all of the data in this problem are convex or linear except the term $\langle x, Qy\rangle$ at the lower level (but it is linear when x is fixed). At the same time, as mentioned above, the problem $(\mathcal{QBP})$–(11.2.1) is implicitly nonconvex due to its bilevel structure, even if $Q \equiv 0 \in \mathbb{R}^{m\times n}$.

Let the following assumptions on the problem $(\mathcal{QBP})$–(11.2.1) hold.

(a) The set $Y_*(x)$ is non-empty and compact for a fixed x, such that

$$x \in Pr(Z) \triangleq \{x \in \mathbb{R}^m \mid \exists y : (x,y) \in Z\}.$$

(b) The function $F(x,y)$ is bounded from below on the non-empty set

$$U := \{(x,y) \in \mathbb{R}^m \times \mathbb{R}^n \mid Ax + By \leq b,\ A_1 x + B_1 y \leq b_1\},$$

so that

$$(\mathcal{H}1)\qquad \exists F_- > -\infty : F(x,y) \geq F_- \ \ \forall (x,y) \in U. \tag{11.2.2}$$

(c) The objective function of the lower level is bounded from below on the set $Y(x)$ for all $x \in Pr(Z)$, so that

$$(\mathcal{H}2)\ \inf_x \inf_y \{\frac{1}{2}\langle y, D_1 y\rangle + \langle d_1, y\rangle + \langle x, Qy\rangle \mid y \in Y(x), x \in Pr(Z)\} > -\infty. \tag{11.2.3}$$

Note that despite the presence of the bilinear term in the objective function, for any fixed $x \in Pr(Z)$ the lower level problem is convex and the Abadie regularity conditions are automatically fulfilled here due to affine constraints [14]. Therefore, the KKT-conditions [14, 15] are necessary and sufficient for the follower problem in $(\mathcal{QBP})$–(11.2.1) $\forall x \in Pr(Z)$.

By replacing the lower level problem in $(\mathcal{QBP})$–(11.2.1) with its KKT-optimality conditions [1, 2, 51, 52], we obtain the following single-level optimization problem with the evident nonconvexity:

$$(\mathcal{DCC}) \qquad \begin{array}{ll} \min\limits_{x,y,v} & F(x,y) \\ \text{s.t.} & Ax + By \le b, \qquad A_1 x + B_1 y \le b_1, \\ & D_1 y + d_1 + x^T Q + v^T B_1 = 0, \\ & v \ge 0, \qquad \langle v, A_1 x + B_1 y - b_1 \rangle = 0. \end{array} \tag{11.2.4}$$

The problem $(\mathcal{DCC})$–(11.2.4) is a problem with a nonconvex feasible set, where the nonconvexity is generated by the latter equality constraint (the complementarity constraint). Let us formulate the main theorem of the KKT-transformation for the problem $(\mathcal{QBP})$–(11.2.1).

Theorem 11.2.1 ([1, 6]) *For the pair (x_*, y_*) to be a global solution to $(\mathcal{QBP})$–(11.2.1), it is necessary and sufficient that there exists a vector $v_* \in \mathbb{R}^q$ such that the triple (x_*, y_*, v_*) is a global solution to $(\mathcal{DCC})$–(11.2.4).* $\triangle$

This theorem reduces the search for an optimistic solution of the bilevel problem $(\mathcal{QBP})$–(11.2.1) to solving the problem $(\mathcal{DCC})$–(11.2.4). Further, we use the exact penalty method [15, 16, 46] to deal with a problem that has a nonconvex objective function and a convex feasible set.

The complementarity constraint can be written in the equivalent form: $\langle v, b_1 - A_1 x - B_1 y \rangle = 0$. Then, obviously, $\forall (x, y, v) \in \mathcal{F}$, where the set

$$\mathcal{F} := \{(x, y, v) | Ax + By \le b, \ A_1 x + B_1 y \le b_1, \ D_1 y + d_1 + x^T Q + v^T B_1 = 0, \ v \ge 0\}$$

is the convex set in $\mathbb{R}^{n+m+q}$, and, in addition, the following inequality takes place: $\langle v, b_1 - A_1 x - B_1 y \rangle \ge 0$. Now the penalized problem [1, 15, 16, 46] can be written as

$$(\mathcal{DC}(\sigma)) \qquad \begin{array}{ll} \min\limits_{x,y,v} & \Phi(x,y,v) := \frac{1}{2}\langle x, Cx \rangle + \langle c, x \rangle + \frac{1}{2}\langle y, Dy \rangle + \\ & \qquad + \langle d, y \rangle + \sigma \, \langle v, b_1 - A_1 x - B_1 y \rangle \\ \text{s.t.} & (x, y, v) \in \mathcal{F}, \end{array} \tag{11.2.5}$$

where $\sigma > 0$ is a penalty parameter. When σ is fixed, this problem belongs to the class of d.c. minimization problems [43, 45] with a convex feasible set , since its cost function is a sum of quadratic and bilinear terms. In what follows (in Sect. 11.4), we demonstrate that the objective function of $(\mathcal{DC}(\sigma))$–(11.2.5) is represented as a difference of two convex functions.

It is well-known that if for some σ the triple $(x(\sigma), y(\sigma), v(\sigma)) \in \text{Sol}(\mathcal{DC}(\sigma))$, and the point $(x(\sigma), y(\sigma), v(\sigma))$ is feasible in the problem ($\mathcal{DCC}$)–(11.2.4), i.e. $(x(\sigma), y(\sigma), v(\sigma)) \in \mathcal{F}$ and $r[\sigma] := \langle v(\sigma), b_1 - A_1x(\sigma) - B_1y(\sigma)\rangle = 0$, then $(x(\sigma), y(\sigma), v(\sigma))$ is a global solution to ($\mathcal{DCC}$)–(11.2.4) [1, 15, 16, 46].

In addition, the following result holds.

Proposition 11.2.2 ([1, 14–16]) *Suppose that for some $\hat{\sigma} > 0$ the equality $r[\hat{\sigma}] = 0$ takes place for a solution $(x(\hat{\sigma}), y(\hat{\sigma}), v(\hat{\sigma}))$ to ($\mathcal{DC}(\sigma)$)–(11.2.5). Then for all values of the parameter $\sigma > \hat{\sigma}$ the function $r[\sigma]$ is equal to zero and the triple $(x(\sigma), y(\sigma), v(\sigma))$ is a solution to ($\mathcal{DCC}$)–(11.2.4).* △

Thus, if the equality $r[\sigma] = 0$ holds, then a solution to ($\mathcal{DC}(\sigma)$) is a solution to ($\mathcal{DCC}$). And when the value of σ grows, this situation remains the same.

Hence, the key point when using the exact penalization theory here is the existence of a threshold value $\hat{\sigma} > 0$ of the parameter σ for which $r[\sigma] = 0 \;\; \forall \sigma \geq \hat{\sigma}$. Due to the fact that the objective function F of the problem ($\mathcal{DCC}$)–(11.2.4) satisfies the Lipschitz property [15, 16] with respect to both variables, the following assertion takes place.

Proposition 11.2.3 ([15, 16, 46]) *Let the triple (x_*, y_*, v_*) be a global solution to the problem ($\mathcal{DCC}$)–(11.2.4). Then, there exists $\hat{\sigma} > 0$ such that (x_*, y_*, v_*) is a global solution to the problem ($\mathcal{DC}(\hat{\sigma})$)–(11.2.5). Moreover, $\forall \sigma > \hat{\sigma}$ any solution $(x(\sigma), y(\sigma), v(\sigma))$ to the problem ($\mathcal{DC}(\sigma)$)–(11.2.5) is feasible in the problem ($\mathcal{DCC}$)–(11.2.4), i.e. $r[\sigma] = 0$, and, therefore, $(x(\sigma), y(\sigma), v(\sigma))$ is a solution to the problem ($\mathcal{DCC}$)–(11.2.4), so that Sol($\mathcal{DCC}$) $\subset$ Sol($\mathcal{DC}(\sigma)$). The latter inclusion implies the equality*

$$\textit{Sol}(\mathcal{DCC}) = \textit{Sol}(\mathcal{DC}(\sigma)) \quad \forall \sigma > \hat{\sigma}, \tag{11.2.6}$$

so that the problems ($\mathcal{DCC}$) and ($\mathcal{DC}(\sigma)$) are equivalent (in the sense of (11.2.6)). △

Therefore, combining Propositions 11.2.2 and 11.2.3 with Theorem 11.2.1, we can conclude that these relations between the problems ($\mathcal{DC}(\sigma)$)–(11.2.5) and ($\mathcal{DCC}$)–(11.2.4) allow us to seek for a global solution to the problem ($\mathcal{DC}(\sigma)$)–(11.2.5), where $\sigma > \hat{\sigma}$, in order to find an optimistic solution to the problem ($\mathcal{QBP}$)–(11.2.1) instead of a solution to the problem ($\mathcal{DCC}$)–(11.2.4). The existence of the threshold value $\hat{\sigma} > 0$ of the penalty parameter enables us to solve a single problem ($\mathcal{DC}(\sigma)$)–(11.2.5) (where $\sigma > \hat{\sigma}$) instead of solving a sequence of problems ($\mathcal{DC}(\sigma)$)–(11.2.5) when the penalty parameter tends to infinity ($\sigma \to +\infty$) (see, for example, [14–16]).

Since the objective function of the problem ($\mathcal{DC}(\sigma)$)–(11.2.5) is a d.c. function for a fixed σ, we will address the problem ($\mathcal{DC}(\sigma)$)–(11.2.5) by means of the GST mentioned above. First, this theory implies construction of a local search procedure that takes into consideration special features of the problem in question.

11.3 Local Search Method

Due to the inoperability of the convex optimization methods [14–16] in nonconvex problems, a number of various local search techniques for nonconvex problems of different classes were invented [33, 43–45, 47, 49, 51]. These methods produce feasible points which are called stationary or critical (with respect to the methods). Generally speaking, a critical point may not be a local solution in the corresponding problem, but it has additional special properties depending on the specific local search method, and, therefore, it is not just only a stationary point (if it is proven) [15, 16, 43].

It can be readily seen that the nonconvexity in the problem ($\mathcal{DC}(\sigma)$)–(11.2.5) is located in the objective function and is generated by the bilinear term $\langle v, b_1 - A_1x - B_1y\rangle$ only. On the basis of this fact, by employing the bilinearity, we suggest to carry out the local search in the problem ($\mathcal{DC}(\sigma)$)–(11.2.5) using the idea of its successive solution by splitting the variables into two groups.

Earlier this idea was successfully applied in several problems with similar bilinear structure in the objective function: finding a Nash equilibrium in polymatrix games [47, 48], problems of bilinear programming [49], and quadratic-linear bilevel problems [51]. We can do it, because the problem ($\mathcal{DC}(\sigma)$)–(11.2.5) turns into a convex quadratic optimization problem for a fixed v and into a linear programming (LP) problem for a fixed pair (x, y) [52].

Denote $\mathcal{F}(v) := \{(x, y) \in \mathbb{R}^{m+n} \mid (x, y, v) \in \mathcal{F}\}$, $\mathcal{F}(x, y) := \{v \in \mathbb{R}^q \mid (x, y, v) \in \mathcal{F}\}$ and describe the following Local Search Method.

Let $(x_0, y_0, v_0) \in \mathcal{F}$ be a starting point.

***XY*-procedure**

Step 0. Set $s := 1$, $(x^s, y^s) := (x_0, y_0)$.

Step 1. Applying a suitable linear programming method, find the approximate $\frac{\rho_s}{2}$-solution v^{s+1} to the following LP problem:

$$(\mathcal{LP}(x^s, y^s)) \qquad \begin{array}{rl} \min\limits_{v} & \langle b_1 - A_1x^s - B_1y^s, v\rangle \\ \text{s.t.} & D_1y^s + d_1 + (x^s)^TQ + v^TB_1 = 0, \quad v \geq 0. \end{array} \tag{11.3.1}$$

Step 2. Find the approximate $\frac{\rho_s}{2}$-solution (x^{s+1}, y^{s+1}) to the following convex problem

$$(\mathcal{QP}(v^{s+1})) \qquad \begin{array}{rl} \min\limits_{x,y} & \frac{1}{2}\langle x, Cx\rangle + \langle c, x\rangle + \frac{1}{2}\langle y, Dy\rangle + \langle d, y\rangle - \\ & -\sigma\left(\langle (v^{s+1})^TA_1, x\rangle + \langle (v^{s+1})^TB_1, y\rangle\right) \\ \text{s.t.} & Ax + By \leq b, \qquad A_1x + B_1y \leq b_1, \\ & D_1y + d_1 + x^TQ + (v^{s+1})^TB_1 = 0. \end{array} \tag{11.3.2}$$

Step 3. Set $s := s + 1$ and move to **Step 1.**

Note that the name of the *XY*-procedure stems from the fact that we do not need all the components of the starting point (x_0, y_0, v_0) to launch the algorithm; using only the pair (x_0, y_0) is sufficient. Herein, it is necessary to choose the components (x_0, y_0, v_0) so that the auxiliary problems of linear and quadratic programming were solvable at the steps of the algorithm. The solvability can be guaranteed, for example, by an appropriate choice of the feasible point $(x_0, y_0, v_0) \in \mathcal{F}$ under the conditions $(\mathcal{H}1)$–(11.2.2) and $(\mathcal{H}2)$–(11.2.3).

The following convergence theorem for the XY-procedure takes place.

Theorem 11.3.1 ([47, 49, 51])

(i) *Let a sequence* $\{\rho_s\}$ *be such that* $\rho_s > 0$ *for all* $s = 1, 2, \ldots,$ *and* $\sum_{s=1}^{\infty} \rho_s < +\infty$. *Then the number sequence* $\{\Phi_s := \Phi(x^s, y^s, v^s)\}$ *generated by the* XY-*procedure converges.*

(ii) *If* $(x^s, y^s, v^s) \to (\hat{x}, \hat{y}, \hat{v})$, *then the limit point* $(\hat{x}, \hat{y}, \hat{v})$ *satisfies the inequalities*

$$\Phi(\hat{x}, \hat{y}, \hat{v}) \leq \Phi(\hat{x}, \hat{y}, v) \quad \forall v \in \mathcal{F}(\hat{x}, \hat{y}), \tag{11.3.3}$$

$$\Phi(\hat{x}, \hat{y}, \hat{v}) \leq \Phi(x, y, \hat{v}) \quad \forall (x, y) \in \mathcal{F}(\hat{v}). \tag{11.3.4}$$

△

Definition 11.3.2 The triple $(\hat{x}, \hat{y}, \hat{v})$ satisfying (11.3.3) and (11.3.4) is said to be *the critical point* to the problem $(\mathcal{DC}(\sigma))$–(11.2.5). △

It can be readily seen that the concept of criticality in the problem $(\mathcal{DC}(\sigma))$–(11.2.5) is due to the structure of the Local Search Method. And any critical point turns out to be partly global solution to the problem $(\mathcal{DC}(\sigma))$–(11.2.5) with respect to two groups of variables (x, y) and v. Such a definition of critical points is quite advantageous when we perform a global search in problems with a bilinear structure (see, for example, [49, 51]). Also, this new concept is tightly connected with the classical definitions of stationarity. In particular, we can prove that for any critical point to the problem $(\mathcal{DC}(\sigma))$–(11.2.5) the classical first-order optimality conditions [14–16] hold.

At the same time, carrying out computational simulations, we have to stop after a finite number of iterations. In this case, we can speak only about an approximate critical point.

Definition 11.3.3 The triple $(\hat{x}, \hat{y}, \hat{v})$ satisfying (11.3.3) and (11.3.4) with the tolerance $\tau \geq 0$, so that

$$\Phi(\hat{x}, \hat{y}, \hat{v}) - \tau \leq \Phi(\hat{x}, \hat{y}, v) \quad \forall v \in \mathcal{F}(\hat{x}, \hat{y}),$$
$$\Phi(\hat{x}, \hat{y}, \hat{v}) - \tau \leq \Phi(x, y, \hat{v}) \quad \forall (x, y) \in \mathcal{F}(\hat{v}),$$

is said to be the *(approximately)* τ*-critical point* to the problem $(\mathcal{DC}(\sigma))$–(11.2.5).

△

We can prove [49, 51] that if at the iteration s $(s > 1)$ of the XY-procedure the following stopping criterion holds (with the tolerance $\tau_1 > 0$):

$$\Phi_s - \widetilde{\Phi}_s \leq \tau_1, \tag{11.3.5}$$

where $\widetilde{\Phi}_s := \Phi(x^s, y^s, v^{s+1})$, then the following inequalities take place:

$$\Phi(x^s, y^s, v^s) - (\tau_1 + \frac{\rho_s}{2}) \leq \inf_v \{\Phi(x^s, y^s, v) \mid v \in \mathcal{F}(x^s, y^s)\},$$

$$\Phi(x^s, y^s, v^s) - (\tau_1 + \frac{\rho_{s-1}}{2} + \frac{\rho_s}{2}) \leq \inf_{(x,y)} \{\Phi(x, y, v^s) \mid v \in \mathcal{F}(v^s)\}.$$

It is clear that if $\tau_1 \leq \frac{\tau}{2}$, $\rho_{s-1} \leq \frac{\tau}{4}$, and $\rho_s \leq \frac{\tau}{4}$, then the point (x^s, y^s, v^s) turns out to be a τ-critical point to the problem $(\mathcal{DC}(\sigma))$–(11.2.5).

Thus, in order for the XY-procedure to be implementable, we should add the following step to the scheme.

Step 2a. If the inequality (11.3.5) is valid for a given $\tau_1 > 0$, then Stop. Hence, the triple (x^s, y^s, v^s) is the approximate critical point to the problem $(\mathcal{DC}(\sigma))$–(11.2.5).

Similarly, we can describe and substantiate other stopping criteria for the XY-procedure (see also [49, 51]). In addition, to implement the local search in the problem $(\mathcal{DC}(\sigma))$–(11.2.5), according to [49, 51], we can consider another variant (V-procedure) of local search, where the auxiliary problems are solved in a different order (initially—with respect to (x, y), and secondly—with respect to v) (see also [52]).

The next section describes the basic elements of the Global Search for the general d.c. minimization problem as well as for the problem $(\mathcal{DC}(\sigma))$–(11.2.5), where σ is fixed. We can find the relevant value of σ at the stage of the Local Search. We start the XY-procedure with the small value of the penalty parameter, and after stopping it we should check the feasibility of the obtained critical point to the problem $(\mathcal{DCC})$–(11.2.4) $(r[\sigma] = 0)$. If that is true, we can move to the Global Search. If not, we should increase the value of the penalty parameter σ in order to obtain a feasible critical point.

Thus, the feasibility is provided by the Local Search Method, whereas the improvement of a current critical point is provided by the Global Search Scheme (GSS) presented below.

11.4 Global Optimality Conditions and a Global Search

It is clear that the above local search method does not provide, in general, a global solution even in small-dimension nonconvex problems, because it can only obtain critical (stationary) points.

First, consider the main elements of a global search in more detail for the following general problem with a d.c. objective function:

$$(\mathcal{P}) \qquad \begin{array}{ll} \min\limits_{x} & \varphi(x) = g(x) - h(x) \\ \text{s.t.} & x \in \mathcal{D}, \end{array} \tag{11.4.1}$$

where $g(\cdot)$, $h(\cdot)$ are convex functions, $\mathcal{D} \subset \mathbb{R}^n$ is a convex set. Moreover, let the function $\varphi(\cdot)$ be bounded below on the set $\mathcal{D}$:

$$\inf(\varphi, \mathcal{D}) > -\infty.$$

For the problem $(\mathcal{P})$, the following Global Optimality Conditions (GOCs) are proved.

Theorem 11.4.1 ([43–46]) *Let a point $z \in \mathcal{D}$ be a global solution to the problem $(\mathcal{P})$ $(\zeta := \varphi(z))$. Then for every pair $(y, \gamma) \in \mathbb{R}^n \times \mathbb{R}$ such that*

$$h(y) = \gamma - \zeta, \tag{11.4.2}$$

the following inequality holds

$$g(x) - \gamma \geq \langle \nabla h(y), x - y \rangle \quad \forall x \in \mathcal{D}. \tag{11.4.3}$$

△

Theorem 11.4.1 represents necessary conditions and reduces the investigation of the nonconvex problem $(\mathcal{P})$ to considering the family of the convex linearized problems

$$(\mathcal{PL}(y)) \qquad \begin{array}{ll} \min\limits_{x} & g(x) - \langle \nabla h(y), x \rangle \\ \text{s.t.} & x \in \mathcal{D}, \end{array} \tag{11.4.4}$$

depending on the parameters $(y, \gamma) \in \mathbb{R}^n \times \mathbb{R}$ which fulfill (11.4.2).

These conditions use the idea of linearization with respect to the basic nonconvexity of the problem $(\mathcal{P})$–(11.4.1), and they are associated with classical optimality conditions. In addition, they possess an algorithmic (constructive) property: if the inequality (11.4.3) is violated, then we can construct a feasible point that is better than the current point z, with respect to the cost function of the problem $(\mathcal{P})$ [43–46].

This constructive property is the basis for building global search algorithms in d.c. minimization problems. In particular, the procedure of escaping the provided by LSM critical point can be represented as the following sequence of operations (at an iteration of k) [43–45] (we will call it the Global Search Scheme (GSS)).

(1) Choose some number $\gamma : \inf(g, \mathcal{D}) \leq \gamma \leq \sup(g, \mathcal{D})$ (for example, we can start with $\gamma_0 = g(z)$).

(2) Construct a finite approximation

$$\mathcal{A}_k(\gamma) = \{v^1, \ldots, v^{N_k} \mid h(v^i) = \gamma - \zeta_k,\ i = 1, \ldots, N_k,\ N_k = N_k(\gamma)\}$$

of the level surface $U_k(\gamma) = \{x \in \mathbb{R}^n \mid h(x) = \gamma - \zeta_k\}$ of the function $h(\cdot)$ that generates a basic nonconvexity in the problem $(\mathcal{P})$. This approximation can be constructed, for example, on the basis of the preliminary given set of vectors Dir (see e.g. [47, 49, 51]).

(3) Find a global δ_k-solution $u^i \in \mathcal{D}$ to the linearized problem $(\mathcal{PL}(v^i))$–(11.4.4), so that:

$$g(u^i) - \langle \nabla h(v_i), u^i \rangle - \delta_k \le \inf_x \{g(x) - \langle \nabla h(v_i), x \rangle \mid x \in \mathcal{D}\}.$$

4) Find a global δ_k-solution w^i $(h(w^i) = \gamma - \zeta_k)$ to the so-called "level problem":

$$\langle \nabla h(w_i), u^i - w^i \rangle + \delta_k \ge \sup_v \{\langle \nabla h(u^i), u^i - v \rangle \mid h(v) = \gamma - \zeta_k\}. \qquad (11.4.5)$$

(5) Compute the value $\eta_k(\gamma) := \eta_k^0(\gamma) + \gamma$, where

$$\eta_k^0(\gamma) := g(u^j) - \langle \nabla h(w_j), u^j - w^j \rangle \stackrel{\triangle}{=} \min_{i \in I_k} \{g(u^i) - \langle \nabla h(w_i), u^i - w^i \rangle\},$$

$i \in I_k \stackrel{\triangle}{=} \{i \in \{1, \ldots, N_k\} \mid g(v^i) \le \gamma\}$. If $\eta_k(\gamma) < 0$, then it can be shown that the point u^j is better than z. Then we can move to the next iteration of the global search and perform a new local search, starting from the point u^j. If $\eta_k(\gamma) \ge 0$, then it is necessary to choose a new value of the numerical parameter γ.

Depending on the nonconvex problem under study, this GSS can be algorithmized in various ways. A number of its stages require some specification, which is carried out on the basis of the statement of the problem in question and its properties (see, for example, [44, 45, 47–49, 51, 52] and the remaining part of this paper).

In this connection, further we describe in detail the procedure of a global search for the problem $(\mathcal{DC}(\sigma))$–(11.2.5) with fixed $\sigma > \hat{\sigma}$:

$$(\mathcal{DC}(\sigma)) \qquad \begin{array}{l} \min\limits_{x,y,v} \ \Phi(x, y, v) \stackrel{\triangle}{=} F(x, y) + \sigma \langle v, b_1 - A_1 x - B_1 y \rangle \\ \text{s.t.} \ (x, y, v) \in \mathcal{F}. \end{array} \qquad (11.2.5)$$

The procedure is based on the Global Optimality Conditions (GOCs) described above. In order to construct the global search procedure, first of all we need to obtain an explicit d.c. representation of the objective function of the problem $(\mathcal{DC}(\sigma))$–(11.2.5).

For this purpose we use the following representation based on the well-known property of the scalar product ($\langle x, y\rangle = \frac{1}{4}\|x+y\|^2 - \frac{1}{4}\|x-y\|^2$):

$$\Phi(x, y, v) = G(x, y, v) - H(x, y, v), \tag{11.4.6}$$

where $G(x, y, v) = \frac{1}{2}\langle x, Cx\rangle + \langle c, x\rangle + \frac{1}{2}\langle y, Dy\rangle + \langle d, y\rangle + \sigma\langle b_1, v\rangle + + \frac{\sigma}{4}\|A_1x - v\|^2 + \frac{\sigma}{4}\|B_1y - v\|^2$, $H(x, y, v) = \frac{\sigma}{4}\|A_1x + v\|^2 + \frac{\sigma}{4}\|B_1y + v\|^2$. Note that the so-called *basic nonconvexity* in the problem ($\mathcal{DC}(\sigma)$)–(11.2.5) is generated by the function $H(\cdot)$ (for more detail, refer to [43–45]).

The necessary GOCs of the so-called contrapositive form have the following statement in terms of the problem ($\mathcal{DC}(\sigma)$)–(11.2.5) (see Theorem 3).

Theorem 11.4.2 ([43–45, 51]) *If the feasible point (x_*, y_*, v_*) is not a (global) solution to the problem ($\mathcal{DC}(\sigma)$)–(11.2.5), then one can find a triple $(z, u, w) \in \mathbb{R}^{m+n+q}$, a feasible vector $(\bar{x}, \bar{y}, \bar{v}) \in \mathcal{F}$, and a scalar γ, such that*

$$H(z, u, w) = \gamma - \zeta, \quad \zeta := \Phi(x_*, y_*, v_*), \tag{11.4.7}$$

$$G(x, y, v) \le \gamma \le \sup_{x,y,v}(g, \mathcal{F}), \tag{11.4.8}$$

and the following inequality takes place:

$$G(\bar{x}, \bar{y}, \bar{v}) - \gamma < \langle \nabla H(z, u, w), (\bar{x}, \bar{y}, \bar{v}) - (z, u, w)\rangle . \tag{11.4.9}$$

△

Theorem 11.4.2 express the algorithmic (constructive) property of the GOCs (11.4.2)–(11.4.3). It means that if one was successful to find the 4-tuple $(\bar{z}, \bar{u}, \bar{w}, \bar{\gamma})$ satisfying (11.4.7)–(11.4.8), and the point $(\bar{x}, \bar{y}, \bar{v}) \in \mathcal{F}$, such that the inequality (11.4.9) holds, then due to the convexity of the function $H(\cdot)$ one obtains $\Phi(\bar{x}, \bar{y}, \bar{v}) < \Phi(x_*, y_*, v_*)$. It means that the triple $(\bar{x}, \bar{y}, \bar{v})$ is better than the triple (x_*, y_*, v_*). One can find such a triple by changing parameters (z, u, w, γ) in (11.4.7)–(11.4.8) for a fixed $\zeta = \zeta_k$ and obtaining approximate solutions $(x(z, u, w, \gamma), y(z, u, w, \gamma), v(z, u, w, \gamma))$ of the linearized problems ($\mathcal{PL}(z, u, w)$) (see ($\mathcal{PL}(y)$)–(11.4.4) and (11.4.9)):

$$(\mathcal{PL}(z, u, w)) \qquad \begin{array}{l} \min\limits_{x,y,v} \; G(x, y, v) - \langle \nabla H(z, u, w), (x, y, v)\rangle \\ \text{s.t.} \;\; (x, y, v) \in \mathcal{F}. \end{array} \tag{11.4.10}$$

Thus, we get a set of starting points to launch the Local Search Method. After that we move to the new level $\zeta_{k+1} := \Phi(x^{k+1}, y^{k+1}, v^{k+1})$, $(x^{k+1}, y^{k+1}, v^{k+1}) := (\bar{x}, \bar{y}, \bar{v})$ and vary the parameters (z, u, w, γ) again.

Hence, according to the GST [43–45], the problem ($\mathcal{DC}(\sigma)$)–(11.2.5) can be split into several simpler problems (linearized problems and problems with respect to other parameters from the global optimality conditions). Taking into account the d.c. representation (11.4.6), on the basis of Theorem 11.4.2 and the above GSS, the Global Search Algorithm (GSA) for quadratic bilevel problems can be formulated in the following way.

Let the following be given: a starting point $(x_0, y_0, v_0) \in \mathcal{F}$, numerical sequences $\{\tau_k\}$, $\{\delta_k\}$ with $(\tau_k, \ \delta_k > 0, \ \ k = 0, 1, 2, \ldots, \tau_k \downarrow 0, \ \delta_k \downarrow 0, (k \to \infty))$, a set $Dir = \{(\bar{z}^1, \bar{u}^1, \bar{w}^1), \ldots, (\bar{z}^N, \bar{u}^N, \bar{w}^N) \in \mathbb{R}^{m+n+q} \ |(\bar{z}^i, \bar{u}^i, \bar{w}^i) \neq 0, i = 1, \ldots, N\}$, the numbers $\gamma_- \stackrel{\triangle}{=} \inf(g, \mathcal{F})$ and $\gamma_+ \stackrel{\triangle}{=} \sup(g, \mathcal{F})$, and the algorithm's parameters M (a positive integer number) and $\mu \in \mathbb{R}_+$.

Step 0. Set $k := 0$, $(\bar{x}^k, \bar{y}^k, \bar{v}^k) := (x_0, y_0, v_0)$, $\gamma := \gamma_-$, $\Delta\gamma = (\gamma_+ - \gamma_-)/M$, $i := 1$.

Step 1. Proceeding from the point $(\bar{x}^k, \bar{y}^k, \bar{v}^k)$ by the XY- or the V-procedure, produce a τ_k-critical point $(x^k, y^k, v^k) \in \mathcal{F}$ to the problem ($\mathcal{DC}(\sigma)$)–(11.2.5). Set $\zeta_k := \Phi(x^k, y^k, v^k)$.

Step 2. Using $(\bar{z}^i, \bar{u}^i, \bar{w}^i) \in Dir$, construct a point (z^i, u^i, w^i) of the approximation $\mathcal{A}_k = \{(z^i, u^i, w^i) \mid H(z^i, u^i, w^i) = \gamma - \zeta_k, \quad i = 1, \ldots, N\}$ of the level surface $\mathcal{U}(\zeta_k) = \{(x, y, v) \mid H(x, y, v) = \gamma - \zeta_k\}$ of the convex function $H(x, y, z)$.

Step 3. If $G(z^i, u^i, w^i) > \gamma + \mu\gamma$, then $i := i + 1$ and return to Step 2.

Step 4. Find a δ_k-solution $(\bar{x}^i, \bar{y}^i, \bar{v}^i)$ of the following linearized problem:

$$(\mathcal{PL}_i) \qquad \begin{array}{ll} \min\limits_{x,y,v} & G(x, y, v) - \langle \nabla H(z^i, u^i, w^i), (x, y, v) \rangle \\ \text{s.t.} & (x, y, v) \in \mathcal{F}. \end{array} \tag{11.4.11}$$

Step 5. Starting at the point $(\bar{x}^i, \bar{y}^i, \bar{v}^i)$, produce a τ_k-critical point $(\hat{x}^i, \hat{y}^i, \hat{v}^i) \in \mathcal{F}$ to the problem ($\mathcal{DC}(\sigma)$)–(11.2.5) by means of the LSM.

Step 6. If $\Phi(\hat{x}^i, \hat{y}^i, \hat{v}^i) \geq \Phi(x^k, y^k, v^k)$, $i < N$, then set $i := i + 1$ and return to Step 2.

Step 7. If $\Phi(\hat{x}^i, \hat{y}^i, \hat{v}^i) \geq \Phi(x^k, y^k, v^k)$, $i = N$ and $\gamma < \gamma_+$, then set $\gamma := \gamma + \Delta\gamma$, $i := 1$ and go to Step 2.

Step 8. If $\Phi(\hat{x}^i, \hat{y}^i, \hat{v}^i) < \Phi(x^k, y^k, v^k)$, then set $\gamma := \gamma_-$ as well as $(\bar{x}^{k+1}, \bar{y}^{k+1}, \bar{v}^{k+1}) := (\hat{x}^i, \hat{y}^i, \hat{v}^i)$, $k := k + 1$, $i := 1$ and return to Step 1.

Step 9. If $\Phi(\hat{x}^i, \hat{y}^i, \hat{v}^i) \geq \Phi(x^k, y^k, v^k)$, $i = N$ and $\gamma = \gamma_+$, then stop. (x^k, y^k, v^k) is the obtained solution of the problem. □

Note that the stage (2) of the GSS is realized at Step 2 of the GSA with the help of the set Dir, stage (3)—at Step 4. Also, we add the extra local search in the GSA at Step 5 to enhance the properties of the current point. As for solving the "level problem" (11.4.5) and calculating the value $\eta_k(\gamma)$ (stages (4)–(5)), we use instead of this the direct comparison of the current value of the objective function with the value of ζ_k. As we could see from our previous computational experience, this variant seems to be more efficient when the objective function is easy to compute [47, 51].

In order to construct a feasible starting point, we used the projection of the chosen infeasible point (x^0, y^0, v^0) onto the feasible set $\mathcal{F}$ by solving the following quadratic programming problem (in this work $(x^0, y^0, v^0) = (0, 0, 0)$ for simplicity):

$$\begin{aligned} \min_{x,y,v}\ & \frac{1}{2}\|(x, y, v) - (x^0, y^0, v^0)\|^2 \\ \text{s.t.}\ & (x, y, v) \in \mathcal{F}. \end{aligned} \tag{11.4.12}$$

The solution to the problem (11.4.12) was taken as a feasible starting point $(x_0, y_0, v_0) \in \mathcal{F}$.

The value of the penalty parameter σ was chosen experimentally at the stage of the preliminary testing of the local search method: $\sigma = 20$.

Furthermore, to carry out the LSM (see Steps 1 and 5), we use both the XY- and V-procedures described in Sect. 11.3. Note that the tolerance for auxiliary problems is $\rho_s = 10^{-7}$. The tolerance of the LSM is $\tau_k = 10^{-5}$.

The selection of the algorithm parameters M and μ can be performed on the basis of our previous experience in solving problems with a bilinear structure [44, 47–49, 51]. The parameter μ is responsible for the tolerance of the inequality (11.4.8) from the GOCs (in order to diminish the computer rounding errors) [44, 47–49, 51] (Step 3). If we set, for example, $\mu = 0.0$ then the algorithm works fast but is not always efficient. When we increase μ, the algorithm's quality improves. However, the run time increases too, because the condition (11.4.8) relaxes and the number of level surface approximation points to be investigated grows. Different values of the parameter M are responsible for splitting the segment $[\gamma_-, \gamma_+]$ into a corresponding number of subsegments to carry out a passive one-dimensional search for γ. When we increase the value of M, the algorithm's accuracy grows, of course, but it happens at the expense of the proportional increase of the run time.

In this work we use rough estimate $\gamma_- := 0.0$ in all cases and the following four sets of parameters: (1) $\gamma_+ := (m + n + q)^2 * \sigma$, $M = 2$, $\mu = 0.0$; (2) $\gamma_+ := (m + n + q) * (n + q) * \sigma$, $M = 5$, $\mu = 0.05$; (3) $\gamma_+ := (m + n + q) * q * \sigma$, $M = 11$, $\mu = 0.02$; (4) $\gamma_+ := (m + n + q) * \sigma$, $M = 50$, $\mu = 0.01$. These values were found experimentally at the stage of the preliminary computational simulation.

Further, we build the approximation $\mathcal{A}_k = \mathcal{A}(\zeta_k)$ for the problem $(\mathcal{DC}(\sigma))$–(11.2.5) with the help of a special sets of directions, based on our previous experience [44, 47–49, 51]:

$$Dir1 = \{(x, y, v) + e^i,\ (x, y, v) - e^i, \quad i = 1, \dots, m + n + q\},$$

where $e^i \in \mathbb{R}^{m+n+q}$ are the Euclidean basic vectors, (x, y, v) is a current critical point of the problem $(\mathcal{DC}(\sigma))$–(11.2.5) (the number of points in the approximation, which is based on the set $Dir1$, is equal to $2(m+n+q)$);

$$Dir2 = \{((x, y) - e^l, v - e^j),\quad l = 1, \ldots, m+n,\quad j = 1, \ldots, q\},$$

where $e^l \in \mathbb{R}^{m+n}$, $e^j \in \mathbb{R}^q$ are the Euclidean basic vectors of the appropriate dimension, (x, y, v) is a current critical point; and

$$Dir3 = \{((x, y) + e^l, v + e^j),\quad l = 1, \ldots, m+n,\quad j = 1, \ldots, q\}.$$

The number of points in the approximation, which is based on the sets $Dir2$ and $Dir3$, is equal to $(m+n) * q$.

Note that $Dir1$ is built in the space of all variables of the problem $(\mathcal{DC}(\sigma))$–(11.2.5). But $Dir2$ and $Dir3$ use different vectors from $(m+n)$-dimensional and q-dimensional spaces. Therefore, the number of points in the two latter sets is rather large (especially when the dimension of the problem grows). So, we apply a special technique for reducing these approximations (see [44, 47]).

Now we are ready to present and discuss the numerical experiment.

11.5 Computational Experiment

For the experimental verification of the efficiency of new methods developed for optimization problems, as well as for their comparison with existing approaches, it is necessary to have a sufficiently wide range of test problems of different complexity and dimension. In the present work, we use the special Calamai-Vicente generation method of bilevel test cases proposed in [53, 54]. The idea of such generation is based on constructing bilevel problems of arbitrary dimension with the help of several classes of the so-called kernel problems, which are one-dimensional bilevel problems with known local and global solutions (see also [49–51]).

At the first stage of the computational experiment, we present the results of solving the bilevel problem $(\mathcal{QBP})$–(11.2.1), where $D_1 = 0$, $Q = 0$. It is a particular case of a quadratic bilevel problem, the so-called quadratic-linear bilevel problem. In [51] one can find the results of solving a wide range of generated problems of such class with different complexity and dimension up to (150×150). Here we describe the comparative numerical results for some complicated test problems obtained by the Global Search Algorithm and the popular KNITRO [55] software package (we use ver. 7.0).

A special series of 15 problems of dimension from 5×5 to 30×30 was generated. When testing, we used a program that implements the Global Search Algorithm and the KNITRO package (with multi-start enabled). Since the KNITRO package does not have the ability to work with bilevel problems directly, the input to this package

Table 11.1 Comparison of the GSA with KNITRO

		KNITRO multistart		Global search algorithm			
Name	F_*	F_{Kms}	T	F_{XY}	T	F_V	T
5×5_1	−21	**−21**	7.7	**−21**	11.4	**−21**	7.3
5×5_2	−9	**−9**	8.7	**−9**	11.3	**−9**	4.7
5×5_3	−5	**−5**	9.9	**−5**	11.7	**−5**	4.6
10×10_1	−38	−30	77.2	**−38**	33.1	**−38**	23.7
10×10_2	−26	**−26**	77.0	**−26**	29.2	**−26**	12.4
10×10_3	−14	**−14**	82.4	**−14**	24.6	**−14**	16.0
15×15_1	−19	**−19**	668.5	**−19**	20.3	**−19**	19.0
15×15_2	−27	−19	343.6	**−27**	38.6	**−27**	31.5
15×15_3	−43	−35	421.2	**−43**	48.1	**−43**	35.5
20×20_1	−24	**−24**	1153.7	**−24**	26.9	**−24**	45.9
20×20_2	−48	**−48**	2000.5	**−48**	72.0	**−48**	70.9
20×20_3	−52	−32	1794.4	**−52**	73.1	**−52**	52.0
30×30_1	−142	−134	5674.9	**−142**	270.3	**−142**	106.8
30×30_2	−58	−38	5483.5	**−58**	86.2	**−58**	111.3
30×30_3	−42	−30	9555.4	**−42**	51.5	**−42**	68.9

was a problem of the form ($\mathcal{DCC}$)–(11.2.4) with the necessary reformulations, equivalent to the corresponding bilevel problem in a global sense. We use here only the third set of algorithm's parameters ($\gamma_+ := (m+n+q)*q*\sigma$, $M = 11$, $\mu = 0.02$) and the approximation based on the set $Dir1$. The test results are presented in Table 11.1, where the following notations were used:

Name is the title of the generated problem ($m \times n\_N$, where m is the dimension of the vector x, n is the dimension of the vector y, N is the number of the problem of this dimension);

F_* is the known global value of the upper level objective function in the test bilevel problem;

F_{Kms} is the value of the upper level objective function in the test bilevel problem at the point found by the KNITRO package;

F_{XY} (F_V) is the value of the upper level objective function at the point obtained using the Global Search Algorithm with the XY-procedure (V-procedure);

T is the running time of the corresponding program (in seconds).

Analyzing the results of the computational experiment, we first note that with the help of the KNITRO package that uses multi-start, global solutions were found only in 53% of the problems with the tolerance $\varepsilon = 10^{-3}$ (marked in bold in the table), while using both variants of the program that implements the GSA, all test problems with this tolerance were solved. Also note the remarkable advantage of the global search in comparison with the KNITRO with respect to maximal computational time for problems of this set: KNITRO—2.66 h, GSA with XY-procedure—4.5 min, GSA with V-procedure—1.85 min).

At this stage of the computational experiment, an approach to the search for optimistic solutions in quadratic-linear bilevel problems, based on the GST, demonstrated an impressive advantage over the KNITRO package for this series of test problems, both in terms of the number of solved problems and calculation time.

In the second stage of the computational experiment, we present the results of the solution of a series of quadratic-quadratic problems generated by the method mentioned above [54]. To carry out the local search, we use both the XY- and V-procedures and choose the best of the two points obtained by them. Here we already use all sets of parameters and all sets of directions which were applied consecutively (see the previous section).

The results of applying the GSA to the generated problems are presented in Table 11.2 with the following notations:

$Cl_{1,2,3,4}$ are the numbers of kernel problems of each class (from 4 classes) in the generated problem;

$Series$ is the number of generated problems of the current structure (in the current series);

$GlobSol$ is the known number of global solutions in the generated problem of the current structure;

$LocSol$ is the number of local solutions which are not global;

Loc_A stands for the average number of start-ups of the LSM required to find the approximate global solution to the problem of such a structure;

St_A is the average number of different critical points obtained (the number of GSA iterations);

T_A is the average operating time of the program (in seconds).

We solved from 1 to 10,000 problems of each type of dimension from 2 to 100 variables at every level. The principal achievement here is that all generated problems were solved globally. It is interesting that the complexity of each type of problems does not directly depend on the number of global and local solutions. Note that if the generation employs the fourth class of kernel problems, we get the complicated type of problems (see the large average number of the LSM executed, large number of critical points obtained, and large average time). If we do not use the fourth class, most of the problems can be easily solved by the GSA, which finds a global solution even to the problem of dimension 100 in 1 h and 7 min.

The GSA can be compared with the results presented in [29, 30], which were obtained for the test quadratic bilevel problems with up to 30 variables at every level generated by the same approach [54]. In [29, 30] the author used a special variant of sequential quadratic programming method. If we compare the new result with our previous findings [51], we can say that for a more complicated bilevel problem with both quadratic objective functions we have reached the dimension 100 at each level. At the same time, as was mentioned above, [51] presents the results of solving quadratic-linear bilevel problems with 150 variables at each level.

We feel confident that our approach can be generalized to more complicated bilevel optimization problems and it can demonstrate competitive efficiency with

Table 11.2 The Global Search in generated quadratic problems

Name	$Cl_{1,2,3,4}$	*Series*	*GlobSol*	*LocSol*	Loc_A	St_A	T_A
2×2_1	(1,0,0,1)	10,000	1	1	77.7	2.0	0.4
3×3_1	(0,1,1,1)	10,000	2	6	126.3	2.0	0.8
4×4_1	(0,4,0,0)	10,000	1	15	163.9	2.0	1.1
5×5_1	(0,2,2,1)	1000	4	28	222.3	2.0	1.6
5×5_2	(2,0,3,0)	1000	8	0	209.1	2.0	1.4
6×6_1	(0,1,0,5)	1000	1	63	2440.0	3.6	19.9
6×6_2	(4,2,0,0)	1000	1	3	251.1	2.0	1.7
8×8_1	(0,0,6,2)	1000	64	192	1708.8	2.7	17.7
8×8_2	(8,0,0,0)	1000	1	0	337.2	2.0	2.4
10×10_1	(2,3,2,3)	1000	4	252	3707.8	10.3	47.7
10×10_2	(0,0,10,0)	1000	1024	0	464.7	2.0	4.7
12×12_1	(5,2,5,0)	1000	32	96	551.4	2.0	6.4
12×12_2	(0,0,0,12)	100	1	4095	15,431.6	7.5	193.0
15×15_1	(4,4,4,3)	100	16	2032	22,018.6	66.6	323.5
15×15_2	(5,5,5,0)	100	32	992	762.3	2.0	12.0
20×20_1	(9,9,0,2)	100	1	2047	71,6211.5	276.9	1558.2
20×20_2	(5,15,0,0)	100	1	32,767	1026.1	2.0	25.7
25×25_1	(8,9,8,0)	100	256	130,816	1403.1	2.0	54.2
25×25_2	(7,6,7,5)	100	128	262,016	95,572.0	268.3	3337.1
30×30_1	(0,15,15,0)	100	32,768	$1.0737 \cdot 10^9$	1831.4	2.0	110.0
30×30_2	(10,5,10,5)	10	1024	1,047,552	162,235.8	406.6	7279.1
40×40_1	(20,0,20,0)	10	1,048,576	0	2426.8	2.0	165.0
40×40_2	(10,10,10,10)	10	1024	$1.0737 \cdot 10^9$	126,645.2	236.7	9689.7
50×50_1	(17,16,17,0)	10	131,072	$8.5898 \cdot 10^9$	3004.0	2.0	622.0
50×50_2	(0,0,45,5)	10	$3.5184 \cdot 10^{13}$	$1.0907 \cdot 10^{15}$	55810.4	17.5	7384.7
60×60_1	(30,30,0,0)	10	1	$1.0737 \cdot 10^9$	3609.4	2.8	712.7
60×60_2	(0,50,0,10)	1	1	$1.1529 \cdot 10^{18}$	269,440.0	459.0	52,730.6
75×75_1	(25,25,25,0)	1	33,554,432	$1.1259 \cdot 10^{15}$	6546.0	4.0	2357.0
75×75_2	(0,0,72,3)	1	$4.7224 \cdot 10^{21}$	$3.3057 \cdot 10^{22}$	136,382.0	41.0	42,485.8
100×100_1	(0,20,80,0)	1	$1.2089 \cdot 10^{24}$	$1.2676 \cdot 10^{30}$	6012.0	2.0	4066.0

respect to any solution methods for bilevel problems. For example, our first attempt to attack bilevel problems with an equilibrium at the lower level can be found in [56].

11.6 Conclusion

This paper presents a new methodology for solving optimistic bilevel optimization problems with a demonstration of its numerical efficiency. This methodology implies their reduction to the single-level but nonconvex optimization problems, which are then solved using the Global Search Theory [43–46]. The numerical results obtained for bilevel test problems of the indicated classes demonstrate the competitiveness of the proposed methodology in comparison with specialized software packages (such as the KNITRO) and the results of publications known at present.

In our future research, we are going to increase the complexity of the bilevel models and apply our approach to practical bilevel problems. Our first attempt to solve a special quadratic bilevel pricing problem in telecommunication networks is presented in [57].

There are still open questions and challenges in the numerical solution of BOPs, in particular, new ideas for handling problems with nonconvex lower level are now at the front edge of the researches in the modern bilevel optimization (see, for example, [22]).

References

1. S. Dempe, *Foundations of Bilevel Programming* (Kluwer Academic Publishers, Dordrecht, 2002)
2. J.F. Bard, *Practical Bilevel Optimization* (Kluwer Academic Publishers, Dordrecht, 1998)
3. S. Dempe, Annotated bibliography on bilevel programming and mathematical programs with equilibrium constraints. Optimization **52**, 333–359 (2003)
4. B. Colson, P. Marcotte, G. Savard, An overview of bilevel optimization. Ann. Oper. Res. **153**, 235–256 (2007)
5. S. Dempe, V.V. Kalashnikov, G.A. Perez-Valdes, N. Kalashnykova, *Bilevel Programming Problems: Theory, Algorithms and Applications to Energy Networks* (Springer, Berlin/Heidelberg, 2015)
6. G.B. Allende, G. Still, Solving bilevel programs with the KKT-approach. Math. Prog. Ser. A. **138**, 309–332 (2013)
7. S. Dempe, A.B. Zemkoho, On the Karush-Kuhn-Tucker reformulation of the bilevel optimization problem. Nonlinear Anal. Theory Methods Appl. **75**, 1202–1218 (2012)
8. Z.-Q. Luo, J.-S. Pang, D. Ralph, *Mathematical Programs with Equilibrium Constraints* (Cambridge University Press, Cambridge, 1996)
9. M. Campelo, S. Dantas, S. Scheimberg, A note on a penalty function approach for solving bilevel linear programs. J. Glob. Optim. **16**, 245–255 (2000)
10. M. Amouzegar, K. Moshirvaziri, A penalty method for linear bilevel programming problems, in *Multilevel Optimization: Algorithms and Applications*, ed. by A. Migdalas, P. Pardalos, P. Varbrand. Nonconvex Optimization and Its Applications, vol. 20, pp. 251–271 (Kluwer, Dordrecht, 1998)
11. G.S. Liu, J.Y. Han, J.Z. Zhang, Exact penalty functions for convex bilevel programming problems. J. Optim. Theory Appl. **110**, 621–643 (2001)
12. L.T. Hoai An, D.C. programming for solving a class of global optimization problems via reformulation by exact penalty. Lect. Notes Comput. Sci. **2861**, 87–101 (2003)
13. S. Jia, Z. Wan, A penalty function method for solving ill-posed bilevel programming problem via weighted summation. J. Syst. Sci. Complex. **26**, 1019–1027 (2013)
14. M.S. Bazara, C.M. Shetty, *Nonlinear Programming. Theory and Algorithms* (Wiley, New York, 1979)
15. J.-F. Bonnans, J.C. Gilbert, C. Lemarechal, C.A. Sagastizabal, *Numerical Optimization: Theoretical and Practical Aspects* (Springer, Berlin/Heidelberg, 2006)

16. J. Nocedal, S.J. Wright, *Numerical Optimization* (Springer, New York/Berlin/Heidelberg, 2006)
17. C. Audet, G. Savard, W. Zghal, New branch-and-cut algorithm for bilevel linear programming. J. Optim. Theory Appl. **134**, 353–370 (2007)
18. J.V. Outrata, M. Kocvara, J. Zowe, *Nonsmooth Approach to Optimization Problems with Equilibrium Constraints: Theory, Applications and Numerical Results* (Kluwer, Boston, 1998)
19. L.D. Muu, N.V. Quy, A global optimization method for solving convex quadratic bilevel programming problems. J. Glob. Optim. **26**, 199–219 (2003)
20. M. Xu, J. Ye, A smoothing augmented Lagrangian method for solving simple bilevel programs. Comput. Optim. Appl. **59**, 353–377 (2014)
21. H. Tuy, A. Migdalas, N.T. Hoai-Phuong, A novel approach to bilevel nonlinear programming. J. Glob. Optim. **38**, 527–554 (2007)
22. A. Mitsos, P. Lemonidis, P.I. Barton, Global solution of bilevel programs with a nonconvex inner program. J. Glob. Optim. **42**, 475–513 (2008)
23. J. Rajesh, K. Gupta, H.S. Kusumakar, V.K. Jayaraman, B.D. Kulkarni, A tabu search based approach for solving a class of bilevel programming problems in chemical engineering. J. Heuristics **9**, 307–319 (2003)
24. L. Vicente, G. Savard, J. Judice, Descent approaches for quadratic bilevel programming. J. Optim. Theory Appl. **81**, 379–399 (1994)
25. A.G. Mersha, S. Dempe, Direct search algorithm for bilevel programming problems. Comput. Optim. Appl. **49**, 1–15 (2011)
26. S. Dempe, J.F. Bard, Bundle trust-region algorithm for bilinear bilevel programming. J. Optim. Theory Appl. **110**, 265–288 (2001)
27. B. Colson, P. Marcotte, G. Savard, A trust-region method for nonlinear bilevel programming: algorithm and computational experience. Comput. Optim. Appl. **30**, 211–227 (2005)
28. G. Liu, S. Xu, J. Han, Trust region algorithm for solving bilevel programming problems. Acta Math. Appl. Sin. Engl. Ser. **29**, 491–498 (2013)
29. J.B. Etoa Etoa, Solving convex quadratic bilevel programming problems using an enumeration sequential quadratic programming algorithm. J. Glob. Optim. **47**, 615–637 (2010)
30. J.B. Etoa Etoa, Solving quadratic convex bilevel programming problems using a smoothing method. Appl. Math. Comput. **217**, 6680–6690 (2011)
31. P.-M. Kleniati, C.S. Adjiman, Branch-and-Sandwich: a deterministic global optimization algorithm for optimistic bilevel programming problems. Part I: Theoretical development. J. Glob. Optim. **60**, 425–458 (2014)
32. P.-M. Kleniati, C.S. Adjiman, Branch-and-Sandwich: a deterministic global optimization algorithm for optimistic bilevel programming problems. Part II: Convergence analysis and numerical results. J. Glob. Optim. **60**, 459–481 (2014)
33. R. Horst, H. Tuy, *Global Optimization. Deterministic Approaches* (Springer, Berlin, 1993)
34. R. Horst, P. Pardalos, N.V. Thoai, *Introduction to Global Optimization* (Kluwer Academic Publishers, Dordrecht/Boston/London, 1995)
35. C.H. Saboia, M. Campelo, S. Scheimberg, A computational study of global algorithms for linear bilevel programming. Numer. Algorithms **35**, 155–173 (2004)
36. H.I. Calvete, C. Gale, S. Dempe, S. Lohse, Bilevel problems over polyhedra with extreme point optimal solutions. J. Glob. Optim. **53**, 573–586 (2012)
37. J. Fliege, L.N. Vicente, Multicriteria approach to bilevel optimization. J. Optim. Theory Appl. **131**, 209–225 (2006)
38. J. Glackin, J.G. Ecker, M. Kupferschmid, Solving bilevel linear programs using multiple objective linear programming. J. Optim. Theory Appl. **140**, 197–212 (2009)
39. N.P. Faisca, V. Dua, B. Rustem, P.M. Saraiva, E. Pistikopoulos, Parametric global optimisation for bilevel programming. J. Glob. Optim. **38**, 609–623 (2007)
40. G.-M. Wang, X.-J. Wang, Z.-P. Wan, Y. Lv, A globally convergent algorithm for a class of bilevel nonlinear programming problem. Appl. Math. Comput. **188**, 166–172 (2007)

41. G.-M. Wang, Z.-P. Wan, X.-J. Wang, Y. Lv, Genetic algorithm based on simplex method for solving linear-quadratic bilevel programming problem. Comput. Math. Appl. **56**, 2550–2555 (2008)
42. J.S. Pang, Three modeling paradigms in mathematical programming. Math. Prog. Ser. B **125**, 297–323 (2010)
43. A.S. Strekalovsky, Elements of nonconvex optimization [in Russian] (Nauka, Novosibirsk, 2003)
44. A.S. Strekalovsky, A.V. Orlov, A new approach to nonconvex optimization. Numer. Methods Prog. (internet-journal: http://num-meth.srcc.msu.su/english/index.html) **8**, 160–176 (2007)
45. A.S. Strekalovsky, On solving optimization problems with hidden nonconvex structures, in *Optimization in Science and Engineering* ed. by T.M. Rassias, C.A. Floudas, S. Butenko (Springer, New York, 2014), pp. 465–502
46. Strekalovsky, A.S., Global optimality conditions and exact penalization. Optim. Lett. **13**, 597–615 (2019)
47. A.V. Orlov, A.S. Strekalovsky, Numerical search for equilibria in bimatrix games. Comput. Math. Math. Phys. **45**, 947–960 (2005)
48. A.V. Orlov, A.S. Strekalovsky, S. Batbileg, On computational search for Nash equilibrium in hexamatrix games. Optim. Lett. **10**, 369–381 (2016)
49. A.V. Orlov, Numerical solution of bilinear programming problems. Comput. Math. Math. Phys. **48**, 225–241 (2008)
50. Gruzdeva, T.V., Petrova, E.G.: Numerical solution of a linear bilevel problem. Comput. Math. Math. Phys. **50**, 1631–1641 (2010)
51. A.S. Strekalovsky, A.V. Orlov, A.V. Malyshev, On computational search for optimistic solution in bilevel problems. J. Glob. Optim. **48**, 159–172 (2010)
52. A.V. Orlov, A nonconvex optimization approach to quadratic bilevel problems, in Learning and Intelligent Optimization, ed. by R. Battiti, D.E. Kvasov, Ya.D. Sergeyev. Lecture Notes in Computer Science, vol. 10556 (Springer International Publishing AG, Beijing, 2017), pp. 222–234
53. P. Calamai, L. Vicente, Generating linear and linear-quadratic bilevel programming problems. SIAM J. Sci. Comput. **14**, 770–782 (1993)
54. P. Calamai, L. Vicente, Generating quadratic bilevel programming test problems. ACM Trans. Math. Softw. **20**, 103–119 (1994)
55. Artelys Knitro - Nonlinear optimization solver. https://www.artelys.com/en/optimization-tools/knitro. Cited 27 Jun 2019
56. A.V. Orlov, T.V. Gruzdeva, The local and global searches in bilevel problems with a matrix game at the lower level, in Mathematical Optimization Theory and Operations Research. MOTOR 2019, ed. by M. Khachay, Y. Kochetov, P. Pardalos. Lecture Notes in Computer Science, vol. 11548 (Springer, Cham, 2019), pp. 172–183
57. A.V. Orlov, The global search theory approach to the bilevel pricing problem in telecommunication networks, in Computational Aspects and Applications in Large Scale Networks, ed. by V.A. Kalyagin et al. (Springer International Publishing AG, Cham, 2018), pp. 57–73

Chapter 12
MPEC Methods for Bilevel Optimization Problems

Youngdae Kim, Sven Leyffer, and Todd Munson

Abstract We study optimistic bilevel optimization problems, where we assume the lower-level problem is convex with a nonempty, compact feasible region and satisfies a constraint qualification for all possible upper-level decisions. Replacing the lower-level optimization problem by its first-order conditions results in a mathematical program with equilibrium constraints (MPEC) that needs to be solved. We review the relationship between the MPEC and bilevel optimization problem and then survey the theory, algorithms, and software environments for solving the MPEC formulations.

Keywords Bilevel optimization · Mathematical program with equilibrium constraints · Stationarity · Algorithms

12.1 Introduction

Bilevel optimization problems model situations in which a sequential set of decisions are made: the leader chooses a decision to optimize its objective function, anticipating the response of the follower, who optimizes its objective function given the leader's decision. Mathematically, we have the following optimization problem:

$$\begin{aligned} \underset{x,y}{\text{minimize}} \quad & f(x, y) \\ \text{subject to} \quad & c(x, y) \geq 0 \\ & y \in \arg\min_{v} \{\, g(x, v) \text{ subject to } d(x, v) \geq 0 \,\}, \end{aligned} \tag{12.1.1}$$

where f and g are the objectives for the leader and follower, respectively, $c(x, y) \geq 0$ is a joint feasibility constraint, and $d(x, y) \geq 0$ defines the feasible actions the

Y. Kim · S. Leyffer · T. Munson (✉)
Argonne National Laboratory, Lemont, IL, USA
e-mail: youngdae@anl.gov; leyffer@anl.gov; tmunson@anl.gov

S. Dempe, A. Zemkoho (eds.), *Bilevel Optimization*, Springer Optimization and Its Applications 161, https://doi.org/10.1007/978-3-030-52119-6_12

follower can take. Throughout this survey, we assume all functions are at least continuously differentiable in all their arguments.

Our survey focuses on optimistic bilevel optimization. Under this assumption, if the solution set to the lower-level optimization problem is not a singleton, then the leader chooses the element of the solution set that benefits it the most.

12.1.1 Applications and Illustrative Example

Applications of bilevel optimization problems arise in economics and engineering domains; see [10, 24, 27–29, 78, 92, 111, 116] and the references therein. As an example bilevel optimization problem, we consider the moral-hazard problem in economics [72, 101] that models a contract between a principal and an agent who works on a project for the principal. The agent exerts effort on the project by taking an action $a \in \mathcal{A}$ that maximizes its utility, where $\mathcal{A} = \{a | d(a) \geq 0\}$ is a convex set, resulting in output value o_q with given probability $p_q(a) > 0$, where $q \in Q$ and Q is a finite set. The principal observes only the output o_q and compensates the agent with c_q defined in the contract. The principal wants to design an optimal contract consisting of a compensation schedule $\{c_q\}_{q \in Q}$ and a recommended action $a \in \mathcal{A}$ to maximize its expected utility.

Since the agent's action is assumed to be neither observable nor forcible by the principal, feasibility of the contract needs to be defined in order to guarantee the desired behavior of the agent. Two types of constraints are specified: a participation constraint and an incentive-compatibility constraint. The participation constraint says that the agent's expected utility from accepting the contract must be at least as good as the utility U it could receive by choosing a different activity. The incentive-compatibility constraint states that the recommended action should provide enough incentive, such as maximizing its utility, so that the agent chooses it.

Mathematically, the problem is defined as follows:

$$
\begin{aligned}
\underset{c,a}{\text{maximize}} \quad & \sum_{q \in Q} p_q(a) w(o_q - c_q) \\
\text{subject to} \quad & \sum_{q \in Q} p_q(a) u(c_q, a) \geq U \\
& a \in \arg\max_{\tilde{a}} \left\{ \sum_{q \in Q} p_q(\tilde{a}) u(c_q, \tilde{a}) \quad \text{subject to } d(\tilde{a}) \geq 0 \right\},
\end{aligned}
$$

where $w(\cdot)$ and $u(\cdot, \cdot)$ are the utility functions of the principal and the agent, respectively. The first constraint is the participation constraint, while the second constraint is the incentive-compatibility constraint. This problem is a bilevel optimization problem with a joint feasibility constraint.

12.1.2 Bilevel Optimization Reformulation as an MPEC

We assume throughout that the lower-level problem is convex with a nonempty, compact feasible region and that it satisfies a constraint qualification for all feasible upper-level decisions, x. Under these conditions, the Karush-Kuhn-Tucker (KKT) conditions for the lower-level optimization problem are both necessary and sufficient, and we can replace the lower-level problem with its KKT conditions to obtain a mathematical program with equilibrium constraints (MPEC) [82, 90, 98]:

$$\begin{aligned} \underset{x,y,\lambda}{\text{minimize}} \quad & f(x, y) \\ \text{subject to} \quad & c(x, y) \geq 0 \\ & \nabla_y g(x, y) - \nabla_y d(x, y)\lambda = 0 \\ & 0 \leq \lambda \perp d(x, y) \geq 0, \end{aligned} \tag{12.1.2}$$

where $\perp$ indicates complementarity, in other words, that either $\lambda_i = 0$ or $d_i(x, y) = 0$ for all i.

This MPEC reformulation is attractive because it results in a single-level optimization problem, and we show in the subsequent sections that this class of problems can be solved successfully. We note that when the lower-level problem is nonconvex, a solution of the MPEC may be infeasible for the bilevel problem [94, Example 1.1]. Thus, we consider only the case where the lower-level problem is convex when using the MPEC formulation.

In the convex case, the relationship between the original bilevel optimization problem (12.1.1) and the MPEC formulation (12.1.2) is nontrivial, as demonstrated in [30, 32] for smooth functions and [33] for nonsmooth functions. These papers show that a global (local) solution to the bilevel optimization problem (12.1.1) corresponds to a global (local) solution to the MPEC (12.1.2) if the Slater constraint qualification is satisfied by the lower-level optimization problem. Under the Slater constraint qualification, a global solution to the MPEC (12.1.2) corresponds to a global solution to the bilevel optimization problem (12.1.1). A local solution to the MPEC (12.1.2) may not correspond to a local solution to the bilevel optimization problem (12.1.1) unless stronger assumptions guaranteeing the uniqueness of multipliers are made, such as MPEC-LICQ, which is described in Sect. 12.2.2; see [30, 32] for details.

By using the Fritz-John conditions, rather than assuming a constraint qualification and applying the KKT conditions, we can ostensibly weaken the requirements to produce an MPEC reformulation [2]. However, this approach has similar theoretical difficulties with the correspondences between local solutions [31], and we have observed computational difficulties when applying MPEC algorithms to solve the Fritz-John reformulation.

12.1.3 MPEC Problems

Generically, we write MPECs such as (12.1.2) as

$$\begin{aligned}
\underset{z}{\text{minimize}} \quad & f(z) \\
\text{subject to} \quad & c(z) \geq 0 \\
& 0 \leq G(z) \ \perp \ H(z) \geq 0,
\end{aligned} \tag{12.1.3}$$

where $z = (x, y, \lambda)$ and where we summarize all nonlinear equality and inequality constraints generically as $c(z) \geq 0$.

MPECs are a challenging class of problem because of the presence of the complementarity constraint. The complementarity constraint can be written equivalently as the nonlinear constraint $G(z)^T H(z) \leq 0$, in which case (12.1.3) becomes a standard nonlinear program (NLP). Unfortunately, the resulting problem violates the Mangasarian-Fromowitz constraint qualification (MFCQ) at any feasible point [108]. Alternatively, we can reformulate the complementarity constraint as a disjunction. Unfortunately, the resulting mixed-integer formulation has very weak relaxations, resulting in large search trees [99].

This survey focuses on practical theory and methods for solving MPECs. In the next section, we discuss the stationarity concepts and constraint qualifications for MPECs. In Sect. 12.3, we discuss algorithms for solving MPECs. In Sect. 12.4, we focus on software environments for specifying and solving bilevel optimization problems and mathematical programs with equilibrium constraints, before providing pointers in Sect. 12.5 to related topics we do not cover in this survey.

12.2 MPEC Theory

We now survey stationarity conditions, constraint qualifications, and regularity for the MPEC (12.1.3). The standard concepts for nonlinear programs need to be rethought because of the complementarity constraint $0 \leq G(z) \ \perp \ H(z) \geq 0$, especially when the solution to (12.1.3) has a nonempty biactive set, that is, when both $G_j(z^*) = H_j(z^*) = 0$ for some indices j at the solution z^*.

We assume that the functions in the MPEC are at least continuously differentiable. When the functions are nonsmooth, enhanced M-stationarity conditions and alternative constraint qualifications have been proposed [119].

12.2.1 Stationarity Conditions

In this section, we define first-order optimality conditions for a local minimizer of an MPEC and several stationarity concepts. The stationarity concepts may involve dual variables, and they collapse to a single stationarity condition when a local minimizer has an empty biactive set. Unlike standard nonlinear programming, these concepts may not correspond to the first-order optimality of the MPEC when there is a nonempty biactive set.

One derivation of stationarity conditions for MPECs is to replace the complementarity condition with a set of nonlinear inequalities, such as $G_j(z)H_j(z) \leq 0$, and then produce the stationarity conditions for the equivalent nonlinear program:

$$\begin{array}{lll} \underset{z}{\text{minimize}} & f(z) & \\ \text{subject to} & c_i(z) \geq 0 & \forall i = 1, \ldots, m \\ & G_j(z) \geq 0,\, H_j(z) \geq 0,\, G_j(z)H_j(z) \leq 0 & \forall j = 1, \ldots, p, \end{array} \tag{12.2.1}$$

where $z \in \mathbb{R}^n$. An alternative formulation as an NLP is obtained by replacing $G_j(z)H_j(z) \leq 0$ by $G(z)^T H(z) \leq 0$. Unfortunately, (12.2.1) violates the MFCQ at *any* feasible point [108] because the constraint $G_j(z)H_j(z) \leq 0$ does not have an interior. Hence, the KKT conditions may not be directly applicable to (12.2.1).

Instead, stationarity concepts are derived from several different approaches: local analysis of the NLPs associated with an MPEC [90, 108]; Clarke's nonsmooth analysis to the complementarity constraints by replacing them with the min function [23, 108]; and Mordukhovich's generalized differential calculus applied to the generalized normal cone [97]. The first method results in B-, weak-, and strong-stationarity concepts. The second and the third lead to C- and M-stationarity, respectively. These stationarity concepts coincide with each other when the solution has an empty biactive set.

We begin by defining the biactive index set $\mathcal{D}$ (or denoted by $\mathcal{D}(z)$ to emphasize its dependency on z) and its partition $\mathcal{D}_{01}$ and $\mathcal{D}_{10}$ such that

$$\mathcal{D} := \{j \mid G_j(z) = H_j(z) = 0\}, \;\; \mathcal{D} = \mathcal{D}_{01} \cup \mathcal{D}_{10}, \mathcal{D}_{01} \cap \mathcal{D}_{10} = \emptyset. \tag{12.2.2}$$

If we define an $\text{NLP}_{\mathcal{D}_{01},\mathcal{D}_{10}}$

$$\begin{array}{lll} \underset{z}{\text{minimize}} & f(z) & \\ \text{subject to} & c_i(z) \geq 0 & \forall i = 1, \ldots, m \\ & G_j(z) = 0 & \forall j : G_j(z) = 0,\, H_j(z) > 0 \\ & H_j(z) = 0 & \forall j : G_j(z) > 0,\, H_j(z) = 0 \\ & G_j(z) = 0,\, H_j(z) \geq 0 & \forall j \in \mathcal{D}_{01} \\ & G_j(z) \geq 0,\, H_j(z) = 0 & \forall j \in \mathcal{D}_{10}, \end{array} \tag{12.2.3}$$

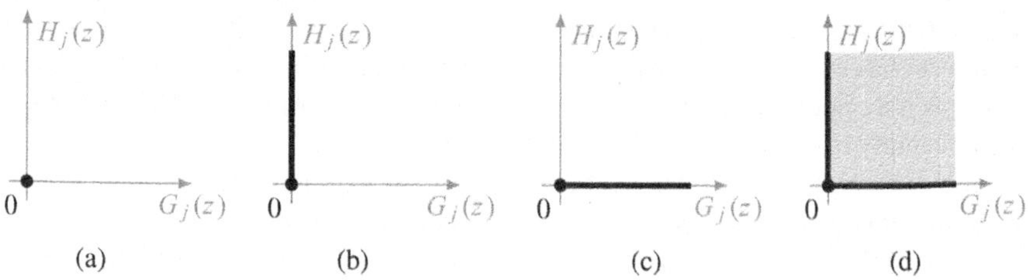

Fig. 12.1 Feasible regions of the NLPs associated with a complementarity constraint $0 \leq G_j(z) \perp H_j(z) \geq 0$ for $j \in \mathcal{D}$. **(a)** Tightened NLP. **(b)** $\mathrm{NLP}_{\mathcal{D}_{01}}$. **(c)** $\mathrm{NLP}_{\mathcal{D}_{10}}$. **(d)** Relaxed NLP

then z^* is a local solution to the MPEC if and only if z^* is a local solution for all the associated NLPs indexed by $(\mathcal{D}_{01}, \mathcal{D}_{10})$. The number of NLPs that need to be checked is exponential in the number of biactive indices, which can be computationally intractable.

Like many other mathematical programs, the geometric first-order optimality conditions are defined in terms of the tangent cone. A feasible point z^* is called a geometric Bouligand- or B-stationary point if $\nabla f(z^*)^T d \geq 0, \forall d \in \mathcal{T}_{\mathrm{MPEC}}(z^*)$, where the tangent cone $\mathcal{T}(z^*)$ to a feasible region $\mathcal{F}$ at $z^* \in \mathcal{F}$ is defined as

$$\mathcal{T}(z^*) := \left\{ d \mid d = \lim_{t_k \downarrow 0} \frac{z^k - z^*}{t_k} \text{ for some } \{z^k\}, z^k \to z^*, z^k \in \mathcal{F}, \forall k \right\}. \tag{12.2.4}$$

To facilitate the analysis of the tangent cone $\mathcal{T}_{\mathrm{MPEC}}(z)$ at $z \in \mathcal{F}_{\mathrm{MPEC}}$, we subdivide it into a set of tangent cones of the associated NLPs:

$$\mathcal{T}_{\mathrm{MPEC}}(z) := \bigcup_{(\mathcal{D}_{01}, \mathcal{D}_{10})} \mathcal{T}_{\mathrm{NLP}_{\mathcal{D}_{01}, \mathcal{D}_{10}}}(z). \tag{12.2.5}$$

Therefore, z^* is a geometric B-stationary point if and only if it is a geometric B-stationary point for all of its associated NLPs indexed by $(\mathcal{D}_{01}, \mathcal{D}_{10})$.

Additionally, two more NLPs, tightened [108] and relaxed [48, 108] NLPs, are defined by replacing the last two conditions of (12.2.3) with a single condition: tightened NLP (TNLP) is defined by setting $G_j(z) = H_j(z) = 0$ for all $j \in \mathcal{D}$, whereas relaxed NLP (RNLP) relaxes the feasible region by defining $G_j(z) \geq 0, H_j(z) \geq 0$ for all $j \in \mathcal{D}$. These NLPs provide a foundation to define constraint qualifications for MPECs and strong stationarity.

Figure 12.1 depicts the feasible regions of the NLPs associated with a complementarity constraint $0 \leq G_j(z) \perp H_j(z) \geq 0$ for $j \in \mathcal{D}$. One can easily verify that

$$\mathcal{T}_{\mathrm{TNLP}}(z) \subseteq \mathcal{T}_{\mathrm{MPEC}}(z) \subseteq \mathcal{T}_{\mathrm{RNLP}}(z). \tag{12.2.6}$$

When $\mathcal{D} = \emptyset$, the equality holds throughout (12.2.6), and *lower-level strict complementarity* [104] holds. From (12.2.6), if z^* is a local minimizer of RNLP(z^*), then z^* is a local minimizer of the MPEC, but not vice versa [108].

Algebraically, B-stationarity is defined by using a linear program with equilibrium constraints (LPEC), an MPEC with all the functions being linear, over a linearized cone. For a feasible z^*, if $d = 0$ is a solution to the following LPEC, then z^* is called a B-stationary (or an algebraic B-stationary) point:

$$\begin{aligned} \underset{z}{\text{minimize}} \quad & \nabla f(z^*)^T d \\ \text{subject to} \quad & d \in \mathcal{T}_{\text{MPEC}}^{\text{lin}}(z^*), \end{aligned} \tag{12.2.7}$$

where

$$\begin{aligned} \mathcal{T}_{\text{MPEC}}^{\text{lin}}(z^*) := \big\{ d \mid \quad & \nabla c_i(z^*)^T d \geq 0, \ \forall i : c_i(z^*) = 0, \\ & \nabla G_j(z^*)^T d = 0, \ \forall j : G_j(z^*) = 0, H_j(z^*) > 0, \\ & \nabla H_j(z^*)^T d = 0, \ \forall j : G_j(z^*) > 0, H_j(z^*) = 0, \\ & 0 \leq \nabla G_j(z^*)^T d \perp \nabla H_j(z^*)^T d \geq 0, \ \forall j \in \mathcal{D}(z^*) \big\}. \end{aligned} \tag{12.2.8}$$

As with geometric B-stationarity, B-stationarity is difficult to check because it involves the solution of an LPEC that may require the solution of an exponential number of linear programs, unless *all* these linear programs share a common multiplier vector. Such a common multiplier vector exists if MPEC-LICQ holds, which we define in Sect. 12.2.2.

Since $\mathcal{T}_{\text{MPEC}}(z^*) \subseteq \mathcal{T}_{\text{MPEC}}^{\text{lin}}(z^*)$ [42], B-stationarity implies geometric B-stationarity, but not vice versa. A similar equivalence (12.2.5) and inclusion relationship (12.2.6) hold between the linearized cones of its associated NLPs [42].

The next important stationarity concept is strong stationarity.

Definition 12.2.1 A point z^* is called *strongly stationary* if there exist multipliers satisfying the stationarity of the RNLP. △

Note that if $\hat{\nu}_{1j}$ and $\hat{\nu}_{2j}$ are multipliers for $G_j(z^*)$ and $H_j(z^*)$, respectively, then $\hat{\nu}_{1j}, \hat{\nu}_{2j} \geq 0$ for $j \in \mathcal{D}(z^*)$. Strong stationarity implies B-stationarity due to (12.2.6) and the inclusion relationship between the tangent and linearized cones. Equivalence of stationarity between (12.2.1) and the RNLP was shown in [5, 48, 103].

Other stationarity conditions differ from strong stationarity in that the conditions on the sign of the multipliers, $\hat{\nu}_{1j}$ and $\hat{\nu}_{2j}$, are relaxed when $G_j(z^*) = H_j(z^*) = 0$. One can easily associate these stationarity concepts with the stationarity conditions for one of its associated NLPs. If we define the biactive set $\mathcal{D}^* := \mathcal{D}(z^*)$, we can

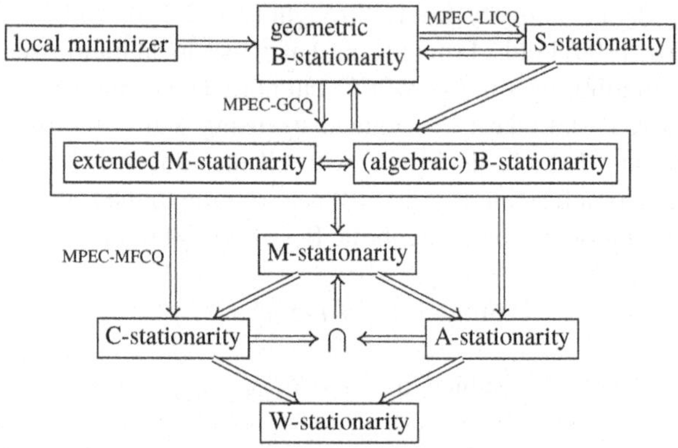

Fig. 12.2 Relationships between stationarity concepts where M-stationarity is the intersection of C- and A-stationarity

state these "stationarity" concepts by replacing the sign of the multipliers in the set $\mathcal{D}^*$ as follows:

- z^* is called weak stationary if there are no sign restrictions on $\hat{\nu}_{1j}$ and $\hat{\nu}_{2j}, \forall j \in \mathcal{D}^*$.
- z^* is called A-stationary if $\hat{\nu}_{1j} \geq 0$ or $\hat{\nu}_{2j} \geq 0, \forall j \in \mathcal{D}^*$.
- z^* is called C-stationary if $\hat{\nu}_{1j}\hat{\nu}_{2j} \geq 0, \forall j \in \mathcal{D}^*$.
- z^* is called M-stationary if either $\hat{\nu}_{1j} > 0$ and $\hat{\nu}_{2j} > 0$ or $\hat{\nu}_{1j}\hat{\nu}_{2j} = 0, \forall j \in \mathcal{D}^*$.

Of particular note is M-stationarity, which implies that if z^* is M-stationary, then z^* satisfies the first-order optimality conditions for at least one of the nonlinear programs (12.2.3). This condition seems to be the best one can achieve without exploring the combinatorial number of possible partitions for the biactive constraints. The extended M-stationarity in [56] extends the notion of M-stationarity in a way that it holds at z^* when M-stationarity is satisfied for each critical direction d defined by $d \in \mathcal{T}^{\text{lin}}_{\text{MPEC}}(z^*)$ and $\nabla f(z^*)^T d \leq 0$, possibly with different multipliers. Thus, if z^* is extended M-stationary, then z^* satisfies the first-order optimality conditions for all the associated NLPs. Hence it is also B-stationary. When a constraint qualification is satisfied, B-stationarity implies extended M-stationarity. Figure 12.2 summarizes relationships between stationarity concepts.

As in the stationarity concepts, the second-order sufficient conditions (SOSCs) of an MPEC are defined in terms of the associated NLPs. In particular, SOSC is defined at a strongly stationary point so its multipliers work for all the associated NLPs. Depending on the underlying critical cone, we have two different SOSCs:

RNLP-SOSC and MPEC-SOSC. In Definition 12.2.2, the Lagrangian $\mathcal{L}$ and the critical cone for the RNLP are defined as follows:

$$\begin{aligned}
\mathcal{L}(z, \lambda, \hat{\nu}_1, \hat{\nu}_2) := f(z) \quad &- \sum_{i \in \mathcal{E} \cup \mathcal{I}} c_i(z)\lambda_i - \sum_{j=1}^{p} G_j(z)\hat{\nu}_{1j} - \sum_{j=1}^{p} H_j(z)\hat{\nu}_{2j}, \\
C_{\mathrm{RNLP}}(z^*) := \{d \mid \quad &\nabla c_i(z^*)^T d = 0, \ \forall i : \lambda_i^* > 0, \\
&\nabla c_i(z^*)^T d \geq 0, \ \forall i : c_i(z^*) = 0, \lambda_i^* = 0, \\
&\nabla G_j(z^*)^T d = 0, \ \forall j : G_j(z^*) = 0, H_j(z^*) > 0, \\
&\nabla H_j(z^*)^T d = 0, \ \forall j : G_j(z^*) > 0, H_j(z^*) = 0, \\
&\nabla G_j(z^*)^T d = 0, \ \forall j \in D(z^*), \hat{\nu}_{1j}^* > 0, \\
&\nabla G_j(z^*)^T d \geq 0, \ \forall j \in D(z^*), \hat{\nu}_{1j}^* = 0, \\
&\nabla H_j(z^*)^T d = 0, \ \forall j \in D(z^*), \hat{\nu}_{2j}^* > 0, \\
&\nabla H_j(z^*)^T d \geq 0, \ \forall j \in D(z^*), \hat{\nu}_{2j}^* = 0 \big\}.
\end{aligned} \tag{12.2.9}$$

Definition 12.2.2 RNLP-SOSC is satisfied at a strongly stationary point z^* with its multipliers $(\lambda^*, \hat{\nu}_1^*, \hat{\nu}_2^*)$ if there exists a constant $\omega > 0$ such that

$$d^T \nabla_{zz}^2 \mathcal{L}(z^*, \lambda^*, \hat{\nu}_1^*, \hat{\nu}_2^*) d \geq \omega \text{ for all } d \neq 0, d \in C_{\mathrm{RNLP}}(z^*).$$

If the conditions hold for $C_{\mathrm{MPEC}}(z^*)$ instead of $C_{\mathrm{RNLP}}(z^*)$, where $C_{\mathrm{MPEC}}(z^*) := C_{\mathrm{RNLP}}(z^*) \cap \{d \mid \min(\nabla G_j(z^*)^T d, \nabla H_j(z^*)^T d) = 0, \forall j \in \mathcal{D}(z^*), \hat{\nu}_{1j}^* = \hat{\nu}_{2j}^* = 0\}$, then we say that MPEC-SOSC is satisfied. △

We note that $d \in C_{\mathrm{MPEC}}(z^*)$ if and only if it is a critical direction of any of the $\mathrm{NLP}_{\mathcal{D}_{01}, \mathcal{D}_{10}}(z^*)$'s at z^* [104, 108]. Thus, MPEC-SOSC holds at z^* if and only if SOSC is satisfied at z^* for all of its associated NLPs, which leads to the conclusion that z^* is a strict local minimizer of the MPEC.

In a similar fashion, we define strong SOSC (SSOSC) for RNLPs. Using RNLP-SSOSC, we can obtain stability results of MPECs by applying the stability theory of nonlinear programs [77, 105] to the RNLP. The stability property can be used to show the uniqueness of a solution of regularized NLPs to solve the MPEC as in Sect. 12.3. A critical cone $C_{\mathrm{RNLP}}^{\mathrm{S}}(z^*)$ is used instead of $C_{\mathrm{RNLP}}(z^*)$, which expands it by removing the feasible directions for inequalities associated with zero multipliers:

$$\begin{aligned}
C_{\mathrm{RNLP}}^{\mathrm{S}}(z^*) := \big\{d \mid \quad &\nabla c_i(z^*)^T d = 0, \ \forall i : \lambda_i^* > 0, \\
&\nabla G_j(z^*)^T d = 0, \ \forall j : \hat{\nu}_{1j}^* \neq 0, \\
&\nabla H_j(z^*)^T d = 0, \ \forall j : \hat{\nu}_{2j}^* \neq 0 \big\}.
\end{aligned} \tag{12.2.10}$$

Definition 12.2.3 RNLP-SSOSC is satisfied at a strongly stationary point z^* with its multipliers $(\lambda^*, \hat{\nu}_1^*, \hat{\nu}_2^*)$ if there exists a constant $\omega > 0$ such that

$$d^T \nabla_{zz}^2 \mathcal{L}(z^*, \lambda^*, \hat{\nu}_1^*, \hat{\nu}_2^*) d \geq \omega \text{ for all } d \neq 0, d \in C_{\mathrm{RNLP}}^{\mathrm{S}}(z^*).$$

△

By definition, we have an inclusion relationship between SOSCs: RNLP-SSOSC ⇒ RNLP-SOSC ⇒ MPEC-SOSC. Reverse directions do not hold in general; an example was presented in [108] showing that MPEC-SOSC ⇏ RNLP-SOSC.

12.2.2 Constraint Qualifications and Regularity Assumptions

Constraint qualifications (CQs) for MPECs guarantee the existence of multipliers for the stationarity conditions to hold. They are an extension of the corresponding CQs for the tightened NLP. Among many CQs in the literature, we have selected five. Three of them are frequently assumed in proving stationarity properties of a limit point of algorithms for solving MPECs described in Sect. 12.3. The remaining two are conceptual and much weaker than the first ones, but they can be used to prove that every local minimizer is at least M-stationary.

We start with the first three CQs. Because of the space limit, we do not specify the definition of the CQs in the context of NLPs. In Definition 12.2.4, LICQ denotes the linear independence constraint qualification, and CPLD represents constant positive linear dependence [102].

Definition 12.2.4 An MPEC (12.1.2) is said to satisfy *MPEC-LICQ (MPEC-MFCQ, MPEC-CPLD)* at a feasible point z if the corresponding tightened NLP satisfies LICQ (MFCQ, CPLD) at z. △

We note that MPEC-LICQ holds at a feasible point z if and only if all of its associated NLPs satisfy LICQ at z. This statement is true because active constraints at any feasible point do not change between the associated NLPs. For example, MPEC-LICQ holds if and only if LICQ holds for the relaxed NLP.

Under MPEC-LICQ, B-stationarity is equivalent to strong stationarity [108] since the multipliers are unique, thus having the same nonnegative signs for a biactive set among the associated NLPs.

The above CQs provide sufficient conditions for the following much weaker CQs to hold. These CQs are in general difficult to verify but provide insight into what stationarity we can expect for a local minimizer. In the following definition, ACQ and GCQ denote Abadie and Guignard constraint qualification, respectively, and for a cone C its polar is defined by $C^\circ := \{y \mid \langle y, x\rangle \leq 0, \forall x \in C\}$.

Definition 12.2.5 The MPEC (12.1.3) is said to satisfy *MPEC-ACQ (MPEC-GCQ)* at a feasible point z if $\mathcal{T}_{\mathrm{MPEC}}(z) = \mathcal{T}_{\mathrm{MPEC}}^{\mathrm{lin}}(z)$ $(\mathcal{T}_{\mathrm{MPEC}}(z)^\circ = \mathcal{T}_{\mathrm{MPEC}}^{\mathrm{lin}}(z)^\circ)$. △

Although MPEC-ACQ or MPEC-GCQ holds, we cannot directly apply the Farkas lemma to show the existence of multipliers because the linearized cone may not be a polyhedral convex set. However, M-stationarity is shown to hold for each local minimizer under MPEC-GCQ by using the limiting normal cones and separating the complementarity constraints from other constraints [43, 56].

Local preservation of constraint qualifications has been studied in [20], which shows that for many MPEC constraint qualifications, if z^* satisfies an MPEC constraint qualification, then all feasible points in a neighborhood of z^* also satisfy that MPEC constraint qualification.

As with standard nonlinear programming, a similar implication holds between CQs for MPECs: MPEC-LICQ $\Rightarrow$ MPEC-MFCQ $\Rightarrow$ MPEC-CPLD $\Rightarrow$ MPEC-ACQ $\Rightarrow$ MPEC-GCQ.

12.3 Solution Approaches

In this section, we classify and outline solution methods for MPECs and summarize their convergence results. Solution methods for MPECs such as (12.1.3) can be categorized into three broad classes:

1. **Nonlinear programming methods** that rewrite the complementarity constraint in (12.1.3) as a nonlinear set of inequalities, such as

$$G(z) \geq 0, \;\; H(z) \geq 0, \;\; \text{and} \;\; G(z)^T H(z) \leq 0, \tag{12.3.1}$$

   and then apply NLP techniques; see, for example, [26, 45, 74, 87, 103, 104, 109]. Unfortunately, convergence properties are generally weak for this class of methods, typically resulting in C-stationary limits unless strong assumptions are made on the limit point.
2. **Combinatorial methods** that tackle the combinatorial nature of the disjunctive complementarity constraint directly. Popular approaches include pivoting methods [40], branch-and-cut methods [8, 9, 11], and active-set methods [57, 84, 88]. This class of methods has the strongest convergence properties.
3. **Implicit methods** that assume that the complementarity constraint has a unique solution for every upper-level choice of variables. For example, if we assume that the lower-level problem has a unique solution, then we can express the lower-level variables $y = y(x)$ in (12.1.2), and use the KKT conditions to eliminate $(y(x), \lambda(x))$, resulting in a reduced nonsmooth problem

$$\underset{x}{\text{minimize}} \;\; f(x, y(x)) \quad \text{subject to} \;\; c(x, y(x)) \geq 0$$

   that can be solved by using methods for nonsmooth optimization. See the monograph [98] and the references [63, 118] for more details.

In this survey, we do not discuss implicit methods further and instead concentrate on the first two classes of methods.

12.3.1 NLP Methods for MPECs

NLP methods are attractive because they allow us to leverage powerful numerical solvers. Unfortunately, the system (12.3.1) violates a standard stability assumption for NLP at any feasible point. In [108], the authors show that (12.3.1) violates MFCQ at any feasible point. Other nonlinear reformulations of the complementarity constraint (12.3.1) are possible. In [46, 83], the authors experiment with a range of reformulations using different nonlinear complementarity functions, but they observe that the formulation (12.3.1) is the most efficient format in the context of sequential quadratic programming (SQP) methods. We also note that reformulations that use nonlinear *equality constraints* such as $G(z)^T H(z) = 0$ are not as efficient, because the redundant lower bound can slow convergence.

Because traditional analyses of NLP solvers rely heavily on a constraint qualification, it is remarkable that convergence results can still be proved. Here, we briefly review how SQP methods can be shown to converge quadratically for MPECs, provided that we reformulate the nonlinear complementarity constraint (12.3.1) using slacks as

$$s_1 = G(z),\ s_2 = H(z) \geq 0,\ s_1 \geq 0,\ s_2 \geq 0, \text{ and } s_1^T s_2 \leq 0. \tag{12.3.2}$$

One can show that close to a strongly stationary point that satisfies MPEC-LICQ and RNLP-SOSC, an SQP method applied to this reformulation converges quadratically to a strongly stationary point, provided that all QP approximations remain feasible. The authors [48] show that these assumptions are difficult to relax, and they give a counterexample that shows that the slacks are necessary for convergence. One undesirable assumption is that all QP approximations must remain feasible, but one can show that this assumption holds if the lower-level problem satisfies a certain mixed-P property; see, for example, [90]. In practice [45], a simple heuristic is implemented that relaxes the linearization of the complementarity constraint.

In general, the failure of MFCQ motivates the use of regularizations within NLP methods, such as penalization or relaxation of the complementarity constraint, and these two classes of methods are discussed next.

12.3.1.1 Relaxation-Based NLP Methods

An attractive way to solve MPECs is to relax the complementarity constraint in (12.3.1) by using a positive relaxation parameter, $t > 0$:

$$\begin{aligned} \underset{z,u,v}{\text{minimize}} \quad & f(z) \\ \text{subject to} \quad & c_i(z) \geq 0 \\ & G_j(z) \geq 0,\ H_j(z) \geq 0,\ G_j(z) H_j(z) \leq t \quad \forall j = 1, \ldots, p. \end{aligned} \tag{12.3.3}$$

This NLP then generally satisfies MFCQ for any $t > 0$, and the main idea is to (approximately) solve these NLPs for a sequence of regularization parameters, $t_k \searrow 0$. This approach has been studied from a theoretical perspective in [104, 109]. Under the unpractical assumption that each regularized NLP is solved exactly, the authors show convergence to C-stationary points.

More recently, interior-point methods based on the relaxation (12.3.3) have been proposed [87, 103] in which the parameter t is chosen to be proportional to the barrier parameter and is updated at the end of each (approximate) barrier subproblem solve. Numerical difficulties may arise when the relaxation parameter becomes small, because the limiting feasible set of the regularized NLP (12.3.3) has no strict interior.

An alternative regularization scheme is proposed in [74], based on the reformulation of (12.3.1) as

$$G_j(z) \geq 0, \; H_j(z) \geq 0, \; \text{ and } \phi(G_j(z), H_j(z), t) \leq 0, \tag{12.3.4}$$

where

$$\phi(a, b, t) = \begin{cases} ab & \text{if } a + b \geq t \\ -\frac{1}{2}(a^2 + b^2) & \text{if } a + b < t. \end{cases}$$

In [74], convergence properties are proved when the nonlinear program is solved inexactly, as one would find in practice. The authors show that typically, but not always, methods converge to M-stationary points with exact NLP solves but converge to only C-stationary points with inexact NLP solves. Other regularization methods have been proposed, and a good theoretical and numerical comparison can be found in [64], with later methods in [73]. Under suitable conditions, these methods can be shown to find M- or C-stationary points for the MPEC when the nonlinear program is solved exactly.

An interesting two-sided relaxation scheme is proposed in [26]. The scheme is motivated by the observation that the complementarity constraint in the slack formulation (12.3.2) can be interpreted as a pair of bounds, $s_{ij} \geq 0$, or, $s_{ij} \leq 0$, and strong stationarity requires a multiplier for at most one of these bounds in each pair. The authors propose a strictly feasible two-sided relaxation of the form

$$s_{1j} \geq -\delta_{1j}, \; s_{2j} \geq -\delta_{2j}, \; s_{1j}s_{2j} \leq \delta_{cj}.$$

By using the multiplier information from inexact subproblem solves, the authors decide which parameter δ_1, δ_2, or δ_c needs to be driven to zero for each complementarity pair. The authors propose an interior-point algorithm and show convergence to C-stationary points, local superlinear convergence, and identification of the optimal active set under MPEC-LICQ and RNLP-SOSC conditions.

12.3.1.2 Penalization-Based NLP Methods

An alternative regularization approach is based on penalty methods. Both exact penalty functions and augmented Lagrangian methods have been proposed for solving MPECs. The penalty approach for MPECs dates back to [41] and penalizes the complementarity constraint in the objective function, after introducing slacks:

$$\begin{aligned} \underset{z,u,v}{\text{minimize}} \quad & f(z) + \pi s_1^T s_2 \\ \text{subject to} \quad & c_i(z) \geq 0 \qquad \forall i = 1, \ldots, m \\ & G_j(z) - s_{1j} = 0,\ H_j(z) - s_{2j} = 0 \quad \forall j = 1, \ldots, p \\ & s_2 \geq 0,\ s_1 \geq 0 \end{aligned} \tag{12.3.5}$$

for positive penalty parameter π. If π is chosen large enough, the solution of the MPEC can be recast as the minimization of a single penalty function problem. The appropriate value of π is unknown in advance, however, and must be estimated during the course of the minimization. In general, if limit points are only B-stationary and not strongly stationary, then the penalty parameter, π, must diverge to infinity, and this penalization is not exact.

This approach was first studied by [3] in the context of active-set SQP methods, although it had been used before to solve engineering problems [41]. It has been adopted as a heuristic to solve MPECs with interior-point methods in LOQO by [15], who present very good numerical results. A more general class of exact penalty functions was analyzed by [66], who derive global convergence results for a sequence of penalty problems that are solved exactly, while the author in [4] derives similar global results in the context of inexact subproblem solves.

In [85], the authors study a general interior-point method for solving (12.3.5). Each barrier subproblem is solved approximately to a tolerance ϵ_k that is related to the barrier parameter, μ_k. Under strict complementarity, MPEC-LICQ, and RNLP-SOSC, the authors show convergence to C-stationary points that are strongly stationary if the product of MPEC-multipliers and primal variables remains bounded. Superlinear convergence can be shown, provided that the tolerance and barrier parameter satisfy the conditions

$$\frac{(\epsilon + \mu)^2}{\epsilon} \to 0, \quad \frac{(\epsilon + \mu)^2}{\mu} \to 0, \text{ and } \frac{\mu}{\epsilon} \to 0.$$

A related approach to penalty functions is the elastic mode that is implemented in SNOPT [58, 59]. The elastic approach [6] combines both a relaxation of the constraints and penalization of the complementarity conditions to solve a sequence of optimization problems

$$\begin{array}{lll} \underset{z,u,v,t}{\text{minimize}} & f(z) + \pi(t + s_1^T s_2) & \\ \text{subject to} & c_i(z) \geq -t & \forall i = 1, \ldots .m \\ & -t \leq G_j(z) - s_{1j} \leq t, -t \leq H_j(z) - s_{2j} \leq t & \forall j = 1, \ldots, p \\ & s_1 \geq 0, s_2 \geq 0, 0 \leq t \leq \bar{t}, & \end{array}$$

where t is a variable with upper bound $\bar{t}$ that relaxes some of the constraints and π is the penalty term on both the constraint relaxation and the complementarity conditions. This problem can be interpreted as a mixture of ℓ_∞ and ℓ_1 penalties. The problem is solved for a sequence of π that may need to converge to infinity. Under suitable conditions such as MPEC-LICQ, the method is shown to converge to M-stationary points, even with inexact NLP solves.

An alternative to the ℓ_1 penalty-function approaches presented above are augmented Lagrangian approaches, which have been adapted to MPECs [51, 69]. These approaches are related to stabilized SQP methods and work by imposing an artificial upper bound on the multiplier. Under MPEC-LICQ one can show that if the sequence of multipliers has a bounded subsequence, then the limit point is strongly stationary. Otherwise, it is only C-stationary.

Remark 12.3.1 NLP methods for MPECs currently provide the most practical approach to solving MPECs, and form the basis for the most efficient and robust software packages; see Sect. 12.4. In general, however, NLP methods may converge only slowly to a solution or fail to converge if the limit point is not strongly stationary. △

The result in [74] seems to show that the best we can hope for from NLP solvers is convergence to C-stationary points. Unfortunately, these points may include limit points at which first-order strict descent directions exist, and which do not correspond to stationary points in the classical sense. To the best of our knowledge, the only way around this issue is to tackle the combinatorial nature of the complementarity constraint directly, and in the next section we describe methods that do so.

12.3.2 *Combinatorial Methods for MPECs*

Despite the success of NLP solvers in tackling a wide range of MPECs, for some classes of problems these solvers still fail. In particular, problems whose stationary points are B-stationary but not strongly stationary can cause NLP solvers to either fail or exhibit slow convergence. Unfortunately, this behavior also occurs when other pathological situations occur (such as convergence to C- or M-stationary points that are not strongly stationary), and it is not easily diagnosed or remedied.

This observation motivates the development of more robust methods for MPECs that guarantee convergence to B-stationary points. Methods with this property must resolve the combinatorial complexity of the complementarity constraint, and we discuss these methods here.

Specialized methods for solving linear programs with linear equilibrium constraints (LPECs), in which all the functions, $f(z), c(z), G(z)$, and $H(z)$, are linear, include methods based on disjunction for computing global solutions [11, 60, 67, 68], pivot-based methods for computing B-stationary points [40], and penalty methods based on a difference of convex functions formulation [70]. Global optimization methods for problems with a convex quadratic objective function and linear complementarity constraints are found in [8]. Algorithms for problems with a nonlinear objective function and linear complementarity constraints include combinatorial [9], active-set [52], sequential quadratic programming [88], and sequential linear programming [57] methods.

One early general method for obtaining B-stationary points is the branch-and-bound method proposed in [11]. It starts by solving a relaxation of (12.1.3) obtained by relaxing the complementarity between $G(z)$ and $H(z)$. If the solution satisfies $G(z)^T H(z) = 0$, then it is a B-stationary point. Otherwise, there exists an index j such that $G_j(z)H_j(z) > 0$, and we branch on this disjunction by creating two new child problems that set $G_j(z) = 0$ and $H_j(z) = 0$, respectively. This process is then repeated, creating a branch-and-bound tree. A branch-and-cut approach is described in [67] for solving LPECs. The algorithm is based on equivalent mixed-integer reformulations and the application of Benders decomposition [14] with linear programming (LP) relaxations. In contrast, the authors in [40] generalize LP pivoting techniques to locally solve LPECs. The authors prove convergence to a B-stationary point, but unlike [67] do not obtain globally optimal solutions.

The SQPEC approach extends SQP methods by taking special care of the complementarity constraint [110]. This method minimizes a quadratic approximation of the Lagrangian subject to a linearized feasible set and a linearized form of the complementarity constraint. Unfortunately, this method can converge to M-stationary points, rather than the desired B-stationary points, as the following counterexample shows [84]:

$$\underset{x,y}{\text{minimize}}\ (x-1)^2 + y^3 + y^2 \quad \text{subject to } 0 \le x \perp y \ge 0. \tag{12.3.6}$$

Starting from $(x_0, y_0) = (0, t)$ for $0 < t < 1$, SQPEC generates iterates

$$(x^{(k+1)}, y^{(k+1)}) = \left(0, \frac{3y^{(k)^2}}{6y^{(k)} + 2}\right)$$

that converge quadratically to the M-stationary point $(0, 0)$, at which we can easily find a descent direction $(1, 0)$ and which is hence not B-stationary.

An alternative class of methods to SQPEC methods that provide convergence to B-stationary points is extensions of SLQP methods to MPECs; see, for example,

[16, 17, 21, 47]. The method is motivated by considering the linearized tangent cone, $\mathcal{T}^{\text{lin}}_{\text{MPEC}}(z^*)$ in (12.2.8) as a direction-finding problem. This method solves a sequence of LPECs inside a trust region [84] with radius Δ around the current point z:

$$\text{LPEC}(z, \Delta) \begin{cases} \underset{d}{\text{minimize}} \ \nabla f(z)^T d \\ \text{subject to } c(z) + \nabla c(z)^T d \geq 0, \\ \qquad\qquad 0 \leq G(z) + \nabla G(z)^T d \ \perp \ H(z) + \nabla H(z)^T d \geq 0, \\ \qquad\qquad \|d\| \leq \Delta. \end{cases}$$

The LPEC need be solved only locally, and the pivoting method of [40] is a practical and efficient method of solving the LPEC. Given a solution $d \neq 0$, we find the active sets that are predicted by the LPEC,

$$\mathcal{A}_c(z+d) := \left\{ i : c_i(z) + \nabla c_i(z)^T d = 0 \right\} \tag{12.3.7}$$

$$\mathcal{A}_G(z+d) := \left\{ j : G_j(z) + \nabla G_j(z)^T d = 0 \right\} \tag{12.3.8}$$

$$\mathcal{A}_H(z+d) := \left\{ j : H_j(z) + \nabla H_j(z)^T d = 0 \right\}, \tag{12.3.9}$$

and solve the corresponding equality-constrained quadratic program (EQP):

$$\text{EQP}(z+d) \begin{cases} \underset{d}{\text{minimize}} \ \nabla f(z)^T d + \frac{1}{2} d^T \nabla^2 \mathcal{L}(z) d \\ \text{subject to} \qquad c_i(z) + a_i(z)^T d = 0, \ \forall i \in \mathcal{A}_c(z+d) \\ \qquad\qquad G_j(z) + \nabla G_j(z)^T d = 0, \ \forall j \in \mathcal{A}_G(z+d) \\ \qquad\qquad H_j(z) + \nabla H_j(z)^T d = 0, \ \forall j \in \mathcal{A}_H(z+d). \end{cases}$$

We note that EQP($z + d$) can be solved as a linear system of equations. The goal of the EQP step is to provide fast convergence near a local minimum. Global convergence is promoted through the use of a three-dimensional filter that separates the complementarity error and the nonlinear infeasibility.

The SLPEC-EQP method has an important advantage over NLP reformulations: the solution of the LPEC matches exactly the definition of B-stationarity, and we therefore always work with the correct tangent cone. In particular, if the zero vector solves the LPEC, then we can conclude that the current point is B-stationary. To our knowledge, this algorithm is the only one that guarantees global convergence to B-stationary points.

12.3.3 Globally Optimal Methods for MPECs

The methods discussed above typically guarantee convergence to a stationary point at best. They do not guarantee convergence to a global minimum, even if the problem functions, f, c, G, and H are convex, because the feasible set of even the simplest complementarity constraint, $0 \leq x_1 \perp x_2 \geq 0$, is nonconvex.

Here, we briefly summarize existing results for obtaining global solutions to certain classes of MPECs. We limit our discussion to cases where the lower-level follower's problems is convex. One approach to obtaining global solutions would be to simply apply global optimization techniques to the nonlinear program (12.2.1). However, state-of-the-art solvers such as BARON [107, 115] or Couenne [12] require finite bounds on all variables to construct valid underestimators, and it is not clear that such bounds are easy to obtain on the multipliers. If we assume that such bounds exist, then we can apply BARON and Couenne. Unfortunately, the effectiveness of these solvers is limited to a few dozen variables at most.

If the MPEC has some special structure, then we can employ formulations and methods that guarantee convergence to a global minimum. One example are LPECs (or QPECs), when $f(z)$ is a linear (or a convex quadratic function), $c(z) = Az - b$ is an affine function, and the complementarity constraint is affine, i.e. $0 \leq G(z) = Mz - c \perp Nz - d = H(z) \geq 0$. Then it is possible to model the complementarity condition using binary variables, $y \in \{0, 1\}^p$, as

$$\Theta y \geq Mz - c \geq 0 \text{ and } (1 - \Theta)y \geq Nz - d \geq 0,$$

where $\Theta > 0$ is an upper bound on $Mz - c$ and $Nz - d$ that can be computed by solving an LP. Unfortunately, the resulting mixed-integer linear program often has a very weak continuous relaxation, resulting in massive branch-and-cut trees. Moreover, numerical issues can cause both $Mz - c$ and $Nz - d$ to be positive. In [8, 67], the authors extend a logical Benders decomposition technique [65] to this class of MPECs that avoids these pitfalls. The approach is based on a minimax principle and derives valid inequalities from LP relaxations. These approaches easily generalize to MPECs with more general (convex) $f(x)$ and $c(x)$.

12.4 Software Environments

Several modeling languages and solvers support bilevel optimization problems and MPECs. Table 12.1 presents a list of them. GAMS/EMP [53] and Pyomo [62] directly support bilevel optimization problems—they provide constructs that allow users to formulate bilevel optimization problems in their natural algebraic form (12.1.1) without applying any reformulations. GAMS introduced an extended mathematical programming (EMP) framework in which users can annotate the variables, functions, and constraints of a model to specify to which level, either upper or lower, they belong. In this case, a single monolithic model is defined, and annotations are provided via a separate text file, called the `empinfo` file. Pyomo takes a different approach by requiring users to define the lower-level problem explicitly as a model using their `Submodel()` construct and to link it to the upper-level model. In this way, not only bilevel optimization problems, but also multilevel

Table 12.1 Modeling languages and solvers that support bilevel or MPECs

Modeling language		Solver	
Bilevel	MPECs	Bilevel	MPECs
GAMS/EMP [53]	AMPL [50]	None	FilterMPEC [44]
Pyomo [62]	AIMMS [106]		KNITRO [7]
	GAMS [55]		NLPEC [55]
	Julia/JuMP [79]		Pyomo [61]
	Pyomo [61]		

problems [92] can be specified by defining submodels recursively and linking them together.

In contrast to bilevel optimization problems, most modeling languages support MPECs by providing a dedicated construct to take complementarity constraints in their natural form. AMPL [50] and Julia [79] provide `complements` and `@complements` keywords, respectively, GAMS [55] has a dot `.` construct, AIMMS [106] defines `ComplementarityVariables`, and Pyomo [61] defines `Complementarity` along with a `complements` expression. All these constructs enable complementarity constraints to be written as first-class expressions so that users can seamlessly use them in their models.

Regarding solvers, to the best of our knowledge, no dedicated, robust, and large-scale solvers are available for bilevel optimization problems at this time. The aforementioned modeling languages for bilevel optimization problems transform these problems into a computationally tractable formulation as an MPEC or generalized disjunctive program and call the associated solvers. Bilevel-specific features, such as optimal value functions, are not exploited in these solution procedures.

Although there are some recent efforts [75, 76, 94] exploiting the value function to globally solve the bilevel problem (12.1.1) using a branch-and-bound scheme, even in the case when the lower-level problem is nonconvex, these methods require repeated global solutions of nonconvex subproblems to compute the lower bounds. This requirement makes these methods unsuitable for large-scale bilevel problems. In particular, the requirement for global optimality may not be relaxed easily as discussed in [75].

In the case of MPECs, a few solvers are available. Most reformulate the MPECs into nonlinear or mixed-integer programming problems by applying relaxation, penalization, or disjunction techniques to the complementarity constraints. FilterMPEC [44] and KNITRO [7] transform the given problem into an NLP and solve it using their own algorithms by taking special care with the complementarity constraints. NLPEC [55] and Pyomo [61] are metasolvers that apply a reformulation and invoke an NLP solver, rather than supplying their own NLP solver. In contrast to the NLP-based approaches, Pyomo additionally provides a disjunctive programming method that formulates the MPEC as a mixed-integer nonlinear program. We believe that a combination of modeling languages for bilevel programs and MPEC solvers is the most promising and viable approach for quickly prototyping and solving bilevel problems when the lower-level problem is convex.

A number of bilevel optimization problems and MPECs are available online. The GAMS/EMP library [54] contains examples of bilevel optimization problems written by using the EMP framework. GAMS also provides MPEC examples [38]. The MacMPEC collection [81] is a set of MPEC examples written in AMPL that are frequently used to test performance of MPEC algorithms. QPECgen [71] generates MPEC examples with quadratic objectives and affine variational inequality constraints for lower-level problems.

12.5 Extensions

Our focus in this survey is on the optimistic bilevel optimization problem where the lower-level optimization problem is convex for all upper-level decisions. This approach contrasts with the pessimistic bilevel optimization problem [34, 35, 89, 117] where the leader plans for the worst by choosing an element of the solution set leading to the worst outcome and results in a problem of the form

$$\begin{aligned} \underset{x,\theta}{\text{minimize}} \quad & \theta \\ \text{subject to} \quad & f(x,y) \leq \theta \quad \forall\, y \in Y(x) \\ & c(x,y) \geq 0 \quad \forall\, y \in Y(x), \end{aligned} \tag{12.5.1}$$

where $Y(x)$ is the solution set of the lower-level optimization problem parameterized by x. This formulation is consistent with robust optimization problems [13] with a complicated uncertainty set $Y(x)$. The two robustness constraints determine the worst objective function value and guarantee satisfaction of the joint feasibility constraint, respectively.

If the lower-level optimization problem is nonconvex, then the global solution of the bilevel optimization problem (12.1.1) may not even be a stationary point for the MPEC reformulation (12.1.2) because the MPEC reformulation is a relaxation of (12.1.1) in these cases; see [93, Example 1]. In the nonconvex case, special algorithms can be devised by a two-layer bounding scheme: bounding the optimal value function of the lower-level problem and computing lower and upper bounds of the upper-level problem, see [75, 76, 94]. These bounds need to be refined as the algorithm progresses, to ensure convergence to a global solution.

Many other extensions of bilevel optimization problems were not covered in this survey, including problems with lower-level second-order cone programs [18, 19], stochastic bilevel optimization [1, 22], discrete bilevel optimization with integer variables in both the upper- and lower-level decisions [36, 96], multiobjective bilevel optimization [25, 39, 91, 100], and multilevel optimization problems [80, 86, 92, 116]. Alternatives to the MPEC reformulations also exist that we did not cover in this survey; see [37, 49, 95, 112–114] for semi-infinite reformulations and [32, 120] for nonsmooth reformulations based on the value function.

Acknowledgements This work was supported by the U.S. Department of Energy, Office of Science, Office of Advanced Scientific Computing Research under Contract No. DE-AC02-06CH11357 at Argonne National Laboratory.

References

1. S.M. Alizadeh, P. Marcotte, G. Savard, Two-stage stochastic bilevel programming over a transportation network. Transp. Res. Part B: Methodol. **58**, 92–105 (2013)
2. G.B. Allende, G. Still, Solving bilevel programs with the KKT-approach. Math. Program. **138**(1), 309–332 (2013)
3. M. Anitescu, On solving mathematical programs with complementarity constraints as nonlinear programs. Preprint ANL/MCS-P864-1200, Mathematics and Computer Science Division, Argonne National Laboratory, Argonne, IL (2000)
4. M. Anitescu, Global convergence of an elastic mode approach for a class of mathematical programs with complementarity constraints. Preprint ANL/MCS-P1143–0404, Mathematics and Computer Science Division, Argonne National Laboratory, Argonne, IL (2004)
5. M. Anitescu, On using the elastic mode in nonlinear programming approaches to mathematical programs with complementarity constraints. SIAM J. Optim. **15**(4), 1203–1236 (2005)
6. M. Anitescu, P. Tseng, S.J. Wright, Elastic-mode algorithms for mathematical programs with equilibrium constraints: global convergence and stationarity properties. Math. Program. **110**(2), 337–371 (2007)
7. Artelys, Artelys KNITRO user's manual. Available at https://www.artelys.com/docs/knitro/index.html
8. L. Bai, J.E. Mitchell, J.-S. Pang, On convex quadratic programs with linear complementarity constraints. Comput. Optim. Appl. **54**(3), 517–554 (2013)
9. J.F. Bard, Convex two-level optimization. Math. Program. **40**(1), 15–27 (1988)
10. J.F. Bard, *Practical Bilevel Optimization: Algorithms and Applications*, vol. 30 (Kluwer Academic Publishers, Dordrecht, 1998)
11. J.F. Bard, J.T. Moore, A branch and bound algorithm for the bilevel programming problem. SIAM J. Sci. Stat. Comput. **11**(2), 281–292 (1990)
12. P. Belotti, J. Lee, L. Liberti, F. Margot, A. Wächter, Branching and bounds tightening techniques for non-convex MINLP. Optim. Methods Softw. **24**(4–5), 597–634 (2009)
13. A. Ben-Tal, L. El Ghaoui, A. Nemirovski, *Robust Optimization* (Princeton University Press, Princeton, 2009)
14. J.F. Benders, Partitioning procedures for solving mixed-variable programming problems. Numerische Mathematik **4**, 238–252 (1962)
15. H. Benson, A. Sen, D.F. Shanno, R. Vanderbei, Interior-point algorithms, penalty methods and equilibrium problems. Comput. Optim. Appl. **34**(2), 155–182 (2006)
16. R.H. Byrd, J. Nocedal, R.A. Waltz, Knitro: an integrated package for nonlinear optimization, in *Large-Scale Nonlinear Optimization* (Springer, New York, 2006), pp. 35–59
17. R.H. Byrd, J. Nocedal, R.A. Waltz, Steering exact penalty methods for nonlinear programming. Optim. Methods Softw. **23**(2), 197–213 (2008)
18. X. Chi, Z. Wan, Z. Hao, The models of bilevel programming with lower level second-order cone programs. J. Inequal. Appl. **2014**(1), 168 (2014)
19. X. Chi, Z. Wan, Z. Hao, Second order sufficient conditions for a class of bilevel programs with lower level second-order cone programming problem. J. Ind. Manage. Optim. **11**(4), 1111–1125 (2015)

20. N.H. Chieu, G.M. Lee, Constraint qualifications for mathematical programs with equilibrium constraints and their local preservation property. J. Optim. Theory Appl. **163**(3), 755–776 (2014)
21. C.M. Chin, R. Fletcher, On the global convergence of an SLP-filter algorithm that takes EQP steps. Math. Program. **96**(1), 161–177 (2003)
22. S. Christiansen, M. Patriksson, L. Wynter, Stochastic bilevel programming in structural optimization. Struct. Multidiscipl. Optim. **21**(5), 361–371 (2001)
23. F.H. Clarke, A new approach to Lagrange multipliers. Math. Oper. Res. **1**(2), 165–174 (1976)
24. B. Colson, P. Marcotte, G. Savard, Bilevel programming: a survey. 4OR **3**(2), 87–107 (2005)
25. F.F. Dedzo, L.P. Fotso, C.O. Pieume, Solution concepts and new optimality conditions in bilevel multiobjective programming. Appl. Math. **3**(10), 1395 (2012)
26. V. DeMiguel, M.P. Friedlander, F.J. Nogales, S. Scholtes, A two-sided relaxation scheme for mathematical programs with equilibrium constraints. SIAM J. Optim. **16**(2), 587–609 (2005)
27. S. Dempe, *Foundations of Bilevel Programming* (Springer Science & Business Media, New York, 2002)
28. S. Dempe, Annotated bibliography on bilevel programming and mathematical programs with equilibrium constraints. Optimization **52**, 333–359 (2003)
29. S. Dempe, *Bilevel Optimization: Theory, Algorithms and Applications* (TU Bergakademie Freiberg, Fakultät für Mathematik und Informatik, 2018)
30. S. Dempe, J. Dutta, Is bilevel programming a special case of a mathematical program with complementarity constraints? Math. Program. **131**(1–2), 37–48 (2012)
31. S. Dempe, S. Franke, On the solution of convex bilevel optimization problems. Comput. Optim. Appl. **63**(3), 685–703 (2016)
32. S. Dempe, A.B. Zemkoho, The bilevel programming problem: reformulations, constraint qualifications and optimality conditions. Math. Program. **138**(1), 447–473 (2013)
33. S. Dempe, A.B. Zemkoho, KKT reformulation and necessary conditions for optimality in nonsmooth bilevel optimization. SIAM J. Optim. **24**(4), 1639–1669 (2014)
34. S. Dempe, B.S. Mordukhovich, A.B. Zemkoho, Necessary optimality conditions in pessimistic bilevel programming. Optimization **63**(4), 505–533 (2014)
35. S. Dempe, G. Luo, S. Franke, *Pessimistic Bilevel Linear Optimization* (Technische Universität Bergakademie Freiberg, Fakultät für Mathematik und Informatik, 2016)
36. S. Dempe, F. Mefo Kue, P. Mehlitz, Optimality conditions for mixed discrete bilevel optimization problems. Optimization **67**(6), 737–756 (2018)
37. M. Diehl, B. Houska, O. Stein, P. Steuermann, A lifting method for generalized semi-infinite programs based on lower level Wolfe duality. Comput. Optim. Appl. **54**(1), 189–210 (2013)
38. S.P. Dirkse, MPEC world, 2001. Available at http://www.gamsworld.org/mpec/mpeclib.htm
39. G. Eichfelder, Multiobjective bilevel optimization. Math. Program. **123**(2), 419–449 (2010)
40. H.-R. Fang, S. Leyffer, T. Munson, A pivoting algorithm for linear programming with linear complementarity constraints. Optim. Methods Softw. **27**(1), 89–114 (2012)
41. M.C. Ferris, F. Tin-Loi, On the solution of a minimum weight elastoplastic problem involving displacement and complementarity constraints. Comput. Methods Appl. Mech. Eng. **174**, 107–120 (1999)
42. M.L. Flegel, C. Kanzow, Abadie-type constraint qualification for mathematical programs with equilibrium constraints. J. Optim. Theory Appl. **124**(3), 595–614 (2005)
43. M.L. Flegel, C. Kanzow, A direct proof of M-stationarity under MPEC-GCQ for mathematical programs with equilibrium constraints, in *Optimization with Multivalued Mappings: Theory, Applications, and Algorithms*, ed. by S. Dempe, V. Kalashnikov (Springer, New York, 2006), pp. 111–122
44. R. Fletcher, S. Leyffer, FilterMPEC. Available at https://neos-server.org/neos/solvers/cp:filterMPEC/AMPL.html
45. R. Fletcher, S. Leyffer, Numerical experience with solving MPECs as NLPs. Numerical Analysis Report NA/210, Department of Mathematics, University of Dundee, Dundee, 2002
46. R. Fletcher, S. Leyffer, Solving mathematical program with complementarity constraints as nonlinear programs. Optim. Methods Softw. **19**(1), 15–40 (2004)

47. R. Fletcher, E. Sainz de la Maza, Nonlinear programming and nonsmooth optimization by successive linear programming. Math. Program. **43**, 235–256 (1989)
48. R. Fletcher, S. Leyffer, D. Ralph, S. Scholtes, Local convergence of SQP methods for mathematical programs with equilibrium constraints. SIAM J. Optim. **17**(1), 259–286 (2006)
49. C. Floudas, O. Stein, The adaptive convexification algorithm: a feasible point method for semi-infinite programming. SIAM J. Optim. **18**(4), 1187–1208 (2008)
50. R. Fourer, D.M. Gay, B.W. Kernighan, AMPL: a modeling language for mathematical programming, 2003. Available at https://ampl.com/resources/the-ampl-book/chapter-downloads
51. M. Fukushima, G.-H. Lin, Smoothing methods for mathematical programs with equilibrium constraints, in *International Conference on Informatics Research for Development of Knowledge Society Infrastructure, 2004. ICKS 2004*. (IEEE, New York, 2004), pp. 206–213
52. M. Fukushima, P. Tseng, An implementable active-set algorithm for computing a B-stationary point of the mathematical program with linear complementarity constraints. SIAM J. Optim. **12**, 724–739 (2002)
53. GAMS Development Corporation, Bilevel programming using GAMS/EMP. Available at https://www.gams.com/latest/docs/UG_EMP_Bilevel.html
54. GAMS Development Corporation, The GAMS EMP library. Available at https://www.gams.com/latest/emplib_ml/libhtml/index.html
55. GAMS Development Corporation, NLPEC (Nonlinear programming with equilibrium constraints). Available at https://www.gams.com/latest/docs/S_NLPEC.html
56. H. Gfrerer, Optimality conditions for disjunctive programs based on generalized differentiation with application to mathematical programs with equilibrium constraints. SIAM J. Optim. **24**(2), 898–931 (2014)
57. G. Giallombardo, D. Ralph, Multiplier convergence in trust-region methods with application to convergence of decomposition methods for MPECs. Math. Program. **112**(2), 335–369 (2008)
58. P.E. Gill, W. Murray, M.A. Saunders, SNOPT: an SQP algorithm for large-scale constrained optimization. SIAM J. Optim. **12**(4), 979–1006 (2002)
59. P.E. Gill, W. Murray, M.A. Saunders, SNOPT: an SQP algorithm for large-scale constrained optimization. SIAM Rev. **47**(1), 99–131 (2005)
60. P. Hansen, B. Jaumard, G. Savard, New branch-and-bound rules for linear bilevel programming. SIAM J. Sci. Stat. Comput. **13**(5), 1194–1217 (1992)
61. W. Hart, J.D. Siirola, Modeling mathematical programs with equilibrium constraints in Pyomo. Technical report, Sandia National Laboratories, Albuquerque, NM, 2015
62. W. Hart, R. Chen, J.D. Siirola, J.-P. Watson, Modeling bilevel programs in Pyomo. Technical report, Sandia National Laboratories, Albuquerque, NM, 2016
63. M. Hintermüller, T. Surowiec, A bundle-free implicit programming approach for a class of elliptic MPECs in function space. Math. Program. **160**(1), 271–305 (2016)
64. T. Hoheisel, C. Kanzow, A. Schwartz, Theoretical and numerical comparison of relaxation methods for mathematical programs with complementarity constraints. Math. Program. **137**(1), 257–288 (2013)
65. J.N. Hooker, G. Ottosson, Logic-based Benders decomposition. Math. Program. **96**(1), 33–60 (2003)
66. X. Hu, D. Ralph, Convergence of a penalty method for mathematical programming with complementarity constraints. J. Optim. Theory Appl. **123**(2), 365–390 (2004)
67. J. Hu, J. Mitchell, J.-S. Pang, K. Bennett, G. Kunapuli, On the global solution of linear programs with linear complementarity constraints. SIAM J. Optim. **19**(1), 445–471 (2008)
68. J. Hu, J. Mitchell, J.-S. Pang, B. Yu, On linear programs with linear complementarity constraints. J. Glob. Optim. **53**(1), 29–51 (2012)
69. A.F. Izmailov, M.V. Solodov, E.I. Uskov, Global convergence of augmented lagrangian methods applied to optimization problems with degenerate constraints, including problems with complementarity constraints. SIAM J. Optim. **22**(4), 1579–1606 (2012)

70. F. Jara-Moroni, J.-S. Pang, A. Wächter, A study of the difference-of-convex approach for solving linear programs with complementarity constraints. Math. Program. **169**(1), 221–254 (2018)
71. H. Jiang, D. Ralph, QPECgen, a MATLAB generator for mathematical programs with quadratic objectives and affine variational inequality constraints. Comput. Optim. Appl. **13**, 25–59 (1999)
72. K.L. Judd, C.-L. Su, Computation of moral-hazard problems, in *Society for Computational Economics, Computing in Economics and Finance*, 2005
73. C. Kanzow, A. Schwartz, Convergence properties of the inexact Lin-Fukushima relaxation method for mathematical programs with complementarity constraints. Comput. Optim. Appl. **59**(1), 249–262 (2014)
74. C. Kanzow, A. Schwartz, The price of inexactness: convergence properties of relaxation methods for mathematical programs with complementarity constraints revisited. Math. Oper. Res. **40**(2), 253–275 (2015)
75. P.-M. Kleniati, C.S. Adjiman, Branch-and-sandwich: a deterministic global optimization algorithm for optimistic bilevel programming problems. Part I: Theoretical development. J. Global Optim. **60**(3), 425–458 (2014)
76. P.-M. Kleniati, C.S. Adjiman, Branch-and-sandwich: a deterministic global optimization algorithm for optimistic bilevel programming problems. Part II: Convergence analysis and numerical results. J. Global Optim. **60**(3), 459–481 (2014)
77. M. Kojima, Strongly stable stationary solutions in nonlinear programming, in *Analysis and Computation of Fixed Points*, ed. by S.M. Robinson, pp. 93–138 (Academic, New York, 1980)
78. C.D. Kolstad, A review of the literature on bi-level mathematical programming. Technical report, Los Alamos National Laboratory Los Alamos, NM, 1985
79. C. Kwon, Complementarity package for Julia/JuMP. Available at https://github.com/chkwon/Complementarity.jl
80. K. Lachhwani, A. Dwivedi, Bi-level and multi-level programming problems: taxonomy of literature review and research issues. Arch. Comput. Methods Eng. **25**(4), 847–877 (2018)
81. S. Leyffer, MacMPEC: AMPL collection of MPECs, 2000. Available at https://wiki.mcs.anl.gov/leyffer/index.php/MacMPEC
82. S. Leyffer, Mathematical programs with complementarity constraints. SIAG/OPT Views-and-News **14**(1), 15–18 (2003)
83. S. Leyffer, Complementarity constraints as nonlinear equations: theory and numerical experience, in emphOptimization and Multivalued Mappings, ed. by S. Dempe, V. Kalashnikov (Springer, Berlin, 2006), pp. 169–208
84. S. Leyffer, T. Munson, A globally convergent filter method for MPECs. Preprint ANL/MCS-P1457-0907, Argonne National Laboratory, Mathematics and Computer Science Division, 2007
85. S. Leyffer, G. Lopez-Calva, J. Nocedal, Interior methods for mathematical programs with complementarity constraints. SIAM J. Optim. **17**(1), 52–77 (2006)
86. G. Li, Z. Wan, J.-W. Chen, X. Zhao, Necessary optimality condition for trilevel optimization problem. J. Ind. Manage. Optim. **41**, 282–290 (2018)
87. X. Liu, J. Sun, Generalized stationary points and an interior-point method for mathematical programs with equilibrium constraints. Math. Program. **101**(1), 231–261 (2004).
88. X. Liu, G. Perakis, J. Sun, A robust SQP method for mathematical programs with linear complementarity constraints. Comput. Optim. Appl. **34**, 5–33 (2006)
89. J. Liu, Y. Fan, Z. Chen, Y. Zheng, Pessimistic bilevel optimization: a survey. Int. J. Comput. Intell. Syst. **11**(1), 725–736 (2018)
90. Z.-Q. Luo, J.-S. Pang, D. Ralph, *Mathematical Programs with Equilibrium Constraints* (Cambridge University Press, Cambridge, 1996)
91. Y. Lv, Z. Wan, Solving linear bilevel multiobjective programming problem via exact penalty function approach. J. Inequal. Appl. **2015**(1), 258 (2015)
92. A. Migdalas, P.M. Pardalos, P. Värbrand (eds.) *Multilevel Optimization: Algorithms and Applications* (Kluwer Academic Publishers, Dordrecht, 1997)

93. J.A. Mirrlees, The theory of moral hazard and unobservable behaviour: part I. Rev. Econ. Stud. **66**(1), 3–21 (1999)
94. A. Mitsos, P. Lemonidis, P.I. Barton, Global solution of bilevel programs with a nonconvex inner program. J. Global Optim. **42**(4), 475–513 (2008)
95. A. Mitsos, P. Lemonidis, C. Lee, P.I. Barton, Relaxation-based bounds for semi-infinite programs. SIAM J. Optim. **19**(1), 77–113 (2008)
96. J.T. Moore, J.F. Bard, The mixed integer linear bilevel programming problem. Oper. Res. **38**(5), 911–921 (1990)
97. J.V. Outrata, Optimality conditions for a class of mathematical programs with equilibrium constraints. Math. Oper. Res. **24**(3), 627–644 (1999)
98. J.V. Outrata, M. Kočvara, J. Zowe, *Nonsmooth Approach to Optimization Problems with Equilibrium Constraints* (Kluwer Academic Publishers, Dordrecht, 1998)
99. J.-S. Pang, Three modeling paradigms in mathematical programming. Math. Program. **125**(2), 297–323 (2010)
100. C.O. Pieume, P. Marcotte, L.P. Fotso, P. Siarry, Solving bilevel linear multiobjective programming problems. Am. J. Oper. Res. **1**(4), 214–219 (2011)
101. E.S. Prescott, A primer on moral-hazard models. Feder. Reserve Bank Richmond Q. Rev. **85**, 47–77 (1999)
102. L. Qi, Z. Wei, On the constant positive linear dependence condition and its application to SQP methods. SIAM J. Optim. **10**(4), 963–981 (2000)
103. A. Raghunathan, L.T. Biegler, An interior point method for mathematical programs with complementarity constraints (MPCCs). SIAM J. Optim. **15**(3), 720–750 (2005)
104. D. Ralph, S.J. Wright, Some properties of regularization and penalization schemes for MPECs. Optim. Methods Softw. **19**(5), 527–556 (2004)
105. S.M. Robinson, Strongly regular generalized equations. Math. Oper. Res. **5**(1), 43–62 (1980)
106. M. Roelofs, J. Bisschop, The AIMMS language reference. Available at https://aimms.com/english/developers/resources/manuals/language-reference
107. N.V. Sahinidis, BARON 17.8.9: Global Optimization of Mixed-Integer Nonlinear Programs, User's Manual, 2017
108. H. Scheel, S. Scholtes, Mathematical program with complementarity constraints: stationarity, optimality and sensitivity. Math. Oper. Res. **25**, 1–22 (2000)
109. S. Scholtes, Convergence properties of a regularization schemes for mathematical programs with complementarity constraints. SIAM J. Optim. **11**(4), 918–936 (2001)
110. S. Scholtes, Nonconvex structures in nonlinear programming. Oper. Res. **52**(3), 368–383 (2004)
111. K. Shimizu, Y. Ishizuka, J.F. Bard, *Nondifferentiable and Two-level Mathematical Programming* (Kluwer Academic Publishers, Dordrecht, 1997)
112. O. Stein, *Bi-Level Strategies in Semi-Infinite Programming*, vol. 71 (Springer Science and Business Media, New York, 2013)
113. O. Stein, P. Steuermann, The adaptive convexification algorithm for semi-infinite programming with arbitrary index sets. Math. Program. **136**(1), 183–207 (2012)
114. O. Stein, A. Winterfeld, Feasible method for generalized semi-infinite programming. J. Optim. Theory Appl. **146**(2), 419–443 (2010)
115. M. Tawarmalani, N.V. Sahinidis, A polyhedral branch-and-cut approach to global optimization. Math. Program. **103**, 225–249 (2005)
116. L.N. Vicente, P.H. Calamai, Bilevel and multilevel programming: a bibliography review. J. Global Optim. **5**(3), 291–306 (1994)
117. W. Wiesemann, A. Tsoukalas, P.-M. Kleniati, B. Rustem, Pessimistic bilevel optimization. SIAM J. Optim. **23**(1), 353–380 (2013)
118. H. Xu, An implicit programming approach for a class of stochastic mathematical programs with complementarity constraints. SIAM J. Optim. **16**(3), 670–696 (2006)

119. J.J. Ye, J. Zhang, Enhanced Karush-Kuhn-Tucker conditions for mathematical programs with equilibrium constraints. J. Optim. Theory Appl. **163**(3), 777–794 (2014)
120. J.J. Ye, D. Zhu, New necessary optimality conditions for bilevel programs by combining the MPEC and value function approaches. SIAM J. Optim. **20**(4), 1885–1905 (2010)

Chapter 13
Approximate Bilevel Optimization with Population-Based Evolutionary Algorithms

Kalyanmoy Deb, Ankur Sinha, Pekka Malo, and Zhichao Lu

Abstract Population-based optimization algorithms, such as evolutionary algorithms, have enjoyed a lot of attention in the past three decades in solving challenging search and optimization problems. In this chapter, we discuss recent population-based evolutionary algorithms for solving different types of bilevel optimization problems, as they pose numerous challenges to an optimization algorithm. Evolutionary bilevel optimization (EBO) algorithms are gaining attention due to their flexibility, implicit parallelism, and ability to customize for specific problem solving tasks. Starting with surrogate-based single-objective bilevel optimization problems, we discuss how EBO methods are designed for solving multi-objective bilevel problems. They show promise for handling various practicalities associated with bilevel problem solving. The chapter concludes with results on an agro-economic bilevel problem. The chapter also presents a number of challenging single and multi-objective bilevel optimization test problems, which should encourage further development of more efficient bilevel optimization algorithms.

Keywords Evolutionary algorithms · Metaheuristics · Evolutionary bilevel optimization · Approximate optimization

---

K. Deb (✉) · Z. Lu
Michigan State University, East Lansing, MI, USA
e-mail: kdeb@egr.msu.edu; luzhicha@msu.edu

A. Sinha
Indian Institute of Management, Ahmedabad, India
e-mail: asinha@iima.ac.in

P. Malo
Aalto University School of Business, Helsinki, Finland
e-mail: pekka.malo@aalto.fi

S. Dempe, A. Zemkoho (eds.), *Bilevel Optimization*, Springer Optimization and Its Applications 161, https://doi.org/10.1007/978-3-030-52119-6_13

13.1 Introduction

Bilevel and multi-level optimization problems are omni-present in practice. This is because in many practical problems there is a hierarchy of two or more problems involving different variables, objectives and constraints, arising mainly from the involvement of different hierarchical stakeholders to the overall problem. Consider an agro-economic problem for which clearly there are at least two sets of stakeholders: policy makers and farmers. Although the overall goal is to maximize food production, minimize cost of cultivation, minimize water usage, minimize environmental impact, maximum sustainable use of the land and others, clearly, the overall solution to the problem involves an optimal design on the following variables which must be settled by using an optimization algorithm: crops to be grown, amount of irrigation water and fertilizers to be used, selling price of crops, fertilizer taxes to be imposed, and others. Constraints associated with the problem are restricted run-off of harmful chemicals, availability of limited budget for agriculture, restricted use of land and other resources, and others. One can attempt to formulate the overall problem as a multi-objective optimization problem involving all the above-mentioned variables, objectives, and constraints agreeable to both stake-holders in a single level. Such a solution procedure has at least two difficulties. First, the number of variables, objectives, and constraints of the resulting problem often becomes large, thereby making the solution of the problem difficult to achieve to any acceptable accuracy. Second, such a single-level process is certainly not followed in practice, simply from an easier managerial point of view. In reality, policy-makers come up with fertilizer taxes and agro-economic regulations so that harmful environmental affects are minimal, crop production is sustainable for generations to come, and revenue generation is adequate for meeting the operational costs. On the other hand, once a set of tax and regulation policies are announced, farmers consider them to decide on the crops to be grown, amount and type of fertilizers to be used, and irrigated water to be utilized to obtain maximum production with minimum cultivation cost. Clearly, policy-makers are at the upper level in this overall agro-economic problem solving task and farmers are at the lower level. However, upper level problem solvers must consider how a solution (tax and regulation policies) must be utilized optimally by the lower level problem solvers (farmers, in this case). Also, it is obvious that the optimal strategy of the lower level problem solvers directly depends on the tax and regulation policies declared by the upper level problem solvers. The two levels of problems are intricately linked in such bilevel problems, but, interestingly, the problem is not symmetric between upper and lower levels; instead the upper level's objectives and constraints control the final optimal solution more than that of the lower level problem.

When both upper and lower level involve a single objective each, the resulting bilevel problem is called a single-objective bilevel problem. The optimal solution in this case is usually a single solution describing both upper and lower level optimal variable values. In many practical problems, each level independently or both levels may involve more than one conflicting objectives. Such problems are termed here as multi-objective bilevel problems. These problems usually have more than one bilevel solution. For implementation purposes, a single bilevel solution must be chosen using the preference information of both upper and lower level problem solvers. Without any coordinated effort by upper level decision makers with lower level decision-makers, a clear choice of the overall bilevel solution is uncertain, which provides another dimension of difficulty to solve multi-objective bilevel problems.

It is clear from the above discussion that bilevel problems are reality in practice, however their practice is uncommon, simply due to the fact that the nested nature of two optimization problems makes the solution procedure computationally expensive. The solution of such problems calls for flexible yet efficient optimization methods, which can handle different practicalities and are capable of finding approximate optimal solutions in a quick computational manner. Recently, evolutionary optimization methods have been shown to be applicable to such problems because of ease of customization that helps in finding near-optimal solution, mainly due to their population approach, use of a direct optimization approach instead of using any gradient, presence of an implicit parallelism mechanism constituting a parallel search of multiple regions simultaneously, and their ability to find multiple optimal solutions in a single application.

In this chapter, we first provide a brief description of metaheuristics in Sect. 13.2 followed by a step-by-step procedure of an evolutionary optimization algorithm as a representative of various approaches belonging to this class of algorithms. Thereafter, in Sect. 13.3, we discuss the bilevel formulation and difficulties involved in solving a bilevel optimization problem. Then, we provide the past studies on population-based methods for solving bilevel problems in Sect. 13.4. Thereafter, we discuss in detail some of the recent efforts in evolutionary bilevel optimization in Sect. 13.5. In Sect. 13.6, we present two surrogate-assisted single-objective evolutionary bilevel optimization (EBO) algorithms—BLEAQ and BLEAQ2. Simulation results of these two algorithms are presented next on a number of challenging standard and scalable test problems provided in the Appendix. Bilevel evolutionary multi-objective optimization (BL-EMO) algorithms are described next in Sect. 13.7. The ideas of optimistic and pessimistic bilevel Pareto-optimal solution sets are differentiated and an overview of research developments on multi-objective bilevel optimization is provided. Thereafter, in Sect. 13.8, a multi-objective agro-economic bilevel problem is described and results using the proposed multi-objective EBO algorithm are presented. Finally, conclusions of this extensive chapter is drawn in Sect. 13.9.

13.2 Metaheuristics Algorithms for Optimization

Heuristics refers to relevant guiding principles, or effective rules, or partial problem information related to the problem class being addressed. When higher level heuristics are utilized in the process of creating new solutions in a search or optimization methodology for solving a larger class of problems, the resulting methodology is called a *metaheuristic*. The use of partial problem information may also be used in metaheuristics to speed up the search process in arriving at a near-optimal solution, but an exact convergence to the optimal solution is usually not guaranteed. For many practical purposes, metaheuristics based optimization algorithms are more pragmatic approaches [13]. These methods are in rise due to the well-known *no-free-lunch* (NFL) theorem.

Theorem 13.2.1 *No single optimization algorithm is most computationally effective for solving all problems.* △

Although somewhat intuitive, a proof of a more specific and formal version of the above theorem was provided in [66]. A corollary to the NFL theorem is that for solving a *specific* problem class, there exists a *customized* algorithm which would perform the best; however the same algorithm may not work so well on another problem class.

It is then important to ask a very important question: 'How does one develop a customized optimization algorithm for a specific problem class?'. The search and optimization literature does not provide a ready-made answer to the above question for every possible problem class, but the literature is full of different application problems and the respective developed customized algorithms for solving the problem class. Most of these algorithms use relevant heuristics derived from the description of the problem class.

We argue here that in order to utilize problem heuristics in an optimization algorithm, the basic structure of the algorithm must allow heuristics to be integrated easily. For example, the well-known steepest-descent optimization method cannot be customized much with available heuristics, as the main search must always take place along the negative of the gradient of the objective functions to allow an improvement or non-deterioration in the objective value from one iteration to the next. In this section, we discuss EAs that are population-based methods belonging to the class of metaheuristics. There exist other population-based metaheuristic methods which have also been used for solving bilevel problems, which we do not elaborate here, but provide a brief description below:

- **Differential Evolution (DE) [59]:** DE is a steady state optimization procedure which uses three population members to create a "mutated" point. It is then compared with the best population member and a recombination of variable exchanges is made to create the final offspring point. An optional selection between offspring and the best population member is used. DE cannot force its population members to stay within specified variable bounds and special operators are needed to handle them as well as other constraints. DE is often considered as a special version of an evolutionary algorithm, described later.
- **Particle Swarm Optimization (PSO) [28, 31]:** PSO is a generational optimization procedure which creates one new offspring point for each parent population member by making a vector operation with parent's previous point, its best point since the start of a run and the best-ever population point. Like DE, PSO cannot

also force its population members to stay within specified variable bounds and special operators are needed to handle them and other constraints.

- **Other Metaheuristics [5, 22]:** There exists more than 100 other metaheuristics-based approaches, which use different natural and physical phenomenon. Variable bounds and constraints are often handled using penalty functions or by using special operators.

These algorithms, along with evolutionary optimization algorithms described below, allow flexibility for customizing the solution procedure by enabling any heuristics or rules to be embedded in their operators.

Evolutionary algorithms (EAs) are mostly synonymous to metaheuristics based optimization methods. EAs work with multiple points (called a *population*) at each iteration (called a *generation*). Here are the usual steps of a generational EA:

1. An initial population P_0 of size N (population size) is created, usually at random within the supplied variable bounds. Every population member is evaluated (objective functions and constraints) and a combined *fitness* or a selection function is evaluated for each population member. Set the generation counter $t = 0$.
2. A termination condition is checked. If not satisfied, continue with Step 3, else report the best point and stop.
3. Select better population members of P_t by comparing them using the fitness function and store them in mating pool M_t.
4. Take pairs of points from M_t at a time and recombine them using a *recombination* operator to create one or more new points.
5. Newly created points are then locally perturbed by using a *mutation* operator. The mutated point is then stored in an offspring population Q_t. Steps 4 and 5 are continued until Q_t grows to a size of N.
6. Two populations P_t and Q_t are combined and N best members are saved in the new parent population P_{t+1}. The generation counter is incremented by one and the algorithm moves to Step 2.

The recombination operator is unique in EAs and is responsible for recombining two different population members to create new solutions. The selection and recombination operators applied on a population constitutes an implicitly parallel search, providing their power. In the above generational EA, $NT_{\max}$ is the total number of evaluations, where $T_{\max}$ is the number of generations needed to terminate the algorithm. Besides the above generational EA, steady-state EAs exist, in which the offspring population Q_t consists of a single new mutated point. In the steady-state EA, the number of generations $T_{\max}$ needed to terminate would be more than that needed for a generational EA, but the overall number of solution evaluations needed for the steady-state EA may be less, depending on the problem being solved.

Each of the steps in the above algorithm description—initialization (Step 1), termination condition (Step 2), selection (Step 3), recombination (Step 4), mutation (Step 5) and survival (Step 6)—can be changed or embedded with problem information to make the overall algorithm customized for a problem class.

13.3 Bilevel Formulation and Challenges

Bilevel optimization problems have two optimization problems staged in a hierarchical manner [55]. The outer problem is often referred to as the upper level problem or the leader's problem, and the inner problem is often referred to as the lower level problem or the follower's problem. Objective and constraint functions of both levels can be functions of all variables of the problem. However, a part of the variable vector, called the upper level variable vector, remains fixed for the lower level optimization problem. For a given upper level variable vector, an accompanying lower level variable set which is optimal to the lower level optimization problem becomes a candidate feasible solution for the upper level optimization problem, subject to satisfaction of other upper level variable bounds and constraint functions. Thus, in such nested problems, the lower level variable vector depends on the upper level variable vector, thereby causing a strong variable linkage between the two variable vectors. Moreover, the upper level problem is usually sensitive to the quality of lower level optimal solution, which makes solving the lower level optimization problem to a high level of accuracy important.

The main challenge in solving bilevel problems is the computational effort needed in solving nested optimization problems, in which for every upper level variable vector, the lower level optimization problem must be solved to a reasonable accuracy. One silver lining is that depending on the complexities involved in two levels, two different optimization algorithms can be utilized, one for upper level and one for lower level, respectively, instead of using the same optimization method for both levels. However, such a nested approach may also be computationally expensive for large scale problems. All these difficulties provide challenges to optimization algorithms while solving a bilevel optimization problem with a reasonable computational complexity.

There can also be situations where the lower level optimization in a bilevel problem has multiple optimal solutions for a given upper level vector. Therefore, it becomes necessary to define, which solution among the multiple optimal solutions at the lower level should be considered. In such cases, one assumes either of the two positions: the *optimistic* position or the *pessimistic* position. In the optimistic position, the follower is assumed to favorable to the leader and chooses the solution that is best for the leader from the set of multiple lower level optimal solutions. In the pessimistic position, it is assumed that the follower may not be favorable to the leader (in fact, the follower is antagonistic to leader's objectives) and may choose the solution that is worst for the leader from the set of multiple lower level optimal solutions. Intermediate positions are also possible and are more pragmatic, which can be defined with the help of selection functions. Most of the literature on bilevel optimization usually focuses on solving optimistic bilevel problems. A general formulation for the bilevel optimization problem is provided below.

Definition 13.3.1 For the upper level objective function $F : \mathbb{R}^n \times \mathbb{R}^m \to \mathbb{R}$ and lower level objective function $f : \mathbb{R}^n \times \mathbb{R}^m \to \mathbb{R}$, the bilevel optimization problem is given by

$$\min_{\mathbf{x}_u, \mathbf{x}_l} \; F(\mathbf{x}_u, \mathbf{x}_l), \tag{13.3.1}$$

$$\text{subject to} \;\; \mathbf{x}_l \in \underset{\mathbf{x}_l}{\text{argmin}}\{f(\mathbf{x}_u, \mathbf{x}_l) : g_j(\mathbf{x}_u, \mathbf{x}_l) \leq 0, \quad j = 1, \ldots, J\}, \tag{13.3.2}$$

$$G_k(\mathbf{x}_u, \mathbf{x}_l) \leq 0, \quad k = 1, \ldots, K, \tag{13.3.3}$$

where $G_k : \mathbb{R}^n \times \mathbb{R}^m \to \mathbb{R}$, $k = 1, \ldots, K$ denotes the upper level constraints, and $g_j : \mathbb{R}^n \times \mathbb{R}^m \to \mathbb{R}$, $j = 1, \ldots, J$ represents the lower level constraints, respectively. Variables $\mathbf{x}_u$ and $\mathbf{x}_l$ are n and m dimensional vectors, respectively. It is important to specify the position one is taking while solving the above formulation. △

13.4 Non-surrogate-Based EBO Approaches

A few early EBO algorithms were primarily either nested approaches or used the KKT conditions of the lower level optimization problem to reduce the bilevel problem to a single level and then a standard algorithm was applied. In this section, we provide a review of such approaches. It is important to note that the implicit parallel search of EAs described before still plays a role in constituting an efficient search within both the upper and lower level optimization tasks.

13.4.1 Nested Methods

Nested EAs are a popular approach to handle bilevel problems, where lower level optimization problem is solved corresponding to each and every upper level member [36, 41, 47]. Though effective, nested strategies are computationally very expensive and not viable for large scale bilevel problems. Nested methods in the area of EAs have been used in primarily two ways. The first approach has been to use an EA at the upper level and a classical algorithm at the lower level, while the second approach has been to utilize EAs at both levels. Of course, the choice between two approaches is determined by the complexity of the lower level optimization problem.

One of the first EAs for solving bilevel optimization problems was proposed in the early 1990s. Mathieu et al. [35] used a nested approach with genetic algorithm at the upper level, and linear programming at the lower level. Another nested approach was proposed in [69], where the upper level was an EA and the lower level was solved using Frank–Wolfe algorithm (reduced gradient method) for every upper level member. The authors demonstrated that the idea can be effectively utilized to solve non-convex bilevel optimization problems. Nested PSO was used in [31] to solve bilevel optimization problems. The effectiveness of the technique was shown on a number of standard test problems with small number of variables, but the computational expense of the nested procedure was not reported. A hybrid approach was proposed in [30], where simplex-based crossover strategy was used at the upper level, and the lower level was solved using one of the classical approaches. The authors report the generations and population sizes required by the algorithm that can be used to compute the upper level function evaluations, but they do not explicitly report the total number of lower level function evaluations, which presumably is high.

DE based approaches have also been used, for instance, in [72], authors used DE at the upper level and relied on the interior point algorithm at the lower level; similarly, in [3] authors have used DE at both levels. Authors have also combined two different specialized EAs to handle the two levels, for example, in [2] authors use an ant colony optimization to handle the upper level and DE to handle the lower level in a transportation routing problem. Another nested approach utilizing ant colony algorithm for solving a bilevel model for production-distribution planning is [9]. Scatter search algorithms have also been employed for solving production-distribution planning problems, for instance [10].

Through a number of approaches involving EAs at one or both levels, the authors have demonstrated the ability of their methods in solving problems that might otherwise be difficult to handle using classical bilevel approaches. However, as already stated, most of these approaches are practically non-scalable. With increasing number of upper level variables, the number of lower level optimization tasks required to be solved increases exponentially. Moreover, if the lower level optimization problem itself is difficult to solve, numerous instances of such a problem cannot be solved, as required by these methods.

13.4.2 Single-Level Reduction Using Lower Level KKT Conditions

Similar to the studies in the area of classical optimization, many studies in the area of evolutionary computation have also used the KKT conditions of the lower level to reduce the bilevel problem into a single-level problem. Most often, such an approach is able to solve problems that adhere to certain regularity conditions at the lower level because of the requirement of the KKT conditions. However, as the reduced single-level problem is solved with an EA, usually the upper level objective function and constraints can be more general and not adhering to such regularities. For instance, one of the earliest papers using such an approach is by Hejazi et al. [24], who reduced the linear bilevel problem to single-level and then used a genetic algorithm, where chromosomes emulate the vertex points, to solve the problem. Another study [8] also proposed a single level reduction for linear bilevel problems. Wang et al. [62] reduced the bilevel problem into a single-level optimization problem using KKT conditions, and then utilized a constraint handling scheme to successfully solve a number of standard test problems. Their algorithm was able to handle non-differentiability at the upper level objective function, but not elsewhere. Later on, Wang et al. [64] introduced an improved algorithm that was able to handle non-convex lower level problem and performed better than the previous approach [62]. However, the number of function evaluations in both approaches remained quite high (requiring function evaluations to the tune of 100,000 for 2–5 variable bilevel problems). In [63], the authors used a simplex-based genetic algorithm to solve linear-quadratic bilevel problems after reducing it to a single level task. Later, Jiang et al. [27] reduced the bilevel optimization problem into a non-linear optimization problem with complementarity constraints, which is sequentially smoothed and solved with a PSO algorithm. Along similar lines of using lower level optimality conditions, Li [29] solved a fractional bilevel optimization problem by utilizing optimality results of the linear fractional lower level problem. In [60], the authors embed the chaos search technique in PSO to solve single-level reduced problem. The search region represented by the KKT conditions can be highly constrained that poses challenges for any optimization algorithm. To address this concern, in a recent study [57], the authors have used approximate KKT-conditions for the lower level problem. One of the theoretical concerns of using the KKT conditions to replace lower level problem directly is that the associated constraint qualification conditions must also be satisfied for every lower level solution.

13.5 Surrogate-Based EBO Approaches

The earlier approaches suffered with some drawbacks like high computational requirements, or reliance on the KKT conditions of the lower level problem. To overcome this, recent research has focused on surrogate-based methods for bilevel problems. Researchers have used surrogates in different ways for solving bilevel problems that we discuss in this section.

Surrogate-based solution methods are commonly used for optimization problems [61], where actual function evaluations are expensive. A meta-model or surrogate model is an approximation of the actual model that is relatively quicker to evaluate. Based on a small sample from the actual model, a surrogate model can be trained and used subsequently for optimization. Given that, for complex problems, it is hard to approximate the entire model with a small set of sample points, researchers often resort to iterative meta modeling techniques, where the actual model is approximated locally during iterations.

Bilevel optimization problems contain an inherent complexity that leads to a requirement of large number of evaluations to solve the problem. Metamodeling of the lower level optimization problem, when used with population-based algorithms,

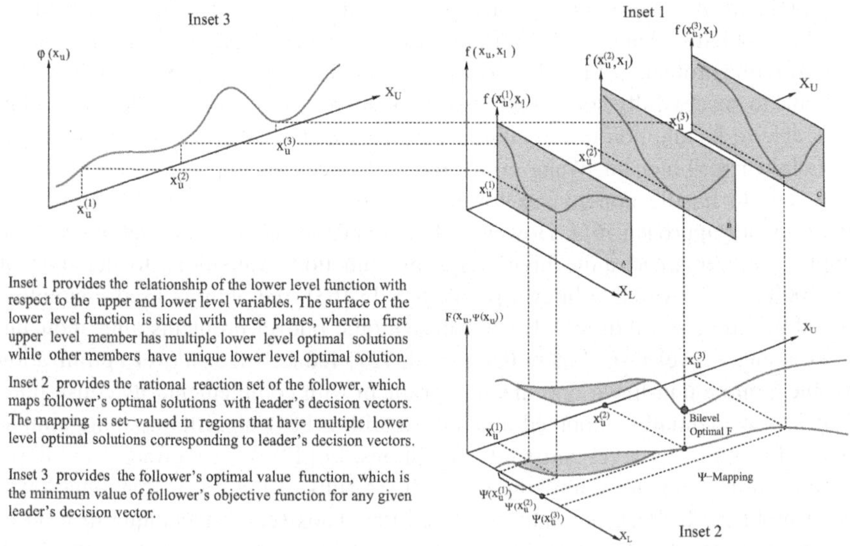

Fig. 13.1 Graphical representation of rational reaction set (Ψ) and lower level optimal value function (φ)

offers a viable means to handle bilevel optimization problems. With good lower level solutions being supplied at the upper level, EA's implicit parallel search power constructs new and good upper level solutions, making the overall search efficient. In this subsection, we discuss three ways in which metamodeling can be applied to bilevel optimization. There are two important mappings in bilevel optimization, referred to as the rational reaction set and lower level optimal value function. We refer the readers to Fig. 13.1, which provides an understanding of these two mappings graphically for a hypothetical bilevel problem.

13.5.1 Reaction Set Mapping

One of the approaches to solve bilevel optimization problems using EAs would be through iterative approximation of the reaction set mapping Ψ. The bilevel formulation in terms of the Ψ-mapping can be written as below.

$$\min_{\mathbf{x}_u, \mathbf{x}_l} \quad F(\mathbf{x}_u, \mathbf{x}_l), \tag{13.5.1}$$

$$\text{subject to} \quad \mathbf{x}_l \in \Psi(\mathbf{x}_u), \tag{13.5.2}$$

$$G_k(\mathbf{x}_u, \mathbf{x}_l) \leq 0, \quad k = 1, \ldots, K. \tag{13.5.3}$$

If the Ψ-mapping in a bilevel optimization problem is known, it effectively reduces the problem to single level optimization. However, this mapping is seldom available; therefore, the approach could be to solve the lower level problem for a few upper level members and then utilize the lower level optimal solutions and corresponding upper level members to generate an approximate mapping $\hat{\Psi}$. It is noteworthy that approximating a set-valued Ψ-mapping offers its own challenges and is not a straightforward task. Assuming that an approximate mapping, $\hat{\Psi}$, can be generated, the following single level optimization problem can be solved for a few generations of the algorithm before deciding to further refine the reaction set.

$$\begin{aligned}
\min_{\mathbf{x}_u, \mathbf{x}_l} \quad & F(\mathbf{x}_u, \mathbf{x}_l), \\
\text{subject to} \quad & \mathbf{x}_l \in \hat{\Psi}(\mathbf{x}_u), \\
& G_k(\mathbf{x}_u, \mathbf{x}_l) \leq 0, \quad k = 1, \ldots, K.
\end{aligned}$$

EAs that rely on this idea to solve bilevel optimization problems are [4, 44, 45, 50]. In some of these studies, authors have used quadratic approximation to approximate the local reaction set. This helps in saving lower level optimization calls when the approximation for the local reaction set is good. In case the approximations generated by the algorithm are not acceptable, the method defaults to a nested approach. It is noteworthy that a bilevel algorithm that uses a surrogate model for reaction set mapping may need not be limited to quadratic models but other models can also be used.

13.5.2 Optimal Lower Level Value Function

Another way to use metamodeling is through the approximation of the optimal value function φ. If the φ-mapping is known, the bilevel problem can once again be reduced to single level optimization problem as follows [68],

$$\begin{aligned}
\min_{\mathbf{x}_u, \mathbf{x}_l} \quad & F(\mathbf{x}_u, \mathbf{x}_l), \\
\text{subject to} \quad & f(\mathbf{x}_u, \mathbf{x}_l) \leq \varphi(\mathbf{x}_u), \\
& g_j(\mathbf{x}_u, \mathbf{x}_l) \leq 0, \quad j = 1, \ldots, J, \\
& G_k(\mathbf{x}_u, \mathbf{x}_l) \leq 0, \quad k = 1, \ldots, K.
\end{aligned}$$

However, since the value function is seldom known, one can attempt to approximate this function using metamodeling techniques. The optimal value function is a single-valued mapping; therefore, approximating this function avoids the complexities associated with set-valued mapping. As described previously, an approximate mapping $\hat{\varphi}$, can be generated with the population members of an EA and the following modification in the first constraint: $f(\mathbf{x}_u, \mathbf{x}_l) \leq \hat{\varphi}(\mathbf{x}_u)$. Evolutionary optimization approaches that rely on this idea can be found in [51, 56, 58].

13.5.3 Bypassing Lower Level Problem

Another way to use a meta-model in bilevel optimization would be to completely by-pass the lower level problem, as follows:

$$\min_{\mathbf{x}_u} \ \hat{F}(\mathbf{x}_u),$$
$$\text{subject to} \ \ \hat{G}_k(\mathbf{x}_u) \leq 0, \quad k = 1, \ldots, K.$$

Given that the optimal $\mathbf{x}_l$ are essentially a function of $\mathbf{x}_u$, it is possible to construct a single level approximation of the bilevel problem by ignoring $\mathbf{x}_l$ completely and writing the objective function and constraints for the resulting single level problem as a function of only $\mathbf{x}_u$. However, the landscape for such a single level problem can be highly non-convex, disconnected, and non-differentiable. Advanced metamodeling techniques might be required to use this approach, which may be beneficial for certain classes of bilevel problems. A training set for the metamodel can be constructed by solving a few lower level problems for different $\mathbf{x}_u$. Both upper level objective F and constraint set (G_k) can then be meta-modeled using $\mathbf{x}_u$ alone. Given the complex structure of such a single-level problem, it might be sensible to create such an approximation locally.

13.5.4 Limitations and Assumptions of Surrogate-Based Approaches

The idea of using function approximations within EAs makes them considerably more powerful than simple random search. However, these algorithms are still constrained by certain background assumptions, and are not applicable to arbitrary problems. In order for the approximations to work, we generally need to impose some constraining assumptions on the bilevel problem to even ensure the existence of the underlying mappings that the algorithms are trying to approximate numerically.

For example, when using function approximations for the reaction set mapping in single objective problems, we need to assume that the lower level problem is convex with respect to the lower level variables. If the lower level problem is not convex, the function could be composed by global as well as local optimal solutions, and if we consider only global optimal solutions in the lower level problem, the respective function could become discontinuous. This challenge can be avoided by imposing the convexity assumption. Furthermore, we generally need to require that the lower level objective function as well as the constraints are twice continuously differentiable. For more insights, we refer to [50], where the assumptions are discussed in greater detail in case of both polyhedral and nonlinear constraints. When these assumptions are not met, there are also no guarantees that the approximation based algorithms can be expected to work any better than a nested search.

In the next section, we describe recent surrogate-based EBO algorithms, but would like to highlight that they are not exempt from the above convexity and differentiability assumptions.

13.6 Single-Objective Surrogate-Based EBO Algorithms

In this section, we discuss the working of two surrogate-based EBO algorithms—BLEAQ and BLEAQ2—that rely on the approximation of the Ψ- and φ-mappings. In a single-objective bilevel problems, both upper and lower level problems have a single objective function: $F(\mathbf{x}_u, \mathbf{x}_l)$ and $f(\mathbf{x}_u, \mathbf{x}_l)$. There can be multiple constraints at each level. We describe these algorithms in the next two subsections.

13.6.1 Ψ-Approach and BLEAQ Algorithm

In the Ψ-approach, the optimal value of each lower level variable is approximated using a surrogate model of the upper level variable set. Thus, a total of $m = |\mathbf{x}_l|$ number of metamodels must be constructed from a few exact lower level optimization results.

After the lower level problem is optimized for a few upper level variable sets, the optimal lower level variable can be modeled using a quadratic function of upper level variables: $x^*_{l,i} = q_i(\mathbf{x}_u)$. In fact, the algorithm creates various local quadratic models, but for simplicity we skip the details here. Thereafter, upper level objective and constraint functions can be expressed in terms of the upper level variables only in an implicit manner, as follows:

$$\begin{aligned} &\min_{\mathbf{x}_u} \; F\left(\mathbf{x}_u, \mathbf{q}(\mathbf{x}_u)\right), \\ &\text{subject to } \; G_k\left(\mathbf{x}_u, \mathbf{q}(\mathbf{x}_u)\right) \le 0, \quad k = 1, 2, \ldots, K. \end{aligned}$$

Note that $\mathbf{q}(\mathbf{x}_u)$ represents the quadratic approximation for the lower level vector. The dataset for creating the surrogate $\mathbf{q}(\mathbf{x}_u)$, is constructed by identifying the optimal lower level vectors corresponding to various upper level vectors by solving the following problem:

$$\mathbf{x}_l^* = \operatorname{argmin}\left\{ f(\mathbf{x}_u, \mathbf{x}_l) \middle| g_j(\mathbf{x}_u, \mathbf{x}_l) \le 0, \quad j = 1, 2, \ldots, J \right\}.$$

The validity of the quadratic approximation is checked after a few iterations. If the existing quadratic model does not provide an accurate approximation, new points are introduced to form a new quadratic model. We call this method BLEAQ [45, 50].

Note that this method will suffer whenever the lower level has multiple optimal solutions at the lower level, and also it requires multiple models $q_i(\mathbf{x}_u),\ \forall\ i \in \{1, \ldots, m\}$, to be created.

13.6.2 *φ-Approach and BLEAQ2 Algorithm*

Next, we discuss the φ-approach that overcomes the various drawbacks of the Ψ-approach. In the φ-approach, an approximation of the optimal lower level objective function value $\varphi(\mathbf{x}_u)$ as a function of $\mathbf{x}_u$ is constructed that we refer to as $\hat{\varphi}(\mathbf{x}_u)$. Thereafter, the following single-level problem is solved:

$$\begin{aligned} &\text{Minimize } F(\mathbf{x}_u, \mathbf{x}_l), \\ &\text{subject to } f(\mathbf{x}_u, \mathbf{x}_l) \le \hat{\varphi}(\mathbf{x}_u), \\ &\qquad\qquad G_k(\mathbf{x}_u, \mathbf{x}_l) \le 0, \quad k = 1, 2, \ldots, K. \end{aligned} \tag{13.6.1}$$

The above formulation allows a single surrogate model to be constructed and the solution of the above problem provides optimal $\mathbf{x}_u$ and $\mathbf{x}_l$ vectors.

In the modified version of BLEAQ, that we called BLEAQ2 [58], both Ψ and φ-approaches are implemented with a check. The model that has higher level of accuracy is used to choose the lower level solution for any given upper level vector. Both Ψ and φ models are approximated locally and updated iteratively.

13.6.3 *Experiments and Results*

Next, we assess the performance of three algorithms, the nested approach, BLEAQ, and BLEAQ2 on a set of bilevel test problems. We perform 31 runs for each test instance. For each run, the upper and lower level function evaluations required until termination are recorded separately. Information about the various parameters and their settings can be found in [50, 58].

13.6.3.1 Results on Non-scalable Test Problems

We first present the empirical results on eight non-scalable test problems selected from the literature (referred to as TP1–TP8). The description for these test problems is provided in the Appendix. Table 13.1 contains the median upper level (UL) function evaluations, lower level (LL) function evaluations and BLEAQ2's overall function evaluation savings as compared to other approaches from 31 runs of the algorithms. The overall function evaluations for any algorithm is simply the sum of upper and lower level function evaluations. For instance, for the median run

Table 13.1 Median function evaluations on non-scalable test problems TP1–TP8

	UL func. evals.			LL func. evals.			BLEAQ2 savings	
	BLEAQ2 Med	BLEAQ Med	Nested Med	BLEAQ2 Med	BLEAQ Med	Nested Med	BLEAQ2 vs BLEAQ	BLEAQ2 vs nested
TP1	136	155	–	242	867	–	63%	Large%
TP2	255	185	436	440	971	5686	40%	89%
TP3	158	155	633	224	894	6867	64%	95%
TP4	198	357	1755	788	1772	19764	54%	95%
TP5	272	243	576	967	1108	6558	8%	83%
TP6	161	155	144	323	687	1984	43%	77%
TP7	112	255	193	287	987	2870	68%	87%
TP8	241	189	403	467	913	7996	36%	92%

While computing savings, we compare the total function evaluations (sum of upper and lower level function evaluations) of one algorithm against the other. Savings for BLEAQ2 when compared against an algorithm A is given as (A − BLEAQ2) /A, where the name of the algorithm denotes the total function evaluations required by the algorithm

with TP1, BLEAQ2 requires 63% less overall function evaluations as compared to BLEAQ, and 98% less overall function evaluations as compared to the nested approach.

All these test problems are bilevel problems with small number of variables, and all the three algorithms were able to solve the eight test instances successfully. A significant computational saving can be observed for both BLEAQ2 and BLEAQ, as compared to the nested approach, as shown in the 'Savings' column of Table 13.1. The performance gain going from BLEAQ to BLEAQ2 is quite significant for these test problems even though none of them leads to multiple lower level optimal solutions. Detailed comparison between BLEAQ and BLEAQ2 in terms of upper and lower level function evaluations is provided through Figs. 13.2 and 13.3, respectively. It can be observed that BLEAQ2 requires comparatively much less number of lower level function evaluations than BLEAQ algorithm, while there is no conclusive argument can be made for the number of upper level function evaluations. However, since the lower level problem is solved more often, as shown Table 13.1, BLEAQ2 requires significantly less number of overall function evaluations to all eight problems.

13.6.3.2 Results on Scalable Test Problems

Next, we compare the performance of the three algorithms on the scalable SMD test suite (presented in the Appendix) which contains 12 test problems [46]. The test suite was later extend to 14 test problems by adding two additional scalable test problems. First we analyze the performance of the algorithms on five variables, and then we provide the comparison results on 10-variable instances of the SMD test problems. For the five-variable version of the SMD test problems, we use $p = 1$,

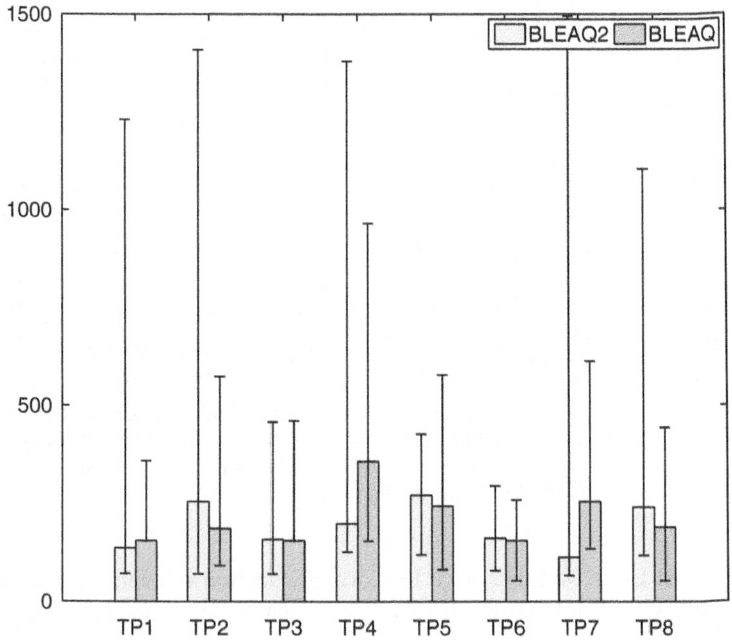

Fig. 13.2 Variation of upper level function evaluations required by BLEAQ and BLEAQ2 algorithms in 31 runs applied to TP1–TP8

$q = 2$ and $r = 1$ for all SMD problems except SMD6 and SMD14. For the five-variable version of SMD6 and SMD14, we use $p = 1$, $q = 0$, $r = 1$ and $s = 2$. For the 10-variable version of the SMD test problems, we use $p = 3$, $q = 3$ and $r = 2$ for all SMD problems except SMD6 and SMD14. For the 10-variable version of SMD6 and SMD14, we use $p = 3$, $q = 1$, $r = 2$ and $s = 2$. In their five-variable versions, there are two variables at the upper level and three variables at the lower level. They also offer a variety of tunable complexities to any algorithm. For instance, the test set contains problems which are multi-modal at the upper and the lower levels, contain multiple optimal solutions at the lower level, contain constraints at the upper and/or lower levels, etc.

Table 13.2 provides the median function evaluations and overall savings for the three algorithms on all 14 five-variable SMD problems. The table indicates that BLEAQ2 is able to solve the entire set of 14 SMD test problems, while BLEAQ fails on two test problems. The overall savings with BLEAQ2 is larger as compared to BLEAQ for all problems. Test problems SMD6 and SMD14 which contain multiple lower level solutions, BLEAQ is unable to handle them. Further details about the required overall function evaluations from 31 runs are provided in Fig. 13.4.

Results for the 10-variable SMD test problems are presented in Table 13.3. BLEAQ2 leads to much higher savings as compared to BLEAQ. BLEAQ is found to fail again on SMD6 and also on SMD7 and SMD8. Both methods outperform

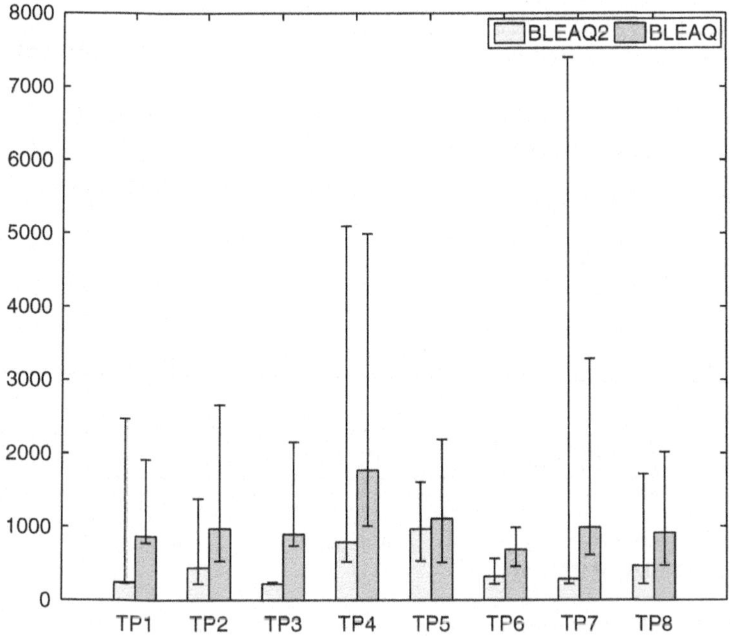

Fig. 13.3 Variation of lower level function evaluations required by BLEAQ and BLEAQ2 algorithms in 31 runs applied to TP1–TP8

Table 13.2 Median function evaluations on five-variable SMD test problems

	UL func. evals.			LL func. evals.			BLEAQ2 savings	
	BLEAQ2	BLEAQ	Nested	BLEAQ2	BLEAQ	Nested	BLEAQ2	BLEAQ2
	Med	Med	Med	Med	Med	Med	vs BLEAQ	vs nested
SMD1	123	98	164	8462	13,425	104,575	37%	92%
SMD2	114	88	106	7264	11,271	74,678	35%	90%
SMD3	264	91	136	12, 452	15,197	101,044	17%	87%
SMD4	272	110	74	8600	12,469	59,208	29%	85%
SMD5	126	80	93	14, 490	19,081	73,500	24%	80%
SMD6	259	–	116	914	–	3074	Large	63%
SMD7	180	98	67	8242	12,580	56,056	34%	85%
SMD8	644	228	274	22, 866	35,835	175,686	35%	87%
SMD9	201	125	127	10, 964	16,672	101,382	34%	89%
SMD10	780	431	–	19, 335	43,720	–	54%	Large
SMD11	1735	258	260	134, 916	158,854	148,520	14%	8%
SMD12	203	557	–	25, 388	135,737	–	81%	Large
SMD13	317	126	211	13, 729	17,752	138,089	21%	90%
SMD14	1014	–	168	12, 364	–	91,197	Large	85%

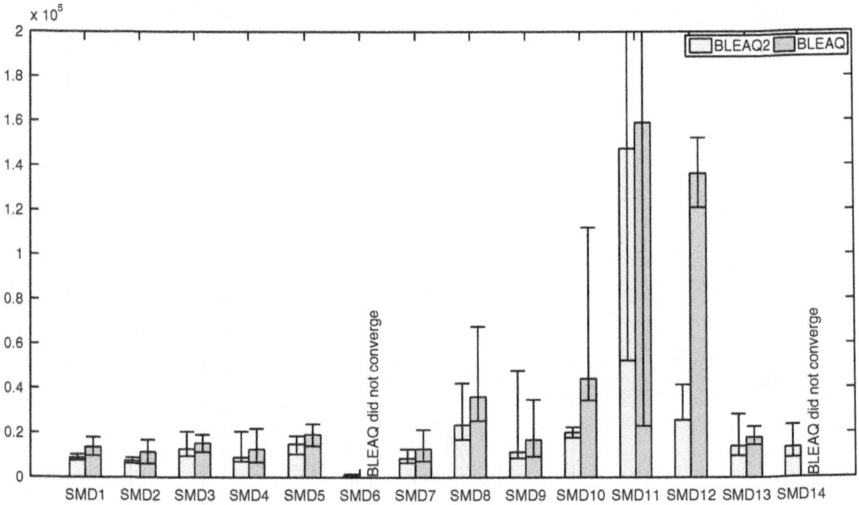

Fig. 13.4 Overall function evaluations needed by BLEAQ and BLEAQ2 for solving five-dimensional SMD1–SMD14 problems

Table 13.3 Median function evaluations on 10-variable SMD test problems

	UL func. evals.			LL func. evals.			BLEAQ2 savings	
	BLEAQ2	BLEAQ	Nested	BLEAQ2	BLEAQ	Nested	BLEAQ2	BLEAQ2
	Med	Med	Med	Med	Med	Med	vs BLEAQ	vs nested
SMD1	670	370	760	52, 866	61,732	1,776,426	14%	97%
SMD2	510	363	652	44, 219	57,074	1,478,530	22%	97%
SMD3	1369	630	820	68, 395	90,390	1,255,015	23%	94%
SMD4	580	461	765	35, 722	59,134	1,028,802	39%	96%
SMD5	534	464	645	65, 873	92,716	1,841,569	29%	96%
SMD6	584	–	824	3950	–	1562,003	Large	99%
SMD7	1486	–	–	83, 221	–	–	Large	Large
SMD8	6551	–	–	231, 040	–	–	Large	Large

the nested method on most of the test problems. We do not provide results for SMD9–SMD14 as none of the algorithms are able to handle these problems. It is noteworthy that SMD9–SMD14 offer difficulties with multi-modalities and having highly constrained search space, which none of the algorithms are able to handle with the parameter setting used here. Details for the 31 runs on each of these test problems are presented in Fig. 13.5.

The advantage of BLEAQ2 algorithm comes from the use of both Ψ and φ-mapping based surrogate approaches. We pick two SMD problems—SMD1 and SMD13—to show that one of the two surrogate approaches perform better depending on their suitability on the function landscape. Figure 13.6 shows that in SMD1 problem, φ-approximation performs better and Fig. 13.7 shows that Ψ-

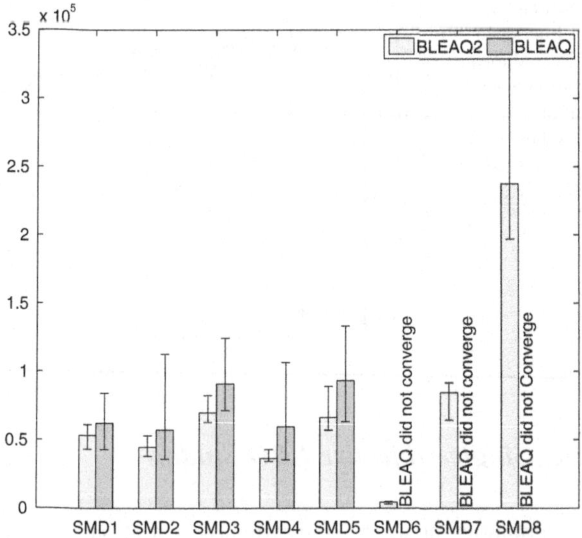

Fig. 13.5 Overall function evaluations needed by BLEAQ and BLEAQ2 for solving 10-dimensional SMD1–SMD14 problems

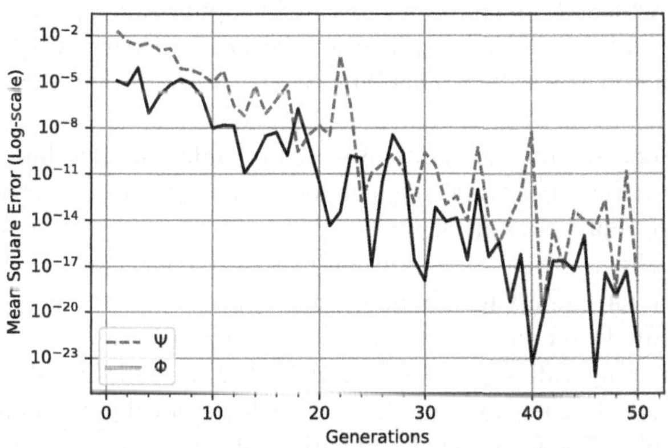

Fig. 13.6 Approximation error (in terms of Euclidean distance) of a predicted lower level optimal solution when using localized Ψ and φ-mapping during the BLEAQ2 algorithm on the five-variable SMD1 test problem

approximation is better on SMD13. In these figures, the variation of the Euclidean distance of lower level solution from exact optimal solution with generations is shown. While both approaches reduce the distance in a noisy manner, BLEAQ2 does it better by using the best of both approaches than BLEAQ which uses only the Ψ-approximation. The two figures show the adaptive nature of the BLEAQ2 algorithm in choosing the right approximation strategy based on the difficulties involved in a bilevel optimization problem.

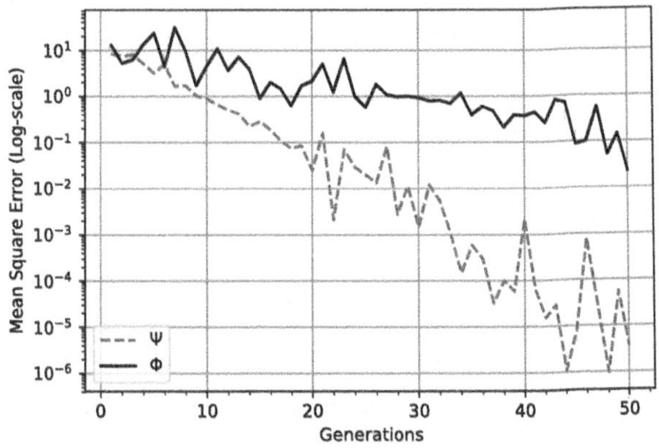

Fig. 13.7 Approximation error (in terms of Euclidean distance) of a predicted lower level optimal solution when using localized Ψ and φ-mapping during the BLEAQ2 algorithm on the five-variable SMD13 test problem

13.6.4 Other Single-Objective EBO Studies

Most often, uncertainties arise from unavoidable variations in implementing an optimized solution. Thus, the issue of uncertainty handling is of great practical significance. Material properties, measurement errors, manufacturing tolerances, interdependence of parameters, environmental conditions, etc. are all sources of uncertainties, which, if not considered during the optimization process, may lead to an optimistic solution without any practical relevance. A recent work [34] introduces the concept of robustness and reliability in bilevel optimization problems arising from uncertainties in both lower and upper level decision variables and parameters. The effect of uncertainties on the final robust/reliable bilevel solution was clearly demonstrated in the study using simple, easy-to-understand test problems, followed by a couple of application problems. The topic of uncertainty handling in bilevel problems is highly practical and timely with the overall growth in research in bilevel methods and in uncertainty handling methods.

Bilevel methods may also be used for achieving an adaptive parameter tuning for optimization algorithms, for instance [48]. The upper level problem considers the algorithmic parameters as variables and the lower level problem uses the actual problem variables. The lower level problem is solved using the algorithmic parameters described by the upper level multiple times and the resulting performance of the algorithm is then used as an objective of the upper level problem. The whole process is able to find optimized parameter values along with the solution of the a number of single-objective test problems.

In certain scenarios, bilevel optimization solution procedures can also be applied to solve single level optimization problems in a more efficient way. For example, the optimal solution of a single-level primal problem can be obtained by solving a dual problem constructed from the Lagranrian function of dual variables. The dual problem formulation is a bilevel (min-max) problem with respect to two sets of variables: upper level involves Lagrangian multipliers or dual variables and lower level involves problem variables or primal variables. A study used evolutionary optimization method to solve the (bilevel) dual problem using a co-evolutionary approach [18]. For zero duality gap problems, the proposed bilevel approach not only finds the optimal solution to the problem, but also produces the Lagrangian multipliers corresponding to the constraints.

13.7 Multi-Objective Bilevel Optimization

Quite often, a decision maker in a practical optimization problem is interested in optimizing multiple conflicting objectives simultaneously. This leads us to the growing literature on multi-objective optimization problem solving [11, 12]. Multiple objectives can also be realized in the context of bilevel problems, where either a leader, or follower, or both might be facing multiple objectives in their own levels [1, 25, 32, 70]. This gives rise to multi-objective bilevel optimization problems that is defined below.

Definition 13.7.1 For the upper level objective function $F : \mathbb{R}^n \times \mathbb{R}^m \to \mathbb{R}^p$ and lower level objective function $f : \mathbb{R}^n \times \mathbb{R}^m \to \mathbb{R}^q$, the multi-objective bilevel problem is given by

$$\begin{aligned} \min_{\mathbf{x}_u, \mathbf{x}_l} \quad & F(\mathbf{x}_u, \mathbf{x}_l) = (F_1(\mathbf{x}_u, \mathbf{x}_l), \ldots, F_p(\mathbf{x}_u, \mathbf{x}_l)), \\ \text{subject to} \quad & \mathbf{x}_l \in \underset{\mathbf{x}_l}{\operatorname{argmin}} \{ f(\mathbf{x}_u, \mathbf{x}_l) = (f_1(\mathbf{x}_u, \mathbf{x}_l), \ldots, f_q(\mathbf{x}_u, \mathbf{x}_l)) : \\ & \qquad g_j(\mathbf{x}_u, \mathbf{x}_l) \le 0, \quad j = 1, \ldots, J \}, \\ & G_k(\mathbf{x}_u, \mathbf{x}_l) \le 0, \quad k = 1, \ldots, K, \end{aligned}$$

where $G_k : \mathbb{R}^n \times \mathbb{R}^m \to \mathbb{R}$, $k = 1, \ldots, K$ denotes the upper level constraints, and $g_j : \mathbb{R}^n \times \mathbb{R}^m \to \mathbb{R}$ represents the lower level constraints, respectively. △

Bilevel problems with multiple objectives at lower level are expected to be considerably more complicated than the single-objective case. Surrogate based methods can once again be used, but even verification of the conditions, for instance, when the use of approximations for the reaction set mapping are applicable, is a challenge and requires quite many tools from variational analysis literature. Some of these conditions are discussed in [53]. As remarked in [52], bilevel optimal solution may not exist for all problems. Therefore, additional regularity and compactness conditions are needed to ensure existence of a solution. This is currently an active area of research. Results have been presented by [21], who established necessary optimality conditions for optimistic multi-objective bilevel problems with the help of Hiriart-Urruty scalarization function.

13.7.1 Optimistic Versus Pessimistic Solutions in Multi-Objective Bilevel Optimization

The optimistic or pessimistic position becomes more prominent in multi-objective bilevel optimization. In the presence of multiple objectives at the lower level, the set-valued mapping $\Psi(\mathbf{x}_u)$ normally represents a set of Pareto-optimal solutions corresponding to any given $\mathbf{x}_u$, which we refer as follower's Pareto-optimal frontier. A solution to the overall problem (with optimistic or pessimistic position) is expected to produce a trade-off frontier for the leader that we refer as the leader's Pareto-optimal frontier. In these problems, the lower level problem produces its own Pareto-optimal set and hence the upper level optimal set depends on which solutions from the lower level would be chosen by the lower level decision-makers. The optimistic and pessimistic fronts at the upper level mark the best and worst possible scenarios at the upper level, given that the lower level solutions are always Pareto-optimal.

Though optimistic position have commonly been studied in classical [20] and evolutionary [17] literature in the context of multi-objective bilevel optimization, it is far from realism to expect that the follower will cooperate (knowingly or unknowingly) to an extent that she chooses any point from her Pareto-optimal frontier that is most suitable for the leader. This relies on the assumption that the follower is indifferent to the entire set of optimal solutions, and therefore decides to cooperate. The situation was entirely different in the single-objective case, where, in case of multiple optimal solutions, all the solutions offered an equal value to the follower. However, this can not be assumed in the multi-objective case. Solution to the optimistic formulation in multi-objective bilevel optimization leads to the best possible Pareto-optimal frontier that can be achieved by the leader. Similarly, solution to the pessimistic formulation leads to the worst possible Pareto-optimal frontier at the upper level.

If the value function or the choice function of the follower is known to the leader, it provides an information as to what kind of trade-off is preferred by the follower. A knowledge of such a function effectively, casually speaking, reduces the lower level optimization problem into a single-objective optimization task, where the value function may be directly optimized. The leader's Pareto-optimal frontier for such intermediate positions lies between the optimistic and the pessimistic frontiers. Figure 13.8 shows the optimistic and pessimistic frontiers for a hypothetical multi-objective bilevel problem with two objectives at upper and lower levels. Follower's frontier corresponding to $\mathbf{x}_u^{(1)}$, $\mathbf{x}_u^{(2)}$ and $\mathbf{x}_u^{(3)}$, and her decisions A_l, B_l and C_l are shown in the insets. The corresponding representations of the follower's frontier and decisions (A_u, B_u and C_u) in the leader's space are also shown.

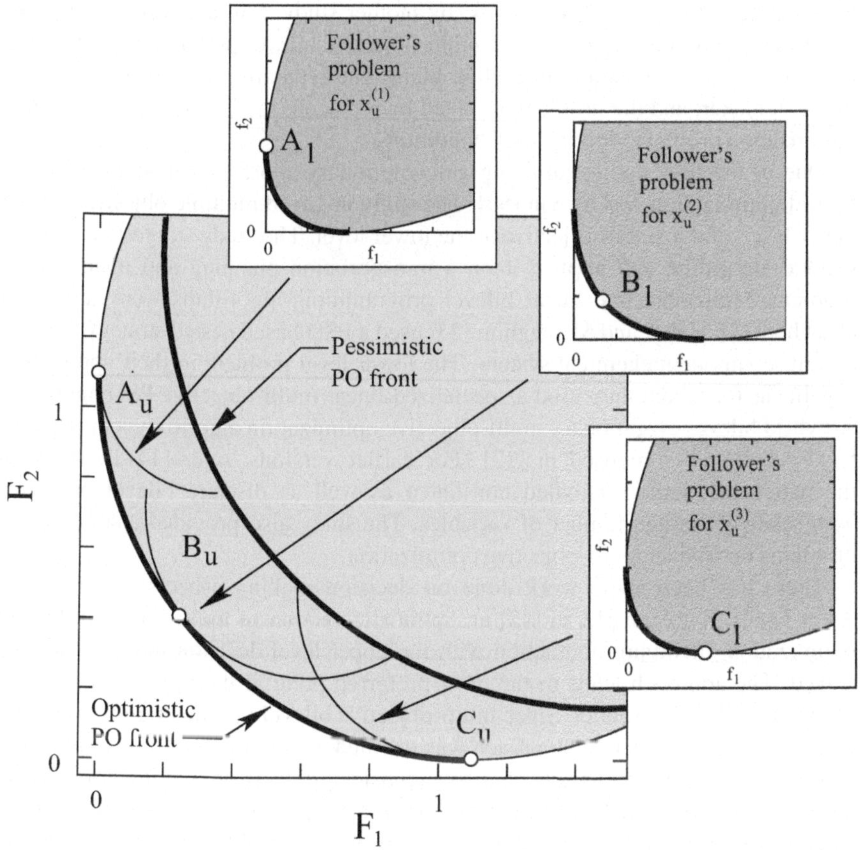

Fig. 13.8 Leader's Pareto-optimal (PO) frontiers for optimistic and pessimistic positions. Few follower's Pareto-optimal (PO) frontiers are shown (in insets) along with their representations in the leader's objective space. Taken from [55]

13.7.2 Bilevel Evolutionary Multi-Objective Optimization Algorithms

There exists a significant amount of work on single objective bilevel optimization; however, little has been done on bilevel multi-objective optimization primarily because of the computational and decision making complexities that these problems offer. For results on optimality conditions in multi-objective bilevel optimization, the readers may refer to [21, 67]. On the methodology side, Eichfelder [19, 20] solved simple multi-objective bilevel problems using a classical approach. The lower level problems in these studies have been solved using a numerical optimization technique, and the upper level problem is handled using an adaptive exhaustive search method. This makes the solution procedure computationally demanding and non-scalable to large-scale problems. In another study, Shi and Xia [40] used ϵ-constraint method at both levels of multi-objective bilevel problem to convert the problem into an ϵ-constraint bilevel problem. The ϵ-parameter is elicited from the decision maker, and the problem is solved by replacing the lower level constrained optimization problem with its KKT conditions.

One of the first studies, utilizing an evolutionary approach for multi-objective bilevel optimization was by Yin [69]. The study involved multiple objectives at the upper lever, and a single objective at the lower level. The study suggested a nested genetic algorithm, and applied it on a transportation planning and management problem. Multi-objective linear bilevel programming algorithms were suggested elsewhere [7]. Halter and Mostaghim [23] used a PSO based nested strategy to solve a multi-component chemical system. The lower level problem in their application was linear for which they used a specialized linear multi-objective PSO approach. A hybrid bilevel evolutionary multi-objective optimization algorithm coupled with local search was proposed in [17] (For earlier versions, refer [14–16, 43]). In the paper, the authors handled non-linear as well as discrete bilevel problems with relatively larger number of variables. The study also provided a suite of test problems for bilevel multi-objective optimization.

There has been some work done on decision making aspects at upper and lower levels. For example, in [42] an optimistic version of multi-objective bilevel optimization, involving interaction with the upper level decision maker, has been solved. The approach leads to the most preferred point at the upper level instead of the entire Pareto-frontier. Since multi-objective bilevel optimization is computationally expensive, such an approach was justified as it led to enormous savings in computational expense. Studies that have considered decision making at the lower level include [49, 52]. In [49], the authors have replaced the lower level with a value function that effectively reduces the lower level problem to single-objective optimization task. In [52], the follower's value function is known with uncertainty, and the authors propose a strategy to handle such problems. Other work related to bilevel multi-objective optimization can be found in [33, 37–39, 71].

13.7.3 BL-EMO for Decision-Making Uncertainty

In most of the practical applications, a departure from the assumption of an indifferent lower level decision maker is necessary [52]. Instead of giving all decision-making power to the leader, the follower is likely to act according to her own interests and choose the most preferred lower level solution herself. As a result, lower level decision making has a substantial impact on the formulation of multi-objective bilevel optimization problems. First, the lower level problem is not a simple constraint that depends only on lower level objectives. Rather, it is more like a selection function that maps a given upper level decision to a corresponding Pareto-optimal lower level solution that it is most preferred by the follower. Second, while solving the bilevel problem, the upper level decision maker now needs to model the follower's behavior by anticipating her preferences towards different objectives. The following formulation of the problem is adapted from [52].

Definition 13.7.2 Let $\xi \in \Xi$ denote a vector of parameters describing the follower's preferences. If the upper level decision maker has complete knowledge of the follower's preferences, the follower's actions can then be modeled via selection mapping

$$\sigma : \mathbf{x}_u \times \Xi \rightarrow \mathbf{x}_l, \quad \sigma(\mathbf{x}_u, \xi) \in \Psi(\mathbf{x}_u), \tag{13.7.1}$$

where Ψ is the set-valued mapping defined earlier. The resulting bilevel problem can be rewritten as follows:

$$\begin{aligned} \min_{\mathbf{x}_u} \quad & F(\mathbf{x}_u, \mathbf{x}_l) = (F_1(\mathbf{x}_u, \mathbf{x}_l), \ldots, F_p(\mathbf{x}_u, \mathbf{x}_l)) \\ \text{subject to} \quad & \mathbf{x}_l = \sigma(\mathbf{x}_u, \xi) \in \Psi(\mathbf{x}_u) \\ & G_k(\mathbf{x}_u, \mathbf{x}_l) \leq 0, \quad k = 1, \ldots, K \end{aligned}$$

△

To model the follower's behavior, one approach is to consider the classical value function framework [52]. We can assume that the follower's preferences are characterized by a function $V : \mathbb{R}^q \times \Xi \rightarrow \mathbb{R}$ that is parameterized by the preference vector ξ. This allows us to write σ as a selection mapping for a value function optimization problem with $\mathbf{x}_u$ and ξ as parameters:

$$\sigma(\mathbf{x}_u, \xi) \in \operatorname*{argmin}_{\mathbf{x}_l}\{V(f(\mathbf{x}_u, \mathbf{x}_l), \xi) : g_j(\mathbf{x}_u, \mathbf{x}_l) \leq 0, \quad j = 1, \ldots, J\}. \tag{13.7.2}$$

When the solution is unique, the above inclusion can be treated as an equality that allows considering σ as a solution mapping for the problem. For most purposes, it is sufficient to assume that V is a linear form where ξ acts as a stochastic weight vector for the different lower level objectives:

$$V(f(\mathbf{x}_u, \mathbf{x}_l), \xi) = \sum_{i=1}^{q} f_i(\mathbf{x}_u, \mathbf{x}_l)\xi_i. \tag{13.7.3}$$

The use of linear value functions to approximate preferences is generally found to be quite effective and works also in situations where the number of objectives is large.

Usually, we have to assume that the follower's preferences are uncertain, i.e. $\xi \sim \mathcal{D}_\xi$, the value function parameterized by ξ is itself a random mapping. To address such problems, the leader can consider the following two-step approach [52]: (1) First, the leader can use her expectation of follower's preferences to

Table 13.4 Two-objective bilevel example problem

Example	Level	Formulation
Variables	Upper level	x_u
	Lower level	$\mathbf{x}_l = (x_{l,1}, \ldots, x_{l,m})$
Objectives	Upper level	$F_1(x_u, \mathbf{x}_l) = (x_{l,1} - 1)^2 + \sum_{i=2}^{m} (x_{l,i})^2 + (x_u)^2$
		$F_2(x_u, \mathbf{x}_l) = (x_{l,1} - 1)^2 + \sum_{i=2}^{m} (x_{l,i})^2 + (x_u - 1)^2$
	Lower level	$f_1(x_u, \mathbf{x}_l) = (x_{l,1})^2 + \sum_{i=2}^{m} (x_{l,i})^2$
		$f_2(x_u, \mathbf{x}_l) = \lvert x_u \rvert (x_{l,1} - x_u)^2 + \sum_{i=2}^{m} (x_{l,i})^2$
Constraints	Upper level	$-1 \leq (x_u, x_{l,1}, \ldots, x_{l,K}) \leq 2$
Preference uncertainty	Lower level	$V(f_1, f_2) = \xi_1 f_1 + \xi_2 f_2,$
		$\xi \sim N_2(\mu_\xi, \Sigma_\xi),\ \mu_\xi = (1, 2),\ \Sigma_\xi = \begin{bmatrix} 0.01 & 0 \\ 0 & 0.01 \end{bmatrix}$

obtain information about the location of the Pareto optimal front by solving the bilevel problem with fixed parameters and value function V. (2) In the second step, the leader can examine the extent of uncertainty by estimating a confidence region around the Pareto optimal frontier corresponding to the expected value function (POF-EVF). Based on the joint evaluation of the expected solutions and the uncertainty observed at different parts of the frontier, the leader can make a better trade-off between her objectives while being aware of the probability of realizing a desired solution. Given the computational complexity of bilevel problems, carrying out these steps requires careful design. One implementation of such algorithm is discussed in detail in [52].

13.7.3.1 An Example

Consider an example which has two objectives at both the levels and the problem is scalable in terms of lower level variables; see Table 13.4. Choosing $m = 14$ will result in a 15 variable bilevel problem with 1 upper level variable and 14 lower level variables. Assume the follower's preferences to follow a linear value function with a bi-variate normal distribution for the weights.

For any x_u, the Pareto-optimal solutions of the lower level optimization problem are given by

$$\Psi(x_u) = \{\mathbf{x}_l \in \mathbb{R}^m \mid x_{l,1} \in [0, x_u],\ x_{l,i} = 0 \text{ for } i = 2, \ldots, m\}.$$

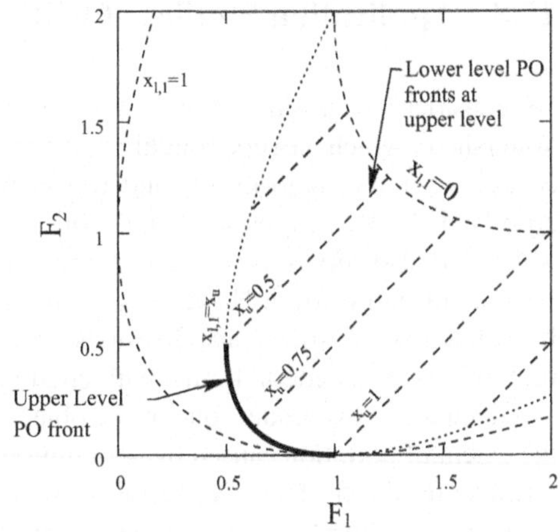

Fig. 13.9 Upper level pareto-optimal front (without lower level decision making) and few representative lower level pareto-optimal fronts in upper level objective space

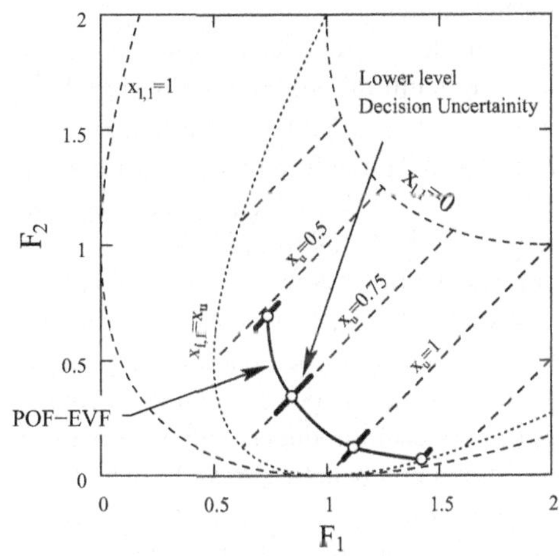

Fig. 13.10 Expected pareto-optimal front at upper level and lower level decision uncertainty

The best possible frontier at the upper level may be obtained when the lower level decision maker does not have any real decision-making power; see Fig. 13.9.

Now let us consider the example with a lower level decision-maker, whose preferences are assumed to follow $V(f_1, f_2) = \xi_1 f_1 + \xi_2 f_2$. The upper level front corresponding to the expected value function is obtained by solving identifying the Pareto optimal front corresponding to the expected value function (POF-EVF). The outcome is shown in Fig. 13.10, where the follower's influence on the bilevel solution is shown as shift of the expected frontier away from the leader's optimal frontier. The extent of decision uncertainty is described using the bold lines around the POF-EVF front. Each line corresponds to the leader's confidence region $C_\alpha(x_u)$ with $\alpha = 0.01$ at different x_u. When examining the confidence regions at different parts of the frontier, substantial variations can be observed.

13.8 Application Studies of EBO

We consider an agri-environmental policy targeting problem for the Raccoon River Watershed, which covers roughly 9400 km^2 in West-Central Iowa. Agriculture accounts for the majority of land use in the study area, with 75.3% of land in crop land, 16.3% in grassland, 4.4% in forests and just 4% in urban use [26]. The watershed also serves as the main source of drinking water for more than 500,000 people in Central Iowa. However, due to its high concentration of nitrate pollution from intensive fertilizer and livestock manure application, nitrate concentrations routinely exceed Federal limits, with citations dating back to the late 1980s.

Given the above issues, one of the objectives the policy maker faces is to reduce the extent of pollution caused by agricultural activities by controlling the amount of fertilizer used [65]. However, at the same time the policy maker does not intend to hamper agricultural activities to an extent of causing significant economic distress for the farmers.

Consider a producer (follower) $i \in \{1, \dots, I\}$ trying to maximize her profits from agricultural production through N inputs $\mathbf{x}^i = \{x_1^i, \dots, x_N^i\}$ and M outputs $\mathbf{y}^i = \{y_1^i, \dots, y_M^i\}$. Out of the N inputs, x_N^i denotes the nitrogen fertilizer input for each farm. The policy maker must choose the optimal spatial allocation of taxes, $\boldsymbol{\tau} = \{\tau^1, \dots, \tau^I\}$ for each farm corresponding to the nitrogen fertilizer usage $\mathbf{x}_N = \{x_N^1, \dots, x_N^I\}$ so as to control the use of fertilizers. Tax vector $\boldsymbol{\tau} = \{\tau^1, \dots, \tau^I\}$ denotes the tax policy for I producers that is expressed as a multiplier on the total cost of fertilizers. Note that the taxes can be applied as a constant for the entire basin, or they can be spatially targeted at semi-aggregated levels or individually at the farm-level. For generality, in our model we have assumed a different tax policy for each producer. The objectives of the upper level are to jointly maximize environmental benefits, $B(\mathbf{x}_N)$—which consists of the total reduction of non-point source emissions of nitrogen runoff from agricultural land—while also maximizing the total basin's profit $\Pi(\boldsymbol{\tau}, \mathbf{x}_N)$. The optimization problem that the policy maker needs to solve in order to identify a Pareto set of efficient policies is given as follows:

$$
\begin{aligned}
\max_{\boldsymbol{\tau}, \mathbf{x}} \ & F(\boldsymbol{\tau}, \mathbf{x}) = \big(\Pi(\boldsymbol{\tau}, \mathbf{x}_N),\ B(\mathbf{x}_N)\big) \qquad (13.8.1) \\
\text{s.t.} \ & x_N^i \in \underset{\mathbf{x}^i}{\text{argmax}}\{\pi^i(\tau^i, x_N^i) : (\tau^i, \mathbf{x}^i) \in \Omega^i\} \\
& x_n^i \geq 0, \ \forall\, i \in \{1, \dots, I\},\ n \in \{1, \dots, N\}, \\
& y_m^i \geq 0, \ \forall\, i \in \{1, \dots, I\},\ m \in \{1, \dots, M\}, \\
& \tau^i \geq 1, \ \forall\, i \in \{1, \dots, I\},
\end{aligned}
$$

The fertilizer tax, $\boldsymbol{\tau}$, serves as a multiplier on the total cost of fertilizer, so $\tau^i \geq 1$. The environmental benefit is the negative of pollution and therefore can be written as the negative of the total pollution caused by the producers, i.e.

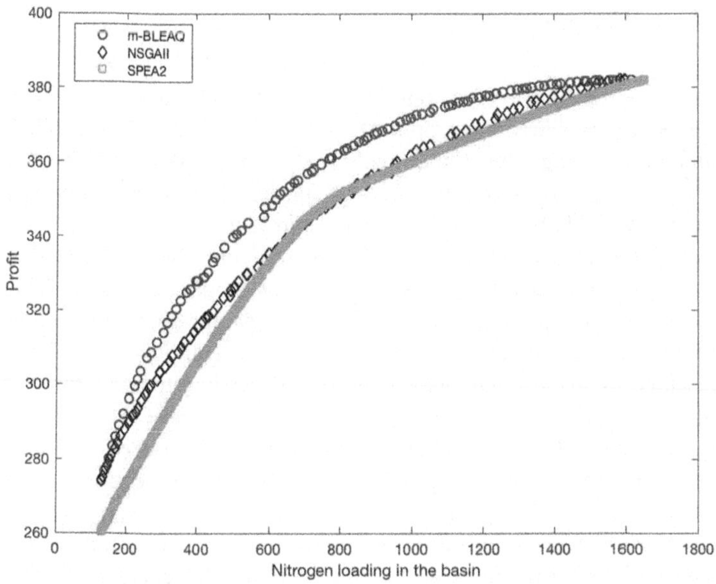

Fig. 13.11 The obtained pareto-optimal frontiers from various methods. The lower level reaction set mapping is analytically defined in this problem. We directly supply the lower level optimal solutions in case of single-level optimization algorithm like NSGA-II and SPEA2 and compare the performance. Taken from [6]

$B(\mathbf{x}_N) = -\sum_{i=1}^{I} p(x_N^i)$. Similarly the total basin profit can be written as the sum total of individual producer's profit, i.e. $\Pi(\boldsymbol{\tau}, \mathbf{x}_N) = \sum_{i=1}^{I} \pi(\tau^i, x_N^i)$. The lower level optimization problem for each agricultural producer can be written as:

$$\max_{x_N^i} \quad \pi^i(\tau^i, x_N^i) = p^i y^i - \sum_{n=1}^{N-1} w_n x_n^i - \tau^i w_N x_N^i, \tag{13.8.2}$$

$$\text{s.t.} \quad y^i \leq P^k(x^i),$$

where w and p are the costs and prices of the fertilizer inputs x and crop yields y, respectively. $P^i(x^i)$ denotes the production frontier for producer i. Heterogeneity across producers, due primarily to differences in soil type, may prevent the use of a common production function that would simplify the solution of (13.8.1). Likewise, the environmental benefits of reduced fertilizer use vary across producers, due to location and hydrologic processes within the basin. This also makes the solution of (13.8.1) more complex.

The simulation results [6] considering all 1175 farms in the Raccoon River Watershed are shown in Fig. 13.11. The Pareto-optimal frontiers trading off the environmental footprints and economic benefits are compared among the three algorithms.

13.9 Conclusions

Bilevel optimization problems are omnipresent in practice. However, these problems are often posed as single-level optimization problems due to the relative ease and availability of single-level optimization algorithms. While this book presents various theoretical and practical aspects of bilevel optimization, this chapter has presented a few viable EAs for solving bilevel problems.

Bilevel problems involve two optimization problems which are nested in nature. Hence, they are computationally expensive to solve to optimality. In this chapter, we have discussed surrogate modeling based approaches for approximating the lower level optimum by a surrogate to speed up the overall computations. Moreover, we have presented multi-objective bilevel algorithms, leading to a number of research and application opportunities. We have also provided a set of systematic, scalable, challenging, single-objective and multi-objective unconstrained and constrained bilevel problems for the bilevel optimization community to develop more efficient algorithms.

Research on bilevel optimization by the evolutionary computation researchers has received a lukewarm interest so far, but hopefully this chapter has provided an overview of some of the existing efforts in the area of evolutionary computation toward solving bilevel optimization problems, so that more efforts are devoted in the near future.

Acknowledgements Some parts of this chapter are adapted from authors' published papers on the topic. Further details can be obtained from the original studies, referenced at the end of this chapter.

Appendix: Bilevel Test Problems

Non-scalable Single-Objective Test Problems from Literature

In this section, we provide some of the standard bilevel test problems used in the evolutionary bilevel optimization studies. Most of these test problems involve only a few fixed number of variables at both upper and lower levels. The test problems (TPs) involve a single objective function at each level. Formulation of the problems are provided in Table 13.5.

Scalable Single-Objective Bilevel Test Problems

Sinha-Malo-Deb (SMD) test problems [46] are a set of 14 scalable single-objective bilevel test problems that offer a variety of controllable difficulties to an algorithm. The SMD test problem suite was originally proposed with eight unconstrained

Table 13.5 Standard test problems TP1–TP8

Problem	Formulation	Best known sol.
TP1 $n = 2$, $m = 2$	$\underset{(\mathbf{x},\mathbf{y})}{\text{Minimize}}\ F(\mathbf{x},\mathbf{y}) = (x_1-30)^2 + (x_2-20)^2 - 20y_1 + 20y_2,$ s.t. $\mathbf{y} \in \underset{(\mathbf{y})}{\text{argmin}}\ \{f(\mathbf{x},\mathbf{y}) = (x_1-y_1)^2 + (x_2-y_2)^2 :$ $0 \le y_i \le 10, \quad i = 1,2\},$ $x_1 + 2x_2 \ge 30, x_1 + x_2 \le 25, x_2 \le 15.$	$F = 225.0$ $f = 100.0$
TP2 $n = 2$, $m = 2$	$\underset{(\mathbf{x},\mathbf{y})}{\text{Minimize}}\ F(\mathbf{x},\mathbf{y}) = 2x_1 + 2x_2 - 3y_1 - 3y_2 - 60,$ s.t. $\mathbf{y} \in \underset{(\mathbf{y})}{\text{argmin}}\ \{f(\mathbf{x},\mathbf{y}) = (y_1 - x_1 + 20)^2 + (y_2 - x_2 + 20)^2 :$ $x_1 - 2y_1 \ge 10, x_2 - 2y_2 \ge 10,$ $-10 \ge y_i \ge 20, \quad i = 1,2\},$ $x_1 + x_2 + y_1 - 2y_2 \le 40,$ $0 \le x_i \le 50, \quad i = 1,2.$	$F = 0.0$ $f = 100.0$
TP3 $n = 2$, $m = 2$	$\underset{(\mathbf{x},\mathbf{y})}{\text{Minimize}}\ F(\mathbf{x},\mathbf{y}) = -(x_1)^2 - 3(x_2)^2 - 4y_1 + (y_2)^2,$ s.t. $\mathbf{y} \in \underset{(\mathbf{y})}{\text{argmin}}\ \{f(\mathbf{x},\mathbf{y}) = 2(x_1)^2 + (y_1)^2 - 5y_2 :$ $(x_1)^2 - 2x_1 + (x_2)^2 - 2y_1 + y_2 \ge -3,$ $x_2 + 3y_1 - 4y_2 \ge 4,$ $0 \le y_i, \quad i = 1,2\},$ $(x_1)^2 + 2x_2 \le 4,$ $0 \le x_i, \quad i = 1,2.$	$F = -18.6787$ $f = -1.0156$
TP4 $n = 2$, $m = 3$	$\underset{(\mathbf{x},\mathbf{y})}{\text{Minimize}}\ F(\mathbf{x},\mathbf{y}) = -8x_1 - 4x_2 + 4y_1 - 40y_2 - 4y_3,$ s.t. $\mathbf{y} \in \underset{(\mathbf{y})}{\text{argmin}}\ \{f(\mathbf{x},\mathbf{y}) = x_1 + 2x_2 + y_1 + y_2 + 2y_3 :$ $y_2 + y_3 - y_1 \le 1,$ $2x_1 - y_1 + 2y_2 - 0.5y_3 \le 1,$ $2x_2 + 2y_1 - y_2 - 0.5y_3 \le 1,$ $0 \le y_i, \quad i = 1,2,3\},$ $0 \le x_i, \quad i = 1,2.$	$F = -29.2$ $f = 3.2$

Table 13.5 Continued.

Problem	Formulation	Best known sol.
TP5 $n = 2$, $m = 2$	$\underset{(\mathbf{x},\mathbf{y})}{\text{Minimize}}\ F(\mathbf{x}, \mathbf{y}) = rt(x)x - 3y_1 - 4y_2 + 0.5t(y)y,$ s.t. $\mathbf{y} \in \underset{(\mathbf{y})}{\text{argmin}}\ \{f(\mathbf{x}, \mathbf{y}) = 0.5t(y)hy - t(b(x))y :$ $-0.333y_1 + y_2 - 2 \leq 0,$ $y_1 - 0.333y_2 - 2 \leq 0,$ $0 \leq y_i, \quad i = 1, 2\},$ where $h = \begin{pmatrix} 1 & 3 \\ 3 & 10 \end{pmatrix}, b(x) = \begin{pmatrix} -1 & 2 \\ 3 & -3 \end{pmatrix} x, r = 0.1,$ $t(\cdot)$ denotes transpose of a vector.	$F = -3.6$ $f = -2.0$
TP6 $n = 1$, $m = 2$	$\underset{(\mathbf{x},\mathbf{y})}{\text{Minimize}}\ F(\mathbf{x}, \mathbf{y}) = (x_1 - 1)^2 + 2y_1 - 2x_1,$ s.t. $\mathbf{y} \in \underset{(\mathbf{y})}{\text{argmin}}\ \{f(\mathbf{x}, \mathbf{y}) = (2y_1 - 4)^2 + (2y_2 - 1)^2 + x_1y_1 :$ $4x_1 + 5y_1 + 4y_2 \leq 12,$ $4y_2 - 4x_1 - 5y_1 \leq -4,$ $4x_1 - 4y_1 + 5y_2 \leq 4,$ $4y_1 - 4x_1 + 5y_2 \leq 4,$ $0 \leq y_i, \quad i = 1, 2\},$ $0 \leq x_1.$	$F = -1.2091$ $f = 7.6145$
TP7 $n = 2$, $m = 2$	$\underset{(\mathbf{x},\mathbf{y})}{\text{Minimize}}\ F(\mathbf{x}, \mathbf{y}) = -\frac{(x_1+y_1)(x_2+y_2)}{1+x_1y_1+x_2y_2},$ s.t. $\mathbf{y} \in \underset{(\mathbf{y})}{\text{argmin}}\ \{f(\mathbf{x}, \mathbf{y}) = \frac{(x_1+y_1)(x_2+y_2)}{1+x_1y_1+x_2y_2} :$ $0 \leq y_i \leq x_i, \quad i = 1, 2\},$ $(x_1)^2 + (x_2)^2 \leq 100,$ $x_1 - x_2 \leq 0,$ $0 \leq x_i, \quad i = 1, 2.$	$F = -1.96$ $f = 1.96$
TP8 $n = 2$, $m = 2$	$\underset{(\mathbf{x},\mathbf{y})}{\text{Minimize}}\ F(\mathbf{x}, \mathbf{y}) = \|2x_1 + 2x_2 - 3y_1 - 3y_2 - 60\|,$ s.t. $\mathbf{y} \in \underset{(\mathbf{y})}{\text{argmin}}\ \{f(\mathbf{x}, \mathbf{y}) = (y_1 - x_1 + 20)^2 + (y_2 - x_2 + 20)^2 :$ $2y_1 - x_1 + 10 \leq 0,$ $2y_2 - x_2 + 10 \leq 0,$ $-10 \leq y_i \leq 20, \quad i = 1, 2\},$ $x_1 + x_2 + y_1 - 2y_2 \leq 40,$ $0 \leq x_i \leq 50, \quad i = 1, 2.$	$F = 0.0$ $f = 100.0$

Note that $\mathbf{x}_u = \mathbf{x}$ and $\mathbf{x}_l = \mathbf{y}$

and four constrained problems [46], it was later extended with two additional unconstrained test problems (SMD13 and SMD14) in [54]. Both these problems contain a difficult φ-mapping, among other difficulties. The upper and lower level functions follow the following structure to induce difficulties due to convergence, interaction, and function dependence between the two levels. The vectors $\mathbf{x}_u = \mathbf{x}$ and $\mathbf{x}_l = \mathbf{y}$ are further divided into two sub-vectors. The φ-mapping is defined by the function f_1. Formulation of SMD test problems are provided in Table 13.6.

$$\begin{aligned} F(\mathbf{x}, \mathbf{y}) &= F_1(\mathbf{x}^1) + F_2(\mathbf{y}^1) + F_3(\mathbf{x}^2, \mathbf{y}^2), \\ f(\mathbf{x}, \mathbf{y}) &= f_1(\mathbf{x}^1, \mathbf{x}^2) + f_2(\mathbf{y}^1) + f_3(\mathbf{x}^2, \mathbf{y}^2), \end{aligned} \tag{13.9.1}$$

where, $\mathbf{x} = (\mathbf{x}^1, \mathbf{x}^2)$ and $\mathbf{y} = (\mathbf{y}^1, \mathbf{y}^2)$.

Bi-objective Bilevel Test Problems

The Deb-Sinha (DS) test suite contains five test problems with two objectives at each level. All problems are scalable with respect to variable dimensions at both levels. Note that $\mathbf{x}_u = \mathbf{x}$ and $\mathbf{x}_l = \mathbf{y}$. The location and shape of respective upper and lower level Pareto-optimal fronts can be found at the original paper [14].

DS 1

Minimize:

$$\begin{cases} F_1(\mathbf{x}, \mathbf{y}) = 1 + r - \cos(\alpha\pi x_1)) + \sum_{j=2}^{K} (x_j - \frac{j-1}{2})^2 + \tau \sum_{i=2}^{K} (y_i - x_i)^2 - r \cos\left(\gamma \frac{pi}{2} \frac{y_1}{x_1}\right) \\ F_2(\mathbf{x}, \mathbf{y}) = 1 + r - \sin(\alpha\pi x_1)) + \sum_{j=2}^{K} (x_j - \frac{j-1}{2})^2 + \tau \sum_{i=2}^{K} (y_i - x_i)^2 - r \sin\left(\gamma \frac{pi}{2} \frac{y_1}{x_1}\right) \end{cases}$$

subject to:

$$\mathbf{y} \in \underset{\mathbf{y}}{\operatorname{argmin}} \begin{cases} f_1(\mathbf{x}, \mathbf{y}) = & y_1^2 + \sum_{i=2}^{K} (y_i - x_i)^2 + \sum_{i=2}^{K} 10(1 - \cos(\frac{\pi}{K}(y_i - x_i))) \\ f_2(\mathbf{x}, \mathbf{y}) = & \sum_{i=1}^{K} (y_i - x_i)^2 + \sum_{i=2}^{K} 10|\sin(\frac{\pi}{K}(y_i - x_i))| \end{cases}$$

$$y_i \in [-K, K],\ i = 1, \ldots, K,\ x_1 \in [1, 4],\ x_j \in [-K, K],\ j = 2, \ldots, K.$$

Recommended parameter setting for this problem, $K = 10$ (overall 20 variables), $r = 0.1, \alpha = 1, \gamma = 1, \tau = 1$. This problem results in a convex upper level Pareto-optimal

Table 13.6 SMD test problems 1–14

Formulation	Variable Bounds
SMD1: $F_1 = \sum_{i=1}^{p} (a_i)^2,$ $F_2 = \sum_{i=1}^{q} (c_i)^2,$ $F_3 = \sum_{i=1}^{r} (b_i)^2 + \sum_{i=1}^{r} (b_i - \tan d_i)^2,$ $f_1 = \sum_{i=1}^{p} (a_i)^2,$ $f_2 = \sum_{i=1}^{q} (c_i)^2,$ $f_3 = \sum_{i=1}^{r} (b_i - \tan d_i)^2.$	$a_i \in [-5, 10],\ \forall i \in \{1, 2, \ldots, p\},$ $b_i \in [-5, 10],\ \forall i \in \{1, 2, \ldots, r\},$ $c_i \in [-5, 10],\ \forall i \in \{1, 2, \ldots, q\},$ $d_i \in (\frac{-\pi}{2}, \frac{\pi}{2}),\ \forall i \in \{1, 2, \ldots, r\}.$
SMD2: $F_1 = \sum_{i=1}^{p} (a_i)^2,$ $F_2 = -\sum_{i=1}^{q} (c_i)^2,$ $F_3 = \sum_{i=1}^{r} (b_i)^2 - \sum_{i=1}^{r} (b_i - \log d_i)^2,$ $f_1 = \sum_{i=1}^{p} (a_i)^2,$ $f_2 = \sum_{i=1}^{q} (c_i)^2,$ $f_3 = \sum_{i=1}^{r} (b_i - \log d_i)^2.$	$a_i \in [-5, 10],\ \forall i \in \{1, 2, \ldots, p\},$ $b_i \in [-5, 1],\ \forall i \in \{1, 2, \ldots, r\},$ $c_i \in [-5, 10],\ \forall i \in \{1, 2, \ldots, q\},$ $d_i \in (0, e],\ \forall i \in \{1, 2, \ldots, r\}.$
SMD3: $F_1 = \sum_{i=1}^{p} (a_i)^2,$ $F_2 = \sum_{i=1}^{q} (c_i)^2,$ $F_3 = \sum_{i=1}^{r} (b_i)^2 + \sum_{i=1}^{r} ((b_i)^2 - \tan d_i)^2,$ $f_1 = \sum_{i=1}^{p} (a_i)^2,$ $f_2 = q + \sum_{i=1}^{q} \left((c_i)^2 - \cos 2\pi c_i\right),$ $f_3 = \sum_{i=1}^{r} ((b_i)^2 - \tan d_i)^2.$	$a_i \in [-5, 10],\ \forall i \in \{1, 2, \ldots, p\},$ $b_i \in [-5, 10],\ \forall i \in \{1, 2, \ldots, r\},$ $c_i \in [-5, 10],\ \forall i \in \{1, 2, \ldots, q\},$ $d_i \in (\frac{-\pi}{2}, \frac{\pi}{2}),\ \forall i \in \{1, 2, \ldots, r\}.$
SMD4: $F_1 = \sum_{i=1}^{p} (a_i)^2,$ $F_2 = -\sum_{i=1}^{q} (c_i)^2,$ $F_3 = \sum_{i=1}^{r} (b_i)^2 - \sum_{i=1}^{r} \left(\lvert b_i \rvert - \log(1 + d_i)\right)^2,$ $f_1 = \sum_{i=1}^{p} (a_i)^2,$ $f_2 = q + \sum_{i=1}^{q} \left((c_i)^2 - \cos 2\pi c_i\right),$ $f_3 = \sum_{i=1}^{r} \left(\lvert b_i \rvert - \log(1 + d_i)\right)^2.$	$a_i \in [-5, 10],\ \forall i \in \{1, 2, \ldots, p\},$ $b_i \in [-1, 1],\ \forall i \in \{1, 2, \ldots, r\},$ $c_i \in [-5, 10],\ \forall i \in \{1, 2, \ldots, q\},$ $d_i \in [0, e],\ \forall i \in \{1, 2, \ldots, r\}.$
SMD5: $F_1 = \sum_{i=1}^{p} (a_i)^2,$ $F_2 = -\sum_{i=1}^{q} \left((c_{i+1} - c_i^2) + (c_i - 1)^2\right),$ $F_3 = \sum_{i=1}^{r} b_i^2 - \sum_{i=1}^{r} \left(\lvert b_i \rvert - d_i^2\right)^2,$ $f_1 = \sum_{i=1}^{p} (a_i)^2,$ $f_2 = \sum_{i=1}^{q} \left((c_{i+1} - c_i^2) + (c_i - 1)^2\right),$ $f_3 = \sum_{i=1}^{r} \left(\lvert b_i \rvert - d_i^2\right)^2.$	$a_i \in [-5, 10],\ \forall i \in \{1, 2, \ldots, p\},$ $b_i \in [-5, 10],\ \forall i \in \{1, 2, \ldots, r\},$ $c_i \in [-5, 10],\ \forall i \in \{1, 2, \ldots, q\},$ $d_i \in [-5, 10],\ \forall i \in \{1, 2, \ldots, r\}.$
SMD6: $F_1 = \sum_{i=1}^{p} (a_i)^2,$ $F_2 = -\sum_{i=1}^{q} c_i^2 + \sum_{i=q+1}^{q+s} c_i^2,$ $F_3 = \sum_{i=1}^{r} b_i^2 - \sum_{i=1}^{r} (b_i - d_i)^2,$ $f_1 = \sum_{i=1}^{p} (a_i)^2,$ $f_2 = \sum_{i=1}^{q} c_i^2 + \sum_{i=q+1, i=i+2}^{q+s-1} (c_{i+1} - c_i)^2,$ $f_3 = \sum_{i=1}^{r} (b_i - d_i)^2.$	$a_i \in [-5, 10],\ \forall i \in \{1, 2, \ldots, p\},$ $b_i \in [-5, 10],\ \forall i \in \{1, 2, \ldots, r\},$ $c_i \in [-5, 10],\ \forall i \in \{1, 2, \ldots, q+s\},$ $d_i \in [-5, 10],\ \forall i \in \{1, 2, \ldots, r\}.$

Table 13.6 Continued.

Formulation	Variable bounds
SMD7: $F_1 = 1 + \frac{1}{400}\sum_{i=1}^{p}(a_i)^2 - \prod_{i=1}^{p}\left(\cos\frac{a_i}{\sqrt{i}}\right)$, $F_2 = -\sum_{i=1}^{q} c_i^2$, $F_3 = \sum_{i=1}^{r} b_i^2 - \sum_{i=1}^{r}(b_i - \log d_i)^2$, $f_1 = \sum_{i=1}^{p} a_i^3$, $f_2 = \sum_{i=1}^{q} c_i^2$, $f_3 = \sum_{i=1}^{r}(b_i - \log d_i)^2$.	$a_i \in [-5, 10],\ \forall\, i \in \{1, 2, \ldots, p\}$, $b_i \in [-5, 10],\ \forall\, i \in \{1, 2, \ldots, r\}$, $c_i \in [-5, 10],\ \forall\, i \in \{1, 2, \ldots, q\}$, $d_i \in (0, e],\ \forall\, i \in \{1, 2, \ldots, r\}$.
SMD8: $F_1 = 20 + e - 20\exp\left(-0.2\sqrt{\frac{1}{p}\sum_{i=1}^{p}(a_i)^2}\right) - \exp\left(\frac{1}{p}\sum_{i=1}^{p}\cos 2\pi a_i\right)$, $F_2 = -\sum_{i=1}^{q}\left((c_{i+1} - c_i^2) + (c_i - 1)^2\right)$, $F_3 = \sum_{i=1}^{r} b_i^2 - \sum_{i=1}^{r}(b_i - d_i^3)^2$, $f_1 = \sum_{i=1}^{p} \lvert a_i \rvert$, $f_2 = \sum_{i=1}^{q}\left((c_{i+1} - c_i^2) + (c_i - 1)^2\right)$, $f_3 = \sum_{i=1}^{r}(b_i - d_i^3)^2$.	$a_i \in [-5, 10],\ \forall\, i \in \{1, 2, \ldots, p\}$, $b_i \in [-5, 10],\ \forall\, i \in \{1, 2, \ldots, r\}$, $c_i \in [-5, 10],\ \forall\, i \in \{1, 2, \ldots, q\}$, $d_i \in [-5, 10],\ \forall\, i \in \{1, 2, \ldots, r\}$.
SMD9: $F_1 = \sum_{i=1}^{p}(a_i)^2$ $F_2 = -\sum_{i=1}^{q}(c_i)^2$, $F_3 = \sum_{i=1}^{r} b_i^2 - \sum_{i=1}^{r}(b_i - \log(1 + d_i))^2$, $f_1 = \sum_{i=1}^{p} a_i^2$, $f_2 = \sum_{i=1}^{q} c_i^2$, $f_3 = \sum_{i=1}^{r}(b_i - \log(1 + d_i))^2$. $G_1 : \sum_{i=1}^{p} a_i^2 + \sum_{i=1}^{r} b_i^2 - \lfloor \sum_{i=1}^{p} a_i^2 + \sum_{i=1}^{r} b_i^2 + 0.5 \rfloor \geq 0$ $g_1 : \sum_{i=1}^{p} c_i^2 + \sum_{i=1}^{r} d_i^2 - \lfloor \sum_{i=1}^{p} c_i^2 + \sum_{i=1}^{r} d_i^2 + 0.5 \rfloor \geq 0$	$a_i \in [-5, 10],\ \forall\, i \in \{1, 2, \ldots, p\}$, $b_i \in [-5, 1],\ \forall\, i \in \{1, 2, \ldots, r\}$, $c_i \in [-5, 10],\ \forall\, i \in \{1, 2, \ldots, q\}$, $d_i \in (-1, -1 + e],\ \forall\, i \in \{1, 2, \ldots, r\}$.
SMD10: $F_1 = \sum_{i=1}^{p}(a_i - 2)^2$ $F_2 = -\sum_{i=1}^{q} c_i^2$, $F_3 = \sum_{i=1}^{r}(b_i^2 - 2)^2 - \sum_{i=1}^{r}(b_i - \tan d_i)^2$, $f_1 = \sum_{i=1}^{p} a_i^2$, $f_2 = \sum_{i=1}^{q}(c_i - 2)^2$, $f_3 = \sum_{i=1}^{r}(b_i - \tan d_i)^2$. $G_j : a_j - \sum_{i=1, i \neq j}^{p} a_i^3 - \sum_{i=1}^{r} b_i^3 \geq 0, \forall j \in \{1, 2, \ldots, p\}$ $G_{p+j} : b_j - \sum_{i=1, i \neq j}^{p} b_i^3 - \sum_{i=1}^{r} b_i^3 \geq 0, \forall j \in \{1, 2, \ldots, r\}$ $g_j : c_j - \sum_{i=1, i \neq j}^{q} c_i^3 \geq 0, \forall j \in \{1, 2, \ldots, q\}$	$a_i \in [-5, 10],\ \forall\, i \in \{1, 2, \ldots, p\}$, $b_i \in [-5, 10],\ \forall\, i \in \{1, 2, \ldots, r\}$, $c_i \in [-5, 10],\ \forall\, i \in \{1, 2, \ldots, q\}$, $d_i \in (\frac{-\pi}{2}, \frac{\pi}{2}),\ \forall\, i \in \{1, 2, \ldots, r\}$.

Table 13.6 Continued.

Formulation	Variable Bounds
SMD11: $F_1 = \sum_{i=1}^{p} a_i^2$ $F_2 = -\sum_{i=1}^{q} c_i^2,$ $F_3 = \sum_{i=1}^{r} b_i^2 - \sum_{i=1}^{r} (b_i - \log d_i)^2,$ $f_1 = \sum_{i=1}^{p} a_i^2,$ $f_2 = \sum_{i=1}^{q} c_i^2,$ $f_3 = \sum_{i=1}^{r} (b_i - \log d_i)^2.$ $G_j : b_j \geq \frac{1}{\sqrt{r}} + \log d_j, \forall j \in \{1, 2, \ldots, r\}$ $g_j : \sum_{i=1}^{r} (b_i - \log d_i) \geq 1$	$a_i \in [-5, 10],\ \forall\, i \in \{1, 2, \ldots, p\},$ $b_i \in [-1, 1],\ \forall\, i \in \{1, 2, \ldots, r\},$ $c_i \in [-5, 10],\ \forall\, i \in \{1, 2, \ldots, q\},$ $d_i \in (\frac{1}{e}, e),\ \forall\, i \in \{1, 2, \ldots, r\}.$
SMD12: $F_1 = \sum_{i=1}^{p} (a_i - 2)^2$ $F_2 = \sum_{i=1}^{q} c_i^2,$ $F_3 = \sum_{i=1}^{r} (b_i^2 - 2)^2 + \sum_{i=1}^{r} \tan \lvert d_i \rvert$ $- \sum_{i=1}^{r} (b_i - \tan d_i)^2,$ $f_1 = \sum_{i=1}^{p} a_i^2,$ $f_2 = \sum_{i=1}^{q} (c_i - 2)^2,$ $f_3 = \sum_{i=1}^{r} (b_i - \tan d_i)^2.$ $G_1 : b_i - \tan d_i \geq 0, \forall i \in \{1, 2, \ldots, r\}$ $G_2 : a_i - \sum_{i=1, i \neq j}^{p} a_i^3 - \sum_{i=1}^{r} b_i^3$ $\geq 0, \forall j \in \{1, 2, \ldots, p\}$ $G_3 : b_i - \sum_{i=1, i \neq j}^{r} b_i^3 - \sum_{i=1}^{p} a_i^3$ $\geq 0, \forall j \in \{1, 2, \ldots, r\}$ $g_1 : \sum_{i=1}^{r} (b_i - \tan d_i)^2 \geq 1$ $g_2 : c_j - \sum_{i=1, i \neq j}^{p} c_i^3, \forall j \in \{1, 2, \ldots, q\}$	$a_i \in [-5, 10],\ \forall\, i \in \{1, 2, \ldots, p\},$ $b_i \in [-14.1, 14.1],\ \forall\, i \in \{1, 2, \ldots, r\},$ $c_i \in [-5, 10],\ \forall\, i \in \{1, 2, \ldots, q\},$ $d_i \in (-1.5, 1.5),\ \forall\, i \in \{1, 2, \ldots, r\}.$
SMD13: $F_1 = \sum_{i=1}^{p-1} (a_i - 1)^2 + (a_{i+1} - (a_i)^2)^2,$ $F_2 = -\sum_{i=1}^{q} \sum_{j=1}^{i} (c_j)^2,$ $F_3 = \sum_{i=1}^{r} \sum_{j=1}^{i} (b_j)^2 - \sum_{i=1}^{r} (b_i - \log d_i)^2,$ $f_1 = \sum_{i=1}^{p} \left(\lvert a_i \rvert + 2 \lvert \sin(a_i) \rvert \right),$ $f_2 = \sum_{i=1}^{q} \sum_{j=1}^{i} (c_j)^2,$ $f_3 = \sum_{i=1}^{r} (b_i - \log d_i)^2.$	$a_i \in [-5, 10],\ \forall\, i \in \{1, 2, \ldots, p\},$ $b_i \in [-5, e],\ \forall\, i \in \{1, 2, \ldots, r\},$ $c_i \in [-5, 10],\ \forall\, i \in \{1, 2, \ldots, q\},$ $d_i \in (0, 10],\ \forall\, i \in \{1, 2, \ldots, r\}.$
SMD14: $F_1 = \sum_{i=1}^{p-1} (a_i - 1)^2 + (a_{i+1} - (a_i)^2)^2,$ $F_2 = -\sum_{i=1}^{q} \lvert c_i \rvert^{i+1} + \sum_{i=q+1}^{q+s} (c_i)^2,$ $F_3 = \sum_{i=1}^{r} i (b_i)^2 - \sum_{i=1}^{r} \lvert d_i \rvert,$ $f_1 = \sum_{i=1}^{p} \lfloor a_i \rfloor,$ $f_2 = \sum_{i=1}^{q} \lvert c_i \rvert^{i+1} + \sum_{i=q+1, i=i+2}^{q+s-1} (c_{i+1} - c_i)^2,$ $f_3 = \sum_{i=1}^{r} \lvert (b_i)^2 - (d_i)^2 \rvert.$	$a_i \in [-5, 10],\ \forall\, i \in \{1, 2, \ldots, p\},$ $b_i \in [-5, 10],\ \forall\, i \in \{1, 2, \ldots, r\},$ $c_i \in [-5, 10],\ \forall\, i \in \{1, 2, \ldots, q+s\},$ $d_i \in [-5, 10],\ \forall\, i \in \{1, 2, \ldots, r\}.$

Note that $(\mathbf{x}^1, \mathbf{x}^2) = (\mathbf{a}, \mathbf{b})$ and $(\mathbf{y}^1, \mathbf{y}^2) = (\mathbf{c}, \mathbf{d})$

front in which one specific solution from each lower level Pareto-optimal front gets associated with each upper level Pareto-optimal solution.

DS 2

Minimize:

$$\begin{cases} F_1(\mathbf{x},\mathbf{y}) = v_1(x_1) + \sum_{j=2}^{K}\left[x_j^2 + 10(1-\cos(\frac{\pi}{K}x_j))\right] + \tau\sum_{i=2}^{K}(y_i - x_i)^2 - r\,\cos\left(\gamma\frac{\pi}{2}\frac{y_1}{x_1}\right) \\ F_2(\mathbf{x},\mathbf{y}) = v_2(x_1) + \sum_{j=2}^{K}\left[x_j^2 + 10(1-\cos(\frac{\pi}{K}x_j))\right] + \tau\sum_{i=2}^{K}(y_i - x_i)^2 - r\,\sin\left(\gamma\frac{\pi}{2}\frac{y_1}{x_1}\right) \\ v_1(x_1) = \begin{cases} \cos(0.2\pi)x_1 + \sin(0.2\pi)\sqrt{|0.02\sin(5\pi x_1)|}, & \text{for } 0 \le x_1 \le 1; \\ x_1 - (1-\cos(0.2\pi)), & \text{for } x_1 > 1. \end{cases} \\ v_2(x_1) = \begin{cases} -\sin(0.2\pi)x_1 + \cos(0.2\pi)\sqrt{|0.02\sin(5\pi x_1)|}, & \text{for } 0 \le x_1 \le 1; \\ 0.1(x_1 - 1) - \sin(0.2\pi), & \text{for } x_1 > 1. \end{cases} \end{cases}$$

subject to:

$$\mathbf{y} \in \underset{\mathbf{y}}{\operatorname{argmin}} \begin{cases} f_1(\mathbf{x},\mathbf{y}) = & y_1^2 + \sum_{i=2}^{K}(y_i - x_i)^2 \\ f_2(\mathbf{x},\mathbf{y}) = & \sum_{i=1}^{K} i(y_i - x_i)^2 \end{cases}$$

$$y_i \in [-K, K],\ i = 1, \ldots, K,\ x_1 \in [0.001, K],\ x_j \in [-K, K],\ j = 2, \ldots, K.$$

Recommended parameter setting for this problem, $K = 10$ (overall 20 variables), $r = 0.25$. Due to the use of periodic terms in v_1 and v_2 functions, the upper level Pareto front corresponds to only six discrete values of $y_1 = [0.001, 0.2, 0.4, 0.6, 0.8, 1]$. Setting $\tau = -1$ will introduces a conflict between upper and lower level problems. For this problem, a number of contiguous lower level Pareto-optimal solutions are Pareto-optimal at the upper level for each upper level Pareto-optimal variable vector.

DS 3

Minimize:

$$\begin{cases} F_1(\mathbf{x},\mathbf{y}) = x_1 + \sum_{j=3}^{K}(x_j - j/2)^2 + \tau\sum_{i=3}^{K}(y_i - x_i)^2 - R(x_1)\cos\left(4\tan^{-1}\left(\frac{x_2 - y_2}{x_1 - y_1}\right)\right), \\ F_2(\mathbf{x},\mathbf{y}) = x_2 + \sum_{j=3}^{K}(x_j - j/2)^2 + \tau\sum_{i=3}^{K}(y_i - x_i)^2 - R(x_1)\sin\left(4\tan^{-1}\left(\frac{x_2 - y_2}{x_1 - y_1}\right)\right), \end{cases}$$

subject to:

$$\mathbf{y} \in \underset{\mathbf{y}}{\operatorname{argmin}} \begin{cases} f_1(\mathbf{x}, \mathbf{y}) = y_1 + \sum_{i=3}^{K} (y_i - x_i)^2, \\ f_2(\mathbf{x}, \mathbf{y}) = y_2 + \sum_{i=3}^{K} (y_i - x_i)^2, \\ \text{subject to :} \quad g_1(y) = (y_1 - x_1)^2 + (y_2 - x_2)^2 \leq r^2, \end{cases}$$
$$G(\mathbf{x}) = x_2 - (1 - x_1^2) \geq 0,$$
$$y_i \in [-K, K], \ i = 1, \ldots, K, \ x_j \in [0, K], \ j = 1, \ldots, K, \ x_1 \text{ is a multiple of } 0.1.$$

In this test problem, the variable x_1 is considered to be discrete, thereby causing only a few x_1 values to represent the upper level Pareto front. Recommended parameter setting for this problem: $R(x_1) = 0.1 + 0.15|\sin(2\pi(x_1 - 0.1)|$ and use $r = 0.2$, $\tau = 1$, and $K = 10$. Like in DS2, in this problem, parts of lower level Pareto-optimal front become upper level Pareto-optimal.

DS 4

Minimize:

$$\begin{cases} F_1(\mathbf{x}, \mathbf{y}) = (1 - y_1)(1 + \sum_{j=2}^{K} y_j^2) x_1, \\ F_2(\mathbf{x}, \mathbf{y}) = y_1 (1 + \sum_{j=2}^{K} y_j^2) x_1, \end{cases} \quad \text{subject to :}$$
$$\mathbf{y} \in \underset{\mathbf{y}}{\operatorname{argmin}} \begin{cases} f_1(\mathbf{x}, \mathbf{y}) = (1 - y_1)(1 + \sum_{j=K+1}^{K+L} y_j^2) x_1, \\ f_2(\mathbf{x}, \mathbf{y}) = y_1 (1 + \sum_{j=K+1}^{K+L} y_j^2) x_1, \end{cases}$$
$$G(\mathbf{x}) = (1 - y_1) x_1 + \tfrac{1}{2} x_1 y_1 - 1 \geq 0,$$
$$1 \leq x_1 \leq 2, \quad -1 \leq y_1 \leq 1, \quad -(K + L) \leq y_i \leq (K + L), \ i = 2, \ldots, (K + L).$$

For this problem, there are a total of $K + L + 1$ variables. The original study recommended $K = 5$ and $L = 4$. This problem has a linear upper level Pareto-optimal front in which a single lower level solution from a linear Pareto-optimal front gets associated with the respective upper level variable vector.

DS 5

This problem exactly the same as DS4, except

$$G(\mathbf{x}) = (1 - y_1) x_1 + \frac{1}{2} x_1 y_1 - 2 + \frac{1}{5} [5(1 - y_1) x_1 + 0.2] \geq 0.$$

This makes a number of lower level Pareto-optimal solutions to be Pareto-optimal at the upper level for each upper level variable vector.

References

1. M.J. Alves, J.P. Costa, An algorithm based on particle swarm optimization for multiobjective bilevel linear problems. Appl. Math. Comput. **247**(C), 547–561 (2014)
2. J.S. Angelo, H.J.C. Barbosa, A study on the use of heuristics to solve a bilevel programming problem. Int. Trans. Oper. Res. **22**(5), 861–882 (2015)
3. J. Angelo, E. Krempser, H. Barbosa, Differential evolution for bilevel programming, in *Proceedings of the 2013 Congress on Evolutionary Computation (CEC-2013)* (IEEE Press, Piscataway, 2013)
4. J.S. Angelo, E. Krempser, H.J.C. Barbosa, Differential evolution assisted by a surrogate model for bilevel programming problems, in *2014 IEEE Congress on Evolutionary Computation (CEC)* (IEEE, Piscataway, 2014), pp. 1784–1791
5. S. Bandaru, K. Deb, Metaheuristic techniques (chapter 11), in *Decision Sciences: Theory and Practice*, ed. by R.N. Sengupta, A. Gupta, J. Dutta (CRC Press, Boca Raton, 2016)
6. B. Barnhart, Z. Lu, M. Bostian, A. Sinha, K. Deb, L. Kurkalova, M. Jha, G. Whittaker, Handling practicalities in agricultural policy optimization for water quality improvements, in *Proceedings of the Genetic and Evolutionary Computation Conference GECCO'17* (ACM, New York, 2017), pp. 1065–1072
7. H.I. Calvete, C. Galé, On linear bilevel problems with multiple objectives at the lower level. Omega **39**(1), 33–40 (2011)
8. H.I. Calvete, C. Gale, P.M. Mateo, A new approach for solving linear bilevel problems using genetic algorithms. Eur. J. Oper. Res. **188**(1), 14–28 (2008)
9. H.I. Calvete, C. Galé, M. Oliveros, Bilevel model for production–distribution planning solved by using ant colony optimization. Comput. Oper. Res. **38**(1), 320–327 (2011)
10. J.-F. Camacho-Vallejo, R. Muñoz-Sánchez, J.L. González-Velarde, A heuristic algorithm for a supply chain's production-distribution planning. Comput. Oper. Res. **61**, 110–121 (2015)
11. C.A.C. Coello, D.A. VanVeldhuizen, G. Lamont, *Evolutionary Algorithms for Solving Multi-Objective Problems* (Kluwer, Boston, 2002)
12. K. Deb, *Multi-Objective Optimization Using Evolutionary Algorithms* (Wiley, Chichester, 2001)
13. K. Deb, C. Myburgh, A population-based fast algorithm for a billion-dimensional resource allocation problem with integer variables. Eur. J. Oper. Res. **261**(2), 460–474 (2017)
14. K. Deb, A. Sinha, Constructing test problems for bilevel evolutionary multi-objective optimization, in *2009 IEEE Congress on Evolutionary Computation (CEC-2009)* (IEEE Press, Piscataway, 2009), pp. 1153–1160
15. K. Deb, A. Sinha, An evolutionary approach for bilevel multi-objective problems, in *Cutting-Edge Research Topics on Multiple Criteria Decision Making, Communications in Computer and Information Science*, vol. 35 (Springer, Berlin, 2009), pp. 17–24
16. K. Deb, A. Sinha, Solving bilevel multi-objective optimization problems using evolutionary algorithms, in *Evolutionary Multi-Criterion Optimization (EMO-2009)* (Springer, Berlin, 2009), pp. 110–124
17. K. Deb, A. Sinha, An efficient and accurate solution methodology for bilevel multi-objective programming problems using a hybrid evolutionary-local-search algorithm. Evol. Comput. J. **18**(3), 403–449 (2010)
18. K. Deb, S. Gupta, J. Dutta, B. Ranjan, Solving dual problems using a coevolutionary optimization algorithm. J. Global Optim. **57**, 891–933 (2013)

19. G. Eichfelder, Solving nonlinear multiobjective bilevel optimization problems with coupled upper level constraints. Technical Report Preprint No. 320, Preprint-Series of the Institute of Applied Mathematics, University of Erlangen-Nornberg (2007)
20. G. Eichfelder, Multiobjective bilevel optimization, Math. Program. **123**(2), 419–449 (2010)
21. N. Gadhi, S. Dempe, Necessary optimality conditions and a new approach to multiobjective bilevel optimization problems. J. Optim. Theory Appl. **155**(1), 100–114 (2012)
22. F. Glover, G.A. Kochenberger, *Handbook of Metaheuristics* (Springer, Berlin, 2003)
23. W. Halter, S. Mostaghim, Bilevel optimization of multi-component chemical systems using particle swarm optimization, in *Proceedings of World Congress on Computational Intelligence (WCCI-2006)* (2006), pp. 1240–1247
24. S.R. Hejazi, A. Memariani, G. Jahanshahloo, M.M. Sepehri, Linear bilevel programming solution by genetic algorithm. Comput. Oper. Res. **29**(13), 1913–1925 (2002)
25. M.M. Islam, H.K. Singh, T. Ray, A nested differential evolution based algorithm for solving multi-objective bilevel optimization problems, in *Proceedings of the Second Australasian Conference on Artificial Life and Computational Intelligence - Volume 9592* (Springer, Berlin, 2016), pp. 101–112
26. M.K. Jha, P.W. Gassman, J.G. Arnold, Water quality modeling for the raccoon river watershed using swat. Trans. ASAE **50**(2), 479–493 (2007)
27. Y. Jiang, X. Li, C. Huang, X. Wu, Application of particle swarm optimization based on CHKS smoothing function for solving nonlinear bilevel programming problem. Appl. Math. Comput. **219**(9), 4332–4339 (2013)
28. J. Kennedy, R. Eberhart, Particle swarm optimization, in *Proceedings of IEEE International Conference on Neural Networks* (IEEE Press, Piscataway, 1995), pp. 1942–1948
29. H. Li, A genetic algorithm using a finite search space for solving nonlinear/linear fractional bilevel programming problems. Ann. Oper. Res. **235**, 543–558 (2015)
30. H. Li, Y. Wand, A hybrid genetic algorithm for solving nonlinear bilevel programming problems based on the simplex method. Int. Conf. Nat. Comput. **4**, 91–95 (2007)
31. X. Li, P. Tian, X. Min, A hierarchical particle swarm optimization for solving bilevel programming problems, in *Proceedings of Artificial Intelligence and Soft Computing (ICAISC'06)* (2006), pp. 1169–1178. Also LNAI 4029
32. H. Li, Q. Zhang, Q. Chen, L. Zhang, Y.-C. Jiao, Multiobjective differential evolution algorithm based on decomposition for a type of multiobjective bilevel programming problems. Knowl. Based Syst. **107**, 271–288 (2016)
33. M. Linnala, E. Madetoja, H. Ruotsalainen, J. Hämäläinen, Bi-level optimization for a dynamic multiobjective problem. Eng. Optim. **44**(2), 195–207 (2012)
34. Z. Lu, K. Deb, A. Sinha, Uncertainty handling in bilevel optimization for robust and reliable solutions. Int. J. Uncertainty, Fuzziness Knowl. Based Syst. **26**(suppl. 2), 1–24 (2018)
35. R. Mathieu, L. Pittard, G. Anandalingam, Genetic algorithm based approach to bi-level linear programming. Oper. Res. **28**(1), 1–21 (1994)
36. J.-A. Mejía-de Dios, E. Mezura-Montes, A physics inspired algorithm for bilevel optimization, in *2018 IEEE International Autumn Meeting on Power, Electronics and Computing (ROPEC)* (2018), pp. 1–6
37. C.O. Pieume, L.P. Fotso, P. Siarry, Solving bilevel programming problems with multicriteria optimization techniques. OPSEARCH **46**(2), 169–183 (2009)
38. S. Pramanik , P.P. Dey, Bi-level multi-objective programming problem with fuzzy parameters. Int. J. Comput. Appl. **30**(10), 13–20 (2011). Published by Foundation of Computer Science, New York, USA
39. S. Ruuska, K. Miettinen, Constructing evolutionary algorithms for bilevel multiobjective optimization, in *2012 IEEE Congress on Evolutionary Computation (CEC)* (2012), pp. 1–7
40. X. Shi, H.S. Xia, Model and interactive algorithm of bi-level multi-objective decision-making with multiple interconnected decision makers. J. Multi-Criteria Decis. Anal. **10**(1), 27–34 (2001)

41. H.K. Singh, M.M. Islam, T. Ray, M. Ryan, Nested evolutionary algorithms for computationally expensive bilevel optimization problems: variants and their systematic analysis. Swarm Evol. Comput. **48**, 329–344 (2019)
42. A. Sinha, Bilevel multi-objective optimization problem solving using progressively interactive evolutionary algorithm, in *Proceedings of the Sixth International Conference on Evolutionary Multi-Criterion Optimization (EMO-2011)* (Springer, Berlin, 2011), pp. 269–284
43. A. Sinha, K. Deb, Towards understanding evolutionary bilevel multi-objective optimization algorithm, in *IFAC Workshop on Control Applications of Optimization (IFAC-2009)*, vol. 7 (Elsevier, Amsterdam, 2009)
44. A. Sinha, P. Malo, K. Deb, Efficient evolutionary algorithm for single-objective bilevel optimization (2013). arXiv:1303.3901
45. A. Sinha, P. Malo, K. Deb, An improved bilevel evolutionary algorithm based on quadratic approximations, in *2014 IEEE Congress on Evolutionary Computation (CEC-2014)* (IEEE Press, Piscataway, 2014), pp. 1870–1877
46. A. Sinha, P. Malo, K. Deb, Test problem construction for single-objective bilevel optimization. Evol. Comput. J. **22**(3), 439–477 (2014)
47. A. Sinha, P. Malo, A. Frantsev, K. Deb, Finding optimal strategies in a multi-period multi-leader-follower stackelberg game using an evolutionary algorithm. Comput. Oper. Res. **41**, 374–385 (2014)
48. A. Sinha, P. Malo, P. Xu, K. Deb, A bilevel optimization approach to automated parameter tuning, in *Proceedings of the 16th Annual Genetic and Evolutionary Computation Conference (GECCO 2014)* (ACM Press, New York, 2014)
49. A. Sinha, P. Malo, K. Deb, Towards understanding bilevel multi-objective optimization with deterministic lower level decisions, in *Proceedings of the Eighth International Conference on Evolutionary Multi-Criterion Optimization (EMO-2015)*. (Springer, Berlin, 2015)
50. A. Sinha, P. Malo, K. Deb, Evolutionary algorithm for bilevel optimization using approximations of the lower level optimal solution mapping. Eur. J. Oper. Res. **257**, 395–411 (2016)
51. A. Sinha, P. Malo, K. Deb, Solving optimistic bilevel programs by iteratively approximating lower level optimal value function, in *2016 IEEE Congress on Evolutionary Computation (CEC-2016)* (IEEE Press, Piscataway, 2016)
52. A. Sinha, P. Malo, K. Deb, P. Korhonen, J. Wallenius, Solving bilevel multi-criterion optimization problems with lower level decision uncertainty. IEEE Trans. Evol. Comput. **20**(2), 199–217 (2016)
53. A. Sinha, P. Malo, K. Deb, Approximated set-valued mapping approach for handling multiobjective bilevel problems. Comput. Oper. Res. **77**, 194–209 (2017)
54. A. Sinha, Z. Lu, K. Deb, P. Malo, Bilevel optimization based on iterative approximation of multiple mappings (2017). arXiv:1702.03394
55. A. Sinha, P. Malo, K. Deb, A review on bilevel optimization: from classical to evolutionary approaches and applications. IEEE Trans. Evol. Comput. **22**(2), 276–295 (2018)
56. A. Sinha, S. Bedi, K. Deb, Bilevel optimization based on kriging approximations of lower level optimal value function, in *2018 IEEE Congress on Evolutionary Computation (CEC)* (IEEE, Piscataway, 2018), pp. 1–8.
57. A. Sinha, T. Soun, K. Deb, Using Karush-Kuhn-Tucker proximity measure for solving bilevel optimization problems. Swarm Evol. Comput. **44**, 496–510 (2019)
58. A. Sinha, Z. Lu, K. Deb, P. Malo, Bilevel optimization based on iterative approximation of multiple mappings. J. Heuristics **26**, 151–185 (2020)
59. R. Storn, K. Price, Differential evolution – a fast and efficient heuristic for global optimization over continuous spaces. J. Global Optim. **11**, 341–359 (1997)
60. Z. Wan, G. Wang, B. Sun, A hybrid intelligent algorithm by combining particle swarm optimization with chaos searching technique for solving nonlinear bilevel programming problems. Swarm Evol. Comput. **8**, 26–32 (2013)
61. G. Wang, S. Shan, Review of metamodeling techniques in support of engineering design optimization. J. Mech. Des. **129**(4), 370–380 (2007)

62. Y. Wang, Y.C. Jiao, H. Li, An evolutionary algorithm for solving nonlinear bilevel programming based on a new constraint-handling scheme. IEEE Trans. Syst. Man Cybern. C Appl. Rev. **32**(2), 221–232 (2005)
63. G. Wang, Z. Wan, X. Wang, Y. Lu, Genetic algorithm based on simplex method for solving linear-quadratic bilevel programming problem. Comput. Math. Appl. **56**(10), 2550–2555 (2008)
64. Y. Wang, H. Li, C. Dang, A new evolutionary algorithm for a class of nonlinear bilevel programming problems and its global convergence. INFORMS J. Comput. **23**(4), 618–629 (2011)
65. G. Whittaker, R. Färe, S. Grosskopf, B. Barnhart, M. Bostian, G. Mueller-Warrant, S. Griffith, Spatial targeting of agri-environmental policy using bilevel evolutionary optimization. Omega **66**(A), 15–27 (2017)
66. D.H. Wolpert, W.G. Macready, No free lunch theorems for optimization. IEEE Trans. Evol. Comput. **1**(1), 67–82 (1997)
67. J.J. Ye, Necessary optimality conditions for multiobjective bilevel programs. Math. Oper. Res. **36**(1), 165–184 (2011)
68. J.J. Ye, D. Zhu, New necessary optimality conditions for bilevel programs by combining the MPEC and value function approaches. SIAM J. Optim. **20**(4), 1885–1905 (2010)
69. Y. Yin, Genetic algorithm based approach for bilevel programming models. J. Transp. Eng. **126**(2), 115–120 (2000)
70. T. Zhang, T. Hu, X. Guo, Z. Chen, Y. Cheng, Solving high dimensional bilevel multiobjective programming problem using a hybrid particle swarm optimization algorithm with crossover operator. Knowl. Based Syst. **53**, 13–19 (2013)
71. T. Zhang, T. Hu, Y. Zheng, X. Guo, An improved particle swarm optimization for solving bilevel multiobjective programming problem. J. Appl. Math. **2012**, 1–13 (2012)
72. X. Zhu, Q. Yu, X. Wang, A hybrid differential evolution algorithm for solving nonlinear bilevel programming with linear constraints, in *5th IEEE International Conference on Cognitive Informatics, 2006. ICCI 2006*, vol. 1 (IEEE, Piscataway, 2006), pp. 126–131

Chapter 14
Methods for Pessimistic Bilevel Optimization

June Liu, Yuxin Fan, Zhong Chen, and Yue Zheng

Abstract Pessimistic bilevel optimization represents an attractive tool to model risk-averse hierarchy problems, and would provide strong ability of analysis for the risk-averse leader. The goal of this chapter is to provide a extensive review on pessimistic bilevel optimization from basic models, definitions and properties to solution approaches. It will directly support researchers in understanding theoretical research results, designing solution algorithms in relation to pessimistic bilevel optimization.

Keywords Pessimistic bilevel optimization · Stackelberg game · Risk-averse

14.1 Introduction

Bilevel decision-making problems [11, 20, 46, 57], motivated by Stackelberg game theory [50], are hierarchical optimization problems in which their constraints are defined in part by another parametric optimization problem. The decision makers at the upper level and the lower level are respectively termed as the leader and the follower, and make their individual decisions in sequence with the aim of optimizing their respective objectives. As is well-known, bilevel optimization plays an exceedingly important role in different application fields, such as transportation, economics, ecology, engineering and others [21]. Note that a linear bilevel programming problem was proved to be strongly NP-hard [25]. Therefore, solving such a problem is not an easy task.

J. Liu · Y. Zheng (✉)
School of Management, Huaibei Normal University, Huaibei, Anhui, P.R. China

Y. Fan
Huazhong University of Science and Technology, Wuhan, P.R. China

Z. Chen
School of Information and Mathematics, Yangtze University, Jingzhou, Hubei, P.R. China
e-mail: chenzhong@yangtzeu.edu.cn

S. Dempe, A. Zemkoho (eds.), *Bilevel Optimization*, Springer Optimization and Its Applications 161, https://doi.org/10.1007/978-3-030-52119-6_14

In general, a bilevel programming problem can be stated as:

$$\begin{aligned} &\text{“}\min_{x}\text{"}\ F(x, y) \\ &\text{s.t.}\quad G(x) \le 0,\ y \in \Psi(x), \end{aligned} \tag{14.1.1}$$

where $\Psi(x)$ is the set of solutions to the lower level problem

$$\begin{aligned} &\min_{y}\ f(x, y) \\ &\text{s.t.}\ g(x, y) \le 0, \end{aligned}$$

Here $x \in R^n$ and $y \in R^m$.

It is worthwhile noting that the lower level problem may have multiple solutions for every (or some) fixed value of the upper level decision making variable. If the solution of the lower level problem is not unique, it is difficult for the leader to predict which point in $\Psi(x)$ the follower will choose. As a result, it is difficult to determine the leader's solution. That is the reason why we use the quotation marks in problem (14.1.1).

To overcome this situation, the majority of authors used optimistic formulation and pessimistic formulation, which represent the two different situations between the leader and the follower. In the optimistic formulation (e.g., see Ref. [20] and the references therein), the follower always selects a strategy in $\Psi(x)$ that suits the leader best. Alternatively, pessimistic formulation (e.g., see Ref. [20] and the references therein) refers to the case where the leader protects himself against the worst possible situation.

There are several survey papers on bilevel/multilevel programming/decision-making in the past 20 years. However, these surveys focus on an optimistic bilevel optimization. For example, Ben-Ayed [12], Wen and Hsu [53] presented the basic models, solution definitions, solution approaches and applications of the optimistic linear bilevel optimization. Colson et al. [17, 18], Vicente and Calamai [52] focused on traditional solution concepts and solution approaches for the optimistic bilevel programming problems. Dempe [21] summarized the main research directions and the main fields of applications of bilevel programming, but mainly fixed attention on the optimistic formulation. Kalashnikov et al. [26] presented a survey of bilevel programming and application area.

Sakawa and Nishizaki [47] reviewed interactive fuzzy programming approaches for the optimistic bilevel/multilevel programming problems which focus on cooperative decision-making in decentralised organizations. Zhang et al. [58] reviewed the fuzzy bilevel decision-making techniques which include models, approaches and systems.

Lu et al. [40] identified nine different kinds of relationships amongst followers by establishing a general framework for bilevel multi-follower decision problems in which the lower level problem has multiple followers. Furthermore, Lu et al. [41] analyzed various kinds of relationships between decision entities in a multi-follower trilevel (MFTL) decision problem, and then proposed an MFTL decision making framework, in which 64 standard MFTL decision situations and their possible combinations are identified. Recently, Lu et al. [42] developed the latest research on multilevel decision-making involving theoretical research results and applications.

Sinha et al. [48] presented an introduction to progress made by evolutionary computation towards handing bilevel optimization. Sinha et al. [49] developed a comprehensive review on bilevel optimization from classical to evolutionary approaches, and then discussed a number of potential application problems.

The above surveys papers have provided good references on optimistic bilevel optimization for researchers. Recently, Liu et al. [32] proposed an extensive review on pessimistic bilevel optimization from basic definitions and properties to solution approaches. This chapter will add some models and the latest newly published articles of pessimistic bilevel optimization on the basis of [32].

The remainder of the chapter is organized as follows. In Sect. 14.2, we describe the classification of pessimistic bilevel optimization. In Sects. 14.3 and 14.4, we review definitions and properties of pessimistic bilevel optimization. In Sect. 14.5, some solution approaches are proposed for pessimistic bilevel optimization. Finally, we conclude this chapter and provide some future directions in Sect. 14.6.

14.2 Classification of Pessimistic Bilevel Optimization

In this section, we describe the classification of pessimistic bilevel optimization into two main categories: pessimistic bilevel programming problems (in their general and linear versions) and further generalizations (semivectorial and multi-follower problems) of pessimistic bilevel optimization.

14.2.1 Pessimistic Bilevel Programming Problems

14.2.1.1 Pessimistic Linear Bilevel Programming Problems

Pessimistic linear bilevel programming problems are pessimistic bilevel optimization problems in which the objective functions of the leader and the follower and their constraint regions are restricted to be affine. Mathematically, pessimistic linear bilevel programming problems are stated as follows:

$$\min_{x \in X} \max_{y \in \Psi(x)} c^\top x + d^\top y \qquad (14.2.1)$$

where X is a closed subset of R^n, and $\Psi(\boldsymbol{x})$ is the set of solutions to the lower level problem

$$\min_{y \geq 0} \boldsymbol{w}^\top \boldsymbol{y}$$
$$\text{s.t. } B\boldsymbol{y} \leq \boldsymbol{b} - A\boldsymbol{x}$$

where $\boldsymbol{c}, \boldsymbol{x} \in R^n$, $\boldsymbol{d}, \boldsymbol{w}, \boldsymbol{y} \in R^m$, $A \in R^{p\times n}$, $B \in R^{p\times m}$, and $\boldsymbol{b} \in R^p$.

Problem (14.2.1) is often called weak linear bilevel programming problem [4]. Note that, when the lower level problem satisfies $A = 0$, i.e. the follower's feasible region does not depend on the upper level decision variables, problem (14.2.1) can be called a independent pessimistic linear bilevel programming problem.

14.2.1.2 General Pessimistic Bilevel Programming Problems

General pessimistic bilevel programming problems can be formulated as follows:

$$\min_{\boldsymbol{x}\in X'} \max_{y\in\Psi(\boldsymbol{x})} F(\boldsymbol{x}, \boldsymbol{y}). \tag{14.2.2}$$

where $X' := \{\boldsymbol{x} : G(\boldsymbol{x}) \leq 0\}$, and $\Psi(\boldsymbol{x})$ is the set of solutions to the lower level problem

$$\min_{y} f(\boldsymbol{x}, \boldsymbol{y})$$
$$\text{s.t. } g(\boldsymbol{x}, \boldsymbol{y}) \leq 0,$$

Here $\boldsymbol{x} \in R^n$ and $\boldsymbol{y} \in R^m$.

When the feasible set of the follower does not depend on the upper level decision variables, i.e., $g(x, y)$ in problem (14.2.2) is reduced to $g(y)$, problem (14.2.2) is referred as an independent pessimistic bilevel problem [54]. Note that it is also called weak Stackelberg problem in Ref. [38].

14.2.2 Further Generalizations of Pessimistic Bilevel

14.2.2.1 Pessimistic Semivectorial Bilevel Programming Problems

Pessimistic semivectorial bilevel programming problems [8] are pessimistic bilevel optimization problems in which the lower level problem is a multiobjective optimization problem. Mathematically, pessimistic semivectorial bilevel programming problems can be formulated as follows:

$$\min_{\boldsymbol{x}\in X'} \max_{y\in\Psi_{Ef}(\boldsymbol{x})} F(\boldsymbol{x}, \boldsymbol{y}). \tag{14.2.3}$$

where $\Psi_{Ef}(\boldsymbol{x})$ is the set of efficient solutions to the lower level problem

$$\begin{aligned}&\min_{\boldsymbol{y}}\ (f_1(\boldsymbol{x},\boldsymbol{y}), f_2(\boldsymbol{x},\boldsymbol{y}),\cdots, f_q(\boldsymbol{x},\boldsymbol{y}))\\&\text{s.t. } g(\boldsymbol{x},\boldsymbol{y})\le 0.\end{aligned}$$

14.2.2.2 Pessimistic Bilevel Multi-Follower Programming Problems

Pessimistic bilevel multi-follower programming problems are pessimistic bilevel optimization problems in which the lower level problem has multiple followers. When multiple followers are involved in a pessimistic bilevel optimization, different relationships among these followers could cause multiple different processes for deriving a solution for the leader. Here three decision models are established and described respectively as follows.

(I) Uncooperative pessimistic linear bilevel multi-follower programming problem [61]. For such a problem, M followers are involved and there is no shared decision variable, objective function or constrain function among them.

$$\min_{\boldsymbol{x}\in X}\ [\boldsymbol{c}^\top \boldsymbol{x} + \sum_{i=1}^{M} \max_{\boldsymbol{y}_i\in\Psi_i(\boldsymbol{x})} \boldsymbol{d}_i^\top \boldsymbol{y}_i] \tag{14.2.4}$$

where $\Psi_i(\boldsymbol{x})$ is the set of solutions to the ith follower's problem

$$\begin{aligned}&\min_{\boldsymbol{y}_i\ge 0}\ \boldsymbol{w}_i^\top \boldsymbol{y}_i\\&\text{s.t. } A_i\boldsymbol{x} + B_i\boldsymbol{y}_i \le \boldsymbol{b}_i.\end{aligned}$$

Here $\boldsymbol{x},\boldsymbol{c}\in R^n$, $\boldsymbol{y}_i,\boldsymbol{d}_i,\boldsymbol{w}_i\in R^{m_i}$, $A_i\in R^{q_i\times n}$, $B_i\in R^{q_i\times m_i}$, $\boldsymbol{b}_i\in R^{q_i}$, and $i=1,2,\ldots,M$.

(II) Pessimistic linear bilevel multi-follower programming problem with partially shared variables among followers [62]. For such a problem, M followers are involved and there is a partially shared decision variable $\boldsymbol{v}$.

$$\min_{\boldsymbol{x}\in X}\ \max_{(\boldsymbol{y}_i,\boldsymbol{v})\in\Psi_i(\boldsymbol{x})}\ [\boldsymbol{c}^\top\boldsymbol{x} + \sum_{i=1}^{M}\boldsymbol{d}_i^\top\boldsymbol{y}_i + \boldsymbol{s}^\top\boldsymbol{v}] \tag{14.2.5}$$

where $\Psi_i(\boldsymbol{x})$ is the set of solutions of the ith follower's problem

$$\begin{aligned}&\min_{\boldsymbol{y}_i,\boldsymbol{v}}\ \boldsymbol{w}_i^\top\boldsymbol{y}_i + z_i^\top\boldsymbol{v}\\&\text{s.t. } A_i\boldsymbol{x} + B_i\boldsymbol{y}_i + C_i\boldsymbol{v}\le\boldsymbol{b}_i,\\&\quad\ \boldsymbol{y}_i, \boldsymbol{z}\ge 0.\end{aligned}$$

Here $\boldsymbol{s},\boldsymbol{v},z_i\in R^l$, $C_i\in R^{q_i\times l}$ and $i=1,2,\ldots,M$.

(III) Pessimistic referential-uncooperative linear bilevel multi-follower programming problem. For such a problem, M followers are involved in a referential-uncooperative situation, i.e. these followers are uncooperative and they do cross reference information by considering other followers' decision results in each of their own decision objectives and constraints.

$$\min_{x} [c^\top x + \sum_{i=1}^{M} \max_{y_i \in \Psi_i(x, y_{-i})} d_i^\top y_i] \tag{14.2.6}$$

where $y_{-i} = (y_1, \cdots, y_{i-1}, y_{i+1}, \cdots, y_M)$ and $\Psi_i(x, y_{-i})$ is the set of solutions of the ith follower's problem

$$\begin{aligned} &\min_{y_i \geq 0} w_i^\top y_i \\ &\text{s.t.} \;\; A_i x + \sum_{j=1}^{M} B_{ij} y_j \leq b_i. \end{aligned}$$

Here $B_{ij} \in R^{q_i \times m_j}$ and $i = 1, 2, \cdots, M$.

14.3 Definitions of Pessimistic Bilevel Optimization

To understand and analyze solution approaches of pessimistic bilevel optimization, this section reviews the notations and definitions of problem (14.2.2).

14.3.1 Definitions

In this subsection, some important definitions of general pessimistic bilevel optimization are itemized below.

Definition 14.3.1

1. Constraint region of problem (14.2.2):

$$S := \{(x, y) : G(x) \leq 0, g(x, y) \leq 0\}.$$

2. Projection of S onto the leader's decision space:

$$S(X) := \{x \in X : \exists y, \text{ such that } (x, y) \in S\}.$$

3. Feasible set for the follower $\forall \boldsymbol{x} \in S(X)$:

$$Y(\boldsymbol{x}) := \{\boldsymbol{y} : g(\boldsymbol{x}, \boldsymbol{y}) \leq 0\}.$$

4. The follower's rational reaction set for $\boldsymbol{x} \in S(X)$:

$$\Psi(\boldsymbol{x}) := \{\boldsymbol{y} : \boldsymbol{y} \in \arg\min[f(\boldsymbol{x}, \boldsymbol{y}) : \boldsymbol{y} \in Y(\boldsymbol{x})]\}.$$

5. Inducible region or feasible region of the leader:

$$\text{IR} := \{(\boldsymbol{x}, \boldsymbol{y}) : (\boldsymbol{x}, \boldsymbol{y}) \in S, \boldsymbol{y} \in \Psi(\boldsymbol{x})\}.$$

△

To introduce the concept of a solution to problem (14.2.2) (also called pessimistic solution), one usually employs the following value function $\varphi(\boldsymbol{x})$:

$$\varphi(\boldsymbol{x}) := \sup_{\boldsymbol{y} \in \Psi(\boldsymbol{x})} F(\boldsymbol{x}, \boldsymbol{y}).$$

Definition 14.3.2 A pair $(\boldsymbol{x}^*, \boldsymbol{y}^*) \in \text{IR}$ is called a solution to problem (14.2.2), if

$$\varphi(\boldsymbol{x}^*) = F(\boldsymbol{x}^*, \boldsymbol{y}^*),$$
$$\varphi(\boldsymbol{x}^*) \leq \varphi(\boldsymbol{x}), \ \forall \, \boldsymbol{x} \in X.$$

△

In the following, some definitions concerning, in particular, approximate solutions (ϵ-Stackelberg solution, ϵ-Stackelberg equilibrium pair) to the weak Stackelberg problem will be described.

Definition 14.3.3 (Loridan and Morgan [35]) Any point $\boldsymbol{x}^*$ verifying $v_1 = \varphi(\boldsymbol{x}^*)$ is called a Stackelberg solution to the weak Stackelberg problem. Here, $v_1 := \inf_{x \in X} \varphi(\boldsymbol{x})$. △

Definition 14.3.4 (Loridan and Morgan [35]) A pair $(\boldsymbol{x}^*, \boldsymbol{y}^*)$ verifying $v_1 = \varphi(\boldsymbol{x}^*)$ and $\boldsymbol{y}^* \in \Psi(\boldsymbol{x}^*)$ is called a Stackelberg equilibrium pair. △

Definition 14.3.5 (Loridan and Morgan [35]) Let $\epsilon > 0$ be a given number. A point $\boldsymbol{x}^*$ is an ϵ-Stackelberg solution if and only if $\boldsymbol{x}^* \in X$ and $\varphi(\boldsymbol{x}^*) \leq v_1 + \epsilon$. △

Definition 14.3.6 (Loridan and Morgan [35]) A pair $(\boldsymbol{x}^*, \boldsymbol{y}^*)$ is an ϵ-Stackelberg equilibrium pair if and only if $\boldsymbol{x}^*$ is an ϵ-Stackelberg solution and $\boldsymbol{y}^* \in \Psi(\boldsymbol{x}^*)$. △

In addition to the above definitions of pessimistic bilevel optimization, Alves and Antunes[8] presented and illustrated a definition of the solution of pessimistic semivectorial bilevel problem (14.2.3), and made comparison of optimistic solution, pessimistic solution, deceiving solution and rewarding solution.

14.4 Properties of Pessimistic Bilevel Optimization

According to the definitions in Sect. 14.3, we categorize the various properties of pessimistic bilevel optimization, and then list some of the well-known properties.

14.4.1 Existence of Solutions

As is well-known, the study of existence of solutions for pessimistic bilevel optimization is a difficult task. An initial step in this direction was developed by Lucchetti et al. [43] who proposed some examples that fail to have a solution. Moreover, the most studies have been devoted to the weak Stackelberg problem. Aboussoror and Loridan [3], Aboussoror [1] have given sufficient conditions to obtain the existence of solutions to the weak Stackelberg problem via a regularized scheme. Any accumulation point of a sequence of regularized solutions is a solution to the weak Stackelberg problem. Aboussoror and Mansouri [5] have deduced the existence of solutions to the weak Stackelberg problem via d.c. problems. Similar results using reverse convex and convex maximization problems are given in Aboussoror et al. [6].

Loridan and Morgan [35] have obtained some results for approximate solutions of the weak Stackelberg problem by using a theoretical approximation scheme. Any accumulation point of a sequence of ϵ-approximate solutions of the approximation problems is an ϵ-approximate solution to the weak Stackelberg problem. The interested reader can refer to the references about the approximate solutions of the weak Stackelberg problem. Furthermore, Loridan and Morgan [36] improved and extended some properties proposed in Ref. [35]. Similar results using approximation scheme were given in Ref. [37].

Lignola and Morgan [28] discussed a more general weak Stackelberg formulation in which the follower's rational reaction set $\Psi(x)$ is replaced by a parameterized constraint $\Psi(t, x)$ (t is a parameter). Marhfour [45] have established existence and stability results for ϵ-mixed solutions of weak Stackelberg problems. In particular, the results are given under general assumptions of minimal character without any convexity assumption.

For the pessimistic linear bilevel programming problems, using the strong dual theorem of linear programming and penalty method, Aboussoror and Mansouri [4] have established the existence results of solutions. Note that, the strong-weak Stackelberg problems have been reduced the weak Stackelberg problem under some conditions. Aboussoror and Loridan [2] have studied the existence of solutions of strong-weak Stackelberg problems. In particular, using the regularization and the notion of variational convergence, Aboussoror [7] have given sufficient conditions to ensure the existence of solutions to such problems. The obtained results in Refs. [2] and [7] can be applied to the weak Stackelberg problem by deleting some variables.

When the pessimistic bilevel optimization problem (14.2.2) does not have a solution which may arise even under strong assumptions, the leader can make do with alternative solution concepts. Based on this, Lignola and Morgan [29] have considered a concept of viscosity solution which can obviate the lack of optimal solutions. In particular, they have given sufficient conditions using regularization families of the solutions map to the lower level problem, ensuring the existence of the corresponding viscosity solutions. More recently, Lignola and Morgan [30] continue this research by considering a new inner regularizations of the lower level problem which not necessarily satisfy the constraints, and by deriving an existence result of related viscosity solutions to pessimistic bilevel optimization.

14.4.2 Optimality Conditions and Complexity

The optimality conditions for pessimistic bilevel optimization and pessimistic semivectorial bilevel optimization have been proposed in the literature. A first attempt was made by Dassanayaka [19] using implicit programming approach, minmax programming approach and duality programming approach respectively. Using the advanced tools of variational analysis and generalized differentiation, Dempe [22] derived several types of necessary optimality conditions via the lower-level value function approach and the Karush-Kuhn-Tucker (KKT) representation of lower-level optimal solution maps. Furthermore, the upper subdifferential necessary optimality conditions are obtained, and the links are also established between the necessary optimality conditions of the pessimistic and optimistic versions in bilevel programming. Using the two-level value function approach, Dempe et al. [24] derived the results on sensitivity analysis and necessary optimality conditions of pessimistic bilevel optimization with nonsmooth data.

In addition, Liu et al. [31] presented the first order necessary optimality conditions of a class of pessimistic semivectorial bilevel optimization. Recently, Lampariello et al. [27] discussed the relations among the perturbed pessimistic bilevel problem, pessimistic bilevel optimization and the two follower game as well as the mathematical program with complementarity constraints with respect to their global minimal points. They also presented the connections between their local minimal points in detail. Aussel and Svensson [9] discussed the relations of global and local solutions of both pessimistic bilevel optimization problem and the associated pessimistic mathematical program with complementarity constraints problem. These various necessary optimality conditions and connections to other optimization problems could be helpful to develop fast algorithms to obtain solutions of pessimistic bilevel optimization.

The computational complexity of pessimistic bilevel optimization is easily confirmed at its simplest version, i.e. pessimistic linear bilevel problem. Under the certain assumptions, Wiesemann et al.[54] obtained that: (a) the independent pessimistic linear bilevel programming problem (14.2.1) can be solved in polynomial time; (b) if m, the number of the lower level decision variables, is constant, the pessimistic linear bilevel programming problem (14.2.1) can be solved in polynomial time. If m is nonconstant, it is strongly NP-hard. These results also imply that a general pessimistic bilevel programming problem is strongly NP-hard if the the number of the lower level decision variables is nonconstant.

14.5 Solution Approaches of Pessimistic Bilevel Optimization

14.5.1 Penalty Method

At present, penalty methods [4, 59] were used to solve the pessimistic linear bilevel programming problems. An initial step in this direction was developed by Aboussoror and Mansouri [4]. Using the strong dual theory of linear programming and penalty methods, they transformed problem (14.2.1) into a single-level optimization problem:

$$\begin{aligned} \min_{\boldsymbol{x},\boldsymbol{t},\boldsymbol{u}} \ & \boldsymbol{c}^\top \boldsymbol{x} + \boldsymbol{w}^\top \boldsymbol{t} + (\boldsymbol{b} - A\boldsymbol{x})^\top \boldsymbol{u} \\ \text{s.t.} \ & -B^\top \boldsymbol{u} \le k\boldsymbol{w} - \boldsymbol{d}, \\ & B\boldsymbol{t} \le k(\boldsymbol{b} - A\boldsymbol{x}), \\ & \boldsymbol{x} \in X, \boldsymbol{t}, \boldsymbol{u} \ge 0, \end{aligned} \tag{14.5.1}$$

where $\boldsymbol{t} \in R^m$, $\boldsymbol{u} \in R^p$ and $k > 0$ is a penalty parameter.

Under some assumptions, Aboussoror and Mansouri [4] proved that there exists a $k^* > 0$ such that for all $k > k^*$, if $(\boldsymbol{x}_k, \boldsymbol{t}_k, \boldsymbol{u}_k)$ is a sequence of solutions of problem (14.5.1), $\boldsymbol{x}_k$ solves problem (14.2.1). Unfortunately, no numerical results were reported.

A more recent contribution by Zheng et al. [59], follows the ideas of Aboussoror and Mansouri [4], who presented a new variant of the penalty method to solve problem (14.2.1). Their method transformed it into the following penalty problem:

$$\begin{aligned} \min_{\boldsymbol{x},\boldsymbol{t},\boldsymbol{u}} \ & \boldsymbol{c}^\top \boldsymbol{x} + k\boldsymbol{w}^\top \boldsymbol{y} + (\boldsymbol{b} - A\boldsymbol{x})^\top \boldsymbol{u} \\ \text{s.t.} \ & -B^\top \boldsymbol{u} \le k\boldsymbol{w} - \boldsymbol{d}, \\ & B\boldsymbol{y} \le \boldsymbol{b} - A\boldsymbol{x}, \\ & \boldsymbol{x} \in X, \boldsymbol{y}, \boldsymbol{u} \ge 0, \end{aligned} \tag{14.5.2}$$

where $\boldsymbol{u} \in R^p$ and $k > 0$ is a penalty parameter. The resulting algorithm involves the minimization of disjoint bilinear programming problem (14.5.2) for a fixed value of k. Two simple examples illustrate the proposed algorithm is feasible.

Replacing the lower level problem with its Kuhn-Tucker optimality condition, Liu et al. [33] transformed problem (14.2.1) into another single-level optimization problem, and investigated the relation of the solution of those problems. In addition, Zheng et al. proposed penalty methods to solve uncooperative pessimistic linear bilevel multi-follower programming problem [61] and pessimistic linear bilevel multi-follower programming problem with partially shared variables among followers [62] respectively.

14.5.2 Kth-Best Algorithm

An important property of the solution to the pessimistic linear bilevel programming problem (14.2.1) is that there exists a solution which occurs at a vertex of the polyhedron W. Here W is the constraint region of problem (14.2.1), i.e.

$$W := \{(\boldsymbol{x}, \boldsymbol{y}) : \boldsymbol{x} \in X, B\boldsymbol{y} \leq \boldsymbol{b} - A\boldsymbol{x}, \boldsymbol{y} \geq 0\}.$$

This property induces the possibility of developing algorithms which search amongst vertices of W in order to solve the pessimistic linear bilevel optimization problems.

Zheng et al. [60] first proposed a modified version of Kth-Best algorithm to the pessimistic linear bilevel optimization problems. After sorting all vertices in ascending order with respect to the value of the upper level objective function, this algorithm selects the first vertex to check if it satisfies the terminate condition, and the current vertex is a solution of problem (14.2.1) if it is yes. Otherwise, the next one will be selected and checked.

14.5.3 Approximation Approach

As is well-known, approximation solution of the independent pessimistic bilevel problems have in fact a long history that dates back at least to papers [28, 37]. For recent surveys and improvements about this topic, see also [15, 29, 30]. These results may provide the reader with an idea to design new algorithms for the pessimistic bilevel programming problems, although the authors in [15, 28–30, 37] did not present the direct algorithms.

In the following, we will consider the perturbed pessimistic bilevel programming problems (PBP_ϵ) [27] which can be formulated as follows:

$$\min_{\boldsymbol{x} \in X'} \max_{\boldsymbol{y} \in \Psi_\epsilon(\boldsymbol{x})} F(\boldsymbol{x}, \boldsymbol{y}). \tag{14.5.3}$$

where $\epsilon > 0$, $\Psi_\epsilon(\boldsymbol{x}) := \{\boldsymbol{y} : f(\boldsymbol{x}, \boldsymbol{y}) \le \phi(\boldsymbol{x}) + \epsilon\}$. Note that $\phi(\boldsymbol{x})$ is the optimal value function of the lower level problem

$$\begin{aligned} &\min_{\boldsymbol{y}} \ f(\boldsymbol{x}, \boldsymbol{y}) \\ &\text{s.t.} \ \ g(\boldsymbol{x}, \boldsymbol{y}) \le 0. \end{aligned}$$

To solve the above perturbed pessimistic bilevel programming problem, Lampariello et al. [27] first presented the perturbed standard pessimistic bilevel problem (SPBP_ϵ):

$$\begin{aligned} &\min_{\boldsymbol{x},\boldsymbol{y}} \ F(\boldsymbol{x}, \boldsymbol{y}) \\ &\text{s.t.} \ \ \boldsymbol{x} \in X', \ \boldsymbol{y} \in R_\epsilon(\boldsymbol{x}), \end{aligned}$$

where $R_\epsilon(\boldsymbol{x})$ is the set of ϵ-solutions of the following optimization problem

$$\begin{aligned} &\max_{\boldsymbol{w}} \ F(\boldsymbol{x}, \boldsymbol{w}) \\ &\text{s.t.} \ \ \boldsymbol{w} \in \Psi_\epsilon(\boldsymbol{x}), \end{aligned}$$

and then by using Multi Follower Game with two followers, proposed an optimistic bilevel problem in which the lower level is a parametric GNEP with two players (MFG_ϵ):

$$\begin{aligned} &\min_{\boldsymbol{x},\boldsymbol{y},\boldsymbol{z}} \ F(\boldsymbol{x}, \boldsymbol{y}) \\ &\text{s.t.} \ \ \boldsymbol{x} \in X', \ (\boldsymbol{y}, \boldsymbol{z}) \in E_\epsilon(\boldsymbol{x}), \end{aligned}$$

where $E_\epsilon(\boldsymbol{x})$ is the equilibrium set of the following GNEP:

$$\begin{aligned} &\min_{\boldsymbol{y}} \ -F(\boldsymbol{x}, \boldsymbol{y}) \\ &\text{s.t.} \ \ \boldsymbol{y} \in Y(\boldsymbol{x}), \\ &\qquad f(\boldsymbol{x}, \boldsymbol{y}) \le f(\boldsymbol{x}, \boldsymbol{z}) + \epsilon, \end{aligned}$$

and

$$\begin{aligned} &\min_{\boldsymbol{z}} \ f(\boldsymbol{x}, \boldsymbol{z}) \\ &\text{s.t.} \ \ \boldsymbol{z} \in Y(\boldsymbol{x}). \end{aligned}$$

Using the KKT conditions, (MFG_ϵ) can be transformed into a single-level Mathematical Program with Complementarity Constraints (MPCC_ϵ) problem. Under some assumptions, Lampariello et al. [27] concluded the relations among (PBP_ϵ),

($SPBP_\epsilon$), (MFG_ϵ) and ($MPCC_\epsilon$). This would provide the readers with a new way to discuss the perturbed pessimistic bilevel programming problems.

When the follower's feasible region does not depend on the upper level decision variables, the above perturbed pessimistic bilevel programming problem also can be called independent perturbed pessimistic bilevel programming problem. It is formulated as follows:

$$\min_{\boldsymbol{x}\in X'} \sup_{\boldsymbol{y}\in\Psi_\epsilon(\boldsymbol{x})} F(\boldsymbol{x},\boldsymbol{y}) \tag{14.5.4}$$

where $\epsilon > 0$ and $\Psi_\epsilon(\boldsymbol{x})$ is the set of ϵ-solutions of the lower level problem

$$\min_{\boldsymbol{y}} f(\boldsymbol{x},\boldsymbol{y})$$
$$\text{s.t. } \boldsymbol{y}\in Y,$$

Several authors presented some approximation approaches. For example, Loridan and Morgan [38] first presented the following strong Stackelberg problem:

$$\min_{\boldsymbol{x}\in X'} \inf_{\boldsymbol{y}\in\Psi(\boldsymbol{x},\beta,\gamma)} F(\boldsymbol{x},\boldsymbol{y}) \tag{14.5.5}$$

where $\beta,\gamma \geq 0$ and $\Psi(\boldsymbol{x},\beta,\gamma)$ is the set of γ-solutions of the parametrized problem

$$\min_{\boldsymbol{y}} l(\boldsymbol{x},\boldsymbol{y},\beta)$$
$$\text{s.t. } \boldsymbol{y}\in Y,$$

where $l(\boldsymbol{x},\boldsymbol{y},\beta) = f(\boldsymbol{x},\boldsymbol{y}) - \beta F(\boldsymbol{x},\boldsymbol{y})$ for any $\boldsymbol{x}\in X'$ and $\boldsymbol{y}\in Y$.

Based on the Molodtsov method, they computed a sequence of solutions of strong Stackelberg problem (14.5.5) and investigated the relations with solutions to the independent pessimistic bilevel problem. Under some assumptions, they have been proved that a sequence of solutions of strong Stackelberg problem convergence to a lower Stackelberg equilibrium for the independent perturbed pessimistic bilevel problem.

On the other hand, Tsoukalas et al. [51], Wiesemann et al. [54] considered the following independent pessimistic bilevel optimization problem:

$$\min_{\boldsymbol{x}\in X} H(\boldsymbol{x}) \tag{14.5.6}$$
$$\text{s.t. } g(\boldsymbol{x},\boldsymbol{y}) \leq 0, \forall \boldsymbol{y}\in \arg\max_{\boldsymbol{y}'\in Y} h(\boldsymbol{x},\boldsymbol{y}').$$

To solve the independent pessimistic bilevel optimization problem (14.5.6), they presented an ϵ-approximation problem:

$$\begin{aligned}
\min_{x \in X'} \ & H(x) \\
\text{s.t.} \ & g(x, y) \leq 0, \forall y \in Y_\epsilon(x), \\
& Y_\epsilon(x) = \{z \in Y : h(x, z) < h(x, z') + \epsilon, \forall z' \in Y\}, \\
& x \in X.
\end{aligned}$$

For a fixed value of ϵ, the above problem was reformulated as a single-level optimization problem:

$$\begin{aligned}
\min_{x,z,\lambda} \ & H(x) \\
\text{s.t.} \ & \lambda(y)[h(x, z) - h(x, y) + \epsilon] \\
& +(1 - \lambda(y))g(x, y) \leq 0, \forall y \in Y, \\
& x \in X', z \in Y, \lambda : Y \longrightarrow [0, 1],
\end{aligned}$$

where the function $\lambda : Y \longrightarrow [0, 1]$ is a decision variable. Furthermore, they developed an iterative solution procedure for the ϵ-approximation problems. Numerical results illustrate the feasibility of the proposed ϵ-approximation method.

In particular, when the lower level corresponds to an equilibrium problem that is represented as a (parametric) variational inequality or, equivalently, a generalized equation, bilevel optimization problem can be called an MPEC (Mathematical Program with Equilibrium Constraints). Červinka et al. [16] proposed a new numerical method, which combines two types of existing codes, a code for derivative-free optimization under box constraints, and a method for solving special parametric MPECs from the interactive system, to compute approximate pessimistic solutions to MPECs.

14.5.4 Reduction Method

Again, to solve problem (14.2.2), Zeng [56] presented its relaxation problem as follows:

$$\begin{aligned}
\min_{x,\bar{y}} \max_{y \in \tilde{Y}(x,\bar{y})} \ & F(x, y) \\
\text{s.t.} \ & x \in X, \\
& g(x, \bar{y}) \leq 0, \\
& \tilde{Y}(x, \bar{y}) := \{y : g(x, y) \leq 0, f(x, y) \leq f(x, \bar{y})\}.
\end{aligned} \tag{14.5.7}$$

Proposition 3 in Ref. [56] means that if $(\boldsymbol{x}^*, \bar{y}', \boldsymbol{y}')$ is a solution to problem (14.5.7), then there exists a point $\boldsymbol{y}^* \in \Psi(\boldsymbol{x}^*)$ such that $(\boldsymbol{x}^*, \boldsymbol{y}^*)$ solves problem (14.2.2). In other words, this result provides the reader with an important idea to compute the pessimistic bilevel optimization via investigating a regular optimistic bilevel programming problem.

Moreover, Zheng et al. [64] proposed a reducibility method for a pessimistic linear bilevel programming problem which is reduced to disjoint bilinear programming problems. For a pessimistic quadratic-linear bilevel optimization problem, Malyshev and Strekalovsky [44] reduced it to a series of optimistic bilevel optimization problems and then developed the global and local search algorithms.

In addition to the above several methods, for the pessimistic linear bilevel programming problem, Dempe et al. [23] proposed the global and local search algorithms via the value method and the strong dual theory of linear programming; Liu et al. [34] presented a new algorithm which embeds a penalty method into a branch and bound algorithm. For the pessimistic bilevel mixed-integer programming problems, Lozano and Smith [39] developed two methods (i.e. two-phase approach and cutting-plane algorithm) based on an optimal-value-function reformulation. Zheng et al. [63] presented a maximum entropy approach to solve a class of the pessimistic bilevel programming problems in which the set of solutions of the lower level problem is discrete. It should be noted that their approach need to obtain the set of solutions of the lower level problem in advance. This is not an easy work, but it may provide the reader with a new way to discuss the pessimistic bilevel optimization.

14.6 Conclusions and Prospective Research Topics

Pessimistic bilevel optimization plays an exceedingly important role in different application fields, such as second best toll pricing [10], production-distribution planning [61], principal-agent [54] and interdiction game [13, 14, 55]. In this chapter, we have addressed pessimistic bilevel optimization. This chapter gives an overview on the state-of-the-art of pessimistic bilevel optimization. Based on the above discussion, we can find that there are some directions that should be discussed for further research.

1. The first order optimality conditions for pessimistic bilevel optimization is proposed, but it has not been organically combined with the algorithm. Furthermore, the higher order optimality conditions also should be studied. In addition, no systematic studies have been conducted on sensitivity analysis.
2. Several existing methods can be used to solve pessimistic linear bilevel optimization and independent pessimistic bilevel optimization problems. It would be interesting to study a general pessimistic bilevel optimization which do not possess the independence property. In particular, it would be instructive to

investigate how pessimistic bilevel optimization can be reduced to a single-level optimization problem and to discuss the relationship between them.

3. The ultimate goal of pessimistic bilevel optimization is to provide strong ability of analysis and decision for the practical problems from the worst-case point of view. In particular, those problems often appear in highly complex and uncertainty environments. This requires further research on how intelligent algorithms can be applied to large-scale pessimistic bilevel optimization in the current age of big data.

Acknowledgement This work was partially supported by the National Natural Science Foundation of China (Nos. 11871383 and 11501233).

References

1. A. Aboussoror, Weak bilevel programming problems: existence of solutions. Adv. Math. Res. **1**, 83–92 (2002)
2. A. Aboussoror, P. Loridan, Strong-weak Stackelberg problems in finite dimensional spaces. Serdica Math. J. **21**, 151–170 (1995)
3. A. Aboussoror, P. Loridan, Existence of solutions to two-level optimization problems with nonunique lower-level solutions. J. Math. Anal. Appl. **254**, 348–357 (2001)
4. A. Aboussoror, A. Mansouri, Weak linear bilevel programming problems: existence of solutions via a penalty method. J. Math. Anal. Appl. **304**, 399–408 (2005)
5. A. Aboussoror, A. Mansouri, Existence of solutions to weak nonlinear bilevel problems via MinSup and d.c. problems. RAIRO Oper. Res. **42**, 87–103 (2008)
6. A. Aboussoror, S. Adly, V. Jalby, Weak nonlinear bilevel problems: existence of solutions via reverse convex and convex maximization problems. J. Ind. Manage. Optim. **7**, 559–571 (2011)
7. A. Aboussoror, S. Adly, F.E. Saissi, Strong-weak nonlinear bilevel problems: existence of solutions in a sequential setting. Set-Valued Var. Anal. **1**, 113–132 (2017)
8. M.J. Alves, C.H. Antunes, An illustration of different concepts of solutions in semivectorial bilevel programming, in IEEE Symposium Series on Computational Intelligence (IEEE, New York, 2016), pp. 1–7
9. D. Aussel, A. Svensson, Is pessimistic bilevel programming a special case of a mathematical program with complementarity constraints? J. Optim. Theory Appl. **181**(2), 504–520 (2019)
10. X.J. Ban, S. Lu, M. Ferris, H.X. Liu, Risk averse second best toll pricing, in *Transportation and Traffic Theory 2009: Golden Jubilee* (Springer, New York, 2009), pp. 197–218
11. J.F. Bard, *Practical Bilevel Optimization: Algorithms and Applications* (Kluwer Academic, Dordrecht, 1998)
12. O. Ben-Ayed, Bilevel linear programming. Comput. Oper. Res. **20**, 485–501 (1993)
13. G. Brown, M. Carlyle, J. Salmeron, K. Wood, Defending critical infrastructure. Interfaces **36**(6), 530–544 (2006)
14. A. Caprara, M. Carvalho, A. Lodi, G.J. Woeginger, Bilevel knapsack with interdiction constraints. INFORMS J. Comput. **28**(2), 319–333 (2016)
15. F. Caruso, M.B. Lignola, J. Morgan, Regularization and approximation methods in Stackelberg Games and bilevel optimization. No. 541. Centre for Studies in Economics and Finance (CSEF), University of Naples, Naples (2019)
16. M. Červinka, C. Matonoha, J.V. Outrata, On the computation of relaxed pessimistic solutions to MPECs. Optim. Methods Softw. **28**, 186–206 (2013)
17. B. Colson, P. Marcotte, G. Savard, Bilevel programming: a survey. 4OR Q. J. Oper. Res. **3**(2), 87–107 (2005)

18. B. Colson, P. Marcotte, G. Savard, An overview of bilevel optimization. Ann. Oper. Res. **153**, 235–256 (2007)
19. S. Dassanayaka, Methods of variational analysis in pessimistic bilevel programming. Wayne State University, PhD Thesis (2010)
20. S. Dempe, *Foundations of Bilevel Programming*. Nonconvex Optimization and its Applications Series (Kluwer Academic, Dordrecht, 2002)
21. S. Dempe, Annotated bibliography on bilevel programming and mathematical problems with equilibrium constraints. Optimization **52**, 333–359 (2003)
22. S. Dempe, B.S. Mordukhovich, A.B. Zemkoho, Necessary optimality conditions in pessimistic bilevel programming. Optimization **63**, 505–533 (2014)
23. S. Dempe, G. Luo, S. Franke, Pessimistic bilevel linear optimization. J. Nepal Math. Soc. **1**, 1–10 (2018)
24. S. Dempe, B.S. Mordukhovich, A.B. Zemkoho, Two-level value function approach to non-smooth optimistic and pessimistic bilevel programs. Optimization **68**(2–3), 433–455 (2019)
25. P. Hansen, B. Jaumard, G. Savard, New branch-and-bound rules for linear bilevel programming. SIAM J. Sci. Stat. Comput. **13**, 1194–1217 (1992)
26. V.V. Kalashnikov, S. Dempe, G.A. Pérez-Valdés, N.I. Kalashnykova, J.F. Camacho-Vallejo, Bilevel programming and applications. Math. Probl. Eng. **2015**, 1–16 (2015)
27. L. Lampariello, S. Sagratella, O. Stein, The standard pessimistic bilevel problem. SIAM J. Optim. **29**(2), 1634–1656 (2019)
28. M.B. Lignola, J. Morgan, Topological existence and stability for Stackelberg problems. J. Optim. Theory Appl. **84**(1), 145–169 (1995)
29. M.B. Lignola, J. Morgan, Inner regularizations and viscosity solutions for pessimistic bilevel optimization problems. J. Optim. Theory Appl. **173**(1), 183–202 (2017)
30. M.B. Lignola, J. Morgan, Further on inner regularizations in bilevel optimization. J. Optim. Theory Appl. **180**(3), 1087–1097 (2019)
31. B. Liu, Z. Wan, J. Chen, G. Wang, Optimality conditions for pessimistic semivectorial bilevel programming problems. J. Inequal. Appl. **41**, 1–26 (2014)
32. J. Liu, Y. Fan, Z. Chen, Y. Zheng, Pessimistic bilevel optimization: a survey. Int. J. Comput. Intell. Syst. **11**(1), 725–736 (2018)
33. J. Liu, Y. Hong, Y. Zheng, A new variant of penalty method for weak linear bilevel programming problems. Wuhan Univ. J. Nat. Sci. **23**(4), 328–332 (2018)
34. J. Liu, Y. Hong, Y. Zheng, A branch and bound-based algorithm for the weak linear bilevel programming problems. Wuhan Univ. J. Nat. Sci. **23**(6), 480–486 (2018)
35. P. Loridan, J. Morgan, Approximate solutions for two-level optimization problems. in *Trends in Mathematical Optimization*, ed. by K. Hoffman, J.-B. Hiriart-Urruty, C. Lemarechal, J. Zowe. International Series of Numerical Mathematics, vol. 84 (Birkhauser Verlag, Basel, 1988), pp. 181–196
36. P. Loridan, J. Morgan, ϵ-regularized two-level optimization problems: approximation and existence results, in *Optimization-Fifth French-German Conference Castel Novel 1988*. Lecture Notes in Mathematics, vol. 1405 (Springer, Berlin, 1989), pp. 99–113
37. P. Loridan, J. Morgan, New results on approximate solution in two-level optimization. Optimization **20**(6), 819–836 (1989)
38. P. Loridan, J. Morgan, Weak via strong Stackelberg problem: new results. J. Glob. Optim. **8**, 263–287 (1996)
39. L. Lozano, J.C. Smith, A value-function-based exact approach for the bilevel mixed-integer programming problem. Oper. Res. **65**(3), 768–786 (2017)
40. J. Lu, C. Shi, G. Zhang, On bilevel multi-follower decision making: general framework and solutions. Inf. Sci. **176**(11), 1607–1627 (2006)
41. J. Lu, G. Zhang, J. Montero, L. Garmendia, Multifollower trilevel decision making models and system. IEEE Trans. Ind. Inf. **8**(4), 974–985 (2012)
42. J. Lu, J. Han, Y. Hu, G. Zhang, Multilevel decision-making: a survey. Inf. Sci. **346–347**, 463–487 (2016)

43. R. Lucchetti, F. Mignanego, G. Pieri, Existence theorems of equilibrium points in Stackelberg games with constraints. Optimization **18**, 857–866 (1987)
44. A.V. Malyshev, A.S. Strekalovskii, Global search for pessimistic solution in bilevel problems, in *Proceedings of the Toulouse Global Optimization Workshop*, ed. by S. Cafieri, B.G. Toth, E.M.T. Hendrix, L. Liberti, F. Messine (2010), pp. 77–80
45. A. Marhfour, Mixed solutions for weak Stackelberg problems: existence and stability results. J. Optim. Theory Appl. **105**, 417–440 (2000)
46. M. Sakawa, I. Nishizaki, *Cooperative and Noncooperative Multi-level Programming* (Springer Science and Business Media, Berlin/Heidelberg, 2009)
47. M. Sakawa, I. Nishizaki, Interactive fuzzy programming for multi-level programming problems: a review. Int. J. Multicrit. Decis. Making **2**, 241–266 (2012)
48. A. Sinha, P. Malo, K. Deb, Evolutionary bilevel optimization: an introduction and recent advances, in *Recent Advances in Evolutionary Multi-objective Optimization* (Springer International Publishing, Cham, 2017), pp. 71–103
49. A. Sinha, P. Malo, K. Deb, A review on bilevel optimization: from classical to evolutionary approaches and applications. IEEE Trans. Evol. Comput. **22**(2), 276–295 (2018)
50. H.V. Stackelberg, *The Theory of Market Economy* (Oxford University Press, Oxford, 1952)
51. A. Tsoukalas, W. Wiesemann, B. Rustem, *Global Optimisation of Pessimistic Bi-level Problems*, ed. by P.M. Pardalos, T.F. Coleman. Lectures on Global Optimization. Fields Institute Communications, vol. 55 (American Mathematical Society, Providence, RI, 2009), pp. 215–243
52. L. Vicente, P. Calamai, Bilevel and multilevel programming: a bibliography review. J. Global Optim. **5**, 291–306 (1994)
53. U.P. Wen, S.T. Hsu, Linear bi-level programming problems-a review. J. Oper. Res. Soc. **42**(2), 125–133 (1991)
54. W. Wiesemann, A. Tsoukalas, P. Kleniati, B. Rustem, Pessimistic bi-level optimisation. SIAM J. Optim. **23**, 353–380 (2013)
55. R.K. Wood, *Bilevel Network Interdiction Models: Formulations and Solutions*. Wiley Encyclopedia of Operations Research and Management Science (Wiley, Hoboken, 2011)
56. B. Zeng, Easier than we thought-a practical scheme to compute pessimistic bilevel optimization problem. Technical report, University of Pittsburgh (2015) Available via optimization-online.org
57. G. Zhang, J. Lu, Y. Gao, *Multi-Level Decision Making: Models, Methods and Applications* (Springer, Berlin, 2015)
58. G. Zhang, J. Han, J. Lu, Fuzzy bi-level decision-making techniques: a survey. Int. J. Comput. Intell. Syst. **9**, 25–34 (2016)
59. Y. Zheng, Z. Wan, K. Sun, T. Zhang, An exact penalty method for weak linear bilevel programming problem. J. Appl. Math. Comput. **42**, 41–49 (2013)
60. Y. Zheng, D. Fang, Z. Wan, A solution approach to the weak linear bilevel programming problems. Optimization **7**, 1437–1449 (2016)
61. Y. Zheng, G. Zhang, J. Han, J. Lu, Pessimistic bilevel optimization model for risk-averse production-distribution planning. Inf. Sci. **372** , 677–689 (2016)
62. Y. Zheng, Z. Zhu, L. Yuan, Partially-shared pessimistic bilevel multi-follower programming: concept, algorithm, and application. J. Inequal. Appl. **2016**, 1–13 (2016)
63. Y. Zheng, X. Zhuo, J. Chen, Maximum entropy approach for sloving pessimistic bilevel programming problems. Wuhan Univ. J. Nat. Sci. **1**, 63–67 (2017)
64. Y. Zheng, G. Zhang, Z. Zhang, J. Lu, A reducibility method for the weak linear bilevel programming problems and a case study in principal-agent. Inf. Sci. **454**, 46–58 (2018)

Part III
Extensions and Uncertainty in Bilevel Optimization

Chapter 15
Methods for Multiobjective Bilevel Optimization

Gabriele Eichfelder

Abstract This chapter is on multiobjective bilevel optimization, i.e. on bilevel optimization problems with multiple objectives on the lower or on the upper level, or even on both levels. We give an overview on the major optimality notions used in multiobjective optimization. We provide characterization results for the set of optimal solutions of multiobjective optimization problems by means of scalarization functionals and optimality conditions. These can be used in theoretical and numerical approaches to multiobjective bilevel optimization.

As multiple objectives arise in multiobjective optimization as well as in bilevel optimization problems, we also point out the results on the connection between these two classes of optimization problems. Finally, we give reference to numerical approaches which have been followed in the literature to solve these kind of problems. We concentrate in this chapter on nonlinear problems, while the results and statements naturally also hold for the linear case.

Keywords Multiobjective bilevel optimization · Semivectorial bilevel problem · Scalarization · Optimistic approach · Numerical methods

15.1 Introduction

In bilevel optimization one typically assumes that one has scalar-valued objective functions in the two levels of the bilevel program. This may be, for instance, costs, or time which has to be minimized. However, many real world problems can only be modeled adequately by considering several, in most cases conflicting, objective functions simultaneously, as weight and stability, or costs and time.

This occurs for instance in transportation planning [51]. We illustrate this with an example, cf. [17]. Let us consider a city bus transportation system financed by

---

G. Eichfelder (✉)
Institute for Mathematics, TU Ilmenau, Ilmenau, Germany
e-mail: gabriele.eichfelder@tu-ilmenau.de

S. Dempe, A. Zemkoho (eds.), *Bilevel Optimization*, Springer Optimization and Its Applications 161, https://doi.org/10.1007/978-3-030-52119-6_15

the public authorities. They have as target the reduction of the money losses in this non-profitable business. As a second target they want to motivate as many people as possible to use the buses instead of their own cars, as it is a public mission to reduce the overall traffic. The public authorities can decide about the bus ticket price, but this will influence the customers in their usage of the buses. The public has several, possibly conflicting, objectives, too, as to minimize their transportation time and costs. Hence, the usage of the public transportation system can be modeled on the lower level with the bus ticket price as parameter and with the solutions influencing the objective values of the public authorities on the upper level. Thus, such a problem can be modeled by multiobjective bilevel optimization.

In [17, 40] also a problem in medical engineering was discussed. There, the task was the configuration of coils. In the original version, this problem was a usual scalar-valued standard optimization problem which had to be reformulated as a bilevel optimization problem due to the need of real-time solutions and because of its structure. It turned out that the engineers would accept—or even prefer—solutions which do not satisfy a previous equality constraint strictly in favor of incorporating some additional objectives. This led to the examination of two objective functions on both levels.

Dempe and Mehlitz consider in [11] a bilevel control problem, where the three objective functions are approximating a given target state as good as possible, minimizing the overall control effort, and the sparcity of the chosen control.

In case we have more than one objective function then we speak of a multiobjective optimization problem. Collecting for instance m objective functions in an m-dimensional vector also leads to the name vector optimization, as now vector-valued maps have to be optimized. In case we have a bilevel program with a vector-valued objective on one or both of the levels we speak of a multiobjective bilevel optimization problem. In case one has multiple functions just on one level one also uses the name semivectorial bilevel problem.

For instance, Pilecka studies in [39] bilevel optimization problems with a single-objective lower level and a vector-valued function on the upper level. The same structure was assumed by Gadhi and Dempe in [23, 24] or Ye in [50]. Bonnel and co-authors [3] study semivectorial bilevel problems with a multiobjective optimization problem on the lower level. For optimality conditions for bilevel problems with such a structure (multiobjective on the lower level, single-objective on the upper level), see also Dempe and co-authors in [12].

In this chapter we will give a short introduction to multiobjective optimization. First, an optimality concept for such optimization problems has to be formulated. By doing that, one can define stricter or weaker concepts, which may have advantages from applications point of view or from the theoretical or numerical point of view, respectively.

We give scalarization results and optimality conditions which allow to characterize the set of optimal solutions of a multiobjective optimization problem. We also point out some of the developments on numerical solvers. Throughout the chapter we emphasize more on nonlinear problems. For the linear case of a multiobjective bilevel optimization problem with multiple functions on each level see for instance

[9] and the references therein. We will also study reformulations of the feasible set of the upper level problem, which can be used to formulate numerical solution approaches.

In bilevel optimization also multiple objectives arise: a function on the upper level and a function on the lower level. Hence, we will also discuss the relationship between multiobjective and bilevel optimization from this perspective.

15.2 Basics of Multiobjective Optimization

As we will consider bilevel optimization problems with a multiobjective optimization problem on the lower or/and on the upper level, we start by giving an overview on the main optimality notions typically used in multiobjective optimization.

In multiobjective optimization one studies optimization problems formally defined by

$$\begin{aligned} \min f(x) = (f_1(x), \ldots, f_m(x))^\top \\ \text{subject to the constraint} \\ x \in \Omega, \end{aligned} \tag{MOP}$$

with a vector-valued objective function $f\colon \mathbb{R}^n \to \mathbb{R}^m$ $(m, n \in \mathbb{N},\ m \geq 2)$ and a nonempty set of feasible points $\Omega \subseteq \mathbb{R}^n$.

For defining minimality for the multiobjective optimization problem (MOP), we need a partial ordering in the image space. Thus one considers partial orderings introduced by closed pointed convex cones $K \subseteq \mathbb{R}^m$. A set K is a convex cone if $\lambda(x + y) \in K$ for all $\lambda \geq 0,\ x, y \in K$, and K is a pointed convex cone if additionally $K \cap (-K) = \{0\}$. A partial ordering introduced by a pointed convex cone is antisymmetric. The partial ordering is then defined by

$$x \leq_K y \qquad \Leftrightarrow \qquad y - x \in K\,.$$

Hence, in the following, let $K \subseteq \mathbb{R}^m$ be a closed pointed convex cone.

For $K = \mathbb{R}^m_+$ we have for $x, y \in \mathbb{R}^m$

$$x \leq_{\mathbb{R}^m_+} y \qquad \Leftrightarrow \qquad x_i \leq y_i,\ i = 1, \ldots, m,$$

i.e. the componentwise ordering. Throughout the chapter, we denote by $\leq$ without a subscript the componentwise ordering defined by $\leq = \leq_{\mathbb{R}^m_+}$. For more details on ordering structures in vector optimization we refer to [21].

By using the concept of a partial ordering, we are now able to define two types of optimal solutions of (MOP)

Definition 15.2.1 A point $\bar{x} \in \Omega$ is called a K-minimal point, or efficient, for (MOP) if

$$(\{f(\bar{x})\} - K) \cap f(\Omega) = \{f(\bar{x})\}.$$

Additionally, for $\mathrm{int}(K) \neq \emptyset$, a point $\bar{x} \in \Omega$ is called a weakly K-minimal point, or weakly efficient, for (MOP) if

$$(\{f(\bar{x})\} - \mathrm{int}(K)) \cap f(\Omega) = \emptyset.$$

△

Of course, any efficient point is also weakly efficient, in case $\mathrm{int}(K) \neq \emptyset$. Whenever we speak of weakly efficient points, we assume that the ordering cone K has a nonempty interior. We denote the set of all K-minimal points, i. e. of all efficient points, as $\mathcal{E}(f(\Omega), K)$, and the set of all weakly efficient points as $\mathcal{E}^w(f(\Omega), K)$. The set $\mathcal{N}(f(\Omega), K) := \{f(x) \in \mathbb{R}^m \mid x \in \mathcal{E}(f(\Omega), K)\}$ is called *nondominated set* and the set $\mathcal{N}^w(f(\Omega), K) := \{f(x) \in \mathbb{R}^m \mid x \in \mathcal{E}^w(f(\Omega), K)\}$ *weakly nondominated set*.

For $K = \mathbb{R}^m_+$ the efficient points are also denoted as *Edgeworth-Pareto (EP)-minimal* points. Then the definition can also be formulated as follows: a point $\bar{x} \in \Omega$ is an EP-minimal point for (MOP) if there is no $x \in \Omega$ with

$$f_i(x) \leq f_i(\bar{x}),\ i = 1, \ldots, m, \text{ and}$$
$$f_j(x) < f_j(\bar{x}) \text{ for at least one } j \in \{1, \ldots, m\}.$$

Similarly, a point is called weakly EP-minimal or weakly efficient for (MOP) w.r.t. $K = \mathbb{R}^m_+$ if there is no $x \in \Omega$ with

$$f_i(x) < f_i(\bar{x}),\ i = 1, \ldots, m.$$

The weakly efficient points are in most cases more of interest from a theoretical or numerical point of view, see, for instance, Theorem 15.3.2, or the results on optimality conditions for a semivectorial bilevel optimization problem as given in [12, 23, 24]. From an application point of view one would in general only be interested in a weakly efficient point which is also efficient, as otherwise one could improve the value of one objective function without deteriorating the others. Nevertheless, in many studies on multiobjective bilevel optimization problems it is assumed that the lower level decision maker would also be satisfied with a weakly efficient solution, see, for instance, [12]. Note that in case Ω is convex and the functions f_i, $i = 1, \ldots, m$ are strictly convex, i.e., if for all $i = 1, \ldots, m$

$$f_i(\lambda x + (1-\lambda)\, y) < \lambda\, f_i(x) + (1-\lambda)\, f_i(y) \text{ for all } x, y \in \Omega,\ x \neq y,\ \lambda \in]0, 1[,$$

then any weakly efficient point for (MOP) w.r.t. $K = \mathbb{R}^m_+$ is also an efficient point of (MOP) w.r.t. $K = \mathbb{R}^m_+$.

For iterative numerical approaches, the following result is useful: in case the feasible set becomes larger and one has already determined the efficient points of the smaller set, then this can be used for determining the efficient points w.r.t. a larger superset.

Lemma 15.2.2 ([17]) *For two sets $A^0, A^1 \subseteq \mathbb{R}^n$, and a vector-valued function $f : \mathbb{R}^n \to \mathbb{R}^m$, consider the sets*

$$A = A^0 \cup A^1 \qquad \text{and}$$
$$\tilde{A} = \mathcal{E}(f(A^0), \mathbb{R}^m_+) \cup A^1.$$

Let $f(A^0)$ be compact. Then it holds $\mathcal{E}(f(A), \mathbb{R}^m_+) = \mathcal{E}(f(\tilde{A}), \mathbb{R}^m_+)$. △

For numerical calculations, also the notion of ε-minimal solutions can be useful. We state here a definition for the ordering cone $K = \mathbb{R}^m_+$, see also for instance [30, 31, 48]:

Definition 15.2.3 Let $\varepsilon \in \mathbb{R}^m$ with $\varepsilon_i > 0,\ i = 1, \ldots, m$, be given. A point $\bar{x} \in \Omega$ is an ε-EP-minimal solution of the multiobjective optimization problem (MOP) if there is no $x \in \Omega$ with

$$\begin{aligned} & f_i(x) + \varepsilon_i \leq f_i(\bar{x}) && \text{for all } i \in \{1, \ldots, m\} \\ \text{and} \quad & f_j(x) + \varepsilon_j < f_j(\bar{x}) && \text{for at least one } j \in \{1, \ldots, m\}. \end{aligned}$$

△

Example to Optimality Notions
Let $\Omega = \{x \in \mathbb{R}^2 \mid x_1 \geq 0,\ x_2 \geq 0,\ x_1 + x_2 \geq 1\}$, $f : \mathbb{R}^2 \to \mathbb{R}^2$ be defined by $f(x) = x$ for all $x \in \mathbb{R}^2$ and $K = \mathbb{R}^2_+$. Then the set of efficient points is

$$\mathcal{E}(f(\Omega), K) = \{x \in \mathbb{R}^2 \mid x_1 \geq 0,\ x_2 \geq 0,\ x_1 + x_2 = 1\}$$

and the set of weakly efficient points is

$$\begin{aligned} \mathcal{E}^w(f(\Omega), K) = \{x \in \mathbb{R}^2 \mid x_1 \geq 0,\ x_2 \geq 0,\ x_1 + x_2 = 1\} \\ \cup \{(0, x_2) \in \mathbb{R}^2 \mid x_2 \geq 1\} \cup \{(x_1, 0) \in \mathbb{R}^2 \mid x_1 \geq 1\}. \end{aligned}$$

As the objective map f is the identity we have

$$\mathcal{N}(f(\Omega), K) = \mathcal{E}(f(\Omega), K) \ \text{ and } \ \mathcal{N}^w(f(\Omega), K) = \mathcal{E}^w(f(\Omega), K).$$

(continued)

For $\varepsilon_i = 1/4$ for $i = 1, 2$, the point $\bar{x} = (1/2 + 1/8, 1/2 + 1/8)$ is an ε-EP-minimal solution of the multiobjective optimization problem $\min_{x \in \Omega} f(x)$ but not an efficient point.

For multiobjective optimization problems also more general optimality notions have been introduced, which use more general ordering structures. For instance, Ye discusses in [50] semivectorial bilevel problems where the preference relation is locally satiated and almost transitive. Moreover, the concept of using a fixed ordering cone can be replaced by using an ordering map which associates to each element y of the space an ordering cone $K(y)$. This was done by Dempe and Gadhi in [10]. They study semivectorial bilevel problems with a multiobjective optimization problem on the upper level and where the cones $K(y)$ are assumed to be Bishop-Phelps-cones. For such multiobjective optimization problems also proper nonlinear scalarizations exist, see [19], which have also been used by Dempe and Gadhi in their approach for formulating necessary optimality conditions. Variable ordering structures have been intensively studied, cf. [18] and, within the context of set-optimization, in [20]. There is a strong relation between bilevel optimization and set optimization, i.e. with optimization with a set-valued objective function. This arises whenever the optimal solution set on the lower level is not unique, see [39].

15.3 Characterization of Optimal Solutions of a Multiobjective Optimization Problem

For solving problem (MOP) several methods are discussed in the literature. For surveys see [13, 29, 35, 42]. An often used approach is to replace the vector optimization problem by a suitable scalar-valued optimization problem. This fact is also used in numerical methods for multiobjective bilevel problems to reformulate the lower level problem by a single-objective problem or by the optimality conditions of this reformulation.

Because of its simplicity, a widely used scalarization for multiobjective optimization is linear scalarization. For that one uses elements of the dual cone

$$K^* := \{w \in \mathbb{R}^m \mid w^\top y \geq 0 \text{ for all } y \in K\}$$

as well as of the quasi-interior of the dual cone

$$K^\# := \{w \in \mathbb{R}^m \mid w^\top y > 0 \text{ for all } y \in K \setminus \{0\}\}.$$

For $K = \mathbb{R}^m_+$ we have $K^* = \mathbb{R}^m_+$ and $K^\# = \mathrm{int}(\mathbb{R}^m_+) = \{y \in \mathbb{R}^m \mid y_i > 0,\ i = 1, \dots, m\}$.

For convex multiobjective optimization problems one can use linear scalarization for a full characterization of the weakly efficient points.

Theorem 15.3.1

(a) Let $w \in K^* \setminus \{0\}$ and $\bar{x} \in \Omega$ with

$$w^\top f(\bar{x}) \leq w^\top f(x) \text{ for all } x \in \Omega.$$

Then $\bar{x}$ is a weakly efficient point for (MOP). If, additionally, $w \in K^\#$, then $\bar{x}$ is even efficient for (MOP).

(b) Let $f(\Omega) + K$ be convex. Then for any weakly efficient point $\bar{x} \in \Omega$ of (MOP) (and thus for any efficient point) there exists some $w \in K^* \setminus \{0\}$ with

$$w^\top f(\bar{x}) \leq w^\top f(x) \text{ for all } x \in \Omega.$$

△

Hence, in the convex case, we have a full characterization of weakly efficient points. A similar result as in (b) does not exist for efficient points and for $K^\#$. This is shown with the next example.

Example on Linear Scalarization

Let $f\colon \mathbb{R}^2 \to \mathbb{R}^2$ with $f(x) = x$, $K = \mathbb{R}^2_+$, and $\Omega = \{x \in \mathbb{R}^2 \mid \|x\|_2 \leq 1\}$. Then

$$\mathcal{E}(f(\Omega), \mathbb{R}^2_+) = \{x \in \mathbb{R}^2 \mid \|x\|_2 = 1, x_1 \leq 0,\ x_2 \leq 0\}.$$

Considering the point $\bar{x} = (0, -1) \in \mathcal{E}(f(\Omega), \mathbb{R}^2_+)$ there exists no $w \in K^\# = \mathrm{int}(\mathbb{R}^m_+)$ with

$$w^\top f(\bar{x}) = -w_2 \leq w^\top f(x) = w_1 x_1 + w_2 x_2 \text{ for all } x \in \Omega.$$

Only for $w = (0, 1)^\top$ the point $\bar{x}$ is a minimal solution of

$$\min_{x \in \Omega} w^\top f(x).$$

This characterization is often used for solving multiobjective bilevel optimization problems. If we have multiple objectives and if the problem is convex, then the set

$$\{\bar{x} \in \Omega \mid \exists w \in K^* \setminus \{0\} :\ w^\top f(\bar{x}) \leq w^\top f(x) \text{ for all } x \in \Omega\} \tag{15.3.1}$$

equals the set of weakly efficient points $\mathcal{E}^w(f(\Omega), K)$. Such linear scalarizations are also called weighted-sum approach, and the components w_i of the vector $w \in K^*$ are interpreted as weights of the objective function.

The approach in (15.3.1) is for instance used in [32, 33]. The authors reformulate a linear semivectorial bilevel programming problem (with a multiobjective optimization problem on the lower level) as a special bilevel programming problem, where the lower level is a parametric linear scalar-valued optimization problem. For characterizing the set in (15.3.1) optimality conditions from scalar-valued optimization can be used. This approach with using the weighted sum is also examined in [11] and it is pointed out that this is a delicate issue in case one studies locally optimal solutions of the parameterized problem only.

Also note that from an application point of view, the weakly efficient points are typically not so much of interest. A point $x \in \Omega$ which is a weakly EP-minimal point of (MOP) but not an EP-minimal point of (MOP) means that there is still some $\tilde{x}$ which satisfies all objective functions equally good and is better with respect to one of the objective functions. By considering instead the set

$$\{\bar{x} \in \Omega \mid \exists w \in K^{\#} : \; w^\top f(\bar{x}) \leq w^\top f(x) \text{ for all } x \in \Omega\}$$

one only has a subset of the set of efficient points of (MOP) as the Example 15.3 has shown. Note that there are numerical approaches for multiobjective bilevel problems which in fact aim on the efficient instead of the weakly efficient points, see for instance [25]

Theorem 15.3.1(b) requires the set $f(\Omega) + K$ to be convex. This holds in case the functions f_i are convex, $K = \mathbb{R}^m_+$, and Ω is a convex set. For nonconvex multiobjective optimization problems one has to use nonlinear scalarizations. A widely used one is a scalarization by Pascoletti and Serafini, [38], also known as Tammer-Weidner-functional [26, 27]. This scalarization approach allows to determine weakly efficient points for an arbitrary ordering cone K while many other well known nonlinear scalarizations as the ε-constraint problem (see [13, 28, 34, 35]) are mainly developed for determining EP-minimal points, only. Moreover, those can also be seen as a special case of the Pascoletti-Serafini-scalarization, see [15]. Thus, results gained for the Pascoletti-Serafini problem can be applied to these scalarization problems, too.

Hence we consider the parameter dependent scalarization problems (according to Pascoletti and Serafini)

$$\begin{aligned} &\min_{(t,x) \in \mathbb{R}^{1+n}} \; t \\ &\text{subject to the constraints} \\ &a + t\,r - f(x) \in K, \\ &x \in \Omega, \\ &t \in \mathbb{R} \end{aligned} \tag{15.3.2}$$

for parameters $a, r \in \mathbb{R}^m$ to the multiobjective optimization problem (MOP). The main properties of this scalarization approach are the following:

Theorem 15.3.2

(a) *Let $\bar{x}$ be an efficient point of* (MOP)*, then* $(0, \bar{x})$ *is a minimal solution of* (15.3.2) *with* $a = f(\bar{x})$*,* $r \in K \setminus \{0\}$*.*
(b) *Let* $(\bar{t}, \bar{x})$ *be a minimal solution of* (15.3.2) *for any* $a, r \in \mathbb{R}^m$*, then* $\bar{x} \in \mathcal{E}^w(f(\Omega), K)$*.*

△

Thus all efficient points of the multiobjective optimization problem can be found even for non-convex problems by choosing suitable parameters. We are even able to restrict the choice of the parameter a to a hyper plane in the image space according to the following theorem ([16], Theorem 3.2).

Theorem 15.3.3 *Let* $\bar{x} \in \mathcal{E}(f(\Omega), K)$ *and* $r \in K$ *be given. We define a hyper plane* H *by* $H = \{y \in \mathbb{R}^m \mid b^\top y = \beta\}$ *with* $b \in \mathbb{R}^m$*,* $b^\top r \neq 0$*,* $\beta \in \mathbb{R}$*. Then there is a parameter* $a \in H$ *and some* $\bar{t} \in \mathbb{R}$ *such that* $(\bar{t}, \bar{x})$ *is a minimal solution of the problem* (15.3.2)*.* △

Also other scalarizations are used for deriving results on multiobjective bilevel optimization problems. For instance, Gadhi and Dempe use in [23] the scalarization known as signed-distance functional for deriving optimality conditions. This functional evaluates the distance of a point to the negative of the ordering cone as well as to the complement of the negative of the ordering cone.

We end this section by giving the KKT-type (i.e., Karush-Kuhn-Tucker-type) optimality conditions for constrained multiobjective optimization problems. KKT-conditions, and optimality conditions in general, are often used to characterize the optimal solutions of the lower level of a bilevel optimization problem. Such an approach was for instance named to be a classical approach by Deb and Sinha in [6] and it was used by Chuong in [4] for a nonsmooth multiobjective bilevel optimization problem. As it is known from single-objective optimization, KKT-conditions typically include complementarity conditions which are difficult to handle mathematically. Moreover, the necessary conditions require some regularity assumptions (cf. (A2) in [33]). For more details on the sufficiency and necessity of the optimality conditions we refer for instance to [13, 29].

We require the feasible set to be given by inequality and equality constraints (also basic convex sets could be handled). For shortness of representation we restrict ourselves here to inequality constraints. Thus, let $g_k : \mathbb{R}^n \to \mathbb{R}$, $k = 1, \ldots, p$ be continuously differentiable functions such that

$$\Omega := \{x \in \mathbb{R}^n \mid g_k(x) \leq 0,\ k = 1, \ldots, p\}. \tag{15.3.3}$$

We give in the following the necessary conditions which also hold for just locally efficient solutions, too. For a proof see for instance [13, Theorem 3.21].

Theorem 15.3.4 (Necessary Optimality Conditions) *Let Ω be given as in (15.3.3) and let f_j, $j = 1, \ldots, m$ and g_k, $k = 1, \ldots, p$ be continuously differentiable. Let $\bar{x}$ be a weakly efficient solution of (MOP) and assume that there exists some $d \in \mathbb{R}^n$ such that*

$$\nabla g_k(\bar{x})^\top d < 0 \text{ for all } k \in I(\bar{x}) := \{k \in \{1, \ldots, p\} \mid g_k(\bar{x}) = 0\},$$

i. e. a constraint qualification is satisfied. Then there exist Lagrange multipliers $\lambda \in \mathbb{R}^m_+ \setminus \{0\}$ and $\mu \in \mathbb{R}^p_+$ such that

$$\begin{aligned} \textstyle\sum_{j=1}^m \lambda_j \nabla f_j(\bar{x}) + \sum_{k=1}^p \mu_k \nabla g_k(\bar{x}) = 0, \\ \mu_k g_k(\bar{x}) = 0,\ k = 1, \ldots, p. \end{aligned} \tag{15.3.4}$$

△

Under appropriate convexity assumptions we can also state sufficient conditions. We state here a simplified version. For weaker assumptions on the convexity of the functions, as quasiconvexity for the constraints, we refer to [29].

Theorem 15.3.5 (Sufficient Optimality Conditions) *Let Ω be given as in (15.3.3) and let f_j, $j = 1, \ldots, m$ and g_k, $k = 1, \ldots, p$ be convex and continuously differentiable. Let $\bar{x} \in \Omega$ and assume that there exist Lagrange multipliers $\lambda \in \mathbb{R}^m_+ \setminus \{0\}$ and $\mu \in \mathbb{R}^p_+$ such that (15.3.4) is satisfied. Then, $\bar{x}$ is a weakly efficient solution of (MOP).* △

15.4 Connection Between Multiobjective Optimization and Bilevel Optimization

In bilevel optimization one has (at least) two objective functions: one on the upper and one on the lower level. In multiobjective optimization one also studies problems with two or more objective functions. This naturally leads to the question on how these problem classes are related, and whether bilevel problems can be reformulated as multiobjective optimization problems. This question was answered for nonlinear problems by Fliege and Vicente in [22]: only by introducing a very special order relation the optimal solutions of the bilevel problem are exactly the efficient points of a multiobjective optimization problem. Only for continuously differentiable problems on the lower level this order relation can be formulated as a more tractable order relation. We give some of the basic ideas in the following, all based on [22]. For a more recent survey and unifying results as well as a more general characterization of the connections between single-level and bilevel multiobjective optimization we refer to [41]. They also provide a thorough literature survey on this topic as well as the historic sources of this approach for bilevel linear problems.

We assume the bilevel problem to be of the form

$$\begin{aligned} &\min\nolimits_{x_u \in \mathbb{R}^{n_u},\, x_\ell \in \mathbb{R}^{n_\ell}} f_u(x_u, x_\ell) \\ \text{s.t.} \quad & x_\ell \in \operatorname{argmin}_{y \in \mathbb{R}^{n_\ell}} \{f_\ell(x_u, y)\}, \end{aligned} \tag{15.4.1}$$

where $f_u, f_\ell\colon\ \mathbb{R}^{n_u} \times \mathbb{R}^{n_\ell} \to \mathbb{R}$ are the upper-level and lower-level objective functions respectively. Hence, we assume that there are no constraints in each level and we assume the optimistic approach. The handling of constraints is also shortly described in [22].

As a related multiobjective optimization problem, it is not enough to study just a problem with the vector-valued function $(f_u(x_u, x_\ell), f_\ell(x_u, x_\ell))$, as counter examples, which can be found in the literature, demonstrate. Instead, a possibility to give a relation is to use a map which includes the vector x_u as a component. But still, it is not enough to use the standard componentwise partial ordering. The authors in [22] present a complete reformulation with an ordering which is not numerically tractable. They also formulate a weaker relation which we present in the following but which gives a sufficient condition only.

The related vector optimization problem, which the authors from [22] finally propose, uses as objective map $F\colon \mathbb{R}^{n_u} \times \mathbb{R}^{n_\ell} \to \mathbb{R}^{n_u} + 3$ defined by

$$F(x_u, x_\ell) = (x_u, f_u(x_u, x_\ell), f_\ell(x_u, x_\ell), \|\nabla_{x_\ell} f_\ell(x_u, x_\ell)\|_2). \tag{15.4.2}$$

In the image space, a cone K is used to define an ordering (but not a partial ordering) by $x \preceq y$ if and only if $y - x \in K$:

$$K = \{(x, f_1, f_2, d) \in \mathbb{R}^{n_u+3} \mid (x = 0 \wedge f_2 > 0) \vee (f_1 > 0 \wedge d \geq 0)\} \cup \{0\}. \tag{15.4.3}$$

Theorem 15.4.1 ([22, Corollary 4.2]) *Let F and K be as in (15.4.2) and (15.4.3). Let a point $\bar{x} = (\bar{x}_u, \bar{x}_\ell) \in \mathbb{R}^{n_u} \times \mathbb{R}^{n_\ell}$ be such that $F(\bar{x})$ is nondominated w.r.t. K, i.e.*

$$(\{F(\bar{x})\} - K \setminus \{0\}) \cap \{F(x) \mid x \in \mathbb{R}^{n_u} \times \mathbb{R}^{n_\ell}\} = \emptyset,$$

then $\bar{x}$ is an optimal solution of the bilevel optimization problem (15.4.1). △

The study about the relation between multiobjective and bilevel optimization problems is also continued and unified in [41] as already mentioned above. The authors also study thereby multiobjective bilevel optimization problems and their connection to multiobjective optimization. A thorough exploration of the use of these relations for numerical approaches for bilevel problems is an open research topic.

15.5 A Multiobjective Bilevel Program

In this section, we introduce the multiobjective bilevel optimization problem. We formulate the general bilevel optimization problem (see (15.5.3)). Then we introduce the so-called optimistic approach (see (15.5.4)) which is a special case of the general bilevel optimization problem. Next, we give a possibility on how to describe the feasible set of such an optimistic bilevel problem as the solution set of a specific multiobjective optimization problem. This will be used in Sect. 15.6 within a numerical method. Then, with an example, we illustrate again what might happen in case the lower level problem has not a unique solution, now in the setting that there is only one objective function on the lower level but several on the upper level.

Moreover, we mention a typical approach from the literature, namely scalarization. We discuss what happens in case one considers weakly efficient points of the lower level problem instead of efficient ones, which is often done when using scalarizations. We end the section by introducing coupling constraints between the upper and the lower level.

As mentioned in the introduction, in bilevel optimization we have a parameter dependent optimization problem on the lower level, which depends on the variable $y \in \mathbb{R}^{n_2}$ of the upper level. In case of multiple objectives, this lower level problems reads as

$$\begin{array}{c} \min\limits_{x \in \mathbb{R}^{n_1}} f(x, y) \\ \text{subject to the constraint} \\ (x, y) \in G \end{array} \tag{15.5.1}$$

with a vector-valued function $f : \mathbb{R}^{n_1} \times \mathbb{R}^{n_2} \to \mathbb{R}^{m_1}$ and a set $G \subseteq \mathbb{R}^{n_1} \times \mathbb{R}^{n_2}$ ($n_1, n_2, m_1 \in \mathbb{N}$). The constraint $(x, y) \in G$ can be replaced by

$$x \in G(y) := \{x \in \mathbb{R}^{n_1} \mid (x, y) \in G\}. \tag{15.5.2}$$

The variables $x \in \mathbb{R}^{n_1}$ of this lower level problem are called lower level variables. For a constant $y \in \mathbb{R}^{n_2}$, let $x(y)$ be a minimal solution of (15.5.1), hence

$$x(y) \in \Psi(y) := \operatorname{argmin}_x \{f(x, y) \mid (x, y) \in G\} \subset \mathbb{R}^{n_1}.$$

For a multiobjective optimization problem on the lower level it has to be clarified what a minimal solution is, i.e. how the set $\Psi(y)$ is defined. An appropriate concept from a practical point of view might be the set of efficient elements w.r.t. a partial ordering, while from a theoretical point of view the set of weakly efficient points might be more suitable.

The optimization problem of the upper level is then given by

$$\begin{aligned} &\text{'}\min_{y \in \mathbb{R}^{n_2}}\text{'}\, F(x(y), y) \\ &\text{subject to the constraints} \\ &x(y) \in \Psi(y), \\ &y \in \tilde{G} \end{aligned} \tag{15.5.3}$$

with a vector-valued function $F: \mathbb{R}^{n_1} \times \mathbb{R}^{n_2} \to \mathbb{R}^{m_2}$, $m_2 \in \mathbb{N}$, and a compact set $\tilde{G} \subset \mathbb{R}^{n_2}$. Here, the constraint $y \in \tilde{G}$ is uncoupled from the lower level variable. An even more general formulation of the bilevel problem is reached by allowing the constraint $y \in \tilde{G}(x)$ to depend on the lower level variable x, what we will discuss shortly later. We speak of a multiobjective bilevel optimization problem if $m_1 \geq 2$ or $m_2 \geq 2$.

If the minimal solution of the lower level problem (15.5.1) is not unique, i. e. the set $\Psi(y)$ consists of more than one point, the objective function $F(x(\cdot), \cdot)$ is not well-defined for $y \in \mathbb{R}^{n_2}$. That is the reason why the word ' min ' is written in quotes in (15.5.3). This difficulty is in some works avoided by just assuming that the solution of the lower level problem is unique. But in the case of a multiobjective optimization problem ($m_1 \geq 2$) on the lower level this is in general not reasonable any more.

Hence, a possible approach is to study the set-valued map

$$y \mapsto \{F(x, y) \mid x \in \Psi(y)\}.$$

In the case of non-uniqueness, another common approach is the optimistic approach. There it is assumed that the decision maker of the lower level chooses among all minimal solutions that one, which is best for the upper level, i. e., which is minimal for the objective function of the upper level. Hence, it is solved

$$\begin{aligned} &\min_{x,y} F(x, y) \\ &\text{subject to the constraints} \\ &x \in \Psi(y), \\ &y \in \tilde{G}. \end{aligned} \tag{15.5.4}$$

Thus, in the optimistic modification (15.5.4), the objective function of the upper level is minimized w. r. t. x and y, while in the general formulation (15.5.3) it is only minimized w. r. t. the upper level variable y. In the following, we consider the optimistic approach only, i. e. the bilevel optimization problem as in (15.5.4).

When assuming this optimistic approach, one assumes at the same time that the decision maker on the lower level allows the decision maker on the upper level to utilize any solution from the set of efficient points of the lower level, which might not be realistic. See also the discussion in [45, 47].

In [45], also the case where the lower level decision maker has sufficient power to make a deterministic decision from the set of efficient points of the lower level is studied. This is formulated by using a value function on the objective functions of the lower level, which needs to be known in this setting. By using a value function the lower level problem is somehow reduced to a single-objective problem. To overcome the difficulty of determining the lower level value function, Sinha and co-authors suggest in [46] to handle uncertainties in the lower level value function.

For the multiobjective optimization problem on the upper level we assume that the partial ordering is given by the closed pointed convex cone $K^2 \subseteq \mathbb{R}^{m_2}$ and for the lower level by the closed pointed convex cone $K^1 \subseteq \mathbb{R}^{m_1}$. For the case $m_1 \geq 2$ the set $\Psi(y)$ is the solution set of a multiobjective optimization problem w. r. t. the ordering cone K^1. Thus, using (15.5.2), we write instead of $x \in \Psi(y)$

$$x \in \mathcal{E}(f(G(y), y), K^1) =: \mathcal{E}_y(f(G), K^1), \tag{15.5.5}$$

with $\mathcal{E}(f(G(y), y), K^1)$ the set of K^1-minimal points of the multiobjective optimization problem (15.5.1) parameterized by y. Hence we study

$$\begin{array}{c} \min_{x,y} F(x, y) \\ \text{subject to the constraints} \\ x \in \mathcal{E}_y(f(G), K^1), \\ y \in \tilde{G}. \end{array} \tag{15.5.6}$$

A helpful result, which can be used for numerical approaches, is that the set of feasible points of the upper level problem of the considered multiobjective bilevel optimization problem can be expressed as the set of efficient points of a multiobjective optimization problem. We will use that in Sect. 15.6. The set of feasible points Ω of the upper level problem in (15.5.6), also called induced set, is given by

$$\Omega = \{(x, y) \in \mathbb{R}^{n_1} \times \mathbb{R}^{n_2} \mid x \in \mathcal{E}_y(f(G), K^1),\ y \in \tilde{G}\}.$$

We show that the set Ω is equivalent to the set of $\hat{K}$-minimal points of the multiobjective optimization problem

$$\begin{array}{c} \min_{x,y} \hat{f}(x, y) := \begin{pmatrix} f(x, y) \\ y \end{pmatrix} \\ \text{subject to the constraints} \\ (x, y) \in G, \\ y \in \tilde{G} \end{array} \tag{15.5.7}$$

w. r. t. the ordering cone

$$\hat{K} := K^1 \times \{0\} \subset \mathbb{R}^{m_1} \times \mathbb{R}^{n_2}.$$

A point $(\bar{x}, \bar{y})$ is thus an element of the set of feasible points Ω if and only if it is a $\hat{K}$-minimal point of the multiobjective optimization problem (15.5.7). This is shown in the following theorem. There is also a close relation to the reformulation using the function F in (15.4.2), see [41].

Theorem 15.5.1 *Let $\hat{\mathcal{E}}$ be the set of $\hat{K}$-minimal points of the multiobjective optimization problem (15.5.7) with $\hat{K} = K^1 \times \{0\}$. Then it holds $\Omega = \hat{\mathcal{E}}$.* △

Proof We have

$$
\begin{aligned}
& (\bar{x}, \bar{y}) \in \Omega \\
\Leftrightarrow\; & \bar{x} \in \mathcal{E}_{\bar{y}}(f(G), K^1) \quad \wedge \quad \bar{y} \in \tilde{G} \\
\Leftrightarrow\; & \big(\nexists \;\; x \in G(\bar{y}) \text{ with } f(\bar{x}, \bar{y}) \in f(x, \bar{y}) + K^1 \setminus \{0\}\big) \\
& \wedge \quad \bar{y} \in \tilde{G} \quad \wedge \quad \bar{x} \in G(\bar{y}) \\
\Leftrightarrow\; & \big(\nexists \;\; (x, y) \in G \text{ with } f(\bar{x}, \bar{y}) \in f(x, y) + K^1 \setminus \{0\} \;\wedge\;\; y = \bar{y}\big) \\
& \wedge \quad \bar{y} \in \tilde{G} \quad \wedge \quad (\bar{x}, \bar{y}) \in G \\
\Leftrightarrow\; & (\nexists \;\; (x, y) \in G \text{ with} \\
& \left.\begin{pmatrix} f(\bar{x}, \bar{y}) \\ \bar{y} \end{pmatrix} \in \begin{pmatrix} f(x, y) \\ y \end{pmatrix} + (K^1 \times \{0\}) \setminus \{0\}\right) \\
& \wedge \quad \bar{y} \in \tilde{G} \quad \wedge \quad (\bar{x}, \bar{y}) \in G \\
\Leftrightarrow\; & \Big(\nexists (x, y) \in G \text{ with } \hat{f}(\bar{x}, \bar{y}) \in \hat{f}(x, y) + \hat{K} \setminus \{0\}\Big) \\
& \wedge \quad \bar{y} \in \tilde{G} \quad \wedge \quad (\bar{x}, \bar{y}) \in G \\
\Leftrightarrow\; & (\bar{x}, \bar{y}) \in \hat{\mathcal{E}}.
\end{aligned}
$$

□

A multiobjective bilevel optimization problem is a quite challenging problem in case there are multiple objectives on each level, as just discussed. But already in case there are only multiple objectives on the upper level, such semivectorial bilevel problems become quite difficult. This is even more the case when the lower level problem does not have unique optimal solutions. We illustrate this with an example taken from [39, Ex. 4.14]. This example also gives insight to the optimistic approach.

Example on Semivectorial Bilevel Problem

We study a bilevel optimization problem as in (15.5.3) with the lower level problem

$$\min_{x \in \mathbb{R}} \{\max\{0, -xy\} \mid x \leq 1\}.$$

(continued)

Then we have for $y \in [0, 1]$

$$\Psi(y) = \begin{cases}]-\infty, 1] & \text{if } y = 0, \\ [0, 1] & \text{if } y \in]0, 1]. \end{cases}$$

For the upper level problem we are now interested in the problem

$$\begin{aligned} &\text{'}\min_{y\in\mathbb{R}}\text{'}\ F(x(y), y) \\ &\text{subject to the contraints} \\ &x(y) \in \Psi(y), \\ &y \in [0, 1] \end{aligned}$$

with

$$F(x, y) = \begin{pmatrix} x - y - 1 \\ y - \max\{0, x\} \end{pmatrix}.$$

As the lower level solutions are not unique, to each $y \in [0, 1]$ we can define the sets

$$\tilde{F}(y) := \bigcup_{x\in\Psi(y)} \{F(x, y)\}$$

where

$$\tilde{F}(y) = \begin{cases} \text{conv}(\{(-1, 0), (0, -1)\}) \cup \{(u, 0) \mid u \in]-\infty, -1]\} & \text{if } y = 0 \\ \text{conv}(\{(-y - 1, y), (-y, y - 1)\}) & \text{if } y =]0, 1] \end{cases}$$

Then $(\hat{x}, \hat{y}) = (\hat{x}, 1)$ with $\hat{x} \in \Psi(1) = [0, 1]$ and $F(\hat{x}, \hat{y}) = (\hat{x} - 2, 1 - \hat{x})$ is not an optimal solution of the optimistic formulation

$$\begin{aligned} &\min_{x,y} F(x(y), y) \\ &\text{subject to the contraints} \\ &x(y) \in \Psi(y), \\ &y \in [0, 1]. \end{aligned}$$

This can be seen by considering $(x, y) = (x, 0)$ with $x \in \Psi(0) \cap]-\infty, -1]$. These points have smaller objective function values: $F(x, 0) = (x - 1, 0)^\top$. By comparing just the nondominated elements of the sets $\tilde{F}(y)$, i.e. by

(continued)

solving a set optimization problem, as proposed in [39, p. 60], there would also be solutions with $\hat{y} = 1$. This is a hint that studying bilevel problems with non-unique lower-level-optimal solutions by using techniques from set optimization might lead to other solutions, which might also be of interest.

Next, we will have a look on a typical approach to multiobjective bilevel problems. This is to scalarize the lower level problem. Thereby, one often considers weakly efficient points of the lower level problem instead of efficient ones. We will also discuss the impact of that. For $K^1 = \mathbb{R}^{m_1}_+$, Lv and Wan use in [32, 33] a weighted-sum approach, as discussed in Sect. 15.3, to characterize the optimal solutions of the lower level in case the lower level problem is convex and some regularity assumptions are satisfied. For the theoretical verification see Theorem 15.3.1. Then, instead of (15.5.6), they study

$$\begin{gathered} \min_{x,y,w} F(x, y) \\ \text{subject to the constraints} \\ y \in \tilde{G}, \\ \textstyle\sum_{i=1}^{m_1} w_i = 1, \\ w \in \mathbb{R}^{m_1}_+, \\ x \in \operatorname{argmin}\left\{w^\top f(x, y) \mid (x, y) \in G\right\}. \end{gathered}$$

Hence, the weighting vector w, which is here the scalarization parameter, is interpreted as a new upper level variable, and thus the original semivectorial bilevel programming problem is transformed into a standard bilevel programming problem (cf. [11, Theorem 3.1]). Thereby it is assumed that one is interested in weakly efficient solutions of the lower level–and not in efficient solutions only. While this approach can be done in view of globally optimal solutions, one has to take care in case one uses locally optimal solutions on the lower level: Dempe and Mehlitz show in [11] that the reformulation using the weighted-sum scalarization and the additional variable w might have locally optimal solutions which are not locally optimal for the original problem.

In a second step, in [32, 33], the constraint

$$x \in \operatorname{argmin}\left\{w^\top f(x, y) \mid (x, y) \in G\right\}$$

is replaced by the KKT-conditions of this scalar-valued optimization problem.

As done in these reformulations, many authors take the weakly efficient elements instead of the efficient, i.e. they replace the condition $x \in \mathcal{E}(f(G(y), y), K^1)$ from the lower level problem by $x \in \mathcal{E}^w(f(G(y), y), K^1)$. The next example shows that this can have a strong impact. We use for this examples ideas from [11].

Example on Weakly Efficient and Efficient Solutions
We study the optimistic semivectorial bilevel optimization problem

$$\begin{gathered}\min_{x,y} x + 2y \\ \text{subject to the constraints} \\ x \in \Psi(y) = \operatorname{argmin}\{(xy, 1-x) \mid x \in [0,1]\}, \\ y \in [0,1].\end{gathered}$$

Then we have

$$\Psi(y) = \begin{cases} \{1\} & \text{if } y = 0, \\ [0,1] & \text{if } y \in (0,1]. \end{cases}$$

The bilevel problem has no optimal solution: for $y = 0$ the objective function value is 1 and for $y > 0$ it is $2y$. In case one takes the set of weakly efficient solutions of the lower level, which we denote by $\Psi^w(y)$, one gets

$$\Psi^w(y) = [0,1] \text{ for } y \in [0,1]$$

and as optimal value of the bilevel problem we obtain 0. The minimal solution is then $(x, y) = (0, 0)$. This shows that it might make a huge difference whether weakly efficient or efficient solutions of the lower level problem are taken into account. In the first case, the bilevel problem is even not solvable.

Instead of enlarging the feasible set by taking the weakly efficient solutions instead of the efficient solutions, one could also use the so-called properly efficient solutions, which are a subset of the efficient solutions. Those can completely be characterized by linear scalarizations, at least in the convex case, by using elements $w \in K^{\#}$. Such an approach was discussed by Gebhardt and Jahn in [25]. The advantage is that the gained optimal values of the bilevel problem are then upper bounds on the optimal values of the original problem, and, what is more, that the gained optimal solutions are in fact feasible for the original problem, as any properly efficient solution is also efficient.

Finally, we would like to draw the attention to additional difficulties which can arise in case of constraints on the upper level which couple the upper and the lower level variables. Thus, in the remaining of this section, we study instead of problem (15.5.6) the more general formulation

$$\begin{gathered}\min_{x,y} F(x,y) \\ \text{subject to the constraints} \\ x \in \mathcal{E}_y(f(G), K^1), \\ (x,y) \in \tilde{G}\end{gathered} \tag{15.5.8}$$

with the constraint set $\tilde{G} \subset \mathbb{R}^{n_1} \times \mathbb{R}^{n_2}$. This results in a coupling of the upper level variable y and the lower level variable x. Then

$$\Omega' := \{(x, y) \in \mathbb{R}^{n_1} \times \mathbb{R}^{n_2} \mid x \in \mathcal{E}_y(f(G), K^1),\ (x, y) \in \tilde{G}\}$$

denotes the induced set of the problem (15.5.8).

First notice that the constraint $(x, y) \in \tilde{G}$ from the upper level cannot just be moved to the lower level as simple examples, see for instance [14, Example 7.4], show. This has also a practical interpretation. The same constraint on the lower level restricting the constraint set on the lower level has a different meaning as on the upper level. There, the feasibility is restricted after the determination of the minimal solution of the lower level and is thus an implicit constraint. For more details see [8, pp. 25].

A generalization of the results from Theorem 15.5.1 is not possible as discussed in [14, Chapter 7]. We also illustrate this with the following example.

Example on Coupled Constraints

We consider the biobjective bilevel optimization problem

$$\min_{y \in \mathbb{R}} F(x, y) = \begin{pmatrix} x_1 - y \\ x_2 \end{pmatrix}$$

subject to the constraints

$$x = x(y) \in \operatorname{argmin}_{x \in \mathbb{R}^2} \left\{ f(x, y) = \begin{pmatrix} x_1 \\ x_2 \end{pmatrix} \middle| (x, y) \in G \right\},$$
$$(x, y) \in \tilde{G}$$

with $n_1 = 2,\ n_2 = 1,\ K^1 = K^2 = \mathbb{R}^2_+, m_1 = m_2 = 2,$

$$G := \{(x, y) \in \mathbb{R}^3 \mid \|x\|_2^2 \le y^2\}$$

and

$$\tilde{G} := \{(x, y) \in \mathbb{R}^3 \mid 0 \le y \le 1,\ x_1 + x_2 \ge -1\}.$$

Then

$$\mathcal{E}_y(f(G), K^1) = \{x = (x_1, x_2) \in \mathbb{R}^2 \mid \|x\|_2^2 = y^2,\ x_1 \le 0,\ x_2 \le 0\}$$

and thus we obtain

$$\Omega' = \{(x, y) \in \mathbb{R}^3 \mid \|x\|_2^2 = y^2,\ x_1 \le 0,\ x_2 \le 0,\ 0 \le y \le 1,\ x_1 + x_2 \ge -1\}.$$

(continued)

Let $\hat{\mathcal{E}}'$ be the set of $(\mathbb{R}^2_+ \times \{0\})$-minimal points of the tricriteria optimization problem

$$\min_{x,y} \begin{pmatrix} x_1 \\ x_2 \\ y \end{pmatrix}$$

subject to the constraints

$$(x, y) \in G \cap \tilde{G} = \{(x, y) \in \mathbb{R}^3 \mid 0 \leq y \leq 1,\ x_1 + x_2 \geq -1,\ \|x\|_2^2 \leq y^2\}$$

similar to (15.5.7). While it holds $\Omega' \subset \hat{\mathcal{E}}'$ (cf. [14, Theorem 7.5]) we show that it does not hold $\hat{\mathcal{E}}' \subset \Omega'$.

For $(\bar{x}, \bar{y}) = (-\frac{1}{2}, -\frac{1}{2}, 1)^\top$ we have $(\bar{x}, \bar{y}) \in \hat{\mathcal{E}}'$ because $(\bar{x}, \bar{y}) \in G \cap \tilde{G}$ and there is no $(x, y) \in G \cap \tilde{G}$ with

$$\begin{pmatrix} \bar{x}_1 \\ \bar{x}_2 \\ \bar{y} \end{pmatrix} \in \left\{ \begin{pmatrix} x_1 \\ x_2 \\ y \end{pmatrix} \right\} + (\mathbb{R}^2_+ \times \{0\}) \setminus \{0\}$$

$$\Leftrightarrow \begin{pmatrix} \bar{x}_1 \\ \bar{x}_2 \end{pmatrix} \in \left\{ \begin{pmatrix} x_1 \\ x_2 \end{pmatrix} \right\} + \mathbb{R}^2_+ \setminus \{0\} \ \wedge\ y = \bar{y}.$$

The set $\{x \in \mathbb{R}^2 \mid (x, y) \in G \cap \tilde{G},\ y = 1\}$ is illustrated in Fig. 15.1. However, $(\bar{x}, \bar{y}) \notin \Omega'$ because $\|\bar{x}\|_2^2 = \frac{1}{2} \neq \bar{y}^2$.

In this example, as the induced set can be determined explicitly, the minimal solution set can be calculated by solving the biobjective optimization problem $\min_{x,y}\{F(x, y) \mid (x, y) \in \Omega'\}$. We get as solution set of the multiobjective bilevel optimization problem:

$$\mathcal{S}^{\min} = \left\{ (x_1, x_2, y) \,\middle|\, x_1 = -1 - x_2,\ x_2 = -\frac{1}{2} \pm \frac{1}{4}\sqrt{8y^2 - 4},\ y \in \left[\frac{\sqrt{2}}{2}, 1\right] \right\}.$$

The image $\left\{F(x_1, x_2, y) \mid (x_1, x_2, y) \in \mathcal{S}^{\min}\right\}$ of this set is

$$\left\{ (z_1, z_2) \in \mathbb{R}^2 \,\middle|\, z_1 = -1 - z_2 - y,\ z_2 = -\frac{1}{2} \pm \frac{1}{4}\sqrt{8y^2 - 4},\ y \in \left[\frac{\sqrt{2}}{2}, 1\right] \right\}.$$

These sets are shown in Fig. 15.2.

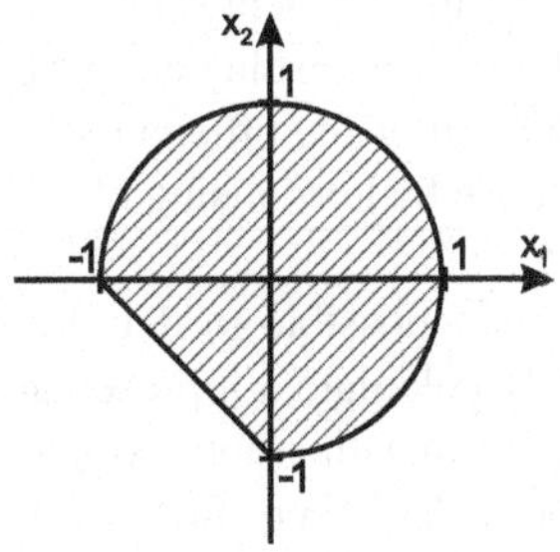

Fig. 15.1 The set $\{x \in \mathbb{R}^2 \mid (x, y) \in G \cap \tilde{G},\ y = 1\}$ of Example 15.5, cf. [14]

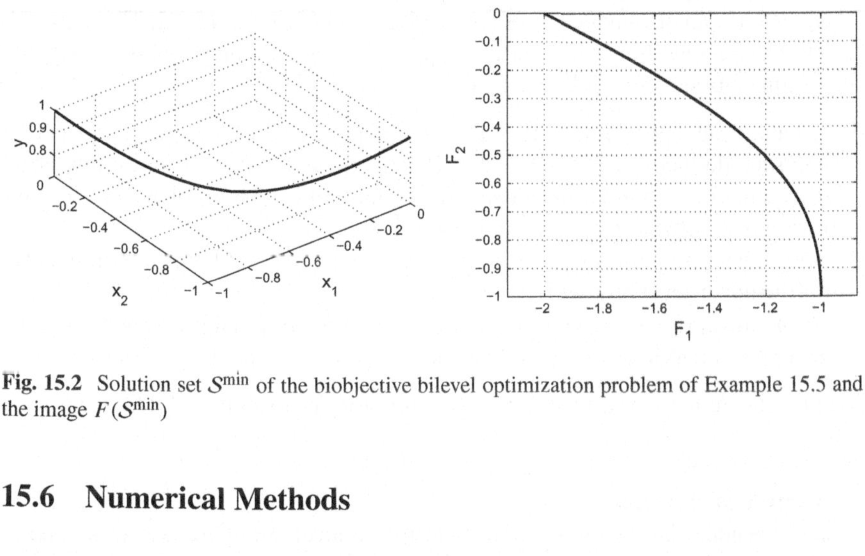

Fig. 15.2 Solution set $\mathcal{S}^{\min}$ of the biobjective bilevel optimization problem of Example 15.5 and the image $F(\mathcal{S}^{\min})$

15.6 Numerical Methods

As stated in the introduction, we concentrate in this chapter on nonlinear problems. Procedures for solving linear multiobjective bilevel problems are presented e. g. in [36].

Much less papers are dealing with nonlinear multiobjective bilevel problems: Shi and Xia [43, 44] present an interactive method, also using a scalarization (the ε-constraint scalarization, cf. [47]). Osman, Abo-Sinna et al. [1, 37] propose the usage of fuzzy set theory for convex problems. Teng et al. [49] give an approach for a convex multiperson multiobjective bilevel problem.

Bonnel and Morgan consider in [2] a semivectorial bilevel optimization problem and propose a solution method based on a penalty approach. However, no numerical results are given.

More recently, in [14, 17] Theorem 15.5.1 was used for developing a numerical algorithm. There, first the efficient set of the multiobjective optimization problem (15.5.7) was calculated (or at least approximated). According to Theorem 15.5.1, this efficient set equals the feasible set of the upper level problem of the original bilevel problem. As the efficient set of (15.5.7) was approximated by a finite set of points, these points can be evaluated and compared easily by the upper level function. The approximation of the efficient set of (15.5.7) is then iteratively refined around points which have had promising objective function values w.r.t. the upper level functions.

Note that in general, it is not possible to determine the whole solution set of problem (15.5.7). We can only calculate a representation or approximation of this set. In [14, 17] it is taken care that this approximation is generated with a high quality in the pre-image space. And then, in a second step, based on sensitivity information, this approximation is refined depending on the behavior of the upper level function. For determining single efficient points of problem (15.5.7) the scalarization according to Pascoletti and Serafini, see Theorem 15.3.2, is used. Thereby, one can also make use of the result given in Lemma 15.2.2. The approach is suitable for low dimensions of the upper level variable only. Summing up, the algorithm consists of the following steps:

1. Calculate an approximation of the efficient set of (15.5.7) w.r.t. the partial ordering defined by the cone $\hat{K}$. This is done by determining equidistant representations of the efficient sets of the multiobjective problems (15.5.1) for various discretization points y in the set $\tilde{G}$.
2. Select all non-dominated points of the lower level problems being non-dominated for the upper level problem.
3. Refine the approximation of the efficient set of the original problem by refining the approximation of the set of feasible points of the upper level problem.

We give the numerical results of this algorithm for one example, cf. [14, Ch. 7].

Numerical Example

We consider the following multiobjective bilevel problem (assuming the optimistic approach) with $n_1 = 2$, $n_2 = 1$, $m_1 = m_2 = 2$ and $K^1 = K^2 = \mathbb{R}^2_+$.

$$\min_{x,y} \begin{pmatrix} F_1(x,y) \\ F_2(x,y) \end{pmatrix} = \begin{pmatrix} x_1 + x_2^2 + y + \sin^2(x_1 + y) \\ \cos(x_2) \cdot (0.1 + y) \cdot \left(\exp(-\frac{x_1}{0.1 + x_2})\right) \end{pmatrix}$$

subject to the constraints

$$x \in \operatorname{argmin}_{x \in \mathbb{R}^2} \left\{ \begin{pmatrix} f_1(x,y) \\ f_2(x,y) \end{pmatrix} \mid (x,y) \in G \right\},$$

$$y \in [0, 10]$$

(continued)

with $f_1, f_2 \colon \mathbb{R}^3 \to \mathbb{R}$,

$$f_1(x,y) = \frac{(x_1-2)^2+(x_2-1)^2}{4} + \frac{x_2 y + (5-y)^2}{16} + \sin\left(\frac{x_2}{10}\right),$$

$$f_2(x,y) = \frac{x_1^2+(x_2-6)^4-2x_1 y-(5-y)^2}{80}$$

and

$$G = \{(x_1,x_2,y) \in \mathbb{R}^3 \mid x_1^2 - x_2 \leq 0,\ 5x_1^2 + x_2 \leq 10,\ x_2 - (5-y/6) \leq 0,\ x_1 \geq 0\}.$$

We apply the algorithm from [14, 17] with distance $\beta = 0.6$ for discretizing the interval $\tilde{G} = [0, 10]$, i.e. we choose $y \in \{0, 0.6, 1.2, 1.8, \ldots, 9.6, 10\}$. For example, for $y = 1.8$ one obtains the representation of the efficient set of problem (15.5.1) as shown in Fig. 15.3a. In the image space of the lower level problem this representation is shown in Fig. 15.3b.

The representation A^0 of the efficient sets of (15.5.1) for all discretization points y of the interval $[0, 10]$ is shown in Fig. 15.4a. The set A^0 is at the same time an approximation of the solution set of problem (15.5.7) and hence of the set of feasible points of the upper level problem, see Theorem 15.5.1. In Fig. 15.4b the set $F(A^0)$ is given. Here the image points under F of the set $\mathcal{E}(F(A^0), \mathbb{R}^2_+)$ are marked with circles and are connected with lines.

In several iterations the discretization of the feasible set of the upper level problem is refined around the efficient points which results in the representation A^f, see Fig. 15.5. Then the algorithm is stopped as only small improvements for the approximation of the efficient set of the multiobjective bilevel optimization problem were gained by the last iteration.

Later, another numerical deterministic approach for nonlinear problems was proposed by Gebhardt and Jahn in [25]. They propose to use a multiobjective search algorithm with a subdivision technique. The studied problems have multiple objectives on both levels. The authors also discuss optimality conditions for replacing the multiobjective problem on the lower level. Instead of taking the weakly efficient solutions of the lower level, they suggest to take the properly efficient solutions. By doing this the feasible set of the upper level gets smaller (instead of larger by using weakly efficient elements) and thus one gets upper bounds, and minimal solutions which are in fact feasible for the original problem. However, the authors do not use this for their numerical approach. Instead, the main ingredient of their algorithm is a subdivision technique for multiobjective problems which uses randomly generated points in boxes, where the size is decreased when the algorithm proceeds. It is assumed that the feasible set of the upper level variable $\tilde{G}$ is such that one can easily work with a discretization of it, for instance an interval. Then the upper level variables are fixed to these values coming from the discretization, and

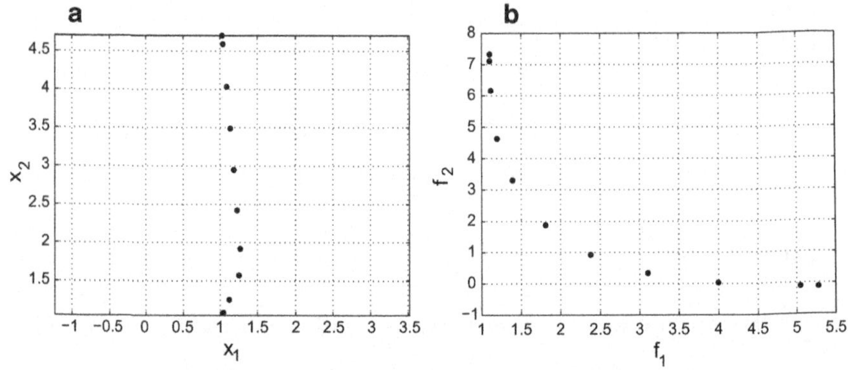

Fig. 15.3 Representation (**a**) of the efficient set and (**b**) the nondominated set of problem (15.5.1) for $y = 1.8$

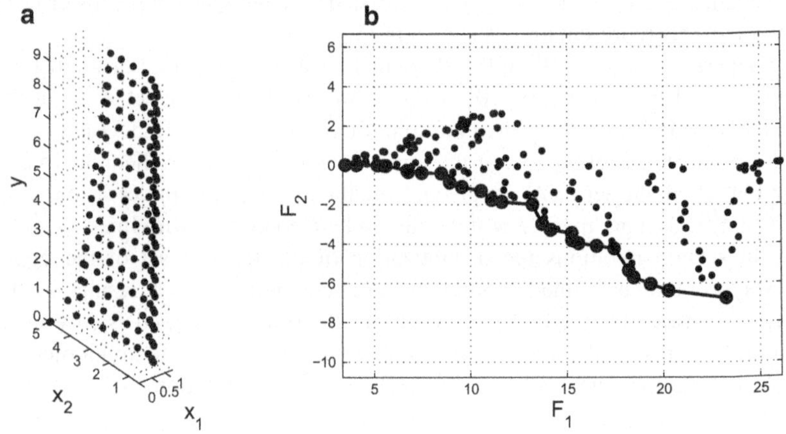

Fig. 15.4 (**a**) Approximation A^0 of the set of feasible points Ω and (**b**) the image $F(A^0)$ of this set under F

the lower level problems are solved with an algorithm from the literature which uses subdivision techniques. While the procedure is appropriate only for small number of variables, it is capable of detecting globally optimal solutions also for highly nonlinear problems.

Also, the construction of numerical test instances is an important aspect. Test instances which are scalable in the number of variables and the number of objectives and where the set of optimal solutions are known are important to evaluate and compare numerical approaches. Deb and Sinha propose five of such test problems in [5]. For the test instances the optimistic approach is chosen.

There have also been some numerical approaches for solving bilevel multiobjective optimization problems by using evolutionary algorithms. An attempt is for

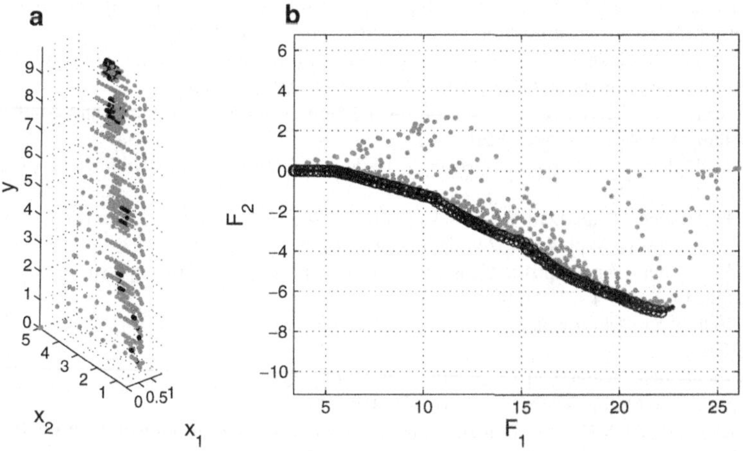

Fig. 15.5 (**a**) Final approximation A^f of the set of feasible points Ω and (**b**) the image $F(A^f)$ of this set under F

instance given by Deb and Sinha in [6]. A hybrid solver combined with a local solver was proposed by the same authors in [7].

References

1. M. Abo-Sinna, A bi-level non-linear multi-objective decision making under fuzziness. J. Oper. Res. Soc. India **38**(5), 484–495 (2001)
2. H. Bonnel, J. Morgan, Semivectorial bilevel optimization problem: penalty approach. J. Optim. Theory Appl. **131**(3), 365–382 (2006)
3. H. Bonnel, I. Todjihoundé, C. Udrişte, Semivectorial bilevel optimization on Riemannian manifolds. J. Optim. Theory Appl. **167**(2), 464–486 (2015)
4. T.D. Chuong, Optimality conditions for nonsmooth multiobjective bilevel optimization problems. Ann. Oper. Res. (2017). https://doi.org/10.1007/s10479-017-2734-6
5. K. Deb, A. Sinha, Constructing test problems for bilevel evolutionary multi-objective optimization, in *IEEE Congress on Evolutionary Computation, Trondheim* (2009), pp. 1153–1160
6. K. Deb, A. Sinha, Solving bilevel multi-objective optimization problems using evolutionary algorithms, in *EMO 2009: Evolutionary Multi-Criterion Optimization* (2009), pp. 110–124
7. K. Deb, A. Sinha, An efficient and accurate solution methodology for bilevel multi-objective programming problems using a hybrid evolutionary-local-search algorithm. Evol. Comput. **18**(3), 403–449 (2010)
8. S. Dempe, *Foundations of Bilevel Programming* (Kluwer Academic Publishers, Dordrecht, 2012)
9. S. Dempe, S. Franke, Bilevel optimization problems with vectorvalued objective functions in both levels Working Paper Department of Mathematics and Computer Science, TU Bergakademie Freiberg, 2012
10. S. Dempe, N. Gadhi, Optimality conditions for bilevel vector optimization problems with a variable ordering structure. Numer. Funct. Anal. Optim. **38**(8), 988–1007 (2017)

11. S. Dempe, P. Mehlitz, Semivectorial bilevel programming versus scalar bilevel programming. Optimization (2019). https://doi.org/10.1080/02331934.2019.1625900
12. S. Dempe, N. Gadhi, A.B. Zemkoho, New optimality conditions for the semivectorial bilevel optimization problem. J. Optim. Theory Appl. **157**(1), 54–74 (2013)
13. M. Ehrgott, *Multicriteria Optimisation*. Lecture Notes in Economics and Mathematical Systems, vol. 491 (Springer, Berlin, 2000)
14. G. Eichfelder, *Adaptive Scalarization Methods in Multiobjective Optimization* (Springer, Heidelberg, 2008)
15. G. Eichfelder, Scalarizations for adaptively solving multi-objective optimization problems. Comput. Optim. Appl. **44**(2), 249–273 (2009)
16. G. Eichfelder, An adaptive scalarization method in multi-objective optimization. SIAM J. Optim. **19**(4), 1694–1718 (2009)
17. G. Eichfelder, Multiobjective bilevel optimization. Math. Program. Ser. A **123**(2), 419–449 (2010)
18. G. Eichfelder, Numerical procedures in multiobjective optimization with variable ordering structures. J. Optim. Theory Appl. **162**(2), 489–514 (2014)
19. G. Eichfelder, T.X.D. Ha, Optimality conditions for vector optimization problems with variable ordering structures. Optimization **62**, 597–627 (2013)
20. G. Eichfelder, M. Pilecka, Set approach for set optimization with variable ordering structures Part II: scalarization approaches. J. Optim. Theory Appl. **171**(3), 947–963 (2016)
21. G. Eichfelder, M. Pilecka, Ordering structures and their applications, in *Applications of Nonlinear Analysis*, ed. by T.M. Rassias (Springer, New York, 2018), pp. 256–304
22. J. Fliege, L.N. Vicente, Multicriteria approach to bilevel optimization. J. Optim. Theory Appl. **131**(2), 209–225 (2006)
23. N. Gadhi, S. Dempe, Necessary optimality conditions and a new approach to multiobjective bilevel optimization problems. J. Optim. Theory Appl. **155**(1), 100–114 (2012)
24. N. Gadhi, S. Dempe, Sufficient optimality conditions for a bilevel semivectorial D.C. problem. Numer. Funct. Anal. Optim. **39**(15), 1622–1634 (2018)
25. E. Gebhardt, J. Jahn, Global solver for nonlinear bilevel vector optimization problems. Pac. J. Optim. **5**(3), 387–401 (2009)
26. C. Gerstewitz (Tammer), Nichtkonvexe Dualität in der Vektoroptimierung. Wissensch. Zeitschr. TH Leuna-Merseburg **25**, 357–364 (1983)
27. C. Gerstewitz (Tammer), E. Iwanow, Dualität für nichtkonvexe Vektoroptimierungsprobleme. Wissensch. Zeitschr. der Techn. Hochschule Ilmenau **31**, 61–81 (1985)
28. Y. Haimes, L. Lasdon, D. Wismer, On a bicriterion formulation of the problems of integrated system identification and system optimization. IEEE Trans. Syst. Man Cybern. **1**, 296–297 (1971)
29. J. Jahn, *Vector Optimization: Theory, Applications and Extensions* (Springer, Berlin, 2004)
30. J. Jahn, Multiobjective search algorithm with subdivision technique. Comput. Optim. Appl. **35**, 161–175 (2006)
31. P. Loridan, ε-solutions in vector minimization problems. J. Optim. Theory Appl. **43**, 265–276 (1984)
32. Y. Lv, Z. Wan, A solution method for the optimistic linear semivectorial bilevel optimization problem. J. Inequal. Appl. (2014). Article number: 164
33. Y. Lv, Z. Wan, A smoothing method for solving bilevel multiobjective programming problems. J. Oper. Res. Soc. China **2**(4), 511–525 (2014)
34. S. Marglin, *Public Investment Criteria* (MIT Press, Cambridge, 1967)
35. K. Miettinen, *Nonlinear Multiobjective Optimization* (Kluwer Academic Publishers, Boston, 1999)
36. I. Nishizaki, M. Sakawa, Stackelberg solutions to multiobjective two-level linear programming problems. J. Optim. Theory Appl. **103**(1), 161–182 (1999)
37. M. Osman, M. Abo-Sinna, A. Amer, O. Emam, A multi-level nonlinear multi-objective decision-making under fuzziness. Appl. Math. Comput. **153**(1), 239–252 (2004)

38. A. Pascoletti, P. Serafini, Scalarizing vector optimization problems. J. Optim. Theory Appl. **42**(4), 499–524 (1984)
39. M. Pilecka, Set-valued optimization and its application to bilevel optimization. Dissertation, Technische Universität Bergakademie Freiberg, 2016
40. J. Prohaska, Optimierung von Spulenkonfigurationen zur Bewegung magnetischer Sonden. Diplomarbeit, Univ. Erlangen-Nürnberg, 2005
41. S. Ruuska, K. Miettine, M.M. Wiecek, Connections between single-level and bilevel multiobjective optimization. J. Optim. Theory Appl. **153**(1), 60–74 (2012)
42. S. Ruzika, M. Wiecek, Approximation methods in multiobjective programming. J. Optim. Theory Appl. **126**(3), 473–501 (2005)
43. X. Shi, H. Xia, Interactive bilevel multi-objective decision making. J. Oper. Res. Soc. **48**(9), 943–949 (1997)
44. X. Shi, H. Xia, Model and interactive algorithm of bi-level multi-objective decision-making with multiple interconnected decision makers. J. Multi-Criteria Decis. Anal. **10**, 27–34 (2001)
45. A. Sinha, P. Maloand, K. Deb, Towards understanding bilevel multi-objective optimization with deterministic lower level decisions, in *Evolutionary Multi-Criterion Optimization*, ed. by A. Gaspar-Cunha, C. Henggeler Antunes, C.C. Coello (2015), pp. 26–443
46. A. Sinha, P. Malo, K. Deb, P. Korhonen, J. Wallenius, Solving bilevel multicriterion optimization problems with lower level decision uncertainty. IEEE Trans. Evol. Comput. **20**(2), 199–217 (2015)
47. A. Sinha, P. Malo, K. Deb, A review on bilevel optimization: from classical to evolutionary approaches and applications. IEEE Trans. Evol. Comput. **22**(2), 276–295 (2018)
48. T. Staib, On two generalizations of Pareto minimality. J. Optim. Theory Appl. **59**(2), 289–306 (1988)
49. C.-X. Teng, L. Li, H.-B. Li, A class of genetic algorithms on bilevel multi-objective decision making problem. J. Syst. Sci. Syst. Eng. **9**(3), 290–296 (2000)
50. J.J. Ye, Necessary optimality conditions for multiobjective bilevel programs. Math. Oper. Res. **36**(1), 165–184 (2011)
51. Y. Yin, Multiobjective bilevel optimization for transportation planning and management problems. J. Adv. Transp. **36**(1), 93–105 (2000)

Chapter 16
Bilevel Optimal Control: Existence Results and Stationarity Conditions

Patrick Mehlitz and Gerd Wachsmuth

Abstract The mathematical modeling of numerous real-world applications results in hierarchical optimization problems with two decision makers where at least one of them has to solve an optimal control problem of ordinary or partial differential equations. Such models are referred to as bilevel optimal control problems. Here, we first review some different features of bilevel optimal control including important applications, existence results, solution approaches, and optimality conditions. Afterwards, we focus on a specific problem class where parameters appearing in the objective functional of an optimal control problem of partial differential equations have to be reconstructed. After verifying the existence of solutions, necessary optimality conditions are derived by exploiting the optimal value function of the underlying parametric optimal control problem in the context of a relaxation approach.

Keywords Bilevel optimal control · Existence results · Inverse optimal control · Stationarity conditions

16.1 What Is Bilevel Optimal Control?

A bilevel programming problem is a hierarchical optimization problem of two decision makers where the objective functional as well as the feasible set of the so-called upper level decision maker (or leader) depend implicitly on the solution set of a second parametric optimization problem which will be called lower level (or follower's) problem. Both decision makers act as follows: First, the leader chooses an instance from his feasible set which then serves as the parameter in the follower's problem. Thus, the follower is in position to solve his problem and passes an optimal solution back to the leader who now may compute the

P. Mehlitz (✉) · G. Wachsmuth
Brandenburgische Technische Universität Cottbus-Senftenberg, Cottbus, Germany
e-mail: mehlitz@b-tu.de; wachsmuth@b-tu.de

S. Dempe, A. Zemkoho (eds.), *Bilevel Optimization*, Springer Optimization and Its Applications 161, https://doi.org/10.1007/978-3-030-52119-6_16

associated value of the objective functional. As soon as the lower level solution set is not a singleton for at least one value of the upper level variable, problems of this type may be ill-posed which is why different solution concepts including the so-called optimistic and pessimistic approach have been developed. Bilevel programming problems generally suffer from inherent lacks of convexity, regularity, and smoothness which makes them theoretically challenging. The overall concept of bilevel optimization dates back to [52] where this problem class is introduced in the context of economical game theory. More than 80 years later, bilevel programming is one of the hottest topics in mathematical optimization since numerous real-world applications can be transferred into models of bilevel structure. A detailed introduction to bilevel programming can be found in the monographs [7, 14, 16, 49] while a satisfying overview of existing literature is given in Appendix: Bilevel Optimization: Bibliography where more than 1350 published books, PhD-theses, and research articles are listed.

Optimal control of ordinary or partial differential equations (ODEs and PDEs, respectively) describes the task of identifying input quantities which control the state function of the underlying differential equation such that a given cost functional is minimized, see [27, 35, 50, 51] for an introduction to this topic. Noting that the decision variables are elements of suitable function spaces, optimal control is a particular field of programming in (infinite-dimensional) Banach spaces, see [10].

In bilevel optimal control, bilevel programming problems are considered where at least one decision maker has to solve an optimal control problem. Thus, we are facing the intrinsic difficulties of bilevel optimization *and* optimal control when investigating this problem class. Naturally, one may subdivide bilevel optimal control problems into three subclasses depending on which decision maker has to perform optimal control. Each of these problem classes appears in practice and has to be tackled with different techniques in order to infer optimality conditions or solution algorithms.

The situation where only the upper level decision maker has to solve an optimal control problem of ordinary differential equations while the lower level problem explicitly depends on the terminal state of the leader's state variable has been considered in [8, 9]. Problems of this type arise from the topic of gas balancing in energy networks, see [32], and can be investigated by combining tools from finite-dimensional parametric optimization and standard optimal control. The situation where parameters within an optimal control problem have to be estimated or reconstructed by certain measurements is a typical example of a bilevel optimal control problem where only the lower level decision maker has to solve an optimal control problem. This particular instance of bilevel optimal control may therefore be also called inverse optimal control. In [2–4, 24, 44], inverse optimal control problems of ODEs are considered in the context of human locomotion. Some more theoretical results for such problems are presented in [25]. First steps regarding the inverse optimal control of PDEs have been done recently in the papers [17, 23, 28]. The paper [46] deals with the scheduling of multiple agents which are controlled at the lower level stage. In [19], the authors discuss a bilevel optimal control problem where airplanes are controlled at multiple lower levels in order to increase the fairness in air racing. Further theoretical results on bilevel optimal control problems where only the lower level decision maker faces an optimal control problem of ODEs are presented in [55, 56]. Finally, it is possible that leader and follower have to solve an optimal control problem. This setting has been discussed theoretically in [11, 36, 40, 45]. Underlying applications arise e.g. when time-dependent coupling of container crane movements is under consideration, see [33, 34].

The optimal control of (quasi-) variational inequalities ((Q)VIs) seems to be closely related to the subject of bilevel optimal control since the underlying variational problem, which assigns to each control the uniquely determined state function, can be modeled as a parametric optimization problem in function spaces. Those problems are of hierarchical structure, but neither leader nor follower has to solve an optimal control problem in the classical meaning. In the seminal work [43], Mignot shows that the control-to-state map of an elliptic VI in the Sobolev space $H_0^1(\Omega)$ is directionally differentiable, and (in the absence of control constraints) this leads to an optimality system of strong-stationarity-type. If control constraints are present, one typically uses a regularization approach for the derivation of optimality conditions. This idea dates back to [6] and we refer to [48] for a modern treatment. Finally, we would like to mention that a comparison of several optimality systems and further references regarding this topic can be found in [21].

16.2 Notation and Preliminaries

Let us briefly recall some essentials of functional analysis we are going to exploit. For a (real) Banach space $\mathcal{X}$, $\|\cdot\|_{\mathcal{X}}\colon \mathcal{X} \to \mathbb{R}$ denotes its norm. Furthermore, $\mathcal{X}^\star$ represents the topological dual of $\mathcal{X}$. We use $\langle\cdot,\cdot\rangle_{\mathcal{X}}\colon \mathcal{X}^\star \times \mathcal{X} \to \mathbb{R}$ in order to denote the associated dual pairing. For a sequence $\{x_k\}_{k\in\mathbb{N}} \subset \mathcal{X}$ and some point $\bar{x} \in \mathcal{X}$, strong and weak convergence of $\{x_k\}_{k\in\mathbb{N}}$ to $\bar{x}$ will be represented by $x_k \to \bar{x}$ and $x_k \rightharpoonup \bar{x}$, respectively. Recall that in a finite-dimensional Banach space $\mathcal{X}$, the concepts of strong and weak convergence coincide. A functional $j\colon \mathcal{X} \to \mathbb{R}$ is said to be weakly sequentially lower (upper) semicontinuous at $\bar{x}$, whenever

$$x_k \rightharpoonup \bar{x} \implies j(\bar{x}) \leq \liminf_{k\to\infty} j(x_k) \qquad \left(x_k \rightharpoonup \bar{x} \implies j(\bar{x}) \geq \limsup_{k\to\infty} j(x_k)\right)$$

holds for all sequences $\{x_k\}_{k\in\mathbb{N}} \subset X$. We say that j is weakly sequentially lower (upper) semicontinuous if it possesses this property at each point from $\mathcal{X}$. It is well known that convex and continuous functionals are weakly sequentially lower semicontinuous. If the canonical embedding $\mathcal{X} \ni x \mapsto \langle\cdot,x\rangle_{\mathcal{X}} \in \mathcal{X}^{\star\star}$ is an isomorphism, then $\mathcal{X}$ is said to be reflexive. The particular Banach space $\mathbb{R}^n$ is equipped with the Euclidean norm $|\cdot|_2$. Furthermore, we use $x \cdot y$ to represent the Euclidean inner product in $\mathbb{R}^n$.

A set $A \subset \mathcal{X}$ is said to be weakly sequentially closed whenever the weak limits of all weakly convergent sequences from A belong to A as well. We note that closed and convex sets are weakly sequentially closed. We call A weakly sequentially compact whenever each sequence from A possesses a weakly convergent subsequence whose limit belongs to A. Each bounded, closed, and convex subset of a reflexive Banach space is weakly sequentially compact.

For a second Banach space $\mathcal{Y}$, $\mathbb{L}[\mathcal{X}, \mathcal{Y}]$ is used to denote the Banach space of all bounded linear operators mapping from $\mathcal{X}$ to $\mathcal{Y}$. For $\mathtt{A} \in \mathbb{L}[\mathcal{X}, \mathcal{Y}]$, $\mathtt{A}^\star \in \mathbb{L}[\mathcal{Y}^\star, \mathcal{X}^\star]$ denotes its adjoint. If $\mathcal{X} \subset \mathcal{Y}$ holds while the associated identity in $\mathbb{L}[\mathcal{X}, \mathcal{Y}]$ is continuous, then $\mathcal{X}$ is said to be continuously embedded into $\mathcal{Y}$ which will be denoted by $\mathcal{X} \hookrightarrow \mathcal{Y}$. Whenever the identity is compact, the embedding $\mathcal{X} \hookrightarrow \mathcal{Y}$ is called compact. For a set-valued mapping $\Gamma\colon \mathcal{X} \rightrightarrows \mathcal{Y}$, the sets $\operatorname{gph}\Gamma := \{(x, y) \in \mathcal{X} \times \mathcal{Y} \mid y \in \Gamma(x)\}$ and $\operatorname{dom}\Gamma := \{x \in \mathcal{X} \mid \Gamma(x) \neq \varnothing\}$ represent the graph and the domain of Γ, respectively.

Let $A \subset \mathcal{X}$ be nonempty and convex. Then, the closed, convex cone

$$A^\circ := \left\{x^\star \in \mathcal{X}^\star \,\middle|\, \forall x \in A\colon\ \langle x^\star, x\rangle_{\mathcal{X}} \leq 0\right\}$$

is called the polar cone of A. For a fixed point $\bar{x} \in A$, $\mathcal{N}_A(\bar{x}) := (A - \{\bar{x}\})^\circ$ is referred to as the normal cone (in the sense of convex analysis) to A at $\bar{x}$. For the purpose of completeness, let us set $\mathcal{N}_A(\hat{x}) := \varnothing$ for all $\hat{x} \in \mathcal{X} \setminus A$. Note that whenever $C \subset \mathcal{X}$ is a closed, convex cone satisfying $\bar{x} \in C$, then we have the relation

$$\mathcal{N}_C(\bar{x}) = C^\circ \cap \{x^\star \in \mathcal{X}^\star \mid \langle x^\star, \bar{x}\rangle_{\mathcal{X}} = 0\}.$$

Detailed information on the function spaces we are going to exploit can be found in the monograph [1].

16.3 Bilevel Programming in Banach Spaces

Let $\mathcal{X}$ and $\mathcal{Z}$ be Banach spaces. We consider the bilevel programming problem

$$\begin{aligned} F(x, z) &\to \min_{x,z} \\ x &\in X_{\mathrm{ad}} \\ z &\in \Psi(x), \end{aligned} \tag{BPP}$$

where $\Psi\colon \mathcal{X} \rightrightarrows \mathcal{Z}$ is the solution mapping of the parametric optimization problem

$$\begin{aligned} f(x, z) &\to \min_{z} \\ z &\in \Gamma(x). \end{aligned} \tag{16.3.1}$$

Note that we minimize the objective functional in (BPP) w.r.t. both variables which is related to the so-called optimistic approach of bilevel programming. In this section, we first want to discuss the existence of optimal solutions associated with (BPP). Afterwards, we briefly present possible approaches which can be used to infer optimality conditions for this problem class.

16.3.1 Existence Theory

In this section, we aim to characterize situations where (BPP) possesses optimal solutions. Noting that compact sets are generally rare in infinite-dimensional spaces, one cannot rely on classical existence results from bilevel programming. Indeed, compactness assumptions on the feasible sets have to be relaxed in order to guarantee applicability of possible results. As a consequence, we need to demand more restrictive properties than (lower semi-) continuity of the appearing objective functionals in order to balance things in a reasonable way. One may check e.g. [30] for a detailed discussion of existence theory for optimization problems in Banach spaces. Particularly, it is presented that each weakly sequentially lower semicontinuous functional achieves its minimum over a nonempty and weakly sequentially compact set. The above remarks justify the subsequently stated general assumptions of this section.

Assumption 16.3.1 *We consider Banach spaces $\mathcal{X}$ and $\mathcal{Z}$. The objective functionals $F, f : \mathcal{X} \times \mathcal{Z} \to \mathbb{R}$ are weakly sequentially lower semicontinuous. The set $X_{\text{ad}} \subset \mathcal{X}$ is assumed to be nonempty and weakly sequentially compact, while $\Gamma : \mathcal{X} \rightrightarrows \mathcal{Z}$ is a set-valued mapping with $X_{\text{ad}} \subset \operatorname{dom} \Gamma$ such that $(X_{\text{ad}} \times \mathcal{Z}) \cap \operatorname{gph} \Gamma$ is weakly sequentially compact.* △

In the setting where $\mathcal{X}$ and $\mathcal{Z}$ are finite-dimensional, e.g. instances of $\mathbb{R}^n$, the above assumptions reduce to the lower semicontinuity of the objective functionals as well as some compactness assumptions on X_{ad} and $\operatorname{gph} \Gamma$ which is rather standard in bilevel programming, see e.g. [14]. For our upcoming analysis, we will exploit the function $\varphi : X_{\text{ad}} \to \overline{\mathbb{R}}$ defined by

$$\forall x \in X_{\text{ad}} : \quad \varphi(x) := \inf_{z}\{f(x, z) \mid z \in \Gamma(x)\}. \tag{16.3.2}$$

By definition, φ assigns to each parameter $x \in X_{\text{ad}}$ the optimal function value of the lower level problem (16.3.1). Assumption 16.3.1 guarantees that the infimal value $\varphi(x)$ is actually attained (i.e. $\Psi(x) \neq \varnothing$) since for all $x \in X_{\text{ad}}$, $f(x, \cdot) : \mathcal{Z} \to \mathbb{R}$ is weakly sequentially lower semicontinuous while $\Gamma(x)$ is nonempty and weakly sequentially compact.

Below, we need to study the (upper) semicontinuity properties of φ. In order to do that, we need to address some continuity properties of the mapping Γ, see [26].

Definition 16.3.2 Fix $(\bar{x}, \bar{z}) \in \operatorname{gph}\Gamma$. Then, Γ is called inner semicontinuous (weakly-weakly inner semicontinuous) at $(\bar{x}, \bar{z})$ if for each sequence $\{x_k\}_{k\in\mathbb{N}} \subset \operatorname{dom}\Gamma$ satisfying $x_k \to \bar{x}$ ($x_k \rightharpoonup \bar{x}$), there exists a sequence $\{z_k\}_{k\in\mathbb{N}} \subset \mathcal{Z}$ satisfying $z_k \in \Gamma(x_k)$ for all $k \in \mathbb{N}$ as well as $z_k \to \bar{z}$ ($z_k \rightharpoonup \bar{z}$). △

It needs to be noted that the concepts of inner and lower semicontinuity of set-valued mappings, see [5], are closely related. Particularly, the lower semicontinuity of Γ at some point $\bar{x} \in \operatorname{dom}\Gamma$ is equivalent to its inner semicontinuity at all points $(\bar{x}, z)$ with $z \in \Gamma(\bar{x})$.

In the particular situation where the mapping Γ is characterized via smooth generalized inequalities, there is an easy criterion which is sufficient for inner semicontinuity.

Remark 16.3.3 We assume that there exists a continuously Fréchet differentiable function $g\colon \mathcal{X} \times \mathcal{Z} \to \mathcal{W}$, where $\mathcal{W}$ is a Banach space, and some nonempty, closed, convex set $C \subset \mathcal{W}$ such that Γ is given by

$$\forall x \in \mathcal{X}\colon \quad \Gamma(x) := \{z \in \mathcal{Z} \mid g(x, z) \in C\}.$$

For fixed $\bar{z} \in \Gamma(\bar{x})$, we assume that the condition

$$g'_z(\bar{x}, \bar{z})\mathcal{Z} - \operatorname{cone}(C - \{g(\bar{x}, \bar{z})\}) = \mathcal{W} \tag{16.3.3}$$

is valid. Then, Γ is inner semicontinuous at $(\bar{x}, \bar{z})$, see e.g. [10, Section 2.3.3]. We note that (16.3.3) often is referred to as Robinson's constraint qualification, see [47], or Kurcyusz–Zowe constraint qualification, see [57]. In the setting of finite-dimensional nonlinear parametric optimization, this condition simply reduces to the Mangasarian–Fromovitz constraint qualification, see [10] for details. Let us note that (16.3.3) trivially holds whenever the operator $g'_z(\bar{x}, \bar{z})$ is surjective. △

We note that weak-weak inner semicontinuity of Γ is inherent whenever this map is actually constant. A nontrivial situation is described in the following example.

Terminal State Dependence of Lower Level

For some time interval $I := (0, T)$ and some natural number n, we consider the Bochner–Sobolev space $\mathcal{X} := H^1(I; \mathbb{R}^n)$, which is a Hilbert space. Clearly, the embedding $\mathcal{X} \hookrightarrow C(\bar{I}; \mathbb{R}^n)$ is compact, see [1]. This means that the evaluation operator $\mathcal{X} \ni x \mapsto x(T) \in \mathbb{R}^n$ is well-defined and compact as well.

For some set-valued mapping $\Upsilon\colon \mathbb{R}^n \rightrightarrows \mathcal{Z}$, we define $\Gamma(x) := \Upsilon(x(T))$ for all $x \in \mathcal{X}$. The above observation implies that Γ is weakly-weakly inner semicontinuous at $(\bar{x}, \bar{z}) \in \operatorname{gph}\Gamma$ whenever Υ is inner semicontinuous at $(\bar{x}(T), \bar{z})$ and the latter can be guaranteed via standard assumptions, see e.g. Remark 16.3.3.

(continued)

The setting in this example reflects the situation of time-dependent coupling between upper and lower level, see [33, Section 5], or a finite-dimensional lower level problem depending only on the terminal value of the leader's state variable, see [8, 9, 32].

It needs to be noted that the analysis of the above situation can be extended to cases where $\mathcal{X}$ is a function space over some domain $\Omega \subset \mathbb{R}^d$ which is embedded compactly into $C(\overline{\Omega})$, and the function of interest is evaluated at finitely many points from $\overline{\Omega}$. This applies to the setting of the Sobolev space $\mathcal{X} := H^2(\Omega)$ where Ω is a bounded Lipschitz domain, see [1].

In the following lemma, which is inspired by [26, Theorem 2.5], we study upper semicontinuity properties of the function φ. This will be useful in order to infer closedness properties of the feasible set associated with (BPP).

Lemma 16.3.4 *Fix $\bar{x} \in X_{\text{ad}}$ and $\bar{z} \in \Psi(\bar{x})$.*

1. *Assume that f is weakly sequentially upper semicontinuous at $(\bar{x}, \bar{z})$ while Γ is weakly-weakly inner semicontinuous at $(\bar{x}, \bar{z})$. Then, φ is weakly sequentially upper semicontinuous at $\bar{x}$.*
2. *Let $\mathcal{X}$ be finite-dimensional. Assume that f is upper semicontinuous at $(\bar{x}, \bar{z})$ while Γ is inner semicontinuous at $(\bar{x}, \bar{z})$. Then, φ is upper semicontinuous at $\bar{x}$.* △

Proof We only verify the first statement of the lemma. The second one can be shown using analogous arguments.

Let $\{x_k\}_{k\in\mathbb{N}} \subset X_{\text{ad}}$ be a sequence satisfying $x_k \rightharpoonup \bar{x}$. Exploiting the weak-weak inner semicontinuity of Γ at $(\bar{x}, \bar{z})$, we find a sequence $\{z_k\}_{k\in\mathbb{N}} \subset \mathcal{Z}$ satisfying $z_k \in \Gamma(x_k)$ for all $k \in \mathbb{N}$ and $z_k \rightharpoonup \bar{z}$. By definition, $\varphi(x_k) \leq f(x_k, z_k)$ holds for all $k \in \mathbb{N}$. Now, the weak sequential upper semicontinuity of f at $(\bar{x}, \bar{z})$ yields

$$\limsup_{k\to\infty} \varphi(x_k) \leq \limsup_{k\to\infty} f(x_k, z_k) \leq f(\bar{x}, \bar{z}) = \varphi(\bar{x}),$$

and this shows the claim. □

Now, we exploit the above lemma in order to infer the existence of optimal solutions to (BPP).

Theorem 16.3.5 *In each of the settings described below,* (BPP) *possesses an optimal solution.*

1. *The mapping f is weakly sequentially upper semicontinuous on $X_{\text{ad}} \times \mathcal{Z}$ while Γ is weakly-weakly inner semicontinuous on $X_{\text{ad}} \times \mathcal{Z}$.*
2. *The Banach space $\mathcal{X}$ is finite-dimensional. The mapping f is upper semicontinuous on $X_{\text{ad}} \times \mathcal{Z}$ while Γ is inner semicontinuous on $X_{\text{ad}} \times \mathcal{Z}$.* △

Proof Again, we only show the theorem's first assertion.

For the proof, we just need to verify that the feasible set $(X_{\mathrm{ad}} \times \mathcal{Z}) \cap \operatorname{gph} \Psi$ of (BPP) is nonempty and weakly sequentially compact since the objective F is supposed to be weakly sequentially lower semicontinuous. Noting that $\Psi(x) \neq \varnothing$ holds true for all $x \in X_{\mathrm{ad}}$, the nonemptiness of $(X_{\mathrm{ad}} \times \mathcal{Z}) \cap \operatorname{gph} \Psi$ is obvious.

Let $\{(x_k, z_k)\}_{k\in\mathbb{N}} \subset (X_{\mathrm{ad}} \times \mathcal{Z}) \cap \operatorname{gph} \Psi$ be an arbitrary sequence. Clearly, we have $\{(x_k, z_k)\}_{k\in\mathbb{N}} \subset (X_{\mathrm{ad}} \times \mathcal{Z}) \cap \operatorname{gph} \Gamma$ and by Assumption 16.3.1, there exists a subsequence (without relabeling) converging weakly to $(\bar{x}, \bar{z}) \in (X_{\mathrm{ad}} \times \mathcal{Z}) \cap \operatorname{gph} \Gamma$. Now, the definition of the function φ, the weak sequential lower semicontinuity of f, see Assumption 16.3.1, and Lemma 16.3.4 yield

$$\varphi(\bar{x}) \leq f(\bar{x}, \bar{z}) \leq \liminf_{k\to\infty} f(x_k, z_k) = \liminf_{k\to\infty} \varphi(x_k) \leq \limsup_{k\to\infty} \varphi(x_k) \leq \varphi(\bar{x}),$$

which shows $\varphi(\bar{x}) = f(\bar{x}, \bar{z})$, i.e. $\bar{z} \in \Psi(\bar{x})$ follows. This yields that the point $(\bar{x}, \bar{z})$ belongs to $(X_{\mathrm{ad}} \times \mathcal{Z}) \cap \operatorname{gph} \Psi$, and this shows the claim. □

Let us apply the above theory to some example problems from bilevel optimal control.

Inverse Nonregularized Control of Poisson's Equation

Let $\Omega \subset \mathbb{R}^d$ be a bounded domain with Lipschitz boundary $\operatorname{bd} \Omega$. For fixed parameters $x^w \in \mathbb{R}^n$ and $x^s \in L^2(\Omega)$, we consider the optimal control of Poisson's equation

$$\begin{aligned} \tfrac{1}{2}\left\| y - \textstyle\sum_{i=1}^n x_i^w f^i \right\|_{L^2(\Omega)}^2 &\to \min_{y,u} \\ -\Delta y &= x^s + u \\ u_a \leq u \leq u_b & \qquad \text{a.e. on } \Omega \end{aligned}$$

where $f^1, \ldots, f^n \in L^2(\Omega)$ are fixed form functions and $u_a, u_b \in L^2(\Omega)$ are given functions satisfying $u_a < u_b$ almost everywhere on Ω. The variables y and u are chosen from the respective spaces $H_0^1(\Omega)$ and $L^2(\Omega)$. The underlying PDE has to be understood in weak sense in $H^{-1}(\Omega) := H_0^1(\Omega)^\star$. In this regard, the source term $x^s + u$ from $L^2(\Omega)$ is embedded into $H^{-1}(\Omega)$, implicitly. Noting that no regularization term w.r.t. the control appears in the objective functional, optimal controls are promoted which take values only at the lower and upper bound u_a and u_b, and such controls are referred to as bang-bang, see [50].

Let $\Psi \colon \mathbb{R}^n \times L^2(\Omega) \rightrightarrows H_0^1(\Omega) \times L^2(\Omega)$ be the solution map associated with the above optimal control problem. In the superordinate upper level problem, we aim to identify the lower level desired state via correct choice

(continued)

of the weights $x^w \in \mathbb{R}^n$ and constant source $x^s \in L^2(\Omega)$ from a nonempty, closed, convex, and bounded set $X_{ad} \subset \mathbb{R}^n \times L^2(\Omega)$ such that a resulting optimal solution is close to observed data functions $y_o, u_o \in L^2(\Omega)$. A suitable model for this program is given by

$$\begin{aligned} \tfrac{1}{2}\,\|y - y_o\|^2_{L^2(\Omega)} + \tfrac{1}{2}\,\|u - u_o\|^2_{L^2(\Omega)} &\to \min_{x,y,u} \\ (x^w, x^s) &\in X_{ad} \\ (y, u) &\in \Psi(x^w, x^s). \end{aligned}$$

Due to continuity and convexity of the appearing objective functionals, they are weakly sequentially lower semicontinuous. Furthermore, the compactness of the embedding $H_0^1(\Omega) \hookrightarrow L^2(\Omega)$ even guarantees that the objective of the lower level problem is weakly sequentially continuous. The set X_{ad} is nonempty and weakly sequentially compact by assumption. Exploiting the linearity and continuity of the solution operator $(-\Delta)^{-1}$ of Poisson's equation, it is not difficult to see that the graph of the lower level feasible set mapping Γ is convex and closed. The boundedness of X_{ad} ensures the boundedness of $(X_{ad} \times H_0^1(\Omega) \times L^2(\Omega)) \cap \operatorname{gph}\Gamma$, and it is not difficult to see that this set is weakly sequentially compact as well. Using the properties of $(-\Delta)^{-1}$, it is easy to see that Γ is weakly-weakly inner semicontinuous at all points of its graph. Now, Theorem 16.3.5 yields the existence of a solution to the bilevel optimal control problem under consideration.

Optimal Control of ODEs with Terminal Penalty Cost

For a fixed given vector $\xi \in \mathbb{R}^n$ of parameters, we consider the parametric optimization problem

$$\begin{aligned} j(\xi, z) &\to \min_{z} \\ g(\xi, z) &\le 0 \end{aligned} \tag{16.3.4}$$

where $j\colon \mathbb{R}^n \times \mathbb{R}^m \to \mathbb{R}$ is continuous and $g\colon \mathbb{R}^n \times \mathbb{R}^m \to \mathbb{R}^k$ is continuously differentiable. Furthermore, we assume that $\Upsilon(\xi) := \{y \in \mathbb{R}^m \mid g(\xi, y) \le 0\}$ is nonempty for each $\xi \in \mathbb{R}^n$, that $\bigcup_{\xi \in \mathbb{R}^n} \Upsilon(\xi)$ is bounded, and that the Mangasarian–Fromovitz constraint qualification holds at all feasible points associated with (16.3.4).

(continued)

The associated upper level problem shall be given by

$$\begin{aligned} \tfrac{1}{2}\|x-x_{\mathrm{d}}\|^2_{L^2(I;\mathbb{R}^n)}+\tfrac{\sigma}{2}\|u\|^2_{L^2(I;\mathbb{R}^p)}+J(x(T),y) &\to \min_{x,u,y}\\ \dot{x}-Ax-Bu&=0\\ x(0)&=0\\ u_a\le u&\le u_b\\ y&\in\Psi(x) \end{aligned} \tag{16.3.5}$$

where $I := (0, T)$ is a time interval, $x_{\mathrm{d}} \in L^2(I; \mathbb{R}^n)$ is a desired state, $\sigma \ge 0$ is a regularization parameter, $J : \mathbb{R}^n \times \mathbb{R}^m \to \mathbb{R}$ is lower semicontinuous, $A \in \mathbb{R}^{n\times n}$ as well as $B \in \mathbb{R}^{n\times p}$ are fixed matrices, $u_a, u_b \in L^2(I; \mathbb{R}^p)$ are fixed functions satisfying $u_a < u_b$ almost everywhere on I, and $\Psi : H^1(I; \mathbb{R}^n) \rightrightarrows \mathbb{R}^m$ assigns to each $x \in H^1(I; \mathbb{R}^n)$ the solution set of (16.3.4) for the fixed parameter $\xi := x(T)$. The controls in (16.3.5) are chosen from $L^2(I; \mathbb{R}^p)$.

Problem (16.3.5) describes the situation where an ODE system has to be controlled in such a way that certain penalty cost resulting from the terminal value of the state function as well as the distance to a desirable state are minimized with minimal control effort. Optimization problems of this kind arise in the context of gas balancing in energy networks and were studied in [8, 9, 32]. Invoking Remark 16.3.3, the subsequently stated example, and Theorem 16.3.5, we obtain the existence of an optimal solution associated with (16.3.5).

Another typical situation arises when the lower level problem (16.3.1) is uniquely solvable for each upper level feasible point.

Theorem 16.3.6 *Assume that there exists a map $\psi : X_{\mathrm{ad}} \to \mathcal{Z}$ sending weakly convergent sequences from X_{ad} to weakly convergent sequences in $\mathcal{Z}$ such that $\Psi(x) = \{\psi(x)\}$ holds for all $x \in X_{\mathrm{ad}}$. Then,* (BPP) *possesses an optimal solution.* $\triangle$

Proof The assumptions of the theorem guarantee that (BPP) is equivalent to

$$\begin{aligned} F(x,\psi(x)) &\to \min_x\\ x&\in X_{\mathrm{ad}}. \end{aligned}$$

Furthermore, $X_{\mathrm{ad}} \ni x \mapsto F(x, \psi(x)) \in \mathbb{R}$ is weakly sequentially lower semicontinuous on X_{ad} since F possesses this property on $\mathcal{X} \times \mathcal{Z}$ while ψ preserves weak convergence of sequences from X_{ad}. Thus, the above problem possesses an optimal solution $\bar{x}$, i.e. (BPP) possesses the optimal solution $(\bar{x}, \psi(\bar{x}))$. $\square$

The above theorem particularly applies to situations where the upper level variable comes from a finite-dimensional Banach space while the solution operator associated to the lower level problem is continuous. This setting has been discussed in [17, 23] and will be of interest in Sect. 16.4.

16.3.2 How to Derive Necessary Optimality Conditions in Bilevel Optimal Control

In order to derive necessary optimality conditions for bilevel programming problems, one generally aims to transfer the hierarchical model into a single-level program first. Therefore, three major approaches are suggested in the literature. First, whenever the lower level problem possesses a uniquely determined solution for each fixed value of the upper level problem, one could use the associated solution operator to eliminate the lower level variable from the model. This approach has been used in [23, 38] in order to derive necessary optimality conditions for bilevel optimal control problems. Second, it is possible to exploit the optimal value function from (16.3.2) in order to replace (BPP) equivalently by the so-called *optimal value reformulation*

$$\begin{aligned} F(x,z) &\to \min_{x,z} \\ x &\in X_{\text{ad}} \\ f(x,z) - \varphi(x) &\leq 0 \\ z &\in \Gamma(x). \end{aligned}$$

In [8, 9, 17, 55, 56], the authors exploited this idea to infer optimality conditions in the context of bilevel optimal control. We will demonstrate in Sect. 16.4, how a relaxation method can be combined with the optimal value approach in order to obtain a satisfactory stationarity condition for a particular problem class from inverse optimal control. Finally, as long as the lower level problem is convex w.r.t. z and regular in the sense that a constraint qualification is satisfied at each feasible point, it is possible to replace the implicit constraint $z \in \Psi(x)$ by suitable necessary and sufficient optimality conditions of Karush–Kuhn–Tucker (KKT) type. In the context of bilevel optimal control, this approach has been discussed in [40]. In this section, we will briefly sketch this last approach. Therefore, we have to fix some assumptions first.

Assumption 16.3.7 *We assume that the mapping Γ is given as stated in Remark 16.3.3 where C is a cone. Furthermore, we suppose that $f(x,\cdot)\colon \mathcal{Z} \to \mathbb{R}$*

is convex and that $g(x,\cdot)\colon \mathcal{Z} \to \mathcal{W}$ *is* C*-convex for each* $x \in X_{ad}$*. The latter means that*

$$\forall z, z' \in \mathcal{Z}\,\forall \gamma \in [0,1]: \quad g(x, \gamma z + (1-\gamma)z') - \gamma g(x,z) - (1-\gamma)g(x,z') \in C$$

holds true. △

Due to the postulated assumptions, for fixed $x \in X_{\text{ad}}$, $z \in \Psi(x)$ holds true if and only if there exists a Lagrange multiplier $\lambda \in \mathcal{W}^\star$ which solves the associated lower level KKT system which is given as stated below:

$$\begin{aligned} 0 &= f'_z(x,z) + g'_z(x,z)^\star\lambda, \\ \lambda &\in C^\circ, \\ 0 &= \langle \lambda, g(x,z)\rangle_{\mathcal{W}}, \end{aligned}$$

see [10, Theorem 3.9] or [57]. Here, it was essential that the lower level problem is convex w.r.t. z while Robinson's constraint qualification (16.3.3) is valid at all lower level feasible points. Due to the above arguments, it is now reasonable to investigate the so-called *KKT reformulation* associated with (BPP) which is given as stated below:

$$\begin{aligned} F(x,z) &\to \min_{x,z,\lambda} \\ x &\in X_{\text{ad}} \\ f'_z(x,z) + g'_z(x,z)^\star\lambda &= 0 \\ g(x,z) &\in C \\ \lambda &\in C^\circ \\ \langle \lambda, g(x,z)\rangle_{\mathcal{W}} &= 0. \end{aligned} \tag{KKT}$$

Let us note that the lower level Lagrange multiplier λ plays the role of a variable in (KKT). This may cause that the problems (BPP) and (KKT) are not equivalent w.r.t. local minimizers as soon as λ is not uniquely determined for each $x \in X_{\text{ad}}$ where $\Psi(x) \neq \varnothing$ holds, see [37]. As reported in [15], this phenomenon is already present in standard finite-dimensional bilevel programming.

In the situation where $C = \{0\}$ holds, the final two constraints in (KKT) are trivial and can be omitted. Then, (KKT) reduces to a standard nonlinear program in Banach spaces which can be tackled via classical arguments. Related considerations can be found in [28]. The subsequently stated example visualizes this approach.

Inverse Control of Poisson's Equation
For a bounded domain $\Omega \subset \mathbb{R}^d$ and a parameter vector $x \in \mathbb{R}^n$, we consider the parametric optimal control problem

$$\begin{aligned} \frac{1}{2}\left\|y - \sum_{i=1}^n x_i f^i\right\|^2_{L^2(\Omega)} + \frac{\sigma}{2}\|u\|^2_{L^2(\Omega)} &\to \min_{y,u} \\ -\Delta y &= u \end{aligned}$$

where $f^1, \ldots, f^n \in L^2(\Omega)$ are given form functions and $\sigma > 0$ is a regularization parameter. For observations $y_o, u_o \in L^2(\Omega)$ and a nonempty, convex, compact set $X_{\text{ad}} \subset \mathbb{R}^n$, we consider the superordinate inverse optimal control problem

$$\begin{aligned} \frac{1}{2}\|y - y_o\|^2_{L^2(\Omega)} + \frac{1}{2}\|u - u_o\|^2_{L^2(\Omega)} &\to \min_{x,y,u} \\ x &\in X_{\text{ad}} \\ (y,u) &\in \Psi(x) \end{aligned} \tag{16.3.6}$$

where $\Psi\colon \mathbb{R}^n \rightrightarrows H_0^1(\Omega) \times L^2(\Omega)$ represents the solution mapping of the aforementioned parametric optimal control problem. We can use Theorem 16.3.6 in order to infer the existence of an optimal solution associated with this bilevel optimal control problem.

Noting that $-\Delta\colon H_0^1(\Omega) \to H^{-1}(\Omega)$ provides an isomorphism, the associated KKT reformulation, given by

$$\begin{aligned} \frac{1}{2}\|y - y_o\|^2_{L^2(\Omega)} + \frac{1}{2}\|u - u_o\|^2_{L^2(\Omega)} &\to \min_{x,y,u,p} \\ x &\in X_{\text{ad}} \\ y - \sum_{i=1}^n x_i f^i - \Delta p &= 0 \\ \sigma u - p &= 0 \\ -\Delta y - u &= 0, \end{aligned}$$

is equivalent to the original hierarchical model. One can easily check that Robinson's constraint qualification is valid at each feasible point of this program which means that its KKT conditions provide a necessary optimality condition for the underlying inverse optimal control problem. Thus, whenever the triplet $(\bar{x}, \bar{y}, \bar{u}) \in \mathbb{R}^n \times H_0^1(\Omega) \times L^2(\Omega)$ is a locally optimal solution of

(continued)

(16.3.6), then we find multipliers $\bar{z} \in \mathbb{R}^n$, $\bar{p}, \bar{\mu}, \bar{\rho} \in H_0^1(\Omega)$, and $\bar{w} \in L^2(\Omega)$ which satisfy

$$\begin{aligned} 0 &= \bar{z} - \left(\left\langle \bar{\mu}, f^i \right\rangle_{L^2(\Omega)}\right)_{i=1}^n, & 0 &= \bar{y} - y_o + \bar{\mu} - \Delta\bar{\rho}, \\ 0 &= \bar{u} - u_o + \sigma\bar{w} - \bar{\rho}, & 0 &= -\Delta\bar{\mu} - \bar{w}, \\ \bar{z} &\in \mathcal{N}_{X_{\mathrm{ad}}}(\bar{x}), & 0 &= \bar{y} - \textstyle\sum_{i=1}^n \bar{x}_i f^i - \Delta\bar{p}, \\ 0 &= \sigma\bar{u} - \bar{p}. \end{aligned}$$

In case where C is a non-trivial cone, the final three constraints of (KKT) form a so-called system of complementarity constraints, i.e. this program is a *mathematical program with complementarity constraints* (MPCC) in Banach spaces. As shown in [40], this results in the violation of Robinson's constraint qualification at all feasible points of (KKT) and, consequently, the KKT conditions of (KKT) may turn out to be too restrictive in order to yield an applicable necessary optimality condition. Instead, weaker problem-tailored stationarity notions and constraint qualifications need to be introduced which respect the specific variational structure, see [37, 40, 53, 54]. In bilevel optimal control, complementarity constraints are typically induced by the cone of nonnegative functions in a suitable function space, e.g. $L^2(\Omega)$, $H_0^1(\Omega)$, or $H^1(\Omega)$. Respective considerations can be found in [13, 20–22, 41, 42].

16.4 Stationarity Conditions in Inverse Optimal Control

In this section, we demonstrate by means of a specific class of parameter reconstruction problems how stationarity conditions in bilevel optimal control can be derived.

For a bounded domain $\Omega \subset \mathbb{R}^d$ and a parameter $x \in \mathbb{R}^n_+$, where $\mathbb{R}^n_+$ denotes the nonnegative orthant in $\mathbb{R}^n$, we study the parametric optimal control problem

$$\begin{aligned} x \cdot j(y) + \tfrac{\sigma}{2}\,\|u\|_{L^2(\Omega)}^2 \;&\rightarrow\; \min_{y,u} \\ \mathtt{A}y - \mathtt{B}u \;&=\; 0 \\ u_a \;\le\; u \;&\le\; u_b \qquad \text{a.e. on } \Omega \end{aligned} \tag{P(x)}$$

as well as the superordinated bilevel optimal control problem

$$\begin{aligned} F(x, y, u) &\to \min_{x,y,u} \\ x &\in X_{\mathrm{ad}} \\ (y, u) &\in \Psi(x) \end{aligned} \tag{IOC}$$

where $\Psi\colon \mathbb{R}^n \rightrightarrows \mathcal{Y} \times L^2(\Omega)$ denotes the solution set mapping of (P(x)). In (P(x)), the state equation $\mathtt{A}y - \mathtt{B}u = 0$ couples the control $u \in L^2(\Omega)$ and the state $y \in \mathcal{Y}$. In this regard, $\mathtt{A}$ can be interpreted as a differential operator. Noting that (IOC) is motivated by underlying applications from parameter reconstruction, it is an inverse optimal control program.

Assumption 16.4.1 *We assume that $\mathcal{Y}$ and $\mathcal{W}$ are reflexive Banach spaces. The functional $F\colon \mathbb{R}^n \times \mathcal{Y} \times L^2(\Omega) \to \mathbb{R}$ is supposed to be continuously Fréchet differentiable and convex. Let $X_{\mathrm{ad}} \subset \mathbb{R}^n_+$ be nonempty and compact. The functional $j\colon \mathcal{Y} \to \mathbb{R}^n$ is assumed to be twice continuously Fréchet differentiable and its n component functions are supposed to be convex. Moreover, we assume that the mapping j satisfies $j(\mathcal{Y}) \subset \mathbb{R}^n_+$. Furthermore, $\sigma > 0$ is fixed. Let linear operators $\mathtt{A} \in \mathbb{L}\left[\mathcal{Y}, \mathcal{W}\right]$ as well as $\mathtt{B} \in \mathbb{L}\left[L^2(\Omega), \mathcal{W}\right]$ be chosen such that $\mathtt{A}$ is continuously invertible while $\mathtt{B}$ is compact. Finally, we assume that $u_a, u_b\colon \Omega \to \overline{\mathbb{R}}$ are measurable functions such that*

$$U_{\mathrm{ad}} := \{u \in L^2(\Omega) \mid u_a \le u \le u_b \text{ a.e. on } \Omega\}$$

is nonempty. △

Below, we present two illustrative examples where all these assumptions hold.

Weighted Lower Level Target-Type Objectives
We choose $\mathcal{Y} := H_0^1(\Omega)$, $\mathcal{W} := H^{-1}(\Omega)$, as well as $\mathtt{A} := -\Delta$ while $\mathtt{B}$ represents the compact embedding $L^2(\Omega) \hookrightarrow H^{-1}(\Omega)$. For fixed functions $y_{\mathrm{d}}^1, \ldots, y_{\mathrm{d}}^n \in L^2(\Omega)$, the lower level objective function is defined by

$$\mathbb{R}^n \times H_0^1(\Omega) \times L^2(\Omega) \ni (x, y, u) \mapsto \textstyle\sum_{i=1}^n x_i \left\| y - y_{\mathrm{d}}^i \right\|_{L^2(\Omega)}^2 + \frac{\sigma}{2} \|u\|_{L^2(\Omega)}^2 \in \mathbb{R}.$$

The upper level feasible set is given by the standard simplex

$$\{x \in \mathbb{R}^n \mid x \ge 0,\ \textstyle\sum_{i=1}^n x_i = 1\}. \tag{16.4.1}$$

(continued)

Such bilevel optimal control problems, where the precise form of the lower level target-type objective mapping has to be reconstructed, have been studied in [23].

Optimal Measuring
Let $\Omega \subset \mathbb{R}^d$, $d \in \{2, 3\}$, be a bounded Lipschitz domain. We fix $p \in (3, 6)$ as in [31, Theorem 0.5]. Let us set $\mathcal{Y} := W_0^{1,p}(\Omega)$ and $\mathcal{W} := W^{-1,p}(\Omega)$. Again, we fix $\mathtt{A} := -\Delta$, and $\mathtt{B}$ represents the embedding $L^2(\Omega) \hookrightarrow W^{-1,p}(\Omega) := W_0^{1,p'}(\Omega)^\star$ where $p' = p/(p-1)$ is the conjugate coefficient associated with p. According to [31, Theorem 0.5], $\mathtt{A}$ is continuously invertible. Due to the Rellich–Kondrachov theorem, the embedding from $W_0^{1,p'}(\Omega)$ to $L^2(\Omega)$ is compact. Since $\mathtt{B}$ is the adjoint of this embedding, Schauder's theorem implies the compactness of $\mathtt{B}$. Furthermore, we note that the embedding $W_0^{1,p}(\Omega) \hookrightarrow C(\overline{\Omega})$ is compact in this setting.

Let $\omega^1, \ldots, \omega^n \in \overline{\Omega}$ be fixed points. We consider the lower level objective function given by

$$\mathbb{R}^n \times W_0^{1,p}(\Omega) \times L^2(\Omega) \ni (x, y, u) \mapsto \sum_{i=1}^n x_i (y(\omega^i) - y_{\mathrm{d}}(\omega^i))^2 + \tfrac{\sigma}{2} \|u\|^2_{L^2(\Omega)} \in \mathbb{R}$$

where $y_{\mathrm{d}} \in C(\overline{\Omega})$ is a given desired state. Noting that the state space $W_0^{1,p}(\Omega)$ is continuously embedded into $C(\overline{\Omega})$, this functional is well-defined. At the upper level stage, we minimize

$$\mathbb{R}^n \times W_0^{1,p}(\Omega) \times L^2(\Omega) \ni (x, y, u) \mapsto \tfrac{1}{2} \|y - y_{\mathrm{d}}\|^2_{L^2(\Omega)} + \tfrac{1}{2} |x|_2^2 \in \mathbb{R}$$

where x comes from the standard simplex given in (16.4.1). The associated bilevel optimal control problem optimizes the *measurement* of the distance between the actual state and the desired state by reduction to pointwise evaluations.

16.4.1 The Lower Level Problem

For brevity, we denote by $f : \mathbb{R}^n \times \mathcal{Y} \times L^2(\Omega) \to \mathbb{R}$ the objective functional of (P(x)). By construction, the map $f(x, \cdot, \cdot) : \mathcal{Y} \times L^2(\Omega) \to \mathbb{R}$ is convex for each $x \in \mathbb{R}^n_+$.

Lemma 16.4.2 *For each $x \in \mathbb{R}^n_+$, (P(x)) possesses a unique optimal solution.* △

Proof Noting that $\mathtt{A}$ is an isomorphism, we may consider the state-reduced problem

$$\min_u\{f(x, \mathtt{S}u, u) \mid u \in U_{\mathrm{ad}}\} \tag{16.4.2}$$

where $\mathtt{S} := \mathtt{A}^{-1} \circ \mathtt{B} \in \mathbb{L}\left[L^2(\Omega), \mathcal{Y}\right]$ is the solution operator of the constraining state equation. Due to the above considerations, the linearity of $\mathtt{S}$, and the continuity of all appearing functions, the objective functional of (16.4.2) is convex and continuous. Observing that $x \cdot j(\mathtt{S}u) \geq 0$ holds for all $u \in L^2(\Omega)$ while $L^2(\Omega) \ni u \mapsto \frac{\sigma}{2}\|u\|^2_{L^2(\Omega)} \in \mathbb{R}$ is coercive, the objective of (16.4.2) is already strongly convex w.r.t. u. Since U_{ad} is weakly sequentially closed, (16.4.2) needs to possess a unique solution $\bar{u}$. Consequently, $(\mathtt{S}\bar{u}, \bar{u})$ is the uniquely determined solution of (P(x)). □

Observing that the lower level problem (P(x)) is regular in the sense that Robinson's constraint qualification is valid at all feasible points, see Remark 16.3.3, its uniquely determined solution for the fixed parameter $x \in \mathbb{R}^n_+$ is characterized by the associated KKT system

$$0 = j'(y)^\star x + \mathtt{A}^\star p, \tag{16.4.3a}$$

$$0 = \sigma u - \mathtt{B}^\star p + \lambda, \tag{16.4.3b}$$

$$\lambda \in \mathcal{N}_{U_{\mathrm{ad}}}(u) \tag{16.4.3c}$$

where $p \in \mathcal{W}^\star$ and $\lambda \in L^2(\Omega)$ are the Lagrange multipliers, see [10, Theorem 3.9] or [57].

The finding of Lemma 16.4.2 allows us to introduce mappings $\psi^y : \mathbb{R}^n_+ \to \mathcal{Y}$ and $\psi^u : \mathbb{R}^n_+ \to L^2(\Omega)$ by $\Psi(x) = \{(\psi^y(x), \psi^u(x))\}$ for all $x \in \mathbb{R}^n_+$. Since $\mathtt{A}^\star$ is continuously invertible, p is uniquely determined by (16.4.3a) and, consequently, the uniqueness of λ follows from (16.4.3b). This gives rise to the mappings $\phi^p : \mathbb{R}^n_+ \to \mathcal{W}^\star$ and $\phi^\lambda : \mathbb{R}^n_+ \to L^2(\Omega)$ that assign to each $x \in \mathbb{R}^n_+$ the lower level Lagrange multipliers p and λ which characterize the unique minimizer $(\psi^y(x), \psi^u(x))$, respectively.

Next, we study the continuity properties of the mappings ψ^y and ψ^u as well as ϕ^p and ϕ^λ.

Lemma 16.4.3 *There are continuous functions $C_y, C_u : \mathbb{R}^n_+ \to \mathbb{R}$ such that the following estimates hold:*

$$\forall x_1, x_2 \in \mathbb{R}^n_+ : \qquad \left\|\psi^y(x_1) - \psi^y(x_2)\right\|_{\mathcal{Y}} \leq C_y(x_1)\,|x_1 - x_2|_2\,,$$

$$\left\|\psi^u(x_1) - \psi^u(x_2)\right\|_{L^2(\Omega)} \leq C_u(x_1)\,|x_1 - x_2|_2\,.$$

Particularly, ψ^y and ψ^u are Lipschitz continuous on X_{ad}. △

Proof Fix $x_1, x_2 \in \mathbb{R}^n_+$ arbitrarily and set $y_i := \psi^y(x_i)$ as well as $u_i := \psi^u(x_i)$ for $i = 1, 2$. Furthermore, let $p_i \in \mathcal{W}^\star$ and $\lambda_i \in L^2(\Omega)$ be the multipliers which solve (16.4.3) for $i = 1, 2$. Testing the associated condition (16.4.3b) with $u_2 - u_1$ and exploiting (16.4.3c), we have

$$\begin{aligned}\langle \sigma u_1 - \mathtt{B}^\star p_1, u_2 - u_1\rangle_{L^2(\Omega)} &= \langle -\lambda_1, u_2 - u_1\rangle_{L^2(\Omega)} \geq 0,\\ \langle \sigma u_2 - \mathtt{B}^\star p_2, u_1 - u_2\rangle_{L^2(\Omega)} &= \langle -\lambda_2, u_1 - u_2\rangle_{L^2(\Omega)} \geq 0.\end{aligned}$$

Adding up these inequalities yields

$$\langle \sigma(u_1 - u_2) - \mathtt{B}^\star(p_1 - p_2), u_2 - u_1\rangle_{L^2(\Omega)} \geq 0.$$

Next, we rearrange this inequality and exploit (16.4.3a), $y_i = (\mathtt{A}^{-1} \circ \mathtt{B})u_i$, $i = 1, 2$, as well as the convexity of the mapping $\mathcal{Y} \ni y \mapsto x_2 \cdot j(y) \in \mathbb{R}$ in order to obtain

$$\begin{aligned}\sigma \|u_1 - u_2\|^2_{L^2(\Omega)} &\leq \langle \mathtt{B}^\star(p_1 - p_2), u_1 - u_2\rangle_{L^2(\Omega)}\\ &= \langle p_1 - p_2, \mathtt{B}(u_1 - u_2)\rangle_{\mathcal{W}} = \langle p_1 - p_2, \mathtt{A}(y_1 - y_2)\rangle_{\mathcal{W}}\\ &= \langle \mathtt{A}^\star(p_1 - p_2), y_1 - y_2\rangle_{\mathcal{Y}} = \langle j'(y_2)^\star x_2 - j'(y_1)^\star x_1, y_1 - y_2\rangle_{\mathcal{Y}}\\ &= \langle j'(y_1)^\star(x_2 - x_1), y_1 - y_2\rangle_{\mathcal{Y}}\\ &\qquad - \langle (j'(y_1) - j'(y_2))^\star x_2, y_1 - y_2\rangle_{\mathcal{Y}}\\ &\leq \langle j'(y_1)^\star(x_2 - x_1), y_1 - y_2\rangle_{\mathcal{Y}}\\ &= \left\langle j'(y_1)^\star(x_2 - x_1), (\mathtt{A}^{-1} \circ \mathtt{B})(u_1 - u_2)\right\rangle_{\mathcal{Y}}\\ &\leq C \left\| j'(y_1)\right\|_{\mathbb{L}[\mathcal{Y},\mathbb{R}^n]} |x_1 - x_2|_2 \|u_1 - u_2\|_{L^2(\Omega)}\end{aligned}$$

for some constant $C > 0$ which does not depend on x_i, y_i, and u_i, $i = 1, 2$. This way, we have

$$\|u_1 - u_2\|_{L^2(\Omega)} \leq (C/\sigma) \left\| j'(\psi^y(x_1))\right\|_{\mathbb{L}[\mathcal{Y},\mathbb{R}^n]} |x_1 - x_2|_2$$

which yields the estimate

$$\forall x_1, x_2 \in \mathbb{R}^n_+\colon\quad \left\|\psi^u(x_1) - \psi^u(x_2)\right\|_{L^2(\Omega)} \leq (C/\sigma) \left\| j'(\psi^y(x_1))\right\|_{\mathbb{L}[\mathcal{Y},\mathbb{R}^n]} |x_1 - x_2|_2 .$$

As a consequence, the map ψ^u is continuous everywhere on $\mathbb{R}^n_+$. Observing that $\psi^y = \mathtt{A}^{-1} \circ \mathtt{B} \circ \psi^u$ holds, ψ^y is continuous on $\mathbb{R}^n_+$ as well. Recalling that j is continuously Fréchet differentiable, the desired estimates follow by setting

$$\begin{aligned}\forall x \in \mathbb{R}^n_+\colon\quad C_u(x) &:= (C/\sigma) \left\| j'(\psi^y(x))\right\|_{\mathbb{L}[\mathcal{Y},\mathbb{R}^n]},\\ C_y(x) &:= \left\|\mathtt{A}^{-1} \circ \mathtt{B}\right\|_{\mathbb{L}[L^2(\Omega),\mathcal{Y}]} C_u(x).\end{aligned}$$

This completes the proof. □

We give an auxiliary result on j.

Lemma 16.4.4 *Let $\hat{X} \subset \mathbb{R}^n_+$ be compact. Then, there exists a constant $C > 0$, such that*

$$\left|j(y_2) - j(y_1) - j'(y_1)(y_2 - y_1)\right|_2 \le \frac{C}{2} \left\|y_2 - y_1\right\|^2_{\mathcal{Y}},$$
$$\left\|j'(y_2) - j'(y_1)\right\|_{\mathbb{L}[\mathcal{Y},\mathbb{R}^n]} \le C \left\|y_2 - y_1\right\|_{\mathcal{Y}}$$

with $y_i := \psi^y(x_i)$, $i = 1, 2$, holds for all $x_1, x_2 \in \hat{X}$. △

Proof First, we define $\hat{\mathcal{Y}} := \operatorname{cl}\operatorname{conv}\{\psi^y(\hat{x}) \,|\, \hat{x} \in \hat{X}\}$. This set is compact due to the compactness of $\hat{X}$ and the continuity of the mapping ψ^y, see Lemma 16.4.3. Since j'' is continuous, we have $C := \sup_{\hat{y} \in \hat{\mathcal{Y}}} \left\|j''(\hat{y})\right\| < \infty$. For $x_1, x_2 \in \hat{X}$, the points $y_i := \psi^y(x_i)$, $i = 1, 2$, belong to the convex set $\hat{\mathcal{Y}}$. Thus, the Taylor estimate follows from [12, Theorem 5.6.2] and the Lipschitz estimate on j' is clear. □

Corollary 16.4.5 *The multiplier mappings ϕ^p and ϕ^λ are continuous on $\mathbb{R}^n_+$ and Lipschitz continuous on X_{ad}.* △

Proof The continuity of ϕ^p and ϕ^λ on $\mathbb{R}^n_+$ follows easily by continuity of ψ^y and ψ^u, see Lemma 16.4.3, exploiting (16.4.3a), (16.4.3b), and the continuity of j'. Since the map $j' \circ \psi^y \colon \mathbb{R}^n_+ \to \mathbb{L}\left[\mathcal{Y}, \mathbb{R}^n\right]$ is continuous on $\mathbb{R}^n_+$ and, by Lemmas 16.4.3 and 16.4.4, Lipschitz continuous on the compact set X_{ad}, we obtain

$$\left\|\phi^p(x_1) - \phi^p(x_2)\right\|_{\mathcal{W}^\star} \le C \left\|j'(\psi^y(x_1))^\star x_1 - j'(\psi^y(x_2))^\star x_2\right\|_{\mathcal{Y}} \le \hat{C} \left|x_1 - x_2\right|_2$$

for all $x_1, x_2 \in X_{\text{ad}}$ and some constants $\hat{C}, C > 0$. The Lipschitz continuity of ϕ^λ on X_{ad} now follows from (16.4.3b). □

We introduce a function $\varphi \colon \mathbb{R}^n_+ \to \mathbb{R}$ by means of

$$\forall x \in \mathbb{R}^n_+ \colon \quad \varphi(x) := f(x, \psi^y(x), \psi^u(x)).$$

Due to the above lemma, φ is continuous and equals the optimal value function associated with (P(x)). Observing that the function f is affine w.r.t. the parameter x, it is easy to see that φ is concave, see [18, Proposition 3.5] as well.

Next, we are going to study the differentiability of φ. We are facing the problem that φ is only defined on the closed set $\mathbb{R}^n_+$. In fact, for $x \in \mathbb{R}^n \setminus \mathbb{R}^n_+$ the objective function of (P(x)) might be non-convex or unbounded from below and (P(x)) might fail to possess a minimizer. To circumvent this difficulty, we first prove that φ admits a first-order Taylor expansion, and then we extend φ to a continuously differentiable function via Whitney's extension theorem.

Lemma 16.4.6 *We define the function $\varphi' \colon \mathbb{R}^n_+ \to \mathbb{R}^n$ via*

$$\forall \bar{x} \in \mathbb{R}^n_+ \colon \qquad \varphi'(\bar{x}) := j(\psi^y(\bar{x})).$$

Then, φ' is continuous and for each compact subset $\hat{X} \subset \mathbb{R}^n_+$ there exists a constant $C > 0$ such that the Taylor-like estimate

$$\forall x, \bar{x} \in \hat{X}: \quad \left|\varphi(x) - \varphi(\bar{x}) - \varphi'(\bar{x}) \cdot (x - \bar{x})\right| \leq C\, |x - \bar{x}|_2^2$$

holds. △

Proof The continuity of φ' follows from Lemma 16.4.3. Now, let $\hat{X} \subset \mathbb{R}^n_+$ be compact. For arbitrary $\bar{x}, x \in \hat{X}$, we define $\bar{y} := \psi^y(\bar{x})$, $\bar{u} := \psi^u(\bar{x})$, $y := \psi^y(x)$, as well as $u := \psi^u(x)$. Then, we have

$$\begin{aligned}
&\varphi(x) - \varphi(\bar{x}) - \varphi'(\bar{x}) \cdot (x - \bar{x}) \\
&\quad = x \cdot j(y) + \tfrac{\sigma}{2} \|u\|^2_{L^2(\Omega)} - \bar{x} \cdot j(\bar{y}) - \tfrac{\sigma}{2} \|\bar{u}\|^2_{L^2(\Omega)} - (x - \bar{x}) \cdot j(\bar{y}) \\
&\quad = x \cdot (j(y) - j(\bar{y})) + \sigma \langle u, u - \bar{u}\rangle_{L^2(\Omega)} - \tfrac{\sigma}{2} \|u - \bar{u}\|^2_{L^2(\Omega)} \cdot
\end{aligned}$$

Next, we are going to employ the optimality condition (16.4.3). To this end, we set $p := \phi^p(x) \in \mathcal{W}^\star$ and $\lambda := \phi^\lambda(x) \in L^2(\Omega)$. Now, (16.4.3) implies

$$\sigma \langle u, u - \bar{u}\rangle_{L^2(\Omega)} = \left\langle \mathtt{B}^\star p, u - \bar{u}\right\rangle_{L^2(\Omega)} - \langle \lambda, u - \bar{u}\rangle_{L^2(\Omega)}$$

and

$$\begin{aligned}
\left\langle \mathtt{B}^\star p, u - \bar{u}\right\rangle_{L^2(\Omega)} &= \langle p, \mathtt{B}(u - \bar{u})\rangle_{\mathcal{W}} = \langle p, \mathtt{A}(y - \bar{y})\rangle_{\mathcal{W}} = \left\langle \mathtt{A}^\star p, y - \bar{y}\right\rangle_{\mathcal{Y}} \\
&= -\left\langle j'(y)^\star x, y - \bar{y}\right\rangle_{\mathcal{Y}} = -x \cdot j'(y)(y - \bar{y}).
\end{aligned} \tag{16.4.4}$$

If we set $\bar{\lambda} := \phi^\lambda(\bar{x})$, (16.4.3c) implies

$$0 \geq -\langle \lambda, u - \bar{u}\rangle_{L^2(\Omega)} \geq \left\langle \bar{\lambda} - \lambda, u - \bar{u}\right\rangle_{L^2(\Omega)} \geq -\left\|\lambda - \bar{\lambda}\right\|_{L^2(\Omega)} \|u - \bar{u}\|_{L^2(\Omega)} \cdot$$

By combining the above estimates, we get

$$\begin{aligned}
\left|\varphi(x) - \varphi(\bar{x}) - \varphi'(\bar{x}) \cdot (x - \bar{x})\right| &\leq |x|_2 \left|j(y) - j(\bar{y}) - j'(y)(y - \bar{y})\right|_2 \\
&\quad + \left\|\lambda - \bar{\lambda}\right\|_{L^2(\Omega)} \|u - \bar{u}\|_{L^2(\Omega)} + \tfrac{\sigma}{2} \|u - \bar{u}\|^2_{L^2(\Omega)} \cdot
\end{aligned}$$

Now, the claim follows from Lemmas 16.4.4 and 16.4.3 as well as Corollary 16.4.5. □

Next, we employ Whitney's extension theorem to extend φ to all of $\mathbb{R}^n$.

Lemma 16.4.7 *There exists a continuously differentiable function $\hat{\varphi}: \mathbb{R}^n \to \mathbb{R}$ such that $\hat{\varphi}(x) = \varphi(x)$ and $\hat{\varphi}'(x) = \varphi'(x)$ for all $x \in X_{\mathrm{ad}}$. Here, φ' is the function defined in Lemma 16.4.6.* △

Proof In order to apply Whitney's extension theorem, see [29, Theorem 2.3.6], we have to show that the function $\eta\colon X_{\mathrm{ad}} \times X_{\mathrm{ad}} \to \mathbb{R}$, defined via

$$\forall x, y \in X_{\mathrm{ad}}\colon \quad \eta(x, y) := \begin{cases} 0 & \text{if } x = y, \\ \frac{|\varphi(x)-\varphi(y)-\varphi'(y)(x-y)|}{|x-y|_2} & \text{if } x \neq y, \end{cases}$$

is continuous on its domain. It is clear that this function is continuous at all points $(x, y) \in X_{\mathrm{ad}} \times X_{\mathrm{ad}}$ satisfying $x \neq y$. Hence, it remains to show

$$\lim_{\substack{x \to a,\, y \to a \\ x, y \in X_{\mathrm{ad}},\, x \neq y}} \frac{|\varphi(x) - \varphi(y) - \varphi'(y)(x - y)|}{|x - y|_2} = 0$$

for all $a \in X_{\mathrm{ad}}$, but this follows from Lemma16.4.6 since X_{ad} is compact. □

Note that we extended φ from X_{ad} to $\mathbb{R}^n$ in Lemma 16.4.7. Technically, this means that the extended function $\hat{\varphi}$ may possess different values than the original optimal value function φ on $\mathbb{R}^n_+ \setminus X_{\mathrm{ad}}$. This, however, does not cause any trouble in the subsequent considerations since we focus on parameters from X_{ad} only.

16.4.2 The Optimal Value Reformulation and Its Relaxation

Based on Lemma 16.4.3, the upcoming result follows from Theorem 16.3.6 while noting that the upper level variables are chosen from a finite-dimensional Banach space.

Theorem 16.4.8 *Problem* (IOC) *possesses an optimal solution.* △

Our aim is to characterize the local minimizers of (IOC) by means of necessary optimality conditions. In order to do so, we want to exploit the continuously differentiable extension $\hat{\varphi}\colon \mathbb{R}^n \to \mathbb{R}$ of the optimal value function φ associated with (P(x)), see Lemma 16.4.7. Observing that $\hat{\varphi}(x) = \varphi(x)$ holds true for all $x \in X_{\mathrm{ad}}$, (IOC) can be transferred into the equivalent problem

$$\begin{aligned} F(x, y, u) &\to \min_{x,y,u} \\ x &\in X_{\mathrm{ad}} \\ f(x, y, u) - \hat{\varphi}(x) &\leq 0 \\ \mathtt{A}y - \mathtt{B}u &= 0 \\ u &\in U_{\mathrm{ad}} \end{aligned} \tag{OVR}$$

where f still denotes the objective of the lower level problem (P(x)). This is a single-level optimization problem with continuously Fréchet differentiable data functions, see Lemma 16.4.7. However, it is easy to check that Robinson's constraint qualification does not hold at the feasible points of (OVR), see e.g. [17, Lemma 5.1]. This failure is mainly caused by the fact that $f(x, y, u) - \hat{\varphi}(x) \leq 0$ is in fact an equality constraint by definition of the optimal value function. Due to this lack of regularity, one cannot expect that the classical KKT conditions provide a necessary optimality condition for (OVR). Furthermore, the nonlinearity of f provokes that the smooth mapping $\mathbb{R}^n \times \mathcal{Y} \times L^2(\Omega) \ni (x, y, u) \mapsto f(x, y, u) - \hat{\varphi}(x) \in \mathbb{R}$ may not serve as an exact penalty function around local minimizers of (OVR). Thus, approaches related to partial penalization w.r.t. the constraint $f(x, y, u) - \hat{\varphi}(x) \leq 0$, see e.g. [8, 9, 55, 56], do not seem to be promising here.

In order to overcome these difficulties, we are going to relax this critical constraint. More precisely, for a sequence $\{\varepsilon_k\}_{k\in\mathbb{N}} \subset \mathbb{R}$ of positive relaxation parameters satisfying $\varepsilon_k \downarrow 0$, we investigate the programs

$$\begin{aligned} F(x, y, u) &\to \min_{x,y,u} \\ x &\in X_{\text{ad}} \\ f(x, y, u) - \hat{\varphi}(x) &\leq \varepsilon_k \\ \mathtt{A}y - \mathtt{B}u &= 0 \\ u &\in U_{\text{ad}}. \end{aligned} \qquad (\text{OVR}(\varepsilon_k))$$

One can easily check that this relaxation provokes regularity of all feasible points. A formal proof of this result parallels the one of [17, Lemma 5.2].

We first want to state an existence result for (OVR(ε_k)).

Lemma 16.4.9 *For each $k \in \mathbb{N}$,* (OVR(ε_k)) *possesses an optimal solution.* △

Proof Let $\{(x_l, y_l, u_l)\}_{l\in\mathbb{N}} \subset \mathbb{R}^n \times \mathcal{Y} \times L^2(\Omega)$ be a minimizing sequence of (OVR(ε_k)), i.e. a sequence of feasible points whose associated objective values tend to the infimal value $\alpha \in \overline{\mathbb{R}}$ of (OVR(ε_k)). The compactness of X_{ad} implies that $\{x_l\}_{l\in\mathbb{N}}$ is bounded and, thus, converges along a subsequence (without relabelling) to $\bar{x} \in X_{\text{ad}}$. By feasibility, we have

$$\tfrac{\sigma}{2} \|u_l\|^2_{L^2(\Omega)} \leq x_l \cdot j(y_l) + \tfrac{\sigma}{2} \|u_l\|^2_{L^2(\Omega)} \leq \varepsilon_k + \hat{\varphi}(x_l)$$

for each $l \in \mathbb{N}$. By boundedness of $\{x_l\}_{l\in\mathbb{N}}$ and continuity of $\hat{\varphi}$ on X_{ad}, we obtain boundedness of $\{u_l\}_{l\in\mathbb{N}}$. Consequently, the latter converges weakly (along a subsequence without relabelling) to $\bar{u} \in L^2(\Omega)$ which belongs to the weakly sequentially closed set U_{ad}. Since $\mathtt{B}$ is compact, we have $y_l \to \bar{y}$ in $\mathcal{Y}$ by validity of the state equation. Here, we used $\bar{y} := (\mathtt{A}^{-1} \circ \mathtt{B})\bar{u}$.

Recall that j and $\hat{\varphi}$ are continuous functions. Exploiting the weak sequential lower semicontinuity of (squared) norms, we obtain

$$\begin{aligned} f(\bar{x}, \bar{y}, \bar{u}) - \hat{\varphi}(\bar{x}) &\leq \lim_{l\to\infty} x_l \cdot j(y_l) + \liminf_{l\to\infty} \tfrac{\sigma}{2} \|u_l\|^2_{L^2(\Omega)} + \lim_{l\to\infty} (-\hat{\varphi})(x_l) \\ &= \liminf_{l\to\infty} \Big(x_l \cdot j(y_l) + \tfrac{\sigma}{2} \|u_l\|^2_{L^2(\Omega)} - \hat{\varphi}(x_l) \Big) \leq \varepsilon_k, \end{aligned}$$

i.e. $(\bar{x}, \bar{y}, \bar{u})$ is feasible to (OVR(ε_k)). Finally, the weak sequential lower semicontinuity of F yields

$$F(\bar{x}, \bar{y}, \bar{u}) \leq \liminf_{l\to\infty} F(x_l, y_l, u_l) \leq \alpha,$$

i.e. $(\bar{x}, \bar{y}, \bar{u})$ is a global minimizer of (OVR(ε_k)). □

Next, we investigate the behavior of a sequence $\{(\bar{x}_k, \bar{y}_k, \bar{u}_k)\}_{k\in\mathbb{N}}$ of global minimizers associated with (OVR(ε_k)) as $k \to \infty$.

Theorem 16.4.10 *For each $k \in \mathbb{N}$, let $(\bar{x}_k, \bar{y}_k, \bar{u}_k) \in \mathbb{R}^n \times \mathcal{Y} \times L^2(\Omega)$ be a global minimizer of* (OVR(ε_k))*. Then, the sequence $\{(\bar{x}_k, \bar{y}_k, \bar{u}_k)\}_{k\in\mathbb{N}}$ possesses a subsequence (without relabelling) such that the convergences $\bar{x}_k \to \bar{x}$, $\bar{y}_k \to \bar{y}$, and $\bar{u}_k \to \bar{u}$ hold where $(\bar{x}, \bar{y}, \bar{u})$ is a global minimizer of* (OVR) *and, thus, of* (IOC). △

Proof Due to compactness of X_{ad}, the sequence $\{\bar{x}_k\}_{k\in\mathbb{N}}$ is bounded and converges along a subsequence (without relabelling) to some $\bar{x} \in X_{\mathrm{ad}}$. We set $\bar{y} := \psi^y(\bar{x})$ and $\bar{u} := \psi^u(\bar{x})$.

Let us set $y_k := \psi^y(\bar{x}_k)$ and $u_k := \psi^u(\bar{x}_k)$ for each $k \in \mathbb{N}$. Due to the componentwise convexity and differentiability of the mapping j, we obtain

$$\begin{aligned} \tfrac{\sigma}{2} \|\bar{u}_k - u_k\|^2_{L^2(\Omega)} &= \tfrac{\sigma}{2} \|\bar{u}_k\|^2_{L^2(\Omega)} - \langle \sigma u_k, \bar{u}_k - u_k \rangle_{L^2(\Omega)} - \tfrac{\sigma}{2} \|u_k\|^2_{L^2(\Omega)} \\ &\leq \bar{x}_k \cdot (j(\bar{y}_k) - j(y_k) - j'(y_k)(\bar{y}_k - y_k)) \\ &\quad + \tfrac{\sigma}{2} \|\bar{u}_k\|^2_{L^2(\Omega)} - \langle \sigma u_k, \bar{u}_k - u_k \rangle_{L^2(\Omega)} - \tfrac{\sigma}{2} \|u_k\|^2_{L^2(\Omega)} \\ &= f(\bar{x}_k, \bar{y}_k, \bar{u}_k) - \hat{\varphi}(\bar{x}_k) \\ &\quad - \bar{x}_k \cdot j'(y_k)(\bar{y}_k - y_k) - \langle \sigma u_k, \bar{u}_k - u_k \rangle_{L^2(\Omega)} \\ &= f(\bar{x}_k, \bar{y}_k, \bar{u}_k) - \hat{\varphi}(\bar{x}_k) + \big\langle \mathtt{B}^\star \phi^p(\bar{x}_k) - \sigma u_k, \bar{u}_k - u_k \big\rangle_{L^2(\Omega)} \\ &= f(\bar{x}_k, \bar{y}_k, \bar{u}_k) - \hat{\varphi}(\bar{x}_k) + \big\langle \phi^\lambda(\bar{x}_k), \bar{u}_k - u_k \big\rangle_{L^2(\Omega)} \\ &\leq f(\bar{x}_k, \bar{y}_k, \bar{u}_k) - \hat{\varphi}(\bar{x}_k) \leq \varepsilon_k. \end{aligned}$$

Here, we used feasibility of $(\bar{y}_k, \bar{u}_k)$ and optimality of (y_k, u_k) for (P(x)) where $x := \bar{x}_k$ holds, as well as (16.4.3), (16.4.4), and the definition of the normal cone.

The above considerations as well as the continuity of ψ^u yield

$$
\begin{aligned}
0 &\le \lim_{k\to\infty} \|\bar{u}_k - \bar{u}\|_{L^2(\Omega)} \le \lim_{k\to\infty} \left(\|\bar{u}_k - u_k\|_{L^2(\Omega)} + \|u_k - \bar{u}\|_{L^2(\Omega)}\right) \\
&= \lim_{k\to\infty} \left(\|\bar{u}_k - u_k\|_{L^2(\Omega)} + \left\|\psi^u(\bar{x}_k) - \psi^u(\bar{x})\right\|_{L^2(\Omega)}\right) \\
&\le \lim_{k\to\infty} \left(\sqrt{2\varepsilon_k/\sigma} + \left\|\psi^u(\bar{x}_k) - \psi^u(\bar{x})\right\|_{L^2(\Omega)}\right) = 0,
\end{aligned}
$$

i.e. $\bar{u}_k \to \bar{u}$ follows. Due to $\bar{y}_k = (\mathtt{A}^{-1} \circ \mathtt{B})\bar{u}_k$ and $\bar{y} = (\mathtt{A}^{-1} \circ \mathtt{B})\bar{u}$, we also have $\bar{y}_k \to \bar{y}$.

Since each feasible point $(x, y, u) \in \mathbb{R}^n \times \mathcal{Y} \times L^2(\Omega)$ of (OVR) is a feasible point of (OVR(ε_k)) for arbitrary $k \in \mathbb{N}$, we have $F(\bar{x}_k, \bar{y}_k, \bar{u}_k) \le F(x, y, u)$ for all $k \in \mathbb{N}$. Taking the limit $k \to \infty$ while observing that F is continuous, we have $F(\bar{x}, \bar{y}, \bar{u}) \le F(x, y, u)$, i.e. $(\bar{x}, \bar{y}, \bar{u})$ is a global minimizer of (OVR) and, thus, of (IOC). □

Clearly, the above theorem is of limited use for the numerical treatment of (IOC) since the programs (OVR(ε_k)) are nonconvex optimal control problems whose constraints comprise the implicitly known function $\hat{\varphi}$. However, we can exploit Theorem 16.4.10 for the derivation of a necessary optimality condition associated with (IOC).

16.4.3 Derivation of Stationarity Conditions

For the derivation of a necessary optimality condition which characterizes the local minimizers of (IOC), we will exploit the relaxation approach described in Sect. 16.4.2. Combining the KKT systems of (OVR(ε_k)) as well as (P(x)) and taking the limit $k \to \infty$ while respecting Theorem 16.4.10 will lead to a stationarity system which characterizes a particular global minimizer of (IOC). Afterwards, this result can be extended to all local minimizers of (IOC). The overall strategy is inspired by theory provided in [17].

As already mentioned in Sect. 16.4.2, we cannot rely on the KKT conditions of (OVR) to be applicable necessary optimality conditions for (IOC). In order to derive a *reasonable* stationarity system, we first observe that for given $x \in X_{\text{ad}}$, we can characterize $(\psi^y(x), \psi^u(x))$ to be the uniquely determined solution of the KKT system (16.4.3) associated with (P(x)). Plugging this system into the constraints of (IOC) in order to eliminate the implicit constraint $(y, u) \in \Psi(x)$, we arrive at the associated KKT reformulation

$$\begin{aligned} F(x,y,u) \to & \min_{x,y,u,p,\lambda} \\ x &\in X_{\mathrm{ad}} \\ \mathtt{A}y - \mathtt{B}u &= 0 \\ j'(y)^\star x + \mathtt{A}^\star p &= 0 \\ \sigma u - \mathtt{B}^\star p + \lambda &= 0 \\ (u,\lambda) &\in \operatorname{gph} \mathcal{N}_{U_{\mathrm{ad}}} \end{aligned} \tag{16.4.5}$$

where $\mathcal{N}_{U_{\mathrm{ad}}}\colon L^2(\Omega) \rightrightarrows L^2(\Omega)$ denotes the normal cone mapping associated with U_{ad}. A simple calculation via pointwise inspection shows

$$\operatorname{gph} \mathcal{N}_{U_{\mathrm{ad}}} = \left\{ (u,\lambda) \in U_{\mathrm{ad}} \times L^2(\Omega) \,\middle|\, \begin{array}{ll} \lambda \geq 0 & \text{a.e. on } \{\omega \in \Omega \,|\, u(\omega) > u_a(\omega)\} \\ \lambda \leq 0 & \text{a.e. on } \{\omega \in \Omega \,|\, u(\omega) < u_b(\omega)\} \end{array} \right\},$$

see [10, Lemma 6.34]. In order to infer a stationarity system for (IOC), we compute the roots of the partial derivatives of the MPCC-Lagrangian associated with (16.4.5). The properties of the multipliers which address the equilibrium condition $(u,\lambda) \in \operatorname{gph} \mathcal{N}_{U_{\mathrm{ad}}}$ are motivated by the pointwise structure of this set and the theory on finite-dimensional complementarity problems. Let us briefly mention that there exist other ways on how to define stationarity notions for optimization problems with pointwise complementarity constraints in function spaces, see e.g. [13, 20].

Definition 16.4.11 A feasible point $(\bar{x},\bar{y},\bar{u}) \in \mathbb{R}^n \times \mathcal{Y} \times L^2(\Omega)$ of (IOC) is said to be weakly stationary (W-stationary) whenever there exist multipliers $\bar{z} \in \mathbb{R}^n$, $\bar{\mu} \in \mathcal{Y}$, $\bar{p}, \bar{\rho} \in \mathcal{W}^\star$, and $\bar{\lambda}, \bar{w}, \bar{\xi} \in L^2(\Omega)$ which satisfy

$$0 = F'_x(\bar{x},\bar{y},\bar{u}) + \bar{z} + j'(\bar{y})\bar{\mu}, \tag{16.4.6a}$$

$$0 = F'_y(\bar{x},\bar{y},\bar{u}) + \mathtt{A}^\star\bar{\rho} + j''(\bar{y})(\bar{\mu})^\star\bar{x}, \tag{16.4.6b}$$

$$0 = F'_u(\bar{x},\bar{y},\bar{u}) + \sigma\bar{w} - \mathtt{B}^\star\bar{\rho} + \bar{\xi}, \tag{16.4.6c}$$

$$0 = \mathtt{A}\bar{\mu} - \mathtt{B}\bar{w}, \tag{16.4.6d}$$

$$\bar{z} \in \mathcal{N}_{X_{\mathrm{ad}}}(\bar{x}), \tag{16.4.6e}$$

$$0 = j'(\bar{y})^\star\bar{x} + \mathtt{A}^\star\bar{p}, \tag{16.4.6f}$$

$$0 = \sigma\bar{u} - \mathtt{B}^\star\bar{p} + \bar{\lambda}, \tag{16.4.6g}$$

$$\bar{\lambda} \geq 0 \quad \text{a.e. on } I^{a+}(\bar{u}), \tag{16.4.6h}$$

$$\bar{\lambda} \leq 0 \quad \text{a.e. on } I^{b-}(\bar{u}), \tag{16.4.6i}$$

$$\bar{\xi} = 0 \quad \text{a.e. on } I^{a+}(\bar{u}) \cap I^{b-}(\bar{u}), \tag{16.4.6j}$$

$$\bar{w} = 0 \quad \text{a.e. on } \{\omega \in \Omega \,|\, \bar{\lambda}(\omega) \neq 0\}. \tag{16.4.6k}$$

Whenever these multipliers additionally satisfy

$$\bar{\xi}\bar{w} \geq 0 \quad \text{a.e. on } \Omega, \tag{16.4.7}$$

$(\bar{x}, \bar{y}, \bar{u})$ is called Clarke-stationary (C-stationary). If (16.4.7) can be strengthened to

$$\bar{\xi} \leq 0 \wedge \bar{w} \leq 0 \quad \text{a.e. on } \{\omega \in \Omega \,|\, \bar{\lambda}(\omega) = 0 \wedge \bar{u}(\omega) = u_a(\omega)\},$$

$$\bar{\xi} \geq 0 \wedge \bar{w} \geq 0 \quad \text{a.e. on } \{\omega \in \Omega \,|\, \bar{\lambda}(\omega) = 0 \wedge \bar{u}(\omega) = u_b(\omega)\},$$

then $(\bar{x}, \bar{y}, \bar{u})$ is referred to as strongly stationary (S-stationary). Here, the measurable sets $I^{a+}(\bar{u})$ and $I^{b-}(\bar{u})$ are given by

$$I^{a+}(\bar{u}) := \{\omega \in \Omega \,|\, \bar{u}(\omega) > u_a(\omega)\}, \qquad I^{b-}(\bar{u}) := \{\omega \in \Omega \,|\, \bar{u}(\omega) < u_b(\omega)\}.$$

Note that all subsets of Ω appearing above are well-defined up to subsets of Ω possessing measure zero. △

Observe that the conditions (16.4.6f) - (16.4.6i) just provide the KKT system (16.4.3) of (P(x)) for $x := \bar{x}$ which characterizes the associated lower level Lagrange multipliers $\bar{p}$ and $\bar{\lambda}$. This way, a feasible point of (16.4.5) is fixed and the actual respective W-, C-, and S-stationarity conditions can be inferred.

In line with the results from [17, 23], we are going to show that the local minimizers of (IOC) are C-stationary. In order to do that, we choose an arbitrary sequence $\{\varepsilon_k\}_{k\in\mathbb{N}} \subset \mathbb{R}$ of positive penalty parameters tending to zero as $k \to \infty$. Due to Lemma 16.4.9, the program (OVR(ε_k)) possesses a global minimizer $(\bar{x}_k, \bar{y}_k, \bar{u}_k) \in \mathbb{R}^n \times \mathcal{Y} \times L^2(\Omega)$. As we mentioned in Sect. 16.4.2, (OVR(ε_k)) is regular as well as smooth at this point and, thus, we find multipliers $z_k \in \mathbb{R}^n$, $\alpha_k \in \mathbb{R}$, $p_k \in \mathcal{W}^\star$, and $\lambda_k \in L^2(\Omega)$ which solve the associated KKT system

$$0 = F'_x(\bar{x}_k, \bar{y}_k, \bar{u}_k) + z_k + \alpha_k(j(\bar{y}_k) - \hat{\varphi}'(\bar{x}_k)), \tag{16.4.8a}$$

$$0 = F'_y(\bar{x}_k, \bar{y}_k, \bar{u}_k) + \alpha_k j'(\bar{y}_k)^\star \bar{x}_k + \mathtt{A}^\star p_k, \tag{16.4.8b}$$

$$0 = F'_u(\bar{x}_k, \bar{y}_k, \bar{u}_k) + \alpha_k \sigma \bar{u}_k - \mathtt{B}^\star p_k + \lambda_k, \tag{16.4.8c}$$

$$z_k \in \mathcal{N}_{X_{\text{ad}}}(\bar{x}_k), \tag{16.4.8d}$$

$$0 \leq \alpha_k \perp f(\bar{x}_k, \bar{y}_k, \bar{u}_k) - \hat{\varphi}(\bar{x}_k) - \varepsilon_k, \tag{16.4.8e}$$

$$\lambda_k \in \mathcal{N}_{U_{\text{ad}}}(\bar{u}_k). \tag{16.4.8f}$$

Furthermore, an evaluation of the lower level KKT system (16.4.3) yields

$$0 = j'(\psi^y(\bar{x}_k))^\star \bar{x}_k + \mathtt{A}^\star \phi^p(\bar{x}_k), \tag{16.4.9a}$$

$$0 = \sigma \psi^u(\bar{x}_k) - \mathtt{B}^\star \phi^p(\bar{x}_k) + \phi^\lambda(\bar{x}_k), \tag{16.4.9b}$$

$$\phi^\lambda(\bar{x}_k) \in \mathcal{N}_{U_{\mathrm{ad}}}(\psi^u(\bar{x}_k)). \tag{16.4.9c}$$

Recall that $\phi^p : \mathbb{R}^n_+ \to \mathcal{W}^\star$ and $\phi^\lambda : \mathbb{R}^n_+ \to L^2(\Omega)$ denote the Lagrange multiplier mappings associated with the lower level problem (P(x)) which are continuous due to Corollary 16.4.5.

Due to Theorem 16.4.10, we may assume that we have $\bar{x}_k \to \bar{x}$, $\bar{y}_k \to \bar{y}$, and $\bar{u}_k \to \bar{u}$ where $(\bar{x}, \bar{y}, \bar{u}) \in \mathbb{R}^n \times \mathcal{Y} \times L^2(\Omega)$ is a global minimizer of (IOC).

Summarizing all these assumptions, we obtain the following results.

Lemma 16.4.12 *There exist* $\bar{z} \in \mathbb{R}^n$, $\bar{\mu} \in \mathcal{Y}$, $\bar{p} \in \mathcal{W}^\star$, *and* $\bar{w}, \bar{\xi} \in L^2(\Omega)$ *such that the convergences*

$$z_k \to \bar{z} \quad \text{in } \mathbb{R}^n, \tag{16.4.10a}$$

$$\alpha_k(\bar{y}_k - \psi^y(\bar{x}_k)) \to \bar{\mu} \quad \text{in } \mathcal{Y}, \tag{16.4.10b}$$

$$\alpha_k(\bar{u}_k - \psi^u(\bar{x}_k)) \rightharpoonup \bar{w} \quad \text{in } L^2(\Omega), \tag{16.4.10c}$$

$$p_k - \alpha_k \phi^p(\bar{x}_k) \to \bar{p} \quad \text{in } \mathcal{W}^\star, \tag{16.4.10d}$$

$$\lambda_k - \alpha_k \phi^\lambda(\bar{x}_k) \rightharpoonup \bar{\xi} \quad \text{in } L^2(\Omega) \tag{16.4.10e}$$

hold at least along a subsequence. Furthermore, the above limits satisfy the conditions (16.4.6a), (16.4.6b), (16.4.6c), (16.4.6d), *and* (16.4.6e). △

Proof We multiply (16.4.9a) by α_k and subtract the resulting equation from (16.4.8b) in order to obtain

$$0 = F'_y(\bar{x}_k, \bar{y}_k, \bar{u}_k) + \alpha_k(j'(\bar{y}_k) - j'(\psi^y(\bar{x}_k)))^\star \bar{x}_k + \mathtt{A}^\star(p_k - \alpha_k \phi^p(\bar{x}_k)). \tag{16.4.11}$$

Testing this equation with $\bar{y}_k - \psi^y(\bar{x}_k)$ while noticing that the first-order derivative of a convex function is a monotone operator, we have

$$\begin{aligned} &\left\langle F'_y(\bar{x}_k, \bar{y}_k, \bar{u}_k) + \mathtt{A}^\star(p_k - \alpha_k \phi^p(\bar{x}_k)), \bar{y}_k - \psi^y(\bar{x}_k)\right\rangle_{\mathcal{Y}} \\ &\quad = -\alpha_k \left\langle (j'(\bar{y}_k) - j'(\psi^y(\bar{x}_k)))^\star \bar{x}_k, \bar{y}_k - \psi^y(\bar{x}_k)\right\rangle_{\mathcal{Y}} \le 0. \end{aligned}$$

This is used to obtain

$$\begin{aligned} &\left\langle \mathtt{B}^\star(p_k - \alpha_k \phi^p(\bar{x}_k)), \bar{u}_k - \psi^u(\bar{x}_k)\right\rangle_{L^2(\Omega)} \\ &\quad = \left\langle p_k - \alpha_k \phi^p(\bar{x}_k), \mathtt{B}(\bar{u}_k - \psi^u(\bar{x}_k))\right\rangle_{\mathcal{W}} \end{aligned}$$

$$
\begin{aligned}
&= \langle p_k - \alpha_k \phi^p(\bar{x}_k), \mathtt{A}(\bar{y}_k - \psi^y(\bar{x}_k))\rangle_{\mathcal{W}} \\
&= \langle \mathtt{A}^\star(p_k - \alpha_k \phi^p(\bar{x}_k)), \bar{y}_k - \psi^y(\bar{x}_k)\rangle_{\mathcal{Y}} \\
&\le \left\langle -F_y'(\bar{x}_k, \bar{y}_k, \bar{u}_k), \bar{y}_k - \psi^y(\bar{x}_k)\right\rangle_{\mathcal{Y}} \\
&= \left\langle -F_y'(\bar{x}_k, \bar{y}_k, \bar{u}_k), (\mathtt{A}^{-1} \circ \mathtt{B})(\bar{u}_k - \psi^u(\bar{x}_k))\right\rangle_{\mathcal{Y}} \\
&\le C \left\| \bar{u}_k - \psi^u(\bar{x}_k)\right\|_{L^2(\Omega)}
\end{aligned}
$$

for some constant $C > 0$ since $\{F_y'(\bar{x}_k, \bar{y}_k, \bar{u}_k)\}_{k\in\mathbb{N}}$ is bounded. Next, we multiply (16.4.9b) by α_k and subtract this from (16.4.8c) in order to obtain

$$
\begin{aligned}
0 = \; & F_u'(\bar{x}_k, \bar{y}_k, \bar{u}_k) + \alpha_k \sigma(\bar{u}_k - \psi^u(\bar{x}_k)) \\
& - \mathtt{B}^\star(p_k - \alpha_k \phi^p(\bar{x}_k)) + \lambda_k - \alpha_k \phi^\lambda(\bar{x}_k).
\end{aligned}
\tag{16.4.12}
$$

Testing this with $\bar{u}_k - \psi^u(\bar{x}_k)$ and exploiting the above estimate as well as the definition of the normal cone, we obtain

$$
\begin{aligned}
&\alpha_k \sigma \left\| \bar{u}_k - \psi^u(\bar{x}_k)\right\|^2_{L^2(\Omega)} \\
&\quad = \langle -F_u'(\bar{x}_k, \bar{y}_k, \bar{u}_k) + \mathtt{B}^\star(p_k - \alpha_k \phi^p(\bar{x}_k)), \bar{u}_k - \psi^u(\bar{x}_k)\rangle_{L^2(\Omega)} \\
&\qquad + \langle \alpha_k \phi^\lambda(\bar{x}_k) - \lambda_k, \bar{u}_k - \psi^u(\bar{x}_k)\rangle_{L^2(\Omega)} \\
&\quad \le \langle -F_u'(\bar{x}_k, \bar{y}_k, \bar{u}_k) + \mathtt{B}^\star(p_k - \alpha_k \phi^p(\bar{x}_k)), \bar{u}_k - \psi^u(\bar{x}_k)\rangle_{L^2(\Omega)} \\
&\quad \le \hat{C} \left\| \bar{u}_k - \psi^u(\bar{x}_k)\right\|_{L^2(\Omega)}
\end{aligned}
$$

for a constant $\hat{C} > 0$. Consequently, the sequence $\{\alpha_k(\bar{u}_k - \psi^u(\bar{x}_k))\}_{k\in\mathbb{N}}$ is bounded and, therefore, possesses a weakly convergent subsequence (without relabelling) whose weak limit will be denoted by $\bar{w}$. Thus, we have shown (16.4.10c). Due to the relation $\bar{y}_k - \psi^y(\bar{x}_k) = (\mathtt{A}^{-1} \circ \mathtt{B})(\bar{u}_k - \psi^u(\bar{x}_k))$ and the compactness of $\mathtt{B}$, we obtain the strong convergence $\alpha_k(\bar{y}_k - \psi^y(\bar{x}_k)) \to \bar{\mu}$ for some $\bar{\mu} \in \mathcal{Y}$ satisfying (16.4.6d). Thus, we have (16.4.10b).

Since j is assumed to be continuously Fréchet differentiable, j is strictly differentiable. Noting that the strong convergences $\bar{y}_k \to \bar{y}$ and $\psi^y(\bar{x}_k) \to \bar{y}$ hold, we have

$$
\frac{j(\bar{y}_k) - j(\psi^y(\bar{x}_k)) - j'(\bar{y})(\bar{y}_k - \psi^y(\bar{x}_k))}{\|\bar{y}_k - \psi^y(\bar{x}_k)\|_{\mathcal{Y}}} \to 0.
$$

Observing that $\{\alpha_k(\bar{y}_k - \psi^y(\bar{x}_k))\}_{k\in\mathbb{N}}$ is particularly bounded, we obtain

$$
\alpha_k \left(j(\bar{y}_k) - j(\psi^y(\bar{x}_k)) - j'(\bar{y})(\bar{y}_k - \psi^y(\bar{x}_k))\right) \to 0.
$$

Since the convergence $j'(\bar{y})(\alpha_k(\bar{y}_k - \psi^y(\bar{x}_k))) \to j'(\bar{y})\bar{\mu}$ is clear from (16.4.10b), we infer

$$\alpha_k(j(\bar{y}_k) - j(\psi^y(\bar{x}_k))) \to j'(\bar{y})\bar{\mu}. \tag{16.4.13}$$

Observing that j is twice continuously Fréchet differentiable, j' is strictly differentiable. Thus, we can reprise the above arguments in order to show the convergence

$$\alpha_k(j'(\bar{y}_k) - j'(\psi^y(\bar{x}_k)))^\star\bar{x}_k \to j''(\bar{y})(\bar{\mu})^\star\bar{x}. \tag{16.4.14}$$

Next, we combine (16.4.11), (16.4.14), and the fact that $\mathtt{A}^\star$ is continuously invertible in order to obtain (16.4.10d) for some $\bar{\rho} \in \mathcal{W}^\star$ (along a subsequence) which satisfies (16.4.6b). Now, we can infer (16.4.10e) for some $\bar{\xi} \in L^2(\Omega)$ from (16.4.12) in a similar way. Taking the weak limit in (16.4.12) yields (16.4.6c). Due to Lemmas 16.4.6 and 16.4.7, (16.4.8a) is equivalent to

$$0 = F'_x(\bar{x}_k, \bar{y}_k, \bar{u}_k) + z_k + \alpha_k(j(\bar{y}_k) - j(\psi^y(\bar{x}_k))).$$

Due to the convergences $F'_x(\bar{x}_k, \bar{y}_k, \bar{u}_k) \to F'_x(\bar{x}, \bar{y}, \bar{u})$ and (16.4.13), we infer (16.4.10a) for some $\bar{z} \in \mathbb{R}^n$ along a subsequence. Particularly, we have (16.4.6a). Finally, (16.4.6e) follows by definition of the normal cone while observing $z_k \to \bar{z}$ and $\bar{x}_k \to \bar{x}$. This completes the proof. □

In the subsequent lemma, we characterize the multipliers $\bar{w}$ and $\bar{\xi}$ from Lemma 16.4.12 in more detail.

Lemma 16.4.13 *Let* $\bar{w}, \bar{\xi} \in L^2(\Omega)$ *be the multipliers characterized in Lemma 16.4.12. Then,* (16.4.6j), (16.4.6k), *and* (16.4.7) *hold.* △

Proof Due to the strong convergence of $\{\bar{u}_k\}_{k\in\mathbb{N}}$ and $\{\psi^u(\bar{x}_k)\}_{k\in\mathbb{N}}$ to $\bar{u}$ in $L^2(\Omega)$, these convergences hold pointwise almost everywhere on Ω along a subsequence (without relabelling). From $\lambda_k \in \mathcal{N}_{U_{\mathrm{ad}}}(\bar{u}_k)$ and $\phi^\lambda(\bar{x}_k) \in \mathcal{N}_{U_{\mathrm{ad}}}(\psi^u(\bar{x}_k))$, we have

$$\lambda_k - \alpha_k\phi^\lambda(\bar{x}_k) = 0 \qquad \text{a.e. on} \qquad \left\{\omega \in \Omega \,\middle|\, \begin{array}{l} u_a(\omega) < \bar{u}_k(\omega) < u_b(\omega) \\ u_a(\omega) < \psi^u(\bar{x}_k)(\omega) < u_b(\omega) \end{array}\right\}$$

by definition of the normal cone. The aforementioned pointwise a.e. convergence yields $\lambda_k(\omega) - \alpha_k\phi^\lambda(\bar{x}_k)(\omega) \to 0$ for almost every $\omega \in I^{a+}(\bar{u}) \cap I^{b-}(\bar{u})$. Since we already have $\lambda_k - \alpha_k\phi^\lambda(\bar{x}_k) \rightharpoonup \bar{\xi}$ from (16.4.10e), (16.4.6j) follows.

Next, we show (16.4.6k). If $\{\alpha_k\}_{k\in\mathbb{N}}$ is bounded, then $\bar{w} = 0$ follows from (16.4.10c) and (16.4.6k) holds trivially. Thus, we assume $\alpha_k \to \infty$. By continuity of ϕ^p and ϕ^λ, see Corollary 16.4.5, we have $\phi^p(\bar{x}_k) \to \phi^p(\bar{x})$ and $\phi^\lambda(\bar{x}_k) \to \phi^\lambda(\bar{x})$. Noting that the lower level Lagrange multipliers are uniquely determined while observing that $\psi^y(\bar{x}_k) \to \bar{y}$ and $\psi^u(\bar{x}_k) \to \bar{u}$ hold, we have $\phi^p(\bar{x}_k) \to \bar{p}$ and $\phi^\lambda(\bar{x}_k) \to \bar{\lambda}$. Here, $\bar{p} \in \mathcal{W}^\star$ and $\bar{\lambda} \in L^2(\Omega)$ satisfy the conditions (16.4.6f), (16.4.6g), (16.4.6h), and (16.4.6i). Thus, (16.4.10e) yields the strong convergence

$\alpha_k^{-1}\lambda_k \to \bar{\lambda}$. Let $G \subset \Omega$ be measurable and $\chi_G \in L^\infty(\Omega)$ be its characteristic function which equals 1 on G and vanishes on $\Omega \setminus G$. By definition of the normal cone, we have

$$\left\langle \alpha_k^{-1}\lambda_k, \alpha_k \chi_G(\bar{u}_k - \psi^u(\bar{x}_k))\right\rangle_{L^2(\Omega)} \geq 0,$$

$$\langle \phi^\lambda(\bar{x}_k), \alpha_k \chi_G(\bar{u}_k - \psi^u(\bar{x}_k))\rangle_{L^2(\Omega)} \leq 0.$$

Taking the limit $k \to \infty$, we thus obtain $\langle \bar{\lambda}, \chi_G \bar{w}\rangle_{L^2(\Omega)} = 0$. Since $G \subset \Omega$ was chosen arbitrarily, (16.4.6k) follows.

Finally, we are going to prove (16.4.7). Therefore, we fix an arbitrary measurable set $G \subset \Omega$. We first observe that due to (16.4.10c) and (16.4.10d), we have

$$\langle \mathtt{B}^\star(p_k - \alpha_k \phi^p(\bar{x}_k)), \alpha_k \chi_G(\bar{u}_k - \psi^u(\bar{x}_k))\rangle_{L^2(\Omega)} \to \langle \mathtt{B}^\star \bar{p}, \chi_G \bar{w}\rangle_{L^2(\Omega)}.$$

Now, we can exploit (16.4.6c), (16.4.12), the weak sequential lower semicontinuity of the map $L^2(\Omega) \ni u \mapsto \langle \sigma u, \chi_G u\rangle_{L^2(\Omega)} \in \mathbb{R}$, and the definition of the normal cone in order to obtain

$$\begin{aligned}
\langle -\bar{\xi}, \chi_G \bar{w}\rangle_{L^2(\Omega)} &= \langle F'_u(\bar{x}, \bar{y}, \bar{u}) - \mathtt{B}^\star \bar{p} \chi_G \bar{w}\rangle_{L^2(\Omega)} + \langle \sigma \bar{w}, \chi_G \bar{w}\rangle_{L^2(\Omega)} \\
&\leq \lim_{k\to\infty} \langle F'_u(\bar{x}_k, \bar{y}_k, \bar{u}_k) - \mathtt{B}^\star(p_k - \alpha_k \phi^\lambda(\bar{x}_k)), \alpha_k \chi_G(\bar{u}_k - \psi^u(\bar{x}_k))\rangle_{L^2(\Omega)} \\
&\qquad + \liminf_{k\to\infty} \langle \sigma \alpha_k(\bar{u}_k - \psi^u(\bar{x}_k)), \alpha_k \chi_G(\bar{u}_k - \psi^u(\bar{x}_k))\rangle_{L^2(\Omega)} \\
&= \liminf_{k\to\infty} \langle \alpha_k \phi^\lambda(\bar{x}_k) - \lambda_k, \alpha_k \chi_G(\bar{u}_k - \psi^u(\bar{x}_k))\rangle_{L^2(\Omega)} \leq 0.
\end{aligned}$$

Noting that $G \subset \Omega$ has been chosen arbitrarily, (16.4.7) follows. □

Above, we have shown that the particular global minimizer $(\bar{x}, \bar{y}, \bar{u})$ which results from the relaxation approach suggested in Sect. 16.4.2 is C-stationary. In order to carry over this analysis to arbitrary local minimizers of (IOC), we exploit a localization argument.

Theorem 16.4.14 *Let $(\bar{x}, \bar{y}, \bar{u}) \in \mathbb{R}^n \times \mathcal{Y} \times L^2(\Omega)$ be a local minimizer of* (IOC). *Then, it is C-stationary.* △

Proof Invoking Lemma 16.4.2, there is some $\varepsilon > 0$ such that $\bar{x}$ is the unique globally optimal solution of

$$F(x, \psi^y(x), \psi^u(x)) + \tfrac{1}{2}|x - \bar{x}|_2^2 \to \min_x$$

$$x \in X_{\text{ad}} \cap \mathbb{B}^\varepsilon(\bar{x})$$

where $\mathbb{B}^\varepsilon(\bar{x})$ denotes the closed ε-ball around $\bar{x}$. Thus, $(\bar{x}, \bar{y}, \bar{u})$ is the unique global minimizer of the bilevel programming problem

$$\begin{aligned} F(x, y, u) + \tfrac{1}{2}\,|x - \bar{x}|_2^2 &\to \min_{x,y,u} \\ x &\in X_{\mathrm{ad}} \cap \mathbb{B}^\varepsilon(\bar{x}) \\ (y, u) &\in \Psi(x). \end{aligned} \tag{16.4.15}$$

Combining Theorem 16.4.10 as well as Lemmas 16.4.12 and 16.4.13, $(\bar{x}, \bar{y}, \bar{u})$ is a C-stationary point of (16.4.15). Noting that the derivative of the functional $\mathbb{R}^n \ni x \mapsto \frac{1}{2}\,|x - \bar{x}|_2^2 \in \mathbb{R}$ vanishes at $\bar{x}$ while $\mathcal{N}_{X_{\mathrm{ad}} \cap \mathbb{B}^\varepsilon(\bar{x})}(\bar{x}) = \mathcal{N}_{X_{\mathrm{ad}}}(\bar{x})$ holds since $\bar{x}$ is an interior point of $\mathbb{B}^\varepsilon(\bar{x})$, the C-stationarity conditions of (16.4.15) and (IOC) coincide at $(\bar{x}, \bar{y}, \bar{u})$. This completes the proof. □

Remark 16.4.15 The counterexample from [23, Section 3.2] shows that the local minimizers of (IOC) are not S-stationary in general. However, it remains an open question whether the multipliers which solve the C-stationarity system associated with a local minimizer of (IOC) additionally satisfy

$$\begin{aligned} \bar{\xi}\bar{w} &= 0 \lor (\bar{\xi} < 0 \land \bar{w} < 0) \quad \text{a.e. on } \{\omega \in \Omega \mid \bar{\lambda}(\omega) = 0 \land \bar{u}(\omega) = u_a(\omega)\}, \\ \bar{\xi}\bar{w} &= 0 \lor (\bar{\xi} > 0 \land \bar{w} > 0) \quad \text{a.e. on } \{\omega \in \Omega \mid \bar{\lambda}(\omega) = 0 \land \bar{u}(\omega) = u_b(\omega)\}. \end{aligned}$$

In line with the terminology of finite-dimensional complementarity programming, the resulting stationarity condition may be referred to as the system of (pointwise) Mordukhovich-stationarity. We would like to briefly note that this system *cannot* be obtained by computing the limiting normal cone to the set gph $\mathcal{N}_{U_{\mathrm{ad}}}$ since the latter turns out to be uncomfortably large, see [41]. More precisely, this strategy results in the W-stationarity system of (IOC) from Definition 16.4.11. Additionally, one cannot rely on the limiting variational calculus in $L^2(\Omega)$ due to an inherent lack of so-called sequential normal compactness, see [39]. Taking into account the outstanding success of variational analysis in the finite-dimensional setting, these observations are quite unexpected. △

References

1. R.A. Adams, J.J.F. Fournier, *Sobolev Spaces* (Elsevier Science, Oxford, 2003)
2. S. Albrecht, M. Ulbrich, Mathematical programs with complementarity constraints in the context of inverse optimal control for locomotion. Optim. Methods Softw. **32**(4), 670–698 (2017)
3. S. Albrecht, C. Passenberg, M. Sobotka, A. Peer, M. Buss, M. Ulbrich, *Optimization Criteria for Human Trajectory Formation in Dynamic Virtual Environments*, Haptics: Generating and Perceiving Tangible Sensations, ed. by A.M.L. Kappers, J.B.F. van Erp, W.M. Bergmann Tiest, F.C.T. van der Helm (Springer, Berlin, 2010), pp. 257–262

4. S. Albrecht, M. Leibold, M. Ulbrich, A bilevel optimization approach to obtain optimal cost functions for human arm movements Numer. Algebra Control Optim. **2**(1), 105–127 (2012)
5. B. Bank, J. Guddat, D. Klatte, B. Kummer, K. Tammer, *Non-Linear Parametric Optimization* (Birkhäuser, Basel, 1983)
6. V. Barbu, *Optimal Control of Variational Inequalities*, Research Notes in Mathematics (Pitman Advanced Pub. Program, Boston, 1984)
7. J.F. Bard, *Practical Bilevel Optimization: Algorithms and Applications* (Kluwer Academic, Dordrecht, 1998)
8. F. Benita, P. Mehlitz, bilevel optimal control with final-state-dependent finite-dimensional lower level. SIAM J. Optim. **26**(1), 718–752 (2016)
9. F. Benita, S. Dempe, P. Mehlitz, Bilevel optimal control problems with pure state constraints and finite-dimensional lower level. SIAM J. Optim. **26**(1), 564–588 (2016)
10. J.F. Bonnans, A. Shapiro, *Perturbation Analysis of Optimization Problems* (Springer, New York, Berlin, Heidelberg, 2000)
11. D.A. Carlson, *Existence of Optimal Controls for a Bi-Level Optimal Control Problem*, Advances in Dynamic Games: Theory, Applications, and Numerical Methods, ed. by V. Křivan, G. Zaccour (Springer, Cham, 2013), pp. 71–84
12. H. Cartan, *Calcul Différentiel* (Hermann, Paris, 1967)
13. C. Clason, Y. Deng, P. Mehlitz, U. Prüfert, Optimal control problems with control complementarity constraints: existence results, optimality conditions, and a penalty method. Optim. Methods Softw. **35**(1), 142–170 (2020). https://doi.org/10.1080/10556788.2019.1604705
14. S. Dempe, *Foundations of Bilevel Programming* (Kluwer, Dordrecht, 2002)
15. S. Dempe, J. Dutta, Is bilevel programming a special case of a mathematical program with complementarity constraints? Math. Program. **131**(1), 37–48 (2012)
16. S. Dempe, V. Kalashnikov, G. Pérez-Valdéz, N. Kalashnykova, *Bilevel Programming Problems - Theory, Algorithms and Applications to Energy Networks* (Springer, Berlin, 2015)
17. S. Dempe, F. Harder, P. Mehlitz, G. Wachsmuth, Solving inverse optimal control problems via value functions to global optimality. J. Global Optim. **74**(2), 297–325 (2019)
18. A.V. Fiacco, J. Kyparisis, Convexity and concavity properties of the optimal value function in parametric nonlinear programming. J. Optim. Theory Appl. **48**(1), 95–126 (1986)
19. F. Fisch, J. Lenz, F. Holzapfel, G. Sachs, *On the solution of bilevel optimal control problems to increase the fairness in air races*. J. Guid. Control Dynam. **35**(4), 1292–1298 (2012)
20. L. Guo, J.J. Ye, Necessary optimality conditions for optimal control problems with equilibrium constraints. SIAM J. Control Optim. **54**(5), 2710–2733 (2016)
21. F. Harder, G. Wachsmuth, Comparison of optimality systems for the optimal control of the obstacle problem. GAMM-Mitteilungen **40**(4), 312–338 (2018)
22. F. Harder, G. Wachsmuth, The limiting normal cone of a complementarity set in Sobolev spaces. Optimization **67**(10), 1579–1603 (2018)
23. F. Harder, G. Wachsmuth, Optimality conditions for a class of inverse optimal control problems with partial differential equations. Optimization **68**(2–3), 615–643 (2019)
24. K. Hatz, Efficient Numerical Methods for Hierarchical Dynamic Optimization with Application to Cerebral Palsy Gait Modeling, Ph.D. thesis, University of Heidelberg, Germany, 2014
25. K. Hatz, J.P. Schlöder, H.G. Bock, Estimating parameters in optimal control problems. SIAM J. Sci. Comput. **34**(3), A1707–A1728 (2012)
26. R. Herzog, F. Schmidt, Weak lower semi-continuity of the optimal value function and applications to worst-case robust optimal control problems. Optimization **61**(6), 685–697 (2012)
27. M. Hinze, R. Pinnau, M. Ulbrich, S. Ulbrich, *Optimization with PDE Constraints* (Springer, Berlin, 2009)
28. G. Holler, K. Kunisch, R.C. Barnard, A bilevel approach for parameter learning in inverse problems. Inverse Probl. **34**(11), 1–28 (2018)
29. L. Hörmander, *The Analysis of Linear Partial Differential Operators I* (Springer, Berlin, 2003)
30. J. Jahn, *Introduction to the Theory of Nonlinear Optimization* (Springer, Berlin, 1996)

31. D. Jerison, C.E. Kenig, The inhomogeneous Dirichlet problem in Lipschitz domains. J. Funct. Anal. **130**(1), 161–219 (1995)
32. V. Kalashnikov, F. Benita, P. Mehlitz, The natural gas cash-out problem: a bilevel optimal control approach. Math. Probl. Eng. 1–17 (2015). https://doi.org/10.1155/2015/286083
33. M. Knauer, Fast and save container cranes as bilevel optimal control problems. Math. Comput. Model. Dynam. Syst. **18**(4), 465–486 (2012)
34. M. Knauer, C. Büskens, Hybrid solution methods for bilevel optimal control problems with time dependent coupling, in *Recent Advances in Optimization and its Applications in Engineering: The 14th Belgian-French-German Conference on Optimization*, ed. by M. Diehl, F. Glineur, E. Jarlebring, W. Michiels (Springer, Berlin, 2010), pp. 237–246
35. F.L. Lewis, D. Vrabie, V.L. Syrmos, *Optimal Control* (Wiley, Hoboken, 2012)
36. P. Mehlitz, Bilevel programming problems with simple convex lower level, Optimization **65**(6), 1203–1227 (2016)
37. P. Mehlitz, Contributions to Complementarity and Bilevel Programming in Banach Spaces, Ph.D. thesis, Technische Universität Bergakademie Freiberg, 2017
38. P. Mehlitz, Necessary optimality conditions for a special class of bilevel programming problems with unique lower level solution. Optimization **66**(10), 1533–1562 (2017)
39. P. Mehlitz, On the sequential normal compactness condition and its restrictiveness in selected function spaces. Set-Valued Var. Anal. **27**(3), 763–782 (2019)
40. P. Mehlitz, G. Wachsmuth, Weak and strong stationarity in generalized bilevel programming and bilevel optimal control. Optimization **65**(5), 907–935 (2016)
41. P. Mehlitz, G. Wachsmuth, The limiting normal cone to pointwise defined sets in Lebesgue spaces. Set-Valued Var. Anal. **26**(3), 449–467 (2018)
42. P. Mehlitz, G. Wachsmuth, The weak sequential closure of decomposable sets in Lebesgue spaces and its application to variational geometry. Set-Valued Var. Anal. **27**(1), 265–294 (2019)
43. F. Mignot, Contrôle dans les inéquations variationelles elliptiques. J. Funct. Anal. **22**(2), 130–185 (1976)
44. K. Mombaur, A. Truong, J. P. Laumond, From human to humanoid locomotion—an inverse optimal control approach. Auton. Robots **28**(3), 369–383 (2010)
45. K.D. Palagachev, M. Gerdts, *Exploitation of the Value Function in a Bilevel Optimal Control Problem*, ed. by L. Bociu, J.-A. Désidéri, A Habbal. System Modeling and Optimization (Springer, Cham, 2016), pp. 410–419
46. K.D. Palagachev, M. Gerdts, *Numerical Approaches Towards Bilevel Optimal Control Problems with Scheduling Tasks*, ed. by L. Ghezzi, D. Hömberg, C. Landry. Math for the Digital Factory (Springer, Cham, 2017), pp. 205–228
47. S.M. Robinson, Stability theory for systems of inequalities, part II: differentiable nonlinear systems. SIAM J. Numer. Anal. **13**(4), 497–513 (1976)
48. A. Schiela, D. Wachsmuth, Convergence analysis of smoothing methods for optimal control of stationary variational inequalities with control constraints. ESAIM Math. Model. Numer. Anal. **47**(3), 771–787 (2013).
49. K. Shimizu, Y. Ishizuka, J. F. Bard, *Nondifferentiable and Two-Level Mathematical Programming* (Kluwer Academic, Dordrecht, 1997)
50. F. Tröltzsch, *Optimal Control of Partial Differential Equations* (Vieweg, Wiesbaden, 2009)
51. J.L. Troutman, *Variational Calculus and Optimal Control* (Springer, New York, 1996)
52. H.V. Stackelberg, *Marktform und Gleichgewicht* (Springer, Berlin, 1934)
53. G. Wachsmuth, Mathematical programs with complementarity constraints in Banach spaces. J. Optim. Theory Appl. **166**(2), 480–507 (2015)
54. G. Wachsmuth, Strong stationarity for optimization problems with complementarity constraints in absence of polyhedricity. Set-Valued Var. Anal. **25**(1), 133–175 (2017)

55. J.J. Ye, Necessary conditions for bilevel dynamic optimization problems. SIAM J. Control Optim. **33**(4), 1208–1223 (1995)
56. J.J. Ye, Optimal strategies for bilevel dynamic problems. SIAM J. Control Optim. **35**(2), 512–531 (1997)
57. J. Zowe, S. Kurcyusz, Regularity and stability for the mathematical programming problem in Banach spaces. Appl. Math. Optim. **5**(1), 49–62 (1979)

Chapter 17
Bilevel Linear Optimization Under Uncertainty

Johanna Burtscheidt and Matthias Claus

Abstract We consider bilevel linear problems, where the right-hand side of the lower level problems is stochastic. The leader has to decide in a here-and-now fashion, while the follower has complete information. In this setting, the leader's outcome can be modeled by a random variable, which gives rise to a broad spectrum of models involving coherent or convex risk measures and stochastic dominance constraints. We outline Lipschitzian properties, conditions for existence and optimality, as well as stability results. Moreover, for finite discrete distributions, we discuss the special structure of equivalent deterministic bilevel programs and its potential use to mitigate the curse of dimensionality.

Keywords Bilevel stochastic programming · Linear · Risk measure · Stability · Differentiability · Deterministic counterpart

17.1 Introduction

In this chapter we consider bilevel optimization models with uncertain parameters. Such models can be classified based on the chronology of decision and observation as well as the nature of the uncertainty involved. A **bilevel stochastic program** arises, if the uncertain parameter is realization of some random vector with known distribution, that can only be observed once the leader has submitted their decision. In contrast, the follower decides under complete information.

If upper and lower level objectives coincide, the bilevel stochastic program collapses to a classical **stochastic optimization problem with recourse** (cf. [1, Chap. 2]). Relations to other mathematical programming problems are explored in the seminal work [2] that also established the existence of solutions, Lipschitzian properties and directional differentiability of a risk-neutral formulation of a bilevel

---

J. Burtscheidt · M. Claus (✉)
Faculty of Mathematics, University of Duisburg-Essen, Essen, Germany
e-mail: johanna.burtscheidt@uni-due.de; matthias.claus@uni-due.de

S. Dempe, A. Zemkoho (eds.), *Bilevel Optimization*, Springer Optimization and Its Applications 161, https://doi.org/10.1007/978-3-030-52119-6_17

stochastic nonlinear model. Moreover, gradient descent and penalization methods were investigated to tackle discretely distributed stochastic mathematical programs with equilibrium constraints (SMPECs).

Reference [3] studies an application to topology optimization problems in structural mechanics. Many other applications are motivated by network related problems that inherit a natural order of successive decision making under uncertainty. Notable examples arise in telecommunications (cf. [4]), grid-based (energy) markets (cf. [5–8]) or transportation science (cf. [9, 10]). An extensive survey on bilevel stochastic programming literature is provided in [11, Chap. 1.4].

In **two-stage stochastic bilevel programming** leader and follower take two decisions: The decision on the respective first-stage variables is made in a **here-and-now** fashion, i.e. without knowledge of the realization of the random parameter. In contrast, the respective second-stage decisions are made in a **wait-and-see** manner, i.e. after observing the parameter (cf. [12]).

This chapter is organized as follows: In Sects. 17.2.1–17.2.5, we outline structural properties, existence and optimality conditions as well as stability results for bilevel stochastic linear problems while paying special attention to the modelling of risk-aversion via coherent or convex risk measures or stochastic dominance constraints. Sections 17.2.6 and 17.2.7 are devoted to the algorithmic treatment of bilevel stochastic linear problems, where the underlying distribution is finite discrete. An application of two-stage stochastic bilevel programming in the context of network pricing is discussed in Sect. 17.3. The chapter concludes with an overview of potential challenges for future research.

17.2 Bilevel Stochastic Linear Optimization

While the analysis in this section is confined to the bilevel stochastic linear problems with random right-hand side, the concepts and underlying principles can be easily transferred to stochastic extensions of more complex bilevel programming models.

17.2.1 Preliminaries

We consider the optimistic formulation of a parametric bilevel linear program

$$\min_x \left\{ c^\top x + \min_y \{ q^\top y \mid y \in \Psi(x, z) \} \mid x \in X \right\}, \qquad \mathrm{P}(z)$$

where $X \subseteq \mathbb{R}^n$ is a nonempty polyhedron, $c \in \mathbb{R}^n$ and $q \in \mathbb{R}^m$ are vectors, $z \in \mathbb{R}^s$ is a parameter, and the lower level optimal solution set mapping $\Psi : \mathbb{R}^n \times \mathbb{R}^s \rightrightarrows \mathbb{R}^m$ is given by

$$\Psi(x, z) := \underset{y}{\operatorname{Argmin}} \, \{ d^\top y \mid Ay \leq Tx + z \}$$

with matrices $A \in \mathbb{R}^{s\times m}$, $T \in \mathbb{R}^{s\times n}$ and a vector $d \in \mathbb{R}^m$. Let $f : \mathbb{R}^n \times \mathbb{R}^s \to \mathbb{R} \cup \{\pm\infty\}$ denote the mapping

$$f(x, z) := c^\top x + \min_y \{q^\top y \mid y \in \Psi(x, z)\}.$$

Lemma 17.2.1 *Assume dom $f \neq \emptyset$, then f is real-valued and Lipschitz continuous on the polyhedron $P = \{(x, z) \in \mathbb{R}^n \times \mathbb{R}^s \mid \exists y \in \mathbb{R}^m : Ay \leq Tx + z\}$.* △

Proof By Eaves [13, Sect. 3], $\emptyset \neq \text{dom } f \subseteq \text{dom } \Psi$ implies dom $\Psi = P$. Consequently, the linear program in the definition of $f(x, z)$ is solvable for any $(x, z) \in P$ by parametric linear programming theory (see [14, Sect. 2.3.2]). Consider any $(x, z), (x', z') \in P$. Without loss of generality, assume that $f(x, z) \geq f(x', z')$ and let $y' \in \Psi(x', z')$ be such that $f(x', z') = c^\top x' + q^\top y'$. Following [15] we obtain

$$\begin{aligned} |f(x, z) - f(x', z')| &= f(x, z) - c^\top x' - q^\top y' \leq c^\top x + q^\top y - c^\top x' - q^\top y' \\ &\leq \|c\| \|x - x'\| + \|q\| \|y - y'\| \end{aligned}$$

for any $y \in \Psi(x, z)$. Let $\mathbb{B}$ denote the Euclidean unit ball and $0 < \Lambda < \infty$ a constant, then [16, Theorem 4.2] yields

$$\Psi(x', z') \subseteq \Psi(x, z) + \Lambda \|(x, z) - (x', z')\| \mathbb{B}$$

and hence $|f(x, z) - f(x', z')| \leq (\|c\| + \Lambda \|q\|) \|(x, z) - (x', z')\|$. □

Remark 17.2.2 An alternate proof for Lemma 17.2.1 is given in [17, Theorem 1]. However, the arguments above can be easily extended to lower level problems with convex quadratic objective function and linear constraints. △

Linear programming theory (cf. [18]) provides verifiable necessary and sufficient condition for dom $f \neq \emptyset$:

Lemma 17.2.3 *dom $f \neq \emptyset$ holds if and only if there exists $(x, z) \in \mathbb{R}^n \times \mathbb{R}^s$ such that*

a. $\{y \mid Ay \leq Tx + z\}$ is nonempty,
b. there is some $u \in \mathbb{R}^s$ satisfying $A^\top u = d$ and $u \leq 0$, and
c. the function $y \mapsto q^\top y$ is bounded from below on $\Psi(x, z)$.

Under these conditions,

$$\min_y \{q^\top y \mid y \in \Psi(x', z')\}$$

is attained for any $(x', z') \in P$. △

17.2.2 Bilevel Stochastic Linear Programming Models

A bilevel stochastic program arises if the parameter $z = Z(\omega)$ in P(z) is the realization of a known random vector Z on some probability space $(\Omega, \mathcal{F}, \mathbb{P})$ and we assume the following chronology of decision and observation:

$$\text{leader decides } x \quad \rightarrow \quad z = Z(\omega) \text{ is revealed} \quad \rightarrow \quad \text{follower decides } y.$$

Throughout the analysis, we assume the stochasticity to be purely exogenous, i.e. the distribution of Z to be independent of x.

Let $\mu_Z := \mathbb{P} \circ Z^{-1} \in \mathcal{P}(\mathbb{R}^s)$ denote the Borel probability measure induced by Z. We shall assume dom $f \neq \emptyset$ and that the lower level problem is feasible for any leader's decision and any realization of the randomness, i.e.

$$X \subseteq P_Z := \{x \in \mathbb{R}^n \mid (x, z) \in P \ \forall z \in \operatorname{supp} \mu_Z\}.$$

In two-stage stochastic programming, a similar assumption is known as **relatively complete recourse** (cf. [1, Sect. 2.1.3]). In this setting, each leader's decision $x \in X$ gives rise to a random variable $f(x, Z(\cdot))$. We thus may fix any mapping $\mathcal{R} : \mathcal{X} \to \mathbb{R}$, where $\mathcal{X}$ is a linear subspace of $L^0(\Omega, \mathcal{F}, \mathbb{P})$ that contains the constants and satisfies

$$\{f(x, Z(\cdot)) \mid x \in X\} \subseteq \mathcal{X},$$

and consider the **bilevel stochastic program**

$$\min_x \{\mathcal{R}[f(x, Z(\cdot))] \mid x \in X\}. \tag{17.2.1}$$

Under suitable moment or boundedness conditions on Z the classical L^p-spaces $L^p(\Omega, \mathcal{F}, \mathbb{P})$ with $p \in [1, \infty]$ are natural choices for the domain $\mathcal{X}$ of $\mathcal{R}$. We define

$$\mathcal{M}_s^p := \left\{\mu \in \mathcal{P}(\mathbb{R}^s) \mid \int_{\mathbb{R}^s} \|z\|^p \, \mu(dz) < \infty\right\},$$

which denotes the set of Borel probability measures on $\mathbb{R}^s$ with finite moments of order $p \in [1, \infty)$, and the set

$$\mathcal{M}_s^\infty := \left\{\mu \in \mathcal{P}(\mathbb{R}^s) \mid \operatorname{supp} \mu_Z \text{ is bounded}\right\}.$$

Lemma 17.2.4 *Assume dom $f \neq \emptyset$ and $\mu_Z \in \mathcal{M}_s^p$ for some $p \in [1, \infty]$. Then the mapping $F : P_Z \to L^0(\Omega, \mathcal{F}, \mathbb{P})$ given by $F(x) := f(x, Z(\cdot))$ takes values in $L^p(\Omega, \mathcal{F}, \mathbb{P})$ and is Lipschitz continuous with respect to the L^p-norm.* △

Proof We first consider the case that p is finite. By $(0, 0) \in P$ and Lemma 17.2.1, there exist a constant L_f such that

$$\begin{aligned} \|F(x)\|_{L^p}^p &\le 2^p|f(0,0)|^p + 2^p \int_{\mathbb{R}^s} |f(x,z) - f(0,0)|^p \,\mu_Z(dz) \\ &\le 2^p|f(0,0)|^p + 2^p L_f^p \|x\|^p + 2^p L_f^p \int_{\mathbb{R}^s} \|z\|^p \,\mu_Z(dz) < \infty \end{aligned}$$

holds for any $x \in P_Z$. Furthermore, for any $x, x' \in P_Z$ we have

$$\|F(x) - F(x')\|_{L^p} = \left(\int_{\mathbb{R}^s} |f(x,z) - f(x',z)|^p \,\mu_Z(dz)\right)^{1/p} \le L_f\|x - x'\|.$$

For $p = \infty$, Lemma 17.2.1 implies that for any fixed $x \in P_Z$, the mapping $f(x, \cdot)$ is continuous on supp μ_Z. Thus, $\mu_Z \in \mathcal{M}_s^\infty$ yields

$$\|F(x)\|_{L^\infty} \le \sup_{z \in \operatorname{supp} \mu_Z} |f(x,z)| < \infty.$$

Moreover, for any $x, x' \in P_Z$ we have

$$\|F(x) - F(x')\|_{L^\infty} \le \sup_{z \in \operatorname{supp} \mu_Z} |f(x,z) - f(x',z)| \le L_f\|x - x'\|. \qquad \square$$

The mapping $\mathcal{R}$ in (17.2.1) can be used to measure the risk associated with the random variable $F(x)$.

Definition 17.2.5 A mapping $\mathcal{R} : \mathcal{X} \to \mathbb{R}$ defined on some linear subspace $\mathcal{X}$ of $L^0(\Omega, \mathcal{F}, \mathbb{P})$ containing the constants is called a **convex risk measure** if the following conditions are fulfilled:

a. (**Convexity**) For any $Y_1, Y_2 \in \mathcal{X}$ and $\lambda \in [0, 1]$ we have

$$\mathcal{R}[\lambda Y_1 + (1 - \lambda)Y_2] \le \lambda\mathcal{R}[Y_1] + (1 - \lambda)\mathcal{R}[Y_2].$$

b. (**Monotonicity**) $\mathcal{R}[Y_1] \le \mathcal{R}[Y_2]$ for all $Y_1, Y_2 \in \mathcal{X}$ satisfying $Y_1 \le Y_2$ with respect to the $\mathbb{P}$-almost sure partial order.

c. (**Translation equivariance**) $\mathcal{R}[Y + t] = \mathcal{R}[Y] + t$ for all $Y \in \mathcal{X}$ and $t \in \mathbb{R}$.

A convex risk measure $\mathcal{R}$ is **coherent** if the following holds true:

d. (**Positive homogeneity**) $\mathcal{R}[tY] = t \cdot \mathcal{R}[Y]$ for all $Y \in \mathcal{X}$ and $t \in [0, \infty)$. △

Definition 17.2.6 A mapping $\mathcal{R} : \mathcal{X} \to \mathbb{R}$ is called **law-invariant** if for all $Y_1, Y_2 \in \mathcal{X}$ with $\mathbb{P} \circ Y_1^{-1} = \mathbb{P} \circ Y_2^{-1}$ we have $\mathcal{R}[Y_1] = \mathcal{R}[Y_2]$. △

Coherent risk measures have been introduced in [19], while the analysis of convex risk measures dates back to [20]. A thorough discussion of their analytical traits is provided in [21]. Below we list some risk measures that are commonly used in stochastic programming (cf. [1, Sect. 6.3.2]).

Examples

(a) The **expectation** $\mathbb{E} : L^1(\Omega, \mathcal{F}, \mathbb{P}) \to \mathbb{R}$,

$$\mathbb{E}[Y] = \int_\Omega Y(\omega)\, \mathbb{P}(d\omega),$$

is a law-invariant and coherent risk measure that turns (17.2.1) into the **risk neutral bilevel stochastic program**

$$\min_x \{\mathbb{E}[F(x)] \mid x \in X\}.$$

(b) The **expected excess of order** $p \in [1, \infty)$ over a predefined level $\eta \in \mathbb{R}$ is the mapping $\mathrm{EE}_\eta^p : L^p(\Omega, \mathcal{F}, \mathbb{P}) \to \mathbb{R}$ given by

$$\mathrm{EE}_\eta^p[Y] := \Big(\mathbb{E}\big[\max\{Y - \eta, 0\}^p\big]\Big)^{1/p}.$$

EE_η^p is law-invariant, convex and nondecreasing, but neither translation-equivariant nor positively homogeneous (cf. [1, Example 6.22]).

(c) The **mean upper semideviation of order** $p \in [1, \infty)$ is the mapping $\mathrm{SD}_\rho^p : L^p(\Omega, \mathcal{F}, \mathbb{P}) \to \mathbb{R}$ defined by

$$\mathrm{SD}_\rho^p[Y] := \mathbb{E}[Y] + \rho \cdot \mathrm{EE}_{\mathbb{E}[Y]}^p[Y] = \mathbb{E}[Y] + \rho \cdot \Big(\mathbb{E}\big[\max\{\mathbb{E}[Y] - \eta, 0\}^p\big]\Big)^{1/p},$$

where $\rho \in (0, 1]$ is a parameter. SD_ρ^p is a law-invariant coherent risk measure (cf. [1, Example 6.20]).

(d) The **excess probability** $\mathrm{EP}_\eta : L^0(\Omega, \mathcal{F}, \mathbb{P}) \to \mathbb{R}$ over a prescribed target level $\eta \in \mathbb{R}$ given by

$$\mathrm{EP}_\eta[Y] = \mathbb{P}[\{\omega \in \Omega \mid Y(\omega) > \eta\}],$$

is nondecreasing and law-invariant. However, it lacks convexity, translation-equivariance and positive homogeneity (cf. [22, Example 2.29]).

(e) The **Value-at-Risk** $\mathrm{VaR}_\alpha : L^0(\Omega, \mathcal{F}, \mathbb{P}) \to \mathbb{R}$ at level $\alpha \in (0, 1)$ defined by

$$\mathrm{VaR}_\alpha[Y] := \inf\{\eta \in \mathbb{R} \mid \mathbb{P}[\{\omega \in \Omega \mid Y(\omega) \leq \eta\}] \geq \alpha\}$$

(continued)

is law-invariant, nondecreasing, translation-equivariant and positively homogeneous, but in general not convex (cf. [23]).

(f) The **Conditional Value-at-Risk** $\mathrm{CVaR}_\alpha : L^1(\Omega, \mathcal{F}, \mathbb{P}) \to \mathbb{R}$ at level $\alpha \in (0, 1)$ given by

$$\mathrm{CVaR}_\alpha[Y] := \inf\{\eta + \frac{1}{1-\alpha}\mathrm{EE}^1_\eta[Y] \mid \eta \in \mathbb{R}\}$$

is a law-invariant coherent risk measure (cf. [23, Proposition 2]). The variational representation above was established in [24, Theorem 10].

(g) The **entropic risk measure** $\mathrm{Entr}_\alpha : L^\infty(\Omega, \mathcal{F}, \mathbb{P}) \to \mathbb{R}$ defined by

$$\mathrm{Entr}_\alpha[Y] := \frac{1}{\alpha}\ln\Big(\mathbb{E}\big[\exp(\alpha Y)\big]\Big),$$

where $\alpha > 0$ is a parameter, is a law-invariant convex (but not coherent) risk measure (cf. [21, Example 4.13, Example 4.34]).

(h) The **worst-case risk measure** $\mathcal{R}_{\max} : L^\infty(\Omega, \mathcal{F}, \mathbb{P}) \to \mathbb{R}$ given by

$$\mathcal{R}_{\max}[Y] := \sup_{\omega \in \Omega} Y(\omega)$$

is law-invariant and coherent (cf. [21, Example 4.8]). This choice of $\mathcal{R}$ in (17.2.1) leads to the **bilevel robust problem**

$$\min_x \{\mathcal{R}_{\max}[F(x)] \mid x \in X\}.$$

Note that $\mathcal{R}_{\max}$ only depends on the so called **uncertainty set** $Z(\Omega) \subseteq \mathbb{R}$. Thus, a bilevel robust problem can be formulated without knowledge of the distribution of the uncertain parameter. In robust optimization, the uncertainty set is often assumed to be finite, polyhedral or ellipsoidal (cf. [25]).

Remark 17.2.7 The set of convex (coherent) risk measures on $L^p(\Omega, \mathcal{F}, \mathbb{P})$ is a convex cone for any fixed $p \in [1, \infty]$. In particular, if $\mathcal{R} : L^p(\Omega, \mathcal{F}, \mathbb{P}) \to \mathbb{R}$ is a convex (coherent) risk measure, then so is $\mathbb{E} + \rho \cdot \mathcal{R}$ for any $\rho \geq 0$. The **mean-risk bilevel stochastic programming model**

$$\min_x \{\mathbb{E}[F(x)] + \rho \cdot \mathcal{R}[F(x)] \mid x \in X\}$$

seeks to minimize a weighted sum of the expected value of the outcome and a quantification of risk. △

Example

Consider the bilevel stochastic problem

$$\min \{\mathcal{R}[\min \Psi(x, Z)] \mid 1 \le x \le 6\},$$

where $\Psi(x,z) := \text{Argmin}_y\{-y \mid y \ge 1,\ y \le x+2+z_1,\ y \le -x+8.5+z_2\}$ and assume that Z is uniformly distributed over the square $[-0.5, 0.5]^2$.

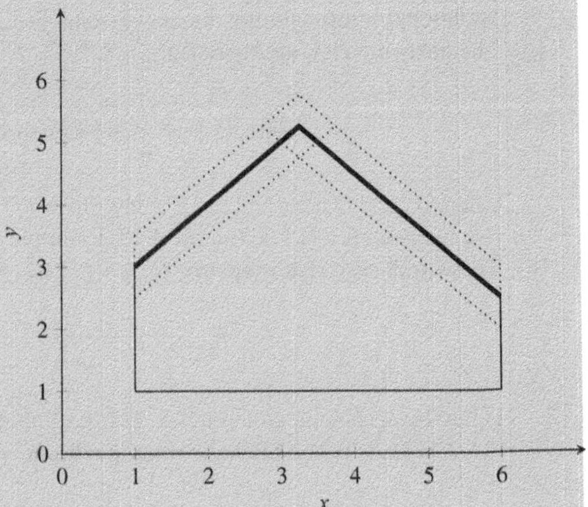

Fig. 17.1 The bold line depicts the graph of $\Psi(\cdot, (0,0))$, while the dotted lines correspond to the graphs of $\Psi(\cdot, (\pm 0.5, \pm 0.5))$ and $\Psi(\cdot, (\mp 0.5, \pm 0.5))$

As it can be seen in Fig. 17.1, we have

$$\Psi(x,z) = \begin{cases} \{x+2+z_1\} & \text{if } x \le 3.25 + 0.5z_2 - 0.5z_1 \\ \{-x+8.5+z_2\} & \text{else} \end{cases}$$

for any $x \in [1,6]$ and $z \in [-0.5, 0.5]^2$. A straightforward calculation shows that

$$\begin{aligned} \mathbb{E}[\min \Psi(x, Z)] &= \int_{-0.5}^{0.5} \int_{-0.5}^{0.5} x + 2 + z_1 \, dz_1 \, dz_2 \\ &= x + 2 \end{aligned}$$

holds for any $x \in [1, 2.75]$. Similarly, we have

$$\begin{aligned} \mathbb{E}[\min \Psi(x, Z)] &= \int_{-0.5}^{2x-6} \int_{-0.5}^{-2x+6.5+z_2} x + 2 + z_1 \, dz_1 \, dz_2 \\ &\quad + \int_{2x-6}^{0.5} \int_{-0.5}^{0.5} x + 2 + z_1 \, dz_1 \, dz_2 \end{aligned}$$

(continued)

$$+\int_{6-2x}^{0.5}\int_{-0.5}^{2x-6.5+z_1} -x+8.5+z_2\,dz_2\,dz_1$$
$$= -\tfrac{4}{3}x^3+11x^2-\tfrac{117}{4}x+\tfrac{1427}{48}$$

for $x \in [2,75,3.25]$ and

$$\begin{aligned}\mathbb{E}[\min\Psi(x,Z)] &= \int_{2x-7}^{0.5}\int_{-0.5}^{-2x+6.5+z_2} x+2+z_1\,dz_1\,dz_2\\ &+\int_{-0.5}^{7-2x}\int_{-0.5}^{2x-6.5+z_1} -x+8.5+z_2\,dz_2\,dz_1\\ &+\int_{7-2x}^{0.5}\int_{-0.5}^{0.5} -x+8.5+z_2\,dz_2\,dz_1\\ &= \tfrac{4}{3}x^3-15x^2+\tfrac{221}{4}x-\tfrac{989}{16}\end{aligned}$$

for $x \in [3.25, 3.75]$. Finally, for $x \in [3.75, 6]$ we calculate

$$\begin{aligned}\mathbb{E}[\min\Psi(x,Z)] &= \int_{-0.5}^{0.5}\int_{-0.5}^{0.5} -x+8.5+z_2\,dz_2\,dz_1\\ &= -x+8.5.\end{aligned}$$

Thus, $\mathbb{E}[\min\Psi(\cdot\,,Z)]$ is piecewise polynomial, non-convex and non-differentiable. It is easy to check that $x^* = 6$ is a global minimizer of the risk-neutral model

$$\min\{\mathbb{E}[\min\Psi(x,Z)] \mid 1 \le x \le 6\}.$$

In this particular example, x^* is also a global minimizer of the bilevel robust problem

$$\min\{\mathcal{R}_{max}[\min\Psi(x,Z)] \mid 1 \le x \le 6\}.$$

17.2.3 Continuity and Differentiability

Continuity properties of $\mathcal{R}$ carry over to Lipschitzian properties of $\mathcal{Q}_{\mathcal{R}} : P_Z \to \mathbb{R}$, $\mathcal{Q}_{\mathcal{R}}(x) := \mathcal{R}[F(x)]$.

Proposition 17.2.8 *Assume dom $f \neq \emptyset$ and $\mu_Z \in \mathcal{M}_s^p$ for some $p \in [1,\infty]$. Then the following statements hold true for any $\mathcal{R} : L^p(\Omega,\mathcal{F},\mathbb{P}) \to \mathbb{R}$:*

a. $\mathcal{Q}_{\mathcal{R}}$ is locally Lipschitz continuous if $\mathcal{R}$ is convex and continuous.
b. $\mathcal{Q}_{\mathcal{R}}$ is locally Lipschitz continuous if $\mathcal{R}$ is convex and nondecreasing.
c. $\mathcal{Q}_{\mathcal{R}}$ is locally Lipschitz continuous if $\mathcal{R}$ is a convex risk measure.

d. $Q_{\mathcal{R}}$ *is Lipschitz continuous if* $\mathcal{R}$ *is Lipschitz continuous.*
e. $Q_{\mathcal{R}}$ *is Lipschitz continuous if* $\mathcal{R}$ *is a coherent risk measure.* △

Proof

a. It is well-known that any real-valued convex and continuous mapping on a normed space is locally Lipschitz continuous (cf. [26]). The result is thus an immediate consequence of Lemma 17.2.4.
b. Any real-valued, convex and nondecreasing functional on the Banach lattice $L^p(\Omega, \mathcal{F}, \mathbb{P})$ is continuous (see e.g. [27, Theorem 4.1]).
c. By definition, any convex risk measure is convex and nondecreasing.
d. This is a straightforward conclusion from Lemma 17.2.4.
e. Any coherent risk measure on $L^p(\Omega, \mathcal{F}, \mathbb{P})$ is Lipschitz continuous by Inoue [28, Lemma 2.1]. □

Remark 17.2.9 Any coherent risk measure $\mathcal{R} : L^\infty(\Omega, \mathcal{F}, \mathbb{P}) \to \mathbb{R}$ is Lipschitz continuous with constant 1 by Föllmer and Schied [21, Lemma 4.3]. Concrete Lipschitz constants for continuous coherent law-invariant risk measures $\mathcal{R}$: $L^p(\Omega, \mathcal{F}, \mathbb{P}) \to \mathbb{R}$ with $p \in [1, \infty)$ may be obtained from representation results (see e.g. [29]). △

Proposition 17.2.8 allows to formulate sufficient conditions for the existence of optimal solutions to the bilevel stochastic linear program (17.2.1):

Corollary 17.2.10 *Assume dom* $f \neq \emptyset$, $\mu_Z \in \mathcal{M}_s^p$ *for some* $p \in [1, \infty]$ *and let* $X \subseteq P_Z$ *be nonempty and compact. Then* (17.2.1) *is solvable for any convex and nondecreasing mapping* $\mathcal{R} : L^p(\Omega, \mathcal{F}, \mathbb{P}) \to \mathbb{R}$. △

Due to the lack of convexity, Proposition 17.2.8 and the subsequent Corollary do not apply to the excess probability and the Value-at-Risk. However, invoking Lemma 17.2.1, the arguments used in the proof of [30, Proposition 3.3] can adapted to the setting of bilevel stochastic linear programming:

Proposition 17.2.11 *Assume dom* $f \neq \emptyset$ *and fix* $\eta \in \mathbb{R}$, *then* Q_{EP_η} *is lower semicontinuous on* P_Z *and continuous at any* $x \in P_Z$ *satisfying*

$$\mu_Z[\{z \in \mathbb{R} \mid f(x, z) = \eta\}] = 0.$$

Furthermore, let $X \subseteq P_Z$ *be nonempty and compact. Then*

$$\min_x \left\{ EP_\eta[F(x)] \mid x \in X \right\}$$

is solvable. △

Q_{VaR_α} has been analyzed in [17, Theorem 2]:

Proposition 17.2.12 *Assume dom* $f \neq \emptyset$ *and* $\alpha \in (0, 1)$*, then* Q_{VaR_α} *is continuous. Moreover, let* $X \subseteq P_Z$ *be nonempty and compact. Then*

$$\min_x \{VaR_\alpha[F(x)] \mid x \in X\}$$

is solvable. △

For specific risk measures, sufficient conditions for differentiability of $Q_\mathcal{R}$ have been investigated in [31].

Proposition 17.2.13 *Assume dom* $f \neq \emptyset$ *and that* $\mu_Z \in \mathcal{M}_s^1$ *is absolutely continuous with respect to the Lebesgue measure. Fix any* $\eta \in \mathbb{R}$*, then* $Q_\mathbb{E}$ *and* $Q_{EE_\eta^1}$ *are continuously differentiable at any* $x_0 \in int\ P_Z$*. Furthermore, for any* $\rho \in [0, 1)$*,* $Q_{SD_\rho^1}$ *is continuously differentiable at any* $x_0 \in int\ P_Z$ *satisfying* $Q_\mathbb{E}(x_0) \neq 0$*.* △

Remark 17.2.14 Theorems 3.7, 3.8 and 3.9 in [31] provide more involved sufficient conditions for continuous differentiability of $Q_\mathbb{E}$, $Q_{\mathrm{EE}_\eta^1}$ and $Q_{\mathrm{SD}_\rho^1}$ that do not require μ_Z to be absolutely continuous. △

Remark 17.2.15 Note that the assumptions of Proposition 17.2.13 are not fulfilled in the example at the end of Sect. 17.2.2: The right-hand side of the restriction system is only **partially random** as the right-hand side of the restriction $y \geq 1$ does not depend on Z. If we extend the system to

$$y \leq x + 2 + z_1', \ \ y \leq -x + 8.5 + z_2', \ \ y \geq 1 + z_3',$$

the third component of the extended random vector Z' has to take the value 0 with probability 1. Thus, $\mathbb{P} \circ Z'^{-1}$ is not absolutely continuous with respect to the Lebesgue measure. △

In the presence of differentiability, necessary optimality conditions for (17.2.1) can be formulated in terms of directional derivatives (cf. [31, Corollary 3.10]).

Proposition 17.2.16 *Assume dom* $f \neq \emptyset$*,* $\mu_Z \in \mathcal{M}_s^p$ *and* $X \subseteq P_Z$*. Furthermore, let* $x_0 \in X$ *be a local minimizer of problem* (17.2.1) *and assume that* $Q_\mathcal{R}$ *is differentiable at* x_0*. Then*

$$Q_\mathcal{R}'(x_0)v \geq 0$$

holds for any feasible direction

$$v \in \{v \in \mathbb{R}^n \mid \exists \epsilon_0 > 0 : \ x_0 + \epsilon v \in X \ \forall \epsilon \in [0, \epsilon_0]\}.$$

△

17.2.4 Stability

While we have only considered $Q_{\mathcal{R}}$ as a functions of the leader's decision x so far, it also depends on the underlying probability measure μ_Z. In stochastic programming, incomplete information about the true underlying distribution or the need for computational efficiency may lead to optimization models that employ an approximation of μ_Z. This section analysis deals with the behaviour of optimal values and (local) optimal solution sets of (17.2.1) under perturbations of the underlying distribution.

Taking into account that the support of the perturbed measure may differ from the original support, we shall assume dom $f \neq \emptyset$ and

$$P = \mathbb{R}^n \times \mathbb{R}^s$$

to ensure that the objective function of (17.2.1) remains well defined. The corresponding assumption in two-stage stochastic programming is called **complete recourse** (cf. [1, Sect. 2.1.3]). Sufficient conditions for dom $f \neq \emptyset$ and $P = \mathbb{R}^n \times \mathbb{R}^s$ are given in [17, Corollary 1] and [17, Corollary 2]. The following characterization is a direct consequence of Gordan's Lemma (cf. [32]):

Lemma 17.2.17 *$P = \mathbb{R}^n \times \mathbb{R}^s$ holds if and only if $u = 0$ is the only non-negative solution to $A^\top u = 0$.* △

Throughout this section, we shall consider the situation that $\mathcal{R} : L^p(\Omega, \mathcal{F}, \mathbb{P}) \to \mathbb{R}$ with $p \in [1, \infty)$ is law-invariant, convex and nondecreasing. Furthermore, for the sake of notational simplicity (cf. [31, Remark 4.1]), we assume that the probability space $(\Omega, \mathcal{F}, \mathbb{P})$ is atomless, i.e. for any $A \in \mathcal{F}$ with $\mathbb{P}[A] > 0$ there exists some $B \in \mathcal{F}$ with $B \subsetneq A$ and $\mathbb{P}[A] > \mathbb{P}[B] > 0$.

Then for any $x \in X$ and $\mu \in \mathcal{M}_s^p$, we have $(\delta_x \otimes \mu) \circ f^{-1} \in \mathcal{M}_1^p$, where $\delta_x \in \mathcal{P}(\mathbb{R}^n)$ denotes the Dirac measure at x. The atomlessness of $(\Omega, \mathcal{F}, \mathbb{P})$ ensures that there exists some $Y_{(x,\mu)} \subset L^p(\Omega, \mathcal{F}, \mathbb{P})$ such that $\mathbb{P} \circ Y_{(x,\mu)}^{-1} = (\delta_x \otimes \mu) \circ f^{-1}$. Thus, we may consider the mapping $Q_{\mathcal{R}} : X \times \mathcal{M}_s^p \to \mathbb{R}$ defined by

$$Q_{\mathcal{R}}(x, \mu) := \mathcal{R}[Y_{(x,\mu)}].$$

Note that the specific choice of $Y_{(x,\mu)}$ does not matter due to the law-invariance of $\mathcal{R}$.

Consider the parametric optimization problem

$$\min_x \{Q_{\mathcal{R}}(x, \mu) \mid x \in X\}. \tag{P$_\mu$}$$

As (P_μ) may be non-convex, we shall pay special attention to sets of locally optimal solutions. For any open set $V \subseteq \mathbb{R}^n$ we introduce the localized optimal value function $\varphi_V : \mathcal{M}_s^p \to \overline{\mathbb{R}}$,

$$\varphi_V(\mu) := \min_x \{Q_\mathcal{R}(x, \mu) \mid x \in X \cap \mathrm{cl}\, V\},$$

as well as the localized optimal solution set mapping $\phi_V : \mathcal{M}_s^p \rightrightarrows \mathbb{R}^n$,

$$\phi_V(\mu) := \operatorname*{Argmin}_x \{Q_\mathcal{R}(x, \mu) \mid x \in X \cap \mathrm{cl}\, V\}.$$

It is well known that additional assumptions are needed when studying stability of local solutions.

Definition 17.2.18 Given $\mu \in \mathcal{M}_s^p$ and an open set $V \subseteq \mathbb{R}^n$, $\phi_V(\mu)$ is called a **complete local minimizing (CLM) set** of (P_μ) w.r.t. V if $\emptyset \neq \phi_V(\mu) \subseteq V$. △

Remark 17.2.19 The set of global optimal solutions $\phi_{\mathbb{R}^n}(\mu)$ and any set of isolated minimizers are CLM sets. However, sets of strict local minimizers may fail to be CLM sets (cf. [33]). △

In the following, we shall equip $\mathcal{P}(\mathbb{R}^s)$ with the topology of weak convergence, i.e. the topology, where a sequence $\{\mu_l\}_{l\in\mathbb{N}} \subset \mathcal{P}(\mathbb{R}^s)$ converges weakly to $\mu \in \mathcal{P}(\mathbb{R}^s)$, written $\mu_l \xrightarrow{w} \mu$, if and only if

$$\lim_{l\to\infty} \int_{\mathbb{R}^s} h(t)\, \mu_l(dt) = \int_{\mathbb{R}^s} h(t)\, \mu(dt)$$

holds for any bounded continuous function $h : \mathbb{R}^s \to \mathbb{R}$ (cf. [34]). The example below (cf. [22, Example 3.2]) shows that even $\varphi_{\mathbb{R}^n}$ may fail to be weakly continuous on the entire space $\mathcal{P}(\mathbb{R}^s)$.

Example
The problem

$$\min_x \left\{x + \int_{\mathbb{R}} z\, \mu(dz) \mid 0 \le x \le 1\right\}$$

arises from a bilevel stochastic linear problem, where $\mathcal{R} = \mathbb{E}$ and $\Psi(x, z) = \{z\} \subsetneq \mathbb{R}$ holds for any (x, z). Assume that μ is the Dirac measure at 0, then the above problem can be rewritten as

$$\min_x \{x \mid 0 \le x \le 1\}$$

(continued)

and its optimal value is 0.

However, while the sequence $\mu_l := (1 - \frac{1}{l})\delta_0 + \frac{1}{l}\delta_l$ converges weakly to δ_0, replacing μ with μ_l yields the problem

$$\min_x \{x + 1 \mid 0 \leq x \leq 1\},$$

whose optimal value is equal to 1 for any $l \in \mathbb{N}$.

We shall follow the approach of [22, 31] and [35] and confine the stability analysis to locally uniformly $\|\cdot\|^p$-integrating sets.

Definition 17.2.20 A set $\mathcal{M} \subseteq \mathcal{M}_s^p$ is said to be **locally uniformly** $\|\cdot\|^p$**-integrating** if for any $\epsilon > 0$ there exists some open neighborhood $\mathcal{N}$ of μ w.r.t. the topology of weak convergence such that

$$\lim_{a\to\infty} \sup_{\nu\in\mathcal{M}\cap\mathcal{N}} \int_{\mathbb{R}^s\setminus a\mathbb{B}} \|z\|^p \, \nu(dz) \leq \epsilon. \qquad \triangle$$

A detailed discussion of locally uniformly $\|\cdot\|^p$-integrating sets and their generalizations is provided in [21, 36, 37], and [38]. The following examples demonstrate the relevance of the concept.

Examples

(a) Fix $\kappa, \epsilon > 0$. Then by Föllmer and Schied [21, Corollary A.47 (c)], the set

$$\mathcal{M}(\kappa, \epsilon) := \left\{\mu \in \mathcal{P}(\mathbb{R}^s) \mid \int_{\mathbb{R}^s} \|z\|^{p+\epsilon} \, \mu(dz) \leq \kappa\right\}$$

of Borel probability measures with uniformly bounded moments of order $p + \epsilon$ is locally uniformly $\|\cdot\|^p$-integrating.

(b) Fix any compact set $\Xi \subset \mathbb{R}^s$. By Föllmer and Schied [21, Corollary A.47, (b)], the set

$$\{\mu \in \mathcal{P}(\mathbb{R}^s) \mid \mu[\Xi] = 1\}$$

of Borel probability measures whose support is contained in Ξ is locally uniformly $\|\cdot\|^p$-integrating.

The following result has been established in [31, Theorem 4.7]:

Theorem 17.2.21 *Assume dom $f \neq \emptyset$ and $P = \mathbb{R}^n \times \mathbb{R}^s$. Let $\mathcal{M} \subseteq \mathcal{M}_s^p$ be locally uniformly $\|\cdot\|^p$-integrating, then*

a. $Q_{\mathcal{R}}|_{\mathbb{R}^n \times \mathcal{M}}$ is real-valued and weakly continuous.
b. $\varphi_{\mathbb{R}^n}|_{\mathcal{M}}$ is weakly upper semicontinuous.

In addition, assume that $\mu_0 \in \mathcal{M}$ is such that $\phi_V(\mu_0)$ is a CLM set of P_{μ_0} w.r.t. some open bounded set $V \subsetneq \mathbb{R}^n$. Then the following statements hold true:

c. $\varphi_V|_{\mathcal{M}}$ is weakly continuous at μ_0.
d. $\phi_V|_{\mathcal{M}}$ is weakly upper semicontinuous at μ_0 in the sense of Berge (cf. [39]), i.e. for any open set $\mathcal{O} \subseteq \mathbb{R}^n$ with $\phi|_V(\mu_0) \subseteq \mathcal{O}$ there exists a weakly open neighborhood $\mathcal{N}$ of μ_0 such that $\phi_V(\mu) \subseteq \mathcal{O}$ for all $\mu \in \mathcal{N} \cap \mathcal{M}$.
e. There exists some weakly open neighborhood $\mathcal{U}$ of μ_0 such that $\phi_V(\mu)$ is a CLM set for (P_μ) *w.r.t. V for any $\mu \in \mathcal{U} \cap \mathcal{M}$.* △

Proof Fix any $x_0 \in \mathbb{R}^n$. By Lemma 17.2.1, f is Lipschitz continuous on $\mathbb{R}^n \times \mathbb{R}^s$. Thus, there exists a constant $L > 0$ such that

$$|f(x,z)| \leq L\|z\| + L\|x - x_0\| + |f(x_0, 0)|$$

and the result follows from [35, Corollary 2.4.]. □

Remark 17.2.22 Under the assumptions of Theorem 17.2.21d., any accumulation point x of a sequence local optimal solutions $x_l \in \phi_V(\mu_l)$ as $\mu_l \overset{w}{\to} \mu$, $\mu_l \in \mathcal{M}$, is a local optimal solution of (P_μ). A detailed discussion of Berge's notion of upper semicontinuity and related concepts is provided in [40, Chap. 5]. △

As any Borel probability measure is the weak limit of a sequence of measures having finite support, Theorem 17.2.21 justifies an approach where the true underlying measure is approximated by a sequence of finite discrete ones. It is well known that approximation schemes based on discretization via empirical estimation [41, 42] or conditional expectations [43, 44] produce weakly converging sequences of discrete probability measures under mild assumptions.

Remark 17.2.23 All results of Sects. 17.2.1–17.2.4 can be easily extended to the pessimistic approach to bilevel stochastic linear programming, where f takes the form

$$f(x,z) = c^\top x - \min_y\{-q^\top y \mid y \in \Psi(x,z)\}$$

(cf. [22, Chap. 4]). △

17.2.5 Stochastic Dominance Constraints

One possibility to model the minimization in

$$\min\{f(x, Z(\cdot)) \mid x \in X\}$$

is doing it w.r.t. some risk measure that maps $f(x, Z(\cdot))$ into the reals, as introduced in Sect. 17.2. In this section, we shall discuss an alternate approach, where a disutility function $g : \mathbb{R}^n \to \mathbb{R}$ is minimized over some subset of random variables of acceptable risk:

$$\min_x \{g(x) \mid x \in X,\ f(x, Z(\cdot)) \in \mathcal{A}\},$$

where $\mathcal{A} \subseteq f(X, Z) := \{f(x, Z(\cdot)) \mid x \in X\}$. The following cases are of particular interest (cf. [1, pp. 90–91]) :

Examples

(a) $\mathcal{A}$ is given by **probabilistic constraints**, i.e.

$$\mathcal{A}_{\text{pc}} = \{h \in f(X, Z) \mid \mathbb{P}[h \leq \beta_j] \geq p_j\ \forall j = 1, \dots, l\}$$

for bounds $\beta_1, \dots, \beta_l \in \mathbb{R}$ and safety levels $p_1, \dots, p_l \in (0, 1)$.

(b) $\mathcal{A}$ is given by **first-order stochastic dominance constraints**, i.e.

$$\mathcal{A}_{\text{fo}} = \{h \in f(X, Z) \mid \mathbb{P}[h \leq \beta] \geq \mathbb{P}[b \leq \beta]\ \forall \beta \in \mathbb{R}\},$$

where $b \in L^0(\Omega, \mathcal{F}, \mathbb{P})$ is a given benchmark variable. If b is discrete with a finite number of realizations, it is sufficient to impose the relation $\mathbb{P}[h \leq \beta] \geq \mathbb{P}[b \leq \beta]$ for any β in a finite subset of $\mathbb{R}$. In this case, $\mathcal{A}$ admits a description by a finite system of probabilistic constraints.

(c) $\mathcal{A}$ is given by **second-order stochastic dominance constraints**, i.e.

$$\mathcal{A}_{\text{so}} = \{h \in f(X, Z) \mid \mathbb{E}[\max\{h - \eta, 0\}] \leq \mathbb{E}[\max\{b - \eta, 0\}]\ \forall \eta \in \mathbb{R}\},$$

where $b \in L^1(\Omega, \mathcal{F}, \mathbb{P})$ is a given benchmark variable.

A discussion of general models involving probabilistic or stochastic dominance constraints can be found in [1, Chap. 8] and [45, Chap. 8.3].

Let $\nu := \mathbb{P} \circ b^{-1} \in \mathcal{P}(\mathbb{R})$ denote the distribution of the benchmark variable b. Then the feasible set under first-order stochastic dominance constraints admits the representation

$$\left\{x \in X \mid \mu_Z\big[\{z \in \mathbb{R}^s \mid f(x,z) \le \beta\}\big] \ge \nu\big[\{b \in \mathbb{R} \mid b \le \beta\}\big] \; \forall \beta \in \mathbb{R}\right\}.$$

Similarly, for second-order stochastic dominance constraints, $\mu \in \mathcal{M}_s^1$ and $\nu \in \mathcal{M}_1^1$, the feasible set takes the form

$$\left\{x \in X \mid \int_{\mathbb{R}^s} \max\{f(x,z) - \eta, 0\} \, \mu_Z(dz) \le \int_{\mathbb{R}} \max\{b - \eta, 0\} \, \nu(db) \; \forall \eta \in \mathbb{N}\right\}.$$

In both cases, the feasibility does only depend on the distribution of the underlying random vector. As in Sect. 17.2.4, we consider situations where μ_Z is replaced with an approximation and study the behaviour of the mappings $C_1 : \mathbb{P}(\mathbb{R}^s) \rightrightarrows \mathbb{R}^n$ defined by

$$C_1(\mu) = \left\{x \in X \mid \mu\big[\{z \in \mathbb{R}^s \mid f(x,z) \le \beta\}\big] \ge \nu\big[\{b \in \mathbb{R} \mid b \le \beta\}\big] \; \forall \beta \in \mathbb{R}\right\}.$$

and $C_2 : \mathcal{M}_s^1 \rightrightarrows \mathbb{R}^n$ given by

$$C_2(\mu) := \left\{x \in X \mid \int_{\mathbb{R}^s} \max\{f(x,z) - \eta, 0\} \, \mu(dz) \le \int_{\mathbb{R}} \max\{b - \eta, 0\} \, \nu(db) \; \forall \eta \in \mathbb{N}\right\}.$$

Invoking Lemma 17.2.1, the following result can be obtained by adapting the proofs of [46, Proposition 2.1] and [47, Proposition 2.2] :

Proposition 17.2.24 *Assume dom $f \neq \emptyset$ and $P = \mathbb{R}^n \times \mathbb{R}^s$. Then the following statements hold true:*

a. *The multifunction C_1 is closed w.r.t. the topology of weak convergence, i.e. for any sequences $\{\mu_l\}_l \subset \mathcal{P}(\mathbb{R}^s)$ and $\{x_l\}_l \subset \mathbb{R}^n$ with $\mu_l \xrightarrow{w} \mu \in \mathcal{P}(\mathbb{R}^s)$, $x_l \to x \in \mathbb{R}^n$ for $l \to \infty$ and $x_l \in C_1(\mu_l)$ for all $l \in \mathbb{N}$ it holds true that $x \in C_1(\mu)$.*
b. *Additionally assume that $\nu \in \mathcal{M}_1^1$, then the multifunction C_2 is closed w.r.t. the topology of weak convergence.* △

By considering the constant sequence $\mu_l = \mu$ for all $l \in \mathbb{N}$ we obtain the closedness of the sets $C_1(\mu)$ and $C_2(\mu)$ under the conditions of Proposition 17.2.24. The closedness of the multifunctions C_1 and C_2 is also the key to proving the following stability result (cf. [46, Proposition 2.5]):

Theorem 17.2.25 *Assume dom $f \neq \emptyset$, $P = \mathbb{R}^n \times \mathbb{R}^s$ and that X is nonempty and compact. Moreover, let g be lower semicontinuous. Then the following statements hold true:*

a. *The optimal value function $\varphi_1 : \mathcal{P}(\mathbb{R}^s) \to \mathbb{R} \cup \{\infty\}$ given by*

$$\varphi_1(\mu) := \inf\{g(x) \mid x \in C_1(\mu)\}$$

is weakly lower semicontinuous on dom C_1.

b. Additionally assume $\nu \in \mathcal{M}_1^1$*, then the function* $\varphi_2 : \mathcal{M}_s^1 \to \mathbb{R} \cup \{\infty\}$ *given by*

$$\varphi_2(\mu) := \inf\{g(x) \mid x \in C_2(\mu)\}$$

is weakly lower semicontinuous on dom C_2*.* △

17.2.6 Finite Discrete Distributions

Throughout this section, we shall assume that the underlying random vector Z is discrete with a finite number of realizations $Z_1, \ldots, Z_K \in \mathbb{R}^s$ and respective probabilities $\pi_1, \ldots, \pi_K \in (0, 1]$. Let I denote the index set $\{1, \ldots, K\}$, then P_Z takes the form

$$P_Z = \{x \in \mathbb{R}^n \mid \forall k \in I \; \exists y \in \mathbb{R}^m : \; Ay \leq Tx + Z_k\}.$$

Suppose that $x_0 \in X$ is such that $\{y \in \mathbb{R}^m \mid Ay \leq Tx_0 + Z_k\} = \emptyset$ holds for some $k \in I$. Then the probability of $f(x_0, Z(\omega)) = \infty$ is a least $\pi_k > 0$, i.e. x_0 should be considered as infeasible for problem (17.2.1). Consequently, $X \subseteq P_Z$ can be understood as an **induced constraint**. Note that $X \cap P_Z$ is a polyhedron if X is a polyhedron.

In this setting, the bilevel stochastic linear problem can be reduced to a standard bilevel program, which allows to adapt optimality conditions and algorithms designed for the deterministic case (cf. [48]).

Proposition 17.2.26 *Assume dom* $f \neq \emptyset$, $\mathcal{R} \in \{\mathbb{E}, EE_\eta^1, SD_\rho^1, EP_\eta, VaR_\alpha, CVaR_\alpha, \mathcal{R}_{\max}\}$ *and let* $X \subseteq P_Z$ *be a polyhedron. If* $\mathcal{R} \in \{EP_\eta, VaR_\alpha\}$*, additionally assume that* X *bounded. Then for any parameter* β*, there exists a constant* $M > 0$ *such that the bilevel stochastic linear problem*

$$\min_x \{\mathcal{R}[F(x)] \mid x \in X\}$$

is equivalent to the standard bilevel program

$$\min_x \Big\{ \underbrace{\inf_{\eta \in \mathbb{R}}}_{\text{if } \mathcal{R}=CVaR_\alpha} \min_w \{a(x, w) \mid w \in \Psi_{\mathcal{R}}(x)\} \mid x \in X \Big\}, \; \text{or}$$

$$\min_x \Big\{ c^\top x + \inf_{\eta \in \mathbb{R}} \{\eta \mid \min_w \{a(x, w) \mid w \in \Psi_{\mathcal{R}}(x)\} \geq \alpha\} \mid x \in X \Big\} \quad \text{if } \mathcal{R} = VaR_\alpha,$$

Table 17.1 Equivalent bilevel linear programs

$\mathcal{R}$	β	w	$a(x,w)$	$b(x,w)$
$\mathbb{E}$		$(y_1,\dots,y_K)\in\mathbb{R}^{Km}$	$c^\top x+\sum_{k\in I}\pi_k q^\top y_k$	
EE^1_η	$\eta\in\mathbb{R}$	$(y_1,\dots,y_K)\in\mathbb{R}^{Km}$ $(v_1,\dots,v_K)\in\mathbb{R}^K$	$\sum_{k\in I}\pi_k v_k$	$\begin{pmatrix} v_k \\ v_k-c^\top x-q^\top y_k+\eta\end{pmatrix}$
SD^1_ρ	$\rho\in$ $(0,1]$	$(y_1,\dots,y_K)\in\mathbb{R}^{Km}$ $(v_1,\dots,v_K)\in\mathbb{R}^K$	$(1-\rho)\sum_{k\in I}\pi_k q^\top y_k$ $+\rho\sum_{k\in I}\pi_k v_k+c^\top x$	$\begin{pmatrix} v_k-q^\top y_k \\ v_k-\sum_{j\in I}\pi_j q^\top y_j\end{pmatrix}$
EP_η	$\eta\in\mathbb{R}$	$(y_1,\dots,y_K)\in\mathbb{R}^{Km}$ $(\theta_1,\dots,\theta_K)\in\{0,1\}^K$	$\sum_{k\in I}\pi_k\theta_k$	$M\theta_k-c^\top x-q^\top y_k+\eta$
VaR_α	$\alpha\in$ $(0,1)$	$(y_1,\dots,y_K)\in\mathbb{R}^{Km}$ $(\theta_1,\dots,\theta_K)\in\{0,1\}^K$	$\sum_{k\in I}\pi_k\theta_k$	$M(1-\theta_k)-c^\top x$ $-q^\top y_k+\eta$
CVaR_α	$\alpha\in$ $(0,1)$	$(y_1,\dots,y_K)\in\mathbb{R}^{Km}$ $(v_1,\dots,v_K)\in\mathbb{R}^K$	$\eta+\frac{1}{1-\alpha}\sum_{k\in I}\pi_k v_k$	cf. EE^1_η
$\mathcal{R}_{\max}$		$(y_1,\dots,y_K)\in\mathbb{R}^{Km}$	$\max_{k\in I} c^\top x+q^\top y_k$	

where $\Psi_{\mathcal{R}}:\mathbb{R}^n\rightrightarrows\mathbb{R}^{dim\, w}$ *with*

$$\Psi_{\mathcal{R}}(x):=\underset{w}{Argmin}\left\{\sum_{k\in I} d^\top y_k \mid Ay_k\le Tx+Z_k,\ b_k(x,w)\ge 0\ \forall k\in I\right\}.$$

The specific formulations can be found in Table 17.1. △

Proof For $\mathcal{R}\in\{\mathbb{E},\mathrm{EE}^1_\eta,\mathrm{SD}^1_\rho,\mathrm{CVaR}_\alpha\}$, we refer to [31, Section 5].

For the excess probability, the first of the considered quantile-based risk measures, we have $\mathrm{EP}_\eta[F(x)]=\mathbb{P}\left[c^\top x+\inf_y\{q^\top y\mid y\in\Psi(x,Z(\cdot))\}>\eta\right]$. Fix $M\in\mathbb{R}$ such that

$$M>\sup\left\{c^\top x+\inf_{y_k}\left\{q^\top y_k\mid y_k\in\Psi(x,Z_k)\right\}\mid x\in X,\ k\in I\right\}-\eta$$

and, for $y_k\in\Psi(x,Z_k)$, let

$$\theta_k:=\begin{cases}0 & \text{if } c^\top x+q^\top y_k-\eta\le 0,\\ 1 & \text{otherwise.}\end{cases}$$

Then the excess probability is equal to

$$\sum_{k\in I}\pi_k\inf_{y_k,\theta_k}\left\{\theta_k\mid M\theta_k\ge c^\top x+q^\top y_k-\eta,\ y_k\in\Psi(x,Z_k),\ \theta_k\in\{0,1\}\right\}$$

$$=\inf_{\substack{y_1,\dots,y_K,\\ \theta_1,\dots,\theta_K}}\left\{\sum_{k\in I}\pi_k\theta_k \;\middle|\; \begin{array}{l} M\theta_k\ge c^\top x+q^\top y_k-\eta,\ y_k\in\Psi(x,Z_k),\ \theta_k\in\{0,1\}\\ \forall k\in I\end{array}\right\}.$$

Similar to the proof for $\mathcal{R} = \mathrm{EP}_\eta$, the expression $\mathbb{P}[f(x, Z(\cdot)) \leq \eta]$ equals

$$\sum_{k\in I} \pi_k \inf_{y_k,\theta_k} \left\{\theta_k \mid M(1-\theta_k) \geq c^\top x + q^\top y_k - \eta,\ y_k \in \Psi(x, Z_k),\ \theta_k \in \{0,1\}\right\}$$

$$= \inf_{\substack{y_1,\dots,y_K,\\ \theta_1,\dots,\theta_K}} \left\{ \begin{array}{l} \sum_{k\in I} \pi_k\theta_k \mid M(1-\theta_i) \geq c^\top x + q^\top y_k - \eta,\ y_k \in \Psi(x, Z_k),\ \theta_k \in \{0,1\} \\ \forall k \in I \end{array} \right\},$$

where

$$\theta_k := \begin{cases} 1 & \text{if } c^\top x + q^\top y_k - \eta \leq 0, \\ 0 & \text{otherwise.} \end{cases}$$

Thereby we get equality of $\mathrm{VaR}_\eta[f(x, Z(\cdot))]$ and

$$\inf \left\{ \eta \in \mathbb{R} \mid \inf_{\substack{y_1,\dots,y_K,\\ \theta_1,\dots,\theta_K}} \left\{ \sum_{k\in I} \pi_k\theta_k \mid (y_1,\dots,y_K,\theta_1,\dots,\theta_K) \in \Psi_{\mathrm{VaR}_\eta}(x) \right\} \geq \alpha \right\}.$$

The worst-case risk measure is equal to $\sup_{k\in I} \left\{c^\top x + \min_{y_k}\{q^\top y \mid y \in \Psi(x, Z_k)\}\right\}$ and the result follows from $\Psi_{\mathcal{R}_{\max}} = \Psi(x, Z_1) \times \ldots \times \Psi(x, Z_K)$. □

Remark 17.2.27

a. The equivalent standard bilevel problem is linear provided that $\mathcal{R} \in \{\mathbb{E}, \mathrm{EE}^1_\eta, \mathrm{SD}^1_\rho\}$.
b. Analogous to [31, Remarks 5.2, 5.4], the inner minimization problems of the standard bilevel linear programs for $\mathcal{R} \in \{\mathbb{E}, \mathrm{EE}^1_\eta, \mathrm{EP}_\eta, \mathcal{R}_{\max}\}$ can be decomposed into K scenario problems that only differ w.r.t. the right-hand side of the constraint system. For the other models, a similar decomposition is possible after Lagrangian relaxation of the coupling constraints involving different scenarios.
c. For $\mathcal{R} = \mathrm{CVaR}_\alpha$, every evaluation of the objective function in the standard bilevel linear program corresponds to solving a bilevel linear problem with scalar upper level variable η.
d. Alternate models for $\mathcal{R} = \mathrm{VaR}_\alpha$ are given in [17] and [49], where the considered bilevel stochastic linear problem is reduced to a mixed-integer nonlinear program and a mathematical programming problem with equilibrium constraints, respectively. A mean-risk model with $\mathcal{R} = \mathrm{CVaR}_\alpha$ is used in [5, Sect. III]. △

Similar reformulations can be obtained for the models discussed in Sect. 17.2.5 if we assume that the disutility function is linear.

Table 17.2 Equivalent programs, notation as in the Example in Sect. 17.2.5

$\mathcal{A}$	γ	w_j	$a(w_j)$	$\mathcal{R}$	δ_j
$\mathcal{A}_{\text{pc}}$	$\beta_j \in \mathbb{R},\ p_j \in (0,1)$ with $j = 1,\dots,l$	$(y_{1j},\dots,y_{Kj}) \in \mathbb{R}^{Km}$ $(\theta_{1j},\dots,\theta_{Kj}) \in \{0,1\}^K$	$\sum_{k\in I} \pi_k \theta_{kj}$	VaR_{β_j}	p_j
$\mathcal{A}_{\text{fo}}$	any finite discrete benchmark variable	$(y_{1j},\dots,y_{Kj}) \in \mathbb{R}^{Km}$ $(\theta_{1j},\dots,\theta_{Kj}) \in \{0,1\}^K$	$\sum_{k\in I} \pi_k \theta_{kj}$	EP_{a_j}	$\bar{a}_j$
$\mathcal{A}_{\text{so}}$	b with realizations $a_1,\dots,a_l$	$(y_{1j},\dots,y_{Kj}) \in \mathbb{R}^{Km}$ $(v_{1j},\dots,v_{Kj}) \in \mathbb{R}^K$	$\sum_{k\in I} \pi_k v_{kj}$	$\text{EE}^1_{a_j}$	$\tilde{a}_j$

Proposition 17.2.28 *Assume dom $f \neq \emptyset$, and let $X \subseteq P_Z$ be a bounded polyhedron. Then for any parameter γ, the problem*

$$\min_x \left\{ g^\top x \mid F(x) \in \mathcal{A},\ x \in X \right\}$$

is equivalent to

$$\min_x \left\{ g^\top x \mid \inf_{w_j} \left\{ a(w_j) \mid w_j \in \Psi_{\mathcal{R}}(x) \right\} \geq \delta_j\ \forall j = 1,\dots,l,\ x \in X \right\}.$$

The specific formulations are listed in Table 17.2, where $\bar{a}_j := 1 - \mathbb{P}[b \leq a_j]$ and $\tilde{a}_j := \int_{\mathbb{R}^s} \max\{b(z) - a_j, 0\}\ \mu_Z(dz)$.

△

17.2.7 Solution Approaches

To solve bilevel problems, it is very common to use a single level reformulation. Often the lower level minimality condition is replaced by its Karush-Kuhn-Tucker or Fritz John conditions and the bilevel problem is reduced to a mathematical programming problem with equilibrium constraints (cf. [5, 17], [48, Chap. 3.5.1]).

For $\mathcal{R} \in \{\mathbb{E}, \text{EE}^1_\eta, \text{SD}^1_\rho, \text{EP}_\eta, \mathcal{R}_{\max}\}$, the equivalent standard bilevel programs in Proposition 17.2.26 can be all restated as

$$\min_u \left\{ g^\top u + \min_w \{h^\top w \mid w \in \Psi(u)\} \mid u \in U \right\}, \tag{17.2.2}$$

where $\Psi : \mathbb{R}^k \rightrightarrows \mathbb{R}^l$ is given by $\Psi(u) = \text{Argmin}_w \{t^\top w \mid Ww \leq Bu + b\}$ for vectors $g \in \mathbb{R}^k$, $h, t \in \mathbb{R}^l$ and $b \in \mathbb{R}^r$, matrices $W \in \mathbb{R}^{r\times l}$ and $B \in \mathbb{R}^{r\times k}$, and $U \subseteq \mathbb{R}^k$ is a nonempty polyhedron. The usage of the KKT conditions of the lower level problem leads to the single-level problem

$$\min_{u,w,v} \left\{ g^\top u + h^\top w \;\middle|\; \begin{array}{l} Ww \leq Bu + b,\ W^\top v = t,\ v \leq 0, \\ v^\top (Ww - Bu - b) = 0,\ u \in U \end{array} \right\}. \tag{17.2.3}$$

More details as well as statements on the coincidence of optimal values and the existence of local and global minimizers are given in [31, Sect. 6]. If the condition $v^\top(Ww - Bu - b) = 0$ is relaxed by $v^\top(Ww - Bu - b) \leq \varepsilon$ (the resulting problem is denoted by P(ε)), the violation of regularity conditions like (MFCQ) and (LICQ) at every feasible point of (17.2.3) can be bypassed. A discussion of other difficulties associated with (17.2.3) is provided in [11, Chap. 3.1.2].

In [31, Sect. 6] it is also shown that $(\bar{u}, \bar{w})$ is a local minimizer of the optimistic formulation, if $(\bar{u}, \bar{w}, \bar{v})$ is an accumulation point of a sequence $\{(u_n, w_n, v_n)\}_{n\in\mathbb{N}}$ of local minimizers of problem P(ε_n) for $\varepsilon_n \downarrow 0$.

In the risk-neutral setting, problem (17.2.2) exhibits a block-structure (cf. Remark 17.2.27 b.). Adapting the solution method for general linear complementarity problems proposed in [50], this special structure has been used in [11, Chap. 6] to construct an efficient algorithm for the global resolution of bilevel stochastic linear problems based on dual decomposition.

Remark 17.2.29 Utilizing the lower level value function, problem (17.2.2) can be reformulated as a single level quasiconcave optimization problem (cf. [48, Chap. 3.6.5]). Solution methods based on a branch-and-bound scheme have been proposed in [51] and [52]. However, without modifications, these algorithms fail to exploit the block structure arising in risk-neutral bilevel stochastic linear optimization models (cf. [11, Chap. 4.2]). △

17.3 Two-Stage Stochastic Bilevel Programs

In two-stage stochastic bilevel programming, both leader and follower have to make their respective first-stage decisions without knowledge of the realization of a stochastic parameter. Afterwards, the second-stage decisions are made under complete information. This leads to the following chronology of decision and observation:

$$\begin{matrix} \text{leader} & & \text{follower} & & z = Z(\omega) & & \text{leader decides} & & \text{follower decides} \\ \text{decides } x_1 & \rightarrow & \text{decides } y_1 & \rightarrow & \text{is revealed} & \rightarrow & x_2(x_1, y_1, z) & \rightarrow & y_2(x_1, y_1, x_2, z) \end{matrix}$$

Remark 17.3.1 The bilevel stochastic linear problems considered in Sect. 17.2 can be understood as special two-stage bilevel programs, where the follower's first-stage and the leader's second stage decision do not influence the outcome. △

In [12], a two-stage stochastic extension of the bilevel network pricing model introduced in [53] is studied. Consider a multicommodity transportation network (N, Λ, K), where (N, Θ) is a directed graph and each commodity $k \in K$ is to be transported from an origin $O(k) \in N$ to a destination $D(k) \in N$ in order to satisfy a demand $n^k \in (0, \infty)$. The set of arcs Θ is partitioned into the subsets θ and $\overline{\theta}$ of tariff and tariff-free arcs, respectively, and the leaders is maximizing the revenue raised from tariffs, knowing that user flows are assigned to cheapest paths. In [53], this situation is modeled as a bilevel program

$$\text{“}\max_{x}\text{”}\left\{\sum_{k\in K} x^\top y^k \mid (y,\bar{y}) \in \Psi(x)\right\},$$

where the lower level is given by

$$\Psi(x,c,d,b) := \underset{y,\bar{y}}{\operatorname{Argmin}}\left\{\sum_{k\in K}\left[(c+x)^\top y^k + \bar{c}^\top \bar{y}^k\right] \;\middle|\; \begin{array}{l} y,\bar{y}\geq 0, \\ Ay^k + \bar{A}\bar{y}^k = b^k \;\forall k\in K \end{array}\right\},$$

and x is the vector of tariffs controlled by the leader, y^k and $\bar{y}^k$ are the flows of commodity k on the tariff and tariff-free arcs, respectively. Moreover, c and $\bar{c}$ are the fixed costs on θ and $\bar{\theta}$, respectively, $(A,\bar{A})$ denotes the node-arc incidence matrix and the vectors b^k defined by

$$b_i^k := \begin{cases} n^k, & \text{if } i = O(k) \\ -n^k, & \text{if } i = D(k) \\ 0, & \text{else} \end{cases}$$

are used to express nodal balance. Reference [12] extends the above model to a two-stage setting including market uncertainties: After deciding on first-stage tariffs, the situation repeats itself on the same network but with different cost and demand parameters. At the first-stage, only the distribution of the second-stage parameter $Z(\omega) = (c_2, d_2, b_2)(\omega)$ is known and the stages are linked by the restriction that the second-stage tariffs should not differ too widely from those set at the first stage. The linking constraint is motivated by policy regulations and competitivity issues. In a risk-neutral setting, this results in the problem

$$\text{“}\max_{x_1}\text{”}\left\{\sum_{k\in K} x_1^\top y_1^k + \mathbb{E}[\Phi(x_1, Z(\cdot))] \mid (y_1,\bar{y}_1) \in \Psi(x_1, c_1, d_1, b_1)\right\}, \tag{17.3.1}$$

where the recourse is given by

$$\Phi(x_1, Z(\omega)) := \text{“}\max_{x_2}\text{”}\left\{\sum_{k\in K} x_2^\top y_2^k \mid (x_1,x_2)\in\Gamma(\delta),\; (y_2,\bar{y}_2)\in\Psi(x_2, Z(\omega))\right\}$$

and the set $\Gamma(\delta)$ is defined as either

$$\Gamma(\delta) := \Gamma_A(\delta) := \{(x_1,x_2) \mid |x_{1,\theta} - x_{2,\theta}| \leq \delta_\theta \;\forall\theta\in\Theta\}$$

if tariff changes are limited in absolute values or

$$\Gamma(\delta) := \Gamma_R(\delta) := \{(x_1, x_2) \mid |x_{1,\theta} - x_{2,\theta}| \leq \delta_\theta |x_{1,\theta}| \; \forall \theta \in \Theta\}$$

if proportional limits are considered. Assuming that the underlying random vector Z is discrete with a finite number of realizations, a reformulation of (17.3.1) as a single-stage bilevel program is established in [12]. Moreover, sensitivity analysis of the optimal value function of (17.3.1) w.r.t. the parameter $\delta \in [0, \infty)^{|\Theta|}$ (cf. [12, Proposition 4.1, Proposition 4.2]) as well as numerical studies are conducted (cf. [12, Sects. 5, 6]).

17.4 Challenges

We shall highlight some aspects of bilevel stochastic programming that are highly deserving of future research:

Going (Further) Beyond the Risk-Neutral Case for Nonlinear Models The first paper on bilevel stochastic programming has already outlined the basic principles as well as existence and sensitivity results for risk neutral models (cf. [2]). Nevertheless, so far, most of the research on bilevel stochastic nonlinear programming is still concerned with the risk-neutral case. Notable exceptions are [5] and [8], where models involving the Conditional Value-at-Risk are considered. In the first paper the problem of maximizing the medium-term revenue of an electricity retailer under uncertain pool prices, demand, and competitor prices is modeled as a bilevel stochastic quadratic problem, while the latter explores links between electricity swing option pricing and bilevel stochastic optimization. However, there exists no systematic analysis of bilevel stochastic nonlinear problems in the broader framework of coherent risk measures or higher stochastic dominance constraints. Future research may also consider distributionally robust models (cf. [54]).

Exploiting (Quasi) Block Structures Arising in Risk-Averse Models Under finite discrete distributions many bilevel stochastic problems can be reformulated as standard bilevel programs. While this reformulation entails a blow-up of the dimension which is usually linear in the number of scenarios, the resulting problems often exhibit (quasi) block structures (cf. Remark 17.2.27b., [2]). For risk-neutral bilevel stochastic linear problems, [11, Chap. 6] utilizes these structures to enhance the mixed integer programming based solution algorithm of [50] resulting in a significant speed-up. Based on the structural similarities an analogous approach should be possible for risk-averse models after Lagrangian relaxation of coupling constraints.

Going Beyond Exogenous Stochasticity While the analysis in the vast majority of papers on stochastic programming is confined to the case of purely exogenous stochasticity, this assumption is known to be unrealistic in economic models, where the decision maker holds market power. Therefore, models with decision dependent distributions are of particular interest in view of stochastic Stackelberg games (cf. [55]).

Acknowledgements The second author thanks the Deutsche Forschungsgemeinschaft for its support via the Collaborative Research Center TRR 154.

References

1. A. Shapiro, D. Dentcheva, A. Ruszczyński, *Lectures on Stochastic Programming: Modeling and Theory*. MPS SIAM Series on Optimization, vol. 9, 2nd edn. (SIAM, Philadelphia, 2014)
2. M. Patriksson, L. Wynter, Stochastic mathematical programs with equilibrium constraints. Oper. Res. Lett. **25**, 159–167 (1999)
3. S. Christiansen, M. Patriksson, L. Wynter, Stochastic bilevel programming in structural optimization. Struct. Multidiscip. Optim. **21**, 361–371 (2001)
4. A. Werner, Bilevel stochastic programming problems: analysis and application to telecommunications. PhD thesis, Norwegian University of Science and Technology, 2005
5. M. Carrión, J.M. Arroyo, A.J. Conejo, A bilevel stochastic programming approach for retailer futures market trading. IEEE Trans. Power Syst. **24**, 1446–1456 (2009)
6. S. Dempe, V.V. Kalashnikov, G.A. Pérez-Valdés, N.I. Kalashnykova, Natural gas bilevel cash-out problem: convergence of a penalty function method. Eur. J. Oper. Res. **215**, 532–538 (2011)
7. S. Kosuch, P. Le Bodic, J. Leung, A. Lisser, On a stochastic bilevel programming problem. Networks **59**, 107–116 (2012)
8. R.M. Kovacevic, G.C. Pflug, Electricity swing option pricing by stochastic bilevel optimization: a survey and new approaches. Eur. J. Oper. Res. **237**, 389–403 (2013)
9. A. Chen, J. Kim, Z. Zhou, P. Chootinan, Alpha reliable network design problem. Transp. Res. Rec. **2029**, 49–57 (2007)
10. M. Patriksson, On the applicability and solution of bilevel optimization models in transportation science: a study on the existence, stability and computation of optimal solutions to stochastic mathematical programs with equilibrium constraints. Transp. Res. Part B Method. **42**, 843–860 (2008)
11. C. Henkel, An algorithm for global resolution of linear stochastic bilevel programs. PhD thesis, University of Duisburg-Essen, 2014
12. S.M. Alizadeh, P. Marcotte, G. Savard, Two-stage stochastic bilevel programming over a transportation network. Transp. Res. B **58**, 92–105 (2013)
13. B.C. Eaves, On quadratic programming. Manag. Sci. **17**, 698–711 (1971)
14. K. Beer, Lösung großer linearer Optimierungsaufgaben (Deutscher Verlag der Wiss., Berlin, 1977)
15. D. Klatte, B. Kummer, Stability properties of infima and optimal solutions of parametric optimization problems, in *Nondifferentiable Optimization: Motivations and Applications, Proceedings of the IIASA Workshop*, Sopron. Lect. Notes Econ. Math. Syst., vol. 255 (Springer, Berlin, 1984), pp. 215–229
16. D. Klatte, G. Thiere, Error bounds for solutions of linear equations and inequalities. ZOR Math. Methods Oper. Res. **41**, 191–214 (1995)
17. S.V. Ivanov, Bilevel stochastic linear programming problems with quantile criterion. Autom. Remote Control **75**, 107–118 (2014)
18. G.B. Dantzig, *Linear Programming and Extensions* (Princeton University Press, Princeton, 1963)
19. P. Artzner, F. Delbaen, J.-M. Eber, D. Heath, Coherent measures of risk. Math. Financ. **9**, 203–228 (1999)

20. H. Föllmer, A. Schied, Convex measures of risk and trading constraints. Finance Stoch. **6**, 429–447 (2002)
21. H. Föllmer, A. Schied, *Stochastic Finance: An Introduction in Discrete Time*, 3rd edn. (de Gruyter, Berlin, 2011)
22. M. Claus, Advancing stability analysis of mean-risk stochastic programs: bilevel and two-stage models. PhD thesis, University of Duisburg-Essen, 2016
23. G.C. Pflug, Some remarks on the value-at-risk and the conditional value-at-risk, in *Probabilistic Constrained Optimization - Methodology and Application*, ed. by S.P. Uryasev (Kluwer Academic, Dordrecht, 2000), pp. 272–281
24. R.T. Rockafellar, S. Uryasev, Conditional value-at-risk for general loss distributions. J. Bank. Financ. **26**, 1443–1471 (2002)
25. A. Ben-Tal, L. El Ghaoui, A. Nemirovski, *Robust Optimization* (Princeton University Press, Princeton, 2009)
26. I. Ekeland, R. Temam, *Analyse convexe et problèmes variationnels* (Dunod, Paris, 1974)
27. P. Cheridito, T. Li, Risk measures on Orlicz hearts. Math. Financ. **18**, 189–214 (2009)
28. A. Inoue, On the worst case conditional expectation. J. Math. Anal. Appl. **286**, 237–247 (2003)
29. D. Belomestny, V. Krätschmer, Central limit theorems for law-invariant coherent risk measures. J. Appl. Prob. **49**, 1–21 (2012)
30. R. Schultz, S. Tiedemann, Risk aversion via excess probabilities in stochastic programs with mixed-integer recourse. SIAM J. Optim. **14**, 115–138 (2003)
31. J. Burtscheidt, M. Claus, S. Dempe, Risk-averse models in bilevel stochastic linear programming. Preprint, arXiv:1901.11349 [math.OC] (2019)
32. P. Gordan, Über die Auflösungen linearer Gleichungen mit reellen Coefficienten. Math. Ann. **6**, 238 (1873)
33. S.M. Robinson, Local epi-continuity and local optimization. Math. Program. **37**, 208–222 (1987)
34. P. Billingsley, *Convergence of Probability Measures* (Wiley, New York, 1968)
35. M. Claus, V. Krätschmer, R. Schultz, Weak continuity of risk functionals with applications to stochastic programming. SIAM J. Optim. **27**, 91–109, S108 (2017)
36. V. Krätschmer, A. Schied, H. Zähle, Qualitative and infinitesimal robustness of tail-dependent statistical functionals. J. Multivar. Anal. **103**, 35–47 (2012)
37. V. Krätschmer, A. Schied, H. Zähle, Comparative and qualitative robustness for law-invariant risk measures. Finance Stoch. **18**, 271–295 (2014)
38. V. Krätschmer, A. Schied, H. Zähle, Domains of weak continuity of statistical functionals with a view on robust statistics. J. Multivar. Anal. **158**, 1–19 (2017)
39. C. Berge, *Espaces topologiques: fonctions multivoques*. Coll. Universitaire de mathématiques, vol. 3 (Dunod, Paris, 1959)
40. R.T. Rockafellar, R.J.-B. Wets, *Variational Analysis* (Springer, Berlin, 2009)
41. D. Pollard, *Convergence of Stochastic Processes* (Springer, New York, 1984)
42. G.R. Shorack, J.A. Wellner, *Empirical Processes with Applications to Statistics* (Wiley, New York, 1986)
43. J.R. Birge, R.J.-B. Wets, Designing approximation schemes for stochastic optimization problems, in particular for stochastic programs with recourse, in *Stochastic Programming 84 Part I*, ed. by A. Prékopa, R.J.B. Wets. Mathematical Programming Studies, vol. 27 (Springer, Berlin, 1986), pp. 54–102
44. P. Kall, A. Ruszczyński, K. Frauendorfer, Approximation techniques in stochastic programming, in *Numerical Techniques for Stochastic Optimization*, ed. by Y. Ermoliev, R.J.-B. Wets (Springer, Berlin, 1988), pp. 33–64
45. A. Prékopa, *Stochastic Programming*. Math. and Its Applications, vol. 324 (Kluwer Academic, Dordrecht, 1995)
46. R. Gollmer, F. Neise, R. Schultz, Stochastic programs with first-order dominance constraints induced by mixed-integer linear recourse. SIAM J. Optim. **19**, 552–571 (2008)
47. R. Gollmer, U. Gotzes, R. Schultz, A note on second-order stochastic dominance constraints induced by mixed-integer linear recourse. Math. Program. A **126**, 179–190 (2011)

48. S. Dempe, *Foundations of Bilevel Programming* (Springer, Berlin, 2002)
49. S. Dempe, S.V. Ivanov, A. Naumov, Reduction of the bilevel stochastic optimization problem with quantile objective function to a mixed-integer problem. Appl. Stoch. Models Bus. Ind. **33**, 544–554 (2017)
50. J. Hu, J.E. Mitchell, J.-S. Pang, K.P. Bennett, G. Kunapuli, On the global solution of linear programs with linear complementarity constraints. SIAM J. Optim. **19**, 445–471 (2008)
51. H. Tuy, Bilevel linear programming, multiobjective programming, and monotonic reverse convex programming, in *Multilevel Optimization: Algorithms and Applications*, ed. by A. Migdalas, P.M. Pardalos, P. Värbrand (Kluwer Academic, Dordrecht, 1998), pp. 295–314
52. H. Tuy, A. Migdalas, P. Värbrand, A quasiconcave minimization method for solving linear two-level programs. J. Glob. Optim. **4**, 243–263 (1994)
53. M. Labbé, P. Marcotte, G. Savard, A bilevel model of taxation and its application to optimal highway pricing. Manag. Sci. **44**, 1608–1622 (1998)
54. J. Zhang, H. Xu, L. Zhang, Quantitative stability analysis for distributionally robust optimization with moment constraints. SIAM J. Optim. **26**, 1855–1882 (2016)
55. V. De Miguel, H. Xu, A stochastic multiple-leader Stackelberg model: analysis, computation, and application. Oper. Res. **57**, 1220–1235 (2009)

Chapter 18
A Unified Framework for Multistage Mixed Integer Linear Optimization

Suresh Bolusani, Stefano Coniglio, Ted K. Ralphs, and Sahar Tahernejad

Abstract We introduce a unified framework for the study of multilevel mixed integer linear optimization problems and multistage stochastic mixed integer linear optimization problems with recourse. The framework highlights the common mathematical structure of the two problems and allows for the development of a common algorithmic framework. Focusing on the two-stage case, we investigate, in particular, the nature of the value function of the second-stage problem, highlighting its connection to dual functions and the theory of duality for mixed integer linear optimization problems, and summarize different reformulations. We then present two main solution techniques, one based on a Benders-like decomposition to approximate either the risk function or the value function, and the other one based on cutting plane generation.

Keywords Multilevel optimization · Multistage stochastic optimization · Discrete optimization · Primal and dual functions · Decomposition methods · Convexification-based methods

18.1 Introduction

This article introduces a unified framework for the study of *multilevel mixed integer linear optimization problems* and *multistage stochastic mixed integer linear optimization problems with recourse*. This unified framework provides insights into

---

S. Bolusani · T. K. Ralphs
Department of Industrial and Systems Engineering, Lehigh University, Bethlehem, PA, USA
e-mail: bsuresh@lehigh.edu; ted@lehigh.edu

S. Coniglio (✉)
Department of Mathematical Sciences, University of Southampton, Southampton, UK
e-mail: s.coniglio@soton.ac.uk

S. Tahernejad
Lindo Systems, Inc., Chicago, IL, USA
e-mail: sahar@lindo.com

S. Dempe, A. Zemkoho (eds.), *Bilevel Optimization*, Springer Optimization and Its Applications 161, https://doi.org/10.1007/978-3-030-52119-6_18

the nature of these two well-known classes of optimization problems, highlights their common mathematical structure, and allows results from the wider literature devoted to both classes to be exploited for the development of a common algorithmic framework.

18.1.1 Motivation

Historically, research in mathematical optimization, which is arguably the most widely applied theoretical and methodological framework for solving optimization problems, has been primarily focused on "idealized" models aimed at informing the decision process of a single *decision maker* (DM) facing the problem of making a single set of decisions at a single point in time under perfect information. Techniques for this idealized case are now well developed, with efficient implementations widely available in off-the-shelf software.

In contrast, most real-world applications involve multiple DMs, and decisions must be made at multiple points in time under uncertainty. To allow for this additional complexity, a number of more sophisticated modeling frameworks have been developed, including multistage and multilevel optimization. In line with the recent optimization literature, we use the term *multistage optimization* to denote the decision process of a single DM over multiple time periods with an objective that factors in the (expected) impact at future stages of the decisions taken at the current stage. With the term *multilevel optimization*, on the other hand, we refer to game-theoretic decision processes in which multiple DMs with selfish objectives make decisions in turn, competing to optimize their own individual outcomes in the context of settings such as, e.g., economic markets.

Because the distinction between multistage and multilevel optimization problems appears substantial from a modeling perspective, their development has been undertaken independently by different research communities. Indeed, multistage problems have arisen out of the necessity to account for stochasticity, which is done by explicitly including multiple decision stages in between each of which the value of a random variable is realized. Knowledge of the precise values of the quantities that were unknown at earlier stages allows for so-called *recourse decisions* to be made in later stages in order to correct possible mis-steps taken due to the lack of precise information. On the other hand, multilevel optimization has been developed primarily to model multi-round games (technically known as *finite, extensive form games with perfect information* in the general case and *Stackelberg games* in the case of two rounds) in which the decision (or strategy) of a given player at a given round must take into account the reactions of other players in future rounds.

Despite these distinctions, these classes of problems share an important common structure from a mathematical and methodological perspective that makes considering them in a single, unifying framework attractive. It is not difficult to understand the source of this common structure—from the standpoint of an individual DM, the complexity of the decision process comes from uncertainty about the future.

From an analytical perspective, the methods we use for dealing with the uncertainty arising from a lack of knowledge of the precise values of input parameters to later-stage decision problems can also be used to address the uncertainty arising from a lack of knowledge of the future actions of another self-interested player. In fact, one way of viewing the outcome of a random variable is as a "decision" made by a DM about whose objective function nothing is known. Both cases require consideration of a set of outcomes arising from either the different ways in which the uncertain future could unfold or from the different possible actions the other players could take. Algorithms for solving these difficult optimization problems must, either explicitly or implicitly, rely on efficient methods for exploring this outcome space. This commonality turns out to be more than a philosophical abstraction. The mathematical connections between multistage and multilevel optimization problems run deep and existing algorithms for the two cases already exhibit common features, as we illustrate in what follows.

18.1.2 Focus

In the rest of this article, we address the broad class of optimization problems that we refer to from here on by the collective name *multistage mixed integer linear optimization problems*. Such problems allow for multiple decision stages, with decisions at each stage made by a different DM and with each stage followed by the revelation of the value of one or more random variables affecting the available actions at the subsequent stages. Each DM is assumed to have their own objective function for the evaluation of the decisions made at all stages following the stage at which they make their first decision, including stages whose decision they do not control. Importantly, the objective functions of different DMs may (but do not necessarily) conflict. Algorithmically, the focus of such problems is usually on determining an optimal decision at the first stage. At the point in time when later-stage decisions must be made, the optimization problem faced by those DMs has the same form as the one faced by early-stage DMs but, in it, the decisions made at the earlier stages act as deterministic inputs. Note that, while we have assumed different DMs at each stage, it is entirely possible to model scenarios in which a single DM makes multiple decisions over time. From a mathematical standpoint, this latter situation is equivalent to the case in which different DMs share the same objective function and we thus do not differentiate these situations in what follows.

Although the general framework we introduce applies more broadly, we focus here on problems with two decision stages and two DMs, as well as stochastic parameters whose values are realized between the two stages. We further restrict our consideration to the case in which we have both continuous and integer variables but the constraints and objective functions are linear. We refer to these problems as *two-stage mixed integer linear optimization problems* (2SMILPs). Despite this restricted setting, the framework can be extended to multiple stages and more general forms of constraints and objective functions in a conceptually straightforward way (see, e.g., Sect. 18.3.3).

18.2 Related and Previous Work

In this section, we give a brief overview of related works. The literature on these topics is vast and the below overview is not intended to be exhaustive by any means, but only to give a general sense of work that has been done to date. The interested reader should consult other articles in this volume for additional background and relevant citations.

18.2.1 Applications

Multilevel and multistage structures, whose two level/stage versions are known as *bilevel optimization problems* and *two-stage stochastic optimization problems with recourse* (2SPRs), arise naturally in a vast array of applications of which we touch on only a small sample here.

In the modeling of large organizations with hierarchical decision-making processes, such as corporations and governments, Bard [8] discusses a bilevel corporate structure in which top management is the leader and subordinate divisions, which may have their own conflicting objectives, are the followers. Similarly, government policy-making can be viewed from a bilevel optimization perspective: [10] models a government encouraging biofuel production through subsidies to the petro-chemical industry, while [5] models a central authority that sets prices and taxes for hazardous waste disposal while polluting firms respond by making location-allocation and recycling decisions.

A large body of work exists on *interdiction problems*, which model competitive games where two players have diametrically opposed goals and the first player has the ability to prevent one or more of the second player's possible activities (variables) from being engaged in at a non-zero level. Most of the existing literature on these problems has focused on variations of the well-studied *network interdiction problem* [47, 65, 76, 80, 81, 105, 139, 141], in which the lower-level DM is an entity operating a network of some sort and the upper-level DM (or interdictor) attempts to *interdict* the network as much as possible via the removal (complete or otherwise) of portions (subsets of arcs or nodes) of the network. A more general case which does not involve networks (the so-called *linear system interdiction problem*) was studied in [79] and later in [52]. A related set of problems involves an attacker disrupting warehouses or other facilities to maximize the resulting transportation costs faced by the firm (the follower) [35, 117, 144]. A *trilevel* version of this problem involves the firm first fortifying the facilities, then the attacker interdicting them, and finally the firm re-allocating customers [34]. More abstract graph-theoretical interdiction problems in which the vertices of a graph are removed in order to reduce the graphs' stability/clique number are studied in [38, 59, 113].

Multilevel problems arise in a wide range of industries. For instance, in the context of the electricity industry, Hobbs and Nelson [78] applies bilevel optimization to demand-side management while [68] formulates a trilevel problem with power network expansion investments in the first level, market clearing in the second, and redispatch in the third. Coniglio et al. [43] and Côté et al. [48] address the capacity allocation and pricing problem for the airline industry. Dempe et al. [51] presents a model for the natural-gas shipping industry. A large amount of work has been carried out in the context of traffic planning problems, including constructing road networks to maximize users' benefits [16], toll revenue maximization [93], and hazardous material transportation [84]. For a general review on these problems and for one specialized to price-setting, the reader is referred to [92, 106]. More applications arise in chemical engineering and bioengineering in the design and control of optimized systems. For example, Clark and Westerberg [36] optimizes a chemical process by controlling temperature and pressure (first-stage actions) where the system (second stage) reaches an equilibrium as it naturally minimizes the Gibbs free energy. Burgard et al. [25] develops gene-deletion strategies (first stage) to allow overproduction of a desired chemical by a cell (second stage). In the area of telecommunication networks, bilevel optimization has been used for modeling the behavior of a networking communication protocol (second-level problem) which the network operator, acting as first-level DM, can influence but not directly control. The case of routing over a TCP/IP network is studied in [1–4, 39].

The literature on game theory features many works on bilevel optimization problems naturally arising from the computation of *Stackelberg equilibria* in different settings. Two main variants of the Stackelberg paradigm are typically considered: one in which the followers can observe the action that the leader draws from its commitment and, therefore, the commitment is in pure strategies [133], and one in which the followers cannot do that directly and, hence, the leader's commitment can be in mixed strategies [45, 134]. While most of the works focus on the case with a single leader and a single follower (which leads to a proper bilevel optimization problem), some work has been done on the case with more than two players: see [12–14, 31, 41, 42, 44, 104] for the *single*-leader *multi*-follower case, [61, 95, 98, 99, 124] for the *multi*-leader *single*-follower case, or [30, 91, 96, 108] for the *multi*-leader *multi*-follower case. Practical applications are often found in *security games*, which correspond to competitive situations where a defender (leader) has to allocate scarce resources to protect valuable targets from an attacker (follower) [6, 87, 109, 128]. Other practical applications are found in, among others, inspection games [7] and mechanism design [116]. The works on the computation of a *correlated equilibrium* [32] as well those on Bayesian persuasion [33], where a leader affects the behavior of the follower(s) by a *signal*, also fall in this category.

Finally, there are deep connections between bilevel optimization and the algorithmic decision framework that drives branch and bound itself, and it is likely that the study of bilevel optimization problems may lead to improved methods for solving single-level optimization problems. For example, the problem of determining the disjunction whose imposition results in the largest bound improvement within a branch-and-bound framework and the problem of determining the maximum bound-improving inequality are themselves bilevel optimization problems [40, 102, 103]. The same applies in n-ary branching when one looks for a branching decision leading to the smallest possible number of child nodes [97, 115].

Multistage problems and, in particular, two-stage stochastic optimization problems with recourse, arise in an equally wide array of application areas, including scheduling, forestry, pollution control, telecommunication and finance. Grass and Fischer [67] surveys literature, applications, and methods for solving disaster-management problems arising in the humanitarian context. Gupta et al. [69] addresses network-design problems where, in the second stage, after one of a finite set of scenarios is realized, additional edges of the network can be bought, and provides constant-factor approximation algorithms. A number of works address the two-stage stochastic optimization with recourse version of classical combinatorial optimization problems: among others, [54] considers the spanning-tree problem, [86] the matching problem, and [64, 66] the vehicle routing problem. For references to other areas of applicability, see the books [18, 83] and, in particular, [135].

18.2.2 Algorithms

The first recognizable formulations for bilevel optimization problems were introduced in the 1970s in [24] and this is when the term was also first coined. Beginning in the early 1980s, these problems attracted increased interest. Vicente and Calamai [131] provides a large bibliography of the early developments.

There is now a burgeoning literature on continuous bilevel linear optimization, but it is only in the past decade that work on the discrete case has been undertaken in earnest by multiple research groups. Moore and Bard [107] was the first to introduce a framework for general integer bilevel linear optimization and to suggest a simple branch-and-bound algorithm. The same authors also proposed a more specialized algorithm for binary bilevel optimization problems in [9]. Following these early works, the focus shifted primarily to various special cases, especially those in which the lower-level problem has the integrality property. Dempe [49] considers a special case characterized by continuous upper-level variables and integer lower-level variables and uses a cutting plane approach to approximate the lower-level feasible region (a somewhat similar approach is adopted in [50] for solving a bilinear mixed integer bilevel problem with integer second-level variables). Wen and Yang [137] considers the opposite case, where the lower-level problem is a linear optimization problem and the upper-level problem is an integer optimization problem, using linear optimization duality to derive exact and heuristic solutions.

The publication of a general algorithm for pure integer problems in [53] (based on the groundwork laid in a later-published dissertation [52]) spurred renewed interest in developing general-purpose algorithms. The evolution of work is summarized in Table 18.1, which indicates the types of variables supported in both the first and

Table 18.1 Evolution of algorithms for bilevel optimization

Citation	Stage 1 variable types	Stage 2 variable types
[137]	B	C
[9]	B	B
[56]	B, C	B, C
[114]	B, C	B, C
[62]	B	C
[52, 53]	G	G
[90]	G or C	G
[11]	B, C	C
[142]	G	G, C
[143]	G, C	G, C
[28]	G	G
[27]	B	B
[77]	B, C	B, C
[127]	G, C	G, C
[136]	G	G
[101]	G	G, C
[57, 58]	G, C	G, C

second stages (C indicates continuous, B indicates binary, and G indicates general integer). The aforementioned network interdiction problem is a special case that continues to receive significant attention, since tractable duality conditions exist for the lower-level problem [47, 65, 76, 80, 81, 105, 139, 141].

As for the case of multistage stochastic optimization, the two-stage linear stochastic optimization problem with recourse in which both the first- and second-stage problems contain only continuous variables has been well studied both theoretically and methodologically. Birge and Louveaux [18] and Kall and Mayer [83] survey the related literature. The integer version of the problem was first considered in the early 1990s by Louveaux and van der Vlerk [100] for the case of two-stage problems with *simple integer recourse*. Combining the methods developed for the linear version with the branch-and-bound procedure, Laporte and Louveaux [94] proposed an algorithm known as the *integer L-shaped method* where the first-stage problem contains only binary variables. Due to the appealing structural properties of a (mixed) binary integer optimization problem, a substantial amount of literature since then has been considering the case of a two-stage stochastic problem with (mixed) binary variables in one or both stages [60, 120, 123]. The case of two-stage problems with a pure integer recourse has also been frequently visited, see [89, 119]. It must be noted that methods such as these, which are typically developed for special cases, often rely on the special structure of the second-stage problem, thus being often not applicable to the two-stage problem with mixed integer restrictions. Algorithms for stochastic optimization problems with integer recourse were proposed by Carøe and Tind [29] and Sherali and Fraticelli [122].

18.3 Setup and Preliminaries

The defining feature of a multistage optimization problem is that the values of the *first-stage* variables (sometimes called *upper-level* or *leader* variables in the bilevel optimization literature) must be (conceptually) fixed without explicit knowledge of future events, the course of which can be influenced by the first-stage decision itself. Due to this influence, the perceived "value" of the first-stage decision must take into account the effect of this decision on the likelihood of occurrence of these future events.

More concretely, the first-stage DM's overall objective is to minimize the sum of two terms, the first term representing the immediate cost of implementation of the first-stage solution and the second term representing the desirability of the first-stage decision in terms of its impact on the decisions taken at later stages. The general form of a *two-stage mixed integer linear optimization problem* is then

$$\min_{x \in \mathcal{P}_1 \cap X} \{cx + \Xi(x)\}, \tag{2SMILP}$$

where

$$\mathcal{P}_1 = \left\{x \in \mathbb{R}^{n_1} \mid A^1 x \geq b^1\right\}$$

is the *first-stage feasible region*, with $A^1 \in \mathbb{Q}^{m_1 \times n_1}$ and $b^1 \in \mathbb{Q}^{m_1}$ defining the associated linear constraints, and $X = \mathbb{Z}_+^{r_1} \times \mathbb{Q}_+^{n_1 - r_1}$ representing integrality, rationality, and non-negativity requirements on the first-stage variables, denoted by x. Note that we require the continuous variables to take on rational values in order to ensure that the second-stage problem defined in (SS) below has an optimal value that is attainable when that value is finite. In practice, solvers always return such solutions, so this is a purely technical detail. The linear function cx with $c \in \mathbb{Q}^{n_1}$ is the aforementioned term that reflects the immediate cost of implementing the first-stage solution. The function $\Xi : \mathbb{Q}^{n_1} \to \mathbb{Q} \cup \{\pm\infty\}$ is the *risk function*, which takes only rational input for technical reasons discussed later. Ξ is the aforementioned term representing the first-stage DM's evaluation of the impact of a given choice for the value of the first-stage variables on future decisions. Similar concepts of risk functions have been employed in many different application domains and will be briefly discussed in Sect. 18.3.3. To enable the development of a practical methodology for the solution of these problems, however, we now define the specific class of functions we consider.

18.3.1 Canonical Risk Function

Our canonical risk function is a generalization of the risk function traditionally used in defining 2SPRs. As usual, let us now introduce a random variable U over an outcome space Ω representing the set of possible future scenarios that could be realized between the making of the first- and second-stage decisions. The values of this random variable will be input parameters to the so-called *second-stage problem* to be defined below.

As is common in the literature on 2SPRs, we assume that U is discrete, i.e., that the outcome space Ω is finite, so that $\omega \in \Omega$ represents which of a finite number of explicitly enumerated scenarios is actually realized. In practice, this assumption is not very restrictive, as one can exploit any algorithm for the case in which Ω is assumed finite to solve cases where Ω is not (necessarily) finite by utilizing a technique for discretization, such as *sample average approximation* (SAA) [121]. As U is discrete in this work, we can associate with it a probability distribution defined by $p \in \mathbb{R}^{|\Omega|}$ such that $0 \leq p_\omega \leq 1$ and $\sum_{\omega\in\Omega} p_\omega = 1$.

With this setup, the canonical risk function for $x \in \mathbb{Q}^{n_1}$ is

$$\Xi(x) = \mathbb{E}\left[\Xi_\omega(x)\right] = \sum_{\omega\in\Omega} p_\omega \Xi_\omega(x), \tag{RF}$$

where $\Xi_\omega(x)$ is the *scenario risk function*, defined as

$$\Xi_\omega(x) = \min\left\{d^1 y^\omega \;\middle|\; y^\omega \in \operatorname{argmin}\{d^2 y \mid y \in \mathcal{P}_2(b^2_\omega - A^2_\omega x) \cap Y\}\right\}; \tag{2SRF}$$

the set

$$\mathcal{P}_2(\beta) = \left\{y \in \mathbb{R}^{n_2} \mid G^2 y \geq \beta\right\}$$

is one member of a family of polyhedra that is parametric w.r.t. the right-hand side vector $\beta \in \mathbb{R}^{m_2}$ and represents the second-stage feasibility conditions; and $Y = \mathbb{Z}^{r_2}_+ \times \mathbb{Q}^{n_2-r_2}_+$ represents the second-stage integrality and non-negativity requirements. The deterministic input data defining Ξ_ω are $d^1, d^2 \in \mathbb{Q}^{n_2}$ and $G^2 \in \mathbb{Q}^{m_2\times n_2}$. $A^2_\omega \in \mathbb{Q}^{m_2\times n_1}$ and $b^2_\omega \in \mathbb{Q}^{m_2}$ represent the realized values of the random input parameters in scenario $\omega \in \Omega$, i.e., $U(\omega) = (A^2_\omega, b^2_\omega)$.

As indicated in (RF) and (2SRF), the inner optimization problem faced by the second-stage DM is parametric only in its right-hand side, which is determined jointly by the value ω of the random variable U and by the chosen first-stage solution. It will be useful in what follows to define a family of *second-stage* optimization problems

$$\inf\left\{d^2 y \;\middle|\; y \in \mathcal{P}_2(\beta) \cap Y\right\} \tag{SS}$$

that are parametric in the right-hand side $\beta \in \mathbb{R}^{m_2}$ (we use "inf" instead of "min" here because, for $\beta \notin \mathbb{Q}^{m_2}$, the minimum may not exist). By further defining

$$\beta^{\omega}(x) = b^2_{\omega} - A^2_{\omega}x$$

to be the parametric right-hand side that arises when the chosen first-stage solution is $x \in X$ and the realized scenario is $\omega \in \Omega$, we can identify the member of the parametric family defined in (SS) in scenario $\omega \in \Omega$ when the chosen first-stage solution is $x \in X$ as that with feasible region $\mathcal{P}_2(\beta^{\omega}(x)) \cap Y$.

Associated with each $x \in \mathbb{Q}^{n_1}$ and $\omega \in \Omega$ is the set of all alternative optimal solutions to the second-stage problem (SS) (we allow for $x \notin X$ here because such solutions arise when solving certain relaxations), called the *rational reaction set* and denoted by

$$\mathcal{R}^{\omega}(x) = \operatorname{argmin}\{d^2 y \mid y \in \mathcal{P}_2(b^2_{\omega} - A^2_{\omega}x) \cap Y\}.$$

For a given $x \in \mathbb{Q}^{n_1}$, $\mathcal{R}^{\omega}(x)$ may be empty if $\mathcal{P}_2(b^2_{\omega} - A^2_{\omega}x) \cap Y$ is itself empty or if the second-stage problem is unbounded (we assume in Sect. 18.3.2 that this cannot happen, however).

When $|\mathcal{R}^{\omega}(x)| > 1$, the second-stage DM can, in principle, choose which alternative optimal solution to implement. We must therefore specify in the definition of the risk function a rule by which to choose one of the alternatives. According to our canonical risk function (RF) and the corresponding scenario risk function (2SRF), the rule is to choose, for each scenario $\omega \in \Omega$, the alternative optimal solution that minimizes $d^1 y^{\omega}$, which corresponds to choosing the collection $\{y^{\omega}\}_{\omega \in \Omega}$ of solutions to the individual scenario subproblems that minimizes

$$d^1\left(\sum_{\omega \in \Omega} p_{\omega} y^{\omega}\right).$$

This is known as the *optimistic* or *semi-cooperative* case in the bilevel optimization literature, since it corresponds to choosing the alternative that is most beneficial to the first-stage DM. Throughout the article, we consider this case unless otherwise specified. In Sect. 18.3.3, we discuss other forms of risk function.

Because of the subtleties introduced above, there are a number of ways one could define the "feasible region" of (2SMILP). We define the *feasible region* for scenario ω (with respect to both first- and second-stage variables) as

$$\mathcal{F}^{\omega} = \{(x, y^{\omega}) \in X \times Y \mid x \in \mathcal{P}_1, y^{\omega} \in \mathcal{R}^{\omega}(x)\} \tag{FR}$$

and members of $\mathcal{F}^{\omega}$ as *feasible solutions* for scenario ω. Note that this definition of feasibility does not prevent having $(x, y^{\omega}) \in \mathcal{F}^{\omega}$ but $d^1 y^{\omega} > \Xi_{\omega}(x)$. This will not cause any serious difficulties, but is something to keep in mind.

We can similarly define the feasible region with respect to just the first-stage variables as

$$\mathcal{F}^1 = \bigcap_{\omega \in \Omega} \operatorname{proj}_x(\mathcal{F}^\omega). \tag{FS-FR}$$

Since $\Xi(x) = \infty$ for $x \in \mathbb{Q}^{n_1}$ if the feasible region $\mathcal{P}_2(\beta^\omega(x)) \cap Y$ of the second-stage problem (SS) is empty for some $\omega \in \Omega$, we have that, for $x \in \mathcal{P}_1 \cap X$, the following are equivalent:

$$x \in \mathcal{F}^1 \Leftrightarrow x \in \bigcap_{\omega \in \Omega} \operatorname{proj}_x(\mathcal{F}^\omega) \Leftrightarrow \mathcal{R}^\omega(x) \neq \emptyset \ \forall \omega \in \Omega \Leftrightarrow \Xi(x) < \infty.$$

Finally, it will be convenient to define $\mathcal{P}^\omega$ to be the feasible region of the relaxation of the deterministic two-stage problem under scenario $\omega \in \Omega$ that is obtained by dropping the optimality requirement for the second-stage variables y^ω, as well as any integrality restrictions. Formally, we have:

$$\mathcal{P}^\omega = \left\{(x, y^\omega) \in \mathbb{R}_+^{n_1+n_2} \mid x \in \mathcal{P}_1,\, y^\omega \in \mathcal{P}_2(b_\omega^2 - A_\omega^2 x)\right\}.$$

Later in Sect. 18.7, we will use these sets to define a relaxation for the entire problem that will be used as the basis for the development of a branch-and-cut algorithm.

18.3.2 Technical Assumptions

We now note the following assumptions made in the remainder of the article.

Assumption 18.3.1 *$\mathcal{P}^\omega$ is bounded for all $\omega \in \Omega$.* △

This assumption, which is made for ease of presentation and can be relaxed, results in the boundedness of (2SMILP).

Assumption 18.3.2 *All first-stage variables with at least one non-zero coefficient in the second-stage problem (the so-called* linking variables*) are integer, i.e.,*

$$L = \left\{i \in \{1, \ldots, n_1\} \;\middle|\; a_i^\omega \neq 0 \textit{ for some } \omega \in \Omega\right\} \subseteq \{1, \ldots, r_1\},$$

where a_i^ω represents the ith column of matrix A_ω^2. △

These two assumptions together guarantee that an optimal solution exists whenever (2SMILP) is feasible [132]. It also guarantees that the convex hull of $\mathcal{F}^\omega$ is a polyhedron, which is important for the algorithms we discuss later. Note that, due to the assumption of optimism, we can assume w.l.o.g. that all first-stage variables are linking variables by simply interpreting the non-linking variables as belonging to the second stage. While this may seem conceptually inconsistent with the intent of the original model, it is not difficult to see that the resulting model is *mathematically* equivalent, since these variables do not affect the second-stage problem and thus, the optimistic selection of values for those variables will be the same in either case.

Before closing this section, we remark that, in this article, we do not allow second-stage variables in the first-stage constraints. While this case can be handled with techniques similar to those we describe in the article from an algorithmic perspective, it does require a more complicated notation which, for the sake of clarity, we prefer not to adopt. Detailed descriptions of algorithms for this more general case in the bilevel setting are provided in [23, 127].

18.3.3 Alternative Models

18.3.3.1 Alternative Form of (2SMILP)

For completeness, we present here an alternative form of (2SMILP) that is closer to the traditional form in which bilevel optimization problems are usually specified in the literature. Adopting the traditional notation, (2SMILP) can be alternatively written as

$$
\begin{aligned}
\min_{x,\{y^{\omega}\}_{\omega\in\Omega}} \quad & cx + \sum_{\omega\in\Omega} p_{\omega} d^1 y^{\omega} \\
\text{s.t.} \quad & A^1 x \geq b^1 \\
& x \in X \\
& \left.\begin{array}{l} y^{\omega} \in \operatorname{argmin}_y \; d^2 y \\ \qquad \text{s.t.} \quad A^2_{\omega} x + G^2 y \geq b^2_{\omega} \\ \qquad\qquad\;\; y \in Y \end{array}\right\} \forall \omega \in \Omega.
\end{aligned}
\tag{2SMILP-Alt}
$$

Note that, in the first stage, the minimization is carried out with respect to both x and y. This again specifies the optimistic case discussed earlier, since the above formulation requires that, for a given $x \in X$, we select $\{y^{\omega}\}_{\omega\in\Omega}$ such that

$$
d^1\left(\sum_{\omega\in\Omega} p_{\omega} y^{\omega}\right)
$$

is minimized.

18.3.3.2 Pessimistic Risk Function

As already pointed out, the canonical risk function defined in (RF) assumes the *optimistic case*, since it encodes selection of the alternative optimal solution to the second-stage problem that is most beneficial to the first-stage DM. This is the case we focus on in the remainder of the article. The pessimistic case, on the other hand, is easily modeled by defining the scenario risk function to be

$$\Xi_\omega(x) = \max\left\{d^1 y^\omega \;\middle|\; y^\omega \in \operatorname{argmin}\{d^2 y \mid y \in \mathcal{P}_2(\beta^\omega(x)) \cap Y\}\right\}.$$

We remark that, while the optimistic and pessimistic cases may coincide in some cases (e.g., when (SS) admits a single optimal solution for every x), this coincidence is rarely observed in practice and would be hard to detect in any case. In general, the pessimistic case is more difficult to solve, though the algorithms discussed in Sect. 18.7 can be modified to handle it.

18.3.3.3 Recursive Risk Functions

Although we limit ourselves to problems with two stages in this article, we briefly mention that more general risk functions can be defined by recursively defining risk functions at earlier stages in terms of later-stage risk functions. This is akin to the recursive definition of the cost-to-go functions that arise in stochastic dynamic programming (see [17]). With such recursive definitions, it is possible to generalize much of the methodology described here in a relatively straightforward way, though the algorithm complexity grows exponentially with the addition of each stage. It is doubtful exact algorithms can be made practical in such cases.

18.3.3.4 Other Risk Functions

Other forms of risk function have been used in the literature, especially in finance. In robust optimization, for example, one might consider a risk function of the form

$$\Xi(x) = \max_{\omega \in \Omega}\{\Xi_\omega(x)\},$$

which models the impact on the first-stage DM of the worst-case second-stage realization of the random variables. A popular alternative in finance applications that is slightly less conservative is the *conditional value at risk*, the expected value taken over the worst α-percentile of outcomes [112, 129]. While it is possible to incorporate such risk functions into the general algorithmic framework we present here, for the purposes of limiting the scope of the discussion, we focus herein only on risk functions in the canonical form (RF).

18.3.4 Related Classes

With Ξ defined as in (RF), the problem (2SMILP) generalizes several well-known classes of optimization problems.

18.3.4.1 Single-Stage Problems

When $d^1 = d^2$ and $|\Omega| = 1$, the two stages of (2SMILP) can be collapsed into a single stage and the problem reduces to a traditional mixed integer linear optimization problem (MILP). It is natural that algorithms for (2SMILP) rely heavily on solving sequences of related single-stage MILPs and we discuss parametric versions of this class in later sections. For continuity, we utilize the notation for the second-stage variables and input data throughout. The case of $r_2 = 0$ (in which there are no integer variables) further reduces to a standard linear optimization problem (LP).

18.3.4.2 Bilevel Problems

When $|\Omega| = 1$ and assuming that we may have $d^1 \neq d^2$, (2SMILP) takes the form of a *mixed integer bilevel linear optimization problem* (MIBLP). Dropping the scenario super/subscript for simplicity, this problem is more traditionally written as

$$\min_{x \in \mathcal{P}_1 \cap X, y \in Y} \left\{ cx + d^1 y \;\middle|\; y \in \underbrace{\arg\min\{d^2 y \mid y \in \mathcal{P}_2(b^2 - A^2 x) \cap Y\}}_{\mathcal{R}(x)} \right\}. \tag{MIBLP}$$

Note that this formulation implicitly specifies the optimistic case since, if $\mathcal{R}(x)$ is not a singleton, it requires that, among the alternative optima, the solution minimizing $d^1 y$ be chosen. In this setting, the *bilevel risk function* can be written as

$$\Xi(x) = \min \left\{ d^1 y \mid y \in \mathcal{R}(x) \right\}.$$

18.3.4.3 Two-Stage Stochastic Optimization Problems with Recourse

When $d^1 = d^2$, either the inner or the outer minimization in (2SRF) is redundant and (2SMILP) takes the form of a *two-stage stochastic mixed integer linear optimization problem with recourse*. In this case, for each scenario $\omega \in \Omega$ we can write the scenario risk function more simply as

$$\Xi_\omega(x) = \min \left\{ d^1 y^\omega \mid y^\omega \in \mathcal{P}_2(b^2_\omega - A^2_\omega x) \cap Y \right\}.$$

The second-stage solution y^ω corresponding to scenario $\omega \in \Omega$ is usually called the *recourse decision*. These problems involve a single DM optimizing a single objective function, but capable of controlling two sets of variables: the first-stage *here-and-now* variables x and the second-stage *wait-and-see* variables y^ω, whose value is set after observing the realization of the random event ω.

18.3.4.4 Zero-Sum and Interdiction Problems

For $d^1 = -d^2$ (and typically, $|\Omega| = 1$), (2SMILP) subsumes the case of *zero-sum problems*, which model competitive games in which two players have exactly opposing goals. An even more specially-structured subclass of zero-sum problems are *interdiction problems*, in which the first-stage variables are in one-to-one correspondence with those of the second stage and represent the ability of the first-stage DM to "interdict" (i.e., forcing to take value zero) individual variables of the second-stage DM. Formally, the effect of interdiction can be modeled using a variable upper-bound constraint

$$y \le u(e - x)$$

in the second-stage problem, where $u \in \mathbb{R}^n$ is a vector of natural upper bounds on the vector of variables y and e is an n-dimensional column vector of ones (here, $n = n_1 = n_2$). Formally, the *mixed integer interdiction problem* is

$$\max_{x \in \mathcal{P}_1 \cap X} \min_{y \in \mathcal{P}_2(x) \cap Y} d^2 y,$$

where (abusing notation slightly), we have

$$\mathcal{P}_2(x) = \left\{ y \in \mathbb{R}^{n_2} \mid G^2 y \ge b^2, \, y \le u(e - x) \right\}.$$

18.4 Computational Complexity

Within the discrete optimization community, the framework typically used for assessing problem complexity is based primarily on the well-known theory of NP-completeness, which has evolved from the foundational work of [46, 63, 85]. This has led to the ubiquitous practice of classifying optimization problems as being either in the class P or the class NP-hard, the latter being an all-encompassing and amorphous class that includes essentially all optimization problems not known to be polynomially solvable. This categorization lacks the refinement necessary for consideration of classes such as those described in this article. It is indeed easy to show that multistage optimization problems are NP-hard in general [15, 26, 73, 82], but this merely tells us that these problems are not in P (assuming P $\neq$ NP), which is not surprising. What we would really like to know is for which complexity class (the decision versions of) these problems are *complete*.

In the presence of a hierarchical structure with k levels (and when Ω is a singleton), the natural complexity class to consider is Σ_k^{P}, i.e., the kth level of the *polynomial hierarchy*. From an optimization perspective, this hierarchy (originally introduced in [125]) is a scheme for classifying multilevel decision problems beyond the usual classes P and NP. The class P (which contains all decision problems that can be solved in polynomial time) occupies the 0th level, also known as Σ_0^{P}. The first level, Σ_1^{P}, is the class also known as NP, which consists of all problems for which there exists a certificate verifiable in polynomial time or, equivalently, all problems that can be solved in non-deterministic polynomial time. The kth level, Σ_k^{P}, contains all problems with certificates that can be verified in polynomial time (equivalently, all problems solvable in non-deterministic polynomial time), assuming the existence of an oracle for solving problems in the class $\Sigma_{k-1}^{\mathsf{P}}$. While it is clear that $\Sigma_k^{\mathsf{P}} \subseteq \Sigma_\ell^{\mathsf{P}}$ for any $k, \ell \in \mathbb{N} \cup \{0\}$ with $k \leq \ell$, $\Sigma_k^{\mathsf{P}} \subset \Sigma_\ell^{\mathsf{P}}$ is conjectured to hold for all $k, \ell \in \mathbb{N} \cup \{0\}$ with $k < \ell$ (the well-known $\mathsf{P} \neq \mathsf{NP}$ conjecture is a special case). It is also known that $\Sigma_k^{\mathsf{P}} = \Sigma_{k+1}^{\mathsf{P}}$ would imply $\Sigma_k^{\mathsf{P}} = \Sigma_\ell^{\mathsf{P}}$ for all $\ell \geq k+1$, which would cause the polynomial hierarchy to collapse to level k (for $k = 0$, we would have P = NP). The notions of completeness and hardness commonly used for NP translate directly to Σ_k^{P}. A proof that k-level optimization problems with binary variables, linear constraints, and linear objective functions are hard for Σ_k^{P} is contained in [82]. Such result suffices to show that the multistage problems with k stages treated in this article are (in their optimization version) Σ_k^{P}-hard (and those with $k = 2$ stages are Σ_2^{P}-hard). A compendium of Σ_2^{P}-complete/hard problems, somewhat similar in spirit to [63], can be found in [118], with more recent updates available online.

For the case of two-stage stochastic optimization problems with recourse with linear constraints, linear objective functions, and mixed integer variables, the assumption of a finite outcome space Ω of either fixed or polynomially bounded size suffices to guarantee that the decision version of such a problem is NP-complete. Indeed, when $|\Omega|$ is considered a constant or is bounded by a polynomial in the total number of variables and constraints, one can directly introduce a block-structured reformulation of the problem with one block per scenario $\omega \in \Omega$ that contains the coefficients of the constraints that y^ω should satisfy (we discuss such a reformulation in Sect. 18.6). As such reformulation is of polynomial size, solutions to the corresponding optimization problem can clearly be certified in polynomial time by checking that they satisfy all the polynomially-many constraints featured in the formulation, which, in turn, implies that the problem belongs to NP. When the outcome space Ω is continuous, the problem becomes #P-hard in general [55, 72]. While a single sample average approximation problem with a finite or polynomially-bounded number of samples can be used to approximate a continuous problem by solving a single discrete optimization problem of polynomial size, Hanasusanto et al. [72] shows that even finding an approximate solution using the SAA method is #P-hard. New results on the complexity of 2SPRs featuring a double-exponential algorithm can be found in [88].

18.5 Duality and the Value Function

Virtually all algorithms for the exact solution of optimization problems produce a proof of optimality that depends on the construction of a solution to a *strong dual*. Although the duality theory for MILPs is not widely known, the most effective algorithms for solving MILPs (which are variants of the well-known branch-and-bound algorithm) *do* produce a solution to a certain dual problem. A natural approach to solving (2SMILP) is therefore to embed the production of the "dual proof" of optimality of the second-stage problem (SS) into the formulation of the first-stage problem, reducing the original two-stage problem to a traditional single-stage optimization problem.

The reformulations and algorithmic approaches that we present in Sects. 18.6 and 18.7 all use some variant of this strategy. In particular, the algorithms we describe are based on iteratively strengthening an initial relaxation in a fashion reminiscent of many iterative optimization algorithms. The strengthening operation essentially consists of the dynamic construction of both a proof of optimality of the second-stage problem and of corresponding first- and second-stage solutions.

In the remainder of the section, we introduce the central concepts of a duality theory for mixed integer linear optimization problems (and more general discrete optimization problems), emphasizing its connection to solution methods for (2SMILP). This introduction is necessarily brief and we refer the reader to [71, 74, 75] for more details specific to the treatment here and to [138, 140] for earlier foundational work on IP duality. Although the "dual problem" is usually a fixed (non-parametric) optimization problem associated with a fixed (non-parametric) "primal problem," the typical concepts of duality employed in constructing dual proofs of optimality and in designing solution algorithms inherently involve parametric families of optimization problems. This makes the tools offered by this theory particularly suitable for employment in this setting. To preserve the connection with the material already introduced, we consider the family of MILPs parameterized on the right-hand side $\beta \in \mathbb{R}^{m_2}$ that was introduced earlier as (SS) and use the same notation. We reproduce it here for convenience:

$$\inf\left\{d^2 y \;\middle|\; y \in \mathcal{P}_2(\beta) \cap Y\right\}, \tag{SS}$$

where

$$\mathcal{P}_2(\beta) = \left\{y \in \mathbb{R}^{n_2} \mid G^2 y \geq \beta\right\}$$

and $\beta \in \mathbb{R}^{m_2}$ is the input parameter. When we want to refer to a (fixed) generic instance in this parametric family, the notation b will be used to indicate a fixed (but arbitrary) right-hand side. We also refer to specific right-hand sides arising in the solution of (2SMILP) using the notation defined earlier.

18.5.1 Value Functions

Among possible notions of duality, the one most relevant to the development of optimization algorithms is one that also has an intuitive interpretation in terms of familiar economic concepts. This theory rests fundamentally on an understanding of the so-called *value function*, which we introduce below. The value function of an MILP has been studied by a number of authors and a great deal is known about its structure and properties. Early work on the value function includes [19–22], while the material here is based on the work in [70, 71, 74, 75].

As a starting point, consider an instance of (SS) with fixed right-hand side b and let us interpret the values of the variables as specifying a numerical "level of engagement" in certain activities in an economic market. Further, let us interpret the constraints as corresponding to limitations imposed on these activities due to available levels of certain scarce resources (it is most natural to think of "$\leq$" constraints in this interpretation). In each row j of the constraint matrix, the coefficient G^2_{ij} associated with activity (variable) i can then be thought of as representing the rate at which resource j is consumed by engagement in activity i. In this interpretation, the optimal primal solution then specifies the level of each activity in which one should engage in order to maximize profits (it is most natural here to think in terms of maximization), given the fixed level of resources b.

Assuming that additional resources were available, how much should one be willing to pay? The intuitive answer is that one should be willing to pay at most the marginal amount by which profits would increase if more of a particular resource were made available. Mathematically, this information can be extracted from the *value function* $\phi : \mathbb{R}^{m_2} \to \mathbb{R} \cup \{\pm\infty\}$ associated with (SS), defined by

$$\phi(\beta) = \inf_{y \in \mathcal{P}_2(\beta) \cap Y} d^2 y, \tag{2SVF}$$

for $\beta \in \mathbb{R}^{m_2}$. Since this function returns the optimal profit for any given basket of resources, its gradient at b (assuming ϕ is differentiable at b) tells us what the marginal change in profit would be if the level of resources available changed in some particular direction. Thus, the gradient specifies a "price" on that basket of additional resources.

The reader familiar with the theory of duality for linear optimization problems should recognize that the solution to the usual dual problem associated with an LP of the form (SS) (i.e., assuming $r_2 = 0$) provides exactly this same information. In fact, we describe below that the set of optimal solutions to the LP dual are precisely the subgradients of its associated value function. This dual solution can hence be interpreted as a linear price on the resources and is sometimes referred to as a vector of "dual prices." The optimal dual prices allow us to easily determine whether it will be profitable to enter into a particular activity i by comparing the profit d^2_i obtained by entering into that activity to the cost uG^2_i of the required resources, where u is a given vector of dual prices and G^2_i is the i-th column of G^2. The difference $d^2_i - uG^2_i$

between the profit and the cost is the *reduced profit/cost* in linear optimization. It is easily proven that the reduced profit of each activity entered into at a non-zero level (i.e., reduced profits of the variables with non-zero value) must be non-negative (again, in the case of maximization) and duality provides an intuitive economic interpretation of this result.

Although the construction of the full value function is challenging even in the simplest case of a linear optimization problem, approximations to the value function in the local area around b can still be used for sensitivity analysis and in optimality conditions. The general dual problem we describe next formalizes this idea by formulating the problem of constructing a function that bounds the value function from below everywhere but yields a strong approximation near a fixed right-hand side b. Such a so-called "dual function" can yield approximate "prices" and its iterative construction can also be used in a more technical way to guide the evolution of an algorithm by providing gradient information helpful in finding the optimal solution, as well as providing a proof of its optimality.

18.5.2 Dual Functions

The above discussion leads to the natural concept of a *dual (price) function* from which we can derive a general notion of a *dual problem.*

Definition 18.5.1 A *dual function* $F : \mathbb{R}^{m_2} \to \mathbb{R}$ is one that satisfies $F(\beta) \leq \phi(\beta)$ for all $\beta \in \mathbb{R}^{m_2}$. We call such a function *strong* at $b \in \mathbb{R}^{m_2}$ if $F(b) = \phi(b)$. △

Dual functions are naturally associated with *relaxations* of the original problem, as the value function of any relaxation yields a feasible dual function. In particular, the value function of the well-known LP relaxation is the best convex under-estimator of the value function.

Also of interest are functions that bound the value function from above, which we refer to as *primal functions*.

Definition 18.5.2 A *primal function* $H : \mathbb{R}^{m_2} \to \mathbb{R}$ is one that satisfies $H(\beta) \geq \phi(\beta)$ for all $\beta \in \mathbb{R}^{m_2}$. We call such a function *strong* at b if $H(b) = \phi(b)$. △

In contrast to dual functions, primal functions are naturally associated with *restrictions* of the original problem and the value function of any such restriction yields a valid primal function.

It is immediately evident that a pair of primal and dual functions yields optimality conditions. If we have a primal function H^* and a dual function F^* such that $F^*(b) = \gamma = H^*(b)$ for some $b \in \mathbb{R}^{m_2}$, then we must also have $\phi(b) = \gamma$. Proofs of optimality of this nature are produced by many optimization algorithms.

18.5.3 Dual Problems

The concepts we have discussed so far further lead us to the definition of a generalized dual problem, originally introduced in [71], for an instance of (SS) with right-hand side $b \in \mathbb{R}^{m_2}$. This problem simply calls for the construction of a dual function that is strong for a particular fixed right-hand side $b \in \mathbb{R}^{m_2}$ by determining

$$\max_{F \in \Upsilon^{m_2}} \{F(b) : F(\beta) \leq \phi(\beta),\ \forall \beta \in \mathbb{R}^{m_2}\}, \tag{MILPD}$$

where $\Upsilon^{m_2} \subseteq \{f \mid f : \mathbb{R}^{m_2} \rightarrow \mathbb{R}\}$. Here, Υ^{m_2} can be taken to be a specific class of functions, such as linear or subadditive, to obtain specialized dual problems for particular classes of optimization problems. It is clear that (MILPD) always has a solution F^* that is strong, provided that the value function is real-valued everywhere (and hence belongs to Υ^{m_2}, however it is defined), since ϕ itself is a solution whenever it is finite everywhere.[1]

Although it may not be obvious, this notion of a dual problem naturally generalizes existing notions for particular problem classes. For example, consider again a parametric family of LPs defined as in (SS) (i.e., assuming $r_2 = 0$). We show informally that the usual LP dual problem with respect to a fixed instance with right-hand side b can be derived by taking Υ^{m_2} to be the set of all non-decreasing linear functions in (MILPD) and simplifying the resulting formulation. First, let a non-decreasing linear function $F : \mathbb{R}^{m_2} \rightarrow \mathbb{R}$ be given. Then, $\exists u \in \mathbb{R}^{m_2}_{+}$ such that $F(\beta) = u\beta$ for all $\beta \in \mathbb{R}^{m_2}$. It follows that

$$F(\beta) = u\beta \leq uG^2 y = \sum_{j=1}^{n_2} uG^2_j y_j \ \forall \beta \in \mathbb{R}^{m_2}.$$

From the above, it then follows that, for any $\beta \in \mathbb{R}^{m_2}$, we have

$$\begin{aligned} uG^2_j \leq d^2_j \ \forall j \in \{1, \ldots, n_2\} \Rightarrow u\beta &\leq uG^2 y \leq d^2 y \ \forall y \in \mathcal{P}_2(\beta) \cap Y \\ \Rightarrow u\beta &\leq \min_{y \in \mathcal{P}_2(\beta) \cap Y} d^2 y \\ \Rightarrow u\beta &\leq \phi(\beta) \\ \Rightarrow F(\beta) &\leq \phi(\beta). \end{aligned}$$

The conditions on the left-hand side above are exactly the feasibility conditions for the usual LP dual and the final condition on the right is the feasibility condition

---

[1] When the value function is not real-valued everywhere, we have to show that there is a real-valued function that coincides with the value function when it is real-valued and is itself real-valued everywhere else, but is still a feasible dual function (see [140]).

for (MILPD). Hence, the usual dual feasibility conditions ensure that u defines a linear function that bounds the value function from below and is a dual function in the sense we have defined. The fact that the epigraph of ϕ is a convex polyhedral cone in this case (it is the max of linear functions associated with extreme points of the feasible region of the dual problem) is enough to show that the dual (MILPD) is strong in the LP case, even when we take Υ^{m_2} to be the set of (non-decreasing) linear functions. Furthermore, it is easy to show that any subgradient of ϕ at b is an optimal solution (and in fact, the set of all dual feasible solutions is precisely the subdifferential of the value function at the origin).

The concepts just discussed can be easily seen in the following small example (note that this example is equality-constrained, in which case most of the above derivation carries through unchanged, but the dual function no longer needs to be non-decreasing).

Dual Function of an LP

$$\begin{aligned} &\min 6y_1 + 7y_2 + 5y_3 \\ &\text{s.t. } 2y_1 - 7y_2 + y_3 = b \\ &\quad y_1, y_2, y_3 \in \mathbb{R}_+. \end{aligned}$$

The solution to the dual of this LP is unique whenever b is non-zero and can be easily obtained by considering the ratios c_j/a_j of objective coefficient to constraint coefficient for $j = 1, 2, 3$, which determine which single primal variable will take a non-zero value in the optimal basic feasible solution. Depending on the sign of b, we obtain one of two possible dual solutions:

$$u^* = \begin{cases} 6/2 = 3 & \text{if } b > 0 \\ 7/(-7) = -1 & \text{if } b < 0. \end{cases}$$

Thus, the value function associated with this linear optimization problem is as shown in Fig. 18.1. Note that, when $b = 0$, the dual solution is not unique and can take any value between -1 and 3. This set of solutions corresponds to the set of subgradients at the single point of non-differentiability of the value function. This function has one of two gradients for all points that are differentiable, and these gradients are equal to one of the two dual solutions derived above.

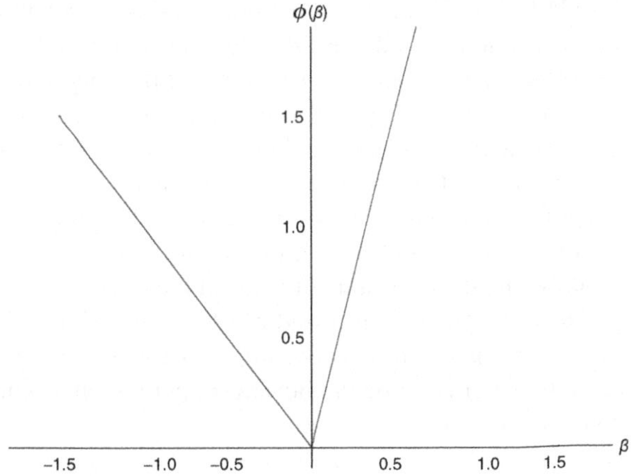

Fig. 18.1 Value function of the LP in the example

18.5.4 Structure of the Value Function

As we mentioned previously, solution methods for (2SMILP) inherently involve the explicit or implicit approximation of several functions, including the value function ϕ in (2SVF) and the risk function Ξ in (RF), which ultimately derives its structure from ϕ. Here, we summarize the results described in the series of papers [70, 71, 75]. Most importantly, the function is piecewise polyhedral, lower semi-continuous, subadditive, and has a discrete structure that is derivative of the structure of two related value functions which we now introduce.

Let y_C, d_C^2, and G_C^2 be the parts of each of the vectors/matrices describing the second-stage problem (SS) that are associated with the continuous variables and let y_I, d_I^2, and G_I^2 be likewise for the integer variables. The *continuous restriction* (CR) is the LP obtained by dropping the integer variables in the second-stage problem (or equivalently, setting them to zero). This problem has its own value function, defined as

$$\phi_C(\beta) = \min_{y_C \in \mathbb{R}_+^{n_2 - r_2}} \{d_C^2 y_C \mid G_C^2 y_C \geq \beta\}. \tag{CRVF}$$

On the other hand, if we instead drop the continuous variables from the problem, we can then consider the *integer restriction* (IR), which has value function

$$\phi_I(\beta) = \min_{y_I \in \mathbb{Z}_+^{r_2}} \{d_I^2 y_I \mid G_I^2 y_I \geq \beta\}. \tag{IRVF}$$

To illustrate how these two functions combine to yield the structure of ϕ and to briefly summarize some of the important results from the study of this function carried out in the aforementioned papers, consider the following simple example of a two-stage stochastic mixed integer linear optimization problem with a single constraint. Note that, in this example, $d^1 = d^2$ and we are again considering the equality-constrained case in order to make the example a bit more interesting.

Value Function of a 2SMILP

We consider the 2SMILP

$$\begin{aligned} \min \quad & \Psi(x_1, x_2) = -3x_1 - 4x_2 + \mathbb{E}[\phi(b^2_\omega - 2x_1 - 0.5x_2)] \\ \text{s.t.} \quad & x_1 \leq 5, x_2 \leq 5 \\ & x_1, x_2 \in \mathbb{R}_+, \end{aligned}$$

where

$$\begin{aligned} \phi(\beta) = \min \quad & 6y_1 + 4y_2 + 3y_3 + 4y_4 + 5y_5 + 7y_6 \\ \text{s.t.} \quad & 2y_1 + 5y_2 - 2y_3 - 2y_4 + 5y_5 + 5y_6 = \beta \\ & y_1, y_2, y_3 \in \mathbb{Z}_+, y_4, y_5, y_6 \in \mathbb{R}_+, \end{aligned}$$

with $\Omega = \{1, 2\}$, $b^2_1 = 6$, and $b^2_2 = 12$. Figure 18.2 shows the objective function Ψ and the second-stage value function ϕ for this example.

Examining Fig. 18.2, it appears that ϕ is comprised of a collection of translations of ϕ_C, each of which has a structure that is similar to the value function of the LP in the Example on page 533. At points where ϕ is differentiable, the gradient always corresponds to one of the two solutions to the dual of the continuous restriction (precisely as in the Example on page 533, dual solutions are the ratios of objective function coefficients to constraint coefficients for the continuous variables), which are in turn also the gradients of ϕ_C. The so-called points of *strict local convexity* are the points of non-differentiability that are the extreme points of the epigraphs of the translations of ϕ_C and are determined by the solutions to the integer restriction. In particular, they correspond to points at which ϕ_I and ϕ are coincident. For instance, observe that, in the example, $\phi_I(5) = \phi(5) = 4$.

Finally, we can observe in Fig. 18.2 how the structure of ϕ translates into that of Ψ.

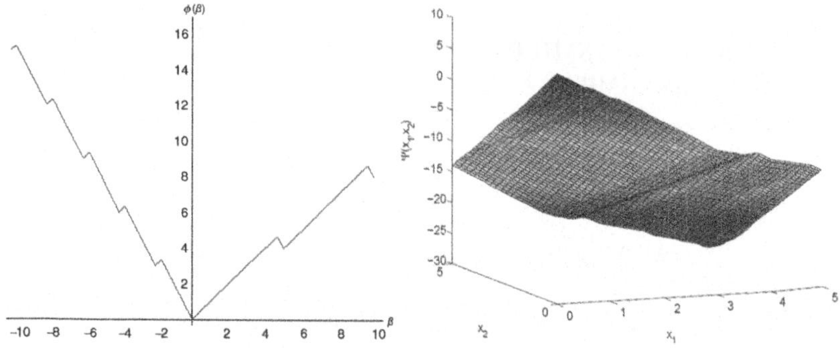

Fig. 18.2 Illustration of the functions from the example, with the second-stage value function $\phi(\beta)$ (left) and the objective function $\Psi(x_1, x_2)$ (right)

Although the case illustrated in the example is of a single constraint, these properties can be made rigorous and do generalize to higher dimension with roughly the same intuition. The general principle is that the value function of an MILP is the minimum of translations of the value function of the continuous restriction ϕ_C, where the points of translation (the aforementioned points of *strict local convexity*) are determined by the value function of the integer restriction ϕ_I.

Theorem 18.5.3 ([75]) *Under the assumption that* $\{\beta \in \mathbb{R}^{m_2} \mid \phi_I(\beta) < \infty\}$ *is compact and* $\text{epi}(\phi_C)$ *is pointed, there exists a finite set* $S \subseteq Y$ *such that*

$$\phi(\beta) = \min_{y_I \in S}\{d_I^2 y_I + \phi_C(\beta - G_I^2 y_I)\}. \qquad \triangle$$

Under the assumptions of the theorem, this result provides a finite description of ϕ.

18.5.5 Approximating the Value Function

Constructing functions that bound the value function of an MILP is an important part of solution methods for both traditional single-stage optimization problems and their multistage counterparts. Functions bounding the value function from below are exactly the *dual functions* defined earlier and arise from *relaxations* of the original problem (such as the LP relaxation, for instance). They are naturally obtained as by-products of common solution algorithms, such as branch and bound, which itself works by iteratively strengthening a given relaxation and produces a dual proof of optimality.

Functions that bound the value function from above are the *primal functions* defined earlier and can be obtained by considering the value function of a *restriction* of the original problem. While it is a little less obvious how to obtain such functions in general (solution algorithms generally do not produce practically useful primal functions), they can be obtained by taking the minimum over the value functions of restrictions obtained by fixing the values of integer variables, as we describe below.

Both primal and dual functions can be iteratively improved by producing new such functions and combining them with existing ones by taking the maximum over several bounding functions in the dual case or the minimum over several bounding functions in the primal case. When such functions are iteratively constructed to be strong at different right-hand side values, such as when solving a sequence of instances with different right-hand sides, such a technique can yield an approximation with good fidelity across a larger part of the domain than any singly constructed function could—this is, in fact, the principle implicitly behind the algorithms we describe in Sect. 18.7.1.

18.5.5.1 Dual Functions from Branch and Bound

Dual functions can be obtained from most practical solution algorithms for solving the MILP associated with (2SVF) with input $\beta = b^2_\omega - A^2_\omega x$, i.e., for computing $\phi(b^2_\omega - A^2_\omega x)$. This is because their existence is (at least implicitly) the very source of the proof of optimality produced by such algorithms. To illustrate, we show how a dual function can be obtained as the by-product of the branch-and-bound algorithm.

Branch and bound is an algorithm that searches the feasible region by partitioning it and then recursively solving the resulting subproblems. Implemented naively, this results in an inefficient complete enumeration, but this potential inefficiency is avoided by utilizing lower and upper bounds computed for each subproblem to intelligently "prune" the search. The recursive partitioning process can be envisioned as a process of searching a rooted tree, each of whose nodes corresponds to a subproblem. Although it is not usually described this way, the branch-and-bound algorithm can be interpreted as constructing a function feasible to (MILPD), thus producing not only a solution but also a dual proof of its optimality.

To understand this interpretation, suppose we evaluate $\phi(b)$ for $b \in \mathbb{R}^{m_2}$ by solving the associated MILP using a standard branch-and-bound algorithm with branching done on elementary (a.k.a. variable) disjunctions of type $y_j \leq \pi_0 \vee y_j \geq \pi_0 + 1$ for some $j \in \{1, \ldots, r_2\}$ and $\pi_0 \in \mathbb{Z}$. Because the individual subproblems in the branch-and-bound tree differ only by the bounds on the variables, it will be convenient to assume that all integer variables have finite initial lower and upper bounds denoted by the vectors $l, u \in \mathbb{Z}^{r_2}$ (such bounds exist by Assumption 1, even if they are not part of the formulation). In this case, we have that

$$\mathcal{P}_2(\beta) = \{y \in \mathbb{R}^{n_2} \mid G^2 y \geq \beta, y_I \geq l, -y_I \geq -u\}.$$

The solution of (SS) by branch and bound for right-hand side b yields a branch-and-bound tree whose leaves, contained in the set $\mathcal{T}$, are each associated with the subproblem

$$\min\{d^2 y \mid y \in \mathcal{P}_2^t(b) \cap Y\},$$

where

$$\mathcal{P}_2^t(\beta) = \{y \in \mathbb{R}^{n_2} \mid G^2 y \geq \beta, y_I \geq l^t, -y_I \geq -u^t\}$$

is the parametric form of the polytope containing the feasible points associated with subproblem $t \in \mathcal{T}$, with $l^t,\ u^t \in \mathbb{Z}_+^{r_2}$ being the modified bounds on the integer variables imposed by branching.

Assuming that no pruning took place, the validity of the overall method comes from the fact that valid methods of branching ensure that

$$\bigcup_{t \in \mathcal{T}} \mathcal{P}_2^t(b) \cap Y = \mathcal{P}_2(b) \cap Y,$$

so that the feasible regions of the leaf nodes constitute a partition of the original feasible region. This partition can be thought of as a single logical disjunction that serves to strengthen the original LP relaxation. The proof of optimality that branch and bound produces derives from global lower and upper bounds derived from local bounds associated with each node $t \in \mathcal{T}$. We denote by L^t and U^t, respectively, the lower and upper bounds on the optimal solution value of the subproblem, where

$$L^t = \phi_{\mathrm{LP}}^t(b), \quad U^t = d^2 y^t,$$

in which ϕ_{LP}^t is as defined in (NVF) below and $y^t \in Y$ is the best solution known for subproblem $t \in \mathcal{T}$ ($U^t = \infty$ if no solution is known). Assuming (SS) is solved to optimality and $y^* \in Y$ is an optimal solution, we must have

$$L := \min_{t \in \mathcal{T}} L^t = d^2 y^* = \min_{t \in \mathcal{T}} U^t =: U,$$

where L and U are the global lower and upper bounds.

From the information encoded in the branch-and-bound tree, the overall dual function can be constructed by deriving a parametric form of the lower bound, combining dual functions for the individual subproblems in set $\mathcal{T}$. For this purpose, we define the value function

$$\phi_{\mathrm{LP}}(\beta, \lambda, \nu) = \min\{d^2 y \mid y \in \mathcal{P}_2(\beta), \lambda \leq y_I \leq \nu, y_C \geq 0\} \qquad \text{(PNVF)}$$

of a generic LP relaxation, which captures the bounds on the integer variables as also being parametric. Using (PNVF), the value function of the LP relaxation associated with a particular node $t \in \mathcal{T}$ (only parametric in the original right-hand side) can be obtained as

$$\phi_{\mathrm{LP}}^t(\beta) = \phi_{\mathrm{LP}}(\beta, l^t, u^t). \qquad \text{(NVF)}$$

For all $t \in \mathcal{T}$ such that $\phi^t_{\mathrm{LP}}(b) < \infty$, let $(v^t, \underline{v}^t, \bar{v}^t)$ be an optimal solution to the dual of the LP relaxation at node $t \in \mathcal{T}$, where v^t, $\underline{v}^t$, and $\bar{v}^t$ are, respectively, the dual variables associated with the original inequality constraints, the lower bounds on integer variables, and the upper bounds on integer variables. Then, by LP duality we have that

$$\phi^t_{\mathrm{LP}}(b) = v^t b + \underline{v}^t l^t - \bar{v}^t u^t.$$

For each node $t \in \mathcal{T}$ for which $\phi^t_{\mathrm{LP}}(b) = \infty$ (the associated subproblem is infeasible), we instead let $(v^t, \underline{v}^t, \bar{v}^t)$ be a dual feasible solution that provides a finite bound exceeding U (such can be found by, e.g., adding some multiple of the dual ray that proves infeasibility to the feasible dual solution found in the final iteration of the simplex algorithm).

Finally, from the above we have that

$$\underline{\phi}^t_{\mathrm{LP}}(\beta) = v^t \beta + \underline{v}^t l^t - \bar{v}^t u^t \tag{NDF}$$

is a dual function w.r.t. the LP relaxation at node t that is strong at b. Finally, we can combine these individual dual functions together to obtain

$$\underline{\phi}^{\mathcal{T}}(\beta) = \min_{t \in \mathcal{T}} \underline{\phi}^t_{\mathrm{LP}}(\beta) = \min_{t \in \mathcal{T}} \{v^t \beta + \underline{v}^t l^t - \bar{v}^t u^t\}, \tag{BB-DF}$$

a dual function for the second-stage problem yielded by the tree $\mathcal{T}$ and which is also *strong* at the right-hand side b, i.e., $\underline{\phi}^{\mathcal{T}}(b) = \phi(b)$.

In principle, a stronger dual function can be obtained by replacing the single linear dual function (which is strong at b for the LP relaxation) associated with each subproblem above by its full value function ϕ^t_{LP} to obtain

$$\underline{\phi}^{\mathcal{T}}_{*}(\beta) = \min_{t \in \mathcal{T}} \phi^t_{\mathrm{LP}}(\beta). \tag{BB-DF-BIS}$$

In practice, constructing a complete description of $\underline{\phi}^{\mathcal{T}}_{*}$ is not practical (even evaluating it for a given β requires the solution of $|\mathcal{T}|$ LPs). We can instead construct a function that bounds *it* from below (and hence is also a dual function for the original problem) by exploiting the entire collection of dual solutions arising during the solution process. For example, let

$$\underline{\phi}_{\mathrm{LP}}(\beta, \lambda, \nu) = \max_{t \in \mathcal{T}} \{v^t \beta + \underline{v}^t \lambda - \bar{v}^t \nu\},$$

which consists of an approximation of the full value function ϕ_{LP} using the optimal dual solutions at each leaf node. Replacing $\underline{\phi}^t_{\mathrm{LP}}(\beta)$ with $\underline{\phi}_{\mathrm{LP}}(\beta, l^t, u^t)$ in (BB-DF) results in a potentially stronger but still practical dual function. Of course, it is also possible to add dual solutions found at non-leaf nodes, as well as other suboptimal dual solutions arising during the solution process, but there is an obvious trade-off between strength and tractability. More details on this methodology are contained in [71, 74].

18.5.5.2 Iterative Refinement

In iterative algorithms such as those we introduce in Sect. 18.7, the single dual function (BB-DF) we get by evaluating the value function for one right-hand side can be iteratively augmented by taking the maximum over a sequence of similarly derived dual functions. Taking this basic idea a step further, [70, 110, 111] developed methods of warm starting the solution process of an MILP. Such methods may serve to enhance tractability, though this is still an active area of research. When evaluating ϕ repeatedly for different values in its domain, we do not need to solve each instance from scratch—it is possible to use the tree resulting from a previous solve as a starting point and simply further refine it to obtain a dual function that remains strong for the previous right-hand side of interest and is made to be strong for a new right-hand side.

Hassanzadeh and Ralphs [74] shows how to use this iterative-refinement approach to construct a lower approximation of the value function of an MILP in the context of a Benders-like algorithm for two-stage stochastic optimization within a single branch-and-bound tree. In fact, with enough sampling this method can be used to construct a single tree whose associated dual function is strong at every right-hand side (provided the set of feasible right-hand sides is finite). The following example illustrates the concept of using this iterative refinement approach in approximating the value function of the example on page 535.

Approximating the Value Function

Consider the value function of the example on page 535, reported in Fig. 18.2. The sequence of evaluations of the value function in this example are the ones arising from first-stage solutions generated by solving the master problem in a generalized Benders algorithm, such as the one described in Sect. 18.7.1. Here, we only illustrate the way in which the dual function is iteratively refined in each step.

We first evaluate $\phi(3.5)$ by branch and bound. Figure 18.3 shows both the tree obtained (far left) and the value function itself (in blue). The dual function arises from the solution to the dual of the LP relaxation in each of the nodes in the branch-and-bound tree. We exhibit the values of the dual solution for each node in the tree in Table 18.2. Explicit upper and lower bounds were added with upper bounds initially taking on a large value M, representing infinity. Note that the dual values associated with the bound constraints are actually nothing more than the reduced costs associated with the variables.

The dual function associated with this first branch-and-bound tree is the minimum of the two linear functions shown in Fig. 18.3 in green and labeled as "Node 1" and "Node 2." Formally, this dual function is

$$\underline{\phi}^{\mathcal{T}_1} = \min\{\underline{\phi}^1_{\mathrm{LP}}, \underline{\phi}^2_{\mathrm{LP}}\},$$

(continued)

where the nodal dual functions for the three nodes are

$$\underline{\phi}^0_{\text{LP}}(\beta) = 0.8\beta$$
$$\underline{\phi}^1_{\text{LP}}(\beta) = \beta$$
$$\underline{\phi}^2_{\text{LP}}(\beta) = -1.5\beta + 11.5.$$

In other words, we have $v_0^1 = 1$ (the value of the dual variable associated with the single equality constraint in Node 1), while $\underline{v}^1 l^1 - \bar{v}^1 u^1 = 0$ (this is the contribution from the dual value corresponding to the bound constraints, which we take to be a constant here, as in (NDF)). Similarly, $v_0^2 = -1.5$ and $\underline{v}^1 l^1 - \bar{v}^1 u^1 = 11.5$. The dual function $\underline{\phi}^{\mathcal{T}_1}_{\text{LP}}$ is strong in the interval [0, 5], but yields a weaker lower bound outside this interval. If we subsequently evaluate the right-hand side 9.5, we see that

$$\underline{\phi}^{\mathcal{T}_1}_{\text{LP}}(9.5) = \min\{9.5, -2.75\} = -2.75 \neq \phi(9.5) = 8.5.$$

To obtain a strong dual function for the new right-hand side, we identify that node 2 is the node whose bound needs to be improved by further refining the tree by branching (this is the linear function yielding the bound in this part of the domain). By further partitioning the subproblem associated with node 2, we obtain the tree pictured to the right of the first tree in Fig. 18.3. We obtain the dual function

$$\underline{\phi}^{\mathcal{T}_2}_{\text{LP}} = \min\{\underline{\phi}^1_{\text{LP}}, \underline{\phi}^3_{\text{LP}}, \underline{\phi}^4_{\text{LP}}\},$$

which is strong at the right-hand side 9.5.

Note that this new function is no longer strong at the initial right-hand side of 3.5. To ensure that this single dual function remains strong for all previously evaluated right-hand sides, we must take the max over the collection of dual functions found at each iteration. This function is still obtained from the single tree, but we are effectively strengthening the leaf node dual functions by taking the max over all dual solutions arising on the path from the root subproblem (this is still a bound on the optimal solution value to the LP relaxation). In this case, we get the strengthened function

$$\min\left\{\max\left\{\underline{\phi}^1_{\text{LP}}, \underline{\phi}^0_{\text{LP}}\right\}, \max\left\{\underline{\phi}^3_{\text{LP}}, \underline{\phi}^2_{\text{LP}}, \underline{\phi}^0_{\text{LP}}\right\}, \max\left\{\underline{\phi}^4_{\text{LP}}, \underline{\phi}^2_{\text{LP}}, \underline{\phi}^0_{\text{LP}}\right\}\right\}.$$

This can be seen as an approximation of $\phi^{\mathcal{T}}_*$ by replacing the full value function at each node with an approximation made of just the dual solutions arising on the path to the root node.

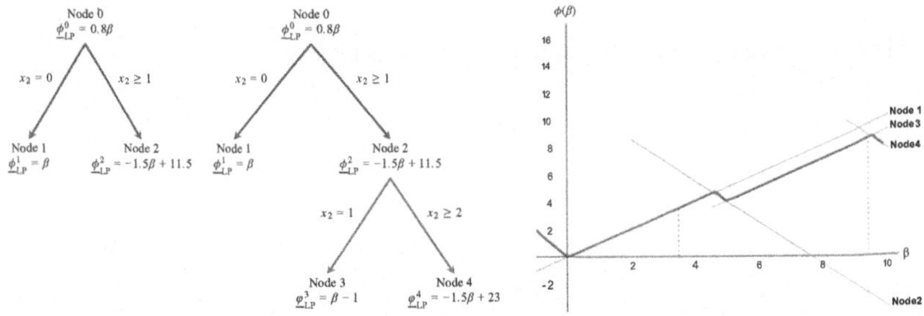

Fig. 18.3 Approximation of the value function of the 2SMILP instance in the example on page 535

Table 18.2 Dual solutions for each node in the branch-and-bound tree

t	v^t	$\underline{v}^t$						$\bar{v}^t$					
0	0.8	4.4	0.0	4.6	5.6	1.0	3.0	0.0	0.0	0.0	0.0	0.0	0.0
1	1.0	4.0	0.0	5.0	6.0	0.0	2.0	0.0	−1.0	0.0	0.0	0.0	0.0
2	−1.5	9.0	11.5	0.0	1.0	12.5	14.5	0.0	0.0	0.0	0.0	0.0	0.0

18.5.5.3 Primal Functions

Upper approximations of ϕ can be obtained by considering the value functions of the second-stage problem (SS) obtained by fixing variables. For example, consider the *integer-fixing value function*

$$\bar{\phi}_{\hat{y}}(\beta) = d_I^2 \hat{y}_I + \phi_C(\beta - A^2_{\omega I} \hat{y}_I) \qquad \text{(IFVF)}$$

obtained by fixing the integer part $\hat{y}_I \in \mathbb{Z}^{r_2}$ to be equal to that from some previously found second-stage solution $\hat{y} \in Y$, where ϕ_C is as defined in (CRVF). Then, we have $\bar{\phi}_{\hat{y}_I}(\beta) \geq \phi(\beta)$ for all $\beta \in \mathbb{R}^{m_2}$. If $\hat{y}$ is an optimal solution to the second-stage problem with respect to a given first-stage solution $x \in \mathcal{P}_1 \cap X$ and a given realized value of ω, then we have $\hat{y} \in \operatorname{argmin}_{y \in \mathcal{P}_2(b^2_\omega - A^2_\omega x) \cap Y} d^2 y$ and $\bar{\phi}_{\hat{y}}$ is strong at $\beta = b^2_\omega - A^2_\omega x$.

In a fashion similar to a cutting plane method, we can iteratively improve the global upper bounding function by taking the minimum of all bounding functions of the form (IFVF) found so far, i.e.,

$$\bar{\phi}(\beta) = \min_{y \in \mathcal{R}} \bar{\phi}_y(\beta),$$

where $\mathcal{R}$ is the set of all second-stage solutions that have been found when evaluating $\phi(\beta)$ for different values of β.

A pictorial example of this type of upper bounding function is shown in Fig. 18.4, where each of the labeled cones shown is the value function of a restriction of the original MILP. The upper bounding function is the minimum over all of these cones.

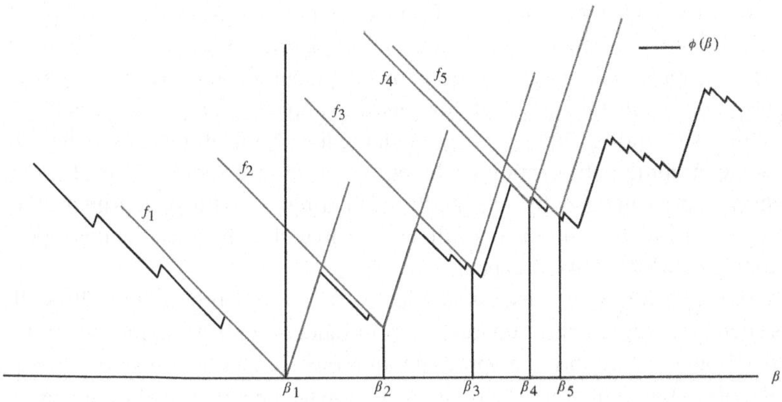

Fig. 18.4 Upper bounding functions obtained at right-hand sides β_i, $i = 1, \ldots, 5$

18.5.6 Reaction and Risk Functions

Because it is particularly relevant in the present context, we also now introduce a function known as the *second-stage (optimistic) reaction function*. This function is closely related to the risk function but its input is a second stage right-hand side, rather than a first-stage solution. Although this function, like the second-stage value function ϕ, takes a right-hand side β as its input, it can nevertheless be used to evaluate a first-stage solution $x \in X$ in scenario $\omega \in \Omega$ by evaluating it with respect to $\beta_\omega(x)$. The function is defined as

$$\rho(\beta) = \inf\left\{d^1 y \mid y \in \operatorname{argmin}\{d^2 y \mid y \in \mathcal{P}_2(\beta) \cap Y\}\right\}, \tag{ReF}$$

for $\beta \in \mathbb{R}^{m_2}$. Note that, although the evaluation of this function appears to require solving a bilevel optimization problem, its evaluation is actually equivalent to solving a lexicographic optimization problem, a somewhat easier computational task.

The importance of the function ρ is to enable us to see that, for (2SMILP), the scenario risk functions Ξ_ω defined in (2SRF) are not in fact completely independent functions but, rather, are connected, since

$$\Xi_\omega(x) = \rho(\beta_\omega(x)).$$

Thus, these functions only differ from each other in the affine map $\beta_\omega(x) = b^2_\omega - A^2_\omega x$ that is applied to x.

The structure of the functions ρ and Ξ derives from that of ϕ and can be understood through a somewhat more involved application of the same principles used to derive the function ϕ. Their structure, though combinatorially much more complex, is nevertheless also piecewise polyhedral. Approximations of Ξ can be derived easily from approximation of ρ in a straightforward way, since $\Xi = \mathbb{E}[\Xi_\omega]$ and the scenario risk functions are themselves defined in terms of ρ, as discussed earlier. The approximation of ρ is quite involved, but it can be obtained by methods that are natural generalizations of those used to approximate ϕ. The main challenge is that the evaluation of $\rho(\beta)$ for a particular value of β itself reduces to the solution of a lexicographic optimization problem, which in turn requires knowledge of ϕ. We may approximate ρ by repeatedly evaluating it, extracting primal and dual information from the solution algorithm as we do with ϕ, but this requires repeatedly evaluating ϕ, which is itself expensive.

In the algorithms we discuss in Sect. 18.7, we approach this difficulty by constructing a single approximation of ϕ in tandem with the approximation of ρ. We need only ensure that the approximation of ϕ is guaranteed to be strong (i.e., equal to the value function) exactly in the region needed to properly evaluate ρ. The result is an approximation of ρ that is strong in the region of a given right-hand side and this is exactly what is needed for a Benders-type algorithm to converge. More details regarding the Benders-type algorithm are contained next in Sect. 18.7. Further details on the structure of and methods for approximating ρ and Ξ can be found in [23], which describes a Benders-type algorithm for solving (2SMILP).

18.5.7 Optimality Conditions

To solidify the connection between the notion of duality described in this section and captured in the dual problem (MILPD), we end this section by formally stating both the weak and strong duality properties arising from this theory. These properties are a proper generalization of the well-known ones from linear optimization and can be used to derive optimality conditions that generalize those from LP duality. These are, in turn, the conditions that can be used to derive the reformulations presented in Sect. 18.6.

Theorem 18.5.4 (*Weak Duality*) *If* $F \in \Upsilon^{m_2}$ *is feasible for* (MILPD) *and* $y \in \mathcal{P}_2(b) \cap Y$, *then* $F(b) \leq d^2 y$. △

Theorem 18.5.5 (*Strong Duality*) *Let* $b \in \mathbb{R}^{m_2}$ *be such that* $\phi(b) < \infty$. *Then, there exists both* $F \in \Upsilon^{m_2}$ *that is feasible for* (MILPD) *and* $y^* \in \mathcal{P}_2(b) \cap Y$ *such that* $F(b) = \phi(b) = d^2 y^*$. △

The form of the dual (MILPD) makes these properties rather self-evident, but Theorem 18.5.5 nevertheless yields optimality conditions that are useful in practice. In particular, the dual functions arising from branch-and-bound algorithms that were described earlier in Sect. 18.5.5.1 are the strong dual functions that provide the optimality certificates for the solutions produced by such algorithms and are the basis on which the algorithms described in Sect. 18.7.1 are developed.

18.6 Reformulations

A crucial step in deriving solution algorithms is to consider some conceptual reformulations that underlie the problems under study, each of which suggests a particular algorithmic strategy. These formulations are heavily based on the duality theory and methodology in the previous section, as we better clarify in the following. In all cases, the goal is to derive, from some initial problem description, a formulation that is, at least in principle, a single-stage mathematical optimization problem that can be tackled with (possibly generalized versions of) the standard algorithmic approaches used for solving mathematical optimization problems. In this case, as in the solution of single-stage MILPs, the main tools are cutting plane methods (branch and cut) and decomposition methods (Benders' algorithm and the exploitation of the block structure using a Dantzig-Wolfe decomposition).

18.6.1 Value-Function (Optimality-Based) Reformulation

The first reformulation we describe is a variation on (2SMILP-Alt), the standard formulation used in most of the bilevel optimization literature. This formulation introduces the second-stage variables explicitly and formulates the problem in the form of a classical mathematical optimization problem using a technique that is standard—replacing the requirement that the solution to the second-stage problem be optimal with explicit optimality conditions. To achieve this, we introduce a constraint involving the value function ϕ, as well as the second-stage feasibility conditions. This is roughly equivalent to imposing primal and dual feasibility along with equality of the primal and dual objectives in the linear optimization case (the constraint involving the value function must be satisfied at equality, though it is stated as an inequality). The formulation is as follows.

$$
\begin{aligned}
\min \quad & cx + \sum_{\omega \in \Omega} p_\omega d^1 y^\omega \\
\text{s.t.} \quad & A^1 x \geq b^1 && && \text{(VFRa)} \\
& G^2 y^\omega \geq b^2_\omega - A^2_\omega x && \forall \omega \in \Omega && \text{(VFRb)} \\
& d^2 y^\omega \leq \phi(b^2_\omega - A^2_\omega x) && \forall \omega \in \Omega && \text{(VFRc)} \\
& x \in X && && \text{(VFRd)} \\
& y^\omega \in Y && \forall \omega \in \Omega. && \text{(VFRe)}
\end{aligned}
$$

It is clear that this formulation cannot be constructed explicitly, but rather, must be solved by the iterative approximation of the constraints involving the value function (which we refer to as the second-stage optimality constraints). This reformulation suggests a family of methods described in Sect. 18.7 in which we replace ϕ with a primal function $\bar{\phi}$, as defined in Definition 18.5.2.

Notice that, when $r_2 = 0$, so that the second-stage problem (SS) is a linear optimization problem, we can exploit the fact that the optimality conditions for this problem involve linear functions. This allows for, in essence, substituting for ϕ the objective function of the classical LP dual of (SS), after introducing the corresponding variables and constraints. This, overall, leads to a tractable *primal-dual* reformulation—the technique is applied, for instance, in [40]. The alternative idea of, rather than the dual of (SS), introducing its KKT conditions, is arguably more popular and has been often exploited in a number of "classical" works on mixed integer bilevel optimization problems, including, among others [93]. Note, however, that while there is an analog of this reformulation that applies in the setting of (2SMILP) (see [52]), it has so far proved not to be practical and, therefore, we will not present any algorithms for its solution in Sect. 18.7.

18.6.2 Risk-Function (Projection-Based) Reformulation

The next reformulation we consider exploits the finiteness of Ω and avoids introducing the second-stage variables explicitly. It reads as follows.

$$
\begin{aligned}
\min \quad & c^1 x + \sum_{\omega \in \Omega} p_\omega z^\omega \\
\text{s.t.} \quad & z^\omega \geq \Xi_\omega(x) && \omega \in \Omega \\
& x \in \mathcal{P}_1 \cap X \\
& z^\omega \in \mathbb{R} && \omega \in \Omega.
\end{aligned}
\tag{RFR}
$$

This reformulation mirrors the original formulation implicitly adopted when we first defined (2SMILP), in which the second-stage variables are not (explicitly) present. However, we can also interpret it as a projection onto the X-space of the value-function reformulation described in the previous section. In fact, it is not hard to see that the set $\{x \in \mathcal{P}_1 \mid \Xi(x) < \infty\}$ is exactly $\mathcal{F}^1$ (the projection of the feasible region onto the space of the first-stage variables) as defined in (FS-FR). As such, this formulation is a natural basis for a Benders-type algorithm that we describe in Sect. 18.7.1, in which we replace Ξ with an under-estimator to obtain a *master problem* which is then iteratively improved until convergence.

18.6.3 Polyhedral (Convexification-Based) Reformulation

An apparently unrelated reformulation generalizes the notion of *convexification* used heavily in the polyhedral theory that underlies the solution methodology of standard MILPs. Convexification considers the following conceptual reformulation:

$$\begin{aligned} \min \quad & cx + \sum_{\omega \in \Omega} p_\omega d^1 y^\omega \\ \text{s.t.} \quad & (x, y^\omega) \in \mathrm{conv}(\mathcal{F}^\omega) \quad \forall \omega \in \Omega, \end{aligned} \tag{POLY-R}$$

where $\mathcal{F}^\omega$ is the feasible region under scenario ω, defined as in (FR). Under our earlier assumptions, the convex hull of $\mathcal{F}^\omega$ is a polyhedron whose extreme points are members of $\mathcal{F}^\omega$. Thus, due to the linearity of the objective function, we can w.l.o.g. replace the requirement that $(x, y^\omega) \in \mathcal{F}^\omega$ with the requirement that $(x, y^\omega) \in \mathrm{conv}(\mathcal{F}^\omega)$, thereby convexifying the feasible region.

With this reformulation, we have transformed the requirement that the second-stage solution be optimal for the parameterized second-stage problem (SS) into a requirement that the combination of first- and second-stage solutions lie in a polyhedral feasible region. This reformulation suggests a different class of algorithms based on the dynamic generation of valid inequalities, such as those so successfully employed in the case of MILPs. We describe an algorithm of such class in Sect. 18.7.2.

18.6.4 Deterministic Equivalent (Decomposition-Based) Reformulation

Finally, we remark that the finiteness of Ω allows for solving the problem via a block-angular reformulation based on the formulation (2SMILP-Alt) presented earlier, which is in the spirit of the so-called deterministic equivalent reformulation used in two-stage stochastic optimization. This renders the stochastic problem as a deterministic MIBLP, which can then be solved via standard methods for that case with the requisite further reformulations (of course, exploiting the block structure of the resulting matrices). For details, see [126].

18.7 Algorithmic Approaches

We now summarize a range of methodologies that arise naturally from the reformulations of the previous section. Any practical method of solving (2SMILP) must have as a fundamental step the evaluation of $\phi(\beta)$ for certain fixed values of $\beta \in \mathbb{R}^{m_2}$, an operation which can be challenging in itself, since the corresponding problem is NP-hard. From the evaluation of ϕ, both primal and dual information is obtained, which can be used to build approximations. While some methods explicitly build such approximations, other methods do it only implicitly. In all cases, information about the value function that is built up through repeated evaluations can be exploited.

Similarly, in the dual methods that we describe below, the function Ξ is also evaluated for various values of $x \in X$ (or rather the function ρ) and, similarly, approximations of this function can be built from primal and dual information obtained during its evaluation. In order to develop computationally tractable methods, a key aspect is to limit the number of such function evaluations as much as possible and to exploit to the maximum extent possible the information generated as a by-product of these single function evaluations.

18.7.1 Decomposition Methods

Decomposition methods are based on generalizations of Benders' original method of decomposing a given mathematical optimization problem by splitting the variables into two subsets and forming a master problem by projecting out one subset. More concretely, we are simply solving a reformulation of the form (RFR).

18.7.1.1 Continuous Problems

For illustration, we consider the simplest case in which we have only continuous variables in both stages ($r_1 = r_2 = 0$) and $d^1 = d^2$. Since the first- and second-stage objectives are the same in this case, the full problem is nothing more than a standard linear optimization problem, but Benders' approach nevertheless applies when either fixing the first stage variables results in a more tractable second-stage LP (such as a min-cost flow problem). In the Benders approach, we (conceptually) rewrite the LP as

$$\min \left\{ cx + \sum_{\omega \in \Omega} p_\omega \phi(\beta^\omega(x)) \;\middle|\; x \in \mathcal{P}_1 \right\},$$

where ϕ is the value function (2SVF). Note that, because $\Xi(x) = \sum_{\omega \in \Omega} p_\omega \phi(\beta^\omega(x))$, this is just a simplification of the original formulation implicitly adopted when we first defined (2SMILP). As we observed earlier, the value function in the LP case is the maximum of linear functions associated with the dual solutions. Recalling that we can restrict the description to only the extreme points of the dual feasible region, we can further rewrite the LP as

$$\min \left\{ cx + \sum_{\omega \in \Omega} p_\omega z^\omega \;\middle|\; \begin{array}{ll} x \in \mathcal{P}_1 & \\ z^\omega \geq u\beta^\omega(x), \; u \in \mathcal{E}, \omega \in \Omega \\ z^\omega \in \mathbb{R}, & \omega \in \Omega \end{array} \right\}, \tag{LP}$$

where $\mathcal{E}$ is the set of such extreme points of the dual of the second-stage LP (which we assumed to be bounded and nonempty). Thus, the linear constraints involving the variable z^{ω} (along with the fact that z^{ω} is minimized) are precisely equivalent to requiring $z^{\omega} = \phi(b_{\omega}^2 - A_{\omega}^2 x)$, so this reformulation is exactly the formulation (RFR) specialized to this case.

A straightforward solution approach is then to solve (LP) by a cutting plane algorithm, which results in the classical L-shaped method for solving (continuous) stochastic linear optimization problems [130]. From the point of view we have taken in this article, this method can be interpreted as one that approximates the value function from below as the maximum of the strong dual functions generated in each iteration. The strong dual functions arise from the solutions to the dual of the second-stage problem and yield what are typically called *Benders cuts* (inequalities of the form imposed in (LP)). The Benders approach is then to iteratively improve this approximation until convergence.

The case $d^1 \neq d^2$ is more complex. The epigraph of the value function of the second-stage problem is no longer necessarily a polyhedral cone, and the function itself is no longer necessarily convex. Formulating the equivalent of (LP) thus requires integer variables. Alternative formulations using the related complementary slackness optimality conditions are also commonly used (see [37]).

18.7.1.2 Discrete Problems

For the case in which there are integer variables, the approach just described can be applied by simply replacing the linear strong dual functions (Benders' cuts) with strong under-estimators of the risk function constructed from dual functions arising from solution algorithms for the second-stage problem, such as those based on branch and bound described in Sect. 18.5.5.1. In this approach, we work directly with the reformulation (RFR), employing the generalized Benders-type algorithm summarized in Fig. 18.5 and a master problem defined as follows.

$$
\begin{aligned}
\min \quad & c^1 x + \sum_{\omega \in \Omega} p_{\omega} z^{\omega} \\
\text{s.t.} \quad & z^{\omega} \geq \Xi_{\omega}(x) && \omega \in \Omega \\
& x \in \mathcal{P}_1 \\
& z^{\omega} \in \mathbb{R} && \omega \in \Omega.
\end{aligned}
\qquad \text{(MASTER)}
$$

When $d^1 = d^2$, the approximation of the scenario risk function and of the risk function itself reduces to the direct approximation of the second-stage value function, and the algorithm can be described rather succinctly. A basic version was originally proposed as the integer L-shaped algorithm for two-stage stochastic optimization problems with integer recourse by Laporte and Louveaux [94] and Carøe and Tind [29]. The version based on dual functions from branch and bound that we describe here is described in [74].

Step 0. Initialize $k \leftarrow 1$, $\underline{\Xi}^0_\omega(x) = -\infty$ for all $x \in \mathbb{Q}^{n_1}$, $\omega \in \Omega$.

Step 1. Solve the Master Problem

a) Set $\underline{\Xi}_\omega = \max_{i=0,\ldots,k-1} \underline{\Xi}^i_\omega$ for $\omega \in \Omega$.
b) Solve (MASTER) to obtain an optimal solution $(x^k, \{z^\omega_k\}_{\omega\in\Omega})$.

Step 2. Solve the Subproblem

a) Evaluate $\Xi_\omega(x^k)$ to obtain an optimal solution $y^{\omega,k}$ for $\omega \in \Omega$ and the strong under-estimator $\underline{\Xi}^k_\omega$.
b) Termination check: Is $z^\omega_k = d^1 y^{\omega,k}$ for $\omega \in \Omega$?

1. If yes, STOP. x^k is an optimal solution to (RFR).
2. If no, set $k \leftarrow k+1$ and go to Step 1.

Fig. 18.5 Generalized Benders algorithm for solving 2SMILPs

To briefly summarize, as in the LP case, we rewrite (2SMILP) as

$$\min \left\{ cx + \sum_{\omega\in\Omega} p_\omega z^\omega \;\middle|\; \begin{array}{ll} x \in \mathcal{P}_1 \cap X & \\ z^\omega \geq \phi(\beta^\omega(x)) & \omega \in \Omega \\ z^\omega \in \mathbb{R} & \omega \in \Omega \end{array} \right\}.$$

By replacing ϕ with the maximum of a set $\mathcal{G}^\omega$ of dual functions associated with scenario $\omega \in \Omega$ (alternatively, we can employ one universal set of dual functions, as indicated in (LP) above), we obtain a convergent Benders-like algorithm based on iteratively solving a master problem of the form

$$\min \left\{ cx + \sum_{\omega\in\Omega} p_\omega z^\omega \;\middle|\; \begin{array}{ll} x \in \mathcal{P}_1 \cap X & \\ z^\omega \geq F(\beta^\omega(x)) & F \in \mathcal{G}^\omega, \omega \in \Omega \\ z^\omega \in \mathbb{R} & \omega \in \Omega \end{array} \right\},$$

which generalizes (LP). The key to making this approach work in practice is that the dual functions we need be easily available as a by-product of evaluating the second-stage value function ϕ for a fixed value of β.

The most general case in which $d^1 \neq d^2$ is conceptually no more complex than that described above, but the details of the methodology for approximating Ξ and in linearizing the master problem are quite involved. The reader is referred to [23] for the details.

18.7.2 Convexification-Based Methods

Primal algorithms are based on the implicit solution of (POLY-R) and generalize the well-known framework of branch and cut that has been so successful in the MILP case. This class of algorithms is based on the iterative approximation of conv($\mathcal{F}^\omega$) beginning with the approximation $\mathcal{P}^\omega$, the feasible region in scenario ω of the following relaxation obtained by dropping both the value-function constraint (VFRc) and the integrality requirements (VFRd) and (VFRe) from (VFR).

$$\begin{aligned}
\min \quad & cx + \sum_{\omega \in \Omega} p_\omega d^1 y^\omega && \\
\text{s.t.} \quad & A^1 x \geq b^1 && \text{(LPRa)}\\
& G^2 y^\omega \geq b^2_\omega - A^2_\omega x \qquad \forall \omega \in \Omega && \text{(LPRb)}\\
& x \in \mathbb{R}^{n_1}_+ && \text{(LPRc)}\\
& y^\omega \in \mathbb{R}^{n_2}_+ \qquad \forall \omega \in \Omega. && \text{(LPRd)}
\end{aligned}$$

Being an LP, this relaxation is easily solved, but it is not difficult to see, however, that it is rather weak (see, e.g., the example on page 552). A straightforward way to strengthen it is simply by including the integrality constraints (VFRd) and (VFRe) from (VFR). This leads to an MILP relaxation with feasible set

$$\mathcal{S}^\omega = \mathcal{P}^\omega \cap (X \times Y)$$

in scenario $\omega \in \Omega$, which, while stronger, is clearly more difficult to solve and also still potentially weak—whether adopting it is a computationally good idea is a purely empirical question. When the number of scenarios is large, dropping constraints (LPRb) from the relaxation may also be advantageous, since this may reduce the size of the relaxation.

As in cutting plane methods for MILPs, the idea is to improve this initial formulation with the addition of inequalities valid for $\mathcal{S}^\omega$, $\mathcal{F}^\omega$, $\bigcup_{\omega\in\Omega} \mathcal{F}^\omega$, or even $\mathcal{F}^1$. In some cases, inequalities may first be derived with respect to $\mathcal{S}^\omega$ or $\mathcal{F}^\omega$ for some particular scenario $\omega \in \Omega$ and then lifted to become valid for a larger set. Inequalities valid for $\mathcal{S}^\omega$ (which can be referred to as *feasibility cuts*) can be generated using any number of well-known procedures associated with cutting plane algorithms for mixed integer linear programming. Inequalities valid for $\mathcal{F}^\omega$ (which can be referred to as *optimality cuts*) are the more interesting case because they can incorporate information about the value function in order to eliminate members of $\mathcal{P}^\omega$ that are not two-stage feasible.

In early work on these methods, the authors of [53] developed inequalities valid for $\mathcal{F}^\omega$ in the case for which Ω is a singleton and the variables must all be integer ($r_1 = n_1$ and $r_2 = n_2$), which illustrate the basic principles. When the input data are integer, a very simple argument can be used to generate an inequality valid for $\mathcal{F}^\omega$ but violated by $(\hat{x}, \hat{y})$, an extreme point of $\mathcal{P}^\omega$ not in $\mathcal{F}^\omega$, by taking advantage of the discrete nature of the feasible set. Assuming the solution of the LP relaxation is an extreme point of $\mathcal{P}^\omega$, there is thus a hyperplane supporting $\mathcal{P}^\omega$ and inducing a face of dimension 0. As such, there exist $f \in \mathbb{R}^{n_1}$, $g \in \mathbb{R}^{n_2}$, and $\gamma \in \mathbb{R}$ such that the hyperplane $\{(x, y) \in \mathbb{R}^{n_1+n_2} \mid fx + gy = \gamma\}$ intersects $\mathcal{P}^\omega$ in the unique point $(\hat{x}, \hat{y})$. Hence, we have that $fx + gy \leq \gamma$ for all $(x, y) \in \mathcal{P}^\omega$. Finally, since the face of $\mathcal{P}^\omega$ induced by this inequality does not contain any other members of $\mathcal{S}^\omega$, we can "push" the hyperplane until it meets the next integer point without separating any additional members of $\mathcal{F}^\omega$. Hence,

$$fx + gy \leq \gamma - 1$$

is valid for $\mathcal{F}^\omega$. This procedure is similar in spirit to the Gomory procedure for standard MILPs. It is used, for instance, in [50]. We next describe the method with a brief example.

Example of Valid Inequalities
Consider the instance

$$\max_x \min_y \{y \mid -x + y \leq 2, -2x - y \leq -2, 3x - y \leq 3, y \leq 3, x, y \in \mathbb{Z}_+\},$$

with $|\Omega| = 1$, whose feasible region is the set $\mathcal{F} = \{(0, 2), (1, 0), (2, 3)\}$ shown in Fig. 18.6. Solving the LP relaxation yields the point (1, 3), which is not feasible. This point is eliminated by the addition of the inequality $x - 2y \geq -4$, which is valid for the feasible region $\mathcal{F}$ and is obtained as a strengthening of the inequality $x - 2y \geq -5$, which is valid for the LP relaxation itself.

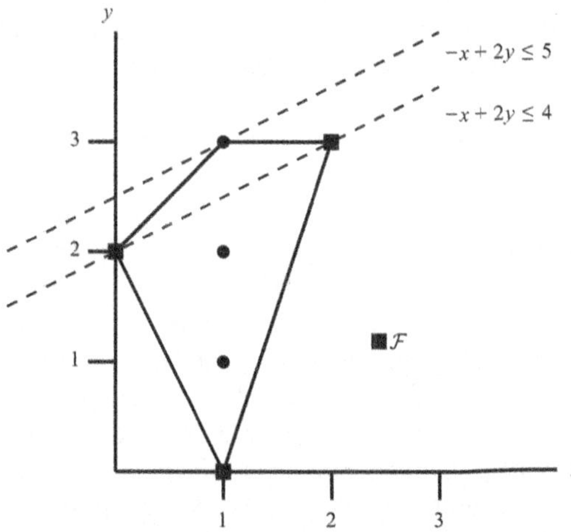

Fig. 18.6 Example of a valid inequality

This cut generation procedure is enough to yield a converging algorithm in this case, but it amounts to removing infeasible points one by one and is not scalable in general. An important observation is that this cut only exploits integrality of the solutions and does not take into account any information about the second-stage value function.

A generalized version of this basic methodology has since been described in [127] and enhanced with additional classes of inequalities, including those valid for the general mixed integer case described in [57, 58]. Inequalities valid for more general discrete probability spaces are derived in [60] for the case $d^1 = d^2$.

Stronger cuts can be obtained by using disjunctive arguments based on knowledge of the value function. In particular, an option is to add constraints of the form

$$d^2 y^\omega \leq \bar{\phi}(b^2_\omega - A^2_\omega x),$$

where $\bar{\phi}$ is a primal function, as defined in Definition 18.5.2. Such primal functions can take many forms and imposing such constraints may be expensive. In general, the form of such functions will be either affine or piecewise polyhedral ("standard" disjunctive programming techniques can be used to obtain a reformulation which only involves linear functions).

18.8 Conclusions

We have introduced a unified framework for multistage mixed integer linear optimization problems which encompasses both multilevel mixed integer linear optimization problems and multistage mixed integer linear optimization problems with recourse. Focusing on the two-stage case, we have investigated the nature of the value function of the second-stage problem and highlighted its connection to dual functions and the theory of duality for mixed integer linear optimization problems. We have summarized different reformulations for this broad class of problems, which rely on either the risk function, the value function, or on their polyhedral nature. We have then presented the main solution techniques for problems of this class, considering both dual- and primal-type methods, the former based on a Benders-like decomposition to approximate either the risk function or the value function, and the latter based on a cutting plane technique that relies on the polyhedral nature of these problems. While much work is still to be done for solving multistage mixed integer linear optimization problems with techniques that are (*mutatis mutandis*, given their intrinsic harder nature) of comparable efficiency to those for solving single-level problems, we believe that the theoretical understanding of multistage mixed integer linear problems is now sufficiently mature to make this an achievable objective.

References

1. E. Amaldi, A. Capone, S. Coniglio, L.G. Gianoli, Energy-aware traffic engineering with elastic demands and MMF bandwidth allocation, in *Proc of 18th IEEE Int. Workshop on Computer Aided Modeling and Design of Communication Links and Networks (CAMAD 2013)* (IEEE, Piscataway, 2013), pp. 169–174
2. E. Amaldi, A. Capone, S. Coniglio, L.G. Gianoli, Network optimization problems subject to max-min fair flow allocation. IEEE Commun. Lett. **17**(7), 1463–1466 (2013)
3. E. Amaldi, S. Coniglio, L.G. Gianoli, C.U. Ileri, On single-path network routing subject to max-min fair flow allocation. Electron Notes Discrete Math. **41**, 543–550 (2013)
4. E. Amaldi, S. Coniglio, L. Taccari, Maximum throughput network routing subject to fair flow allocation, in *Proc. of Int. Symp. on Combinatorial Optimization (ISCO 2014)* (Springer, Berlin, 2014), pp. 1–12
5. M.A. Amouzegar, K. Moshirvaziri, Determining optimal pollution control policies: an application of bilevel programming. Eur. J. Oper. Res. **119**(1), 100–120 (1999)
6. B. An, J. Pita, E. Shieh, M. Tambe, C. Kiekintveld, J. Marecki, Guards and protect: next generation applications of security games. ACM SIGecom Exch. **10**(1), 31–34 (2011)
7. R. Avenhaus, A. Okada, S. Zamir, Inspector leadership with incomplete information, in *Game Equilibrium Models IV* (Springer, Berlin, 1991), pp. 319–361
8. J. Bard, An algorithm for solving the general bilevel programming problem. Math. Oper. Res. **8**(2), 260–272 (1983)
9. J. Bard, J.T. Moore, An algorithm for the discrete bilevel programming problem. Nav. Res. Logist. **39**(3), 419–435 (1992)
10. J.F. Bard, J. Plummer, J.C. Sourie, A bilevel programming approach to determining tax credits for biofuel production. Eur. J. Oper. Res. **120**, 30–46 (2000)
11. L. Baringo, A.J. Conejo, Transmission and wind power investment. IEEE Trans. Power Syst. **27**(2), 885–893 (2012)
12. N. Basilico, S. Coniglio, N. Gatti, Methods for finding leader-follower equilibria with multiple followers (extended abstract), in *Proc. of 2016 Int. Conf. on Autonomous Agents and Multiagent Systems (AAMAS 2016)* (2016), pp. 1363–1364
13. N. Basilico, S. Coniglio, N. Gatti, A. Marchesi, Bilevel programming approaches to the computation of optimistic and pessimistic single-leader-multi-follower equilibria, in *Proc. of 16th Int. Symp. on Experimental Algorithms (SEA 2017)* (Schloss Dagstuhl-Leibniz-Zentrum fuer Informatik, Wadern, 2017)
14. N. Basilico, S. Coniglio, N. Gatti, A. Marchesi, Bilevel programming methods for computing single-leader-multi-follower equilibria in normal-form and polymatrix games. EURO J. Comput. Optim. **8**, 3–31 (2020)
15. O. Ben-Ayed, C. Blair, Computational difficulties of bilevel linear programming. Oper. Res. **38**, 556–560 (1990)
16. O. Ben-Ayed, C. Blair, D. Boyce, L. LeBlanc, Construction of a real-world bilevel linear programming model of the highway network design problem. Ann. Oper. Res. **34**(1), 219–254 (1992)
17. D.P. Bertsekas, *Dynamic Programming and Optimal Control* (Athena Scientific, Belmont, 2017)
18. J.R. Birge, F. Louveaux, *Introduction to Stochastic Programming* (Springer Science & Business Media, New York, 2011)
19. C.E. Blair, A closed-form representation of mixed-integer program value functions. Math. Program. **71**(2), 127–136 (1995)
20. C.E. Blair, R.G. Jeroslow, The value function of a mixed integer program: I. Discret. Math. **19**(2), 121–138 (1977)
21. C.E. Blair, R.G. Jeroslow, The value function of a mixed integer program: II. Discret. Math. **25**(1), 7–19 (1979)

22. C.E. Blair, R.G. Jeroslow, Constructive characterizations of the value-function of a mixed-integer program I. Discret. Appl. Math. **9**(3), 217–233 (1984)
23. S. Bolusani, T.K. Ralphs, A framework for generalized Benders' decomposition and its application to multilevel optimization. Technical report 20T-004, COR@L Laboratory, Lehigh University, 2020. http://coral.ie.lehigh.edu/~ted/files/papers/MultilevelBenders20.pdf
24. J. Bracken, J.T. McGill, Mathematical programs with optimization problems in the constraints. Oper. Res. **21**(1), 37–44 (1973)
25. A.P. Burgard, P. Pharkya, C.D. Maranas, OptKnock: a bilevel programming framework for identifying gene knockout strategies for microbial strain optimization. Biotechnol. Bioeng. **84**, 647–657 (2003)
26. P. Calamai, L. Vicente, Generating quadratic bilevel programming problems. ACM Trans. Math. Softw. **20**, 103–119 (1994)
27. A. Caprara, M. Carvalho, A. Lodi, G. Woeginger, Bilevel knapsack with interdiction constraints. INFORMS J. Comput. **28**(2), 319–333 (2016)
28. M. Caramia, R. Mari, Enhanced exact algorithms for discrete bilevel linear problems. Optim. Lett. **9**(7), 1447–1468 (2015)
29. C.C. Carøe, J. Tind, L-shaped decomposition of two-stage stochastic programs with integer recourse. Math. Program. **83**(1), 451–464 (1998)
30. M. Castiglioni, A. Marchesi, N. Gatti, Be a leader or become a follower: the strategy to commit to with multiple leaders, in *Proc. of 28th Int. Joint Conf. on Artificial Intelligence (IJCAI 2019)* (2019)
31. M. Castiglioni, A. Marchesi, N. Gatti, S. Coniglio, Leadership in singleton congestion games: what is hard and what is easy. Artif. Intell. **277**, 103–177 (2019)
32. A. Celli, S. Coniglio, N. Gatti, Computing optimal ex ante correlated equilibria in two-player sequential games, in *Proc. of 18th Int. Conf. on Autonomous Agents and MultiAgent Systems (AAMAS 2019)* (International Foundation for Autonomous Agents and Multiagent Systems, Richland, 2019), pp. 909–917
33. A. Celli, S. Coniglio, N. Gatti, Private Bayesian persuasion with sequential games, in *Proc. of 34th AAAI Conf. on Artificial Intelligence (AAAI 2020)* (AAAI Press, New York, 2020), pp. 1–8
34. R.L. Church, M.P. Scaparra, Protecting critical assets: the r-interdiction median problem with fortification. Geogr. Anal. **39**(2), 129–146 (2006)
35. R.L. Church, M.P. Scaparra, R.S. Middleton, Identifying critical infrastructure: the median and covering facility interdiction problems. Ann. Assoc. Am. Geogr. **94**(3), 491–502 (2004)
36. P.A. Clark, A.W. Westerberg, Bilevel programming for steady-state chemical process design I. Fundamentals and algorithms. Comput. Chem. Eng. **14**(1), 87–97 (1990)
37. B. Colson, P. Marcotte, G. Savard, Bilevel programming: a survey. 4OR **3**(2), 87–107 (2005)
38. S. Coniglio, S. Gualandi, On the separation of topology-free rank inequalities for the max stable set problem, in *Proc. of 16th Int. Symp. on Experimental Algorithms (SEA 2017)* (Schloss Dagstuhl-Leibniz-Zentrum fuer Informatik, Wadern, 2017)
39. S. Coniglio, L.G. Gianoli, E. Amaldi, A. Capone, Elastic Traffic Engineering Subject to a Fair Bandwidth Allocation via Bilevel Programming, in *IEEE/ACM Transactions on Networking*, https://doi.org/10.1109/TNET.2020.3007572
40. S. Coniglio, M. Tieves, On the generation of cutting planes which maximize the bound improvement, in *Proc. of 14th Int. Symp. on Experimental Algorithms (SEA 2015)* (Springer, Berlin, 2015), pp. 97–109
41. S. Coniglio, N. Gatti, A. Marchesi, Pessimistic leader-follower equilibria with multiple followers, in *Proc. of 26th Int. Joint Conf. on Artificial Intelligence (IJCAI 2017)* (AAAI Press, New York, 2017), pp. 171–177
42. S. Coniglio, N. Gatti, A. Marchesi, Computing a pessimistic Stackelberg equilibrium with multiple followers: the mixed-pure case. Algorithmica **82**, 1189–1238 (2020)
43. S. Coniglio, M. Sirvent, M. Weibelzahl, Airport capacity extension, fleet investment, and optimal aircraft scheduling in a multi-level market model: quantifying the costs of imperfect markets (2020, under review). http://www.optimization-online.org/DB_HTML/2017/05/5989.html

44. V. Conitzer, D. Korzhyk, Commitment to correlated strategies, in *Proc. of 25th AAAI Conf. on Artificial Intelligence (AAAI 2011)* (2011), pp. 632–637
45. V. Conitzer, T. Sandholm, Computing the optimal strategy to commit to, in *Proc. of 7th ACM Conf. on Electronic Commerce (EC 2006)* (2006), pp. 82–90
46. S.A. Cook, The complexity of theorem-proving procedures, in *Proc. of 3rd Annual ACM Symposium on Theory of Computing* (ACM, New York, 1971), pp. 151–158
47. K.J. Cormican, D.P. Morton, R.K. Wood, Stochastic network interdiction. Oper. Res. **46**(2), 184–197 (1998)
48. J.-P. Côté, P Marcotte, G. Savard, A bilevel modelling approach to pricing and fare optimisation in the airline industry. J. Revenue Pricing Manag. **2**(1), 23–36 (2003)
49. S. Dempe, Discrete bilevel optimization problems. Technical Report D-04109, Institut fur Wirtschaftsinformatik, Universitat Leipzig, Leipzig, 2001
50. S. Dempe, F. Mefo Kue, Solving discrete linear bilevel optimization problems using the optimal value reformulation. J. Glob. Optim. **68**(2), 255–277 (2017)
51. S. Dempe, V. Kalashnikov, R.Z. Rios-Mercado, Discrete bilevel programming: application to a natural gas cash-out problem. Eur. J. Oper. Res. **166**(2), 469–488 (2005)
52. S. DeNegre, Interdiction and discrete bilevel linear programming. PhD, Lehigh University, 2011 http://coral.ie.lehigh.edu/~ted/files/papers/ScottDeNegreDissertation11.pdf
53. S. DeNegre, T.K. Ralphs, A branch-and-cut algorithm for bilevel integer programming, in *Proc. of 11th INFORMS Computing Society Meeting* (2009), pp. 65–78. http://coral.ie.lehigh.edu/~ted/files/papers/BILEVEL08.pdf
54. K. Dhamdhere, R. Ravi, M. Singh, On two-stage stochastic minimum spanning trees, in *International Conference on Integer Programming and Combinatorial Optimization* (Springer, Berlin, 2005), pp. 321–334
55. M. Dyer, L. Stougie, Computational complexity of stochastic programming problems. Math. Program. **106**(3), 423–432 (2006)
56. N.P. Faísca, V. Dua, B. Rustem, P.M. Saraiva, E.N. Pistikopoulos, Parametric global optimisation for bilevel programming. J. Glob. Optim. **38**, 609–623 (2007)
57. M. Fischetti, I. Ljubić, M. Monaci, M. Sinnl, A new general-purpose algorithm for mixed-integer bilevel linear programs. Oper. Res. **65**(6), 1615–1637 (2017)
58. M. Fischetti, I. Ljubić, M. Monaci, M. Sinnl, On the use of intersection cuts for bilevel optimization. Math. Program. **172**, 77–103 (2018)
59. F. Furini, I. Ljubic, S. Martin, P. San Segundo, The maximum clique interdiction game. Eur. J. Oper. Res. **277**(1), 112–127 (2019)
60. D. Gade, S. Küçükyavuz, S. Sen, Decomposition algorithms with parametric gomory cuts for two-stage stochastic integer programs. Math. Program. **144**, 1–26 (2012)
61. J. Gan, E. Elkind, M. Wooldridge, Stackelberg security games with multiple uncoordinated defenders, in *Proc. of 17th Int. Conf. on Autonomous Agents and Multiagent Systems (AAMAS 2008)* (2018)
62. L.P. Garcés, A.J. Conejo, R. García-Bertrand, R. Romero, A bilevel approach to transmission expansion planning within a market environment. IEEE Trans. Power Syst. **24**(3), 1513–1522 (2009)
63. M.R. Garey, D.S. Johnson, *Computers and Intractability: A Guide to the Theory of NP-Completeness* (W.H. Freeman and Company, New York, 1979)
64. M. Gendreau, G. Laporte, R. Séguin, Stochastic vehicle routing. Eur. J. Oper. Res. **88**(1), 3–12 (1996)
65. P.M. Ghare, D.C. Montgomery, W.C. Turner, Optimal interdiction policy for a flow network. Nav. Res. Logist. Q. **18**, 27–45 (1971)
66. I.L. Gørtz, V. Nagarajan, R. Saket, Stochastic vehicle routing with recourse, in *International Colloquium on Automata, Languages, and Programming* (Springer, Berlin, 2012), pp. 411–423
67. E. Grass, K. Fischer, Two-stage stochastic programming in disaster management: a literature survey. Surv. Oper. Res. Manag. Sci. **21**(2), 85–100 (2016)

68. V. Grimm, A. Martin, M. Martin, M. Weibelzahl, G. Zöttl, Transmission and generation investment in electricity markets: the effects of market splitting and network fee regimes. Eur. J. Oper. Res. **254**(2), 493–509 (2016)
69. A. Gupta, R. Ravi, A. Sinha, Lp rounding approximation algorithms for stochastic network design. Math. Oper. Res. **32**(2), 345–364 (2007)
70. M. Güzelsoy, Dual methods in mixed integer linear programming. PhD, Lehigh University, 2009. http://coral.ie.lehigh.edu/~ted/files/papers/MenalGuzelsoyDissertation09.pdf
71. M. Güzelsoy, T.K. Ralphs, Duality for mixed-integer linear programs. Int. J. Oper. Res. **4**, 118–137 (2007). http://coral.ie.lehigh.edu/~ted/files/papers/MILPD06.pdf
72. G.A. Hanasusanto, D. Kuhn, W. Wiesemann, A comment on "computational complexity of stochastic programming problems". Math. Program. **159**(1), 557–569 (2016)
73. P. Hansen, B. Jaumard, G. Savard, New branch-and-bound rules for linear bilevel programming. SIAM J. Sci. Stat. Comput. **13**(5), 1194–1217 (1992)
74. A. Hassanzadeh, T.K. Ralphs, A generalized Benders' algorithm for two-stage stochastic program with mixed integer recourse. Technical Report COR@L Laboratory Report 14T-005, Lehigh University, 2014. http://coral.ie.lehigh.edu/~ted/files/papers/SMILPGenBenders14.pdf
75. A. Hassanzadeh, T.K. Ralphs, On the value function of a mixed integer linear optimization problem and an algorithm for its construction. Technical report, COR@L Laboratory Report 14T-004, Lehigh University, 2014. http://coral.ie.lehigh.edu/~ted/files/papers/MILPValueFunction14.pdf
76. H. Held, D.L. Woodruff, Heuristics for multi-stage interdiction of stochastic networks. J. Heuristics **11**(5–6), 483–500 (2005)
77. M. Hemmati, J.C. Smith, A mixed integer bilevel programming approach for a competitive set covering problem. Technical report, Clemson University, 2016
78. B.F. Hobbs, S.K. Nelson, A nonlinear bilevel model for analysis of electric utility demand-side planning issues. Ann. Oper. Res. **34**(1), 255–274 (1992)
79. E. Israeli, System interdiction and defense. PhD thesis, Naval Postgraduate School, 1999
80. E. Israeli, R.K. Wood, Shortest path network interdiction. Networks **40**(2), 97–111 (2002)
81. U. Janjarassuk, J. Linderoth, Reformulation and sampling to solve a stochastic network interdiction problem. Networks **52**, 120–132 (2008)
82. R.G. Jeroslow, The polynomial hierarchy and a simple model for competitive analysis. Math. Program. **32**(2), 146–164 (1985)
83. P. Kall, J. Mayer, *Stochastic Linear Programming: Models, Theory, and Computation* (Springer, Berlin, 2010)
84. B. Kara, V. Verter, Designing a road network for hazardous materials transportation. Transp. Sci. **38**(2), 188–196 (2004)
85. R.M. Karp, On the computational complexity of combinatorial problems. Networks **5**, 45–68 (1975)
86. I. Katriel, C. Kenyon-Mathieu, E. Upfal, Commitment under uncertainty: two-stage stochastic matching problems, in *International Colloquium on Automata, Languages, and Programming* (Springer, Berlin, 2007), pp. 171–182
87. C. Kiekintveld, M. Jain, J. Tsai, J. Pita, F. Ordóñez, M. Tambe, Computing optimal randomized resource allocations for massive security games, in *Proceedings of AAMAS* (2009), pp. 689–696
88. K.-M. Klein, About the complexity of two-stage stochastic IPs (2019)
89. N. Kong, A.J. Schaefer, B. Hunsaker, Two-stage integer programs with stochastic right-hand sides: a superadditive dual approach. Math. Program. **108**(2), 275–296 (2006)
90. M. Köppe, M. Queyranne, C.T. Ryan, Parametric integer programming algorithm for bilevel mixed integer programs. J. Optim. Theory Appl. **146**(1), 137–150 (2010)
91. A.A. Kulkarni, U.V. Shanbhag, A shared-constraint approach to multi-leader multi-follower games. Set-Valued Var. Anal. **22**(4), 691–720 (2014)
92. M. Labbé, A. Violin, Bilevel programming and price setting problems. 4OR **11**(1), 1–30 (2013)

93. M. Labbé, P. Marcotte, G. Savard, A bilevel model of taxation and its application to optimal highway pricing. Manag. Sci. **44**, 1608–1622 (1998)
94. G. Laporte, F.V. Louveaux, The integer l-shaped method for stochastic integer programs with complete recourse. Oper. Res. Lett. **13**(3), 133–142 (1993)
95. A. Laszka, J. Lou, Y. Vorobeychik, Multi-defender strategic filtering against spear-phishing attacks, in *Proc. of 30th AAAI Conf. on Artificial Intelligence (AAAI 2016)* (2016)
96. S. Leyffer, T. Munson, Solving multi-leader-common-follower games. Optim. Methods Softw. **25**(4), 601–623 (2010)
97. A. Lodi, T.K. Ralphs, F. Rossi, S. Smriglio, Interdiction branching. Technical Report OR/09/10, DEIS-Università di Bologna, 2009
98. J. Lou, Y. Vorobeychik, Equilibrium analysis of multi-defender security games, in *Proc. of 24th Int. Joint Conf. on Artificial Intelligence (IJCAI 2019)* (2015)
99. J. Lou, A.M. Smith, Y. Vorobeychik, Multidefender security games. IEEE Intell. Syst. **32**(1), 50–60 (2017)
100. F.V. Louveaux, M.H. van der Vlerk, Stochastic programming with simple integer recourse. Math. Program. **61**(1), 301–325 (1993)
101. L. Lozano, J.C. Smith, A value-function-based exact approach for the bilevel mixed-integer programming problem. Oper. Res. **65**(3), 768–786 (2017)
102. A. Mahajan, On selecting disjunctions in mixed integer linear programming. PhD, Lehigh University, 2009. http://coral.ie.lehigh.edu/~ted/files/papers/AshutoshMahajanDissertation09.pdf
103. A. Mahajan, T.K. Ralphs, On the complexity of selecting disjunctions in integer programming. SIAM J. Optim. **20**(5), 2181–2198 (2010). http://coral.ie.lehigh.edu/~ted/files/papers/Branching08.pdf
104. A. Marchesi, S. Coniglio, N. Gatti, Leadership in singleton congestion games, in *Proc. of 27th Int. Joint Conf. on Artificial Intelligence (IJCAI 2018)* (AAAI Press, New York, 2018), pp. 447–453
105. A.W. McMasters, T.M. Mustin, Optimal interdiction of a supply network. Nav. Res. Logist. Q. **17**, 261–268 (1970)
106. A. Migdalas, Bilevel programming in traffic planning: models, methods and challenge. J. Glob. Optim. **7**, 381–405 (1995)
107. J.T. Moore, J.F. Bard, The mixed integer linear bilevel programming problem. Oper. Res. **38**(5), 911–921 (1990)
108. J-S. Pang, M. Fukushima, Quasi-variational inequalities, generalized Nash equilibria, and multi-leader-follower games. Comput. Manag. Sci. **2**(1), 21–56 (2005)
109. P. Paruchuri, J.P. Pearce, J. Marecki, M. Tambe, F. Ordonez, S. Kraus, Playing games for security: an efficient exact algorithm for solving Bayesian Stackelberg games, in *Proc. of 7th Int. Conf. on Autonomous Agents and Multiagent Systems (AAMAS 2008)* (2008), pp. 895–902
110. T.K. Ralphs, M. Güzelsoy, The SYMPHONY callable library for mixed integer programming, in *Proceedings of the Ninth INFORMS Computing Society Conference* (2005), pp. 61–76. http://coral.ie.lehigh.edu/~ted/files/papers/SYMPHONY04.pdf
111. T.K. Ralphs, M. Güzelsoy, Duality and warm starting in integer programming, in *The Proceedings of the 2006 NSF Design, Service, and Manufacturing Grantees and Research Conference* (2006). http://coral.ie.lehigh.edu/~ted/files/papers/DMII06.pdf
112. R.T. Rockafellar, S. Uryasev, Optimization of conditional value-at-risk. J. Risk **2**, 21–42 (2000)
113. V. Rutenburg, Propositional truth maintenance systems: classification and complexity analysis. Ann. Math. Artif. Intell. **10**(3), 207–231 (1994)
114. G.K. Saharidis, M.G. Ierapetritou, Resolution method for mixed integer bi-level linear problems based on decomposition technique. J. Glob. Optim. **44**(1), 29–51 (2008)
115. P. San Segundo, S. Coniglio, F. Furini, I. Ljubić, A new branch-and-bound algorithm for the maximum edge-weighted clique problem. Eur. J. Oper. Res. **278**(1), 76–90 (2019)
116. W.H. Sandholm, Evolutionary implementation and congestion pricing. Rev. Econ. Stud. **69**(3), 667–689 (2002)

117. M.P. Scaparra, R.L. Church, A bilevel mixed-integer program for critical infrastructure protection planning. Comput. Oper. Res. **35**(6), 1905–1923 (2008)
118. M. Schaefer, C. Umans, Completeness in the polynomial-time hierarchy: a compendium. SIGACT News **33**(3), 32–49 (2002)
119. R. Schultz, L. Stougie, M.H. Van Der Vlerk, Solving stochastic programs with integer recourse by enumeration: a framework using Gröbner basis. Math. Program. **83**(1), 229–252 (1998)
120. S. Sen, J.L. Higle, The C^3 theorem and a D^2 algorithm for large scale stochastic mixed-integer programming: set convexification. Math. Program. **104**(1), 1–20 (2005)
121. A. Shapiro, Monte Carlo sampling methods, in *Handbooks in Operations Research and Management Science*, vol. 10 (Elsevier, Amsterdam, 2003), pp. 353–425
122. H.D. Sherali, B.M.P. Fraticelli, A modification of Benders' decomposition algorithm for discrete subproblems: an approach for stochastic programs with integer recourse. J. Glob. Optim. **22**(1), 319–342 (2002)
123. H.D. Sherali, X. Zhu, On solving discrete two-stage stochastic programs having mixed-integer first-and second-stage variables. Math. Program. **108**(2), 597–616 (2006)
124. A. Smith, Y. Vorobeychik, J. Letchford, Multidefender security games on networks. ACM SIGMETRICS Perform. Eval. Rev. **41**(4), 4–7 (2014)
125. L.J. Stockmeyer, The polynomial-time hierarchy. Theor. Comput. Sci. **3**, 1–22 (1976)
126. S. Tahernejad, Two-stage mixed integer stochastic bilevel linear optimization. PhD, Lehigh University, 2019
127. S. Tahernejad, T.K. Ralphs, S.T. DeNegre, A branch-and-cut algorithm for mixed integer bilevel linear optimization problems and its implementation. Math. Program. Comput. (2020). http://coral.ie.lehigh.edu/~ted/files/papers/MIBLP16.pdf
128. M. Tambe, *Security and Game Theory: Algorithms, Deployed Systems, Lessons Learned* (Cambridge University Press, Cambridge, 2011)
129. S. Uryasev, Conditional value-at-risk: optimization algorithms and applications, in *Proc. of the IEEE/IAFE/INFORMS 2000 Conference on Computational Intelligence for Financial Engineering (CIFEr)* (IEEE, Piscataway, 2000), pp. 49–57
130. R.M. Van Slyke, R. Wets, L-shaped linear programs with applications to optimal control and stochastic programming. SIAM J. Appl. Math. **17**, 638–663 (1969)
131. L.N. Vicente, P.H. Calamai, Bilevel and multilevel programming: a bibliography review. J. Glob. Optim. **5**(3), 291–306 (1994)
132. L. Vicente, G. Savard, J. Júdice, Discrete linear bilevel programming problem. J. Optim. Theory Appl. **89**(3), 597–614 (1996)
133. H. Von Stackelberg, *Market Structure and Equilibrium* (Springer Science & Business Media, Berlin, 2010)
134. B. von Stengel, S. Zamir, Leadership games with convex strategy sets. Games Econom. Behav. **69**(2), 446–457 (2010)
135. S.W. Wallace, W.T. Ziemba, *Applications of Stochastic Programming* (SIAM, Philadelphia, 2005)
136. L. Wang, P. Xu, The watermelon algorithm for the bilevel integer linear programming problem. SIAM J. Optim. **27**(3), 1403–1430 (2017)
137. U.P. Wen, Y.H. Yang, Algorithms for solving the mixed integer two-level linear programming problem. Comput. Oper. Res. **17**(2), 133–142 (1990)
138. H.P. Williams, Duality in mathematics and linear and integer programming. J. Optim. Theory Appl. **90**(2), 257–278 (1996)
139. R. Wollmer, Removing arcs from a network. Oper. Res. **12**(6), 934–940 (1964)
140. L.A. Wolsey, Integer programming duality: price functions and sensitivity analysis. Math. Program. **20**(1), 173–195 (1981)
141. R.K. Wood, Deterministic network interdiction. Math. Comput. Model. **17**(2), 1–18 (1993)
142. P. Xu, L. Wang, An exact algorithm for the bilevel mixed integer linear programming problem under three simplifying assumptions. Comput. Oper. Res. **41**, 309–318 (2014)

143. B. Zeng, Y. An, Solving bilevel mixed integer program by reformulations and decomposition. Technical report, University of South Florida, 2014
144. Y. Zhang, L.V. Snyder, T.K. Ralphs, Z. Xue, The competitive facility location problem under disruption risks. Transport. Res. Part E Log. Transport. Rev. **93** (2016). http://coral.ie.lehigh.edu/~ted/files/papers/CFLPD16.pdf

Part IV
Numerical and Research Tools for Bilevel Optimization

Chapter 19
BOLIB: Bilevel Optimization LIBrary of Test Problems

Shenglong Zhou, Alain B. Zemkoho, and Andrey Tin

Abstract This chapter presents the Bilevel Optimization LIBrary of the test problems (BOLIB–for short), which contains a collection of test problems, with continuous variables, to help support the development of numerical solvers for bilevel optimization. The library contains 173 examples with 138 nonlinear, 24 linear, and 11 simple bilevel optimization problems. This BOLIB collection is probably the largest bilevel optimization library of test problems. Moreover, as the library is computation-enabled with the MATLAB m-files of all the examples, it provides a uniform basis for testing and comparing algorithms. The library, together with all the related codes, is freely available at https://biopt.github.io/bolib/.

Keywords Bilevel optimization · Test problems · Numerical methods · Library of examples · MATLAB codes

19.1 Introduction

The bilevel optimization problem can take the form

$$\begin{aligned} \min_{x,y}\ & F(x,y) \\ \text{s.t. } & G(x,y) \leq 0, \\ & y \in S(x) := \arg\min_{y} \left\{ f(x,y) \mid g(x,y) \leq 0 \right\}, \end{aligned} \tag{19.1.1}$$

where the functions $G : \mathbb{R}^{n_x} \times \mathbb{R}^{n_y} \to \mathbb{R}^{n_G}$ and $g : \mathbb{R}^{n_x} \times \mathbb{R}^{n_y} \to \mathbb{R}^{n_g}$ define the upper-level and lower-level constraints, respectively. As for $F : \mathbb{R}^{n_x} \times \mathbb{R}^{n_y} \to \mathbb{R}$ and $f : \mathbb{R}^{n_x} \times \mathbb{R}^{n_y} \to \mathbb{R}$, they denote the upper-level and lower-level objective functions, respectively. The set-valued map $S : \mathbb{R}^{n_x} \rightrightarrows \mathbb{R}^{n_y}$ represents the

---

S. Zhou · A. B. Zemkoho (✉) · A. Tin
School of Mathematics, University of Southampton, Southampton, United Kingdom
e-mail: shenglong.zhou@soton.ac.uk; a.b.zemkoho@soton.ac.uk; a.tin@soton.ac.uk

S. Dempe, A. Zemkoho (eds.), *Bilevel Optimization*, Springer Optimization and Its Applications 161, https://doi.org/10.1007/978-3-030-52119-6_19

optimal solution/argminimum mapping of the lower-level problem. Further recall that problem (19.1.1) as a whole is often called upper-level problem.

Our aim is to propose a computation-enabled library of test problems to help accelerate the development of numerical methods for bilevel programs in the form (19.1.1). Note that the bilevel optimization problems in this library, that we classify into the following three categories, only involve continuous variables:

- *Nonlinear bilevel programs*, which are problems in the form (19.1.1) with at least one of the functions involved being nonlinear.
- *Linear bilevel programs* are problems in the form (19.1.1) with functions F, f, and all the components of G and g being linear.
- *Simple bilevel programs* (term coined in [20]) are optimization problems where the feasible set is partly defined by the optimal solution set of a second optimization problem. But unlike in (19.1.1), the lower-level problem is not a parametric optimization problem. More precisely, a simple bilevel optimization has the form:

$$\begin{aligned} &\min_{y} F(y) \\ &\text{s.t. } G(y) \leq 0, \\ &\quad\ y \in S := \arg\min_{y} \{f(y) \mid g(y) \leq 0\}, \end{aligned} \tag{19.1.2}$$

where, similarly to (19.1.1), $G : \mathbb{R}^{n_y} \to \mathbb{R}^{n_G}$ and $g : \mathbb{R}^{n_y} \to \mathbb{R}^{n_g}$ describe the upper-level and lower-level constraints, respectively, while the real-valued function F (resp. f), defined $\mathbb{R}^{n_y}$, represents the upper-level (resp. lower-level) objective function. The expression "simple bilevel program" is used in [45] to refer to bilevel optimization problems of the form (19.1.1), where y (resp. x) is not involved in the upper-level (resp. lower-level) constraints.

The main contributions of the library are three-fold. First, BOLIB provides MATLAB codes for 173 examples, including 138 nonlinear, 24 linear, and 11 simple bilevel programs, ready to be used to test numerical algorithms. Secondly, it puts together the true or best known solutions and the corresponding references for all the examples included. Hence, can serve as a benchmark platform for numerical accuracy evaluation for methods designed to solve problem (19.1.1). Thirdly, all examples as well as their gradients and Hessians are programmed and stored in the MATLAB m-files. Thus, facilitating the use of the examples and corresponding derivatives in the implementation of numerical methods, where such information is necessary.

To the best of our knowledge, this is the largest library of test examples for bilevel optimization, especially for the nonlinear class of the problem. It includes bilevel optimization problems from Colson's BIPA [17], Leyffer's MacMPEC [44], as well as from Mitsos and Barton's technical report [53]. We would like to emphasize that the fundamental objective that we hope to achieve with BOLIB is the acceleration of numerical software development for bilevel optimization, as it is our opinion that the level of expansion of applications of the problem has outpaced the development rate for numerical solvers, especially for the nonlinear class of the problem.

In the next section, we describe the library with details on the inputs and outputs of the codes, as well as some useful insights on the examples. In the subsequent section, a guideline is given on how to access the library.

19.2 Description of the Library

This section describes the structure of the library, while focusing on the inputs and outputs of each example, as well as the list of all examples together with their true or best known solutions and corresponding references. Before we proceed, note that each `m-file` contains information about the corresponding example, which include the first and second order derivatives of the input functions. For the upper-level objective function $F : \mathbb{R}^{n_x} \times \mathbb{R}^{n_y} \to \mathbb{R}$, these derivatives are defined as follows

$$
\begin{aligned}
\nabla_x F(x, y) &= \begin{bmatrix} \nabla_{x_1} F \\ \vdots \\ \nabla_{x_{n_x}} F \end{bmatrix} \in \mathbb{R}^{n_x}, \\
\nabla^2_{xx} F(x, y) &= \begin{bmatrix} \nabla^2_{x_1 x_1} F & \cdots & \nabla^2_{x_{n_x} x_1} F \\ \vdots & \ddots & \vdots \\ \nabla^2_{x_1 x_{n_x}} F & \cdots & \nabla^2_{x_{n_x} x_{n_x}} F \end{bmatrix} \in \mathbb{R}^{n_x \times n_x}, \\
\nabla^2_{xy} F(x, y) &= \begin{bmatrix} \nabla^2_{x_1 y_1} F & \cdots & \nabla^2_{x_{n_x} y_1} F \\ \vdots & \ddots & \vdots \\ \nabla^2_{x_1 y_{n_x}} F & \cdots & \nabla^2_{x_{n_x} y_{n_x}} F \end{bmatrix} \in \mathbb{R}^{n_y \times n_x}.
\end{aligned}
\tag{19.2.1}
$$

Similar expressions are valid for $\nabla_y F(x, y) \in \mathbb{R}^{n_y}$, $\nabla^2_{yy} F(x, y) \in \mathbb{R}^{n_y \times n_y}$, and the lower-level objective function f. As the constraint functions are vector-valued, we use the following notations to refer to derivative information in the context of the upper-level constraint function $G : \mathbb{R}^{n_x} \times \mathbb{R}^{n_y} \to \mathbb{R}^{n_G}$, for instance:

$$
\nabla_x G(x, y) = \begin{bmatrix} \nabla_x G_1 \\ \vdots \\ \nabla_x G_{n_G} \end{bmatrix} = \begin{bmatrix} \nabla_{x_1} G_1 & \cdots & \nabla_{x_{n_x}} G_1 \\ \vdots & \ddots & \vdots \\ \nabla_{x_1} G_{n_G} & \cdots & \nabla_{x_{n_x}} G_{n_G} \end{bmatrix} \in \mathbb{R}^{n_G \times n_x},
$$

$$
\nabla^2_{xx} G(x, y) = \begin{bmatrix} \nabla^2_{xx} G_1 \\ \vdots \\ \nabla^2_{xx} G_{n_G} \end{bmatrix} = \begin{bmatrix} \nabla^2_{x_1 x_1} G_1 & \cdots & \nabla^2_{x_{n_x} x_1} G_1 \\ \vdots & \ddots & \vdots \\ \nabla^2_{x_1 x_{n_x}} G_1 & \cdots & \nabla^2_{x_{n_x} x_{n_x}} G_1 \\ \vdots & \ddots & \vdots \\ \nabla^2_{x_1 x_1} G_{n_G} & \cdots & \nabla^2_{x_{n_x} x_1} G_{n_G} \\ \vdots & \ddots & \vdots \\ \nabla^2_{x_1 x_{n_x}} G_{n_G} & \cdots & \nabla^2_{x_{n_x} x_{n_x}} G_{n_G} \end{bmatrix} \in \mathbb{R}^{(n_G n_x) \times n_x},
$$

$$\nabla^2_{xy}G(x,y) = \begin{bmatrix} \nabla^2_{xy}G_1 \\ \vdots \\ \nabla^2_{xy}G_{n_G} \end{bmatrix} = \begin{bmatrix} \nabla^2_{x_1y_1}G_1 & \cdots & \nabla^2_{x_{n_x}y_1}G_1 \\ \vdots & \ddots & \vdots \\ \nabla^2_{x_1y_{n_y}}G_1 & \cdots & \nabla^2_{x_{n_x}y_{n_y}}G_1 \\ \vdots & \ddots & \vdots \\ \nabla^2_{x_1y_1}G_{n_G} & \cdots & \nabla^2_{x_{n_x}y_1}G_{n_G} \\ \vdots & \ddots & \vdots \\ \nabla^2_{x_1y_{n_y}}G_{n_G} & \cdots & \nabla^2_{x_{n_x}y_{n_y}}G_{n_G} \end{bmatrix} \in \mathbb{R}^{(n_G n_y)\times n_x}. \tag{19.2.2}$$

Similar formulas are also valid for $\nabla_y G(x, y) \in \mathbb{R}^{n_G \times n_y}$, $\nabla^2_{yy}G(x, y) \in \mathbb{R}^{n_G n_y \times n_y}$, and the lower-level constraint g. It is important to emphasize that in the context of the constraints, $\nabla_x G(x, y) \in \mathbb{R}^{1\times n_x}$, for example, is a row vector when $n_G = 1$. However, $\nabla_x F(x, y) \in \mathbb{R}^{n_x}$ and $\nabla_x f(x, y) \in \mathbb{R}^{n_x}$ are column vectors.

19.2.1 Inputs and Outputs

The `BOLIBver2` folder (see Sect. 19.3 on how to access the library), which contains all the library material, includes 3 sub-folders named `Nonlinear`, `Linear`, and `Simple`. In the `Nonlinear` subfolder, there are 138 MATLAB m-files. Each one specifies a nonlinear bilevel optimization test example, named by a combination of authors' surnames, year of publication, and when necessary, the order of the example in the corresponding reference. For example, as in following figure (showing a partial list of the examples), `AiyoshiShimizu1984Ex2.m` stands for Example 2 in the paper by Aiyoshi and Shimizu published in 1984 [1]. However, for a few examples (`DesignCentringP1`, `NetworkDesignP1`, etc.), the problem naming is based on previous use in the literature and therefore could help to easily recognize them.

In folder `Linear`, there are 24 MATLAB m-files defining 24 liner bilevel optimization test examples. The rule of naming each example is same as in the nonlinear case. Similarly, folder `Simple` contains 11 simple bilevel optimization test examples.

Now we describe the inputs and outputs of the m-file of a given example. Each file has the function handle named in the following way:

$$\texttt{w = example\_name(x, y, keyf, keyxy)}. \tag{19.2.3}$$

For the inputs, we have

$$\begin{aligned} x &\in \mathbb{R}^{n_x}, \quad y \in \mathbb{R}^{n_y}, \\ \texttt{keyf} &\in \{\texttt{'F'}, \ \texttt{'G'}, \ \texttt{'f'}, \ \texttt{'g'}\}, \\ \texttt{keyxy} &\in \{[\,], \texttt{'x'}, \ \texttt{'y'}, \ \texttt{'xx'}, \ \texttt{'xy'}, \ \texttt{'yy'}\}, \end{aligned}$$

where `'F'`, `'G'`, `'f'`, and `'g'`, respectively stand for the four functions involved in (19.1.1). `'x'` and `'y'` represent the first order derivative with respect to x and y, respectively. Finally, `'xx'`, `'xy'`, and `'yy'` correspond to the second order derivative of the function F, G, f, and g, with respect to xx, xy, and yy, respectively.

For the outputs, `w=example_name(x,y,keyf)` or `w = example_name(x,y, keyf, [])` returns the function value of `keyf` while `w=example_name(x,y,keyf, keyxy)` can additionally return the first or second order derivative of `keyf` w.r.t. the choice of `keyxy` as described above. We summarize all the scenarios in Table 19.1:

For the dimension of w in each scenario, see (19.2.1)–(19.2.2). If $n_G = 0$ (or $n_g = 0$), all outputs related to G (or g) should be empty, namely, $w = [\,]$. To further clarify the outputs, let us look at some specific usage:

- $w = \texttt{example\_name}(x, y, \texttt{'F'})$ or $w = \texttt{example\_name}(x, y, \texttt{'F'}, [\,])$ returns the function value of F, i.e., $w = F(x, y)$; this is similar for G, f, and g;
- $w = \texttt{example\_name}(x, y, \texttt{'F'}, \texttt{'x'})$ returns the partial derivative of F with respect to x, i.e., $w = \nabla_x F(x, y)$;
- $w = \texttt{example\_name}(x, y, \texttt{'G'}, \texttt{'y'})$ returns the Jacobian matrix of G with respect to y, i.e., $w = \nabla_y G(x, y)$;

Table 19.1 Input–output scenarios from the m-files containing the examples

keyf/keyxy	[ ]	'x'	'y'	'xx'	'xy'	'yy'
'F'	$F(x, y)$	$\nabla_x F(x, y)$	$\nabla_y F(x, y)$	$\nabla^2_{xx} F(x, y)$	$\nabla^2_{xy} F(x, y)$	$\nabla^2_{yy} F(x, y)$
'G'	$G(x, y)$	$\nabla_x G(x, y)$	$\nabla_y G(x, y)$	$\nabla^2_{xx} G(x, y)$	$\nabla^2_{xy} G(x, y)$	$\nabla^2_{yy} G(x, y)$
'f'	$f(x, y)$	$\nabla_x f(x, y)$	$\nabla_y f(x, y)$	$\nabla^2_{xx} f(x, y)$	$\nabla^2_{xy} f(x, y)$	$\nabla^2_{yy} f(x, y)$
'g'	$g(x, y)$	$\nabla_x g(x, y)$	$\nabla_y g(x, y)$	$\nabla^2_{xx} g(x, y)$	$\nabla^2_{xy} g(x, y)$	$\nabla^2_{yy} g(x, y)$

- w = `example_name`(x, y, `'f'`, `'xy'`) returns the Hessian matrix of f with respect to xy, i.e., $w = \nabla^2_{xy} f(x, y)$;
- w = `example_name`(x, y, `'g'`, `'yy'`) returns the second order derivative of g with respect to yy, i.e., $w = \nabla^2_{yy} g(x, y)$.

We now use two examples to illustrate the definitions above. The first one is nonlinear while the second one is a simple bilevel program.

Example
Shimizu et al. (1997), see [65], considered the bilevel program (19.1.1) with

$$
\begin{aligned}
F(x, y) &:= (x-5)^2 + (2y+1)^2, \\
f(x, y) &:= (y-1)^2 - 1.5xy, \\
g(x, y) &:= \begin{bmatrix} -3x + y + 3 \\ x - 0.5y - 4 \\ x + y - 7 \end{bmatrix}.
\end{aligned}
$$

Here, we have dimensions $n_x = 1$, $n_y = 1$, $n_G = 0$, and $n_g = 3$. The `m-file` is named by `ShimizuEtal1997a` (i.e., `exmaple_name` = `ShimizuEtal1997a`) and was coded in MATLAB as it can be seen in Table 19.2. If we are given some inputs (as in left column of the table below), then `ShimizuEtal1997a` will return us corresponding results as in the right column:

Inputs	**Outputs**
`x   = 4`	`x   = 4`
`y   = 0`	`y   = 0`
`F   = ShimizuEtal1997a(x,y,'F')`	`F   = 2`
`Fx  = ShimizuEtal1997a(x,y,'F','x')`	`Fx  = -2`
`Gy  = ShimizuEtal1997a(x,y,'G','y')`	`Gy  = []`
`fxy = ShimizuEtal1997a(x,y,'f','xy')`	`fxy = -1.5`
`gyy = ShimizuEtal1997a(x,y,'g','yy')`	`gyy = [0;0;0]`

Table 19.2 Matlab code for `ShimizuEtal1997a`

```
function w=ShimizuEtal1997a(x,y,keyf,keyxy)
if nargin<4 || isempty(keyxy)
    switch keyf
    case 'F'; w = (x-5)^2+(2*y+1)^2;
    case 'G'; w = [];
    case 'f'; w = (y-1)^2-1.5*x*y;
    case 'g'; w = [-3*x+y+3; x-0.5*y-4; x+y-7];
    end
else
    switch keyf
    case 'F'
        switch keyxy
        case 'x' ; w = 2*(x-5);
        case 'y' ; w = 4*(2*y+1);
        case 'xx'; w = 2;
        case 'xy'; w = 0;
        case 'yy'; w = 8;
        end
    case 'G'
        switch keyxy
        case 'x' ; w = [];
        case 'y' ; w = [];
        case 'xx'; w = [];
        case 'xy'; w = [];
        case 'yy'; w = [];
        end
    case 'f'
        switch keyxy
        case 'x' ; w = -1.5*y;
        case 'y' ; w = 2*(y-1)-1.5*x;
        case 'xx'; w = 0;
        case 'xy'; w = -1.5;
        case 'yy'; w = 2;
        end
    case 'g'
        switch keyxy
        case 'x' ; w = [-3;  1;   1];
        case 'y' ; w = [ 1;-0.5;  1];
        case 'xx'; w = [ 0;  0;   0];
        case 'xy'; w = [ 0;  0;   0];
        case 'yy'; w = [ 0;  0;   0];
        end
    end
end
end
```

Example
Franke et al. (2018), see [31], considered the bilevel program (19.1.1) with

$$F(y) := -y_2,$$
$$f(y) := y_3,$$
$$g(y) := \begin{bmatrix} y_1^2 - y_3 \\ y_1^2 + y_2^2 - 1 \\ -y_3 \end{bmatrix}.$$

Here, we have dimensions $n_x = 0$, $n_y = 3$, $n_G = 0$, and $n_g = 3$. The `m-file` is named `FrankeEtal2018Ex513` (i.e., `exmaple_name = FrankeEtal2018Ex513`) and is equally coded in MATLAB as described in Table 19.3.

Remark 19.2.1 It is worth mentioning that despite the lack of variable x in the latter example, we still treat it as an input, for the sake of unifying the inputs of the function handle as in (19.2.3). Hence, for all the simple bilevel optimization examples, we input `x` as a scalar. In this way, `x` has no impact on the example itself. △

19.2.2 *Useful Details on the Examples*

The details related to each example presented in the BOLIB library are in a column of Table 19.4 below. As we mentioned before, those examples are classified into 3 categories: nonlinear, linear and simple bilevel optimisation test examples. The first column of the table provides the list of problems, as they appear in the `Examples` subfolder of the `BOLIBver2` folder. The second column gives the reference in the literature where the example might have first appeared. The third column combines the labels corresponding to the nature of the functions involved in (19.1.1). Precisely, "N" and "L" will be used to indicate whether the functions F, G, f, and g are nonlinear (N) or linear (L), while "O" is used to symbolize that there is either no function G or g present in problem (19.1.1). Then follows the column with n_x and n_y for the upper and lower-level variable dimensions, as well as n_G (resp. n_g) to denote the number of components of the upper (resp. lower)-level constraint function. On the other hand, F^* and f^* denote the best known optimal upper and lower-level objective function values, respectively, according to the reference that is listed in the last column.

Table 19.3 Matlab code for FrankeEtal2018Ex513

```
function w=FrankeEtal2018Ex513(x,y,keyf,keyxy)
if nargin<4 || isempty(keyxy)
    switch keyf
    case 'F'; w = -y(2);
    case 'G'; w = [];
    case 'f'; w = y(3);
    case 'g'; w = [y(1)^2-y(3); y(1)^2+y(2)^2-1; -y(3)];
    end
else
    switch keyf
    case 'F'
        switch keyxy
        case 'x' ; w = 0;
        case 'y' ; w = [0; -1; 0];
        case 'xx'; w = 0;
        case 'xy'; w = zeros(3,1);
        case 'yy'; w = zeros(3,3);
        end
    case 'G'
       switch keyxy
        case 'x' ; w = [];
        case 'y' ; w = [];
        case 'xx'; w = [];
        case 'xy'; w = [];
        case 'yy'; w = [];
        end
    case 'f'
        switch keyxy
        case 'x' ; w = 0;
        case 'y' ; w = [0; 0; 1];
        case 'xx'; w = 0;
        case 'xy'; w = zeros(3,1);
        case 'yy'; w = zeros(3,3);
        end
    case 'g'
        switch keyxy
        case 'x' ; w = zeros(3,1);
        case 'y' ; w = [2*y(1) 0 -1; 2*y(1) 2*y(2) 0; 0 0 -1];
        case 'xx'; w = zeros(3,1);
        case 'xy'; w = zeros(9,1);
        case 'yy'; w = [2 0 0;0 0 0;0 0 0;2 0 0; 0 2 0;zeros(4,3)];
        end
    end
end
end
```

Table 19.4 List of bilevel programs with related labels and known solutions

Example name	RefI	F-G-f-g	n_x	n_y	n_G	n_g	F^*	f^*	RefII
Nonlinear bilevel programs									
AiyoshiShimizu1984Ex2	[1]	L-L-N-L	2	2	5	6	5	0	[1]
AllendeStill2013	[2]	N-L-N-N	2	2	5	2	1	−0.5	[2]
AnEtal2009	[3]	N-L-N-L	2	2	6	4	2251.6	565.8	[3]
Bard1988Ex1	[6]	N-L-N-L	1	1	1	4	17	1	[6]
Bard1988Ex2	[6]	N-L-N-L	4	4	9	12	−6600	54	[17]
Bard1988Ex3	[6]	N-N-N-N	2	2	3	4	−12.68	−1.02	[18]
Bard1991Ex1	[7]	L-L-N-L	1	2	2	3	2	12	[7]
BardBook1998Ex832	[8]	N-L-L-L	2	2	4	7	0	5	
CalamaiVicente1994a	[12]	N-O-N-L	1	1	0	3	0	0	[12]
CalamaiVicente1994b	[12]	N-O-N-L	4	2	0	6	0.3125	−0.4063	[12]
CalamaiVicente1994c	[12]	N-O-N-L	4	2	0	6	0.3125	−0.4063	[12]
CalveteGale1999P1	[13]	L-L-L-N	2	3	2	6	−29.2	0.31	[13, 33]
ClarkWesterberg1990a	[15]	N-L-N-L	1	1	2	3	5	4	[61]
Colson2002BIPA1	[17]	N-L-N-L	1	1	3	3	250	0	
Colson2002BIPA2	[17]	N-L-N-L	1	1	1	4	17	2	[18]
Colson2002BIPA3	[17]	N-L-N-L	1	1	2	2	2	24.02	[18]
Colson2002BIPA4	[17]	N-L-N-L	1	1	2	2	88.79	−0.77	[18]
Colson2002BIPA5	[17]	N-L-N-N	1	2	1	6	2.75	0.57	[18]
Dempe1992a	[19]	L-N-N-N	2	2	1	2	×	×	
Dempe1992b	[19]	N-O-N-N	1	1	0	1	31.25	4	[18]
DempeDutta2012Ex24	[21]	N-O-N-N	1	1	0	1	0	0	[21]
DempeDutta2012Ex31	[21]	L-N-N-N	2	2	4	2	−1	4	[21]
DempeEtal2012	[25]	L-L-N-L	1	1	2	2	−1	−1	[25]
DempeFranke2011Ex41	[22]	N-L-N-L	2	2	4	4	5	−2	[22]
DempeFranke2011Ex42	[22]	N-L-N-L	2	2	4	3	2.13	−3.5	[22]
DempeFranke2014Ex38	[23]	L-L-N-L	2	2	4	4	−1	−4	[23]
DempeLohse2011Ex31a	[24]	N-O-N-L	2	2	0	4	−5.5	0	[24]
DempeLohse2011Ex31b	[24]	N-O-N-L	3	3	0	5	−12	0	
DeSilva1978	[26]	N-O-N-L	2	2	0	4	−1	0	[18]
FalkLiu1995	[28]	N-O-N-L	2	2	0	4	−2.1962	0	[18]
FloudasEtal2013	[29]	L-L-N-L	2	2	4	7	0	200	[69]
FloudasZlobec1998	[30]	N-L-L-N	1	2	2	6	1	−1	[33, 53]
GumusFloudas2001Ex1	[33]	N-L-N-L	1	1	3	3	2250	197.75	[53]
GumusFloudas2001Ex3	[33]	L-L-N-L	2	3	4	9	−29.2	0.31	[53]
GumusFloudas2001Ex4	[33]	N-L-N-L	1	1	5	2	9	0	[53]
GumusFloudas2001Ex5	[33]	L-L-N-N	1	2	2	6	0.19	−7.23	[53]
HatzEtal2013	[34]	L-O-N-L	1	2	0	2	0	0	[34]
HendersonQuandt1958	[36]	N-L-N-L	1	1	2	1	−3266.7	−711.11	[36]
HenrionSurowiec2011	[37]	N-O-N-O	1	1	0	0	$-c^2/4$	$-c^2/8$	[27]

(continued)

Table 19.4 (continued)

Example name	RefI	*F*-*G*-*f*-*g*	n_x	n_y	n_G	n_g	F^*	f^*	RefII
IshizukaAiyoshi1992a	[39]	N-L-L-L	1	2	1	5	0	−M	[39]
KleniatiAdjiman2014Ex3	[40]	L-L-N-L	1	1	2	2	−1	0	[40]
KleniatiAdjiman2014Ex4	[40]	N-N-N-N	5	5	13	11	−10	−3.1	[40]
LamparSagrat2017Ex23	[41]	L-L-N-L	1	2	2	2	−1	1	[41]
LamparSagrat2017Ex31	[42]	N-L-L-L	1	1	1	1	1	0	[42]
LamparSagrat2017Ex32	[42]	N-O-N-O	1	1	0	0	0.5	0	[42]
LamparSagrat2017Ex33	[42]	N-L-L-L	1	2	1	3	0.5	0	[42]
LamparSagrat2017Ex35	[42]	N-L-L-L	1	1	2	3	0.8	−0.4	[42]
LucchettiEtal1987	[48]	N-L-N-L	1	1	2	2	0	0	[48]
LuDebSinha2016a	[47]	N-L-N-O	1	1	4	0	1.14	1.69	[47]
LuDebSinha2016b	[47]	N-L-N-O	1	1	4	0	0	1.66	[47]
LuDebSinha2016c	[47]	N-L-N-O	1	1	4	0	1.12	0.06	[47]
LuDebSinha2016d	[47]	L-N-L-N	2	2	11	3	×	×	
LuDebSinha2016e	[47]	N-L-L-N	1	2	6	3	×	×	
LuDebSinha2016f	[47]	L-N-N-O	2	1	9	0	×	×	
MacalHurter1997	[49]	N-O-N-O	1	1	0	0	81.33	−0.33	[49]
Mirrlees1999	[52]	N-O-N-O	1	1	0	0	1	0.06	[52]
MitsosBarton2006Ex38	[53]	N-L-N-L	1	1	4	2	0	0	[53]
MitsosBarton2006Ex39	[53]	L-L-N-L	1	1	3	2	−1	−1	[53]
MitsosBarton2006Ex310	[53]	L-L-N-L	1	1	2	2	0.5	−0.1	[53]
MitsosBarton2006Ex311	[53]	L-L-N-L	1	1	2	2	−0.8	0	[53]
MitsosBarton2006Ex312	[53]	N-L-N-L	1	1	2	2	0	0	[53]
MitsosBarton2006Ex313	[53]	L-L-N-L	1	1	2	2	−1	0	[53]
MitsosBarton2006Ex314	[53]	N-L-N-L	1	1	2	2	0.25	−0.08	[53]
MitsosBarton2006Ex315	[53]	L-L-N-L	1	1	2	2	0	−0.83	[53]
MitsosBarton2006Ex316	[53]	L-L-N-L	1	1	2	2	−2	0	[53]
MitsosBarton2006Ex317	[53]	N-L-N-L	1	1	2	2	0.19	−0.02	[53]
MitsosBarton2006Ex318	[53]	N-L-N-L	1	1	2	2	−0.25	0	[53]
MitsosBarton2006Ex319	[53]	N-L-N-L	1	1	2	2	−0.26	0	[53]
MitsosBarton2006Ex320	[53]	N-L-N-L	1	1	2	2	0.31	−0.08	[53]
MitsosBarton2006Ex321	[53]	N-L-N-L	1	1	2	2	0.21	−0.07	[53]
MitsosBarton2006Ex322	[53]	N-L-N-N	1	1	2	3	0.21	−0.07	[53]
MitsosBarton2006Ex323	[53]	N-N-L-N	1	1	3	3	0.18	−1	[53]
MitsosBarton2006Ex324	[53]	N-L-N-L	1	1	2	2	−1.75	0	[53]
MitsosBarton2006Ex325	[53]	N-N-N-N	2	3	6	9	−1	−2	[53]
MitsosBarton2006Ex326	[53]	N-N-N-L	2	3	7	6	−2.35	−2	[53]
MitsosBarton2006Ex327	[53]	N-N-N-N	5	5	13	13	2	−1.1	[53]
MitsosBarton2006Ex328	[53]	N-N-N-N	5	5	13	13	−10	−3.1	[53]
MorganPatrone2006a	[54]	L-L-N-L	1	1	2	2	−1	0	[54]
MorganPatrone2006b	[54]	L-O-N-L	1	1	0	4	−1.25	0	[54]

(continued)

Table 19.4 (continued)

Example name	RefI	F-G-f-g	n_x	n_y	n_G	n_g	F^*	f^*	RefII
MorganPatrone2006c	[54]	L-O-N-L	1	1	0	4	−1	−0.25	[54]
MuuQuy2003Ex1	[55]	N-L-N-L	1	2	2	3	−2.08	−0.59	[55]
MuuQuy2003Ex2	[55]	N-L-N-L	2	3	3	4	0.64	1.67	[55]
NieEtal2017Ex34	[56]	L-L-N-N	1	2	2	2	2	0	[56]
NieEtal2017Ex52	[56]	N-N-N-N	2	3	5	2	−1.71	−2.23	[56]
NieEtal2017Ex54	[56]	N-N-N-N	4	4	3	2	−0.44	−1.19	[56]
NieEtal2017Ex57	[56]	N-N-N-N	2	3	5	2	−2	−1	[56]
NieEtal2017Ex58	[56]	N-N-N-N	4	4	3	2	−3.49	−0.86	[56]
NieEtal2017Ex61	[56]	N-N-N-N	2	2	5	1	−1.02	−1.08	[56]
Outrata1990Ex1a	[57]	N-O-N-L	2	2	0	4	−8.92	−6.05	[57]
Outrata1990Ex1b	[57]	N-O-N-L	2	2	0	4	−7.56	−0.58	[57]
Outrata1990Ex1c	[57]	N-O-N-L	2	2	0	4	−12	−112.71	[57]
Outrata1990Ex1d	[57]	N-O-N-L	2	2	0	4	−3.6	−2	[57]
Outrata1990Ex1e	[57]	N-O-N-L	2	2	0	4	−3.15	−16.29	[57]
Outrata1990Ex2a	[57]	N-L-N-L	1	2	1	4	0.5	−14.53	[57]
Outrata1990Ex2b	[57]	N-L-N-L	1	2	1	4	0.5	−4.5	[57]
Outrata1990Ex2c	[57]	N-L-N-L	1	2	1	4	1.86	−10.93	[57]
Outrata1990Ex2d	[57]	N-L-N-N	1	2	1	4	0.92	−19.47	[57]
Outrata1990Ex2e	[57]	N-L-N-N	1	2	1	4	0.90	−14.94	[57]
Outrata1993Ex31	[58]	N-L-N-N	1	2	1	4	1.56	−11.67	[58]
Outrata1993Ex32	[58]	N-L-N-N	1	2	1	4	3.21	−20.53	[58]
Outrata1994Ex31	[59]	N-L-N-N	1	2	2	4	3.21	−20.53	[59]
OutrataCervinka2009	[60]	L-L-N-L	2	2	1	3	0	0	[60]
PaulaviciusEtal2017a	[61]	N-L-N-L	1	1	4	2	0.25	0	[61]
PaulaviciusEtal2017b	[61]	L-L-N-L	1	1	4	2	−2	−1.5	[61]
SahinCiric1998Ex2	[62]	N-L-N-L	1	1	2	3	5	4	[62]
ShimizuAiyoshi1981Ex1	[64]	N-L-N-L	1	1	3	3	100	0	[64]
ShimizuAiyoshi1981Ex2	[64]	N-L-N-L	2	2	3	4	225	100	[64]
ShimizuEtal1997a	[65]	N-O-N-L	1	1	0	3	×	×	
ShimizuEtal1997b	[65]	N-L-N-L	1	1	2	2	2250	197.75	[65]
SinhaMaloDeb2014TP3	[66]	N-N-N-N	2	2	3	4	−18.68	−1.02	[66]
SinhaMaloDeb2014TP6	[66]	N-L-N-L	1	2	1	6	−1.21	7.62	[66]
SinhaMaloDeb2014TP7	[66]	N-N-N-L	2	2	4	4	−1.96	1.96	[66]
SinhaMaloDeb2014TP8	[66]	N-L-N-L	2	2	5	6	0	100	[66]
SinhaMaloDeb2014TP9	[66]	N-O-N-L	10	10	0	20	0	1	[66]
SinhaMaloDeb2014TP10	[66]	N-O-N-L	10	10	0	20	0	1	[66]
TuyEtal2007	[69]	N-L-L-L	1	1	2	3	22.5	−1.52	[69]
Vogel2002	[72]	N-L-N-L	1	1	2	1	1	−2	[72]
WanWangLv2011	[73]	N-O-L-L	2	3	0	8	10.63	−0.5	[73]
YeZhu2010Ex42	[75]	N-L-N-L	1	1	2	1	1	−2	[75]
YeZhu2010Ex43	[75]	N-L-N-L	1	1	2	1	1.25	−2	[75]
Yezza1996Ex31	[76]	N-L-N-L	1	1	2	2	1.5	−2.5	[76]

(continued)

Table 19.4 (continued)

Example name	RefI	F-G-f-g	n_x	n_y	n_G	n_g	F^*	f^*	RefII
Yezza1996Ex41	[76]	N-O-N-L	1	2	0	2	0.5	2.5	[76]
Zlobec2001a	[77]	N-O-L-L	1	2	0	3	−1	−1	[77]
Zlobec2001b	[77]	L-L-L-N	1	1	2	4	No	Solution	[77]
DesignCentringP1	[67]	N-N-N-N	3	6	3	3	×	×	
DesignCentringP2	[67]	N-N-N-N	4	6	5	3	×	×	
DesignCentringP3	[67]	N-N-N-N	6	6	3	3	×	×	
DesignCentringP4	[67]	N-N-N-N	4	6	3	12	×	×	
NetworkDesignP1	[17]	N-L-N-L	5	5	5	6	300.5	419.8	[18]
NetworkDesignP2	[17]	N-L-N-L	5	5	5	6	142.9	81.95	[18]
OptimalControl	[50]	N-N-N-L	2	n_y	3	$2n_y$	×	×	
RobustPortfolioP1	[67]	L-N-N-N	N+1	N	N+3	N+1	1.15	0	[67]
RobustPortfolioP2	[67]	L-N-N-N	N+1	N	N+3	N+1	1.15	0	[67]
TollSettingP1	[17]	N-L-N-L	3	8	3	18	−7	12	[18]
TollSettingP2	[17]	N-L-N-L	3	18	3	38	−4.5	32	[18]
TollSettingP3	[17]	N-L-N-L	3	18	3	38	−3.5	32	[18]
TollSettingP4	[17]	N-O-N-L	2	4	0	8	−4	14	[18]
TollSettingP5	[17]	N-O-N-L	1	4	0	8	−2.5	14	[18]
Linear bilevel programs									
AnandalinghamWhite1990	[4]	L-L-L-L	1	1	1	6	−49	15	[4]
Bard1984a	[5]	L-L-L-L	1	1	1	5	28/9	−60/9	[5]
Bard1984b	[5]	L-L-L-L	1	1	1	5	−37.6	1.6	[5]
Bard1991Ex2	[7]	L-L-L-L	1	2	1	5	−1	−1	[7]
BardFalk1982Ex2	[9]	L-L-L-L	2	2	2	5	−3.25	−4	[9]
Ben-AyedBlair1990a	[10]	L-L-L-L	1	2	2	4	−2.5	−5	[10]
Ben-AyedBlair1990b	[10]	L-L-L-L	1	1	1	4	−6	5	[10]
BialasKarwan1984a	[11]	L-L-L-L	1	2	1	7	−2	−0.5	[11]
BialasKarwan1984b	[11]	L-L-L-L	1	1	1	6	−11	11	[11]
CandlerTownsley1982	[14]	L-L-L-L	2	3	2	6	−29.2	3.2	[14]
ClarkWesterberg1988	[16]	L-O-L-L	1	1	0	3	−37	14	[16]
ClarkWesterberg1990b	[15]	L-L-L-L	1	2	2	5	−13	−4	[15]
GlackinEtal2009	[32]	L-L-L-L	2	1	3	3	6	0	[32]
HaurieSavardWhite1990	[35]	L-O-L-L	1	1	0	4	27	−3	[35]
HuHuangZhang2009	[38]	L-L-L-L	1	2	1	5	−76/9	−41/9	[38]
LanWenShihLee2007	[43]	L-L-L-L	1	1	1	7	−85.09	50.17	[43]
LiuHart1994	[46]	L-L-L-L	1	1	1	4	−16	4	[46]
MershaDempe2006Ex1	[51]	L-L-L-L	1	1	1	5	×	×	[51]
MershaDempe2006Ex2	[51]	L-L-L-L	1	1	2	2	−20	−6	[51]
TuyEtal1993	[68]	L-L-L-L	2	2	3	4	−3.25	−6	[68]
TuyEtal1994	[70]	L-L-L-L	2	2	3	3	6	0	[70]
TuyEtal2007Ex3	[69]	L-L-L-L	10	6	12	13	−467.46	−11.62	[69]
VisweswaranEtal1996	[71]	L-L-L-L	1	1	1	5	28/9	−60/9	[71]
WangJiaoLi2005	[74]	L-L-L-L	1	2	2	2	−1000	−1	[74]

(continued)

Table 19.4 (continued)

Example name	RefI	F-G-f-g	n_x	n_y	n_G	n_g	F^*	f^*	RefII
Simple bilevel programs									
`FrankeEtal2018Ex53`	[31]	`N-L-N-L`	0	2	4	4	1	1	[31]
`FrankeEtal2018Ex511`	[31]	`N-O-L-L`	0	3	0	4	3	0	[31]
`FrankeEtal2018Ex513`	[31]	`L-O-L-N`	0	3	0	3	−1	0	[31]
`FrankeEtal2018Ex521`	[31]	`L-O-L-N`	0	2	0	3	−1	0	[31]
`MitsosBarton2006Ex31`	[53]	`L-L-L-L`	0	1	2	2	1	−1	[53]
`MitsosBarton2006Ex32`	[53]	`L-L-L-L`	0	1	3	2	No	Solution	[53]
`MitsosBarton2006Ex33`	[53]	`L-L-N-N`	0	1	2	3	−1	1	[53]
`MitsosBarton2006Ex34`	[53]	`L-L-N-L`	0	1	2	2	1	−1	[53]
`MitsosBarton2006Ex35`	[53]	`L-L-N-L`	0	1	2	2	0.5	−1	[53]
`MitsosBarton2006Ex36`	[53]	`L-L-N-L`	0	1	2	2	−1	−1	[53]
`ShehuEtal2019Ex42`	[63]	`N-O-N-O`	0	n_y	0	0	×	×	

Note that examples `Zlobec2001b` and `MitsosBartonEx32` have no optimal solutions. There are 4 examples involving parameters; i.e., `CalamaiVicente1994a` with $\rho \geq 1$ (its F^* and f^* listed in the table are under $\rho = 1$, other cases can be found in [12]), `HenrionSurowiec2011` with $c \in \mathbb{R}$, `IshizukaAiyoshi1992a` with $M > 1$ and `RobustPortfolioP1` with $\delta \in [1, +\infty]$ (its F^* and f^* listed in the table are under $\delta = 2$). Dimensions n_x, n_y, n_G or n_g of examples `OptimalControl`, `RobustPortfolioP1`, `RobustPortfolioP2`, and `ShehuEtal2019Ex42` can be altered to get problems of different sizes, as necessary.

Remark 19.2.2 It is worth pointing out that some examples involve equality constraints in the upper or lower-level problems. As only 8% of the BOLIB problems have such constraints, we preserve the uniformity in the structure of the codes by converting equalities constraints into inequalities. For the sake of clarity, we list all the examples with equality constraints below (Table 19.5). △

19.3 How to Access the Library?

The library can be accessed through the dedicated website https://biopt.github.io/bolib/. Under this link, you will find the zipped folder named `BOLIBver2` containing all the relevant files for the version of the library presented in this paper. The folder contains the subfolder named `Examples`, which contains all the m-files with the codes of the examples, as described in the previous section. The pdf file named `Formulas` collects all the mathematical formulas of all the examples in this library. To start with the library, it is advised to consult the `readme` file for some further useful instructions on how to use it.

Table 19.5 List of bilevel programs with equalities constraints

Example name	Ref.	F-G-H-f-g-h	n_x	n_y	n_G	n_H	n_g	n_h
`DempeDutta2012Ex31`	[21]	`L-L-N-N-N-O`	2	2	2	1	2	0
`DempeFranke2011Ex41`	[22]	`N-L-L-N-L-O`	2	2	2	1	4	0
`DempeFranke2011Ex42`	[22]	`N-L-L-N-L-O`	2	2	2	1	3	0
`Zlobec2001b`	[77]	`L-L-O-L-L-N`	1	1	2	0	2	1
`NetworkDesignP1`	[17]	`N-L-O-N-O-L`	5	5	5	0	0	3
`NetworkDesignP2`	[17]	`N-L-O-N-O-L`	5	5	5	0	0	3
`OptimalControl`	[50]	`N-N-O-N-L-L`	2	n_y	3	0	n_y	$\frac{1}{2}n_y$
`RobustPortfolioP1`	[67]	`L-N-L-N-N-O`	N+1	N	N+1	1	N+1	0
`RobustPortfolioP2`	[67]	`L-N-L-N-N-O`	N+1	N	N+1	1	N+1	0
`TollSettingP1`	[17]	`N-L-O-N-L-L`	3	8	3	0	8	5
`TollSettingP2`	[17]	`N-L-O-N-L-L`	3	18	3	0	18	10
`TollSettingP3`	[17]	`N-L-O-N-L-L`	3	18	3	0	18	10
`TollSettingP4`	[17]	`N-O-O-N-L-L`	2	4	0	0	2	4
`TollSettingP5`	[17]	`N-O-O-N-L-L`	1	4	0	0	2	4

Acknowledgements The work of the first and second authors is partly funded by the EPSRC Grant EP/P022553/1. The third author's work is partly funded by the University of Southampton's Presidential Scholarship. We thank Dr Patrick Mehlitz (Brandenburgische Technische Universität Cottbus-Senftenberg) for the `OptimalControl` example and related MATLAB files.

References

1. E. Aiyoshi, K. Shimizu, A solution method for the static constrained Stackelberg problem via penalty method. IEEE Trans. Autom. Control **29**, 1111–1114 (1984)
2. G.B. Allende, G. Still, Solving bilevel programs with the KKT-approach. Math. Program. **138**(1–2), 309–332 (2013)
3. L.T.H. An, P.D. Tao, N.N. Canh, N.V. Thoai, DC programming techniques for solving a class of nonlinear bilevel programs. J. Global Optim. **44**(3), 313–337 (2009)
4. A. Anandalingham, D.J. White, A solution method for the linear static Stackelberg problem using penalty functions. IEEE Trans. Autom. Control **35**(10), 1170–1173 (1990)
5. J. Bard, Optimality conditions for the biievel programming problem. Nav. Res. Logist. Quarterly **31**, 13–26 (1984)
6. J.F. Bard, Convex two-level optimization. Math. Program. **40**, 15–27 (1988)
7. J.F. Bard, Some properties of the bilevel programming problem. J. Optim. Theory Appl. **68**(2), 371–378 (1991)
8. J.F. Bard, *Practical Bilevel Optimization: Algorithms and Applications* (Kluwer Academic Publishers, Dordrecht, 1998)
9. J.F. Bard, J.E. Falk, An explicit solution to the multi-level programming problem. Comput. Oper. Res. **9**, 77–100 (1982)
10. O. Ben-Ayed, C. Blair, Computational difficulties of bilevel linear programming. Oper. Res. **38**, 556–560 (1990)
11. W. Bialas, M. Karwan, Two-level linear programming. Management Science **30**, 1004–1020 (1984)

12. P.H. Calamai, L.N. Vicente, Generating quadratic bilevel programming test problems. ACM Trans. Math. Softw. **20**(1), 103–119 (1994)
13. H.I. Calvete, C. Galé, The bilevel linear/linear fractional programming problem. Eur. J. Oper. Res. **114**(1), 188–197 (1999)
14. W. Candler, R. Townsley, A linear two-level programming problem. Comput. Oper. Res. **9**, 59–76 (1982)
15. P.A. Clark, A.W. Westerberg, Bilevel programming for steady-state chemical process design–I. Fundamentals and algorithms. Comput. Chem. Eng. **14**(1), 87–97 (1990)
16. P.A. Clark, A.W. Westerberg, A note on the optimality conditions for the bilevel programming problem. Nav. Res. Logist. Q. **35**(5), 413–418 (1988)
17. B. Colson, BIPA (Bilevel Programming with Approximation Methods): Software guide and test problems, Technical report, 2002
18. B. Colson, P. Marcotte, G. Savard, A trust-region method for nonlinear bilevel programming: algorithm and computational experience. Comput. Optim. Appl. **30**(3), 211–227 (2005)
19. S. Dempe, A necessary and a sufficient optimality condition for bilevel programming problems. Optimization **25**, 341–354 (1992)
20. S. Dempe, N. Dinh, J. Dutta, Optimality conditions for a simple convex bilevel programming problem, in *Variational Analysis and Generalized Differentiation in Optimization and Control* (Springer, New York, NY, 2010), pp. 149–161
21. S. Dempe, J. Dutta, Is bilevel programming a special case of a mathematical program with complementarity constraints? Math. Program. **131**(1–2), 37–48 (2012)
22. S. Dempe, S. Franke, *An Algorithm for Solving a Class of Bilevel Programming Problems*. Preprint 2011–04 (TU Bergakademie, 2011)
23. S. Dempe, S. Franke, Solution algorithm for an optimistic linear Stackelberg problem. Comput. Oper. Res. **41**, 277–281 (2014)
24. S. Dempe, S. Lohse, *Dependence of Bilevel Programming on Irrelevant Data*. Preprint (TU Bergakademie, 2011). http://www.optimization-online.org/DB_HTML/2011/05/3038.html
25. S. Dempe, B. Mordukhovich, A.B. Zemkoho, Sensitivity analysis for two-level value functions with applications to bilevel programming. SIAM J. Optim. **22**(4), 1309–1343 (2012)
26. A.H. De Silva, Sensitivity formulas for nonlinear factorable programming and their application to the solution of an implicitly defined optimization model of US crude oil production, Ph.D. thesis, George Washington University, 1978
27. F. Facchinei, H. Jiang, L. Qi, A smoothing method for mathematical programs with equilibrium constraints. Math. Program. **85**(1), 107–134 (1999)
28. J.E. Falk, J. Liu, On bilevel programming, Part I: general nonlinear cases. Math. Program. **70**(1), 47–72 (1995)
29. C.A. Floudas, P.M. Pardalos, C. Adjiman, W.R. Esposito, Z.H. Gümüs, S.T. Harding, J.L. Klepeis, C.A. Meyer, C.A. Schweiger, *Handbook of Test Problems in Local and Global Optimization* (Springer, New York, 1999)
30. C.A. Floudas, S. Zlobec, Optimality and duality in parametric convex lexicographic programming, in *Multilevel Optimization: Algorithms and Applications* (Springer, New York, 1998), pp. 359–379
31. S. Franke, P. Mehlitz, M. Pilecka, Optimality conditions for the simple convex bilevel programming problem in Banach spaces. Optimization **67**(2), 237–268 (2018)
32. J. Glackin, J.G. Ecker, M. Kupferschmid, Solving bilevel linear programs using multiple objective linear programming. J. Optim. Theory Appl. **140**(2), 197–212 (2009)
33. Z.H. Gümüş, C.A. Floudas, Global optimization of nonlinear bilevel programming problems. J. Global Optim. **20**(1), 1–31 (2001)
34. K. Hatz, S. Leyffer, J.P. Schlöder, H.G. Bock, Regularizing bilevel nonlinear programs by lifting, Technical Report, Argonne National Laboratory, USA, 2013
35. A. Haurie, G. Savard, D. White, A note on: An efficient point algorithm for a linear two-stage optimization problem. Oper. Res. **38**, 553–555 (1990)
36. J.M. Henderson, R.E. Quandt, *Microeconomic Theory: A Mathematical Approach* (McGraw-Hill, New York, 1958)

37. R. Henrion, T. Surowiec, On calmness conditions in convex bilevel programming. Applicable Analysis **90**(6), 951–970 (2011)
38. T. Hu, B. Huang, X. Zhang, A neural network approach for solving linear bilevel programming problem, in *Proceedinds of the 6th ISNN Conference, AISC*, vol. 56, pp. 649–658 (2009)
39. Y. Ishizuka, E. Aiyoshi, Double penalty method for bilevel optimization problems. Ann. Oper. Res. **34**(1), 73–88 (1992)
40. P.-M. Kleniati, C.S. Adjiman, Branch-and-sandwich: a deterministic global optimization algorithm for optimistic bilevel programming problems, Part II: Convergence analysis and numerical results. J. Global Optim. **60**(3), 459–481 (2014)
41. L. Lampariello, S. Sagratella, Numerically tractable optimistic bilevel problems. Optimization-Online (2017)
42. L. Lampariello, S. Sagratella, A bridge between bilevel programs and nash games. J. Optim. Theory Appl. **174**(2), 613–635 (2017)
43. K.M. Lan, U.P. Wen, H.S. Shih, E.S. Lee, A hybrid neural network approach to bilevel programming problems. Appl. Math. Lett. **20**(8), 880–884 (2007)
44. S. Leyffer, *MacMPEC: AMPL Collection of MPECs* (Argonne National Laboratory, 2000). Available at https://wiki.mcs.anl.gov/leyffer/index.php/MacMPEC
45. G.H. Lin, M. Xu, J.Y. Jane, On solving simple bilevel programs with a nonconvex lower level program. Math. Program. **144**(1–2), 277–305 (2014)
46. Y.-H. Liu, S.M. Hart, Characterizing an optimal solution to the linear bilevel programming problem. Eur. J. Oper. Res. **73**(1), 164–166 (1994)
47. Z.C. Lu, K. Deb, A. Sinha, Robust and reliable solutions in bilevel optimization problems under uncertainties. COIN Report 2016026. Retrived on 19 November 2017 from http://www.egr.msu.edu/kdeb/papers/c2016026.pdf
48. R. Lucchetti, F. Mignanego, G. Pieri, Existence theorems of equilibrium points in Stackelberg games with contraints. Optimization **18**(6), 857–866 (1987)
49. C.M. Macal, A.P. Hurter, Dependence of bilevel mathematical programs on irrelevant constraints. Comput. Oper. Res. **24**(12), 1129–1140 (1997)
50. P. Mehlitz, G. Wachsmuth, Weak and strong stationarity in generalized bilevel programming and bilevel optimal control. Optimization **65**(5), 907–935 (2016)
51. A.G. Mersha, S. Dempe, Linear bilevel programming with upper level constraints depending on the lower level solution. Appl. Math. Comput. **180**, 247–254 (2006)
52. J.A. Mirrlees, The theory of moral hazard and unobservable behaviour: Part I. Rev. Econ. Stud. **66**(1), 3–21 (1999)
53. A. Mitsos, P.I. Barton, A Test Set for Bilevel Programs. Technical report (MIT, 2006). http://www.avt.rwthaachen.de/cms/AVT/Forschung/Software/~dfyqp
54. J. Morgan, F. Patrone, Stackelberg problems: Subgame perfect equilibria via tikhonov regularization. Adv. Dynamic Games 209–221 (2006)
55. L.D. Muu, N.V. Quy, A global optimization method for solving convex quadratic bilevel programming problems. J. Global Optim. **26**(2), 199–219 (2003)
56. J.W. Nie, L. Wang, J.J. Ye, Bilevel polynomial programs and semidefinite relaxation methods. SIAM J. Optim. **27**(3), 1728–1757 (2017)
57. J.V. Outrata, On the numerical solution of a class of Stackelberg problems. Z. Oper. Res. **34**(4), 255–277 (1990)
58. J.V. Outrata, Necessary optimality conditions for Stackelberg problems. J. Optim. Theory Appl. **76**(2), 305–320 (1993)
59. J.V. Outrata, On optimization problems with variational inequality constraints. SIAM J. Optim. **4**(2), 340–357 (1994)
60. J.V. Outrata, M. Červinka, On the implicit programming approach in a class of mathematical programs with equilibrium constraints. Control Cybern. **38**(4B), 1557–1574 (2009)
61. R. Paulavicius, P.M. Kleniati, C.S. Adjiman, BBASBL: Branch-And-Sandwich BiLevel solver. Implementation and computational study with the BASBLib test set. Comput. Chem. Eng. **132**, 106609 (2020)

62. K.H. Sahin, A.R. Ciric, A dual temperature simulated annealing approach for solving bilevel programming problems. Comput. Chem. Eng. **23**(1), 11–25 (1998)
63. Y. Shehu, P.T. Vuong, A.B. Zemkoho, An inertial extrapolation method for convex simple bilevel optimization. Optim. Methods Softw. (in press). Available online at https://www.tandfonline.com/doi/full/10.1080/10556788.2019.1619729
64. K. Shimizu, E. Aiyoshi, A new computational method for Stackelberg and min-max problems by use of a penalty method. IEEE Trans. Autom. Control **26**(2), 460–466 (1981)
65. K. Shimizu, Y. Ishizuka, J.F. Bard, *Nondifferentiable and Two-Level Mathematical Programming* (Kluwer Academic Publishers, 1997)
66. A. Sinha, P. Malo, K. Deb, An improved bilevel evolutionary algorithm based on quadratic approximations, in *2014 IEEE Congress Evolutionary Computation (CEC)*, pp. 1870–1877 (2014)
67. O. Stein, G. Still, Solving semi-infinite optimization problems with interior point techniques. SIAM J. Control Optim. **42**(3), 769–688 (2003)
68. H. Tuy, A. Migdalas, P. Värbrand, A global optimization approach for the linear two-level program. J. Global Optim. **3**(1), 1–23 (1993)
69. H. Tuy, A. Migdalas, N.T. Hoai-Phuong, A novel approach to bilevel nonlinear programming. J. Global Optim. **38**(4), 527–554 (2007)
70. H. Tuy, A. Migdalas, P. Värbrand, A quasiconcave minimization method for solving linear two-level programs. J. Global Optim. **4**, 243–263 (1994)
71. V. Visweswaran, C.A. Floudas, M.G. Ierapetritou, E.N. Pistikopoulos, A decomposition-based global optimization approach for solving bi-level linear and quadratic programs, in *State of the Art in Global Optimization*, ed. by C.A. Floudas, P.M. Pardalos. Nonconvex Optimization and Its Applications, vol. 7, pp. 139–162 (Springer, New York, 1996)
72. S. Vogel, Zwei-Ebenen-Optimierungsaufgaben mit nichtkonvexer Zielfunktion in der unteren Ebene, PhD thesis, Department of Mathematics and Computer Science, TU Bergakademie Freiberg, Freiberg, Germany, 2012
73. Z.P. Wan, G.M. Wang, Y.B. Lv, A dual-relax penalty function approach for solving nonlinear bilevel programming with linear lower level problem. Acta Math. Sci. **31**(2), 652–660 (2011)
74. Y. Wang, Y.C. Jiao, H. Li, A evolutionary algorithm for solving nonlinear bilevel programming based on a new contraint-handling scheme. IEEE Trans. Syst. Man Cybern. C **35**(2), 221–232 (2005)
75. J.J. Ye, D.L. Zhu, New necessary optimality conditions for bilevel programs by combining the mpec and value function approaches. SIAM J. Optim. **20**(4), 1885–1905 (2010)
76. A. Yezza, First-order necessary optimality conditions for general bilevel programming problems. J. Optim. Theory Appl. **89**(1), 189–219 (1996)
77. S. Zlobec, Bilevel programming: optimality conditions and duality, in *Encyclopedia of Optimization*, pp. 180–185, ed. by C. Floudas, P. Pardalos (Springer, New York, 2001)

Chapter 20
Bilevel Optimization: Theory, Algorithms, Applications and a Bibliography

Stephan Dempe

Abstract Bilevel optimization problems are hierarchical optimization problems where the feasible region of the so-called upper level problem is restricted by the graph of the solution set mapping of the lower level problem. Aim of this article is to collect a large number of references on this topic, to show the diversity of contributions and to support young colleagues who try to start research in this challenging and interesting field.

Keywords Bilevel optimization · Mathematical programs with complementarity constraints · Optimality conditions · Applications · Necessary optimality conditions · Solution algorithms · Metaheuristics · Optimistic and pessimistic bilevel optimization problems

20.1 Introduction

Bilevel optimization problems are hierarchical optimization problems of two or more players. For defining them consider first a parametric optimization problem

$$\min_y \{f(x, y) : g(x, y) \leq 0, \ y \in Y\}, \tag{20.1.1}$$

where $f, g_i : \mathbb{R}^m \times \mathbb{R}^n \to \mathbb{R}, i = 1, \ldots, p$ and $Y \subseteq \mathbb{R}^n$. Here, equality constraints can be added if the regularity conditions are adapted accordingly. This is the problem of the *lower-level decision maker*, sometimes called the *follower's*

S. Dempe (✉)
TU Bergakademie Freiberg, Institute of Numerical Mathematics and Optimization, Freiberg, Germany
e-mail: dempe@tu-freiberg.de

S. Dempe, A. Zemkoho (eds.), *Bilevel Optimization*, Springer Optimization and Its Applications 161, https://doi.org/10.1007/978-3-030-52119-6_20

problem. If problems with more than one decision maker in the lower level are considered, then e.g. a Nash equilibrium is searched for between them. Let

$$\varphi(x) := \min_y \{f(x, y) : g(x, y) \leq 0, \ y \in Y\} \tag{20.1.2}$$

denote the optimal value function of problem (20.1.1) and

$$\Psi(x) := \{y \in Y : g(x, y) \leq 0, \ f(x, y) \leq \varphi(x)\} \tag{20.1.3}$$

the solution set mapping of problem (20.1.1). If $\operatorname{gph}\Psi := \{(x, y) : y \in \Psi(x)\}$ is used to abbreviate the graph of the solution set mapping Ψ, the *bilevel optimization problem*

$$\text{'}\min_x\text{'}\{F(x, y) : G(x) \leq 0, \ (x, y) \in \operatorname{gph}\Psi, \ x \in X\} \tag{20.1.4}$$

can be formulated with $X \subseteq \mathbb{R}^m$, $F : \mathbb{R}^m \times \mathbb{R}^n \to \mathbb{R}$, $G_j : \mathbb{R}^m \to \mathbb{R}$, $j = 1, \ldots, q$. Sometimes, this problem is called the *upper level* optimization problem or the *problem of the leader*. Here we used quotation marks to indicate that this problem is not well-defined in case of multiple lower-level optimal solutions.

For simplicity we assume that $\Psi(x) \neq \emptyset$ for all $x \in X$ with $G(x) \leq 0$. This assumption can be weakened and is only used to guarantee that the optimal value function (20.1.2) and the solution set mapping defined in (20.1.3) are well defined.

20.2 History

Problem (20.1.1), (20.1.4) has been formulated for the first time in an economic context by v. Stackelberg [1245]. Many economic articles investigate related principal-agency problems, see some references below. Hence, it is sometimes called *Stackelberg game* and its solution a Stackelberg equilibrium. About 40 years later, this problem has been introduced into the mathematical community [208, 273, 274, 522, 761, 981]. Since then a large number of articles illustrating different views on the topic, investigating various questions both from theoretical or numerical point-of-view or numerous applications of the problem appeared. It is our aim here to give the reader some insight into the topic and its investigations. Citations in this article refer to items in the bibliography given at the end of this chapter. Without any doubt, this bibliography cannot be complete. Also, due to space limitations, not all items in the large list of references can be mentioned in the text. This, of course, does not mean that the items not used in the text are of less importance.

Bilevel optimization problems are nonconvex and nondifferentiable optimization problems even if all their defining functions are convex and smooth, see [326].

20.3 Overviews and Introductions

Well formulated introductory texts on bilevel optimization can be found in [66, 311, 336, 646, 755, 782, 1311]. Reference [1220] gives an overview over evolutionary methods. In [1084], the authors compare Nash, Cournot, Bertrand and Stackelberg games, describe ideas for solving them as well as some applications, see also [171, 1410]. An overview over the investigations at the Montreal school in the years before 2008 is given in [218]. An overview over solution algorithms for (mixed-integer) linear bilevel optimization problems can be found in [1143], and for general bilevel optimization problems in [1076]. Structural properties of the feasible set of mixed-integer bilevel optimization problems can be found in [138].

Monographs, textbooks and edited volumes on the topic are Bard [131], Dempe [385], Dempe et al. [413], Dempe and Kalashnikov [397], Kalashnikov et al. [710], Migdalas et al. [961], Mesanovic et al. [955], Sakawa [1146], Sakawa and Nishizaki [1151], Shimizu et al. [1200], Stein [1249], Talbi [1270], Xu et al. [1388], Zhang et al. [1458]. Bilevel optimization problems are the topic of a chapter in the monograph [479].

Bilevel optimization problems with many followers and the three-level optimization problem have been investigated in [40, 64, 600–602, 604, 817, 889, 894, 1035]. A comment to [64] can be found in [1227].

We have used constraints in the upper-level problem of the form $G(x) \leq 0$. Sometimes upper-level constraints of the form $G(x, y) \leq 0$, so-called joint constraints, are investigated. These problems are not easy to interpret in a game theoretic context since then, the leader has to select first his / her decision and gives it to the follower. The latter then computes one optimal reply on the leader's choice and gives it back to the leader, who only now is able to check if his / her initial selection was a feasible one. If the follower's selection was not unique, feasibility of the leader's selection also depends on the response of the follower. This was the motivation of the authors in [1192] to suggest to move joint upper-level constraints into the lower level to derive a "correct" definition. This approach is shown to be not correct in [101, 952] since it changes the problem seriously. Joint constraints can make the feasible set of the bilevel optimization problem empty even if $\Psi(x) \neq \emptyset$ for all $x \in X$, or disconnected even if gph Ψ is connected, see [952].

The three-level problem is investigated in [126, 362, 517]. The articles [128, 158, 1131] explain the geometry of bilevel and multilevel optimization problems. A global optimal solution of multilevel optimization problems is approximated in [726]. In [517], three different types of optimistic formulations of three-level optimization problems are suggested and compared. The electrical network defense is formulated as a three-level mixed-integer linear programming model [1411]. Necessary optimality conditions and assumptions guaranteeing the existence of an optimal solution for the three-level optimization problem can be found in [817].

Examples showing nonexistence of optimal solutions can be found in [130].

Survey papers are [148, 335, 384, 387, 435, 707, 713, 884, 934, 935, 960, 1360]. A survey for the pessimistic bilevel optimization problems is [858].

First bibliographies can be found in [386, 1312, 1333].

The formulation of problems with an infinite number of lower-level decision makers as a stochastic bilevel optimization problem is given in [895].

20.4 Theoretical Properties and Relations to Other Optimization Problems

20.4.1 Formulation of the Bilevel Optimization Problem

20.4.1.1 Optimistic vs. Pessimistic Formulation

The formulation of the bilevel optimization problem as given in (20.1.1), (20.1.4) is not clear in case of multiple lower-level optimal solutions for some of the selections of the upper-level decision maker. In that case, the leader may assume that the follower can be motivated to select a best optimal solution in $\Psi(x)$ with respect to the leader's objective function. This is the so-called optimistic or weak formulation of the bilevel optimization problem, investigated in most of the references:

$$\min\{\varphi_o(x) : G(x) \leq 0,\ x \in X\},$$

where

$$\varphi_o(x) = \min_y\{F(x, y) : y \in \Psi(x)\}. \tag{20.4.1}$$

This problem is almost equivalent to

$$\min_{x,y}\{F(x, y) : G(x) \leq 0,\ x \in X,\ (x, y) \in \operatorname{gph} \Psi\}, \tag{20.4.2}$$

see [385]. If the upper-level objective function is of a special type the optimistic bilevel optimization problem can be interpreted as an inverse optimization problem [23, 670, 1470].

Relations to generalized semi-infinite optimization problems can be found in [1250].

If this optimistic assumption is not possible or even not allowed the leader is forced to bound the damage resulting from an unwelcome selection of the follower resulting in the pessimistic or strong formulation of the bilevel optimization problem:

$$\min\{\varphi_p(x) : G(x) \leq 0,\ x \in X\}, \tag{20.4.3}$$

where

$$\varphi_p(x) = \max_y \{F(x, y) : y \in \Psi(x)\}. \qquad (20.4.4)$$

The formulation of the pessimistic Stackelberg game is given in [805, 986]. The existence of a pessimistic (strong) or an optimistic (weak) optimal solution is considered in [5, 14, 15, 893], the same based on d.c. optimization is investigated in [8]. For the existence and stability of pessimistic solutions in general spaces, the reader is referred to [833, 834, 836, 840]. Topic of the article [1068] is the existence of solutions in Banach spaces if the solution of the lower-level problem is strongly stable. The possible nonexistence of pessimistic optimal solutions is shown in [893].

In [1370], the pessimistic bilevel optimization problem with an objective function not depending on the lower-level variable is formulated as

$$\min\{F(x) : x \in X,\ G(x, y) \le 0\ \forall\ y \in \Psi(x)\}. \qquad (20.4.5)$$

The relations between this formulation and pessimistic bilevel optimization as given in (20.4.3) are investigated in [1370]. In the same vein, the optimistic bilevel optimization problem reads as

$$\min\{F(x) : x \in X,\ G(x, y) \le 0\ \exists\ y \in \Psi(x)\}.$$

Using an idea in [787], the pessimistic bilevel optimization is formulated as an optimistic one with a two follower Nash game in the lower level [789]. Since the pessimistic bilevel optimization problem does not have an optimal solution in general, the authors replace the solution set mapping of the lower-level problem with the ε-optimal solutions. They show that the resulting perturbed pessimistic bilevel optimization problem and the optimistic bilevel problem with the two follower Nash game in the lower level are equivalent with respect to global optimal solutions.

20.4.1.2 Optimization over the Efficient Set

The search for a "best" efficient solution of a multicriterial optimization problem can be formulated as a bilevel optimization problem [159, 160, 188, 189, 194, 472, 644, 645, 695, 738, 843, 993, 994, 1241, 1281, 1283, 1296], see also the overview in [1399].

Properties of the problem and the replacement of the upper-level objective function $\langle d, x\rangle$ by a constraint $\langle d, x\rangle \le t$ for an unknown t in the lower-level problem can be found in [157]. The "best" efficient solution is obtained for the smallest t for which the resulting problem has an optimal solution.

A special model is the simple bilevel optimization problem in [392], where a "best" solution of an optimization problem is searched for. Optimality conditions for that problem can be found in [392], a solution algorithm is given in [1135, 1235, 1236].

Stochastic problems of this type are topic of [193].

20.4.1.3 Semivectorial Bilevel Optimization: Vector-Valued Lower-Level Problems and Problems with Multiobjective Upper-Level Problems

Bilevel optimization problems where the lower-level problem is a multiobjective optimization problem are often called semivectorial bilevel optimization problems [36, 87, 192, 196, 198, 259, 409, 533, 818, 904, 1011, 1105].

Here, the scalarization approach is often used to transform the semivectorial bilevel optimization problem into an optimistic bilevel problem, see e.g. [409]. For that, the scalarization vector appears as new variable in the upper-level problem. In the case that local optimal solutions are computed for the resulting bilevel problem, this is a delicate issue [424].

Using the scalarization approach and indicator functions as terms in the upper-level objective function the semivectorial bilevel optimization problem is transformed into a single-level one for which the nonsmooth Mangasarian-Fromovitz constraint qualification can be satisfied, see [534].

The use of utility functions as well as optimistic and pessimistic approaches to investigate linear bilevel problems with multiobjective functions in both the leader's and the follower's problems can be found in [1015], for the same in the case of stochastic data see [1017].

Multiobjective optimization in the upper level of a linear bilevel optimization problem is the topic of [890, 905], fuzzy optimization approaches are applied in this case in [202]. Problems with multiobjective upper-level problems are investigated in [1418] using a combination of the KKT and the optimal value function approach to transform the bilevel optimization problem into a single-level one.

Optimality conditions for nonlinear bilevel vector optimization problems and a global solver can be found in [551].

20.4.1.4 Fuzzy Bilevel Optimization Problems

The investigation of bilevel optimization problems with fuzzy lower-level problems can be found in [432, 657, 850, 1453, 1457].

The fuzzy linear bilevel optimization problem is transformed into a crisp problem and then solved using a k-th best algorithm in [1100, 1147, 1148, 1454] or using the KKT approach [1449, 1452]. Solving this problem using an interactive approach has been the topic of [1109, 1148, 1152, 1154–1156]. A solution algorithm for fuzzy bilevel optimization problems using α-cuts is given in [225, 542, 544].

For the computation of a satisfactory solution see [786, 1085, 1086, 1127, 1140]. Here, in some sense, an approach is used which is related to multiobjective optimization, see remarks in [388]. The transformation of a bilevel optimization problem using ideas from fuzzy optimization into the problem of maximizing membership functions related to both the objective functions of the leader's and the follower's problem at the same time is not a possible approach for solving the bilevel optimization problem, see [388].

Fuzzy (random) bilevel optimization has been applied

1. in the Shuibuya hydropower project [1385],
2. in logistics planning [702],
3. to model the water exchange in eco-industrial parks [112].

20.4.1.5 Stochastic Problems

The model of stochastic bilevel optimization and solution algorithms can be found in [41, 285, 320, 368, 412, 536, 639, 668, 669, 717, 762, 892, 1052, 1054, 1055, 1078]. Stochastic bilevel multi-leader multi-follower problems are investigated in [367, 1384]. In [367] the authors show uniqueness of the stochastic multi-leader Stackelberg-Nash-Cournot equilibrium and suggest an algorithm for computing it.

The transformation of two-stage stochastic bilevel optimization problems into mixed-discrete optimization problems can be found in [1409].

20.4.1.6 Bilevel Optimization Problems with Fixed-Point Constraints

Existence theorems for bilevel optimization problems with fixed-point constraints are the topic of [847]. Special cases are MPECs and semi-infinite optimization problems.

Robust polynomial bilevel optimization problems are investigated in [322]. Robust Stackelberg problems are the topic of [864].

20.4.1.7 Bilevel Equilibrium Problems

A bilevel equilibrium problem has been formulated in [990] which has been the topic of investigations of many articles since then, see e.g. [888, 1308]. This problem is an hierarchical problem where both the upper and the lower-level problems are formulated as variational inequalities. As in bilevel optimization, the solution of the upper-level problem is a parameter of the lower-level problem and the solution of the lower-level problem is used to formulate the constraint in the upper-level one.

20.4.2 Dependence on Data Perturbations

The dependence of optimal solutions of bilevel optimization problems on data perturbations has been investigated by some authors, see e.g. [13, 1341]. The authors of [12] replace the lower-level problem using ε-optimal solutions and consider convergence of the solutions for $\varepsilon \downarrow 0$. Stability considerations can be found in [399, 511, 681, 833–836, 919], the same using the transformation by an inclusion constraint is given in [448, 449].

A surprising fact is that, global optimal solutions of the bilevel optimization problem need not to remain globally optimal if a constraint is added to the lower-level problem which is inactive at the optimal solution [421, 910].

The structure of the feasible set of bilevel optimization problems has been the topic of [411, 457, 694], see also [1440]. The Mangasarian-Fromovitz constraint qualification is not generically satisfied in the lower-level problem at optimal solutions of (optimistic) bilevel optimization problems. If it is violated for some values of the upper-level variable, the feasible set of the KKT transformation is in general no longer closed. That encouraged the authors of [43] to replace the lower-level problem using the F.-John necessary optimality conditions. Generic properties of an optimal solution of the resulting problem can be found in [43], see [402] with a comment to that approach.

Upper and lower bounds for the optimal objective function value of bilevel problems with interval coefficients are computed in [254]. Bilevel linear programming problems with interval coefficients have also been considered in [1102, 1110, 1162].

An important result related to this is well-posedness of bilevel optimization problems [78].

20.4.3 *Possible Transformations*

To investigate properties, for the formulation of optimality conditions and solution algorithms, the bilevel optimization problem needs to be transformed into a single-level problem. For this, different approaches are possible. Assume that $Y = \mathbb{R}^n$ in this subsection.

20.4.3.1 Use of the Karush-Kuhn-Tucker Conditions of the Lower-Level Problem

If the functions $y \mapsto f(x, y)$, $y \mapsto g_i(x, y)$, $i = 1, \dots, p$, are differentiable and a regularity condition is satisfied for the lower-level problem for all $(x, y) \in \operatorname{gph} \Psi$, problem (20.4.2) can be replaced by

$$\begin{aligned}\min_{x,y,u} \{F(x, y) : G(x) \le 0,\ x \in X,&\\ \nabla_y \{f(x, y) + u^\top g(x, y)\} = 0,&\\ u \ge 0,\ g(x, y) \le 0,\ u^\top g(x, y) = 0\}.&\end{aligned} \tag{20.4.6}$$

It is shown in [967] that this approach is only possible if the lower-level problem is a convex one. Problem (20.4.6) is a so-called mathematical program with equilibrium (or complementarity) constraints (MPEC), see [897]. This problem is a nonconvex optimization problem for which the Mangasarian-Fromovitz constraint qualification is violated at every feasible point [1169]. This transformation is the most often used one, MPECs have intensively been investigated. In [103, 522, 532] the complementarity constraint in the Karush-Kuhn-Tucker (KKT) transformation (20.4.6) is replaced using Boolean variables. Problem (20.4.3) is compared with its transformation using the KKT conditions of the lower-level problem in [110].

Relations between the KKT and the optimal value transformations as well as between the KKT transformation and the original bilevel optimization problem are highlighted in [393]. A similar question is investigated in [1248] for global optimal solutions of the bilevel optimization problem and its KKT transformation formulated as a mixed Boolean optimization problem. The combination of the KKT and the optimal value function approaches for the transformation of bilevel optimization problems (with nonconvex lower-level problems) has been suggested in [1073, 1420]. The MPEC-LICQ is generically satisfied for MPECs [1172]. An example showing that the MPEC-LICQ can be violated for the KKT transformation of a bilevel optimization problem, even if LICQ and a strong sufficient optimality condition of second order are satisfied at the optimal solution of the lower-level problem, can be found in [614] together with a lifting approach to satisfy this regularity condition for the perturbed problems.

20.4.3.2 Use of Necessary Optimality Conditions Without Lagrange Multipliers

Let, for $x \in X$,

$$M(x) := \{y : g(x, y) \leq 0\}$$

denote the feasible set of the lower-level problem, assume that $y \mapsto f(x, y)$ is a convex function and, for arbitrary, fixed $x \in X$, $M(x) \subseteq \mathbb{R}^m$ is a convex set. Then, $y \in \Psi(x)$ if and only if $0 \in \partial_y f(x, y) + N_{M(x)}(y)$, where

$$N_A(z) = \{d : d^\top (w - z) \leq 0 \; \forall \; w \in A\}$$

denotes the normal cone in the sense of convex analysis to a closed set A and it is assumed that $N_A(z) = \emptyset$ if $z \notin A$. Thus, (20.4.2) can be replaced by

$$\min\{F(x, y) : G(x) \leq 0, \; x \in X, \; 0 \in \partial_y f(x, y) + N_{M(x)}(y)\}. \qquad (20.4.7)$$

Problem (20.4.7) is fully equivalent to (20.4.2). Problem (20.4.7) is also called an optimization problem with a generalized equation or with a variational inequality, it has been studied in [435, 982, 983, 985, 1419].

20.4.3.3 Use of the Optimal Value Function

Problem (20.4.2) can be equivalently replaced by

$$\min\{F(x,y) : G(x) \leq 0,\ x \in X,\ g(x,y) \leq 0,\ f(x,y) \leq \varphi(x)\}. \tag{20.4.8}$$

This transformation has first been used in [1037, 1038]. Problem (20.4.8) is a nonsmooth optimization problem since the optimal value function $\varphi(x)$ is, even under restrictive assumptions, in general not differentiable. Moreover, the nonsmooth Mangasarian-Fromovitz constraint qualification is violated at every feasible point, see [1072, 1422].

20.4.3.4 Transformation Using a Variational Inequality

If the definition of the normal cone is used, problem (20.4.7) is

$$\begin{aligned} \min\{F(x,y) : G(x) \leq 0,\ x \in X,\ y \in M(x), \\ \langle \nabla_y f(x,y), z - y \rangle \geq 0\ \forall\, z \in M(y)\}, \end{aligned} \tag{20.4.9}$$

where the problem

$$\text{find } y \in M(x) \text{ such that } \langle \nabla_y f(x,y), z - y \rangle \geq 0\ \forall z \in M(x)$$

is often called a generalized variational inequality. Here, f is assumed to be differentiable, see [1424].

The problem of minimizing some objective function subject to the solution of a parametric variational inequality is called a generalized bilevel optimization problem in [835].

Bilevel variational inequalities have been introduced in [714, 1320] and solved e.g. in [164, 832].

The existence of solutions for bilevel variational inequalities, where both the upper- and lower-level problems are formulated using variational inequalities, is topic in [835, 837, 838].

The investigation of bilevel variational inequalities under invexity assumptions can be found in [298].

The application of bilevel optimization to solve fuzzy variational inequalities is suggested in [500].

20.4.3.5 Formulation as a Set-Valued Optimization Problem

Using

$$\mathcal{F}(x) := \bigcup_{y \in \Psi(x)} F(x,y),$$

problem (20.4.2) can be replaced with

$$\text{" min "}\{\mathcal{F}(x) : G(x) \leq 0, \ x \in X\}. \tag{20.4.10}$$

This formulation has been the topic of [428, 1074]. For this approach, the notion of an optimal solution needs to be defined first.

20.4.4 General Properties

Optimal solutions of certain bilevel optimization problems can be found at vertices of the feasible set [128, 177, 241–243, 250, 255, 865, 942, 943].

Since the bilevel optimization problem can be interpreted as a hierarchical game it is interesting to ask if it is beneficial to act as leader, see [53, 701, 704, 1290, 1441].

Optimal solutions of the bilevel optimization problem are in general not Pareto optimal if optimization is done w.r.t. the objective functions of both levels at the same time [388, 1234]. The relationship between the bilevel problem and bicriterial optimization is illustrated in [271, 932, 959, 1359]. The correct formulation of multicriterial optimization problems which are equivalent to the bilevel problem can be found in [516, 531, 667, 1070, 1132]. An application of these results to solve linear bilevel optimization problems is given in [562].

In [1361, 1362, 1366], a possible transformation of an optimal solution of the bilevel optimization problem into a Pareto optimal solution is investigated.

For problems with multiple followers see [692, 1450], problems with multiple leaders have been investigated in [109, 368, 1186]. Nine different kinds of relationships between followers in bilevel optimization problems with multiple followers can be found in [885]. For a variational inequality formulation of problems with multiple leaders we refer to [649]. The existence of optimal solutions for problems with multiple leaders as well as the relation of this problem to its KKT transformation (both with respect to global and local optimal solutions) has been investigated in [109].

Phenomena of inverse Stackelberg problems are described in [1027, 1028].

$\mathcal{NP}$-hardness of bilevel optimization problems has been shown in [130, 185, 439, 676]. Computational complexity of the bilevel knapsack problems is the topic of [278].

Relations to mixed-integer optimization problems are topic of [523]. The investigation of a special problem which is polynomially solvable can be found in [748, 749, 1081].

20.4.5 Problems in Infinite Dimensional Spaces

Bilevel optimization problems in general spaces are considered in [450, 757, 944, 946].

For the investigation of semivectorial bilevel optimization on Riemannian manifolds the reader is referred to [200]. The existence of Stackelberg equilibrium points on strategy sets which are geodesic convex in certain Riemannian manifolds has been shown in [766].

The KKT transformation is applied to the bilevel optimization problem in infinite dimensional spaces in [944]. For the optimal value transformation, regularity conditions and optimality conditions see [1416]

Necessary optimality conditions in form of M-stationarity conditions for problems with second order cone programming problems in the lower-level can be found in [310, 1469]. The robust optimization method for bilevel optimization problems with uncertain data in the lower-level problem can lead to a bilevel programming problem with a second-order cone program in the lower-level [309].

The existence of optimal solutions has been investigated for problems in Banach spaces [837] and for problems in locally convex Hausdorff topological vector spaces [846].

Stackelberg games go back to the original definition by H.v. Stackelberg [1245] and refer to problems where the feasible set of the lower level does not depend on the upper-level variable. Recently investigated related problems are:

1. closed-loop Stackelberg games [734, 976, 1205, 1289],
2. dynamic Stackelberg problems [962, 974, 1012, 1013],
3. reverse Stackelberg problems [578–582]: a special point is the computation of optimal incentive functions resulting in realization of the leader's aim by the follower.
4. Stackelberg differential games [1041]
5. Stackelberg equilibria in an oligopoly [139].

Bilevel optimal control problems are another rather recent area of research. Here, two optimal control problems are combined in a hierarchical sense. Such problems have been considered in [152, 154, 162, 745, 757, 948]. An application of Stackelberg games to optimal control problems can be found in [987]. Pursuit-evasion games are discretized and transformed into a bilevel programming problem in [478].

20.4.6 Discrete Bilevel Optimization Problems

A general introduction into discrete bilevel optimization problems can be found in [1315]. For the special case, when the lower level problem is a parametric knapsack problem see [214, 215, 429, 1089] or a matroid problem [503]. For nonlinear integer bilevel optimization problems see [455, 675].

A transformation of bilevel optimization problems with mixed-integer lower-level problems into a single-level problem using minima of optimal value functions of (partial) lower-level problems is suggested in [813].

Using a special penalization approach, the mixed-discrete bilevel optimization problem can be transformed into a continuous one. Assuming partial calmness, optimality conditions for both the optimistic and the pessimistic problems are derived in [418].

Complexity of a bilevel perfect matching problem is the topic of [549], for the bilevel minimum spanning tree problem see [282, 283].

Optimality conditions for problems with discrete parametric lower-level problems using the radial subdifferential can be found in [501] or, in case of a parametric matroid problem in the lower-level, in [503].

Multi-leader-follower games are investigated in [769, 812, 1048, 1292]. The existence for equilibria in such problems is investigated in [1433]. A Gauss–Seidel method for solving the EPEC transformation of such problems has been developed in [643].

20.5 Optimality Conditions

20.5.1 Strongly Stable Lower-Level Optimal Solution

For necessary optimality conditions and solution algorithms using strongly stationary solutions in the lower-level problem see [370, 371, 373–375, 381, 382, 854, 1038, 1039]. Necessary optimality conditions using the implicit function theorem and variational analysis can be found in [1440]. Here the author verifies that the posed necessary optimality conditions are generically satisfied at local minima of smooth bilevel optimization problems and that partial calmness is often violated. Reference [635] describes an idea to formulate necessary (and sufficient) optimality conditions after deleting some inequality constraints in (20.4.6).

20.5.2 Use of the KKT Transformation

Necessary optimality conditions using the KKT transformation are formulated in [434, 436, 1340] and using a generalized equation in [17, 434].

The transformation using the F.-John conditions applied to the lower-level problem in place of the KKT conditions is investigated in [43], see [402] for an example with a nonconvex lower-level problem where the global optimum cannot be computed with this approach.

Using Boolean variables, the resulting mathematical program with complementarity constraints is transformed into a mixed integer optimization problem in [103], bounds for the sufficiently large constants (big-M) can be found in [445]. Verifying that a given big-M does not cut off any feasible vertex of the linear bilevel optimization problem cannot be done in polynomial time unless $\mathcal{P} = \mathcal{NP}$, see [741].

20.5.3 Transformation Using the Optimal Value Function

The generalized derivative of the optimal value function of the lower-level problem, violation of the MFCQ, and necessary optimality conditions can be found in [302]. Calmness properties of the transformed problem can be found in [633], see also [1037]. Optimality conditions using the optimal value transformation for the optimistic bilevel optimization problem are derived in [394–396, 425, 433, 469, 854, 984, 1422, 1427]. For the case when the lower-level problem is an optimal control problem, see [1413, 1415]. For optimality conditions for infinite and semi-infinite bilevel optimization problems see [452].

Necessary and sufficient optimality conditions based on Fenchel-Lagrange duality are investigated in [6].

Optimality conditions using variational analysis for the pessimistic problem are the topic of [426].

Problems with multiobjective upper-level problems have been investigated in [405, 408, 480]. Convexificators and exhausters can be used to study extremum problems involving functions which are not convex, quasidifferentiable or locally Lipschitz continuous. They are used to derive necessary optimality conditions in [783].

20.5.4 Set-Valued Optimization Approach

Optimality conditions using the formulation of the bilevel optimization problem as a set-valued optimization problem can be found in [428, 1444]. Optimality conditions are derived applying the coderivative of Mordukhovich [982] to the graph of the solution set mapping Ψ, see [1474, 1475].

20.5.5 Transformation into a Semi-infinite Optimization Problem

Optimality conditions using a semi-infinite transformation of the bilevel optimization problem can be found in [127]. A counterexample to this result is given in [323].

20.5.6 *Optimality Conditions for Semivectorial Bilevel Problems*

Optimality conditions for the pessimistic semivectorial bilevel optimization problem using variational analysis and the transformation using the optimal value function of the lower-level problem can be found in [853]. Application of the scalarization approach to the lower-level problem followed by the KKT or the optimal value transformation of the resulting problem leads to necessary optimality conditions, see [818].

20.5.7 *Optimality Conditions for the Simple Bilevel Optimization Problem*

The problem of finding a special point in the set of optimal solutions of a convex optimization problem is called a simple bilevel optimization problem. In some sense, this problem parallels the aim to find a "best" (weak) Pareto optimal solution of a multiobjective optimization problem. Optimality conditions for a simple bilevel optimization problem are given in [26, 392].

20.5.8 *Optimality Conditions for the Three-Level Optimization Problem*

In [832] the lower level is first transformed using the KKT conditions and then, optimality conditions for the resulting problem are formulated using variational analysis.

20.5.9 *Other Approaches*

The formulation of necessary optimality conditions exploiting convexificators can be found in [115, 404, 1265].

The extremal principle [982] is used for describing necessary optimality conditions in [140, 407]. For problems with set-valued optimization problems in both levels, optimality conditions using the variational principle can be found in [410].

Different optimality conditions and transformations are collected in [415].

Necessary and sufficient optimality conditions using a linearization of the inducible region [296] and by using a description of the tangent cone to the feasible set [1313] have been derived.

Input optimization is used in [1293].

For necessary and sufficient optimality conditions under generalized invexity assumptions see [205].

Necessary conditions for a global optimal solution using a bilevel Farkas lemma can be found in [679].

20.5.10 Second-Order Optimality Conditions

Second order necessary and sufficient optimality conditions for the optimistic bilevel optimization problem are obtained in [406].

20.6 Solution Algorithms

20.6.1 Pessimistic Problem

Properties, existence and stability of the pessimistic bilevel problem are investigated in [423, 770, 873–882].

Using ε-optimal solutions in the lower-level problem, convergence to a pessimistic solution for $\varepsilon \downarrow 0$ is shown in [918].

In [16] the duality gap of the linear lower-level problem is penalized in the upper-level objective function to solve the pessimistic bilevel optimization problem. Some incorrectness in this article is found and corrected in [1483]. The pessimistic linear bilevel optimization problem with multiple followers is solved using a penalization of the duality gap in [1495]. Here, the different followers share some of the resources.

An algorithm for solving the pessimistic bilevel optimization problem using a regularization approach can be found in [170]. For using an entropy approach see [1496].

For use of the k-th best algorithm to solve the pessimistic linear bilevel problem see [1483].

Partial cooperation between the leader and the follower (i.e. weighted sum of the optimistic and the pessimistic approaches) for linear bilevel optimization problems is the topic of [276, 1487].

20.6.2 Optimistic Problem

20.6.2.1 Enumeration

Properties and an algorithm for linear bilevel optimization problems can be found in [1129, 1130].

Enumeration of the basic matrices of the lower-level problem in linear bilevel optimization is suggested in [275, 1000]. Using this idea it is shown in [866] that the algorithm is of polynomial time if the number of variables in the lower-level problem is fixed.

Vertex enumeration plus a descent algorithm is used in [599]. Convergence to a local optimum for linear bilevel optimization problems by investigating adjacent extreme points is shown in [177, 179, 369]. A simplex-type algorithm applied to an exactly penalized problem is suggested in [269]. A descent algorithm computing a local optimal solution for linear-quadratic bilevel optimization problems can be found in [1257, 1260].

The k-th best algorithm has originally been published in [1356], see also [247, 1483]. The same algorithm for bilevel problems with partially shared variables between followers is given in [1193]. For an application of the k-th best algorithm to three-level optimization problems the reader is referred to [600, 1459].

The use of the solution package Pyomo [1087] for solving the optimistic bilevel optimization problem is described in [610].

20.6.2.2 Use of KKT Transformation

Solution algorithms using the Karush-Kuhn-Tucker transformation can be found in [122, 123, 125, 129, 132, 697–699, 886]. Solving the KKT transformation using branch-and-bound is suggested in [133, 522]. Gomory-like cuts in a branch-and-cut algorithm solving the KKT transformation of the bilevel problem are applied in [105].

A penalty function algorithm for problems with quadratic lower-level problems can be found in [916].

Solution algorithm for the problem (20.4.6) after replacing the complementarity constraint using Boolean variables can be found in [104, 608].

Branch-and-bound algorithm is applied in [179, 474, 487, 608], and for problems with multiple followers in [887].

Penalization of the duality gap for bilevel problems with linear lower-level problem is done in [68, 245, 265, 266, 902, 1368, 1397, 1488–1490], for nonlinear lower level problems see [938].

A penalty function approach to the lower-level problem is applied in [24, 25, 1367] to derive an unconstrained optimization problem which can be replaced by its gradient. Similar ideas for problems with connecting upper-level constraints can be found in [659, 950].

Penalization of the complementarity constraint for the linear bilevel optimization problem is done in [1030], the results of this article are corrected in [267], see also [268, 270]. Exact penalization of the complementarity constraint under partial calmness is used in [856].

An approximate global optimal solution is searched for in [1484].

The authors of [634] solve the KKT-transformation by implicit use of the inequality constraints. The complementarity constraints are replaced by concave inequalities in [30].

A successive approximation of the feasible set of the KKT transformation is obtained in [1269].

The use of approximations for both the complementarity constraint and the y-gradient of the Lagrange function of the lower-level problem make the computation of local or global optimal solutions of the bilevel optimization problem possible, see [403] and [953] for an earlier approach. A similar approximation has been used in [1225] in combination with an evolutionary algorithm.

Application of disjunctive cuts to the KKT transformation of the bilevel optimization problem can be found in [102].

Benders decomposition algorithm is realized in [521] for problems with discrete upper and linear lower level problems.

20.6.3 If the Lower-Level Problem Has a Strongly Stationary Solution

Solution algorithms using strong stability of the lower-level optimal solution can be found in [378, 441, 496, 497]. Investigation of large problems is done in [756]. Comparison of different solution algorithms (Hooke-Jeeves algorithm, bilevel descent algorithm, MINOS and others) can be found in [1263].

After inserting the (optimal) solution function of the lower-level problem into the upper-level problem, a nonsmooth optimization problem arises which can be solved using various solution algorithms.

1. An interior point algorithm is applied in [807],
2. a trust region algorithm is the choice in [855],
3. an inexact restoration algorithm where the lower-level problem is solved at each iteration, can be found in [71],
4. a bundle algorithm is suggested in [383, 384, 391]. This approach is generalized to the case of nonunique optimal solutions in the lower-level problem in [430].
5. A feasible direction method is used in [954],
6. a steepest descent algorithm can be found in [1166, 1314],
7. Reference [451] use an extragradient cutting-plane-method,
8. a pattern search method has been described in [456].

20.6.4 Use of the Optimal Value Function of the Lower-Level Problem

Solution algorithms using the optimal value function transformation can be found in [400, 402, 975, 1199].

Smooth upper approximations of the optimal value function of the lower-level problem are used in [100] and in [970, 971] for solving mixed-discrete bilevel problems. If the optimal value function of the lower level problem is convex or concave these properties can be applied for an upper estimation of the feasible set of problem (20.4.8). This has been done for the computation of global and local optimal solutions in [400, 402].

Kriging is used in [1210] for an interpolation of the optimal value function.

The optimal value transformation is used to find relations between Stackelberg and Nash equilibria in [787]. A related algorithm solving bilevel optimization problems where the lower-level problem is fully convex with a parameter independent feasible set is suggested in [788].

A cutting plane approach is applied to a reverse-convex transformation of the problem in [11, 57, 60, 995, 1297–1301].

An algorithm for the computation of an approximate global optimal solution of the optimal value transformation for bilevel optimization problems with nonconvex lower-level problems and a global optimal solution for ones with convex lower-level problems using semidefinite optimization is presented in [322, 678]. A smoothing SQP method for solving these problems is suggested in [1392], a smoothing augmented Lagrangian method in [1390, 1391] and a smoothing projected gradient method in [845].

If all functions describing the bilevel optimization problem are polynomials and the bilevel optimization problem is transformed equivalently into a semi-infinite optimization problem, a combination of the exchange technique with Lasserre-type semidefinite relaxations [1009] can be used to solve the problem. If the constraints in the lower-level problem do not depend on the leader's variables, this algorithm converges to a global optimal solution [1010].

20.6.5 Global Optimization

In [586], focus is on global optimization using the αBB approach.

A branch-and-sandwich algorithm can be found in [742–744, 1057].

Global optimization of the KKT-transformation is done in [1318].

For global optimization using sensitivity analysis in the lower-level problem see [493, 495, 1079].

If the lower-level problem is replaced by a variational inequality, an active set algorithm is suggested in [1342].

An algorithm for the computation of a global optimal solution for bilevel problems with quadratic upper and linear lower-level problems can be found in [1258, 1261, 1262]. Using a transformation with d.c. constraints, the same problems can be solved globally, see [61, 585].

20.6.6 Metaheuristics

Different metaheuristics have been applied to bilevel optimization problems:

1. Different genetic algorithms [67, 256–258, 263, 537, 597, 627, 759, 819, 822, 827, 852, 940, 941, 1016, 1149, 1150, 1153, 1214, 1223, 1346, 1403, 1471].
2. Memetic algorithm [37, 660, 663].
3. Ant colony systems [259].
4. Tabu search algorithm [552].
5. Particle swarm optimization [46, 536, 541, 543, 602, 603, 725, 775, 908, 1324, 1462, 1463, 1477, 1478]. This algorithm is used for solving bilevel linear optimization problems with multiple objective functions in the upper-level problem [48]. In [46] the algorithm is used to approximate the set of Pareto optimal solutions of the multiobjective, nonlinear bilevel optimization problem with linear optimization problems in the lower-level problem which are solved exactly for each particle in the swarm.
6. Evolutionary algorithm [261, 355–357, 820, 914, 1218, 1221, 1347]. Evolutionary algorithm applied to multiobjective bilevel optimization problems using quadratic fibres to approximate the set of Pareto optimal solutions of the lower-level problem [1217].
7. Differential evolution algorithm for problems with multiobjective upper-level problem [826] and for problems with linear equality constraints [784]. The differential evolution algorithm for general bilevel optimization problems is formulated in[74, 75].
8. Simulated annealing [690, 1060, 1144, 1375].
9. Estimation of distribution algorithm [1322].
10. Neural network algorithm [619, 650, 790, 900, 903, 1119, 1195].
11. Fruit fly algorithm [911, 1330].

20.6.7 Special Algorithms

Solution algorithms for special problems can be found in [141, 233, 1135, 1236].

1. Direct search algorithm [953, 1446].
2. Combination of the simplex algorithm with projected gradients [1252].
3. A trust-region algorithm [331, 334, 937].
4. Application of ideas from bicriterial optimization for solving bilevel optimization problems [1304], comment on this algorithm in [271, 1359].
5. Transformation into multicriterial optimization problem using certain membership functions [483]. References [149, 617] show that the algorithms in [125] (Grid search algorithm) and in [180] (parametric complementary pivot algorithm) fail in general.

6. Use of fuzzy optimization to compute a satisfactory solution [95, 654, 969, 1157, 1194, 1485].
7. Use of derivative-free solution algorithms [337].

20.6.8 Integer Bilevel Problems

Solution algorithms for mixed-integer bilevel optimization problems are suggested in [280, 438, 455, 475, 587, 760, 980, 1142, 1363, 1395]. The watermelon algorithm proposed in [1335] is an exact algorithm solving discrete linear bilevel optimization problems using disjunction cuts to remove infeasible solutions for the bilevel problem from the search space. An efficient cutting plane algorithm can be found in [512–514]. A cutting plane algorithm for special discrete bilevel optimization problem is given in [629].

The solution of Boolean bilevel optimization problems using the optimal value reformulation and a cutting plane algorithm has been investigated in [417], see also [883]. An interactive approach to integer bilevel optimization problems can be found in [484]. An extended version of the k-th best algorithm can be found in [1188, 1189]. For a mixture of cutting plane and k-th best algorithm for integer fractional bilevel optimization problems, see [201, 1282].

The exact solution of bilevel assignment problems is topic of [145], special cases of this $\mathcal{NP}$-hard problem can be solved in polynomial time.

For solving integer bilevel optimization problems different solution algorithms have been suggested:

1. Some kind of a k-th best algorithm [1282], see the remarks to that algorithm in [244].
2. A cutting plane approach [380, 587, 1183].
3. A branch and bound algorithm [134].
4. Other approaches can be found in [504].

A polynomial-time algorithm for the bilevel time minimization (or bottleneck) transportation problem can be found in [1182, 1238, 1239, 1380].

20.6.9 Related Problems

Problems where the upper-level constraints and objective function depend on the optimal value function of the lower-level problem can be found in [1199].

20.7 Bilevel Problems with Multiobjective Functions in the Lower or Upper Level, or with Multiple Followers

Problems with vector-valued objective function in the upper-level problem are considered in [252]. For problems with multiple followers the reader is referred to [249, 357]. Semivectorial bilevel optimization problems, i.e. bilevel optimization problems where the lower-level problem has a vector-valued objective function are topic of [192, 196–198, 409, 899, 1486].

In [1491], multiobjective (linear) problems in both levels are considered. The lower-level problem is replaced using Benson's approach. The authors compute a satisfactory solution applying certain k-th best approach.

Application of the idea in [531] to problems with multiobjective linear optimization problems in both levels is realized in [1071].

20.8 Applications

Here, we give only topics of diverse applications.

1. Agricultural economics [272, 274], support of biofuel production [135, 136].
2. Agricultural credit distribution to improve rural income [1031].
3. Aid distribution after the occurrence of a disaster [262].
4. Airline revenue management [339].
5. Aircraft structural design [607].
6. Aluminium production [1006–1008].
7. Analysis of the possible mechanisms of optimization of biodiversity [38].
8. Autonomous cars driving [1480].
9. Bioengineering and biotechnology [229, 1067], optimization of bioprocess productivity based on metabolic-genetic network models [671], optimization of low-carbon product family and its manufacturing process [1379].
10. Capacity (expansion) planning [517, 547].
11. Chemical equilibria [324, 1233].
12. Contact shape optimization [634].
13. Control of container cranes in high rack warehouses [745].
14. Credit allocation [1137].
15. Critical infrastructure protection planning [1167].
16. Deception in games [814].
17. Defense applications [59, 209, 808]. Interdiction problems below describe also applications related to defense problems. Electric grid defense planning [40, 1411].
18. Discount decisions for the retailer [729].
19. Dynamic storage pricing strategy in supply hub in industrial park [1088].

20. Ecological problems: Greenhouse gas emissions [618, 637, 803, 891, 1369], water exchange in eco-industrial parks [112, 1273].
21. Electron tomography [1504].
22. Electricity markets and networks [20, 106–108, 120, 172, 184, 498, 546, 615, 616, 636, 639–641, 652, 794, 898, 1256, 1307, 1326, 1354, 1438].
    a. Control of renewable energy generation [1267].
    b. Optimal location and size of storage devices in transmission networks [470].
    c. Bids of wind power producers in the day-ahead market with stochastic market [804]. Real-time pricing schemes in electricity markets [1505]. Optimal strategic bidding of energy producers [632, 765].
    d. Electricity swing option pricing [763].
    e. Local electricity markets under decentralized and centralized designs [799].
    f. Power system vulnerability analysis [96].
    g. Pay-as-clear electricity market with demand elasticity [44].
    h. Transmission-Distribution System Operator Coordination [234].
    i. Three-level model to optimize the operating costs by the system operator [561].
23. Environmental policy [379, 589].
24. Evacuation planning [89, 1114, 1381, 1428, 1481].
25. Facility location and production problem [1, 165, 167–169, 187, 286, 442, 630, 746, 767, 939, 963, 1016, 1080, 1117, 1145, 1264, 1465], facility location and freight transportation problem [350, 596], production planning problem [896]. Best location of stone industrial parks which pollute the environment [536]. Location-allocation problem [917]. Facility location problem with customer's patronization [287]. Competitive facility location problems [758].
26. Fisheries management [1044].
27. Flow shop scheduling problems [2].
28. Gas cash-out problem [414, 716, 720], entry-exit gas market [575].
29. Global investment strategies with financial intermediation [153].
30. Hazardous materials transportation [264, 486].
31. Health insurance problem [1446].
32. Human arm movements [32, 33, 977].
33. Identification of enzymatic capacity constraints [1407].
34. Image segmentation [1096], image reconstruction [446, 1499].
35. Incentive systems [377, 1482], Principal-agent problems [291, 535, 583, 584, 677, 735, 793, 1003, 1097, 1098, 1125, 1128, 1207, 1492].
36. Inferring oncoenzymes [1377].
37. Interdiction problems [29, 40, 116, 119, 219–222, 277, 554, 871, 1004, 1020, 1061, 1118, 1138, 1168, 1232, 1373, 1374, 1378]. Many interdiction problems are formulated as three-level optimization problems, some of the references describe especially tailored solution algorithms. Heuristic algorithms for generalized interdiction problems, where the assumption that the objective function of the leader is opposite to that of the follower is removed, can be found in [515]. In this article the authors report also very extensive computational

results. Optimal resource allocation for the critical infrastructure protection [1019].
38. Inverse optimization [464].
39. Local access networks (LAN) [263].
40. Machine learning problems [156, 306, 307, 773, 1066], statistical learning methods [155], parameter learning in image applications [352, 1023, 1024].
41. Material transportation at the Lancang River Hydropower Base [909].
42. Maximally violated valid inequality generation often has a natural interpretation as a bilevel program [868].
43. Misclassification minimization [924–926, 933].
44. Mechanics [1046].
45. Network problems:

    a. Highway network design [99, 147, 150, 151, 315, 566, 778, 779, 800, 927, 930, 1339, 1342]. Complexity of the highway pricing problem [444, 624]. Network design problem [521, 527–529, 545, 703, 844, 1159, 1266], the same with uncertain travel demand [290, 317]. Sensitivity analysis is used to solve the network design problem in [696]. The algorithm in [800] has been shown not to converge in [929]. Network design problem with congestion effects [928].
    b. The mathematical structure of the strategic pricing problem is investigated in [936].
    c. O-D adjustment problem [303, 328, 458, 519, 1181, 1405], O-D demands estimation [1115], optimal tolls in transportation networks [204, 206, 216, 217, 223, 398, 401, 422, 431, 445, 447, 467, 574, 798, 931]. The same with a real application in Hong Kong [1400].
    d. Solution algorithms for an application in a traffic network [327] with some comments in [329].
    e. Traffic network signal control [605, 724, 1244, 1303, 1406]. Use of traffic flow guidance systems [1202].
    f. Hierarchical transportation network with transshipment facilities [341, 342]. Expansion of a highway network [77].
    g. An overview over pricing problems in transportation and marketing is given in [625]. Multiobjective pricing problems are considered in [1334].
    h. Interaction of public and private sections using the example of Korea [737]. Model for public-private partnerships [795–797].
    i. Review of related problems [958], bilevel traffic management [1053]. Investigation of an approximation algorithm for the toll setting problem [574, 1123]. Different models for traffic assignment problems [870]. A comparison of algorithms for solving a bi-level toll setting problem can be found in [703].
    j. Public Rail Transportation under Incomplete Data [1059].
    k. Computational complexity of the problem is investigated and a cutting plane approach is suggested in [626].
    l. Pricing toll roads under uncertainty [453].

m. In [991] the problem is transformed using the KKT transformation, the complementarity conditions are replaced using the Fischer-Burmeister function and the resulting problems are solved globally.
n. Load balancing in mobile networks [491].
o. Rank pricing problem [240].
p. Transportation of hazardous materials [50, 181, 461, 486, 721, 732].
q. Two-level stochastic optimization problem over transportation network [41].
r. Trajectory planning for a robot [949].
s. Location of hydrogen filling stations to promote the use of electric cars [21, 965].
t. Vehicle routing problem [940].
u. Hub arc location model [1163], the same under competition [343, 344, 912, 1164].
v. Railway transport hub planning [733].
w. School bus routing [1051].

46. Newsboy problem [346, 680, 1500].
47. New product design [1251].
48. Optimal drug combination causing minimal side effects in biomedicine [492].
49. Optimal partial hedging in discrete markets [988].
50. Optimal standardization [567–569].
51. Optimizing bus-size and headway in transit networks [338, 359].
52. Parameter estimation in chemical engineering [191, 973].
53. Physical layer security in cognitive radio networks [499].
54. Pipe network design [1472].
55. Predatory pricing in a multiperiod Stackelberg game [998].
56. Prediction of underground blast-induced ground vibration amplitudes [1047].
57. Price-based market clearing under marginal pricing [506, 507, 509].
58. Price setting problems [780, 781], in part related to toll setting problems in transportation networks. Price setting problems on a network [559] and on an oligopolistic market [1120].
59. Process design problem [324, 325].
60. Product selection with partial exterior financing [750].
61. Profitability of merger in Stackelberg markets [655].
62. Quantitative policy analysis [212].
63. Real-time path planning approach for unmanned aerial vehicles [862].
64. Relations between central economic units and subunits: [4, 121, 124, 128, 289, 740, 806], hazardous waste management [56, 58], applications in economics [211, 227, 228, 251].
65. Resource allocation model [289, 598, 1001, 1002, 1338], special model for HIV control [728] and in wireless networks [233]. Problems with resource allocation constraints lead to minimization problems over the efficient set [1279].
66. Scheduling problems [723].
67. Set invariance characterization of dynamical systems affected by time-delay [792].

68. Stackelberg-Nash-Cournot equilibria [511, 1187, 1287, 1288, 1434]. A stochastic problem of this type is investigated in [354, 1384]. A critical comment to some of the results in [1187] can be found in [477]. Under some conditions, the Stackelberg equilibrium is also a Cournot equilibrium [701]. Stackelberg Equilibria of Linear Strategies in Risk-Seeking Insider Trading [564]. Stackelberg solution in static and dynamic nonzero-sum two-player games (open-loop Stackelberg solution) [576, 1205].
69. Supply chain configuration [256, 774, 1139, 1158, 1403], corporate social responsibility in a supply chain [54, 731], supply chain management [672, 849, 864, 1161, 1336, 1396]. Different metaheuristics are applied to a location-allocation problem related to a supply chain problem. Timberlands supply chain model is investigated in [1425, 1426].
70. Support vector machines are solved as bilevel optimization problem [772].
71. Truss topology optimization [526].
72. Uncapacitated lot-sizing problem [739].
73. Virtual power plants [722, 1448].
74. Water conflict problem between India and Bangladesh [65, 183], water allocation issues [653, 1387, 1389], water distribution system [1231], water rights trading [1349]. Water integration in eco-industrial parks [1095].

20.9 Test Problems

Methods to generate test problems can be found for linear bilevel optimization problems in [989], for more general problems in [236–239, 1034, 1213], see also Chapter 9 of [520]. A bilevel optimization problem library can be found on the internet page *http://coral.ise.lehigh.edu/data-sets/bilevel-instances/.* For another test set see [972]. The seemingly most comprehensive set of test problems can be found in [1497], see also [1498] in this volume.

20.10 Master, PhD and Habilitation Theses

S. Addoune [19], G.B. Allende [42], M. Andersson [69], T. J. Becker[142], O. Ben-Ayed [146], Z. Bi [173], H.C. Bylling [232], W.D. Cai [235], L.M. Case [288], M. Červinka [1309], Y. Chen [299], B. Colson [330, 332], S.M. Dassanayaka [347], S. Dempe [372], S. DeNegre [437], deSilva [440], J. Deuerlein [443], S. Dewez [444], J. Eckardt [471], T. Edmunds [473], A. Ehrenmann [476], D. Fanghänel [502], S. Franke [524], Y. Gao [540], N. Groot [577], F. Harder [609], K. Hatz [613], C. Henkel [631], X. Hu [651], E. Israeli [664], D. Joksimocic [693], F.M. Kue [768], S. Lohse [869], J. Lžičař [907], P. Mehlitz [947], A.G. Mersha [951], G.M. Moore [978], J. Moore [979], S.Nagy, [997], A. Nwosu [1021], W. Oeder [1026], F. Parraga [1050], T. Petersen [1062], A. G. Petoussis [1063], O. Pieume [1069], M. Pilecka

[1072, 1074], P. Pisciella [1077], R. Rog [1124], A. Ruziyeva [1133], R. Saboiev [1136], G. Savard [1165], G. Schenk [1170], H. Schmidt [1171], J. Shaw [1185], S.A Siddiqui [1203], Z.C. Taskin [1276], L. Vicente [1310], S. Vogel [1316], A. Werner [1365], U. Wen [1355], R. Winter [1371], P. Xu [1394], J. Zhang [1466], A.B. Zemkoho [1442, 1443]

Edited volumes are Anandalingam and Friesz [66], Dempe and Kalashnikov [397], Migdalas et al. [961].

Acknowledgements The author's work has been supported by Deutsche Forschungsgemeinschaft, Projects GZ DE650/7 and DE650/10.

The author is indebted to an anonymous referee for very careful reading of the original manuscript and making many useful suggestions and remarks, which resulted in essential improvements of the paper.

Bibliography

1. K. Aardal, M. Labbé, J. Leung, M. Queranne, On the two-level uncapacitated facility location problem. INFORMS J. Comput. **8**, 289–301 (1996)
2. S.A. Abass, Bilevel programming approach applied to the flow shop scheduling problem under fuzziness. Comput. Manag. Sci. **2**(4), 279–293 (2005)
3. S.A. Abass, An interval number programming approach for bilevel linear programming problem. Int. J. Manag. Sci. Eng. Manag. **5**(6), 461–464 (2010)
4. H. Abou-Kandil, P. Bertrand, Government—private sector relations as a Stackelberg game: a degenerate case. J. Econom. Dyn. Control **11**, 513–517 (1987)
5. A. Aboussoror, Weak bilevel programming problems: existence of solutions. Adv. Math. Res. **1**, 83–92 (2002)
6. A. Aboussoror, S. Adly, A Fenchel-Lagrange duality approach for a bilevel programming problem with extremal-value function. J. Optim. Theory Appl. **149**(2), 254–268 (2011)
7. A. Aboussoror, S. Adly, New necessary and sufficient optimality conditions for strong bilevel programming problems. J. Global Optim. **70**(2), 309–327 (2018)
8. A. Aboussoror, S. Adly, V. Jalby, Weak nonlinear bilevel problems: existence of solutions via reverse convex and convex maximization problems. J. Ind. Manag. Optim. **7**(3), 559–571 (2011)
9. A. Aboussoror, S. Adly, F.E. Saissi, Strong-weak nonlinear bilevel problems: existence of solutions in a sequential setting. Set-Valued Var. Anal. **25**(1), 113–132 (2017)
10. A. Aboussoror, S. Adly, F.E. Saissi, A duality approach for a class of semivectorial bilevel programming problems. Vietnam J. Math. **46**(1), 197–214 (2018)
11. A. Aboussoror, Z. Ankhili, A. Mansour, Bilevel programs: approximation results involving reverse convex programs. Pac. J. Optim. **4**, 279–291 (2008)
12. A. Aboussoror, P. Loridan, Existence and approximation results involving regularized constrained Stackelberg problems. J. Math. Anal. Appl. **188**(1), 101–117 (1994)
13. A. Aboussoror, P. Loridan, Sequential stability of regularized constrained Stackelberg problems. Optimization **33**(3), 251–270 (1995)
14. A. Aboussoror, P. Loridan, Strong-weak Stackelberg problems in finite dimensional spaces. Serdica Math. J. **21**, 151–170 (1995)
15. A. Aboussoror, P. Loridan, Existence of solutions to two-level optimization problems with nonunique lower-level solutions. J Math. Anal. Appl. **254**(2), 348–357 (2001)
16. A. Aboussoror, A. Mansouri, Weak linear bilevel programming problems: existence of solutions via a penalty method. J. Math. Anal. Appl. **304**, 399–408 (2005)

17. L. Adam, R. Henrion, J. Outrata, On M-stationarity conditions in MPECs and the associated qualification conditions. Math. Program. **168**(1–2), 229–259 (2018)
18. P. Adasme, A. Lisser, A computational study for bilevel quadratic programs using semidefinite relaxations. Eur. J. Oper. Res. **254**(1), 9–18 (2016)
19. S. Addoune, Optimisation à deux niveaux : Conditions d'optimalité, approximation et stabilité, Ph.D. thesis (Université de Bourgogne/Département de Mathématique, Erasme, 1994)
20. S. Afşar, L. Brotcorne, P. Marcotte, G. Savard, Achieving an optimal trade-off between revenue and energy peak within a smart grid environment. Renew. Energy **91**, 293–301 (2016)
21. S. Aghajani, M. Kalantar, Operational scheduling of electric vehicles parking lot integrated with renewable generation based on bilevel programming approach. Energy **139**, 422–432 (2017)
22. J. Agor, O.Y. Özaltın, Feature selection for classification models via bilevel optimization. Comput. Oper. Res. **106**, 156–168 (2019)
23. R.K. Ahuja, J.B. Orlin, Inverse optimization. Oper. Res. **49**(5), 771–783 (2001)
24. E. Aiyoshi, K. Shimizu, Hierarchical decentralized systems and its new solution by a barrier method. IEEE Trans. Syst. Man Cybern. **11**, 444–449 (1981)
25. E. Aiyoshi, K. Shimizu, A solution method for the static constrained Stackelberg problem via penalty method. IEEE Trans. Autom. Control **29**, 1111–1114 (1984)
26. M.A. Aizerman, A.V. Malishevski, Conditions for universal reducibility of a two-stage extremization problem to a one-stage problem. J. Math. Anal. Appl. **119**, 361–388 (1986)
27. T. Akbari, S.Z. Moghaddam, E. Poorghanaat, F. Azimi, Coordinated planning of generation capacity and transmission network expansion: a game approach with multi-leader-follower. Int. Trans. Electr. Energy Syst. **27**(7), e2339 (2017)
28. H.G. Akdemir, F. Tiryaki, Bilevel stochastic transportation problem with exponentially distributed demand. Bitlis Eren Univer. J. Sci. Technol. **2**(1), 32–37 (2012)
29. D. Aksen, S.S. Akca, N. Aras, A bilevel partial interdiction problem with capacitated facilities and demand outsourcing. Comput. Oper. Res. **41**, 346–358 (2014)
30. F. Al-Khayyal, R. Horst, P. Pardalos, Global optimization of concave functions subject to quadratic constraints: an application in nonlinear bilevel programming. Annal. Oper. Res. **34**, 125–147 (1992)
31. S. Albaek, Stackelberg leadership as a natural solution under cost uncertainty. J. Ind. Econ. **38**, 335–347 (1990)
32. S. Albrecht, M. Leibold, M. Ulbrich, A bilevel optimization approach to obtain optimal cost functions for human arm-movements. Numer. Algebra Control Optim. **2**(1), 105–127 (2012)
33. S. Albrecht, K. Ramirez-Amaro, F. Ruiz-Ugalde, D. Weikersdorfer, M. Leibold, M. Ulbrich, M. Beetz, Imitating human reaching motions using physically inspired optimization principles, in *Proceedings of the 11th IEEE-RAS International Conference on Humanoid Robots (Humanoids) 2011* (IEEE, New York, 2011), pp. 602–607
34. E. Alekseeva, Y. Kochetov, Matheuristics and exact methods for the discrete (r| p)-centroid problem, in *Metaheuristics for Bi-level Optimization* (Springer, Berlin, 2013), pp. 189–219
35. E. Alekseeva, Y. Kochetov, A. Plyasunov, An exact method for the discrete $(r|p)$-centroid problem. J. Global Optim. **63**(3), 445–460 (2015)
36. E. Alekseeva, Y. Kochetov, E.-G. Talbi, A matheuristic for the discrete bilevel problem with multiple objectives at the lower level. Int. Trans. Oper. Res. **24**(5), 959–981 (2017)
37. E. Alekseeva, N. Kochetova, Y. Kochetov, A. Plyasunov, A hybrid memetic algorithm for the competitive p-median problem. IFAC Proc. **42**(4), 1533–1537 (2009)
38. G.M. Aleshchenko, E.N. Bukvareva, Two-level hierarchical model of optimal biological diversity. Biol. Bull. **37**(1), 1–9 (2010)
39. N. Alexandrov, J.E. Dennis, Algorithms for bilevel optimization, in *Institute for Computer Applications in Science and Engineering* (NASA Langley Research Center, Hampton, 1994)
40. N. Alguacil, A. Delgadillo, J.M. Arroyo, A trilevel programming approach for electric grid defense planning. Comput. Oper. Res. **41**, 282–290 (2014)

41. S.M. Alizadeh, P. Marcotte, G. Savard, Two-stage stochastic bilevel programming over a transportation network. Transp. Res. B Methodol. **58**, 92–105 (2013)
42. G.B. Allende, Mathematical programs with equilibrium constraints: solution techniques from parametric optimization, Ph.D. thesis (University of Twente/EEMCS Faculty, Drienerlolaan, 2006)
43. G.B. Allende, G. Still, Solving bilevel programs with the KKT-approach. Math. Program. **138**, 309–332 (2013)
44. E. Allevi, D. Aussel, R. Riccardi, On an equilibrium problem with complementarity constraints formulation of pay-as-clear electricity market with demand elasticity. J. Global Optim. **70**, 329–346 (2018)
45. H. Almutairi,S. Elhedhli, Carbon tax based on the emission factor: a bilevel programming approach. J. Global Optim. **58**(4), 795–815 (2014)
46. M.J. Alves, Using MOPSO to solve multiobjective bilevel linear problems, in *International Conference on Swarm Intelligence* (Springer, Berlin, 2012), pp. 332–339
47. M.J. Alves, C.H. Antunes, A differential evolution algorithm to semivectorial bilevel problems, in *International Workshop on Machine Learning, Optimization, and Big Data* (Springer, New York, 2017), pp. 172–185
48. M.J. Alves, J.P. Costa, An algorithm based on particle swarm optimization for multiobjective bilevel linear problems. Appl. Math. Comput. **247**, 547–561 (2014)
49. M.J. Alves, S. Dempe, J.J. Júdice, Computing the Pareto frontier of a bi-objective bi-level linear problem using a multiobjective mixed-integer programming algorithm. Optimization **61**(3), 335–358 (2012)
50. E. Amaldi, M. Bruglieri, B. Fortz, On the hazmat transport network design problem, in *International Conference on Network Optimization* (Springer, New York, 2011), pp. 327–338
51. A.H. Amer, Implementation of the constraint method in special class of multi-objective fuzzy bi-level nonlinear problems. Pak. J. Stat. Oper. Res. **13**(4), 739–756 (2017)
52. M. Amini, F. Yousefian, An iterative regularized incremental projected subgradient method for a class of bilevel optimization problems (2018). arXiv preprint:1809.10050
53. R. Amir, A. Stepanova, Second-mover advantage and price leadership in Bertrand duopoly. Games Econ. Behav. **55**(1), 1–20 (2006)
54. O. Amirtaheri, M. Zandieh, B. Dorri, A.R. Motameni, A bi-level programming approach for production-distribution supply chain problem. Comput. Ind. Eng. **110**, 527–537 (2017)
55. M.A. Amouzegar, A global optimization method for nonlinear bilevel programming problems. IEEE Trans. Syst. Man Cybern. Part B Cybern. **29**(6), 771–777 (1999)
56. M.A. Amouzegar, S.E. Jacobsen, A decision support system for regional hazardous waste management alternatives. J. Appl. Math. Decis. Sci. **2**, 23–50 (1998)
57. M.A. Amouzegar, K. Moshirvaziri, A penalty method for linear bilevel programming problems, in *Multilevel Optimization: Algorithms and Applications*, ed. by A. Migdalas, P.M. Pardalos, P. Värbrand (Kluwer Academic Publishers, Dordrecht, 1998), pp. 251–271
58. M.A. Amouzegar, K. Moshirvaziri, Determining optimal pollution control policies: an application of bilevel programming. Eur. J. Oper. Res. **119**(1), 100–120 (1999)
59. B. An, F. Ordóñez, M. Tambe, E. Shieh, R. Yang, C. Baldwin, J. DiRenzo III, K. Moretti, B. Maule, G. Meyer, A deployed quantal response-based patrol planning system for the US coast guard. Interfaces **43**(5), 400–420 (2013)
60. L.T.H. An, P.D. Tao, N.N. Canh, N.V. Thoai, DC programming techniques for solving a class of nonlinear bilevel programs. J. Global Optim. **44**(3), 313–337 (2009)
61. L.T.H. An, P.D. Tao, N.N. Canh, N.V. Thoai, DC programming techniques for solving a class of nonlinear bilevel programs. J. Global Optim. **44**(3), 313–337 (2009)
62. L.T.H. An, P.D. Tao, L.D. Muu, Numerical solution for optimization over the efficient set by DC optimization algorithms. Oper. Res. Lett. **19**(3), 117–128 (1996)
63. G. Anandalingam, An analysis of information and incentives in bi-level programming, in *IEEE 1985 Proceedings of the International Conference on Cybernetics and Society* (1985), pp. 925–929

64. G. Anandalingam, A mathematical programming model of decentralized multi-level systems. J. Oper. Res. Soc. **39**(11), 1021–1033 (1988)
65. G. Anandalingam, V. Apprey, Multi-level programming and conflict resolution. Eur. J. Oper. Res. **51**(2), 233–247 (1991)
66. G. Anandalingam, T.L. Friesz, Hierarchical optimization: an introduction. Ann. Oper. Res. **34**, 1–11 (1992)
67. G. Anandalingam, R. Mathieu, L. Pittard, N. Sinha, Artificial intelligence based approaches for solving hierarchical optimization problems, in *Impacts of Recent Computer Advances on Operations Research*, ed. by R. Sharda, B. Golden, E. Wasil, O. Balci, W. Stewart (Elsevier, Amsterdam, 1983), pp. 289–301
68. G. Anandalingam, D.J. White, A solution method for the linear static Stackelberg problem using penalty functions. IEEE Trans. Autom. Control **35**(10), 1170–1173 (1990)
69. M. Andersson, A bilevel approach to parameter tuning of optimization algorithms using evolutionary computing: understanding optimization algorithms through optimization, Ph.D. thesis (University of Skövde, Skövde, 2018)
70. M. Andersson, S. Bandaru, A. Ng, A. Syberfeldt, Parameter tuning of MOEAs using a bilevel optimization approach, in *Evolutionary Multi-Criterion Optimization*, ed. by A. Gaspar-Cunha, A. Carlos Henggeler, C. Coello Coello. Lecture Notes in Computer Science, vol. 9018 (Springer, New York, 2015), pp. 233–247
71. R. Andreani, S.L.C. Castro, J.L. Chela, A. Friedlander, S.A. Santos, An inexact-restoration method for nonlinear bilevel programming problems. Comput. Optim. Appl. **43**(3), 307–328 (2009)
72. R. Andreani, J.M. Martinez, On the solution of mathematical programs with equilibrium constraints. Z. Oper. Res. **54**, 345–358 (2001)
73. R. Andreani, V.A. Ramirez, S.A. Santos, L.D. Secchin, Bilevel optimization with a multiobjective problem in the lower level. Numer. Algorithms **81**(3), 915–946 (2019)
74. J.S. Angelo, H.J.C. Barbosa, A study on the use of heuristics to solve a bilevel programming problem. Int. Trans. Oper. Res. **22**(5), 861–882 (2015)
75. J.S. Angelo, E. Krempser, H.J.C. Barbosa, Differential evolution for bilevel programming, in *IEEE Congress on Evolutionary Computation (CEC)* (IEEE, New York, 2013), pp. 470–477
76. J.S. Angelo, E. Krempser, H.J.C. Barbosa, Differential evolution assisted by a surrogate model for bilevel programming problems, in *IEEE Congress on Evolutionary Computation (CEC)* (IEEE, New York, 2014), pp. 1784–1791
77. E. Angulo, E. Castillo, R. García-Ródenas, J. Sánchez-Vizcaíno, A continuous bi-level model for the expansion of highway networks. Comput. Oper. Res. **41**, 262–276 (2014)
78. L.Q. Anh, P.Q. Khanh, D.T.M. Van, Well-posedness under relaxed semicontinuity for bilevel equilibrium and optimization problems with equilibrium constraints. J. Optim. Theory Appl. **153**(1), 42–59 (2012)
79. P.N. Anh, A new extragradient iteration algorithm for bilevel variational inequalities. Acta Math. Vietnam **37**, 95–107 (2012)
80. P.N. Anh, J.K. Kim, L.D. Muu, An extragradient algorithm for solving bilevel pseudomonotone variational inequalities. J. Global Optim. **52**(3), 627–639 (2012)
81. T.T.H. Anh, L.B. Long, T.V. Anh, A projection method for bilevel variational inequalities. J. Inequal. Appl. **2014**(1), 205 (2014)
82. T.V. Anh, A strongly convergent subgradient extragradient-halpern method for solving a class of bilevel pseudomonotone variational inequalities. Vietnam J. Math. **45**(3), 317–332 (2017)
83. T.V. Anh, L.D. Muu, A projection-fixed point method for a class of bilevel variational inequalities with split fixed point constraints. Optimization **65**(6), 1229–1243 (2016)
84. M. Anitescu, *On Solving Mathematical Programs with Complementarity Constraints as Nonlinear Programs*. Technical Report ANL/NCS-P864–1200 (Department of Mathematics, University of Pittsburgh, Pittsburgh, 2002)
85. M. Anitescu, Global convergence of an elastic mode approach for a class of mathematical programs with equilibrium constraints. SIAM J. Optim. **16**, 120–145 (2005)

86. Z. Ankhili, Multiobjective bilevel optimization problem: Penalty method, in *Proceedings of the International Conference on Learning and Optimization Algorithms: Theory and Applications* (ACM, New York, 2018), p. 10
87. Z. Ankhili, A. Mansouri, An exact penalty on bilevel programs with linear vector optimization lower level. Eur. J. Oper. Res. **197**(1), 36–41 (2009)
88. T. Aonuma, A facet-following coordination for linear bilevel planning process, Technical Report 86 (Kobe University of Commerce, Institute of Economic Research, Kobe, 1985)
89. P. Apivatanagul, R.A. Davidson, L.K. Nozick, Bi-level optimization for risk-based regional hurricane evacuation planning. Nat. Hazards **60**(2), 567–588 (2012)
90. C. Arbib, M. Tonelli, A non-metric bilevel location problem, Technical report (Università degli Studi dell'Aquila, L'Aquila, 2015)
91. J. Arica, S. Scheimberg, A necessary optimality condition for bilevel programming problem. Technical report, in *Programa de Engenharia de Sistemas e Comutacao* (Universidade Federal do Rio de Janeiro, Brazil, 1993)
92. J. Arica, S. Scheimberg, The bilevel programming problem: optimality conditions, Technical report (Universidade Estadual do Norte Fluminense, Rio de Janeiro, Brasil, 1995). Publicacão Técnica Interna No. 03/95
93. A. Arizti, A. Mauttone, M.E. Urquhart, A bilevel approach to frequency optimization in public transportation systems, in *OASIcs-OpenAccess Series in Informatics*, vol. 65 (Schloss Dagstuhl-Leibniz-Zentrum fuer Informatik, Wadern, 2018)
94. R. Arora, S.R. Arora, An algorithm for solving an integer linear fractional/quadratic bi-level programming problem. Adv. Model. Optim. **14**, 57–78 (2012)
95. S.R. Arora, R. Gupta, Interactive fuzzy goal programming approach for bilevel programming problem. Eur. J. Oper. Res. **194**, 368–376 (2009)
96. J.M. Arroyo, Bilevel programming applied to power system vulnerability analysis under multiple contingencies. IET Gener. Transm. Distrib. **4**(2), 178–190 (2010)
97. J.M. Arroyo, F.J. Fernández, A genetic algorithm approach for the analysis of electric grid interdiction with line switching, in *Proceedings of the 15th International Conference on Intelligent System Applications to Power Systems, 2009 (ISAP'09)* (IEEE, New York, 2009), pp. 1–6
98. M.G. Ashtiani, A. Makui, R. Ramezanian, A robust model for a leader—follower competitive facility location problem in a discrete space. Appl. Math. Model. **37**(1–2), 62–71 (2013)
99. R. Askin, F. Camacho, V. Kalashnikov, N. Kalashnykova, Comparison of algorithms for solving a bi-level toll setting problem. Int. J. Innovative Comput. Inf. Control **6**(8), 3529–3549 (2010)
100. A. Aswani, A. Ouattara, Duality approach to bilevel programs with a convex lower level (2016). arXiv preprint:1608.03260
101. C. Audet, J. Haddad, G. Savard, A note on the definition of a linear bilevel programming problem. Appl. Math. Comput. **181**, 351–355 (2006)
102. C. Audet, J. Haddad, G. Savard, Disjunctive cuts for continuous linear bilevel programming. Optim. Lett. **1**(3), 259–267 (2007)
103. C. Audet, P. Hansen, B. Jaumard, G. Savard, Links between linear bilevel and mixed 0–1 programming problems. J. Optim. Theory Appl. **93**, 273–300 (1997)
104. C. Audet, P. Hansen, B. Jaumard, G. Savard, On the linear maxmin and related programming problems, in *Multilevel Optimization: Algorithms and Applications* ed. by A. Migdalas, P.M. Pardalos, P. Värbrand (Kluwer Academic, Dordrecht, 1998), pp. 181–208
105. C. Audet, G. Savard, W. Zghal, New branch-and-cut algorithm for bilevel linear programming. J. Optim. Theory Appl. **134**(2), 353–370 (2007)
106. D. Aussel, P. Bendotti, M. Pištěk, Nash equilibrium in a pay-as-bid electricity market: Part 1—existence and characterization. Optimization **66**, 1013–1025 (2017)
107. D. Aussel, P. Bendotti, M. Pištěk, Nash equilibrium in a pay-as-bid electricity market: Part 2—best response of a producer. Optimization **66**, 1027–1053 (2017)
108. D. Aussel, R. Correa, M. Marechal, Electricity spot market with transmission losses. J. Ind. Manag. Optim. **9**(2), 275–290 (2013)

109. D. Aussel, A. Svensson, Some remarks about existence of equilibria, and the validity of the EPCC reformulation for multi-leader-follower games. J. Nonlinear Convex Anal. **19**(7), 1141–1162 (2018)
110. D. Aussel, A. Svensson, Is pessimistic bilevel programming a special case of a mathematical program with complementarity constraints?. J. Optim. Theory Appl., **181**(2), 504–520 (2019). Online first publication
111. Y. Averboukh, A. Baklanov, Stackelberg solutions of differential games in the class of nonanticipative strategies. Dynam. Games Appl. **4**(1), 1–9 (2014)
112. K.B. Aviso, R.R. Tan, A.B. Culaba, J.B. Cruz, Bi-level fuzzy optimization approach for water exchange in eco-industrial parks. Process. Saf. Environ. Prot. **88**(1), 31–40 (2010)
113. S. Avraamidou, E.N. Pistikopoulos, A multi-parametric optimization approach for bilevel mixed-integer linear and quadratic programming problems. Comput. Chem. Eng. **125**, 98–113 (2019)
114. N. Azarmir, M. Zohrehbandian, A lexicographic approach for solving multiobjective bilevel programming problems. Caspian J. Appl. Sci. Res. **5**(4), 1–4 (2016)
115. H. Babahadda, N. Gadhi, Necessary optimality conditions for bilevel optimization problems using convexificators. J. Global Optim. **34**(4), 535–549 (2006)
116. M. Backhaus, G. Schaefer, Towards optimally resilient topologies against optimal attacks, in *Proceedings of the IFIP/IEEE Symposium on Integrated Network and Service Management (IM), 2017* (IEEE, New York, 2017), pp. 1065–1070
117. S.A. Bagloee, M. Asadi, M. Sarvi, M. Patriksson, A hybrid machine-learning and optimization method to solve bi-level problems. Expert Syst. Appl. **95**(Supplement C), 142–152 (2018)
118. B. Bahmani-Firouzi, S. Sharifinia, R. Azizipanah-Abarghooee, T. Niknam, Scenario-based optimal bidding strategies of GENCOs in the incomplete information electricity market using a new improved prey—predator optimization algorithm. IEEE Syst. J. **9**(4), 1485–1495 (2015).
119. N.O. Bakır, A Stackelberg game model for resource allocation in cargo container security. Ann. Oper. Res. **187**(1), 5–22 (2011)
120. A.G. Bakirtzis, N.P. Ziogos, A.C. Tellidou, G.A. Bakirtzis, Electricity producer offering strategies in day-ahead energy market with step-wise offers. IEEE Trans. Power Syst. **22**(4), 1804–1818 (2007)
121. K.R. Balachandran, J. Ronen, Incentive contracts when production is subcontracted. Eur. J. Oper. Res. **40**, 169–185 (1989)
122. J.F. Bard, A grid search algorithm for the linear bilevel programming problem, in *Proceedings of the 14th Annual Meeting of the American Institute for Decision Science* (1982), pp. 256–258
123. J.F. Bard, An algorithm for the general bilevel programming problem. Math. Oper. Res. **8**, 260–272 (1983)
124. J.F. Bard, Coordination of a multidivisional organization through two levels of management. OMEGA **11**, 457–468 (1983)
125. J.F. Bard, An efficient point algorithm for a linear two-stage optimization problem. Oper. Res. **31**, 670–684 (1983)
126. J.F. Bard, An investigation of the linear three level programming problem. IEEE Trans. Syst. Man Cybern. **14**, 711–717 (1984)
127. J.F. Bard, Optimality conditions for the bilevel programming problem. Naval Res. Logistics Q. **31**, 13–26 (1984)
128. J.F. Bard, Geometric and algorithm developments for a hierarchical planning problem. Eur. J. Oper. Res. **19**, 372–383 (1985)
129. J.F. Bard, Convex two-level optimization. Math. Program. **40**, 15–27 (1988)
130. J.F. Bard, Some properties of the bilevel programming problem. J. Optim. Theory Appl. **68**, 371–378 (1991)
131. J.F. Bard, *Practical Bilevel Optimization: Algorithms and Applications* (Kluwer Academic, Dordrecht, 1998)

132. J.F. Bard, J. Falk, An explicit solution to the multi-level programming problem. Comput. Oper. Res. **9**, 77–100 (1982)
133. J.F. Bard, J. Moore, A branch and bound algorithm for the bilevel programming problem. SIAM J. Sci. Stat. Comput. **11**, 281–292 (1990)
134. J.F. Bard, J. Moore, An algorithm for the discrete bilevel programming problem. Nav. Res. Logist. **39**, 419–435 (1992)
135. J.F. Bard, J.C. Plummer, J.C. Sourie, Determining tax credits for converting nonfood crops to biofuels: an application of bilevel programming, in *Multilevel Optimization: Algorithms and Applications* ed. by A. Migdalas, P.M. Pardalos, P. Värbrand (Kluwer Academic, Dordrecht, 1998), pp. 23–50
136. J.F. Bard, J.C. Plummer, J.C. Sourie, A bilevel programming approach to determining tax credits for biofuel production. Eur. J. Oper. Res. **120**, 30–46 (2000)
137. B. Barnhart, Z. Lu, M. Bostian, A. Sinha, K. Deb, L. Kurkalova, M. Jha, G. Whittaker, Handling practicalities in agricultural policy optimization for water quality improvements, in *Proceedings of the Genetic and Evolutionary Computation Conference* (ACM, New York, 2017), pp. 1065–1072
138. A. Basu, C.T. Ryan, S. Sankaranarayanan, *Mixed-Integer Bilevel Representability*, Technical report (Johns Hopkins University, Baltimore, 2018). www.optimization-online.org
139. K. Basu, Stackelberg equilibrium in oligopoly: an explanation based on managerial incentives. Econ. Lett. **49**(4), 459–464 (1995)
140. M. Bazine, A. Bennani, N. Gadhi, Fuzzy optimality conditions for fractional multiobjective bilevel problems under fractional constraints. Numer. Funct. Anal. Optim. **32**(2), 126–141 (2011)
141. A. Beck, S. Sabach, A first order method for finding minimal norm-like solutions of convex optimization problems. Math. Program. **147**(1), 25–46 (2014)
142. T.J. Becker, *Bilevel Clique Interdiction and Related Problems*, Ph.D. Thesis (Rice University, Houston, 2017)
143. K. Bedhrinath, J.R.J. Rao, *Bilevel Models for Optimum Designs which are Insensitive to Perturbations in Variables and Parameters*, Technical report (University of Houston, USA, 2003)
144. B. Beheshti, O.Y. Özaltın, M.H. Zare, O.A. Prokopyev, Exact solution approach for a class of nonlinear bilevel knapsack problems. J. Global Optim. **61**(2), 291–310 (2015)
145. B. Beheshti, O.A. Prokopyev, E.L. Pasiliao, Exact solution approaches for bilevel assignment problems. Comput. Optim. Appl. **64**(1), 215–242 (2016)
146. O. Ben-Ayed, *Bilevel Linear Programming: Analysis and Application to the Network Design Problem*, Ph.D. thesis (University of Illinois, Urbana-Champaign, 1988)
147. O. Ben-Ayed, A bilevel linear programming model applied to Tunisian interegional High way network design problem. Revue Tunesienne d'Economie et de Gestion **V**, 234–277 (1990)
148. O. Ben-Ayed, Bilevel linear programming. Comput. Oper. Res. **20**, 485–501 (1993)
149. O. Ben-Ayed, C. Blair, Computational difficulties of bilevel linear programming. Oper. Res. **38**, 556–560 (1990)
150. O. Ben-Ayed, C. Blair, D. Boyce, L. LeBlanc, Construction of a real-world bilevel linear programming model of the highway design problem. Ann. Oper. Res. **34**, 219–254 (1992)
151. O. Ben-Ayed, D. Boyce, C. Blair, A general bilevel linear programming formulation of the network design problem. Transp. Res. **22 B**, 311–318 (1988)
152. F. Benita, S. Dempe, P. Mehlitz, Bilevel optimal control problems with pure state constraints and finite-dimensional lower level. SIAM J. Optim. **26**(1), 564–588 (2016)
153. F. Benita, F. López-Ramos, S. Nasini, A bi-level programming approach for global investment strategies with financial intermediation. Eur. J. Oper. Res. **274**(1), 375–390 (2019)
154. F. Benita, P. Mehlitz, Bilevel optimal control with final-state-dependent finite-dimensional lower level. SIAM J. Optim. **26**(1), 718–752 (2016)

155. K.P. Bennett, J. Hu, X. Ji, G. Kunapuli, J.-S. Pang, Model selection via bilevel optimization, in *Proceedings of the 2006 IEEE International Joint Conference on Neural Network* (IEEE, New York, 2006), pp. 1922–1929
156. K.P. Bennett, G. Kunapuli, J. Hu, J.-S. Pang, Bilevel optimization and machine learning, in *Computational Intelligence: Research Frontiers* (Springer, New York, 2008), pp. 25–47
157. H.P. Benson, Optimization over the efficient set. J. Math. Anal. Appl. **98**, 562–580 (1984)
158. H.P. Benson, On the structure and properties of a linear multilevel programming problem. J. Optim. Theory Appl. **60**, 353–373 (1989)
159. H.P. Benson, An all-linear programming relaxation algorithm for optimizing over the efficient set. J. Global Optim. **1**(1), 83–104 (1991)
160. H.P. Benson, A finite, nonadjacent extreme-point search algorithm for optimization over the efficient set. J. Optim. Theory Appl. **73**(1), 47–64 (1992)
161. H.Y. Benson, D.F. Shanno, R.J. Vanderbei, Interior-point methods for nonconvex programming: complementarity constraints, Technical report (Operations Research and Financial Engineering Department, Princeton University, Princeton, 2002)
162. A. Bensoussan, M.H.M. Chau, Y. Lai, S.C.P. Yam, Linear-quadratic mean field Stackelberg games with state and control delays. SIAM J. Control Optim. **55**(4), 2748–2781 (2017)
163. A. Bensoussan, M.H.M. Chau, S.C.P. Yam, Mean field Stackelberg games: aggregation of delayed instructions. SIAM J. Control Optim. **53**(4), 2237–2266 (2015)
164. G.C. Bento, J.X. Cruz Neto, J.O. Lopes, P.A. Soares Jr, A. Soubeyran, Generalized proximal distances for bilevel equilibrium problems. SIAM J. Optim. **26**(1), 810–830 (2016)
165. V. Beresnev, Branch-and-bound algorithm for a competitive facility location problem. Comput. Oper. Res. **40**(8), 2062–2070 (2013)
166. V.L. Beresnev, I.A. Davydov, P.A. Kononova, A.A. Melnikov, Bilevel "defender–attacker" model with multiple attack scenarios. J. Appl. Ind. Math. **12**(3), 417–425 (2018)
167. V.L. Beresnev, A.A. Melnikov, Approximate algorithms for the competitive facility location problem. J. Appl. Ind. Math. **5**(2), 180–190 (2011)
168. V.L. Beresnev, A.A. Melnikov, The branch-and-bound algorithm for a competitive facility location problem with the prescribed choice of suppliers. J. Appl. Ind. Math. **8**(2), 177–189 (2014)
169. V.L. Beresnev, A.A. Melnikov, Approximation of the competitive facility location problem with MIPs. Comput. Oper. Res. **104**, 139–148 (2019)
170. M. Bergouniuox, M. Haddou, A regularization method for ill-posed bilevel optimization problems. RAIRO Oper. Res. **40**, 19–35 (2006)
171. F. Bernstein, A. Federgruen, Pricing and replenishment strategies in a distribution system with competing retailers. Oper. Res. **51**(3), 409–426 (2003)
172. C.A. Berry, B.F. Hobbs, W.A. Meroney, R.P. O'Neill, W.R. Stewart Jr., Analyzing strategic bidding behavior in transmission networks. Utility Policy **8**, 139–158 (1999)
173. Z. Bi, *Numerical Methods for Bilevel Programming Problems*, Ph.D. thesis (Department of Systems Design Engineering, University of Waterloo, Waterloo, 1992)
174. Z. Bi, P. Calamai, Optimality conditions for a class of bilevel programming problems, Technical report #191-O-191291 (Department of Systems Design Engineering, University of Waterloo, Waterloo, 1991)
175. Z. Bi, P. Calamai, A. Conn, An exact penalty function approach for the linear bilevel programming problem, Technical Report #167-O-310789 (Department of Systems Design Engineering, University of Waterloo, Waterloo, 1989)
176. Z. Bi, P. Calamai, A. Conn, An exact penalty function approach for the nonlinear bilevel programming problem, Technical Report #180-O-170591 (Department of Systems Design Engineering, University of Waterloo, Waterloo, 1991)
177. W. Bialas, M. Karwan, Multilevel linear programming, Technical report 78–1 (Operations Research Program, State University of New York at Buffalo, Buffalo, 1978)
178. W. Bialas, M. Karwan, On two-level optimization. IEEE Trans. Autom. Control **27**, 211–214 (1982)
179. W. Bialas, M. Karwan, Two-level linear programming. Manag. Sci. **30**, 1004–1020 (1984)

180. W. Bialas, M. Karwan, J. Shaw, A parametric complementary pivot approach for two-level linear programming, Technical Report 80–82 (Operations Research Program, State University of New York, Buffalo, 1980)
181. L. Bianco, M. Caramia, S. Giordani, A bilevel flow model for hazmat transportation network design. Transp. Res. C Emerg. Technol. **17**(2), 175–196 (2009)
182. R. Birla, V.K. Agarwal, I.A. Khan, V.N. Mishra, An alternative approach for solving bi-level programming problems. Am. J. Oper. Res. **7**(03), 239 (2017)
183. J. Bisschop, W. Candler, J. Duloy, G. O'Mara, The indus basin model: a special application of two-level linear programming. Math. Program. Study **20**, 30–38 (1982)
184. M. Bjørndal, K. Jørnsten, The deregulated electricity market viewed as a bilevel programming problem. J. Global Optim. **33**(3), 465–475 (2005)
185. C. Blair, The computational complexity of multi-level linear programs. Ann. Oper. Res. **34**, 13–19 (1992)
186. R.I. Boţ, D.-K. Nguyen, A forward—backward penalty scheme with inertial effects for monotone inclusions. Applications to convex bilevel programming. Optimization **68**(10), 1855–1880 (2019)
187. G. Boglárka, K. Kovács, Solving a huff-like Stackelberg location problem on networks. J. Global Optim. **64**(2), 233–247 (2016)
188. S. Bolintinéanu, Minimization of a quasi-concave function over an efficient set. Math. Program. **61**(1–3), 89–110 (1993)
189. S. Bolintinéanu, Necessary conditions for nonlinear suboptimization over the weakly-efficient set. J. Optim. Theory Appl. **78**(3), 579–598 (1993)
190. S. Bolintinéanu, Optimality conditions for minimization over the (weakly or properly) efficient set. J. Math. Anal. Appl. **173**, 523–523 (1993)
191. G.M. Bollas, P.I. Barton, A. Mitsos, Bilevel optimization formulation for parameter estimation in vapor—liquid (-liquid) phase equilibrium problems. Chem. Eng. Sci. **64**(8), 1768–1783 (2009)
192. H. Bonnel, Optimality conditions for the semivectorial bilevel optimization problem. Pac. J. Optim. **2**(3), 447–467 (2006)
193. H. Bonnel, J. Collonge, Stochastic optimization over a Pareto set associated with a stochastic multi-objective optimization problem. J. Optim. Theory Appl. **162**(2), 405–427 (2014)
194. H. Bonnel, J. Collonge, Optimization over the Pareto outcome set associated with a convex bi-objective optimization problem: theoretical results, deterministic algorithm and application to the stochastic case. J.Global Optim. **62**(3), 481–505 (2015)
195. H. Bonnel, C.Y. Kaya, Optimization over the efficient set of multi-objective convex optimal control problems. J. Optim. Theory Appl. **147**(1), 93–112 (2010)
196. H. Bonnel, J. Morgan, Semivectorial bilevel optimization problem: penalty approach. J. Optim. Theory Appl. **131**, 365–382 (2006)
197. H. Bonnel, J. Morgan, Semivectorial bilevel convex optimal control problems: existence results. SIAM J. Control Optim. **50**(6), 3224–3241 (2012)
198. H. Bonnel, J. Morgan, Optimality conditions for semivectorial bilevel convex optimal control problems, in *Computational and Analytical Mathematics*, ed. by H. Bauschke, M. Théera (Springer, Berlin, 2013), pp. 45–78
199. H. Bonnel, N.S. Pham, Non-smooth optimization over the (weakly or properly) Pareto set of a linear-quadratic multiobjective control problem: explicit optimality conditions. J. Ind. Manag. Optim. **7**(4), 789–809 (2011)
200. H. Bonnel, L. Todjihoundé, C. Udrişte, Semivectorial bilevel optimization on Riemannian manifolds. J. Optim. Theory Appl. **167**, 464–486 (2015)
201. M. Borza, A.S. Rambely, M. Saraj, A Stackelberg solution to a two-level linear fractional programming problem with interval coefficients in the objective functions. Sains Malaysiana **41**, 1651–1656 (2012)
202. M. Borza, A.S. Rambely, M. Saraj, Two-level linear programming problems with two decision-makers at the upper level: An interactive fuzzy approach. Mod. Appl. Sci. **8**, 211–222 (2014)

203. M. Bostian, G. Whittaker, B. Barnhart, R. Färe, S. Grosskopf, Valuing water quality tradeoffs at different spatial scales: an integrated approach using bilevel optimization. Water Resour. Econ. **11**, 1–12 (2015)
204. M. Bouhtou, S. van Hoesel, A.F van der Kraaij, J.-L. Lutton, Tariff optimization in networks. INFORMS J. Comput. **19**(3), 458–469 (2007)
205. K. Bouibed, H. Slimani, M.S. Radjef, Global efficiency for multiobjective bilevel programming problems under generalized invexity. J. Appl. Math. Comput. **53**(1-2), 507–530 (2017)
206. D. Boyce, L. Mattsson, Modeling residential location choice in relation to housing location and road tolls on congested urban highway networks. Transp. Res. Part B Methodol. **33**(8), 581–591 (1999)
207. J. Bracken, J. Falk, J. McGill, Equivalence of two mathematical programs with optimization problems in the constraints. Oper. Res. **22**, 1102–1104 (1974)
208. J. Bracken, J. McGill, Mathematical programs with optimization problems in the constraints. Oper. Res. **21**, 37–44 (1973)
209. J. Bracken, J. McGill, Defense applications of mathematical programs with optimization problems in the constraints. Oper. Res. **22**, 1086–1096 (1974)
210. J. Bracken, J. McGill, A method for solving mathematical programs with nonlinear programs in the constraints. Oper. Res. **22**, 1097–1101 (1974)
211. J. Bracken, J. McGill, Production and marketing decisions with multiple objectives in a competitive environment. J. Optim. Theory Appl. **24**, 449–458 (1978)
212. A. Breiner, M. Avriel, Two-stage approach for quantitative policy analysis using bilevel programming. J. Optim. Theory Appl. **100**, 15–27 (1999)
213. M. Breton, A. Alj, A. Haurie, Sequential Stackelberg equilibria in two-person games. J. Optim. Theory Appl. **59**, 71–97 (1988)
214. L. Brotcorne, S. Hanafi, R. Mansi, A dynamic programming algorithm for the bilevel knapsack problem. Oper. Res. Lett. **37**(3), 215–218 (2009)
215. L. Brotcorne, S. Hanafi, R. Mansi, One-level reformulation of the bilevel knapsack problem using dynamic programming. Discrete Optim. **10**(1), 1–10 (2013)
216. L. Brotcorne, M. Labbé, P. Marcotte, G. Savard, A bilevel model and solution algorithm for a freight tariff setting problem. Transp. Sci. **34**, 289–302 (2000)
217. L. Brotcorne, M. Labbé, P. Marcotte, G. Savard, A bilevel model for toll optimization on a multicommodity transportation network. Transp. Sci. **35**(4), 345–358 (2001)
218. L. Brotcorne, P. Marcotte, G. Savard, Bilevel programming: the Montreal school. INFOR **46**(4), 231–246 (2008)
219. G. Brown, M. Carlyle, D. Diehl, J. Kline, K. Wood, A two-sided optimization for theater ballistic missile defense. Oper. Res. **53**(5), 745–763 (2005)
220. G. Brown, M. Carlyle, J. Salmerón, K. Wood, Defending critical infrastructure. Interfaces **36**(6), 530–544 (2006)
221. G.G. Brown, W.M. Carlyle, R.C. Harney, E.M. Skroch, R.K. Wood, Interdicting a nuclear-weapons project. Oper. Res. **57**(4), 866–877 (2009)
222. G.G. Brown, W.M. Carlyle, J. Salmeron, K. Wood, Analyzing the vulnerability of critical infrastructure to attack and planning defenses, in *Emerging Theory, Methods, and Applications, INFORMS* (2005), pp. 102–123
223. A. Budnitzki, Computation of the optimal tolls on the traffic network. Eur. J. Oper. Res. **235**(1), 247–251 (2014)
224. A. Budnitzki, k-th best algorithm for fuzzy bilevel optimization problem, in *6th German-Polish Conference on Optimization, Book of Abstracts* (2014), pp. 21–23
225. A. Budnitzki, The solution approach to linear fuzzy bilevel optimization problems. Optimization **64**(5), 1195–1209 (2015)
226. L.F. Bueno, G. Haeser, J.M. Martínez, An inexact restoration approach to optimization problems with multiobjective constraints under weighted-sum scalarization. Optim. Lett. **10**(6), 1315–1325 (2016)
227. V.A. Bulavski, V.V. Kalashnikov, Equilibrium in generalized Cournot and Stackelberg models. Economica i Matematicheskie Metody **31**(3), 151–163 (1995)

228. V.A. Bulavski, V.V. Kalashnikov, Equilibrium in generalized Cournot and Stackelberg models. Zeitschrift für Angewandte Mathematik und Mechanik **76**, 387–388 (1996)
229. A.P. Burgard, P. Pharkya, C.D. Maranas, Optknock: a bilevel programming framework for identifying gene knockout strategies for microbial strain optimization. Biotechnol. Bioeng. **84**(6), 647–657 (2003)
230. J. Burtscheidt, M. Claus, S. Dempe, Risk-averse models in bilevel stochastic linear programming. SIAM J. Optim. **30**(1), 377–406 (2020)
231. J.A. Bustos, S.H. Olavarria, V.M. Albornoz, S.V. Rodríguez, M.A. Jiménez-Lizárraga, A Stackelberg game model between manufacturer and wholesaler in a food supply chain, in *Proceedings of the ICORES* (2017), pp. 409–415
232. H.C. Bylling, Bilevel optimization with applications in energy, Ph.D. thesis (University of Copenhagen, Faculty of Science, Copenhagen, 2018)
233. A. Cabot, Proximal point algorithm controlled by a slowly vanishing term: applications to hierarchical minimization. SIAM J. Optim. **15**(2), 555–572 (2005)
234. H. Le Cadre, I. Mezghani, A. Papavasiliou, A game-theoretic analysis of transmission-distribution system operator coordination. Eur. J. Oper. Res. **274**(1), 317–339 (2018)
235. W.D. Cai, Electricity markets for the smart grid: Networks, timescales, and integration with control, Ph.D. thesis (California Institute of Technology, California, 2016)
236. P. Calamai, L. Vicente, Generating linear and linear-quadratic bilevel programming problems. SIAM J. Sci. Stat. Comput. **14**, 770–782 (1993)
237. P. Calamai, L. Vicente, Algorithm 728: Fortran subroutines for generating quadratic bilevel programming test methods. ACM Trans. Math. Soft. **20**, 120–123 (1994)
238. P. Calamai, L. Vicente, Generating quadratic bilevel programming test problems. ACM Trans. Math. Softw. **20**, 103–119 (1994)
239. P.H. Calamai, L.N. Vicente, J.J. Júdice, A new technique for generating quadratic programming test problems. Math. Program. **61**(1–3), 215–231 (1993)
240. H.I. Calvete, C. Domínguez, C. Galé, M. Labbé, A. Marín, The rank pricing problem: Models and branch-and-cut algorithms. Comput. Oper. Res. **105**, 12–31 (2019)
241. H.I. Calvete, C. Galé, On the quasiconcave bilevel programming problem. J. Optim. Theory Appl. **98**, 613–622 (1998)
242. H.I. Calvete, C. Galé, The bilevel linear/linear fractional programming problem. Eur. J. Oper. Res. **114**, 188–197 (1999)
243. H.I. Calvete, C. Galé, Local optimality in quasiconcave bilevel programming, in *Proceedings of the 7th Zaragoza-Pau Conference on Applied and Statistical Mathematics, Jaca (Huesca), September 17–18, 2001* ed. by Madaune-Tort, M., et al. University of de Zaragoza, Zaragoza, Seminario Matemático "García de Galdeano". Monographia Seminario Matematico "García de Galdeano", vol. 27 (2003), pp. 153–160
244. H.I. Calvete, C. Galé, A note on 'Bilevel linear fractional programming problem'. Eur. J. Oper. Res. **152**(1), 296–299 (2004)
245. H.I. Calvete, C. Galé, A penalty method for solving bilevel linear fractional/linear programming problems. Asia-Pacific J. Oper. Res. **21**, 207–224 (2004)
246. H.I. Calvete, C. Galé, Optimality conditions for the linear fractional/quadratic bilevel problem, in *VIII Journées Zaragoza-Pau de Mathématiques Appliquées et de Statistiques*. Monographia Seminario Matematico García Galdeano, vol. 31 (Prensas University Zaragoza, Zaragoza, 2004), pp. 285–294
247. H.I. Calvete, C. Galé, Solving linear fractional bilevel programs. Oper. Res. Lett. **32**(2), 143–151 (2004)
248. H.I. Calvete, C. Galé, Note on the 'Optimality conditions for linear fractional bilevel programs'. Indian J. Pure Appl. Math. **36**(1), 23–34 (2005)
249. H.I. Calvete, C. Galé, Linear bilevel multi-follower programming with independent followers. J. Global Optim. **39**(3), 409–417 (2007)
250. H.I. Calvete, C. Galé, Bilevel multiplicative problems: a penalty approach to optimality and a cutting plane based algorithm. J. Comput. Appl. Math. **218**(2), 259–269 (2008)

251. H.I. Calvete, C. Galé, A multiobjective bilevel program for production-distribution planning in a supply chain, in *Multiple Criteria Decision Making for Sustainable Energy and Transportation Systems* ed. by Ehrgott, M., et al. Proceedings of the 19th International Conference on Multiple Criteria Decision Making, Auckland, New Zealand, 7th–12th January 2008. Lecture Notes in Economics and Mathematical Systems, vol. 634 (Springer, Berlin, 2010), pp. 155–165
252. H.I. Calvete, C. Galé, Linear bilevel programs with multiple objectives at the upper level. J. Comput. Appl. Math. **234**(4), 950–959 (2010)
253. H.I. Calvete, C. Galé, On linear bilevel problems with multiple objectives at the lower level. Omega **39**, 33–40 (2011)
254. H.I. Calvete, C. Galé, Linear bilevel programming with interval coefficients. J. Comput. Appl. Math. **236**(15), 3751–3762 (2012)
255. H.I. Calvete, C. Galé, S. Dempe, S. Lohse, Bilevel problems over polyhedra with extreme point optimal solutions. J. Global Optim. **53**(3), 573–586 (2012)
256. H.I. Calvete, C. Galé, J.A. Iranzo, Planning of a decentralized distribution network using bilevel optimization. Omega **49**, 30–41 (2014)
257. H.I. Calvete, C. Galé, P.M. Mateo, A new approach for solving linear bilevel problems using genetic algorithms. Eur. J. Oper. Res. **188**(1), 14–28 (2008)
258. H.I. Calvete, C. Galé, P.M. Mateo, A genetic algorithm for solving linear fractional bilevel problems. Ann. Oper. Res. **166**, 39–56 (2009)
259. H.I. Calvete, C. Galé, M.-J. Oliveros, Bilevel model for production-distribution planning solved by using ant colony optimization. Comput. Oper. Res. **38**(1), 320–327 (2011)
260. F. Camacho, Two examples of a bilevel toll setting problem, in *Proceedings, International Business and Economics Research Conference, Las Vegas* (2006)
261. J.-F. Camacho-Vallejo, Á.E. Cordero-Franco, R.G. González-Ramírez, Solving the bilevel facility location problem under preferences by a Stackelberg-evolutionary algorithm. Math. Prob. Eng. **2014**, 14 (2014)
262. J.-F. Camacho-Vallejo, E. González-Rodríguez, F.-J. Almaguer, R.G. González-Ramírez, A bi-level optimization model for aid distribution after the occurrence of a disaster. J. Cleaner Prod. **105**, 134–145 (2014)
263. J.-F. Camacho-Vallejo, J. Mar-Ortiz, F. López-Ramos, R.P. Rodríguez, A genetic algorithm for the bi-level topological design of local area networks. PLoS ONE **10**(6), 21 (2015)
264. J.F. Camacho-Vallejo, R. Muñoz S.ánchez, A path based algorithm for solve the hazardous materials transportation bilevel problem, in *Applied Mechanics and Materials* ed. by G. Li, C. Chen. Translations Technology Publication, vol. 253 (2013), pp. 1082–1088
265. M. Campêlo, S. Dantas, S. Scheimberg, A note on a penalty function approach for solving bilevel linear programs. J. Global Optim. **16**, 245–255 (2000)
266. M. Campêlo, S. Scheimberg, An analysis of the bilevel linear problem by a penalty approach, Technical report (Universidade Federal do Rio de Janeiro, Brazil, 1998)
267. H.I. Calvete, C. Galé, P.M. Mateo, A note on a modified simplex approach for solving bilevel linear programming problems. Eur. J. Oper. Res. **126**, 454–458 (2000)
268. H.I. Calvete, C. Galé, P.M. Mateo, Theoretical and computational results for a linear bilevel problem, in *Advances in Convex Analysis and Global Optimization* (Springer, Berlin, 2001), pp. 269–281
269. H.I. Calvete, C. Galé, P.M. Mateo, A simplex approach for finding local solutions of a linear bilevel program by equilibrium points. Ann. Oper. Res. **138**(1), 143–157 (2005)
270. H.I. Calvete, C. Galé, P.M. Mateo, A study of local solutions in linear bilevel programming. J. Optim. Theory Appl. **125**(1), 63–84 (2005)
271. W. Candler, A linear bilevel programming algorithm: a comment. Comput. Oper. Res. **15**, 297–298 (1988)
272. W. Candler, J. Fortuny-Amat, B. McCarl, The potential role of multilevel programming in agricultural economics. Am. J. Agric. Econ. **63**, 521–531 (1981)
273. W. Candler, R. Norton, Multilevel programming, Technical report 20 (World Bank Development Research Center, Washington, 1977)

274. W. Candler, R. Norton, Multilevel programming and development policy, Technical report 258 (World Bank Development Research Center, Washington D.C., 1977)
275. W. Candler, R. Townsley, A linear two-level programming problem. Comput. Oper. Res. **9**, 59–76 (1982)
276. D. Cao, L.C. Leung, A partial cooperation model for non-unique linear two-level decision problems. Eur. J. Oper. Res. **140**, 134–141 (2002)
277. P. Cappanera, M.P. Scaparra, Optimal allocation of protective resources in shortest-path networks. Transp. Sci. **45**(1), 64–80 (2011)
278. A. Caprara, M. Carvalho, A. Lodi, G.J. Woeginger, A study on the computational complexity of the bilevel knapsack problem. SIAM J. Optim. **24**(2), 823–838 (2014)
279. A. Caprara, M. Carvalho, A. Lodi, G.J. Woeginger, Bilevel knapsack with interdiction constraints. INFORMS J. Comput. **28**(2), 319–333 (2016)
280. M. Caramia, R. Mari, Enhanced exact algorithms for discrete bilevel linear problems. Optim. Lett. **9**(7), 1447–1468 (2015)
281. M. Caramia, R. Mari, A decomposition approach to solve a bilevel capacitated facility location problem with equity constraints. Optim. Lett. **10**(5), 997–1019 (2016)
282. J. Cardinal, E.D. Demaine, S. Fiorini, G. Joret, S. Langerman, I. Newman, O. Weimann, The Stackelberg minimum spanning tree game, in *Proceedings of the 10th International Workshop Algorithms and Data Structures, WADS 2007, Halifax, Canada, August 15–17, 2007*, ed. by D. Frank, et al. Lecture Notes in Computer Science, vol. 4619 (Springer, Berlin, 2007), pp. 64–76
283. J. Cardinal, E.D. Demaine, S. Fiorini, G. Joret, S. Langerman, I. Newman, O. Weimann, The Stackelberg minimum spanning tree game. Algorithmica **59**(2), 129–144 (2011)
284. R. Carli, M. Dotoli, Bi-level programming for the energy retrofit planning of street lighting systems, in *Proceeding of the IEEE 14th International Conference on Networking, Sensing and Control (ICNSC), 2017* (IEEE, New York, 2017), pp. 543–548
285. M. Carrion, J.M. Arroyo, A.J. Conejo, A bilevel stochastic programming approach for retailer futures market trading. IEEE Trans. Power Syst. **24**(3), 1446–1456 (2009)
286. M.-S. Casas-Ramírez, J.-F. Camacho-Vallejo, Solving the p-median bilevel problem with order through a hybrid heuristic. Appl. Soft Comput. **60**, 73–86 (2017)
287. M.-S. Casas-Ramírez, J.-F. Camacho-Vallejo, I.-A. Martínez-Salazar, Approximating solutions to a bilevel capacitated facility location problem with customer's patronization toward a list of preferences. Appl. Math. Comput. **319**(Supplement C), 369–386 (2018)
288. L.M. Case, An l_1 penalty function approach to the nonlinear bilevel programming problem, Ph.D. thesis (University of Waterloo, Canada, 1999)
289. R. Cassidy, M. Kirby, W. Raike, Efficient distribution of resources through three levels of government. Manag. Sci. **17**, 462–473 (1971)
290. M. Catalano, M. Migliore, A Stackelberg-game approach to support the design of logistic terminals. J. Transp. Geogr. **41**, 63–73 (2014)
291. M. Cecchini, J. Ecker, M. Kupferschmid, R. Leitch, Solving nonlinear principal-agent problems using bilevel programming. Eur. J. Oper. Res. **230**(2), 364–373 (2013)
292. L.-C. Ceng, Y.-C. Liou, C.-F. Wen, A hybrid extragradient method for bilevel pseudomonotone variational inequalities with multiple solutions. J. Nonlinear Sci. Appl. **9**(6), 4052–4069 (2016)
293. L.-C. Ceng, Y.-C. Liou, C.-F. Wen, A. Latif, Hybrid steepest-descent viscosity methods for triple hierarchical variational inequalities with constraints of mixed equilibria and bilevel variational inequalities. J. Nonlinear Sci. Appl. **10**(3), 1126–1147 (2017)
294. A. Chaabani, S. Bechikh, L.B. Said, A new co-evolutionary decomposition-based algorithm for bi-level combinatorial optimization. Appl. Intell. **48**(9), 2847–2872 (2018)
295. O. Chadli, Q.H. Ansari, S. Al-Homidan, Existence of solutions and algorithms for bilevel vector equilibrium problems: an auxiliary principle technique. J. Optim. Theory Appl. **172**(3), 726–758 (2017)

296. T.-S. Chang, P.B. Luh, Derivation of necessary and sufficient conditions for single-stage Stackelberg games via the inducible region concept. IEEE Trans. Autom. Control **AC-29**, 63–66 (1984)
297. H. Chen, B. An, D. Niyato, Y. Soh, C. Miao, Workload factoring and resource sharing via joint vertical and horizontal cloud federation networks. IEEE J. Sel. Areas Commun. **35**(3), 557-570 (2017)
298. J. Chen, Z. Wan, Y. Zou, Bilevel invex equilibrium problems with applications. Optim. Lett. **8**(2), 447–461 (2014)
299. Y. Chen, Bilevel programming problems: analysis, algorithms and applications, Ph.D. thesis (Université de Montréal, École Polytechnique, 1993)
300. Y. Chen, M. Florian, The nonlinear bilevel programming problem: a general formulation and optimality conditions, Technical Report CRT-794 (Centre de Recherche sur les Transports, East Liberty, 1991)
301. Y. Chen, M. Florian, On the geometry structure of linear bilevel programs: a dual approach, Technical Report CRT-867 (Centre de Recherche sur les Transports, East Liberty, 1992)
302. Y. Chen, M. Florian, The nonlinear bilevel programming problem: formulations, regularity and optimality conditions. Optimization **32**, 193–209 (1995)
303. Y. Chen, M. Florian, Congested O-D trip demand adjustment problem: bilevel programming formulation and optimality conditions, in *Multilevel Optimization: Algorithms and Applications* ed. by A. Migdalas, P.M. Pardalos, P. Värbrand (Kluwer Academic Publishers, Dordrecht, 1998), pp. 1–22
304. Y. Chen, M. Florian, S. Wu, A descent dual approach for linear bilevel programs, Technical Report CRT-866 (Centre de Recherche sur les Transports, East Liberty, 1992)
305. Y. Chen, H. Lu, J. Li, L. Ren, L. He, A leader-follower-interactive method for regional water resources management with considering multiple water demands and eco-environmental constraints. J. Hydrol. **548**, 121–134 (2017)
306. Y. Chen, T. Pock, R. Ranftl, H. Bischof, Revisiting loss-specific training of filter-based MRFs for image restoration, in *Pattern Recognition* (Springer, Berlin, 2013), pp. 271–281
307. Y. Chen, R. Ranftl, T. Pock, Insights into analysis operator learning: from patch-based sparse models to higher order MRFs. IEEE Trans. Image Process. **23**(3), 1060–1072 (2014)
308. C.-B. Cheng, H.-S. Shih, B. Chen, Subsidy rate decisions for the printer recycling industry by bi-level optimization techniques. Oper. Res. **17**(3), 901–919 (2017)
309. X. Chi, Z. Wan, Z. Hao, The models of bilevel programming with lower level second-order cone programs. J. Inequal. Appl. **2014**(1), 168 (2014)
310. X. Chi, Z. Wan, Z. Hao, Second order sufficient conditions for a class of bilevel programs with lower level second-order cone programming problem. J. Ind. Manag. Optim. **11**(4), 1111–1125 (2015)
311. A. Chinchuluun, P.M. Pardalos, H.-X. Huang, Multilevel (hierarchical) optimization: complexity issues, optimality conditions, algorithms. Adv. Appl. Math. Global Optim. **17**, 197–221 (2009)
312. S.-W. Chiou, Optimization of area traffic control for equilibrium network flows. Transp. Sci. **33**(3), 279–289 (1999)
313. S.-W. Chiou, TRANSYT derivatives for area traffic control optimisation with network equilibrium flows. Transp. Res. B Methodol. **37**(3), 263–290 (2003)
314. S.-W. Chiou, Bilevel programming for the continuous transport network design problem. Transp. Res. B Methodol. **39**(4), 361–383 (2005)
315. S.-W. Chiou, A bi-level programming for logistics network design with system-optimized flows. Inf. Sci. **179**(14), 2434–2441 (2009)
316. S.-W. Chiou, Optimization of robust area traffic control with equilibrium flow under demand uncertainty. Comput. Oper. Res. **41**, 399–411 (2014)
317. S.-W. Chiou, A bi-level decision support system for uncertain network design with equilibrium flow. Decis. Support Syst. **69**, 50–58 (2015)
318. S.-W. Chiou, A cutting plane projection method for bi-level area traffic control optimization with uncertain travel demand. Appl. Math. Comput. **266**, 390–403 (2015)

319. A. Chowdhury, A.R. Zomorrodi, C.D. Maranas, Bilevel optimization techniques in computational strain design. Comput.Chem. Eng. **72**, 363–372 (2015)
320. S. Christiansen, M. Patriksson, L. Wynter, Stochastic bilevel programming in structural optimization. Struct. Multidiscip. Optim. **21**, 361–371 (2001)
321. T.D. Chuong, Optimality conditions for nonsmooth multiobjective bilevel optimization problems. Ann. Oper. Res. **287**(2), 617–642 (2020)
322. T.D. Chuong, V. Jeyakumar, Finding robust global optimal values of bilevel polynomial programs with uncertain linear constraints. J. Optim. Theory Appl. **173**(2), 683–703 (2017)
323. P. Clarke, A. Westerberg, A note on the optimality conditions for the bilevel programming problem. Nav. Res. Logist. **35**, 413–418 (1988)
324. P. Clarke, A. Westerberg, Bilevel programming for steady-state chemical process design—I. Fundamentals and algorithms. Comput.Chem. Eng. **14**, 87–98 (1990)
325. P. Clarke, A. Westerberg, Bilevel programming for steady-state chemical process design—II. Performance study for nondegenerate problems. Comput.Chem. Eng. **14**, 99–110 (1990)
326. P.A. Clarke, A.W. Westerberg, Optimization for design problems having more than one objective. Comput.Chem. Eng. **7**, 259–278 (1983)
327. J. Clegg, M.J. Smith, Cone projection versus half-space projection for the bilevel optimization of transportation networks. Transp. Res. B **35**, 71–82 (2001)
328. E. Codina, L. Montero, Approximation of the steepest descent direction for the OD matrix adjustment problem. Ann. Oper. Res. **144**(1), 329–362 (2006)
329. G. Cohen, J.-P. Quadrat, L. Wynter, Technical note: On the halfplane and cone algorithms for bilevel programming problems by Clegg and Smith, Technical Report (INRIA, France, 2001)
330. B. Colson, Mathematical programs with equilibrium constraints and nonlinear bilevel programming problems, Master's thesis (Department of Mathematics, FUNDP, Namur, Belgium, 1999)
331. B. Colson, BIPA (BIlevel Programming with Approximate methods) software guide and test problems, Technical Report (Département de Mathématique, Facultés Universitaires Notre-Dame de la Paix, Namur, Belgique, 2002)
332. B. Colson, Trust-region algorithms for derivative-free optimization and nonlinear bilevel programming, Ph.D. thesis (Department of Mathematics, The University of Namur, Belgium, 2003)
333. B. Colson, Trust-region algorithms for derivative-free optimization and nonlinear bilevel programming. 4OR, Q. J. Belgian French Ital. Oper. Res. Soc. **2**(1), 85–88 (2004)
334. B. Colson, P. Marcotte, G. Savard, A trust-region method for nonlinear bilevel programming: algorithm and computational experience. Comput. Optim. Appl. **30**(3), 211–227 (2005)
335. B. Colson, P. Marcotte, G. Savard, Bilevel programming: a survey, 4OR **3**, 87–107 (2005)
336. B. Colson, P. Marcotte, G. Savard, An overview of bilevel optimization. Ann. Oper. Res. **153**, 235–256 (2007)
337. A.R. Conn, L.N. Vicente, Bilevel derivative-free optimization and its application to robust optimization. Optim. Methods Softw. **27**(3), 561–577 (2012)
338. I. Constantin, M. Florian, Optimizing frequencies in a transit network: a nonlinear bi-level programming approach. Int. Tran. Oper. Res. **2**(2), 149–164 (1995)
339. J.-P. Côté, P. Marcotte, G. Savard, A bilevel modelling approach to pricing and fare optimisation in the airline industry. J. Revenue Pricing Manag. **2**(1), 23–36 (2003)
340. J.B.Jr. Cruz, Leader-follower strategies for multilevel systems. IEEE Trans. Autom. Control **AC-23**, 244–255 (1978)
341. J. Current, H. Pirkul, The hierarchical network design problem with transshipment facilities. Eur. J. Oper. Res. **52**, 338–347 (1991)
342. J.R. Current, The design of a hierarchical transportation network with transshipment facilities. Transp. Sci. **22**(4), 270–277 (1988)
343. D.D. Čvokić, Y.A. Kochetov, A.V. Plyasunov, A leader-follower hub location problem under fixed markups, in *International Conference on Discrete Optimization and Operations Research* (Springer, Berlin, 2016), pp. 350–363

344. D.D. Čvokić, Y.A. Kochetov, A.V. Plyasunov, The existence of equilibria in the leader-follower hub location and pricing problem, in *Proceedings of the 2015 Operations Research* (Springer, Berlin, 2017), pp. 539–544
345. P. Daniele, Evolutionary variational inequalities and applications to complex dynamic multi-level models. Transp. Res. Logist. Transp. Rev. **46**(6), 855–880 (2010)
346. B. Das, M. Maiti, An application of bi-level newsboy problem in two substitutable items under capital cost. Appl. Math. Comput. **190**(1), 410–422 (2007)
347. S.M. Dassanayaka, Methods of variational analysis in pessimistic bilevel programming, Ph.D. thesis (Wayne University Detroit, Michigan, 2010)
348. J.P. Dauer, Optimization over the efficient set using an active constraint approach. Zeitschrift für Oper. Res. **35**(3), 185–195 (1991)
349. J.P. Dauer, T.A. Fosnaugh, Optimization over the efficient set. J. Global Optim. **7**(3), 261–277 (1995)
350. I. Davydov, Y. Kochetov, S. Dempe, Local search approach for the competitive facility location problem in mobile networks. Int. J. Artif. Intell. Educ. **16**(1), 130–143 (2018)
351. I.A. Davydov, Y.A. Kochetov, N. Mladenovic, D. Urosevic, Fast metaheuristics for the discrete (rlp)-centroid problem. Autom. Remote Control **75**(4), 677–687 (2014)
352. J.C. De los Reyes, C.B. Schönlieb, T. Valkonen, Bilevel parameter learning for higher-order total variation regularisation models. J. Math. Imaging Vision **57**(1), 1–25 (2017)
353. C.H.M. de Sabóia, M. Campêlo, S. Scheimberg, A computational study of global algorithms for linear bilevel programming. Numer. Algorithms **35**(2–4), 155–173 (2004)
354. D. De Wolf, Y. Smeers, A stochastic version of a Stackelberg-Nash-Cournot equilibrium model. Manag. Sci. **43**(2), 190–197 (1997)
355. K. Deb, A. Sinha, Constructing test problems for bilevel evolutionary multi-objective optimization, in *IEEE Congress on Evolutionary Computation (CEC'09)* (IEEE, New York, 2009), pp. 1153–1160
356. K. Deb, A. Sinha, Solving bilevel multi-objective optimization problems using evolutionary algorithms, in *Evolutionary Multi-Criterion Optimization* (Springer, Berlin, 2009), pp. 110–124
357. K. Deb, A. Sinha, An efficient and accurate solution methodology for bilevel multi-objective programming problems using a hybrid evolutionary-local-search algorithm. Evol. Comput. **18**(3), 403–449 (2010)
358. A. Dekdouk, A. Azzouz, H. Yahyaoui, S. Krichen, Solving energy ordering problem with multiple supply-demand using bilevel optimization approach. Procedia Comput. Sci. **130**, 753–759 (2018). The 9th International Conference on Ambient Systems, Networks and Technologies (ANT 2018)/The 8th International Conference on Sustainable Energy Information Technology (SEIT-2018)/Affiliated Workshops
359. L. dell' Olio, A. Ibeas, F. Ruisánchez, Optimizing bus-size and headway in transit networks. Transportation **39**(2), 449–464 (2012)
360. V.T. Dement'ev, A.I. Erzin, R.M. Larin, Yu.V. Shamardin, in *Problems of the optimization of hierarchical structures (Russian)* (Izdatel'stvo Novosibirskogo Universiteta, Novosibirsk, 1996)
361. V.T. Dement'ev, A.V. Pyatkin, On a decentralized transportation problem. (Russian). Diskretn. Anal. Issled. Oper. **15**(3), 22–30, 95–96 (2008). translation in J. Appl. Ind. Math. 3(1), 32–37 (2009)
362. V.T. Dement'ev, Y.V. Shamardin, A three-level model for the choice of nomenclature of products. Diskret. Anal. Issled. Oper. **8**(1), 40–46 (2001)
363. V.T. Dement'ev, Y.V. Shamardin, The problem of price selection for production under the condition of obligatory satisfaction of demand (Russian). Diskretn. Anal. Issled. Oper. Ser. 2 **9**(2), 31–40 (2002)
364. V.T. Dement'ev, Y.V. Shamardin, A two-level assignment problem with a generalized Monge condition (Russian). Diskretn. Anal. Issled. Oper. Ser. 2 **10**(2), 19–28 (2003)
365. V.T. Dement'ev, Y.V. Shamardin, On a polynomially solvable case of a decentralized transportation problem. (Russian). Diskretn. Anal. Issled. Oper. **18**(1), 20–26, 102 (2011)

366. V. DeMiguel, W. Murray, A local convergence analysis of bilevel decomposition algorithms. Optim. Eng. **7**(2), 99–133 (2006)
367. V. DeMiguel, H. Xu, A stochastic multiple-leader Stackelberg model: analysis, computation, and application. Oper. Res. **57**(5), 1220–1235 (2009)
368. V. DeMiguel, H. Xu, A stochastic multiple-leader Stackelberg model: analysis, computation, and application. Oper. Res. **57**(5), 1220–1235 (2009)
369. S. Dempe, A simple algorithm for the linear bilevel programming problem. Optimization **18**, 373–385 (1987)
370. S. Dempe, On an optimality condition for a two-level optimization problem. Vestn. Leningr. Univ. Ser. I **1989**(3), 10–14 (1989, in Russian)
371. S. Dempe, Optimality condition for bilevel programming problems. Vestn. Leningr. Univ. Math. **22**(3), 11–16 (1989)
372. S. Dempe, Richtungsdifferenzierbarkeit der Lösung parametrischer Optimierungsaufgaben und ihre Anwendung bei der Untersuchung von Zwei-Ebenen-Problemen, Ph.D. thesis (Technische Universität Karl-Marx-Stadt, Sektion Mathematik, 1991). Habilitation thesis
373. S. Dempe, A necessary and a sufficient optimality condition for bilevel programming problems. Optimization **25**, 341–354 (1992)
374. S. Dempe, Optimality conditions for bilevel programming problems, in *System Modelling and Optimization*, ed. by P. Kall, et al. Lecture Notes in Control and Information Science (180) (Springer, Berlin, 1992), pp. 17–24
375. S. Dempe, On the directional derivative of a locally upper Lipschitz continuous point–to–set mapping and its application to optimization problems, in *Parametric Optimization and Related Topics, III, P. Lang*, ed. by J. Guddat, H.Th. Jongen, B. Kummer, F. Nožička (1993)
376. S. Dempe, On the leader's dilemma and a new idea for attacking bilevel programming problems, Technical Report (Technische Universität Chemnitz, Fachbereich Mathematik, Chemnitz, 1993)
377. S. Dempe, Computing optimal incentives via bilevel programming. Optimization **33**, 29–42 (1995)
378. S. Dempe, On generalized differentiability of optimal solutions and its application to an algorithm for solving bilevel optimization problems, in *Recent Advances in Nonsmooth Optimization*, ed. by D.-Z. Du, L. Qi, R.S. Womersley (World Scientific, Singapore, 1995), pp. 36–56
379. S. Dempe, Applicability of two-level optimization to issues of environmental policy, in *Modelling the Environmental Concerns of Production*, ed. by K. Richter. Discussion Paper, vol. 62 (Europa-Universität Viadrina Frankfurt (Oder), Fakultät für Wirtschaftswissenschaften, 1996), pp. 41–50
380. S. Dempe, Discrete bilevel optimization problems, Technical Report 12 (Universität Leipzig, Wirtschaftswissenschaftliche Fakultät, Leipzig, 1996)
381. S. Dempe, First-order necessary optimality conditions for general bilevel programming problems. J. Optim. Theory Appl. **95**, 735–739 (1997)
382. S. Dempe, An implicit function approach to bilevel programming problems, in *Multilevel Optimization: Algorithms and Applications*, ed. by A. Migdalas, P.M. Pardalos, P. Värbrand (Kluwer Academic Publishers, Dordrecht, 1998), pp. 273–294
383. S. Dempe, A bundle algorithm applied to bilevel programming problems with non-unique lower level solutions. Comput. Optim. Appl. **15**, 145–166 (2000)
384. S. Dempe, Bilevel programming: the implicit function approach, in *Encyclopedia of Optimization* (Kluwer Academic Publishers, Dordrecht, 2001), pp. 167–173
385. S. Dempe, *Foundations of Bilevel Programming* (Kluwer Academic Publishers, Dordrecht, 2002)
386. S. Dempe, Annotated bibliography on bilevel programming and mathematical programs with equilibrium constraints. Optimization **52**, 333–359 (2003)
387. S. Dempe, Bilevel programming, in *Essays and Surveys in Global Optimization*, ed. by C. Audet, P. Hansen, G. Savard (Kluwer Academic Publishers, Boston, 2005), pp. 165–194

388. S. Dempe, Comment to Interactive fuzzy goal programming approach for bilevel programming problem by S.R. Arora and R. Gupta. Eur. J. Oper. Res. **212**(2), 429–431 (2011)
389. S. Dempe, Bilevel optimization: Reformulation and first optimality conditions, in *Generalized Nash Equilibrium Problems, Bilevel Programming and MPEC*, ed. by D. Aussel, C.S. Lalitha (Springer, Berlin, 2017), pp. 1–20
390. S. Dempe, J.F. Bard, A bundle trust region algorithm for bilinear bilevel programming, in *Operations Research Proceedings 1999* (Springer, Berlin, 2000), pp. 7–12
391. S. Dempe, J.F. Bard, Bundle trust-region algorithm for bilinear bilevel programming. J. Optim. Theory Appl. **110**(2), 265–288 (2001)
392. S. Dempe, N. Dinh, J. Dutta, Optimality conditions for a simple convex bilevel programming problem, in *Variational Analysis and Generalized Differentiation in Optimization and Control*, ed. by R.S. Burachik, J.-C. Yao. Springer Optimization and Its Applications, vol. 47 (Springer, Berlin, 2010), pp. 149–162
393. S. Dempe, J. Dutta, Is bilevel programming a special case of a mathematical program with complementarity constraints?. Math. Program. **131**, 37–48 (2012)
394. S. Dempe, J. Dutta, S. Lohse, Optimality conditions for bilevel programming problems. Optimization **55**, 505–524 (2006)
395. S. Dempe, J. Dutta, B.S. Mordukhovich, New necessary optimality conditions in optimistic bilevel programming. Optimization **56**, 577–604 (2007)
396. S. Dempe, J. Dutta, B.S. Mordukhovich, Variational analysis in bilevel programming, in *Mathematical Programming and Game Theory for Decision Making*, ed. by S.K. Neogy, et al. (World Scientific, Singapore, 2008), pp. 257–277
397. S. Dempe, V. Kalashnikov *Optimization with Multivalued Mappings: Theory, Applications and Algorithms* (Springer/LLC, New York, 2006)
398. S. Dempe, D. Fanghänel, T. Starostina, Optimal toll charges: fuzzy optimization approach, in *Methods of Multicriteria Decision—Theory and Applications*, ed. by F. Heyde, A. Löhne, C. Tammer (Shaker, Aachen, 2009), pp. 29–45
399. S. Dempe, S. Franke, Bilevel programming: Stationarity and stability. Pac. J. Optim. **9**(2), 183–199 (2013)
400. S. Dempe, S. Franke, Solution algorithm for an optimistic linear Stackelberg problem. Comput. Oper. Res. **41**, 277–281 (2014)
401. S. Dempe, S. Franke, The bilevel road pricing problem. Int. J. Comput. Optim. **2**, 71–92 (2015)
402. S. Dempe, S. Franke, On the solution of convex bilevel optimization problems. Comput. Optim. Appl. **63**, 685–703 (2016)
403. S. Dempe, S. Franke, Solution of bilevel optimization problems using the KKT approach. Optimization **68**, 1471–1489 (2019)
404. S. Dempe, N. Gadhi, Necessary optimality conditions for bilevel set optimization problem. J. Global Optim. **39**(4), 529–542 (2007)
405. S. Dempe, N. Gadhi, Necessary optimality conditions of a D.C. set-valued bilevel optimization problem. Optimization **57**, 777–793 (2008)
406. S. Dempe, N. Gadhi, Second order optimality conditions for bilevel set optimization problems. J. Global Optim. **47**(2), 233–245 (2010)
407. S. Dempe, N. Gadhi, Optimality results for a specific bilevel optimization problem. Optimization **60**(7–9), 813–822 (2011)
408. S. Dempe, N. Gadhi, A new equivalent single-level problem for bilevel problems. Optimization **63**(5), 789–798 (2014)
409. S. Dempe, N. Gadhi, A.B. Zemkoho, New optimality conditions for the semivectorial bilevel optimization problem. J. Optim. Theory Appl. **157**(1), 54–74 (2013)
410. S. Dempe, N.A. Gadhi, L. Lafhim, Fuzzy and exact optimality conditions for a bilevel set-valued problem via extremal principles. Numer. Funct. Anal. Optim. **31**(8), 907–920 (2010)
411. S. Dempe, H. Günzel, H.Th. Jongen, On reducibility in bilevel problems. SIAM J. Optim. **20**, 718–727 (2009)

412. S. Dempe, S. Ivanov, A. Naumov, Reduction of the bilevel stochastic optimization problem with quantile objective function to a mixed-integer problem. Appl. Stochastic Models Bus. Ind. **33**(5), 544–554 (2017)
413. S. Dempe, V. Kalashnikov, G.A. Pérez-Valdés, N. Kalashnykova, *Bilevel Programming Problems: Theory, Algorithms and Application to Energy Networks* (Springer, Berlin, 2015)
414. S. Dempe, V. Kalashnikov, R.Z. Rios-Mercado, Discrete bilevel programming: application to a natural gas cash-out problem. Eur. J. Oper. Res. **166**, 469–488 (2005)
415. S. Dempe, V.V. Kalashnikov, N. Kalashnykova, Optimality conditions for bilevel programming problems, in *Optimization with Multivalued Mappings: Theory, Applications and Algorithms*, ed. by S. Dempe, V. Kalashnikov (Springer/LLC, New York, 2006), pp. 3–28
416. S. Dempe, V.V. Kalashnikov, N.I. Kalashnykova, A.A. Franco, A new approach to solving bi-level programming problems with integer upper level variables. ICIC Express Lett. **3**(4), 1281–1286 (2009)
417. S. Dempe, F.M. Kue, Solving discrete linear bilevel optimization problems using the optimal value reformulation. J. Global Optim. **68**(2), 255–277 (2017)
418. S. Dempe, F.M. Kue, P. Mehlitz, Optimality conditions for mixed discrete bilevel optimization problems. Optimization **67**(6), 737–756 (2018)
419. S. Dempe, F.M. Kue, P. Mehlitz, Optimality conditions for special semidefinite bilevel optimization problems. SIAM J. Optim. **28**(2), 1564–1587 (2018)
420. S. Dempe, S. Lohse, Inverse linear programming, Recent Advances in Optimization, in *Proceedings of the 12th French-German-Spanish Conference on Optimization held in Avignon, September 20-24, 2004*, ed. by A. Seeger. Lectures Notes in Economics and Mathematical Systems, vol. 563 (Springer, Berlin, 2006), pp. 19–28
421. S. Dempe, S. Lohse, Dependence of bilevel programming on irrelevant data, Technical Report 2011-01 (TU Bergakademie Freiberg, Department of Mathematics and Computer Science, Freiberg, 2011). www.optimization-online.org
422. S. Dempe, S. Lohse, Optimale Mautgebühren—Ein Modell und ein Optimalitätstest. at—Automatisierungstechnik **60**(4), 225–232 (2012)
423. S. Dempe, G. Luo, S. Franke, Pessimistic bilevel linear optimization. J. Nepal Math.Soc. **1**, 1–10 (2018)
424. S. Dempe, P. Mehlitz, Semivectorial bilevel programming versus scalar bilevel programming. Optimization, **69**(4), 657–679 (2020)
425. S. Dempe, B.S. Mordukhovich, A.B. Zemkoho, Sensitivity analysis for two-level value functions with applications to bilevel programming. SIAM J. Optim. **22**, 1309–1343 (2012)
426. S. Dempe, B.S. Mordukhovich, A.B. Zemkoho, Necessary optimality conditions in pessimistic bilevel programming. Optimization **63**(4), 505–533 (2014)
427. S. Dempe, B.S. Mordukhovich, A.B. Zemkoho, Two-level value function approach to non-smooth optimistic and pessimistic bilevel programs. Optimization **68**(2–3), 433–455 (2019)
428. S. Dempe, M. Pilecka, Necessary optimality conditions for optimistic bilevel programming problems using set-valued programming. J. Global Optim. **61**(4), 769–788 (2015)
429. S. Dempe, K. Richter, Bilevel programming with knapsack constraints. Cent. Eur. J. Oper. Res. **8**, 93–107 (2000)
430. S. Dempe, H. Schmidt, On an algorithm solving two-level programming problems with nonunique lower level solutions. Comput. Optim. Appl. **6**, 227–249 (1996)
431. S. Dempe, T. Starostina, Optimal toll charges in a fuzzy flow problem, in *Computational Intelligence, Theory and Applications. Advances in Soft Computing*, ed. by B. Reusch (Springer, Berlin, 2006), pp. 405–413
432. S. Dempe, T. Starostina, On the solution of fuzzy bilevel programming problems, Technical report (Department of Mathematics and Computer Science, TU Bergakademie Freiberg, 2007)
433. S. Dempe, A.B. Zemkoho, The generalized Mangasarian-Fromowitz constraint qualification and optimality conditions for bilevel programs. J. Optim. Theory Appl. **148**(1), 46–68 (2011)
434. S. Dempe, A.B. Zemkoho, On the Karush-Kuhn-Tucker reformulation of the bilevel optimization problem. Nonlinear Anal. Theory Methods Appl. **75**, 1202–1218 (2012)

435. S. Dempe, A.B. Zemkoho, The bilevel programming problem: reformulations, constraint qualifications and optimality conditions. Math. Program. **2013**, 447–473 (138)
436. S. Dempe, A.B. Zemkoho, KKT reformulation and necessary conditions for optimality in nonsmooth bilevel optimization. SIAM J. Optim. **24**(4), 1639–1669 (2014)
437. S. DeNegre, Interdiction and discrete bilevel linear programming, Ph.D. thesis (Lehigh University, Lehigh, 2011)
438. S.T. DeNegre, T.K. Ralphs, A branch-and-cut algorithm for integer bilevel linear programs, in *Operations Research and Cyber-Infrastructure*, ed. by J.W. Chinneck, B. Kristjansson, M. Saltzman. Operations Research/Computer Science Interfaces, vol. 47 (Springer, Berlin, 2009), pp. 65–78
439. X. Deng, Complexity issues in bilevel linear programming, in *Multilevel Optimization: Algorithms and Applications*, ed. by A. Migdalas, P.M. Pardalos, P. Värbrand (Kluwer Academic, Dordrecht, 1998), pp. 149–164
440. A. deSilva, Sensitivity formulas for nonlinear factorable programming and their application to the solution of an implicitly defined optimization model of us crude oil production, Ph.D. thesis (George Washington University, Washington, 1978)
441. A. deSilva, G. McCormick, Implicitly defined optimization problems. Ann. Oper. Res. **34**, 107–124 (1992)
442. M. Desrochers, P. Marcotte, M. Stan, The congested facility location problem, in *Proceedings of the 14th International Symposium on Mathematical Programming, Amsterdam, August 5–9* (1991)
443. J. Deuerlein, Hydraulische Systemanalyse von Wasserversorgungsnetzen, Ph.D. thesis (Universität Karlsruhe, Karlsruhe, 2002)
444. S. Dewez, On the toll setting problem, Ph.D. thesis (Université Libre de Bruxelles, Bruxelles, 2004)
445. S. Dewez, M. Labbé, P. Marcotte, G. Savard, New formulations and valid inequalities for a bilevel pricing problem. Oper. Res. Lett. **36**(2), 141–149 (2008)
446. S. Diamond, V. Sitzmann, S. Boyd, G. Wetzstein, F. Heide, Dirty pixels: optimizing image classification architectures for raw sensor data (2017). arXiv preprint:1701.06487
447. M. Didi-Biha, P. Marcotte, G. Savard, Path-based formulations of a bilevel toll setting problem, in *Optimization with Multivalued Mappings: Theory, Applications and Algorithms*, ed. by S. Dempe, V. Kalashnikov. Optimization and its Applications, vol. 2 (Springer/LLC, New York, 2006), pp. 29–50
448. P.H. Dien, N.D. Yen, On implicit function theorems for set-valued maps and their application to mathematical programming under inclusion constraints. Appl. Math. Optim. **24**(1), 35–54 (1991)
449. P.H. Dien, N.D. Yen, Correction: On implicit function theorems for set-valued maps and their application to mathematical programming under inclusion constraints. Appl. Math. Optim. **26**(1), 111–111 (1992)
450. X. P. Ding, Y.-C. Liou, Bilevel optimization problems in topological spaces. Taiwanese J. Math. **10**(1), 173–179 (2006)
451. B.V. Dinh, P.G. Hung, L.D. Muu, Bilevel optimization as a regularization approach to pseudomonotone equilibrium problems. Numer. Funct. Anal. Optim. **35**(5), 539–563 (2014)
452. N Dinh, B Mordukhovich, T.T.A. Nghia, Subdifferentials of value functions and optimality conditions for DC and bilevel infinite and semi-infinite programs. Math. Program. **123**(1), 101–138 (2010)
453. T. Dokka, A. Zemkoho, S.S. Gupta, F.T. Nobibon, Pricing toll roads under uncertainty, in *OASIcs-OpenAccess Series in Informatics*, vol. 54. Schloss Dagstuhl-Leibniz-Zentrum fuer Informatik (2016)
454. V.F. Dökmeci, Optimum location of hierarchical production units with respect to price-elastic demand. Environ. Plann. A **23**(11), 1671–1678 (1991)
455. L.F. Domínguez, E.N. Pistikopoulos, Multiparametric programming based algorithms for pure integer and mixed-integer bilevel programming problems. Comput.Chem. Eng. **34**(12), 2097–2106 (2010)

456. Y. Dong, Z. Wan, A pattern search filter method for bilevel programming problems, in *2009 WRI World Congress on Computer Science and Information Engineering*, vol. 6 (IEEE, New York, 2009), pp. 53–59
457. D. Dorsch, H.Th. Jongen, V. Shikhman, On intrinsic complexity of Nash equilibrium problems and bilevel optimization. J. Optim. Theory Appl. **159**(3), 606–634 (2013)
458. O. Drissi-Kaitouni, J.T. Lundgren, Bilevel origin-destination matrix estimation using a descent approach, Technical Report LiTH-MAT-R-1992-49 (Linköping Institute of Technology, Department of Mathematics, Sweden, 1992)
459. G. Du, Y. Xia, R.J. Jiao, X. Liu, Leader-follower joint optimization problems in product family design. J. Intell. Manuf. **30**(3), 1387–1405 (2019)
460. G. Du, Y. Zhang, X. Liu, R.J. Jiao, Y. Xia, Y. Li, A review of leader-follower joint optimization problems and mathematical models for product design and development. Int. J. Adv. Manuf. Technol. **103**(9–12), 3405-3424 (2019)
461. J. Du, X. Li, L. Yu, R. Dan, J. Zhou, Multi-depot vehicle routing problem for hazardous materials transportation: a fuzzy bilevel programming. Inf. Sci. **399**, 201–218 (2017)
462. Q. Duan, M. Xu, Y. Lu, L. Zhang, A smoothing augmented Lagrangian method for nonconvex, nonsmooth constrained programs and its applications to bilevel problems. J. Ind. Manag. Optim., **15**(3), 1241–1261 (2019)
463. X. Duan, S. Song, J. Zhao, Emergency vehicle dispatching and redistribution in highway network based on bilevel programming. Math. Prob. Eng. **2015**, 12 (2015)
464. Z. Duan, L. Wang, Heuristic algorithms for the inverse mixed integer linear programming problem. J. Global Optim. **51**(3), 463–471 (2011)
465. P.M. Duc, L.D. Muu, A splitting algorithm for a class of bilevel equilibrium problems involving nonexpansive mappings. Optimization **65**, 1855–1866 (2016)
466. T. Dudas, B. Klinz, G.J. Woeginger, The computational complexity of multi-level bottleneck programming problems, in *Multilevel Optimization: Algorithms and Applications*, ed. by A. Migdalas, P.M. Pardalos, P. Värbrand (Kluwer Academic, Dordrecht, 1998), pp. 165–179
467. J.-P. Dussault, P. Marcotte, S. Roch, G. Savard, A smoothing heuristic for a bilevel pricing problem. Eur. J. Oper. Res. **174**(3), 1396–1413 (2006)
468. J. Dutta, Optimality conditions for bilevel programming: an approach through variational analysis, in *Generalized Nash Equilibrium Problems, Bilevel Programming and MPEC*, ed. by D. Aussel, C.S. Lalitha (Springer, Singapore, 2017), pp. 43–64
469. J. Dutta, S. Dempe, Bilevel programming with convex lower level problems, in *Optimization with Multivalued Mappings: Theory, Applications and Algorithms*, ed. by S. Dempe, V. Kalashnikov (Springer/LLC, New York, 2006)
470. Y. Dvorkin, R. Fernández-Blanco, D.S. Kirschen, H. Pandžić, J.P. Watson, C.A. Silva-Monroy, Ensuring profitability of energy storage. IEEE Trans. Power Syst. **32**(1), 611–623 (2017)
471. J. Eckardt, Zwei-Ebenen-Optimierung mit diskreten Aufgaben in der unteren Ebene, Master's thesis (TU Bergakademie Freiberg, Fakultät für Mathematik und Informatik, Freiberg, 1998)
472. J.G. Ecker, J.H. Song, Optimizing a linear function over an efficient set. J. Optim. Theory Appl. **83**(3), 541–563 (1994)
473. T. Edmunds, Algorithms for nonlinear bilevel mathematical programs, Ph.D. thesis (Department of Mechanical Engineering, University of Texas, Austin, 1988)
474. T. Edmunds, J.F. Bard, Algorithms for nonlinear bilevel mathematical programming. IEEE Trans. Syst. Man Cybern. **21**, 83–89 (1991)
475. T. Edmunds, J.F. Bard, An algorithm for the mixed-integer nonlinear bilevel programming problem. Ann. Oper. Res. **34**, 149–162 (1992)
476. A. Ehrenmann, Equilibrium problems with equilibrium constraints and their application to electricity markets, Ph.D. thesis (University of Cambridge, Cambridge, 2004)
477. A. Ehrenmann, Manifolds of multi-leader Cournot equilibria. Oper. Res. Lett. **32**(2), 121–125 (2004)

478. H. Ehtamo, T. Raivio, On applied nonlinear and bilevel programming for pursuit-evasion games. J. Optim. Theory Appl. **108**, 65–96 (2001)
479. G. Eichfelder, *Adaptive Scalarization Methods in Multiobjective Optimization* (Springer, Berlin, 2008)
480. G. Eichfelder, Multiobjective bilevel optimization. Math. Program. **123**, 419–449 (2010)
481. H.A. Eiselt, G. Laporte, J.-F. Thisse, Competitive location models: a framework and bibliography. Transp. Sci. **27**(1), 44–54 (1993)
482. B. El-Sobky, Y. Abo-Elnaga, A penalty method with trust-region mechanism for nonlinear bilevel optimization problem. J. Comput. Appl. Math. **340**, 360–374 (2018)
483. O.E. Emam, A fuzzy approach for bi-level integer non-linear programming problem. Appl. Math. Comput. **172**, 62–71 (2006)
484. O.E. Emam, Interactive approach to bi-level integer multi-objective fractional programming problem. Appl. Math. Comput. **223**, 17–24 (2013)
485. E. Erkut, O. Alp, Designing a road network for hazardous materials shipments. Comput. Oper. Res. **34**(5), 1389–1405 (2007)
486. E. Erkut, F. Gzara, Solving the hazmat transport network design problem. Comput. Oper. Res. **35**(7), 2234–2247 (2008)
487. M.S. Ershova, The branch and bound method for a quadratic problem of bilevel programming. Diskret. Anal. Issled. Oper. **13**(1), 40–56 (2006, in Russian)
488. M. Esmaeili,H. Sadeghi, An investigation of the optimistic solution to the linear trilevel programming problem. Mathematics **6**(10), 179 (2018)
489. J.B.E. Etoa, Solving convex quadratic bilevel programming problems using an enumeration sequential quadratic programming algorithm. J. Global Optim. **47**(4), 615–637 (2010)
490. J.B.E. Etoa, Solving quadratic convex bilevel programming problems using a smoothing method. Appl. Math. Comput. **217**(15), 6680–6690 (2011)
491. J.B. Eytard, M. Akian, M. Bouhtou, S. Gaubert, A bilevel optimization model for load balancing in mobile networks through price incentives, in *Proceedings of the 15th International Symposium on Modeling and Optimization in Mobile, Ad Hoc, and Wireless Networks (WiOpt), 2017* (IEEE, New York, 2017), pp. 1–8
492. G Facchetti, C Altafini, Partial inhibition and bilevel optimization in flux balance analysis. BMC Bioinf. **14**(1), 344 (2013)
493. N.P. Faísca, V. Dua, B. Rustem, P.M. Saraiva, E.N. Pistikopoulos, Parametric global optimisation for bilevel programming. J. Global Optim. **38**(4), 609–623 (2007)
494. N.P. Faísca, V. Dua, P.M. Saraiva, B. Rustem, E.N. Pistikopoulos, A global parametric programming optimisation strategy for multilevel problems, in *Proceedings of the 16th European Symposium on Computer Aided Process Engineering and 9th International Symposium on Process Systems Engineering*, ed. by W. Marquardt, C. Pantelides. Computer Aided Chemical Engineering, vol. 21 (Elsevier, Amsterdam, 2006), pp. 215–220
495. N.P. Faísca, P.M. Saraiva, B. Rustem, E.N. Pistikopoulos, A multi-parametric programming approach for multilevel hierarchical and decentralised optimisation problems. Comput. Manag. Sci. **6**(4), 377–397 (2009)
496. J.E. Falk, J. Liu, Algorithms for general nonlinear bilevel programs. Cent. Eur. J. Oper. Res. **2**, 101–117 (1993)
497. J.E. Falk, J. Liu, On bilevel programming, Part I: General nonlinear cases. Math. Program. **70**, 47–72 (1995)
498. M. Fampa, L.A. Barroso, D. Candal, L. Simonetti, Bilevel optimization applied to strategic pricing in competitive electricity markets. Comput. Optim. Appl. **39**(2), 121–142 (2008)
499. H. Fang, L. Xu, K.-K.R. Choo, Stackelberg game based relay selection for physical layer security and energy efficiency enhancement in cognitive radio networks. Appl. Math. Comput. **296**, 153–167 (2017)
500. S.-C. Fang, C.-F. Hu, Solving fuzzy variational inequalities. Fuzzy Optim. Decis. Making **1**(1), 113–133 (2002)
501. D. Fanghänel, Optimality criteria for bilevel programing problems using the radial subdifferential, in *Optimization with Multivalued Mappings: Theory, Applications and Algorithms*,

ed. by S. Dempe, V. Kalashnikov. Optimization and its Applications, vol. 2 (Springer/LLC, New York, 2006), pp. 73–95
502. D. Fanghänel, Zwei-Ebenen-Optimierung mit diskreter unterer Ebene und stetiger oberer Ebene, Ph.D. thesis (TU Bergakademie Freiberg, Germany, 2006)
503. D. Fanghänel, Optimality conditions for a bilevel matroid problem. J. Comb. Optim. **22**(4), 594–608 (2011)
504. D. Fanghänel, S. Dempe, Bilevel programming with discrete lower level problems. Optimization **58**, 1029–1047 (2009)
505. A.M.F. Fard, M. Hajaghaei-Keshteli, A tri-level location-allocation model for forward/reverse supply chain. Appl. Soft Comput. **62**, 328–346 (2018)
506. R. Fernández-Blanco, J.M. Arroyo, N. Alguacil, A unified bilevel programming framework for price-based market clearing under marginal pricing. IEEE Trans. Power Syst. **27**(1), 517–525 (2012)
507. R. Fernández-Blanco, J.M. Arroyo, N. Alguacil, Network-constrained day-ahead auction for consumer payment minimization. IEEE Trans. Power Syst. **29**(2), 526–536 (2014)
508. R. Fernández-Blanco, J.M. Arroyo, N. Alguacil, Bilevel programming for price-based electricity auctions: a revenue-constrained case. EURO J. Comput. Optim. **3**(3), 163–195 (2015)
509. R. Fernández-Blanco, J.M. Arroyo, N. Alguacil, On the solution of revenue-and network-constrained day-ahead market clearing under marginal pricing. Part i: An exact bilevel programming approach. IEEE Trans. Power Syst. **32**(1), 208–219 (2017)
510. B. Fernando, S. Gould, Discriminatively learned hierarchical rank pooling networks. Int. J. Comput. Vision **124**(3), 335–355 (2017)
511. F.A. Ferreira, F. Ferreira, M. Ferreira, A.A. Pinto, Flexibility in a Stackelberg leadership with differentiated goods. Optimization **64**(4), 877–893 (2015)
512. M. Fischetti, I. Ljubić, M. Monaci, M. Sinnl, Intersection cuts for bilevel optimization, in *Proceedings of the 18th International Conference on Integer Programming and Combinatorial Optimization*, ed. by Q. Louveaux, M. Skutella (Springer, Berlin, 2016), pp. 77–88
513. M. Fischetti, I. Ljubić, M. Monaci, M. Sinnl, A new general-purpose algorithm for mixed-integer bilevel linear programs. Oper. Res. **65**(6), 1615–1637 (2017)
514. M. Fischetti, I. Ljubić, M. Monaci, M. Sinnl, On the use of intersection cuts for bilevel optimization. Math. Program. **172**(1), 77–103 (2018)
515. M. Fischetti, M. Monaci, M. Sinnl, A dynamic reformulation heuristic for generalized interdiction problems. Eur. J. Oper. Res. **267**(1), 40–51 (2018)
516. J. Fliege, L.N. Vicente, Multicriteria approach to bilevel optimization. J. Optim. Theory Appl. **131**(2), 209–225 (2006)
517. C. Florensa, P. Garcia-Herreros, P. Misra, E. Arslan, S. Mehta, I.E. Grossmann, Capacity planning with competitive decision-makers: trilevel MILP formulation, degeneracy, and solution approaches. Eur. J. Oper. Res. **262**, 449–463 (2017)
518. M. Florian, Y. Chen, A bilevel programming approach to estimating O-D matrix by traffic counts, Technical Report CRT-750 (Centre de Recherche sur les Transports, East Liberty, 1991)
519. M. Florian, Y. Chen, A coordinate descent method for bilevel O-D matrix estimation problems. Int. Trans. Oper. Res. **2**, 165–179 (1995)
520. C.A. Floudas, P.M. Pardalos, C. Adjiman, W.R. Esposito, Z.H. Gümüs, S.T. Harding, J.L. Klepeis, C.A. Meyer, C.A. Schweiger, *Handbook of Test Problems in Local and Global Optimization*, vol. 33 (Springer, Berlin, 2013)
521. P. Fontaine, S. Minner, Benders decomposition for discrete–continuous linear bilevel problems with application to traffic network design. Transp. Res. B Methodol. **70**, 163–172 (2014)
522. J. Fortuny-Amat, B. McCarl, A representation and economic interpretation of a two-level programming problem. J. Oper. Res. Soc. **32**, 783–792 (1981)
523. A. Frangioni, On a new class of bilevel programming problems and its use for reformulating mixed integer problems. Eur. J. Oper. Res. **82**, 615–646 (1995)

524. S. Franke, Bilevel programming: optimal value and Karush-Kuhn-Tucker reformulation, Ph.D. thesis (TU Bergakademie Freiberg, Freiberg, 2014)
525. S. Franke, P. Mehlitz, M. Pilecka, Optimality conditions for the simple convex bilevel programming problem in Banach spaces. Optimization **67**(2), 237–268 (2018)
526. A. Friedlander, F.A.M. Gomes, Solution of a truss topology bilevel programming problem by means of an inexact restoration method. Comput. Appl. Math. **30**(1), 109–125 (2011)
527. T. Friesz, C. Suwansirikul, R. Tobin, Equilibrium decomposition optimization: a heuristic for the continuous equilibrium network design problem. Transp. Sci. **21**, 254–263 (1987)
528. T.L. Friesz, G. Anandalingam, N.J. Mehta, K. Nam, S.J. Shah, R.L. Tobin, The multiobjective equilibrium network design problem revisited: a simulated annealing approach. Eur. J. Oper. Res.. **65**(1), 44–57 (1993)
529. T.L. Friesz, H.-J. Cho, N.J. Mehta, R.L. Tobin, G. Anandalingam, A simmulated annealing approach to the network design problem with variational inequality constraints. Transp. Sci. **26**, 18–26 (1992)
530. T.L. Friesz, R.L. Tobin, H.-J. Cho, N.J. Mehta, Sensitivity analysis based heuristic algorithms for mathematical programs with variational inequality constraints. Math. Program. **48**(1–3), 265–284 (1990)
531. J. Fülöp, On the equivalence between a linear bilevel programming problem and linear optimization over the efficient set, Technical Report WP 93–1, in *Laboratory of Operations Research and Decision Systems, Computer and Automation Institute, Hungarian Academy of Sciences* (1993)
532. S.A. Gabriel, F.U. Leuthold, Solving discretely-constrained MPEC problems with applications in electric power markets. Energy Econ. **32**(1), 3–14 (2010)
533. N. Gadhi, S. Dempe, Necessary optimality conditions and a new approach to multiobjective bilevel optimization problems. J. Optim. Theory Appl. **155**, 100–114 (2012)
534. N. Gadhi, M. El idrissi, An equivalent one level optimization problem to a semivectorial bilevel problem. Positivity **22**(1), 261–274 (2018)
535. A.A. Gaivoronski, A. Werner, Stochastic programming perspective on the agency problems under uncertainty, in *Managing Safety of Heterogeneous Systems* (Springer, New York, 2012), pp. 137–167
536. J. Gang, Y. Tu, B. Lev, J. Xu, W. Shen, L. Yao, A multi-objective bi-level location planning problem for stone industrial parks. Comput. Oper. Res. **56**, 8–21 (2015)
537. J. Gao, B. Liu, Fuzzy multilevel programming with a hybrid intelligent algorithm. Comput. Math. Appl. **49**(9), 1539–1548 (2005)
538. J. Gao, F. You, Economic and environmental life cycle optimization of noncooperative supply chains and product systems: modeling framework, mixed-integer bilevel fractional programming algorithm, and shale gas application. ACS Sustainable Chem. Eng. **5**(4), 3362–3381 (2017)
539. J. Gao, F. You, Game theory approach to optimal design of shale gas supply chains with consideration of economics and life cycle greenhouse gas emissions. AIChE J. **63**(7), 2671–2693 (2017)
540. Y. Gao, Bi-level decision making with fuzzy sets and particle swarm optimisation, Ph.D. thesis, in *Faculty of Engineering and Information Technology* (University of Technology, Sydney, 2010)
541. Y. Gao, G. Zhang, J. Lu, A particle swarm optimization based algorithm for fuzzy bilevel decision making with constraints-shared followers, in *Proceedings of the 2009 ACM Symposium on Applied Computing* (ACM, New York, 2009), pp. 1075–1079
542. Y. Gao, G. Zhang, J. Lu, T. Dillon, X. Zeng, A λ-cut approximate algorithm for goal-based bilevel risk management systems. Int. J. Inf. Technol. Decis. Making **7**(04), 589–610 (2008)
543. Y. Gao, G. Zhang, J. Lu, H.-M. Wee, Particle swarm optimization for bi-level pricing problems in supply chains. J. Global Optim. **51**(2), 245–254 (2011)
544. Y. Gao, G. Zhang, J. Ma, J. Lu, A-cut and goal-programming-based algorithm for fuzzy-linear multiple-objective bilevel optimization. IEEE Trans. Fuzzy Syst. **18**(1), 1–13 (2010)

545. Z. Gao, H. Sun, H. Zhang, A globally convergent algorithm for transportation continuous network design problem. Optim. Eng. **8**(3), 241–257 (2007)
546. L.P. Garcés, A.J. Conejo, R. García-Bertrand, R. Romero, A bilevel approach to transmission expansion planning within a market environment. IEEE Trans. Power Syst. **24**(3), 1513–1522 (2009)
547. P. Garcia-Herreros, L. Zhang, P. Misra, E. Arslan, S. Mehta, I.E. Grossmann, Mixed-integer bilevel optimization for capacity planning with rational markets. Comput. Chem. Eng. **86**, 33–47 (2016)
548. I. Gaspar, J. Benavente, M. Bordagaray, B. Alonso, J.L. Moura, Á. Ibeas, A bilevel mathematical programming model to optimize the design of cycle paths. Transp. Res. Procedia **10**, 423–432 (2015)
549. E. Gassner, B. Klinz, The computational complexity of bilevel assignment problems. 4OR **7**, 379–394 (2009)
550. X. Ge, Y. Chen, W. Wang, Model and algorithm for inventory-transportation integrated optimization based on bi-level programming. Int. J. Adv. Comput. Technol **5**, 460–468 (2013)
551. E. Gebhardt, J. Jahn, Global solver for nonlinear bilevel vector optimization problems. Pac. J. Optim. **5**(3), 387–401 (2009)
552. M. Gendreau, P. Marcotte, G. Savard, A hybrid tabu-ascent algorithm for the linear bilevel programming problem. J. Global Optim. **8**, 217–233 (1996)
553. R. Gessing, Optimal control laws for two-level hierarchical ressource allocation. Large Scale Syst. **12**, 69–82 (1987)
554. N. Ghaffarinasab, R. Atayi, An implicit enumeration algorithm for the hub interdiction median problem with fortification. Eur. J. Oper. Res. **267**(1), 23–39 (2018)
555. N. Ghaffarinasab, A. Motallebzadeh, Hub interdiction problem variants: models and metaheuristic solution algorithms. Eur. J. Oper. Res. **267**(2), 496–512 (2018)
556. M. Ghamkhari, A. Sadeghi-Mobarakeh, H. Mohsenian-Rad, Strategic bidding for producers in nodal electricity markets: a convex relaxation approach. IEEE Trans. Power Syst. **32**(3), 2324–2336 (2017)
557. E. Ghotbi, A.K. Dhingra, A bilevel game theoretic approach to optimum design of flywheels. Eng. Optim. **44**(11), 1337–1350 (2012)
558. A. Gibali, K.-H. Küfer, P. Süss, Reformulating the Pascoletti-Serafini problem as a bi-level optimization problem. Contemp. Math. **636**, 121–129 (2015)
559. F. Gilbert, P. Marcotte, G. Savard, A numerical study of the logit network pricing problem. Transp. Sci. **49**(3), 706–719 (2015)
560. E.Kh. Gimadi, E.N. Goncharov, A two-level choice problem for a system of machines and nodes with a nonlinear production function. Sibirskii Zhurnal Industrial'noi Matematiki **9**(2), 44–54 (2006, in Russian)
561. L. Gkatzikis, I. Koutsopoulos, T. Salonidis, The role of aggregators in smart grid demand response markets. IEEE J. Sel. Areas Commun. **31**(7), 1247–1257 (2013)
562. J. Glackin, J.G. Ecker, M. Kupferschmid, Solving bilevel linear programs using multiple objective linear programming. J. Optim. Theory Appl. **140**(2), 197–212 (2009)
563. A.I. Gladyshev, V.T. Dement'ev, A.I. Erzin, Models and problems of the optimal synthesis of homogeneous hierarchical systems (Russian), in *Models and methods of optimization (Russian)*. Trudy Instituta Matematiki, 28, Izdat. Ross. Akad. Nauk Sib. Otd. Inst. Mat., Novosibirsk, **149**, 63–76 (1994)
564. F. Gong, Y. Zhou, Sequential fair Stackelberg equilibria of linear strategies in risk-seeking insider trading. J. Syst. Sci. Complexity **31**(5), 1302–1328 (2018)
565. P.H. Gonzalez, L. Simonetti, P. Michelon, C. Martinhon, E. Santos, A variable fixing heuristic with local branching for the fixed charge uncapacitated network design problem with user-optimal flow. Comput. Oper. Res. **76**, 134–146 (2016)
566. V.J.L. González, J.F. Camacho Vallejo, G. Pinto Serrano, A scatter search algorithm for solving a bilevel optimization model for determining highway tolls. Computación y Sistemas **19**(1), 3529–3549 (2015)

567. L.E. Gorbachevskaya, Algorithms and complexity of the bilevel standardization problems with profit correction. Diskretnij Analiz i Issledovanie Operazij, Seriya 2 **5**, 20–33 (1998)
568. L.E. Gorbachevskaya, V.T. Dement'ev, Y.V. Shamardin, Two-level extremal problems of selecting the nomenclature of products, Technical Report 41 (Russian Academy of Sciences, Siberian Branch, Insitut of Mathemetics, Novosibirsk, 1997, in Russian)
569. L.E. Gorbachevskaya, V.T. Dement'ev, Y.V. Shamardin, The bilevel standardization problem with uniqueness condition for an optimal customer choice. Diskretnij Analiz i Issledovanie Operazij, Seriya 2 **6**, 3–11 (1999, in Russian)
570. V.A. Gorelik, Approximate search for the maximin with constraints connecting the variables. Zhurnal Vychislitelnoi Matematiki i Matematicheskoi Fiziki **12**, 510–519 (1972, in Russian)
571. V.A. Gorelik, Dynamic systems with hierarchical control structure. Cybernetics **14**(3), 427–430 (1978)
572. V.A. Gorelik, Hierarchical optimization-coordination systems. Kibernetika **1**, 87–94 (1978, in Russian)
573. V.A. Gorelik, M.S. Shtil'man, On one class of two-level models for the regularization of economic-ecologic processes. Economica i Matematicheskie Metody **XIII**, 1251–1263 (1977, in Russian)
574. A. Grigoriev, S. Van Hoesel, A.F. Van Der Kraaij, M. Uetz, M. Bouhtou, Pricing network edges to cross a river, in *Lecture Notes in Computer Science*, vol. 3351 (Springer, Berlin, 2004), pp. 140–153
575. V. Grimm, L. Schewe, M. Schmidt, G. Zöttl, A multilevel model of the European entry-exit gas market. Math. Methods Oper. Res. **89**, 223–256 (2019)
576. F. Groot, C. Withagen, A. De Zeeuw, Note on the open-loop von Stackelberg equilibrium in the cartel versus fringe model. Econ. J. **102**(415), 1478–1484 (1992)
577. N. Groot, Reverse Stackelberg games: theory and applications in traffic control, Ph.D. thesis (Delft Center for Systems and Control, Delft, 2013)
578. N. Groot, B. De Schutter, H. Hellendoorn, A full characterization of the set of optimal affine leader functions in the reverse Stackelberg game, in *Proceedings of the 51st IEEE Conference on Decision and Control* (2012), pp. 6484–6488
579. N. Groot, B. De Schutter, H. Hellendoorn, Reverse Stackelberg games, part II: Results and open issues, *Proceedings of the IEEE International Conference on Control Applications (CCA), 2012* (IEEE, New York, 2012), pp. 427–432
580. N. Groot, B. De Schutter, H. Hellendoorn, Optimal leader functions for the reverse Stackelberg game: splines and basis functions, in *European Control Conference (ECC), 2013* (IEEE, New York, 2013), pp. 696–701
581. N. Groot, B. De Schutter, H. Hellendoorn, On systematic computation of optimal nonlinear solutions for the reverse Stackelberg game. IEEE Trans. Syst. Man Cybern. Syst. **44**(10), 1315–1327 (2014)
582. N. Groot, B. Schutter, H. Hellendoorn, Optimal affine leader functions in reverse Stackelberg games. J. Optim. Theory Appl. **168**(1), 348–374 (2014)
583. S.J. Grossman, O.D. Hart, An analysis of the principal-agent problem. Econometrica **51**, 7–45 (1983)
584. S.J. Grossman, O.D. Hart, An analysis of the principal-agent problem, in *Foundations of Insurance Economics (Georges Dionne and ScottE. Harrington)*. Huebner International Series on Risk, Insurance and Economic Security, vol. 14 (Springer, Netherlands, 1992), pp. 302–340
585. T.V. Gruzdeva, E.G. Petrova, Numerical solution of a linear bilevel problem. Comput. Math. Math. Phys. **50**(10), 1631–1641 (2010)
586. Z.H. Gümüs, C.A. Floudas, Global optimization of nonlinear bilevel programming problems. J. Global Optim. **20**, 1–31 (2001)
587. Z.H. Gümüs, C.A. Floudas, Global optimization of mixed-integer bilevel programming problems. Comput. Manag. Sci. **2**, 181–212 (2005)
588. P. Guo, X. Zhu, Focus programming: a fundamental alternative for stochastic optimization problems (2019), p. 15. Available at SSRN: https://ssrn.com/abstract=3334211

589. Z. Guo, J. Chang, Q. Huang, L. Xu, C. Da, H. Wu, Bi-level optimization allocation model of water resources for different water industries. Water Sci. Technol. Water Supply **14**(3), 470–477 (2014)
590. A. Gupta, C.D. Maranas, A two-stage modeling and solution framework for multisite midterm planning under demand uncertainty. Ind. Eng. Chem. Res. **39**(10), 3799–3813 (2000)
591. W.J. Gutjahr, N. Dzubur, Bi-objective bilevel optimization of distribution center locations considering user equilibria. Transp. Res. Logist. Transp. Rev. **85**, 1–22 (2016)
592. F. Gzara, A cutting plane approach for bilevel hazardous material transport network design. Oper. Res. Lett. **41**(1), 40–46 (2013)
593. M. Haan, H. Maks, Stackelberg and Cournot competition under equilibrium limit pricing. J. Econ. Stud. **23**(5/6), 110–127 (1996)
594. A. Hafezalkotob, Competition of domestic manufacturer and foreign supplier under sustainable development objectives of government. Appl. Math. Comput. **292**, 294–308 (2017)
595. M. Hajiaghaei-Keshteli, A.M. Fathollahi-Fard, A set of efficient heuristics and metaheuristics to solve a two-stage stochastic bi-level decision-making model for the distribution network problem. Comput. Ind. Eng. **123**, 378–395 (2018)
596. L. Hajibabai, Y. Bai, Y. Ouyang, Joint optimization of freight facility location and pavement infrastructure rehabilitation under network traffic equilibrium. Transp. Res. B Methodol. **63**, 38–52 (2014)
597. M. Hajinassiry, N. Amjady, H. Sharifzadeh, Hydrothermal coordination by bi-level optimization and composite constraint handling method. Int. J. Electr. Power Energy Syst. **62**, 476–489 (2014)
598. S. Hakim, A. Seifi, A. Ghaemi, A bi-level formulation for DEA-based centralized resource allocation under efficiency constraints. Comput. Ind. Eng. **93**, 28–35 (2015)
599. J. Han, G. Liu, S. Wang, A new descent algorithm for solving quadratic bilevel programming problems. Acta Math. Appl. Sin. Engl. Ser. **16**, 235–244 (2000)
600. J. Han, J. Lu, Y. Hu, G. Zhang, Tri-level decision-making with multiple followers: model, algorithm and case study. Inf. Sci. **311**, 182–204 (2015)
601. J. Han, J. Lu, G. Zhang, S. Ma, Multi-follower tri-level decision making with uncooperative followers, in *Proceedings of the 11th International FLINS Conference, Brazil* (2014), pp. 524–529
602. J. Han, G. Zhang, Y. Hu, J. Lu, Solving tri-level programming problems using a particle swarm optimization algorithm, in *Proceedings of the IEEE 10th Conference on Industrial Electronics and Applications (ICIEA), 2015* (IEEE, New York, 2015), pp. 569–574
603. J. Han, G. Zhang, Y. Hu, J. Lu, A solution to bi/tri-level programming problems using particle swarm optimization. Inf. Sci. **370**, 519–537 (2016)
604. J. Han, G. Zhang, J. Lu, Y. Hu, S. Ma, Model and algorithm for multi-follower tri-level hierarchical decision-making, in *Proceedings of the Neural Information Processing* (Springer, Berlin, 2014), pp. 398–406
605. K. Han, Y. Sun, H. Liu, T.L. Friesz, T. Yao, A bi-level model of dynamic traffic signal control with continuum approximation. Transp. Res. C Emerg. Technol. **55**, 409–431 (2015)
606. S.D. Handoko, L.H. Chuin, A. Gupta, O.Y. Soon, H.C. Kim, T.P. Siew, Solving multi-vehicle profitable tour problem via knowledge adoption in evolutionary bi-level programming, in *Proceedings of the IEEE Congress on Evolutionary Computation (CEC), 2015* (IEEE, New York, 2015), pp. 2713–2720
607. L.U. Hansen, P. Horst, Multilevel optimization in aircraft structural design evaluation. Comput. Struct. **86**(1), 104–118 (2008)
608. P. Hansen, B. Jaumard, G. Savard, New branch-and-bound rules for linear bilevel programming. SIAM J. Sci. Stat. Comput. **13**, 1194–1217 (1992)
609. F. Harder, Optimal control of the obstacle problem using the value function, Master's thesis (TU Chemnitz, Department of Mathematics, Chemnitz, 2016)

610. W.E. Hart, R.L.-Y. Chen, J.D. Siirola, J.-P. Watson, Modeling bilevel programs in Pyomo, Technical Report (Sandia National Laboratories (SNL-NM), Albuquerque; Sandia National Laboratories, Livermore, 2015)
611. A. Hassanpour, J. Bagherinejad, M. Bashiri, A robust bi-level programming model to design a closed loop supply chain considering government collection's policy. Scientia Iranica **26**(6), 3737–3764 (2019)
612. B. Hassanzadeh, J. Liu, J.F. Forbes, A bilevel optimization approach to coordination of distributed model predictive control systems. Ind. Eng. Chem. Res. **57**(5), 1516–1530 (2018)
613. K. Hatz, Efficient numerical methods for hierarchical dynamic optimization with application to cerebral palsy gait modeling, Ph.D. thesis (Universität Heidelberg, Heidelberg, 2014)
614. K. Hatz, S. Leyffer, J.P. Schlöder, H.G. Bock, Regularizing bilevel nonlinear programs by lifting, Technical Report (Argonne National Laboratory, USA, 2013). Preprint ANL/MCS-P4076-0613
615. A. Haurie, R. Loulou, G. Savard, A two-level systems analysis model of power cogeneration under asymmetric pricing, in *Proceedings of IEEE Automatic Control Conference (San Diego)* (1990)
616. A. Haurie, R. Loulou, G. Savard, A two player game model of power cogeneration in New England. IEEE Trans. Autom. Control **37**, 1451–1456 (1992)
617. A. Haurie, G. Savard, D. White, A note on: an efficient point algorithm for a linear two-stage optimization problem. Oper. Res. **38**, 553–555 (1990)
618. L. He, G.H. Huang, H. Lu, Greenhouse gas emissions control in integrated municipal solid waste management through mixed integer bilevel decision-making. J. Hazard. Mater. **193**, 112–119 (2011)
619. X. He, C. Li, T. Huang, C. Li, Neural network for solving convex quadratic bilevel programming problems. Neural Netw. **51**, 17–25 (2014)
620. X. He, C. Li, T. Huang, C. Li, J. Huang, A recurrent neural network for solving bilevel linear programming problem. IEEE Trans. Neural Networks Learn. Syst. **25**(4), 824–830 (2014)
621. X. He, Y. Zhou, Z. Chen, Evolutionary bilevel optimization based on covariance matrix adaptation. IEEE Trans. Evol. Comput. **23**(2), 258–272 (2018)
622. D.W. Hearn, M.V. Ramana, Solving congestion toll pricing models, in *Equilibrium and advanced transportation modelling*, ed. by P. Marcotte, S. Nguyen (Springer, Berlin, 1998), pp. 109–124
623. L. Hecheng, W. Yuping, Exponential distribution-based genetic algorithm for solving mixed-integer bilevel programming problems. J. Syst. Eng. Electron. **19**(6), 1157–1164 (2008)
624. G. Heilporn, M. Labbé, P. Marcotte, G. Savard, *The Highway Problem: Models, Complexity and Valid Inequalities*, Technical Report (Université Libre de Bruxelles, Bruxelles, Belgique. 2006)
625. G. Heilporn, M. Labbé, P. Marcotte, G. Savard, A parallel between two classes of pricing problems in transportation and marketing. J. Revenue Pricing Manag. **9**(1-2), 110–125 (2010)
626. G. Heilporn, M. Labbé, P. Marcotte, G. Savard, A polyhedral study of the network pricing problem with connected toll arcs. Networks **55**(3), 234–246 (2010)
627. S.R. Hejazi, A. Memariani, G. Jahanshaloo, M.M. Sepehri, Linear bilevel programming solution by genetic algorithm. Comput. Oper. Res. **29**, 1913–1925 (2002)
628. H. Held, D.L. Woodruff, Heuristics for multi-stage interdiction of stochastic networks. J. Heuristics **11**(5-6), 483–500 (2005)
629. M. Hemmati, J.C. Smith, A mixed-integer bilevel programming approach for a competitive prioritized set covering problem. Discrete Optim. **20**, 105–134 (2016)
630. E.M.T. Hendrix, On competition in a Stackelberg location-design model with deterministic supplier choice. Ann. Oper. Res. **246**(1-2), 19–30 (2016)
631. C. Henkel, An algorithm for the global resolution of linear stochastic bilevel programs, Ph.D. thesis (Universität Duisburg-Essen, Fakultät für Mathematik, 2014)

632. R. Henrion, J. Outrata, T. Surowiec, Analysis of M-stationary points to an EPEC modeling oligopolistic competition in an electricity spot market. ESAIM Control Optim. Calc. Var. **18**(2), 295–317 (2012)
633. R. Henrion, T. Surowiec, On calmness conditions in convex bilevel programming. Appl. Anal. **90**(5–6), 951–970 (2011)
634. J. Herskovits, A. Leontiev, G. Dias, G. Santos, Contact shape optimization: a bilevel programming approach. Int. J. of Struc. Multidisc. Optim. **20**, 214–221 (2000)
635. J. Herskovits, M. Tanaka Filho, A. Leontiev, An interior point technique for solving bilevel programming problems. Optim. Eng. **14**(3), 381–394 (2013)
636. M.R. Hesamzadeh, M. Yazdani, Transmission capacity expansion in imperfectly competitive power markets. IEEE Trans. Power Syst. **29**(1), 62–71 (2014)
637. G. Hibino, M. Kainuma, Y. Matsuoka, Two-level mathematical programming for analyzing subsidy options to reduce greenhouse-gas emissions, Technical Report WP-96-129 (IIASA, Laxenburg, 1996)
638. M. Hintermüller, T. Wu, Bilevel optimization for calibrating point spread functions in blind deconvolution. Inverse Prob. Imaging **9**(4), 1139–1169 (2015)
639. Y.-C. Ho, P.B. Luh, R. Muralidharan, Information structure, Stackelberg games, and incentive controllability. IEEE Trans. Autom. Control **26**(2), 454–460 (1981)
640. B. Hobbs, S. Nelson, A nonlinear bilevel model for analysis of electric utility demand-side planning issues. Ann. Oper. Res. **34**, 255–274 (1992)
641. B.F. Hobbs, C.B. Metzler, J.-S. Pang, Strategic gaming analysis for electric power systems: an MPEC approach. IEEE Trans. Power Syst. **15**(2), 638–645 (2000)
642. F. Hooshmand, S.A. MirHassani, An effective bilevel programming approach for the evasive flow capturing location problem. Netw. Spatial Econ., **18**(4), 909–935 (2018)
643. A. Hori, M. Fukushima, Gauss–Seidel method for multi-leader–follower games. J. Optim. Theory Appl. **180**(2), 651–670 (2019)
644. R. Horst, N.V. Thoai, Maximizing a concave function over the efficient set or weakly-efficient set. Eur. J. Oper. Res. **117**, 239–252 (1999)
645. R. Horst, N.V. Thoai, Y. Yamamoto, D. Zenke, On optimization over the efficient set in linear multicriteria programming. J. Optim. Theory Appl. **134**(3), 433–443 (2007)
646. S. Hsu, U. Wen, A review of linear bilevel programming problems, in *Proceedings of the National Science Council, Republic of China, Part A: Physical Science and Engineering*, vol. 13 (1989), pp. 53–61
647. C.-F. Hsueh, A bilevel programming model for corporate social responsibility collaboration in sustainable supply chain management. Transp. Res. Logist. Transp. Rev. **73**, 84–95 (2015)
648. C.-F. Hu, F.-B. Liu, Solving mathematical programs with fuzzy equilibrium constraints. Comput. Math. Appl. **58**(9), 1844–1851 (2009)
649. M. Hu, M. Fukushima, Variational inequality formulation of a class of multi-leader-follower games. J. Optim. Theory Appl. **151**(3), 455–473 (2011)
650. T. Hu, X. Guo, X. Fu, Y. Lv, A neural network approach for solving linear bilevel programming problem. Knowledge-Based Syst. **23**(3), 239–242 (2010)
651. X. Hu, Mathematical Programs with Complementarity Constraints and Game Theory Models in Electricity Markets, Ph.D. thesis (University of Melbourne, Melbourne, 2002)
652. X. Hu, D. Ralph, Using EPECs to model bilevel games in restructured electricity markets with locational prices. Oper. Res. **55**(5), 809–827 (2007)
653. Z. Hu, C. Wei, L. Yao, C. Li, Z. Zeng, Integrating equality and stability to resolve water allocation issues with a multiobjective bilevel programming model. J. Water Resour. Plann. Manage. **142**(7), 04016013 (2016)
654. C. Huang, D. Fang, Z. Wan, An interactive intuitionistic fuzzy method for multilevel linear programming problems. Wuhan Univer. J. Nat. Sci. **20**(2), 113–118 (2015)
655. S. Huck, K.A. Konrad, W. Müller, Big fish eat small fish: on merger in Stackelberg markets. Econ. Lett. **73**(2), 213–217 (2001)
656. Y. Huo, The upper semi-convergence of optimal solution sets of approximation problems for bilevel stochastic programming. J. Syst. Sci. Math. Sci. **34**, 674–681 (2014, in Chinese)

657. M. Inuiguchi, P. Sariddichainunta, Bilevel linear programming with ambiguous objective function of the follower. Fuzzy Optim. Decis. Making **15**(4), 415–434 (2016)
658. Y. Ishizuka, Optimality conditions for quasi-differentiable programs with applications to two-level optimization. SIAM J. Control Optim. **26**, 1388–1398 (1988)
659. Y. Ishizuka, E. Aiyoshi, Double penalty method for bilevel optimization problems. Ann. Oper. Res. **34**, 73–88 (1992)
660. M.M. Islam, H.K. Singh, T. Ray, A memetic algorithm for solving bilevel optimization problems with multiple followers, in *Proceedings of the 2016 IEEE Congress on Evolutionary Computation (CEC)* (2016), pp. 1901–1908
661. M.M. Islam, H.K. Singh, T. Ray, A surrogate assisted approach for single-objective bilevel optimization. IEEE Trans. Evol. Comput. **21**(5), 681–696 (2017)
662. M.M. Islam, H.K. Singh, T. Ray, Use of a non-nested formulation to improve search for bilevel optimization, in *Proceedings of the Australasian Joint Conference on Artificial Intelligence* (Springer, Berlin, 2017), pp. 106–118
663. M.M. Islam, H.K. Singh, T. Ray, A. Sinha, An enhanced memetic algorithm for single-objective bilevel optimization problems. Evol. Comput. **25**(4), 607–642 (2017)
664. E. Israeli, System interdiction and defense, Ph.D. thesis (Naval Postgraduate School Monterey, USA, 1999)
665. E. Israeli, R.K. Wood, Shortest-path network interdiction. Networks **40**(2), 97–111 (2002)
666. D. Ivanenko, A. Plyasunov, Lower and upper bounds for the bilevel capacitated facility location problem, Technical Report (Sobolev Institute of Mathematics, Novosibirsk, 2003)
667. D.S. Ivanenko, A.V. Plyasunov, Reducibility of bilevel programming problems to vector optimization problems. J. Appl. Ind. Math. **2**(2), 179–195 (2008)
668. S.V. Ivanov, Bilevel stochastic linear programming problems with quantile criterion. Autom. Remote Control **75**(1), 107–118 (2014)
669. S.V. Ivanov, A bilevel stochastic programming problem with random parameters in the follower's objective function. J. Appl. Ind. Math. **12**(4), 658–667 (2018), Original Russian Text published in Diskret. Anal. Issled. Oper. **25**(4), 27–45 (2018)
670. G. Iyengar, W. Kang, Inverse conic programming with applications. Oper. Res. Lett. **33**(3), 319–330 (2005)
671. B. Jabarivelisdeh, S. Waldherr, Optimization of bioprocess productivity based on metabolic-genetic network models with bilevel dynamic programming. Biotechnol. Bioeng. **115**(7), 1829–1841 (2018)
672. C.K. Jaggi, M. Gupta, A. Kausar, S. Tiwari, Inventory and credit decisions for deteriorating items with displayed stock dependent demand in two-echelon supply chain using Stackelberg and Nash equilibrium solution. Ann. Oper. Res. **274**(1), 309–329 (2019)
673. A. Jahanshahloo, M. Zohrehbandian, A cutting plane approach for solving linear bilevel programming problems, in *Proceedings of the Advanced Computational Methods for Knowledge Engineering* (Springer, Berlin, 2015), pp. 3–13
674. M.Z. Jamaludin, C.L.E. Swartz, A bilevel programming formulation for dynamic real-time optimization. IFAC-PapersOnLine **48**(8), 906–911 (2015)
675. R.-H. Jan, M.-S. Chern, Nonlinear integer bilevel programming. Eur. J. Oper. Res. **72**, 574–587 (1994)
676. R.G. Jeroslow, The polynomial hierarchy and a simple model for competitive analysis. Math. Program. **32** 146–164 (1985)
677. I. Jewitt, Justifying the first-order approach to principal-agent problems. Econom. J. Econom. Soc. **56**, 1177–1190 (1988)
678. V. Jeyakumar, J.-B. Lasserre, G. Li, T.S. Pham, Convergent semidefinite programming relaxations for global bilevel polynomial optimization problems. SIAM J. Optim. **26**(1), 753–780 (2016)
679. V. Jeyakumar, G. Li, A bilevel Farkas lemma to characterizing global solutions of a class of bilevel polynomial programs. Oper. Res. Lett. **43**(4), 405–410 (2015)
680. X. Ji, Z. Shao, Model and algorithm for bilevel newsboy problem with fuzzy demands and discounts. Appl. Math. Comput. **172**, 163–174 (2006)

681. F. Jia, F. Yang, S.-Y. Wang, Sensitivity analysis in bilevel linear programming. Syst. Sci. Math. Sci. **11**, 359–366 (1998)
682. L. Jia, Z. Li, An ameliorated teaching-learning based optimization algorithm for nonlinear bilevel programming, in *Proceedings of the 12th International Conference on Computational Intelligence and Security (CIS), 2016* (IEEE, New York, 2016), pp. 52–56
683. L. Jia, Y. Wang, L. Fan, Multiobjective bilevel optimization for production-distribution planning problems using hybrid genetic algorithm. Integr. Comput. Aided Eng. **21**(1), 77–90 (2014)
684. L. Jia, Y. Wang, L. Fan, An improved uniform design-based genetic algorithm for multi-objective bilevel convex programming. Int. J. Comput. Sci. Eng. **12**(1), 38–46 (2016)
685. L. Jia, G. Zou, Z. Li, Target-vector based particle swarm optimization for multi-objective bilevel programming problem, in *Proceedings of the 11th International Conference on Computational Intelligence and Security (CIS), 2015* (IEEE, New York, 2015), pp. 295–298
686. M. Jiang, Z. Meng, R. Shen, X. Xu, A quadratic objective penalty function for bilevel programming. J. Syst. Sci. Complexity **27**(2), 327–337 (2014)
687. Y. Jiang, X. Li, C. Huang, X. Wu, Application of particle swarm optimization based on CHKS smoothing function for solving nonlinear bilevel programming problem. Appl. Math. Comput. **219**(9), 4332–4339 (2013)
688. Y. Jiang, X. Li, C. Huang, X. Wu, An augmented Lagrangian multiplier method based on a CHKS smoothing function for solving nonlinear bilevel programming problems. Knowledge-Based Syst. **55**, 9–14 (2014)
689. Z. Jiang, J. Yuan, E. Feng, Robust identification and its properties of nonlinear bilevel multistage dynamic system. Appl. Math. Comput. **219**(12), 6979–6985 (2013)
690. Q. Jin, S. Feng, Bi-level simulated annealing algorithm for facility location. Syst. Eng. **2**(02), 36–40 (2007)
691. S. Jin, R. Fan, G. Wang, X. Bu, Network utility maximization in wireless networks over fading channels with uncertain distribution. IEEE Commun. Lett. **21**(5), 1107–1110 (2017)
692. Y.W. Jing, S.Y. Zhang, The solution to a kind of Stackelberg game system with multi-follower: coordinative and incentive, in *Proceedings of the Analysis and Optimization of Systems (Antibes, 1988)*. Lecture Notes in Control and Information in Science, vol. 111 (Springer, Berlin, 1988), pp. 593–602
693. D. Joksimocic, Dynamic bi-level optimal toll design approach for dynamic traffic networks, Ph.D. thesis (Delft University of Technology, Delft, 2007)
694. H.T. Jongen, V. Shikhman, Bilevel optimization: on the structure of the feasible set. Math. Program. **136**, 65–90 (2012)
695. J.M. Jorge, A bilinear algorithm for optimizing a linear function over the efficient set of a multiple objective linear programming problem. J. Global Optim. **31**(1), 1–16 (2005)
696. M. Josefsson, M. Patriksson, Sensitivity analysis of separable traffic equilibrium equilibria with application to bilevel optimization in network design. Transp. Res. B Methodol. **41**(1), 4–31 (2007)
697. J. Júdice, A. Faustino, The solution of the linear bilevel programming problem by using the linear complementarity problem. Investigação Operacional **8**, 77–95 (1988)
698. J. Júdice, A. Faustino, A sequential LCP method for bilevel linear programming. Ann. Oper. Res. **34**, 89–106 (1992)
699. J. Júdice, A. Faustino, The linear-quadratic bilevel programming problem. INFOR **32**, 87–98 (1994)
700. J.J. Júdice, A.M. Faustino, I.M. Ribeiroa, A.S. Neves, On the use of bilevel programming for solving a structural optimization problem with discrete variables, in *Optimization with Multivalued Mappings: Theory, Applications and Algorithms* ed. by S. Dempe, V. Kalashnikov. Optimization and its Applications, vol. 2 (Springer/LLC, New York, 2006), pp. 123–142
701. L.A. Julien, A note on Stackelberg competition. J. Econom. **103**(2), 171–187 (2011)
702. C. Kahraman, G. Zhang, J. Lu, Model and approach of fuzzy bilevel decision making for logistics planning problem. J. Enterp. Inf. Manag. **20**(2), 178–197 (2007)

703. V. Kalashnikov, F. Camacho, R. Askin, N. Kalashnykova, Comparison of algorithms for solving a bi-level toll setting problem. Int. J. Innovative Comput. Inf. Control **6**(8), 3529–3549 (2010)
704. V. Kalashnikov, A.E. Cordero, V. Kalashnikov, Cournot and Stackelberg equilibrium in mixed duopoly models, Optimization **59**(5), 689–706 (2010)
705. V. Kalashnikov, S. Dempe, B. Mordukhovich, S.V. Kavun, Bilevel optimal control, equilibrium, and combinatorial problems with applications to engineering. Math. Prob. Eng. **2017**, 3 (2017)
706. V. Kalashnikov, N. Kalashnykova, J.G. Flores-Muñiz, Solution of the portfolio optimization model as a fuzzy bilevel programming problem, in *Proceedings of the International Forum for Interdisciplinary Mathematics* (Springer, Berlin, 2015), pp. 164–178
707. V. Kalashnikov, T.I. Matis, J.F. Camacho Vallejo, S.V. Kavun, Bilevel programming, equilibrium, and combinatorial problems with applications to engineering. Math. Prob. Eng. **2016**, 3 (2016)
708. V.V. Kalashnikov, Actuality of the portfolio optimization model as a bilevel programming problem, in *Proceedings of the International forum for safety Rezensenty (INFOS-2017)* (2017), pp. 211–214
709. V.V. Kalashnikov, F. Benita, P. Mehlitz, The natural gas cash-out problem: a bilevel optimal control approach. Math. Prob. Eng. **2015**, 17 (2015)
710. V.V. Kalashnikov, S. Dempe, N.I. Kalashnykova, *Operations Research and Bilevel Programming* (Editorial Digital del Tecnológico de Monterrey, 2013)
711. V.V. Kalashnikov, S. Dempe, G.A. Pérez-Valdés, N.I. Kalashnykova, *Reduction of Dimension of the Upper Level Problem in a Bilevel Programming Model, Part 1* (Intelligent Decision Technologies, Springer, Berlin, 2011), pp. 255–264
712. V.V. Kalashnikov, S. Dempe, G.A. Pérez-Valdés, N.I. Kalashnykova, *Reduction of Dimension of the Upper Level Problem in a Bilevel Programming Model, Part 2* (Intelligent Decision Technologies, Springer, Berlin, 2011), pp. 265–272
713. V.V. Kalashnikov, S. Dempe, G.A. Pérez-Valdés, N.I. Kalashnykova, J.-F. Camacho-Vallejo, Bilevel programming and applications. Math. Prob. Eng. **2015**, 16 (2015)
714. V.V. Kalashnikov, N.I. Kalashnikova, Solving two-level variational inequality. J. Global Optim. **17**, 289–294 (1991)
715. V.V. Kalashnikov, N.I. Kalashnykova, M.A. Leal-Coronado, Solution of the portfolio optimization model as a bilevel programming problem, in *Proceedings of the Cherkasy University Bulletin: Economics Sciences*, vol. 1 (2017)
716. V.V. Kalashnikov, G.A. Pérez-Valdés, N.I. Kalashnykova, A linearization approach to solve the natural gas cash-out bilevel problem. Ann. Oper. Res. **181**(1), 423–442 (2010)
717. V.V. Kalashnikov, G.A. Pérez-Valdés, A. Tomasgard, N.I. Kalashnykova, Natural gas cash-out problem: bilevel stochastic optimization approach. Eur. J. Oper. Res. **206**(1), 18–33 (2010)
718. V.V. Kalashnikov, R.Z. Ríos-Mercado, An algorithm to solve a gas cash out problem, in *Proceedings of the International Business and Economic Research Conference (IBERC2002)* (Puerto Vallarta, Mexico, 2002), p. 9
719. V.V. Kalashnikov, R.Z. Ríos-Mercado, A penalty-function approach to a mixed-integer bilevel programming problem, Technical Report (Universidad Autónoma de Nuevo León, Mexico, 2002)
720. V.V. Kalashnikov, R.Z. Ríos-Mercado, A natural gas cash-out problem: A bilevel programming framework and a penalty function method. Optim. Eng. **7**, 403–420 (2006)
721. B.Y. Kara, V. Verter, Designing a road network for hazardous materials transportation. Transp. Sci. **38**(2), 188–196 (2004)
722. E.G. Kardakos, C.K. Simoglou, A.G. Bakirtzis, Optimal offering strategy of a virtual power plant: a stochastic bi-level approach. IEEE Trans. Smart Grid **7**(2), 794–806 (2016)
723. J.K. Karlof, W. Wang, Bilevel programming applied to the flow shop scheduling problem. Comput. Oper. Res. **23**, 443–451 (1996)

724. A. Karoonsoontawong, S.T. Waller, Integrated network capacity expansion and traffic signal optimization problem: robust bi-level dynamic formulation. Netw. Spatial Econom. **10**(4), 525–550 (2010)
725. C. Kasemset, V. Kachitvichyanukul, A PSO-based procedure for a bi-level multi-objective TOC-based job-shop scheduling problem. Int. J.Oper. Res. **14**(1), 50–69 (2012)
726. A.M. Kassa, S.M. Kassa, A branch-and-bound multi-parametric programming approach for non-convex multilevel optimization with polyhedral constraints. J. Global Optim. **64**(4), 745–764 (2016)
727. A.M. Kassa, S.M. Kassa, Deterministic solution approach for some classes of nonlinear multilevel programs with multiple followers. J. Global Optim. **68**(4), 729–747 (2017)
728. S.M. Kassa, Three-level global resource allocation model for HIV control: a hierarchical decision system approach. Math. Biosci. Eng. **15**(1), 255–273 (2018)
729. G.Y. Ke, J.H. Bookbinder, Coordinating the discount policies for retailer, wholesaler, and less-than-truckload carrier under price-sensitive demand: a tri-level optimization approach. Int. J. Prod. Econ. **196**, 82–100 (2018)
730. H. Ke, H. Huang, D.A. Ralescu, L. Wang, Fuzzy bilevel programming with multiple non-cooperative followers: model, algorithm and application. Int. J. General Syst. **45**(3), 336–351 (2016)
731. M. Khademi, M. Ferrara, M. Salimi, S. Sharifi, A dynamic Stackelberg game for green supply chain management (2015). arXiv preprint:1506.06408
732. A. Kheirkhah, H.-R. Navidi, M. Messi Bidgoli, A bi-level network interdiction model for solving the hazmat routing problem. Int. J. Prod. Res. **54**(2), 459–471 (2016)
733. A.I. Kibzun, A.V. Naumov, S.V. Ivanov, Bilevel optimization problem for railway transport hub planning. Upravlenie Bol'shimi Sistemami **38**, 140–160 (2012)
734. R. Kicsiny, Z. Varga, A. Scarelli, Backward induction algorithm for a class of closed-loop Stackelberg games. Eur. J. Oper. Res. **237**, 1021–1036 (2014)
735. S. Kiener, *Die Prinzipal-Agenten-Theorie aus informationsökonomischer Sicht* (Physica, Heidelberg, 1990)
736. N.T.B. Kim, T.N. Thang, Optimization over the efficient set of a bicriteria convex programming problem. Pac. J. Optim. **9**, 103–115 (2013)
737. T. Kim, S. Suh, Toward developing a national transportation planning model: a bilevel programming approach for Korea. Ann. Reg. Sci. **22**, 65–80 (1988)
738. G. Kirlik, S. Sayın, Bilevel programming for generating discrete representations in multiobjective optimization. Math. Program. **169**, 585–604 (2018)
739. T. Kis, A. Kovács, Exact solution approaches for bilevel lot-sizing. Eur. J. Oper. Res. **226**(2), 237–245 (2013)
740. K.-P. Kistner, M. Switalski, Hierarchical production planning: necessity, problems, and methods. Zeitschrift für Oper. Res. **33**, 199–212 (1989)
741. T. Kleinert, M. Labbé, F. Plein, M. Schmidt, There's no free lunch: on the hardness of choosing a correct big-M in bilevel optimization, Technical Report (Friedrich-Alexander-Universität Erlangen-Nürnberg, Erlangen-Nürnberg, 2019)
742. P.-M. Kleniati, C.S. Adjiman, Branch-and-sandwich: a deterministic global optimization algorithm for optimistic bilevel programming problems. Part I: Theoretical development. J. Global Optim. **60**(3), 425–458 (2014)
743. P.-M. Kleniati, C.S. Adjiman, Branch-and-sandwich: a deterministic global optimization algorithm for optimistic bilevel programming problems. Part II: Convergence analysis and numerical results. J. Global Optim. **60**(3), 459–481 (2014)
744. P.-M. Kleniati, C.S. Adjiman, A generalization of the branch-and-sandwich algorithm: From continuous to mixed-integer nonlinear bilevel problems. Comput. Chem. Eng. **72**, 373–386 (2015)
745. M. Knauer, Fast and save container cranes as bilevel optimal control problems. Math. Comput. Model. Dyn. Syst. **18**(4), 465–486 (2012)

746. Y. Kochetov, N. Kochetova, A. Plyasunov, A matheuristic for the leader-follower facility location and design problem, in *Proceedings of the 10th Metaheuristics International Conference (MIC 2013)*, vol. 32 (2013), p. 3
747. Y.A. Kochetov, A.A. Panin, A.V. Plyasunov, Comparison of metaheuristics for the bilevel facility location and mill pricing problem. J. Appl. Ind. Math. **9**(3), 392–401 (2015)
748. Y.A. Kochetov, A.V. Pljasunov, Efficient algorithm for a class of bilevel linear programming problems, in *Operations Research Proceedings 1996* (Springer, Berlin, 1997), pp. 10–13
749. Y.A. Kochetov, A.V. Pljasunov, A polynomially solvable class of two-level linear programming problems. Diskret. Anal. Issled. Oper. **4**(2), 23–33 (1997)
750. Y.A. Kochetov, A.V. Pljasunov, The problem of selecting a number of products with partial exterior financing. Diskret. Anal. Issled. Oper., Serija 2 **9**(2), 78–96 (2002, in Russian)
751. A. Koh, Solving transportation bi-level programs with differential evolution, in *Proceedings of the IEEE Congress on Evolutionary Computation, 2007 (CEC 2007)* (IEEE, New York, 2007), pp. 2243–2250
752. A. Koh, A metaheuristic framework for bi-level programming problems with multi-disciplinary applications, in *Proceedings of the Metaheuristics for Bi-level Optimization* (Springer, Berlin, 2013), pp. 153–187
753. B. Kohli, Variational inequalities and optimistic bilevel programming problem via convexifactors, in *Topics in Nonconvex Optimization: Theory and Applications* ed. by S.K. Mishra (Springer, New York, 2011), pp. 243–255
754. B. Kohli, Optimality conditions for optimistic bilevel programming problem using convexifactors. J. Optim. Theory Appl. **152**(3), 632–651 (2012)
755. C. Kolstad, A review of the literature on bi-level mathematical programming, Technical Report LA-10284-MS, US-32 (Los Alamos National Laboratory, New York, 1985)
756. C. Kolstad, L. Lasdon, Derivative evaluation and computational experience with large bilevel mathematical programs. J. Optim. Theory Appl. **65**, 485–499 (1990)
757. A.F. Kononenko, V.V. Chumakov, Decision making in a two-level hierarchical control system in the presence of exogeneous noncontrollable factors. Avtomat. i Telech., 1, 92–101 (1988, in Russian), Autom. Remote Control, **49**(1), 73–80 (1988)
758. A.V. Kononov, Y.A. Kochetov, A.V. Plyasunov, Competitive facility location models. Comput. Math. Math. Phys. **49**(6), 994–1009 (2009)
759. D. Konur, M.M. Golias, Analysis of different approaches to cross-dock truck scheduling with truck arrival time uncertainty. Comput. Ind. Eng. **65**(4), 663–672 (2013)
760. M. Köppe, M. Queyranne, C.T. Ryan, Parametric integer programming algorithm for bilevel mixed integer programs. J. Optim. Theory Appl. **146**(1), 137–150 (2010)
761. J. Kornaj, T. Liptak, Two-level planning. Econometrica **33**, 141–169 (1965)
762. S. Kosuch, P. Le Bodic, J. Leung, A. Lisser, On a stochastic bilevel programming problem. Networks **59**(1), 107–116 (2012)
763. R.M. Kovacevic, G.C. Pflug, Electricity swing option pricing by stochastic bilevel optimization: a survey and new approaches. Eur. J. Oper. Res. **237**(2), 389–403 (2014)
764. A. Kovács, Bilevel programming approach to optimizing a time-variant electricity tariff for demand response, in *IEEE International Conference on Smart Grid Communications (SmartGridComm), 2016* (IEEE, New York, 2016), pp. 674–679
765. G. Kozanidis, E. Kostarelou, P. Andrianesis, G. Liberopoulos, Mixed integer parametric bilevel programming for optimal strategic bidding of energy producers in day-ahead electricity markets with indivisibilities. Optimization **62**(8), 1045–1068 (2013)
766. A. Kristály, S. Nagy, Followers' strategy in Stackelberg equilibrium problems on curved strategy sets. Acta Polytech. Hungarica **10**(7), 69–80 (2013)
767. H. Küçükaydin, N. Aras, I.K. Altınel, Competitive facility location problem with attractiveness adjustment of the follower: a bilevel programming model and its solution. Eur. J. Oper. Res. **208**(3), 206–220 (2011)
768. F.M. Kue, Mixed integer bilevel programming problems, Ph.D. thesis (TU Bergakademie, Freiberg, 2017)

769. A.A. Kulkarni, U.V. Shanbhag, A shared-constraint approach to multi-leader multi-follower games. Set-Valued Variational Anal. **22**(4), 691–720 (2014)
770. A.A. Kulkarni, U.V. Shanbhag, An existence result for hierarchical Stackelberg v/s Stackelberg games. IEEE Trans. Autom. Control **60**(12), 3379–3384 (2015)
771. G. Kunapuli, K. Bennett, J. Hu, J.-S. Pang, Bilevel model selection for support vector machines, in *CRM Proceedings and Lecture Notes*, vol. 45 (2008), pp. 129–158
772. G. Kunapuli, K.P. Bennett, Jing Hu, J.-S. Pang, Classification model selection via bilevel programming. Optim. Methods Softw. **23**(4), 475–489 (2008)
773. K. Kunisch, T. Pock, A bilevel optimization approach for parameter learning in variational models. SIAM J. Imag. Sci. **6**(2), 938–983 (2013)
774. R.J. Kuo, Y.S. Han, A hybrid of genetic algorithm and particle swarm optimization for solving bi-level linear programming problem–a case study on supply chain model. Appl. Math. Model. **35**(8), 3905–3917 (2011)
775. R.J. Kuo, C.C. Huang, Application of particle swarm optimization algorithm for solving bi-level linear programming problem. Comput. Math. Appl. **58**(4), 678–685 (2009)
776. R.J. Kuo, Y.H. Lee, F.E. Zulvia, F.C. Tien, Solving bi-level linear programming problem through hybrid of immune genetic algorithm and particle swarm optimization algorithm. Appl. Math. Comput. **266**, 1013–1026 (2015)
777. M.A. Laamim, A. Makrizi, E.H. Essoufi, Application of genetic algorithm for solving bilevel linear programming. Bioinspired Heuristics Optim. **774**, 123–136 (2018)
778. M. Labbé, P. Marcotte, G. Savard, A bilevel model of taxation and its application to optimal highway pricing. Manag. Sci. **44**, 1608–1622 (1998)
779. M. Labbé, P. Marcotte, G. Savard, On a class of bilevel programs, in *Nonlinear Optimization and Related Topics*, ed. by D. di Pillo, F. Gianessi, (Springer, Berlin, 2000), pp. 183–206
780. M. Labbé, A. Violin, Bilevel programming and price setting problems, 4OR **11**(1), 1–30 (2013)
781. M. Labbé, A. Violin, Bilevel programming and price setting problems. Ann. Oper. Res. **240**, 141–169 (2016)
782. K. Lachhwani, A. Dwivedi, Bi-level and multi-level programming problems: taxonomy of literature review and research issues. Arch. Comput. Methods Eng. **25**(4), 847–877 (2018)
783. L. Lafhim, N. Gadhi, K. Hamdaoui, F. Rahou, Necessary optimality conditions for a bilevel multiobjective programming problem via a ψ-reformulation. Optimization **67**, 2179–2189 (2018)
784. K.A.P. Lagares, J.S. Angelo, H.S. Bernardino, H.J.C. Barbosa, A differential evolution algorithm for bilevel problems including linear equality constraints, in *Proceedings of the 2016 IEEE Congress on Evolutionary Computation (CEC)* (2016), pp. 1885–1892
785. F. Lagos, F. Ordóñez, M. Labbé, A branch and price algorithm for a Stackelberg security game. Comput. Ind. Eng. **111**, 216–227 (2017)
786. Y.-J. Lai, Hierarchical optimization: a satisfactory solution. Fuzzy Sets Syst. **77**, 321–335 (1996)
787. L. Lampariello, S. Sagratella, A bridge between bilevel programs and Nash games. J. Optim. Theory Appl. **174**(2), 613–635 (2017)
788. L. Lampariello, S. Sagratella, Numerically tractable optimistic bilevel problems, Technical Report (Roma Tre University, Department of Business Studies, Rome, 2017)
789. L. Lampariello, S. Sagratella, O. Stein, The standard pessimistic bilevel problem. SIAM J. Optim. **29**(2), 1634–1656 (2019)
790. K.-M. Lan, U.-P. Wen, H.-S. Shih, E.S. Lee, A hybrid neural network approach to bilevel programming problems. Appl. Math. Lett. **20**(8), 880–884 (2007)
791. Y. Lan, R. Zhao, W. Tang, A bilevel fuzzy principal-agent model for optimal nonlinear taxation problems. Fuzzy Optim. Decis. Making **10**, 211–232 (2011)
792. M.-T. Laraba, M. Hovd, S. Olaru, S.-I. Niculescu, A bilevel optimization approach for D-invariant set design. IFAC-PapersOnLine **49**(10), 235–240 (2016)
793. H. Laux, H.Y. Schenk-Mathes, *Lineare und nichtlineare Anreizsysteme* (Physica, Netherlands, 1992)

794. D. Lavigne, R. Loulou, G. Savard, Pure competition, regulated and Stackelberg equilibria: application to the energy system of Quebec. Eur. J. Oper. Res. **125**, 1–17 (2000)
795. S. Lavlinskii, A.A. Panin, A.V. Plyasunov, Public-private partnership models with tax incentives: Numerical analysis of solutions, in *Proceedings of the International Conference on Optimization Problems and Their Applications* (Springer, Berlin, 2018), pp. 220–234
796. S.M. Lavlinskii, A.A. Panin, A.V. Plyasunov, A bilevel planning model for public–private partnership. Autom. Remote Control **76**(11), 1976–1987 (2015)
797. S.M. Lavlinskii, A.A. Panin, A.V. Plyasunov, Comparison of models of planning public-private partnership. J. Appl. Ind. Math. **10**(3), 356–369 (2016)
798. S. Lawphongpanich, D.W. Hearn, An MPEC approach to second-best toll pricing. Math. Program. **101**, 33–55 (2004)
799. H. Le Cadre, On the efficiency of local electricity markets under decentralized and centralized designs: a multi-leader Stackelberg game analysis. Cent. Eur. J. Oper. Res., **27**(4), 953–984 (2019)
800. L. LeBlanc, D. Boyce, A bilevel programming algorithm for exact solution of the network design problem with user-optimal flows. Transp. Res. **20B**, 259–265 (1986)
801. E.S. Lee, Fuzzy multiple level programming. Appl. Math. Comput. **120**(1–3), 79–90 (2001)
802. F. Legillon, A. Liefooghe, E.-G. Talbi, Cobra: a coevolutionary metaheuristic for bi-level optimization, in *Proceedings of the Metaheuristics for Bi-level Optimization*, ed. by E.-G. Talbi (Springer, Berlin, 2013), pp. 95–114
803. L. Lei, Y. Wei, Research of leader-follower problem to tradable emission permits, in *Proceedings of the International Conference on Management Science and Engineering, 2007 (ICMSE 2007)* (IEEE, New York, 2007), pp. 2184–2189
804. M. Lei, J. Zhang, X. Dong, J.J. Ye, Modeling the bids of wind power producers in the day-ahead market with stochastic market clearing. Sustainable Energy Technol. Assess. **16**, 151–161 (2016)
805. G. Leitman, On generalized Stackelberg strategies. J. Optim. Theory Appl. **26**, 637–643 (1978)
806. J.M. Leleno, H.D. Sherali, A leader-follower model and analysis for a two-stage network of oligopolies. Ann. Oper. Res. **34**, 37–72 (1992)
807. A. Leontiev, J. Herskovits, An interior point technique for solving bilevel programming problems. Optim. Eng. **14**, 381–394 (2013)
808. A.M. Lessin, B.J. Lunday, R.R. Hill, A bilevel exposure-oriented sensor location problem for border security. Comput. Oper. Res. **98**, 56–68 (2018)
809. E.S. Levitin, Optimization problems with extremal constraints. Part I: General concepts, formulation, and main problems. Avtomatika i Telemekhanika **12**, 1–15 (1995, in Russian)
810. E.S. Levitin, Optimization problems with extremal constraints. Part II: Description as mathematical problem of systems analysis. Avtomatika i Telemekhanika **12**, 16–31 (1995, in Russian)
811. E.S. Levitin, Two-stage models of optimization. Matematiceskoje Modelirovanie **8**, 45–54 (1996, in Russian)
812. S. Leyffer, T. Munson, Solving multi-leader–common-follower games. Optim. Method. Softw. **25**(4), 601–623 (2010)
813. C. Li, L. Guo, A single-level reformulation of mixed integer bilevel programming problems. Oper. Res. Lett. **45**(1), 1–5 (2017)
814. D. Li, J.B. Cruz, Information, decision-making and deception in games. Decision Support Syst. **47**(4), 518–527 (2009)
815. G. Li, Z. Wan, On bilevel programs with a convex lower-level problem violating Slater's constraint qualification. J. Optim. Theory Appl. **179**(3), 820–837 (2018)
816. G. Li, Z. Wan, J.-W. Chen, X. Zhao, Existence of solution and algorithms for a class of bilevel variational inequalities with hierarchical nesting structure. Fixed Point Theory Appl. **2016**(1), 41 (2016)
817. G. Li, Z. Wan, J.-W. Chen, X. Zhao, Necessary optimality condition for trilevel optimization problem. J. Ind. Manag. Optim. **13**(5), 282–290 (2018)

818. G. Li, Z. Wan, X. Zhao, Optimality conditions for bilevel optimization programs. Pac. J. Optim. **13**, 421–441 (2017)
819. H. Li, A genetic algorithm using a finite search space for solving nonlinear/linear fractional bilevel programming problems. Ann. Oper. Res. **235**(1), 543–558 (2015)
820. H. Li, L. Fang, An evolutionary algorithm for solving bilevel programming problems using duality conditions. Math. Prob. Eng. **2012**, 14 (2012)
821. H. Li, L. Fang, Co-evolutionary algorithm: an efficient approach for bilevel programming problems. Eng. Optim. **46**(3), 361–376 (2014)
822. H. Li, Y. Wang, A hybrid genetic algorithm for solving nonlinear bilevel programming problems based on the simplex method, in *Proceedings of the 3rd International Conference on Natural Computation*, vol. 4 (IEEE, New York, 2007), pp. 91–95
823. H. Li, Y. Wang, Exponential distribution-based genetic algorithm for solving mixed-integer bilevel programming problems. J. Syst. Eng. Electron. **19**(6), 1157–1164 (2008)
824. H. Li, L. Zhang, Solving linear bilevel programming problems using a binary differential evolution, in *Proceedings of the 11th International Conference on Computational Intelligence and Security (CIS)* (IEEE, New York, 2015), pp. 38–42
825. H. Li, L. Zhang, Y. Jiao, Solution for integer linear bilevel programming problems using orthogonal genetic algorithm. J. Syst. Eng. Electron. **25**(3), 443–451 (2014)
826. H. Li, Q. Zhang, Q. Chen, L. Zhang, Y.-C. Jiao, Multiobjective differential evolution algorithm based on decomposition for a type of multiobjective bilevel programming problems. Knowledge-Based Syst. **107**, 271–288 (2016)
827. M. Li, D. Lin, S. Wang, Solving a type of biobjective bilevel programming problem using NSGA-II. Comput. Math. Appl. **59**(2), 706–715 (2010)
828. N. Li, Z. Yu, Forward-backward stochastic differential equations and linear-quadratic generalized Stackelberg games. SIAM J. Control Optim. **56**(6), 4148–4180 (2018)
829. X. Li, P. Tian, X. Min, A hierarchical particle swarm optimization for solving bilevel programming problems, in *Proceedings of the ICAISC 2006*, ed. by L. Rutkowski, Lecture Notes in Artificial Intelligence, vol. 4029 (Springer, Berlin, 2006), pp. 1169–1178
830. X.-Y. Li, X.-M. Li, X.-W. Li, H.-T. Qiu, Multi-agent fare optimization model of two modes problem and its analysis based on edge of chaos. Phys. A Stat. Mech. Appl. **469**, 405–419 (2017)
831. Z. Li, W. Shen, J. Xu, B. Lev, Bilevel and multi-objective dynamic construction site layout and security planning. Autom. Constr. **57**, 1–16 (2015)
832. G. Lia, Z. Wana, J.-W. Chenb, X. Zhaoa, Optimality conditions for pessimistic trilevel optimization problem with middle-level problem being pessimistic. J. Nonlinear Sci. Appl.(JNSA) **9**(6), 3864–3878 (2016)
833. M.B. Lignola, J. Morgan, Topological existence and stability for Stackelberg problems. J. Optim. Theory Appl. **84**, 145–169 (1995)
834. M.B. Lignola, J. Morgan, Stability of regularized bilevel programming problems. J. Optim. Theory Appl. **93**, 575–596 (1997)
835. M.B. Lignola, J. Morgan, Existence of solutions to generalized bilevel programming problem. *Multilevel Optimization: Algorithms and Applications*, ed. by A. Migdalas, P.M. Pardalos, P. Värbrand (Kluwer Academic, Dordrecht, 1998), pp. 315–332
836. M.B. Lignola, J. Morgan, Well-posedness for optimization problems with constraints defined by variational inequalities having a unique solution. J. Global Optim. **16**, 57–67 (2000)
837. M.B. Lignola, J. Morgan, Existence for optimization problems with equilibrium constraints in reflexive Banach spaces, in *Proceedings of the Optimization in Economics, Finance and Industry, Datanova, Milano 2002* (2002), pp. 15–36
838. M.B. Lignola, J. Morgan, Existence of solutions to bilevel variational problems in Banach spaces, in *Equilibrium Problems: Nonsmooth Optimization and Variational Inequality Models* ed. by F. Giannessi, A. Maugeri, P.M. Pardalos (Kluwer Academic, Dordrecht, 2002), p. 161–174
839. M.B. Lignola, J. Morgan, Asymptotic behavior of semi-quasivariational optimistic bilevel problems in Banach spaces. J. Math. Anal. Appl. **424**(1), 1–20 (2015)

840. M.B. Lignola, J. Morgan, Inner regularizations and viscosity solutions for pessimistic bilevel optimization problems. J. Optim. Theory Appl. **173**, 183–202 (2017)
841. M.B. Lignola, J. Morgan, Further on inner regularizations in bilevel optimization. J. Optim. Theory Appl. **180**(3), 1087–1097 (2019)
842. C. Lim, J.C. Smith, Algorithms for discrete and continuous multicommodity flow network interdiction problems. IIE Trans. **39**(1), 15–26 (2007)
843. P. Limleamthong, G. Guillén-Gosálbez, Rigorous analysis of Pareto fronts in sustainability studies based on bilevel optimization: application to the redesign of the UK electricity mix. J. Cleaner Prod. **164**, 1602–1613 (2017)
844. D.-Y. Lin, A. Karoonsoontawong, S.T. Waller, A Dantzig-Wolfe decomposition based heuristic scheme for bi-level dynamic network design problem. Netw. Spatial Econ. **11**(1), 101–126 (2011)
845. G.-H. Lin, M. Xu, J.J. Ye, On solving simple bilevel programs with a nonconvex lower level program. Math. Program. **144**(1–2), 277–305 (2014)
846. L.-J. Lin, H.J. Shie, Existence theorems of quasivariational inclusion problems with applications to bilevel problems and mathematical programs with equilibrium constraint. J. Optim. Theory Appl. **138**(3), 445–457 (2008)
847. L.J. Lin, Existence theorems for bilevel problem with applications to mathematical program with equilibrium constraint and semi-infinite problem. J. Optim. Theory Appl. **137**(1), 27–40 (2008)
848. M. Linnala, E. Madetoja, H. Ruotsalainen, J. Hämäläinen, Bi-level optimization for a dynamic multiobjective problem. Eng. Optim. **44**(2), 195–207 (2012)
849. Y.-C. Liou, S. Schaible, J.-C. Yao, Supply chain inventory management via a Stackelberg equilibrium. J. Ind. Manag. Optim. **2**(1), 81–94 (2006)
850. Y.-C. Liou, S.-Y. Wu, J.-C. Yao, Bilevel decision with generalized semi-infinite optimization for fuzzy mappings as lower level problems. Fuzzy Optim. Decis. Making **4**, 41–50 (2005)
851. Y.-C. Liou, J.-C. Yao, Bilevel decision via variational inequalities. Comput. Math. Appl. **49**(7), 1243–1253 (2005)
852. B. Liu, Stackelberg-Nash equilibrium for multilevel programming with multiple followers using genetic algorithms. Comput. Math. Appl. **36**(7), 79–89 (1998)
853. B. Liu, Z. Wan, J. Chen, G. Wang, Optimality conditions for pessimistic semivectorial bilevel programming problems. J. Inequalities Appl. **2014**(1), 41 (2014)
854. G. Liu, J. Han, Optimality conditions for nonconvex bilevel programming problems. Syst. Sci. Math. Sci. **10**, 183–192 (1997)
855. G. Liu, J. Han, S. Wang, A trust region algorithm for bilevel programming problems. Chin. Sci. Bull. **43**, 820–824 (1998)
856. G.S. Liu, J.Y. Han, J.Z. Zhang, Exact penalty functions for convex bilevel programming problems. J. Optim. Theory Appl. **110**, 621–643 (2001)
857. G.S. Liu, J.Y. Han, J.Z. Zhang, A trust region algorithm for solving bilevel programming problems. Acta Math. Appl. Sin. English Ser. **29**(3), 491–498 (2013)
858. J. Liu, Y. Fan, Z. Chen, Y. Zheng, Pessimistic bilevel optimization: A survey. Int. J. Comput. Int. Syst. **11**(1), 725–736 (2018)
859. J. Liu, Y. Hong, Y. Zheng, A branch and bound-based algorithm for the weak linear bilevel programming problems. Wuhan Univer. J. Nat. Sci. **23**(6), 480–486 (2018)
860. J. Liu, Y. Hong, Y. Zheng, A new variant of penalty method for weak linear bilevel programming problems. Wuhan Univer. J. Nat. Sci. **23**(4), 328–332 (2018)
861. J. Liu, T. Zhang, Y.-X. Fan, B. Han, Y. Zheng, An objective penalty method for optimistic bilevel programming problems. J. Oper. Res. Soc. China **6**(1), 177–187 (2020)
862. W. Liu, K.-Y. Zheng, Z. Cai, Bi-level programming based real-time path planning for unmanned aerial vehicles. Knowledge-Based Syst. **44**, 34–47 (2013)
863. X. Liu, G. Du, R.J. Jiao, Bilevel joint optimisation for product family architecting considering make-or-buy decisions. Int. J. Prod. Res. **55**(20), 5916–5941 (2017)
864. Y. Liu, H. Xu, S.-J.S. Yang, J. Zhang, Distributionally robust equilibrium for continuous games: Nash and Stackelberg models. Eur. J. Oper. Res. **265**(2), 631–643 (2018)

865. Y.-H. Liu, S.M. Hart, Characterizing an optimal solution to the linear bilevel programming problem. Eur. J. Oper. Res. **73**, 164–166 (1994)
866. Y.-H. Liu, T.H. Spencer, Solving a bilevel linear program when the inner decision maker controls few variables. Eur. J. Oper. Res. **81**, 644–651 (1995)
867. Z. Liu, M. Ehrgott, Primal and dual algorithms for optimization over the efficient set. Optimization **67**, 1661–1686 (2018)
868. A. Lodi, T.K. Ralphs, G.J. Woeginger, Bilevel programming and the separation problem. Math. Program. **146**(1-2), 437–458 (2014)
869. S. Lohse, Eine spezielle Klasse von Zwei-Ebenen-Optimierungsaufgaben, Ph.D. thesis (TU Bergakademie, Freiberg, 2011)
870. G. Londono, A. Lozano, A bilevel optimization program with equilibrium constraints for an urban network dependent on time. Transp. Res. Procedia **3**, 905–914 (2014)
871. J.M. López-Lezama, J. Cortina-Gómez, N. Muñoz-Galeano, Assessment of the electric grid interdiction problem using a nonlinear modeling approach. Electr. Power Syst. Res. **144**, 243–254 (2017)
872. F. López-Ramos, S. Nasini, A. Guarnaschelli, Road network pricing and design for ordinary and hazmat vehicles: integrated model and specialized local search. Comput. Oper. Res. **109**, 170–187 (2019)
873. P. Loridan, J. Morgan, Approximate solutions for two-level optimization problems, in *Trends in Mathematical Optimization*, ed. by K. Hoffman, J. Hiriart-Urruty, C. Lemarechal, J. Zowe. International Series of Numerical Mathematics, vol. 84 (Birkhäuser, Basel, 1988), pp. 181–196
874. P. Loridan, J. Morgan, A sequential stability result for constrained Stackelberg problems. Richerche di Matematica **38**, 19–32 (1989)
875. P. Loridan, J. Morgan, A theoretical approximation scheme for Stackelberg problems. J. Optim. Theory Appl. **61**, 95–110 (1989)
876. P. Loridan, J. Morgan, New results on approximate solutions in two-level optimization. Optimization **20**, 819–836 (1989)
877. P. Loridan, J. Morgan, ε-regularized two-level optimzation problems: approximation and existence results, in *Proceedings of the Optimization—Fifth French-German Conference (Varez)*. Lecture Notes in Mathematics (Springer, Berlin, 1989), pp. 99–113. Nov. 1405
878. P. Loridan, J. Morgan, *Quasi Convex Lower Level Problem and Applications in Two Level Optimization*. Lecture Notes in Economics and Mathematical Systems, vol. 345, pp. 325–341 (Springer, Berlin, 1990)
879. P. Loridan, J. Morgan, Regularization for two-level optimization problems, in *Proceedings of the 6th French-German Conference on Optimization, Lambrecht, Advances in Optimization* (Springer, Berlin, 1991), pp. 239–255
880. P. Loridan, J. Morgan, Least-norm regularization for weak two-level optimization problems, in *Proceedings of the Optimization, Optimal Control and Partial Differential Equations, International Series of Numerical Mathematics*, vol. 107 (Birkhäuser, Basel, 1992), pp. 307–318
881. P. Loridan, J. Morgan, On strict ε-solutions for a two-level optimization problem, in *Proceedings of the International Conference on Operations Research*, vol. 90 (Springer, Berlin, 1992), pp. 165–172
882. P. Loridan, J. Morgan, Weak via strong Stackelberg problem: New results. J. Global Optim. **8**, 263–287 (1996)
883. L. Lozano, J. Cole Smith, A value-function-based exact approach for the bilevel mixed-integer programming problem. Oper. Res. **65**(3), 768–786 (2017)
884. J. Lu, J. Han, Y. Hu, G. Zhang, Multilevel decision-making: a survey. Inf. Sci. **346–347**, 463–487 (2016)
885. J. Lu, C. Shi, G. Zhang, On bilevel multi-follower decision making: general framework and solutions. Inf. Sci. **176**(11), 1607–1627 (2006)

886. J. Lu, C. Shi, G. Zhang, T. Dillon, Model and extended Kuhn–Tucker approach for bilevel multi-follower decision making in a referential-uncooperative situation. J. Global Optim. **38**(4), 597–608 (2007)
887. J. Lu, C. Shi, G. Zhang, D. Ruan, An extended branch and bound algorithm for bilevel multi-follower decision making in a referential-uncooperative situation. Int. J. Inf. Technol. Decis. Making **6**(02), 371–388 (2007)
888. J. Lu, Y.-B. Xiao, N.-J. Huang, A Stackelberg quasi-equilibrium problem via quasi-variational inequalities. Carpathian J. Math. **34**(3), 355–362 (2018)
889. J. Lu, G. Zhang, J. Montero, L. Garmendia, Multifollower trilevel decision making models and system. IEEE Trans. Ind. Inf. **8**(4), 974–985 (2012)
890. Y.-B. Lü, Z.-P. Wan, A smoothing method for solving bilevel multiobjective programming problems. J. Oper. Res. Soc. China **2**(4), 511–525 (2014)
891. Y.-B. Lü, Z.-P. Wan, X.-N. Guo, Bilevel model of emission permits market trading. Xitong Gongcheng Lilun yu Shijian/Syst. Eng. Theory Pract. **34**(2), 343–348 (2014)
892. Z. Lu, K. Deb, A. Sinha, Finding reliable solutions in bilevel optimization problems under uncertainties, in *Proceedings of the 2016 on Genetic and Evolutionary Computation Conference* (ACM, New York, 2016), pp. 941–948
893. R. Lucchetti, F. Mignanego, G. Pieri, Existence theorem of equilibrium points in Stackelberg games with constraints. Optimization **18**, 857–866 (1987)
894. P.B. Luh, T.-S. Chang, T. Ning, Three-level Stackelberg decision problems. IEEE Trans. Autom. Control **AC-29**, 280–282 (1984)
895. P.B. Luh, T.-S. Chang, T. Ning, Pricing problems with a continuum of customers as stochastic Stackelberg games. J. Optim. Theory Appl. **55**, 119–131 (1987)
896. Z. Lukač, K. Šorić, V.V. Rosenzweig, Production planning problem with sequence dependent setups as a bilevel programming problem. Eur. J. Oper. Res. **187**(3), 1504–1512 (2008)
897. Z.-Q. Luo, J.-S. Pang, D. Ralph, *Mathematical Programs with Equilibrium Constraints* (Cambridge University, Cambridge, 1996)
898. T. Lv, Q. Ai, Y. Zhao, A bi-level multi-objective optimal operation of grid-connected microgrids. Electr. Power Syst. Res. **131**, 60–70 (2016)
899. Y. Lv, J. Chen, A discretization iteration approach for solving a class of semivectorial bilevel programming problem. J. Nonlinear Sci. Appl. **9**(5), 2888–2899 (2016)
900. Y. Lv, Z. Chen, Z. Wan, A neural network for solving a convex quadratic bilevel programming problem. J. Comput. Appl. Math. **234**(2), 505–511 (2010)
901. Y. Lv, T. Hu, Z. Wan, A penalty function method for solving weak price control problem. Appl. Math. Comput. **186**(2), 1520–1525 (2007)
902. Y. Lv, T. Hu, G. Wang, Z. Wan, A penalty function method based on Kuhn–Tucker condition for solving linear bilevel programming. Appl. Math. Comput. **188**(1), 808–813 (2007)
903. Y. Lv, T. Hu, G. Wang, Z. Wan, A neural network approach for solving nonlinear bilevel programming problem. Comput. Math. Appl. **55**(12), 2823–2829 (2008)
904. Y. Lv, Z. Wan, A solution method for the optimistic linear semivectorial bilevel optimization problem. J. Inequalities Appl. **2014**(1), 164 (2014)
905. Y. Lv, Z. Wan, Solving linear bilevel multiobjective programming problem via exact penalty function approach. J. Inequalities Appl. **2015**(1), 258 (2015)
906. Y. Lv, Z. Wan, Linear bilevel multiobjective optimization problem: Penalty approach. J. Ind. Manag. Optim., **15**(3), 1213–1223 (2019)
907. J. Lžičař, Solving methods for bilevel optimization problems, Master's thesis (Univerzita Karlova, Matematicko-fyzikální fakulta, 2019)
908. W. Ma, M. Wang, X. Zhu, Improved particle swarm optimization based approach for bilevel programming problem-an application on supply chain model. Int. J. Mach. Learn. Cybern. **5**(2), 281–292 (2014)
909. Y. Ma, F. Yan, K. Kang, X. Wei, A novel integrated production-distribution planning model with conflict and coordination in a supply chain network. Knowledge-Based Syst. **105**, 119–133 (2016)

910. C.M. Macal, A.P. Hurter, Dependence of bilevel mathematical programs on irrelevant constraints. Comput. Oper. Res. **24**, 1129–1140 (1997)
911. G. Mahapatra, S. Banerjee, Bilevel optimization using firefly algorithm, in *Proceedings of the 1st International Science and Technology Congress* (2014), p. 10
912. A.I. Mahmutogullari, B.Y. Kara, Hub location under competition. Eur. J. Oper. Res. **250**(1), 214–225 (2016)
913. C. Makasu, A bilevel programming approach to double optimal stopping. Appl. Math. Comput. **238**, 393–396 (2014)
914. S. Maldonado-Pinto, M.-S. Casas-Ramírez, J.-F. Camacho-Vallejo, Analyzing the performance of a hybrid heuristic for solving a bilevel location problem under different approaches to tackle the lower level. Math. Prob. Eng. **2016**, 10 (2016)
915. N. Malhotra, S.R. Arora, Optimality conditions for linear fractional bilevel programs. Indian J. Pure Appl. Math. **30**, 373–384 (1999)
916. N. Malhotra, S.R. Arora, Optimality conditions and an algorithm for linear-quadratic bilevel programming. Manag. Sci. Financial Eng. **7**(1), 41–56 (2001)
917. L. Mallozzi, A.P. di Napoli, Optimal transport and a bilevel location-allocation problem. J. Global Optim. **67**(1–2), 207–221 (2017)
918. L. Mallozzi, J. Morgan, ε-mixed strategies for static continuous-kernel Stackelberg problems. J. Optim. Theory Appl. **78**, 303–316 (1993)
919. L. Mallozzi, J. Morgan, Weak Stackelberg problem and mixed solutions under data perturbations. Optimization **32**, 269–290 (1995)
920. L. Mallozzi, J. Morgan, On approximate mixed Nash equilibria and average marginal functions for a two-stage three-players games, in *Optimization with Multivalued Mappings: Theory, Applications and Algorithms*, ed. by S. Dempe, V. Kalashnikov. Optimization and its Applications, vol. 2 (Springer/LLC, New York, 2006), pp. 97–107
921. A.V. Malyshev, A.S. Strekalovsky, Global search for pessimistic solution in bilevel problems, in *Proceedings of the Toulouse global optimization workshop* (2010), pp. 77–80
922. A.V. Malyshev, A.S. Strekalovsky, Global search for guaranteed solutions in quadratic-linear bilevel optimization problems. Izvestiya Irkutskogo Gosudarstvennogo Universiteta. Seriya "Matematika" **4**(1), 73–82 (2011)
923. A.V. Malyshev, A.S. Strekalovsky, On global search for pessimistic solution in bilevel problems (Special Issue: Bilevel programming, optimization methods, and applications to economics). Int. J. Biomed. Soft Comput. Human Sci. Off. J Biomed. Fuzzy Syst. Assoc. **18**(1), 57–61 (2013)
924. O.L. Mangasarian, Misclassification minimization. J. Global Optim. **5**, 309–323 (1994)
925. O.L. Mangasarian, Regularized linear programs with equilibrium constraints, in *Reformulation—Nonsmooth, Piecewise Smooth, Semismooth and Smoothing Methods*, ed. by M. Fukushima, L. Qi (Kluwer Academic, Dordrecht, 1998), pp. 259–268
926. O.L. Mangasarian, J.-S. Pang, Exact penalty functions for mathematical programs with linear complemantarity constraints. Optimization **42**, 1–8 (1997)
927. P. Marcotte, Network optimization with continuous control parameters. Transp. Sci. **17**, 181–197 (1983)
928. P. Marcotte, Network design problem with congestion effects: a case of bilevel programming. Math. Program. **34**, 142–162 (1986)
929. P. Marcotte, A note on a bilevel programming algorithm by LeBlanc and Boyce. Transp. Res. **22B**, 233–237 (1988)
930. P. Marcotte, G. Marquis, Efficient implementation of heuristics for the continuous network design problem. Ann. Oper. Res. **34**, 163–176 (1992)
931. P. Marcotte, A. Mercier, G. Savard, V. Verter, Toll policies for mitigating hazardous materials transport risk. Transp. Sci. **43**(2), 228–243 (2009)
932. P. Marcotte, G. Savard, A note on the Pareto optimality of solutions to the linear bilevel programming problem. Comput. Oper. Res. **18**, 355–359 (1991)
933. P. Marcotte, G. Savard, Novel approaches to the discrimination problem. Zeitschrift für Oper. Res. **36**, 517–545 (1992)

934. P. Marcotte, G. Savard, Bilevel programming: applications, in *Encyclopedia of Optimization* (Kluwer Academic, Dordrecht, 2001)
935. P. Marcotte, G. Savard, Bilevel programming: a combinatorial perspective, in *Proceedings of the Graph Theory and Combinatorial Optimization, GERAD 25th Anniversary Series*, vol. 8 (Springer, New York, 2005), pp. 191–217
936. P. Marcotte, G. Savard, D. Zhu, Mathematical structure of a bilevel strategic pricing model. Eur. J. Oper. Res. **193**(2), 552–566 (2009)
937. P. Marcotte, G. Savard, D.L. Zhu, A trust region algorithm for nonlinear bilevel programming. Oper. Res. Lett. **29**, 171–179 (2001)
938. P. Marcotte, D.L. Zhu, Exact and inexact penalty methods for the generalized bilevel programming problem. Math. Program. **74**, 141–157 (1996)
939. V. Marianov, D. Serra, Hierarchical location–allocation models for congested systems. Eur. J. Oper. Res. **135**(1), 195–208 (2001)
940. Y. Marinakis, A. Migdalas, P.M. Pardalos, A new bilevel formulation for the vehicle routing problem and a solution method using a genetic algorithm. J. Global Optim. **38**(4), 555–580 (2007)
941. R. Mathieu, L. Pittard, G. Anandalingam, Genetic algorithm based approach to bi-level linear programming. RAIRO. Recherche Opérationnelle **28**, 1–21 (1994)
942. K. Mathur, M.C. Puri, A bilevel linear programming problem with bottleneck objectives. Opsearch **31**, 177–201 (1994)
943. K. Mathur, M.C. Puri, A bilevel bottleneck programming problem. Eur. J. Oper. Res. **86**, 337–344 (1995)
944. A. Maugeri, L. Scrimali, A new approach to solve convex infinite-dimensional bilevel problems: application to the pollution emission price problem. J. Optim. Theory Appl. **169**(2), 370–387 (2016)
945. A. Mauttone, M. Labbé, R. Figueiredo, A tabu search approach to solve a network design problem with user-optimal flows, in *Proceedings of the ALIO/EURO Workshop on Applied Combinatorial Optimization, Buenos Aires, Argentina, 2008* (2007)
946. P. Mehlitz, Bilevel programming problems with simple convex lower level. Optimization **65**(6), 1203–1227 (2016)
947. P. Mehlitz, Contributions to complementarity and bilevel programming in Banach spaces, Ph.D. thesis (TU Bergakademie Freiberg, Freiberg, 2017)
948. P. Mehlitz, G. Wachsmuth, Weak and strong stationarity in generalized bilevel programming and bilevel optimal control. Optimization **65**(5), 907–935 (2016)
949. R. Menasri, A. Nakib, B. Daachi, H. Oulhadj, P. Siarry, A trajectory planning of redundant manipulators based on bilevel optimization. Appl. Math. Comput. **250**, 934–947 (2015)
950. Z. Meng, C. Dang, R. Shen, M. Jiang, An objective penalty function of bilevel programming. J. Optim. Theory Appl. **153**(2), 377–387 (2012)
951. A.G. Mersha, Solution methods for bilevel programming problems, Ph.D. thesis (TU Bergakademie Freiberg, Freiberg, 2008)
952. A.G. Mersha, S. Dempe, Linear bilevel programming with upper level constraints depending on the lower level solution. Appl. Math. Comput. **180**(1), 247–254 (2006)
953. A.G. Mersha, S. Dempe, Direct search algorithm for bilevel programming problems. Comput. Optim. Appl. **49**(1), 1–15 (2011)
954. A.G. Mersha, S. Dempe, Feasible direction method for bilevel programming problem. Optimization **61**(4–6), 597–616 (2012)
955. M. Mesanovic, D. Macko, Y. Takahara, *Theory of Hierarchical, Multilevel Systems* (Academic Press, New York, 1970)
956. B. Metev, Multiobjective optimization methods help to minimize a function over the efficient set. Cybern. Inf. Technol. **7**(2), 22–28 (2007)
957. C. Miao, G. Du, R.J. Jiao, T. Zhang, Coordinated optimisation of platform-driven product line planning by bilevel programming. Int. J. Prod. Res. **55**(13), 3808–3831 (2017)
958. A. Migdalas, Bilevel programming in traffic planning: Models, methods and challenge. J. Global Optim. **7**, 381–405 (1995)

959. A. Migdalas, When is Stackelberg equilibrium Pareto optimum?, in *Advances in Multicriteria Analysis*, ed. by P. Pardalos et al. (Kluwer Academic, Dordrecht, 1995)
960. A. Migdalas, P. Pardalos, Editorial: Hierarchical and bilevel programming. J. Global Optim. **8**, 209–215 (1996)
961. A. Migdalas, P.M. Pardalos, P. Värbrand, *Multilevel Optimization: Algorithms and Applications* (Kluwer Academic Publishers, Dordrecht, 1998)
962. F. Mignanego, A. Sciomachen, Incentive strategies with threats in dynamic constrained-Stackelberg problems, a bilevel programming approach. Optimization **38**, 263–276 (1996)
963. T. Miller, T. Friesz, R. Tobin, Heuristic algorithms for delivered price spatially competitive network facility location problems. Ann. Oper. Res. **34**, 177–202 (1992)
964. T.C. Miller, R.L. Tobin, T.L. Friesz, Stackelberg games on a network with Cournot-Nash oligopolistic competitors. J. Reg. Sci. **31**(4), 435–454 (1991)
965. M. Miralinaghi, Y. Lou, B.B. Keskin, Y.-T. Hsu, R. Shabanpour, Hydrogen refueling station location problem with traffic deviation considering route choice and demand uncertainty. Int. J. Hydrogen Energy **42**, 3335–3351 (2017)
966. S.A. MirHassani, S. Raeisi, A. Rahmani, Quantum binary particle swarm optimization-based algorithm for solving a class of bi-level competitive facility location problems. Optim. Method. Softw. **30**(4), 756–768 (2015)
967. J.A. Mirrlees, The theory of moral hazard and unobservable bevaviour: part I. Rev. Econ. Studies **66**, 3–21 (1999)
968. S. Mishra, Weighting method for bi-level linear fractional programming problems. Eur. J. Oper. Res. **183**(1), 296–302 (2007)
969. S. Mishra, A. Ghosh, Interactive fuzzy programming approach to bi-level quadratic fractional programming problems. Ann. Oper. Res. **143**, 251–263 (2006)
970. A. Mitsos, Global solution of bilevel mixed-integer nonlinear programs, in *Proceedings of the 2008 Annual Meeting Computing and Systems Technology Division, Philadelphia, 2008* (2008)
971. A. Mitsos, Global solution of nonlinear mixed-integer bilevel programs. J. Global Optim. **47**(4), 557–582 (2010)
972. A. Mitsos, P.I. Barton, A test set for bilevel programs, Technical Report (Massachusetts Institute of Technology, Cambridge, 2006)
973. A. Mitsos, G.M. Bollas, P.I. Barton, Bilevel optimization formulation for parameter estimation in liquid—liquid phase equilibrium problems. Chem. Eng. Sci. **64**(3), 548–559 (2009)
974. A. Mitsos, B. Chachuat, P.I. Barton, Towards global bilevel dynamic optimization. J. Global Optim. **45**(1), 63–93 (2009)
975. A. Mitsos, P. Lemonidis, P.I. Barton, Global solution of bilevel programs with a nonconvex inner program. J. Global Optim. **42**(4), 475–513 (2008)
976. K. Mizukami, H. Xu, Closed-loop Stackelberg strategies for linear-quadratic descriptor systems. J. Optim. Theory Appl. **74**, 151–170 (1992)
977. K. Mombaur, A. Truong, J.-P. Laumond, From human to humanoid locomotion—an inverse optimal control approach. Auton. Robots **28**(3), 369–383 (2010)
978. G.M. Moore, Bilevel programming algorithms for machine learning model selection, Ph.D. thesis (Rensselaer Polytechnic Institute Troy, New York, 2010)
979. J. Moore, Extensions to the multilevel linear programming problem, Ph.D. thesis (Department of Mechanical Engineering, University of Texas, Austin, 1988)
980. J. Moore, J.F. Bard, The mixed integer linear bilevel programming problem. Oper. Res. **38**, 911–921 (1990)
981. M. Moraal, Stackelberg solutions in linear programming problems, in *Proceedings of the Symposium on Operations Research*, vol. 6 (University of Augsburg, Augsburg, 1981). Part II, Methods of Operations Research (1983), pp. 375–383
982. B.S. Mordukhovich, Variational Analysis and Generalized Differentiation Vol. 1: Basic Theory (Springer, Berlin, 2006)
983. B.S. Mordukhovich, Variational Analysis and Generalized Differentiation Vol. 2: Applications (Springer, Berlin, 2006)

984. B.S. Mordukhovich, N.M. Nam, H.M. Phan, Variational analysis of marginal functions with applications to bilevel programming. J. Optim. Theory Appl. **152**(3), 557–586 (2012)
985. B.S. Mordukhovich, J.V. Outrata, Coderivative analysis of quasi-variational inequalities with applications to stability and optimization. SIAM J. Optim. **18**(2), 389–412 (2007)
986. J. Morgan, Constrained well-posed two-level optimization problems, in *Proceedings of the Nonsmooth Optimization and Related Topics*, ed. by F.H. Clarke et al. (Plenum Press, New York, 1989), pp. 307–325
987. J. Morgan, P. Loridan, Approximation of the Stackelberg problem and applications in control theory, in *Control Application of Nonlinear Programming and Optimization: Proceedings of the Fifth IFAC Workshop, Capri, Italy 11–14 June*, ed. by G. Di Pillo (1985), pp. 121–124
988. V.V. Morozov, A.I. Soloviev, On optimal partial hedging in discrete markets. Optimization **62**(11), 1403–1418 (2013)
989. K. Moshirvaziri, M.A. Amouzegar, S. E. Jacobsen, Test problem construction for linear bilevel programming problems. J. Global Optim. **8**(3), 235–243 (1996)
990. A. Moudafi, Proximal methods for a class of bilevel monotone equilibrium problems. J. Global Optim. **47**(2), 287–292 (2010)
991. R.E. Msigwa, Y. Lu, Y. Ge, L. Zhang, A smoothing approach for solving transportation problem with road toll pricing and capacity expansions. J. Inequalities Appl. **2015**, 237 (2015)
992. A. Mukherjee, L. Zhao, Profit raising entry. J. Ind. Econ. **57**(4), 870 (2009)
993. L.D. Muu, On the construction of initial polyhedral convex set for optimization problems over the efficient set and bilevel linear programs. Vietnam J. Math. **28**, 177–182 (2000)
994. L.D. Muu, W. Oettli, Optimization over equilibrium sets. Optimization **49**, 179–189 (2001)
995. L.D. Muu, N.V. Quy, A global optimization method for solving convex quadratic bilevel programming problems. J. Global Optim. **26**, 199–219 (2003)
996. S. Nagy, Stackelberg equilibria via variational inequalities and projections. J. Global Optim. **57**(3), 821–828 (2013)
997. S. Nagy, Variational approach to Stackelberg equilibria, Ph.D. thesis (Babeş-Bolyai University, Romania, 2015)
998. J. Naoum-Sawaya, S. Elhedhli, Controlled predatory pricing in a multiperiod Stackelberg game: an MPEC approach. J. Global Optim. **50**(2), 345–362 (2011)
999. S. Narula, A. Nwosu, A dynamic programming solution for the hierarchical linear programming problem, Technical Report 37–82 (Department of Operations Research and Statistics, Rensselaer Polytechnic Institute, New York, 1982)
1000. S. Narula, A. Nwosu, Two-level hierarchical programming problems, in *Essays and surveys on multiple criteria decision making*, ed. by P. Hansen (Springer, Berlin, 1983), pp. 290–299
1001. S. Narula, A. Nwosu, An algorithm to solve a two-level resource control pre-emptive hierarchical programming problem, in *Mathematics of multiple-objective programming*, ed. by P. Serafini (Springer, Berlin, 1985)
1002. S. Narula, A. Nwosu, Two-level resource control pre-emptive hierarchical linear programming problem: a review, in *Recent Developments in Mathematical Programming*, ed. by S. Kumar (Gordon and Breach Science Publication, Philadelphia, 1991), pp. 29–43
1003. M. Nasri, Characterizing optimal wages in principal-agent problems without using the first-order approach. Optimization **65**(2), 467–478 (2016)
1004. N. Nezamoddini, S. Mousavian, M. Erol-Kantarci, A risk optimization model for enhanced power grid resilience against physical attacks. Electr. Power Syst. Res. **143**, 329–338 (2017)
1005. T.Q. Nguyen, M. Bouhtou, J.-L. Lutton, DC approach to bilevel bilinear programming problem: application in telecommunication pricing, in *Optimization and Optimal Control*, ed. by P. Pardalos, I, Tseveendorj, R. Enkhbat (World Scientific, Singapore, 2003), pp. 211–231
1006. M.G. Nicholls, Aluminium production modelling—a non-linear bi-level programming approach. Oper. Res. **43**, 208–218 (1995)
1007. M.G. Nicholls, The application of nonlinear bilevel programming to the aluminium industry. J. Global Optim. **8**, 245–261 (1996)

1008. M.G. Nicholls, Developing an integrated model of an aluminium smelter incorporating submodels with different time bases and levels of aggregation. Eur. J. Oper. Res. **99**, 477–490 (1997)
1009. J. Nie, Optimality conditions and finite convergence of Lasserre's hierarchy. Math. Program. **146**(1–2), 97–121 (2014)
1010. J. Nie, L. Wang, J. Ye, Bilevel polynomial programs and semidefinite relaxation methods. SIAM J. Optim. **27**, 1728–1757 (2017)
1011. P.-Y. Nie, A note on bilevel optimization problems. Int. J. Appl. Math. Sci. **2**, 31–28 (2005)
1012. P.-Y. Nie, Dynamic discrete-time multi-leader–follower games with leaders in turn. Comput. Math. Appl. **61**(8), 2039–2043 (2011)
1013. P.-Y. Nie, M.-Y. Lai, S.-J. Zhu, Dynamic feedback Stackelberg games with non-unique solutions. Nonlinear Anal. Theory Methods Appl. **69**(7), 1904–1913 (2008)
1014. T. Nishi, O. Yoshida, Optimization of multi-period bilevel supply chains under demand uncertainty. Procedia CIRP **41**, 508–513 (2016)
1015. I. Nishizaki, M. Sakawa, Stackelberg solutions to multiobjective two-level linear programming problems. J. Optim. Theory Appl. **103**, 161–182 (1999)
1016. I. Nishizaki, M. Sakawa, Computational methods through genetic algorithms for obtaining Stackelberg solutions to two-level mixed zero-one programming problems. Cybern. Syst. **31**(2), 203–221 (2000)
1017. I. Nishizaki, M. Sakawa, H. Katagiri, Stackelberg solutions to multiobjective two-level linear programming problems with random variable coefficients. Cent. Eur. J. Oper. Res. **11**(3), 281–296 (2003)
1018. V.I. Norkin, Optimization models of anti-terrorist protection. Cybern. Syst. Anal. **54**(6), 918–929 (2018)
1019. V.I. Norkin, A.A. Gaivoronski, V.A. Zaslavsky, P.S. Knopov, Models of the optimal resource allocation for the critical infrastructure protection. Cybern. Syst. Anal. **54**(5), 696–706 (2018)
1020. A.J. Novak, G. Feichtinger, G. Leitmann, A differential game related to terrorism: Nash and Stackelberg strategies. J. Optim. Theory Appl. **144**(3), 533–555 (2010)
1021. A. Nwosu, Pre-emptive hierarchical programming problem: a decentralized decision model, Ph.D. thesis (Department of Operations Research and Statistics, Rensselaer Polytechnic Institute, New York, 1983)
1022. R. Oberdieck, N.A Diangelakis, S. Avraamidou, E.N. Pistikopoulos, On unbounded and binary parameters in multi-parametric programming: applications to mixed-integer bilevel optimization and duality theory. J. Global Optim. **69**, 587–606 (2017)
1023. P. Ochs, R. Ranftl, T. Brox, T. Pock, Bilevel optimization with nonsmooth lower level problems, in *Proceedings of the International Conference on Scale Space and Variational Methods in Computer Vision* (Springer, Berlin, 2015), pp. 654–665
1024. P. Ochs, R. Ranftl, T. Brox, T. Pock, Techniques for gradient-based bilevel optimization with non-smooth lower level problems. J. Math. Imaging Vision **56**(2), 175–194 (2016)
1025. V. Oduguwa, R. Roy, Bi-level optimisation using genetic algorithm, in *Proceedings of the IEEE International Conference on Artificial Intelligence Systems, 2002* (IEEE, New York, 2002), pp. 322–327
1026. W. Oeder, Ein Verfahren zur Lösung von Zwei-Ebenen-Optimierungsaufgaben in Verbindung mit der Untersuchung von chemischen Gleichgewichten, Ph.D. thesis (Technische Universität Karl-Marx-Stadt, 1988)
1027. G.L. Olsder, Phenomena in inverse Stackelberg games, part 1: Static problems. J. Optim. Theory Appl. **143**, 589–600 (2009)
1028. P. Ochs, R. Ranftl, T. Brox, T. Pock, Phenomena in inverse Stackelberg games, part 2: Dynamic problems. J. Optim. Theory Appl. **143**, 601–618 (2009)
1029. H. Önal, Computational experience with a mixed solution method for bilevel linear/quadratic programs, Technical Report (University of Illinois, Urbana-Champaign, 1992)
1030. H. Önal, A modified simplex approach for solving bilevel linear programming problems. Eur. J. Oper. Res. **67**, 126–135 (1993)

1031. H. Önal, D.H. Darmawan, S.H. Johnson III, A multilevel analysis of agricultural credit distribution in East Java, Indonesia. Comput. Oper. Res. **22**, 227–236 (1995)
1032. A.V. Orlov, Numerical solution of bilinear programming problems. Comput. Math. Math. Phys. **48**(2), 225–241 (2008)
1033. A.V. Orlov, Global search for optimistic solutions in bilevel problem of optimal tariff choice by telecommunication company. Izvestiya Irkutskogo Gosudarstvennogo Universiteta. Seriya "Matematika" **6**(1), 57–71 (2013)
1034. A.V. Orlov, A.V. Malyshev, Test problem generation for quadratic-linear pessimistic bilevel optimization. Numer. Anal. Appl. **7**(3), 204–214 (2014)
1035. M.S. Osman, M.A. Abo-Sinna, A.H. Amer, O.E. Emam, A multi-level non-linear multi-objective decision-making under fuzziness. Appl. Math. Comput. **153**(1), 239–252 (2004)
1036. A. Ouattara, A. Aswani, Duality approach to bilevel programs with a convex lower level, in *Proceedings of the 2018 Annual American Control Conference (ACC)* (IEEE, New York, 2018), pp. 1388–1395
1037. J.V. Outrata, A note on the usage of nondifferentiable exact penalties in some special optimization problems. Kybernetika **24**(4), 251–258 (1988)
1038. J.V. Outrata, On the numerical solution of a class of Stackelberg problems. ZOR—Math. Methods Oper. Res. **34**, 255–277 (1990)
1039. J.V. Outrata, Necessary optimality conditions for Stackelberg problems. J. Optim. Theory Appl. **76**, 305–320 (1993)
1040. O.Y. Özaltın, O.A. Prokopyev, A.J. Schaefer, The bilevel knapsack problem with stochastic right-hand sides. Oper. Res. Lett. **38**(4), 328–333 (2010)
1041. M. Pachter, Linear-quadratic reversed Stackelberg differential games with incentives. IEEE Trans. Autom. Control **AC-29**, 644–647 (1984)
1042. B.B. Pal, B.N. Moitra, A fuzzy goal programming procedure for solving quadratic bilevel programming problems. Int. J. Int. Syst. **18**(5), 529–540 (2003)
1043. K.D. Palagachev, M. Gerdts, Numerical approaches towards bilevel optimal control problems with scheduling tasks, in *Math for the Digital Factory*, ed. by L. Ghezzi, D. Hömberg, C. Landry (Springer, New York, 2017), pp. 205–228
1044. M. Pan, P.S. Leung, S.G. Pooley, A decision support model for fisheries management in Hawaii: a multilevel and multiobjective programming approach. North Am. J. Fish. Manage. **21**(2), 293–309 (2001)
1045. Q. Pan, Z. An, H. Qi, Exact penalty method for the nonlinear bilevel programming problem. Wuhan Univer. J. Nat. Sci. **15**(6), 471–475 (2010)
1046. P.D. Panagiotopoulos, E.S. Mistakidis, G.E. Stavroulakis, O.K. Panagouli, Multilevel optimization methods in mechanics, in *Multilevel Optimization: Algorithms and Applications*, ed. by A. Migdalas, P. Pardalos, P. Värbrand (Kluwer Academic, Dordrecht, 1998)
1047. G. Paneiro, F.O. Durão, M.C. e Silva, P.A. Bernardo, Neural network approach based on a bilevel optimization for the prediction of underground blast-induced ground vibration amplitudes. Neural Comput. Appl. **32**, 5975–5987 (2020)
1048. J.-S. Pang, M. Fukushima, Quasi-variational inequalities, generalized Nash equilibria, and multi-leader-follower games. Comput. Manag. Sci. **2**(1), 21–56 (2005)
1049. G. Papavassilopoulos, Algorithms for static Stackelberg games with linear costs and polyhedral constraints, in *Proceedings of the 21st IEEE Conference on Decisions and Control* (1982), pp. 647–652
1050. F. Parraga, Hierarchical programming and applications to economic policy, Ph.D. thesis (Systems and Industrial Engineering Department, University of Arizona, Arizona, 1981)
1051. S.P. Parvasi, M. Mahmoodjanloo, M. Setak, A bi-level school bus routing problem with bus stops selection and possibility of demand outsourcing. Applied Soft Comput. **61**, 222–238 (2017)
1052. M. Patriksson, On the applicability and solution of bilevel optimization models in transportation science: A study on the existence, stability and computation of optimal solutions to stochastic mathematical programs with equilibrium constraints. Transp. Res. B Methodol. **42**(10), 843–860 (2008)

1053. M. Patriksson, R.T. Rockafellar, A mathematical model and descent algorithm for bilevel traffic management. Transp. Sci. **36**(3), 271–291 (2002)
1054. M. Patriksson, L. Wynter, Stochastic nonlinear bilevel programming, Technical Report (PRISM, Université de Versailles—Saint Quentin en Yvelines, Versailles, France, 1997)
1055. M. Patriksson, L. Wynter, Stochastic mathematical programs with equilibrium constraints. OR Lett. **25**, 159–167 (1999)
1056. R. Paulavicius, C.S. Adjiman, BASBL: branch-and-sandwich bilevel solver. i. theoretical advances and algorithmic improvements, Technical Report (Imperial College London, 2017)
1057. R. Paulavicius, P.M. Kleniati, C.S. Adjiman, Global optimization of nonconvex bilevel problems: implementation and computational study of the branch-and-sandwich algorithm. Comput. Aided Chem. Eng. **38**, 1977–1982 (2016)
1058. R. Paulavicius, P.M. Kleniati, C.S. Adjiman, BASBL: branch-and-sandwich bilevel solver. ii. implementation and computational study with the basbl test set, Technical Report (Imperial College London, 2017)
1059. K. Pavlova, T. Stoilov, K. Stoilova, Bi-level model for public rail transportation under incomplete data. Cybern. Inf. Technol. **17**(3), 75–91 (2017)
1060. R. Peng, X. Rui-hua, Q. Jin, *Bi-level simulated annealing algorithm for facility location problem*, in *Proceedings of the International Conference on Information Management, Innovation Management and Industrial Engineering (ICIII'08)*, vol. 3 (IEEE, New York, 2008), pp. 17–22
1061. A.A. Pessoa, M. Poss, M.C. Roboredo, L. Aizemberg, Solving bilevel combinatorial optimization as bilinear min-max optimization via a branch-and-cut algorithm, in *Proceedings of the Anais do XLV Simpósio Brasileiro de Pesquisa Operacional* (2013)
1062. T. Petersen, *Optimale Anreizsysteme* (Gabler, Wiesbaden, 1989)
1063. A.G. Petoussis, Supply function equilibrium analysis for electricity markets, Ph.D. thesis (University of Warwick, Warwick, 2009)
1064. E.G. Petrova, T.V. Gruzdeva, The linear bilevel problems via nonconvex constraint problems, in *Proceedings of the Toulouse Global Optimization workshop TOGO10, Toulouse, France, August-September 2010*, ed. by C. Cafieri, E.M.T. Hendrix, L. Liberti, F. Messine (2010), pp. 123–126
1065. E.G. Petrova, A.S. Strekalovsky, The quadratic-linear bilevel problems solving via nonconvex constraint problems (Special Issue: Bilevel programming, optimization methods, and applications to economics). Int. J. Biomed. Soft Comput. Human Sci. Off. J. Biomed. Fuzzy Syst. Assoc. **18**(1), 63–67 (2013)
1066. G. Peyré, J.M. Fadili, Learning Analysis Sparsity Priors, in *Proceedings of the Sampta'11* (2011)
1067. P. Pharkya, A.P. Burgard, C.D. Maranas, Exploring the overproduction of amino acids using the bilevel optimization framework OptKnock. Biotechnol. Bioeng. **84**(7), 887–899 (2003)
1068. G. Pieri, Sufficient conditions for the existence of the solution for bilevel minimization problems with constraints in Banach spaces. Rivista di Matematica Pura ed Applicata **5**, 41–48 (1989)
1069. C.O. Pieume, Multiobjective optimization approaches in bilevel optimization, Ph.D. thesis (Université de Yaounde I, Yaounde, 2011)
1070. C.O. Pieume, L.P. Fotso, P. Siarry, Solving bilevel programming problems with multicriteria optimization techniques. Opsearch **46**, 169–183 (2009)
1071. C.O. Pieume, P. Marcotte, L.P. Fotso, P. Siarry, Solving bilevel linear multiobjective programming problems. Am. J. Oper. Res. **1**, 214–219 (2011)
1072. M. Pilecka, Combined reformulation of bilevel programming problems, Master's thesis (TU Bergakademie Freiberg, Fakultät für Mathematrik und Informatik, Freiberg, 2011)
1073. M. Pilecka, Combined reformulation of bilevel programming problems. Schedae Inf. **2012 21**, 65–79 (2012)
1074. M. Pilecka, Set—valued optimization and its application to bilevel optimization, Ph.D. thesis (TU Bergakademie Freiberg, Freiberg, 2016)

1075. E.A. Pilotta, G.A. Torres, An inexact restoration package for bilevel programming problems. Appl. Math. **2012**(3), 1252–1259 (2012)
1076. S. Pineda, H. Bylling, J.M. Morales, Efficiently solving linear bilevel programming problems using off-the-shelf optimization software. Optim. Eng. **19**(1), 187–211 (2018)
1077. P. Pisciella, Methods for evaluation of business models for provision of advanced mobile services under uncertainty, Ph.D. thesis (Norvegian University of Science and Technology, Norvegian, 2012)
1078. P. Pisciella, A.A. Gaivoronski, Stochastic programming bilevel models for service provision with a balancing coordinator. IMA J. Manag. Math. **28**, 131–152 (2017)
1079. E.N. Pistikopoulos, V. Dua, J.-H. Ryu, Global optimization of bilevel programming problems via parametric programming, in *Frontiers in Global Optimization*, ed. by Floudas, A. Christodoulos, Pardalos, Panos (Springer, New York, 2004), pp. 457–476
1080. F. Plastria, L. Vanhaverbeke, Discrete models for competitive location with foresight. Comput. Oper. Res. **35**(3), 683–700 (2008)
1081. A.V. Plyasunov, A polynomially solvable class of two-level nonlinear programming problems. Diskretnyj Analiz i Issledovanie Operatsii Seriya 2 **7**, 89–113 (2000, in Russian)
1082. A.V. Plyasunov, A two-level linear programming problem with a multivariant knapsack at the lower level. Diskret. Anal. Issled. Oper. **10**(1), 44–52 (2003)
1083. P.-L. Poirion, S. Toubaline, C. D' Ambrosio, L. Liberti, Bilevel mixed-integer linear programs and the zero forcing set, Technical Report (École Polytechnique, Palaiseau, France, 2016)
1084. D. Pozo, E. Sauma, J. Contreras, Basic theoretical foundations and insights on bilevel models and their applications to power systems. Ann. Oper. Res. **254**(1–2), 303–334 (2017)
1085. S. Pramanik, Bilevel programming problem with fuzzy parameters: A fuzzy goal programing approach. J. Appl. Quant. Methods **1**(7), 9–24 (2011)
1086. S. Pramanik, T.K. Roy, Fuzzy goal programming approach to multilevel programming problems. Eur. J. Oper. Res. **176**(2), 1151–1166 (2007)
1087. Pyomo, *Installation, Documentation and Examples*. http://www.pyomo.org/
1088. X. Qiu, G.Q. Huang, Storage pricing, replenishment, and delivery schedules in a supply hub in industrial park: a bilevel programming approach. Int. J. Prod. Res. **51**(23–24), 6950–6971 (2013)
1089. X. Qiu, W. Kern, Improved approximation algorithms for a bilevel knapsack problem. Theor. Comput. Sci. **595**, 120–129 (2015)
1090. N.V. Quy, An algorithm for a class of bilevel split equilibrium problems: application to a differentiated Nash-Cournot model with environmental constraints. Optimization **68**(4), 753–771 (2019)
1091. A. Rahmani, S.A. MirHassani, Lagrangean relaxation-based algorithm for bi-level problems. Optim. Methods Softw. **30**(1), 1–14 (2015)
1092. A. Rahmani, M. Yousefikhoshbakht, An effective branch-and-cut algorithm in order to solve the mixed integer bi-level programming. Int. J. Prod. Manag. Eng. **5**(1), 1–10 (2017)
1093. J. Rajesh, K. Gupta, H.S. Kusumakar, V.K. Jayaraman, B.D. Kulkarni, A tabu search based approach for solving a class of bilevel programming problems in chemical engineering. J. Heuristics **9**(4), 307–319 (2003)
1094. T. Ralphs, E. Adams, Computational optimization research at Lehigh: Bilevel optimization problem library, Technical Report. COR @ L (2005). http://coral.ise.lehigh.edu/data-sets/bilevel-instances/
1095. M.A. Ramos, M. Boix, D. Aussel, L. Montastruc, S. Domenech, Water integration in eco-industrial parks using a multi-leader-follower approach. Comput.Chem. Eng. **87**, 190–207 (2016)
1096. R. Ranftl, T. Pock, A deep variational model for image segmentation, in *Proceedings of the Pattern Recognition* (Springer, Berlin, 2014), pp. 107–118
1097. R. Rees, The theory of principal and agent. part 1. Bull. Econ. Res. **37**, 3–26 (1985)
1098. R. Rees, The theory of principal and agent. part 2. Bull. Econ. Res. **37**, 75–95 (1985)

1099. M. Reisi, S.A. Gabriel, B. Fahimnia, Supply chain competition on shelf space and pricing for soft drinks: a bilevel optimization approach. Int. J. Prod. Econ. **211**, 237–250 (2019)
1100. A. Ren, A novel method for solving the fully fuzzy bilevel linear programming problem. Math. Prob. Eng. **2015**, 11 (2015)
1101. A. Ren, Solving the fully fuzzy bilevel linear programming problem through deviation degree measures and a ranking function method. Math. Prob. Eng. **2016**, 11 (2016)
1102. A. Ren, Y. Wang, A cutting plane method for bilevel linear programming with interval coefficients. Ann. Oper. Res. **223**(1), 355–378 (2014)
1103. A. Ren, Y. Wang, Optimistic Stackelberg solutions to bilevel linear programming with fuzzy random variable coefficients. Knowledge-Based Syst. **67**, 206–217 (2014)
1104. A. Ren, Y. Wang, An interval approach based on expectation optimization for fuzzy random bilevel linear programming problems. J. Oper. Res. Soc. **66**(12), 2075–2085 (2015)
1105. A. Ren, Y. Wang, A novel penalty function method for semivectorial bilevel programming problem. Appl. Math. Model. **40**(1), 135–149 (2016)
1106. A. Ren, Y. Wang, An approach based on reliability-based possibility degree of interval for solving general interval bilevel linear programming problem. Soft Comput. **23**, 997–1006 (2019)
1107. A. Ren, Y. Wang, An approach for solving a fuzzy bilevel programming problem through nearest interval approximation approach and KKT optimality conditions. Soft Comput. **21**(18), 5515–5526 (2017)
1108. A. Ren, Y. Wang, A new approach based on possibilistic programming technique and fractile optimization for bilevel programming in a hybrid uncertain circumstance. Appl. Intell. **48**(10), 3782–3796 (2018)
1109. A. Ren, Y. Wang, X. Xue, Interactive programming approach for solving the fully fuzzy bilevel linear programming problem. Knowledge-Based Syst. **99**, 103–111 (2016)
1110. A. Ren, Y. Wang, X. Xue, A novel approach based on preference-based index for interval bilevel linear programming problem. J. Inequalities Appl. **2017**(1), 112 (2017)
1111. A. Ren, X. Xue, A new solution method for a class of fuzzy random bilevel programming problems, in *Proceedings of the International Conference on Intelligent Information Hiding and Multimedia Signal Processing* (Springer, Berlin, 2017), pp. 233–241
1112. A. Ren, X. Xue, Solution strategy for bilevel linear programming in fuzzy random circumstances, in *Proceedings 13th International Conference on Computational Intelligence and Security (CIS), 2017* (IEEE, New York, 2017), pp. 508–511
1113. A. Ren, X. Xue, A new solving method for fuzzy bilevel optimization with triangular fuzzy coefficients, in *Proceedings 2018 14th International Conference on Computational Intelligence and Security (CIS)* (2018), pp. 50–53
1114. G. Ren, Z. Huang, Y. Cheng, X. Zhao, Y. Zhang, An integrated model for evacuation routing and traffic signal optimization with background demand uncertainty. J. Adv. Transp. **47**(1), 4–27 (2013)
1115. H.-L. Ren, Origin-destination demands estimation in congested dynamic transit networks, in *Proceedings of the International Conference on Management Science and Engineering, 2007 (ICMSE 2007)* (IEEE, New York, 2007), pp. 2247–2252
1116. H. Riahi, Z. Chbani, M.-T. Loumi, Weak and strong convergences of the generalized penalty forward—forward and forward—backward splitting algorithms for solving bilevel hierarchical pseudomonotone equilibrium problems. Optimization **67**, 1745–1767 (2018)
1117. M.J. Rider, J.M. López-Lezama, J. Contreras, A. Padilha-Feltrin, Bilevel approach for optimal location and contract pricing of distributed generation in radial distribution systems using mixed-integer linear programming, IET Generation. Transm. Distrib. **7**(7), 724–734 (2013)
1118. G. Ridinger, R.S. John, M. McBride, N. Scurich, Attacker deterrence and perceived risk in a Stackelberg security game. Risk Anal. **36**(8), 1666–1681 (2016)
1119. R.M. Rizk-Allah, M.A. Abo-Sinna, Integrating reference point, Kuhn–Tucker conditions and neural network approach for multi-objective and multi-level programming problems. OPSEARCH **54**(4), 663–683 (2017)

1120. M.J. Robbins, B.J. Lunday, A bilevel formulation of the pediatric vaccine pricing problem. Eur. J. Oper. Res. **248**(2), 634–645 (2016)
1121. A.J. Robson, Stackelberg and marshall, in *The American Economic Review* (1990), 69–82
1122. S. Roch, P. Marcotte, G. Savard, *Design and Analysis of an Approximation Algorithm for Stackelberg Network Pricing*, Technical report (École Polytechnique de Montréal, (Québec, Canada, 2003)
1123. S. Roch, G. Savard, P. Marcotte, An approximation algorithm for Stackelberg network pricing. Networks **46**(1), 57–67 (2005)
1124. R. Rog, Lösungsalgorithmen für die KKT-Transformation von Zwei-Ebenen-Optimierungsaufgaben, Master's thesis (TU Bergakademie Freiberg, Fakultät für Mathematik und Informatik, Freiberg, 2017)
1125. W.P. Rogerson, The first-order approach to principal-agent problems. Econom. J. Econom. Soc., 1357–1367 (1985)
1126. E. Roghanian, M.B. Aryanezhad, S.J. Sadjadi, Integrating goal programming, Kuhn-Tucker conditions, and penalty function approaches to solve linear bi-level programming problems. Appl. Math. Comput. **195**(2), 585–590 (2008)
1127. E. Roghanian, S.J. Sadjadi, M.-B. Aryanezhad, A probabilistic bi-level linear multi-objective programming problem to supply chain planning. Appl. Math. Comput. **188**(1), 786–800 (2007)
1128. S.A. Ross, The economic theory of agency: the principal's problem. AER **63**, 134–139 (1973)
1129. G. Ruan, The properties for the linear bilevel programming problem. Nat. Sci. J. Xiangtan Univ. **15**, 5–9 (1993, in Chinese)
1130. G. Ruan, An algorithm for the linear bilevel programming problem. Nat. Sci. J. Xiangtan Univ. **16**, 1–5 (1994, in Chinese)
1131. G.Z. Ruan, S.Y. Wang, Y. Yamamoto, S.S. Zhu, Optimality conditions and geometric properties of a linear multilevel programming problem with dominated objective functions. J. Optim. Theory Appl. **123**(2), 409–429 (2004)
1132. S. Ruuska, K. Miettinen, M.M. Wiecek, Connections between single-level and bilevel multiobjective optimization. J. Optim. Theory Appl. **153**(1), 60–74 (2012)
1133. A. Ruziyeva, Fuzzy bilevel programming, Ph.D. thesis (TU Bergakademie Freiberg, Freiberg, 2013)
1134. J.-H. Ryu, V. Dua, E.N. Pistikopoulos, A bilevel programming framework for enterprise-wide process networks under uncertainty. Comput. Chem. Eng. **28**(6–7), 1121–1129 (2004)
1135. S. Sabach, S. Shtern, A first order method for solving convex bilevel optimization problems. SIAM J. Optim. **27**(2), 640–660 (2017)
1136. R. Saboiev, Solution methods for linear bilevel optimization problems, Ph.D. thesis (TU Bergakademie Freiberg, Freiberg, 2016)
1137. S.M. Sadatrasou, M.R. Gholamian, K. Shahanaghi, An application of data mining classification and bi-level programming for optimal credit allocation. Decis. Sci. Lett. **4**, 35–50 (2015)
1138. S. Sadeghi, A. Seifi, E. Azizi, Trilevel shortest path network interdiction with partial fortification. Comput. Ind. Eng. **106**, 400–411 (2017)
1139. A.S. Safaei, S. Farsad, M.M. Paydar, Robust bi-level optimization of relief logistics operations. Appl. Math. Model. **56**, 359–380 (2018)
1140. N. Safaei, M. Saraj, A new method for solving fully fuzzy linear bilevel programming problems. Int. J. Appl. Oper. Res. **4**(1), 39–46 (2014)
1141. K.S. Sagyngaliev, Coordinated resource allocation in a three-level active system. Avtomatika i Telemechanika **10**, 81–88 (1986, in Russian)
1142. G.K. Saharidis, M.G. Ierapetritou, Resolution method for mixed integer bi-level linear problems based on decomposition technique. J. Global Optim. **44**(1), 29–51 (2009)
1143. G.K.D. Saharidis, A.J. Conejo, G. Kozanidis, Exact solution methodologies for linear and (mixed) integer bilevel programming, in *Metaheuristics for Bi-level Optimization*, ed. by E.-G. Talbi, (Springer, Berlin, 2013), pp. 221–245

1144. K.H. Sahin, A.R. Ciric, A dual temperature simulated annealing approach for solving bilevel programming problems. Comput.Chem. Eng. **23**, 11–25 (1998)
1145. M.E. Sáiz, E.M.T. Hendrix, J. Fernández, B. Pelegrín, On a branch-and-bound approach for a huff-like Stackelberg location problem. OR Spectr. **31**(3), 679–705 (2009)
1146. M. Sakawa, *Genetic algorithms and fuzzy multiobjective optimization*, vol. 14 (Springer, Berlin, 2012)
1147. M. Sakawa, H. Katagiri, Stackelberg solutions for fuzzy random two-level linear programming through level sets and fractile criterion optimization. Cent. Eur. J. Oper. Res. **20**(1), 101–117 (2012)
1148. M. Sakawa, H. Katagiri, T. Matsui, Stackelberg solutions for fuzzy random two-level linear programming through probability maximization with possibility. Fuzzy Sets Syst. **188**(1), 45–57 (2012)
1149. M. Sakawa, I. Nishizaki, Interactive fuzzy programming for multi-level nonconvex nonlinear programming problems through genetic algorithms, in *Dynamical Aspects in Fuzzy Decision Making*, ed. by Y. Yoshida, (Springer, Berlin, 2001), pp. 99–116
1150. M. Sakawa, I. Nishizaki, Interactive fuzzy programming for two-level nonconvex programming problems with fuzzy parameters through genetic algorithms. Fuzzy Sets Syst. **127**(2), 185–197 (2002)
1151. M. Sakawa, I. Nishizaki, *Cooperative and Noncooperative Multi-Level Programming*, vol. 48 (Springer, Berlin, 2009)
1152. M. Sakawa, I. Nishizaki, Interactive fuzzy programming for multi-level programming problems: a review. Int. J. Multicriteria Decis. Making **2**(3), 241–266 (2012)
1153. M. Sakawa, I. Nishizaki, M. Hitaka, Interactive fuzzy programming for multi-level 0–1 programming problems through genetic algorithms. Eur. J. Oper. Res. **114**(3), 580–588 (1999)
1154. M. Sakawa, I. Nishizaki, M. Hitaka, Interactive fuzzy programming for multi-level 0–1 programming problems with fuzzy parameters through genetic algorithms. Fuzzy Sets Syst. **117**(1), 95 111 (2001)
1155. M. Sakawa, I. Nishizaki, Y. Uemura, Interactive fuzzy programming for multilevel linear programming problems. Comput. Math. Appl. **36**(2), 71–86 (1998)
1156. M. Sakawa, I. Nishizaki, Y. Uemura, Interactive fuzzy programming for multi-level linear programming problems with fuzzy parameters. Fuzzy Sets Syst. **109**(1), 3–19 (2000)
1157. M. Sakawa, I. Nishizaki, Y. Uemura, Interactive fuzzy programming for two-level linear and linear fractional production and assignment problems: a case study. Eur. J. Oper. Res. **135**, 142–157 (2001)
1158. S.S. Sana, A production-inventory model of imperfect quality products in a three-layer supply chain. Decis. Support Syst. **50**(2), 539–547 (2011)
1159. N.G.F. Sancho, A suboptimal solution to a hierarchical network design problem using dynamic programming. Eur. J. Oper. Res. **83**, 237–244 (1995)
1160. M. Saraj, S. Sadeghi, Quadratic bi-level programming problems: a fuzzy goal programming approach. Int. J. Appl. Oper. Res. **4**(2), 83–88 (2014)
1161. S. Saranwong, C. Likasiri, Bi-level programming model for solving distribution center problem: a case study in Northern Thailand's sugarcane management. Comput. Ind. Eng. **103**, 26–39 (2017)
1162. P. Sariddichainunta, M. Inuiguchi, Global optimality test for maximin solution of bilevel linear programming with ambiguous lower-level objective function. Ann. Oper. Res. **256**(2), 285–304 (2017)
1163. M. Sasaki, J.F. Campbell, M. Krishnamoorthy, A.T. Ernst, A Stackelberg hub ARC location model for a competitive environment. Comput. Oper. Res. **47**, 27–41 (2014)
1164. M. Sasaki, M. Fukushima, Stackelberg hub location problem. J. Oper. Res. Soc. Jpn. **44**(4), 390–402 (2001)
1165. G. Savard, Contribuitions à la programmation mathématique à deux niveaux, Ph.D. thesis, École Polytechnique (Université de Montréal, Montréal, 1989)

1166. G. Savard, J. Gauvin, The steepest descent direction for the nonlinear bilevel programming problem. Oper. Res. Lett. **15**, 265–272 (1994)
1167. M.P. Scaparra, R.L. Church, A bilevel mixed-integer program for critical infrastructure protection planning. Comput. Oper. Res. **35**(6), 1905–1923 (2008)
1168. M.P. Scaparra, R.L. Church, Protecting supply systems to mitigate potential disaster: a model to fortify capacitated facilities. Int. Reg. Sci. Rev. **35**(2), 188–210 (2012)
1169. H. Scheel, S. Scholtes, Mathematical programs with equilibrium constraints: stationarity, optimality, and sensitivity. Math. Oper. Res. **25**, 1–22 (2000)
1170. G. Schenk, A multilevel programming model for determining regional effluent charges, Master's thesis (Department of Industrial Engineering, State University of New York, Buffalo, 1980)
1171. H. Schmidt, Zwei-Ebenen-Opitmierungsaufgaben mit mehrelementiger Lösung der unteren Ebenen, Ph.D. thesis (TU Chemnitz, 1996)
1172. S. Scholtes, M. Stöhr, How stringent is the linear independence assumption for mathematical programs with stationarity constraints?. Math. Oper. Res. **26**, 851–863 (2001)
1173. J. Schulte, N. Feldkamp, S. Bergmann, V. Nissen, Bilevel innovization: knowledge discovery in scheduling systems using evolutionary bilevel optimization and visual analytics, in *Proceedings of the Genetic and Evolutionary Computation Conference Companion* (ACM, New York, 2018), pp. 197–198
1174. J. Schulte, N. Feldkamp, S. Bergmann, V. Nissen, Knowledge discovery in scheduling systems using evolutionary bilevel optimization and visual analytics, in *Proceedings of the International Conference on Evolutionary Multi-Criterion Optimization* (Springer, New York, 2019), pp. 439–450
1175. R. Segall, Bi-level geometric programming: a new optimization model, Technical Report (Department of Mathematics, University of Lowell Olsen Hall, Lowell Olsen Hall, 1989)
1176. R.S. Segall, Using branch-and-bound to solve bi-level geometric programming problems: a new optimization model. Appl. Math. Model. **14**(5), 271–274 (1990)
1177. R.S. Segall, An update on bi-level geometric programming: a new optimization model. Appl. Math. Model. **17**(4), 219–222 (1993)
1178. S.P. Sethi, Q. Zhang, Multilevel hierarchical open-loop and feedback controls in stochastic marketing-production systems. IEEE Trans. Rob. Autom. **10**(6), 831–839 (1994)
1179. Y.V. Shamardin, Three-level problems of allocation of the production, Technical Report 47 (Russian Academy of Sciences, Siberian Branch, Insitut of Mathemetics, Novosibirsk, 1998, in Russian)
1180. Y.V. Shamardin, On a two-level location problem with constraints on the volume of production. Diskret. Anal. Issled. Oper. **7**(2), 114–118 (2000)
1181. H. Shao, W.H.K. Lam, A. Sumalee, A. Chen, M.L. Hazelton, Estimation of mean and covariance of peak hour origin–destination demands from day-to-day traffic counts. Transp. Res. B Methodol. **68**, 52–75 (2014)
1182. A. Sharma, V. Verma, P. Kaur, K. Dahiya, An iterative algorithm for two level hierarchical time minimization transportation problem. Eur. J. Oper. Res. **246**(3), 700–707 (2015)
1183. V. Sharma, K. Dahiya, V. Verma, A class of integer linear fractional bilevel programming problems. Optimization **63**(10), 1565–1581
1184. Y. Sharma, D.P. Williamson, Stackelberg thresholds in network routing games or the value of altruism. Games Econ. Behav. **67**(1), 174–190 (2009)
1185. J. Shaw, A parametric complementary pivot approach to multilevel programming, Master's thesis (Department of Industrial Engineering, State University of New York, Buffalo, 1980)
1186. H. Sherali, A multiple leader Stackelberg model and analysis. Oper. Res. **32**, 390–404 (1984)
1187. H.D. Sherali, A.L. Soyster, F.H. Murphy, Stackelberg-Nash-Cournot equilibria: characterizations and Computations. Oper. Res. **31**, 253–276 (1983)
1188. C. Shi, J. Lu, G. Zhang, An extended Kuhn-Tucker approach for linear bilevel programming. Appl. Math. Comput. **162**, 51–63 (2005)
1189. C. Shi, J. Lu, G. Zhang, An extended Kth-best approach for linear bilevel programming. Appl. Math. Comput. **164**(3), 843–855 (2005)

1190. C. Shi, J. Lu, G. Zhang, H. Zhou, An extended branch and bound algorithm for linear bilevel programming. Appl. Math. Comput. **180**(2), 529–537 (2006)
1191. C. Shi, G. Zhang, J. Lu, The k-th-best approach for linear bilevel multi-follower programming. J. Global Optim. **33**(4), 563–578 (2005)
1192. C. Shi, G. Zhang, J. Lu, On the definition of linear bilevel programming solution. Appl. Math. Comput. **160**, 169–176 (2005)
1193. C. Shi, H. Zhou, J. Lu, G. Zhang, Z. Zhang, The kth-best approach for linear bilevel multifollower programming with partial shared variables among followers. Appl. Math. Comput. **188**(2), 1686–1698 (2007)
1194. H.-S. Shih, Y.-L. Lai, E. S. Lee, Fuzzy approach for multilevel programming problems. Comput. Oper. Res. **23**, 73–91 (1996)
1195. H.-S. Shih, U.-P. Wen, S. Lee, K.-M. Lan, H.-C. Hsiao, A neural network approach to multiobjective and multilevel programming problems. Comput. Math. Appl. **48**(1), 95–108 (2004)
1196. H.S. Shih, C.B. Cheng, U.P. Wen, Y.C. Huang, M.Y. Peng, Determining a subsidy rate for Taiwan's recycling glass industry: an application of bi-level programming. J. Oper. Res. Soc. **63**(1), 28–37 (2012)
1197. K. Shimizu, Two-level decision problems and their new solution methods by a penalty method, in *Proceedings of the Control Science and Technology for the Progress of Society*, vol. 2 (IFAC, New York, 1982), pp. 1303–1308
1198. K. Shimizu, E. Aiyoshi, A new computational method for Stackelberg and min-max problems by use of a penalty method. IEEE Trans. Autom. Control **26**, 460–466 (1981)
1199. K. Shimizu, Y. Ishizuka, Optimality conditions and algorithms for parameter design problems with two-level structure. IEEE Trans. Autom. Control **30**(10), 986–993 (1985)
1200. K. Shimizu, Y. Ishizuka, J.F. Bard, *Nondifferentiable and Two-Level Mathematical Programming* (Kluwer Academic Publishers, Dordrecht, 1997)
1201. K. Shimizu, M. Lu, A global optimization method for the Stackelberg problem with convex functions via problem transformations and concave programming. IEEE Trans. Syst. Man Cybern. **25**, 1635–1640 (1995)
1202. C. Shouhua, Y. Zhenzhou, L. Yanhong, W. Xianyu, Model for road network stochastic user equilibrium based on bi-level programming under the action of the traffic flow guidance system. J. Transp. Syst. Eng. Inf. Technol. **7**(4), 36–42 (2007)
1203. S.A. Siddiqui, Solving two-level optimization problems with applications to robust design and energy markets, Ph.D. thesis (University of Maryland, Maryland, 2011)
1204. M. Simaan, Stackelberg optimization of two-level systems. IEEE Trans. Syst. Man Cybern. **7**, 554–557 (1977)
1205. M. Simaan, J.B. Cruz, On the Stackelberg strategy in nonzero-sum games. J. Optim. Theory Appl. **11**, 533–555 (1973)
1206. M. Simaan, J.B. Cruz Jr., On the Stackelberg Strategy in Nonzero-Sum Games, in *Multicriteria Decision Making and Differential Games*, ed. by G. Leitmann. Mathematical Concepts and Methods in Science and Engineering (Springer, New York, 1976), pp. 173–195
1207. B. Sinclair-Desagne, The first-order approach to multi-signal principal-agent systems. Econometrica **62**, 459–465 (1994)
1208. S. Singh, An approach to solve bilevel quadratic-linear programming problems, in *Proceedings of the International MultiConference of Engineers and Computer Scientists*, Lecture Notes in Engineering and Computer Science, vol. 2196 (2012), pp. 1473–1476
1209. V.P. Singh, D. Chakraborty, Solving bi-level programming problem with fuzzy random variable coefficients. J. Int. Fuzzy Syst. **32**(1), 521–528 (2017)
1210. A. Sinha, S. Bedi, K. Deb, Bilevel optimization based on kriging approximations of lower level optimal value function, in *Proceedings of the 2018 IEEE Congress on Evolutionary Computation (CEC)* (IEEE, New York, 2018), pp. 1–8
1211. A. Sinha, K. Deb, Towards understanding evolutionary bilevel multi-objective optimization algorithm. IFAC Proc. Volumes **42**(2), 338–343 (2009)

1212. A. Sinha, P. Malo, K. Deb, An improved bilevel evolutionary algorithm based on quadratic approximations, in *Proceedings of the IEEE Congress on Evolutionary Computation (CEC)* (IEEE, New York, 2014), pp. 1870–1877
1213. A. Sinha, P. Malo, K. Deb, Test problem construction for single-objective bilevel optimization. Evol. Comput. **22**(3), 439–477 (2014)
1214. A. Sinha, P. Malo, K. Deb, Towards understanding bilevel multi-objective optimization with deterministic lower level decisions, in *Proceedings of the International Conference on Evolutionary Multi-Criterion Optimization* (Springer, Berlin, 2015), pp. 426–443
1215. A. Sinha, P. Malo, K. Deb, Transportation policy formulation as a multi-objective bilevel optimization problem, *IEEE Congress on Evolutionary Computation (CEC)* (IEEE, New York, 2015), pp. 1651–1658
1216. A. Sinha, P. Malo, K. Deb, Solving optimistic bilevel programs by iteratively approximating lower level optimal value function, in *Proceedings of the IEEE Congress on Evolutionary Computation (CEC)* (IEEE, New York, 2016), pp. 1877–1884
1217. A. Sinha, P. Malo, K. Deb, Approximated set-valued mapping approach for handling multiobjective bilevel problems. Comput. Oper. Res. **77**, 194–209 (2017)
1218. A. Sinha, P. Malo, K. Deb, Evolutionary algorithm for bilevel optimization using approximations of the lower level optimal solution mapping. Eur. J. Oper. Res. **257**(2), 395–411 (2017)
1219. A. Sinha, P. Malo, K. Deb, Evolutionary bilevel optimization: an introduction and recent advances, in *Recent Advances in Evolutionary Multi-objective Optimization*, ed. by S. Bechikh, R. Datta, A. Gupta (Springer, New York, 2017), pp. 71–103
1220. A. Sinha, P. Malo, K. Deb, A review on bilevel optimization: from classical to evolutionary approaches and applications. IEEE Trans. Evol. Comput. **22**(2), 276–295 (2018)
1221. A. Sinha, P. Malo, K. Deb, P. Korhonen, J. Wallenius, Solving bilevel multicriterion optimization problems with lower level decision uncertainty. IEEE Trans. Evol. Comput. **20**(2), 199–217 (2016)
1222. A. Sinha, P. Malo, A. Frantsev, K. Deb, Multi-objective Stackelberg game between a regulating authority and a mining company: a case study in environmental economics, in *Proceedings of the IEEE Congress on Evolutionary Computation (CEC)* (IEEE, New York, 2013), pp. 478–485
1223. A. Sinha, P. Malo, A. Frantsev, K. Deb, Finding optimal strategies in a multi-period multi-leader–follower Stackelberg game using an evolutionary algorithm. Comput. Oper. Res. **41**, 374–385 (2014)
1224. A. Sinha, P. Malo, P. Xu, K. Deb, A bilevel optimization approach to automated parameter tuning, in *Proceedings of the 2014 Annual Conference on Genetic and Evolutionary Computation* (ACM, New York, 2014), pp. 847–854
1225. A. Sinha, T. Soun, K. Deb, Evolutionary bilevel optimization using KKT proximity measure, in *Proceedings of the IEEE Congress on Evolutionary Computation (CEC), 2017* (IEEE, New York, 2017), pp. 2412–2419
1226. A. Sinha, T. Soun, K. Deb, Using Karush-Kuhn-Tucker proximity measure for solving bilevel optimization problems. Swarm Evol. Comput. **44**, 496–510 (2019)
1227. S. Sinha, A comment on Anandalingam (1988). A mathematical programming model of decentralized multi-level systems. J. Oper. Res. Soc. **52**(5), 594–596 (2001)
1228. S. Sinha, Fuzzy mathematical programming applied to multi-level programming problems. Comput. Oper. Res. **30**(9), 1259–1268 (2003)
1229. S. Sinha, Fuzzy programming approach to multi-level programming problems. Fuzzy Sets Syst. **136**, 189–202 (2003)
1230. B. Sixou, Y. Li, F. Peyrin, Determination of blur kernel for HR-pQCT with bilevel optimization, in *Journal of Physics: Conference Series*, vol. 1131 (IOP Publishing, Bristol, 2018)
1231. O. Skulovich, L. Perelman, A. Ostfeld, Bi-level optimization of closed surge tanks placement and sizing in water distribution system subjected to transient events. Procedia Eng. **89**, 1329–1335 (2014)

1232. J.C. Smith, C. Lim, Algorithms for network interdiction and fortification games, in *Pareto Optimality Game Theory Equilibria*, ed. by P. Pardalos, A. Migdalas, L. Pitsoulis (Springer, Berlin, 2008), pp. 609–644
1233. W.R. Smith, R.W. Missen, *Chemical Reaction Equilibrium Analysis: Theory and Algorithms* (Wiley, New York, 1982)
1234. M. Soismaa, A note on efficient solutions for the linear bilevel programming problem. Eur. J. Oper. Res. **112**, 427–431 (1999)
1235. M.V. Solodov, A bundle method for a class of bilevel nonsmooth convex minimization problems. SIAM J. Optim. **18**, 242–259 (2007)
1236. M.V. Solodov, An explicit descent method for bilevel convex optimization. J. Convex Anal. **14**(2), 227–237 (2007)
1237. H.-M. Song, H. Yang, A. Bensoussan, Optimizing production and inventory decisions in a supply chain with lot size, production rate and lead time interactions. Appl. Math. Comput. **224**, 150–165 (2013)
1238. K.A. Sonia, A. Khandelwal, M.C. Puri, Bilevel time minimizing transportation problem. Discrete Optim. **5**(4), 714–723 (2008)
1239. K.A. Sonia, M.C. Puri, Two level hierarchical time minimizing transportation problem. Top **12**(2), 301–330 (2004)
1240. K.A. Sonia, M.C. Puri, Bilevel time minimizing assignment problem. Appl. Math. Comput. **183**(2), 990–999 (2006)
1241. W. Sosa, F. Raupp, On optimization over weakly efficient sets. Optimization **56**, 207–219 (2007)
1242. R. Sousa, N. Shah, L.G. Papageorgiou, Supply chain design and multilevel planning—an industrial case. Comput.Chem. Eng. **32**(11), 2643–2663 (2008)
1243. P. Sprechmann, A.M. Bronstein, G. Sapiro, Supervised non-euclidean sparse NMF via bilevel optimization with applications to speech enhancement, in *Proceedings of the 4th Joint Workshop on Hands-free Speech Communication and Microphone Arrays (HSCMA), 2014* (IEEE, New York, 2014), pp. 11–15
1244. S. Srivastava, S.K. Sahana, Nested hybrid evolutionary model for traffic signal optimization. Appl. Intell. **46**(1), 113–123 (2017)
1245. H.v. Stackelberg, *Marktform und Gleichgewicht* (Springer, Wien, 1934). Engl. transl.: *The Theory of the Market Economy* (Oxford University, Oxford, 1952)
1246. W. Stanford, Pure strategy Nash equilibria and the probabilistic prospects of Stackelberg players. Oper. Res. Lett. **38**(2), 94–96 (2010)
1247. T. Starostina, S. Dempe, Sensitivity analysis for fuzzy shortest path problem, in *Computational Intelligence, Theory and Applications*, ed. by B. Reusch (Springer, New York, 2005), pp. 695–702
1248. S. Steffensen, Global solution of bilevel programming problems, in *Operations Research Proceedings 2014* (Springer, New York, 2016), pp. 575–580
1249. O. Stein, *Bi-level Strategies in Semi-Infinite Programming* (Kluwer Akademic Publishers, Boston, 2003)
1250. O. Stein, G. Still, On generalized semi-infinite optimization and bilevel optimization. Eur. J. Oper. Res. **142**(3), 444–462 (2002)
1251. W.J. Steiner, A Stackelberg-Nash model for new product design. OR Spectr. **32**(1), 21–48 (2010)
1252. G. Still, Linear bilevel problems: genericity results and an efficient method for computing local minima. Math. Methods Oper. Res. **55**(3), 383–400 (2002)
1253. K. Stoilova, Fast resource allocation by bilevel programming problem, in *Proceedings of the International IFAC workshop DECOM-TT 2004* (Bansko, Bulgaria, 2004), pp. 249–254
1254. K. Stoilova, T. Stoilov, V. Ivanov, Bi-level optimization as a tool for implementation of intelligent transportation systems. Cybern. Inf. Technol. **17**(2), 97–105 (2017)
1255. K. Stoilova, T.Stoilov, Predictive coordination in two level hierarchical systems, in *Proceedings of the IEEE Symposium Intelligent Systems, 10–12 September 2002, Varna*, vol. I (2002), pp. 332–337

1256. A. Street, A. Moreira, J.M. Arroyo, Energy and reserve scheduling under a joint generation and transmission security criterion: an adjustable robust optimization approach. IEEE Trans. Power Syst. **29**(1), 3–14 (2014)
1257. A.S. Strekalovskii, A.V. Orlov, A.V. Malyshev, A local search for the quadratic-linear bilevel programming problem. Sibirskii Zhurnal Vychislitel'noi Matematiki **13**(1), 75–88 (2010)
1258. A.S. Strekalovskii, A.V. Orlov, A.V. Malyshev, Numerical solution of a class of bilevel programming problems. Sibirskii Zhurnal Vychislitel'noi Matematiki **13**(2), 201–212 (2010)
1259. A.S. Strekalovsky, Methods for solving the bilevel optimization problems, in *Proceedings of the II International Conference OPTIMA-2011 (Petrovac, Montenegro)* (2011), pp. 205–208
1260. A.S. Strekalovsky, A.V. Orlov, A.V. Malyshev, Local search in a quadratic-linear bilevel programming problem. Numer. Anal. Appl. **3**(1), 59–70 (2010)
1261. A.S. Strekalovsky, A.V. Orlov, A.V. Malyshev, Numerical solution of a class of bilevel programming problems. Numer. Anal. Appl. **3**(2), 165–173 (2010)
1262. A.S. Strekalovsky, A.V. Orlov, A.V. Malyshev, On computational search for optimistic solutions in bilevel problems. J. Global Optim. **48**(1), 159–172 (2010)
1263. S. Suh, T. Kim, Solving nonlinear bilevel programming models of the equilibrium network design problem: a comparative review. Ann. Oper. Res. **34**, 203–218 (1992)
1264. H. Sun, Z. Gao, J. Wu, A bi-level programming model and solution algorithm for the location of logistics distribution centers. Appl. Math. Model. **32**(4), 610–616 (2008)
1265. S.K. Suneja, B. Kohli, Optimality and duality results for bilevel programming problem using convexifactors. J. Optim. Theory Appl. **150**(1), 1–19 (2011)
1266. C. Suwansirikul, T. Friesz, R. Tobin, Equilibrium decomposed optimization: a heuristic for the continuous equilibrium network design problem. Transp. Sci. **21**, 254–263 (1987)
1267. A.F. Taha, N.A. Hachem, J.H. Panchal, A quasi-feed-in-tariff policy formulation in micro-grids: a bi-level multi-period approach. Energy Policy **71**, 63–75 (2014)
1268. S. Tahernejad, T.K. Ralphs, S.T. DeNegre, A branch-and-cut algorithm for mixed integer bilevel linear optimization problems and its implementation. Math. Program. Comput. 1–40 (2020)
1269. A. Takeda, M. Kojima, Successive convex relaxation approach to bilevel quadratic optimization problems, in *Complementarity: applications, algorithms and extensions*, ed. by M.C. Ferris, O.L. Mangasarian, J.-S. Pang (Kluwer, Dordrecht, 2001), pp. 317–340
1270. E.-G. Talbi, *Metaheuristics for Bi-level Optimization*. Studies in Computational Intelligence (482), ed. by E.-G. Talbi (Springer, Berlin, 2013)
1271. E.-G. Talbi, A taxonomy of metaheuristics for bi-level optimization, in *Metaheuristics for Bi-level Optimization*, ed. by E.-G. Talbi (Springer, Berlin, 2013), pp. 1–39
1272. M.L. Tam, W.H.K. Lam, Balance of car ownership under user demand and road network supply conditions. case study in Hong Kong. J. Urban Plann. Dev. **130**(1), 24–36 (2004)
1273. R.R. Tan, K.B. Aviso, J.B. Cruz, A.B. Culaba, A note on an extended fuzzy bi-level optimization approach for water exchange in ECO-industrial parks with hub topology. Process. Saf. Environ. Prot. **89**(2), 106–111 (2011)
1274. Y. Tang, J.-P.P. Richard, J.C. Smith, A class of algorithms for mixed-integer bilevel min–max optimization. J. Global Optim. **66**(2), 225–262 (2016)
1275. Z. Tao, A stochastic bilevel programming model for the iron and steel production optimization problem under carbon trading mechanism, in *Proceedings of the 10th International Conference on Management Science and Engineering Management* (Springer, Berlin, 2017), pp. 699–710
1276. Z.C. Taskin, Algorithms for solving multi-level optimization problems with discrete variables at multiple levels, Ph.D. thesis (University of Florida, Florida, 2009)
1277. C. Tawfik, S. Limbourg, Bilevel optimization in the context of intermodal pricing: state of art. Transp. Res. Procedia **10**, 634–643 (2015)
1278. A. Tesoriere, Stackelberg equilibrium with multiple firms and setup costs. J. Math. Econ. **73**, 86–102 (2017)
1279. P.T. Thach, T.V. Thang, Problems with resource allocation constraints and optimization over the efficient set. J. Global Optim. **58**(3), 481–495 (2014)

1280. J. Thai, R. Hariss, A. Bayen, A multi-convex approach to latency inference and control in traffic equilibria from sparse data, in *Proceedings of the American Control Conference (ACC), 2015* (IEEE, New York, 2015), pp. 689–695
1281. H.A. Le Thi, T.P. Dinh, L.D. Muu, Simplicially-constrained D.C. optimization over efficient and weakly efficient sets. J. Optim. Theory Appl. **117**, 503–531 (2003)
1282. D. Thirwani, S.R. Arora, An algorithm for the integer linear fractional bilevel programming problem. Optimization **39**(1), 53–67 (1997)
1283. N.V. Thoai, Reverse convex programming approach in the space of extreme criteria for optimization over efficient sets. J. Optim. Theory Appl. **147**(2), 263–277 (2010)
1284. L.Q. Thuy, T.N. Hai, A projected subgradient algorithm for bilevel equilibrium problems and applications. J. Optim. Theory Appl. **175**, 411–431 (2017)
1285. S.L. Tilahun, S.M. Kassa, H.C. Ong, A new algorithm for multilevel optimization problems using evolutionary strategy, inspired by natural adaptation, in *Proceedings of the PRICAI 2012: Trends in Artificial Intelligence (Berlin, Heidelberg)*, ed. by P. Anthony, M. Ishizuka, D. Lukose (Springer, Berlin, 2012), pp. 577–588
1286. F. Tiryaki, Interactive compensatory fuzzy programming for decentralized mult-level linear programming (DMLLP) problems. Fuzzy Sets Syst. **157**, 3072–3090 (2006)
1287. R. Tobin, T. Friesz, Spatial competition facility location models: definition, formulation and solution approach. Ann. Oper. Res. **6**, 49–74 (1986)
1288. R.L. Tobin, Uniqueness results and algorithms for Stackelberg-Cournot-Nash equilibrium. Ann. Oper. Res. **34**, 21–36 (1992)
1289. B. Tolwinski, Closed-loop Stackelberg solution to a multistage linear-quadratic game. J. Optim. Theory Appl. **34**, 485–501 (1981)
1290. C.A. Tovey, Asymmetric probabilistic prospects of Stackelberg players. J. Optim. Theory Appl. **68**, 139–159 (1991)
1291. F. Tramontana, L. Gardini, T. Puu, Mathematical properties of a discontinuous Cournot—Stackelberg model. Chaos Solitons Fractals **44**(1), 58–70 (2011)
1292. K.K. Trejo, J.B. Clempner, A.S. Poznyak, An optimal strong equilibrium solution for cooperative multi-leader-follower Stackelberg Markov chains games. Kybernetika **52**(2), 258–279 (2016)
1293. R. Trujillo-Cortez, S. Zlobec, Bilevel convex programming models. Optimization **58**(8), 1009–1028 (2009)
1294. A. Tsoukalas, B. Rustem, E.N. Pistikopoulos, A global optimization algorithm for generalized semi-infinite, continuous minimax with coupled constraints and bi-level problems. J. Global Optim. **44**(2), 235–250 (2009)
1295. A. Tsoukalas, W. Wiesemann, B. Rustem, Global optimisation of pessimistic bi-level problems. Lect. Global Optim. **55**, 215–243 (2009)
1296. T.V. Tu, Optimization over the efficient set of a parametric multiple objective linear programming problem. Eur. J. Oper. Res. **122**, 570–583 (2000)
1297. H. Tuy, Bilevel linear programming, multiobjective programming, and monotonic reverse convex programming, in *Multilevel Optimization: Algorithms and Applications*, ed. by A. Migdalas, P.M. Pardalos, P. Värbrand (Kluwer Academic, Dordrecht, 1998), pp. 295–314
1298. H. Tuy, S. Ghannadan, A new branch and bound method for bilevel linear programs, in *Multilevel Optimization: Algorithms and Applications*, ed. by A. Migdalas, P.M. Pardalos, P. Värbrand (Kluwer Academic, Dordrecht, 1998), pp. 231–249
1299. H. Tuy, A. Migdalas, N.T. Hoai-Phuong, A novel approach to bilevel nonlinear programming. J. Global Optim. **38**(4), 527–554 (2007)
1300. H. Tuy, A. Migdalas, P. Värbrand, A global optimization approach for the linear two-level program. J. Global Optim. **3** , 1–23 (1993)
1301. H. Tuy, A. Migdalas, P. Värbrand, A quasiconcave minimization method for solving linear two-level programs. J. Global Optim. **4**, 243–263 (1994)
1302. F. Ugranli, E. Karatepe, A.H. Nielsen, MILP approach for bilevel transmission and reactive power planning considering wind curtailment. IEEE Trans. Power Syst. **32**(1), 652–661 (2017)

1303. S. Ukkusuri, K. Doan, H.M.A. Aziz, A bi-level formulation for the combined dynamic equilibrium based traffic signal control. Procedia Soc. Behav. Sci. **80**, 729–752 (2013)
1304. G. Ünlü, A linear bilevel programming algorithm based on bicriteria programming. Comput. Oper. Res. **14**, 173–179 (1987)
1305. T. Uno, H. Katagiri, K. Kato, An evolutionary multi-agent based search method for Stackelberg solutions of bilevel facility location problems. Int. J. Innovative Comput. Inf. Control **4**(5), 1033–1042 (2008)
1306. B. Vahdani, M. Soltani, M. Yazdani, S.M. Mousavi, A three level joint location-inventory problem with correlated demand, shortages and periodic review system: robust meta-heuristics. Comput. Ind. Eng. **109**, 113–129 (2017)
1307. M. Vahid-Ghavidel, N. Mahmoudi, B. Mohammadi-Ivatloo, Self-scheduling of demand response aggregators in short-term markets based on information gap decision theory. IEEE Trans. Smart Grid **10**(2), 2115–2126 (2018)
1308. B. Van Dinh, L.D. Muu, On penalty and gap function methods for bilevel equilibrium problems. J. Appl. Math. **2011** (2011)
1309. M. Červinka, Oligopolistic markets in terms of equilibrium problems with equilibrium constraints, Technical Report (Charles University in Prague, Faculty of Social Sciences, Prague, 2006). Bachelor Thesis
1310. L. Vicente, Bilevel programming, Master's thesis (Department of Mathematics, University of Coimbra, Coimbra, 1992)
1311. L.N. Vicente, Bilevel programming: introduction, history, and overviev, in *Encyclopedia of Optimization*, ed. by P.M. Pardalos et al. (Kluwer Academic, Dordrecht, 2001), pp. 178–180
1312. L.N. Vicente, P.H. Calamai, Bilevel and multilevel programming: a bibliography review. J. Global Optim. **5**(3), 291–306 (1994)
1313. L.N. Vicente, P.H. Calamai, Geometry and local optimality conditions for bilevel programs with quadratic strictly convex lower levels, in *Minimax and Applications*, ed. by D.-Z. Du, P.M.Pardalos (Kluwer Academic, Dordrecht, 1995), pp. 141–151
1314. L.N. Vicente, G. Savard, J.J. Júdice, Descent approaches for quadratic bilevel programming. J. Optim. Theory Appl. **81**(2), 379–399 (1994)
1315. L.N. Vicente, G. Savard, J.J. Júdice, Discrete linear bilevel programming problem. J. Optim. Theory Appl. **89**(3), 597–614 (1996)
1316. S. Vogel, Zwei-Ebenen-Optimierungsaufgaben mit nichtkonvexer Zielfunktion in der unteren Ebene: Pfadverfolgung und Spruenge, Ph.D. thesis (TU Bergakademie Freiberg, Freiberg, 2002)
1317. S. Vogel, S. Dempe, Pathfollowing and jumps in bilevel programming, in *Operations Research Proceedings 1999*, ed. by K. Inderfurth et al. (Springer, Berlin, 2000), pp. 30–35
1318. V.Visweswaran, C.A. Floudas, M.G. Ierapetritou, E.N. Pistikopoulos, A decomposition-based global optimization approach for solving bilevel linear and quadratic programs, in *State of the Art in Global Optimization: Computational Methods and Applications*, ed. by C.A. Floudas, P.M. Pardalos (Kluwer Academic Publishers, Dordrecht, 1996)
1319. Z. Wan, Some approximating results on bilevel programming problems. J. Syst. Sci. Syst. Eng. **20** , 289–294 (2000)
1320. Z. Wan, J.-W. Chen, On bilevel variational inequalities. J. Oper. Res. Soc. China **1**(4), 483–510 (2013)
1321. Z. Wan, M. Jiang, T. Hu, Approximate decomposition algorithm for solving the bilevel programming with the minimum risk. J. Eng. Math. (Xi'an) **17**, 25–30 (2000)
1322. Z. Wan, L. Mao, G. Wang, Estimation of distribution algorithm for a class of nonlinear bilevel programming problems. Inf. Sci. **256**, 184–196 (2014)
1323. Z. Wan, G. Wang, Y. Lv, A dual-relax penalty function approach for solving nonlinear bilevel programming with linear lower level problem. Acta Math. Sci. **31**(2), 652–660 (2011)
1324. Z. Wan, G. Wang, B. Sun, A hybrid intelligent algorithm by combining particle swarm optimization with chaos searching technique for solving nonlinear bilevel programming problems. Swarm Evol. Comput. **8**, 26–32 (2013)

1325. Z. Wan, S. Zhou, The convergence of approach penalty function method for approximate bilevel programming problem. Acta Math. Sci. Ser. B (English Edition) **21**, 69–76 (2001)
1326. B. Wang, X.-Z. Zhou, J. Watada, A unit commitment-based fuzzy bilevel electricity trading model under load uncertainty. Fuzzy Optim. Decis. Making **15**(1), 103–128 (2016)
1327. C.-Y. Wang, K.-C. Yen, S.-R. Hu, C.-P. Chu, Y.-T. Jhuang, A network signal timing design bilevel optimization model with traveler trip-chain route choice behavior consideration. J. Traffic Transp. Eng. **5**, 203–216 (2017)
1328. F.-S. Wang, Nested differential evolution for mixed-integer bi-level optimization for genome-scale metabolic networks. Differ. Evol. Chem. Eng. Dev. Appl. **6**, 352 (2017)
1329. G. Wang, Z. Gao, M. Xu, H. Sun, Models and a relaxation algorithm for continuous network design problem with a tradable credit scheme and equity constraints. Comput. Oper. Res. **41**, 252–261 (2014)
1330. G. Wang, L. Ma, J. Chen, A bilevel improved fruit fly optimization algorithm for the nonlinear bilevel programming problem. Knowledge-Based Syst. **138**(Supplement C), 113–123 (2017)
1331. G. Wang, Z. Wan, X. Wang, Y. Lv, Genetic algorithm based on simplex method for solving linear-quadratic bilevel programming problem. Comput. Math. Appl. **56**(10), 2550–2555 (2008)
1332. G. Wang, X. Wang, Z. Wan, Y. Lv, A globally convergent algorithm for a class of bilevel nonlinear programming problem. Appl. Math. Comput. **188**(1), 166–172 (2007)
1333. G.-M. Wang, Z. Wan, X.-J. Wang, Bibliography on bilevel programming. Adv. Math. **36**(5), 513–529 (2007)
1334. J.Y.T. Wang, M. Ehrgott, K.N. Dirks, A. Gupta, A bilevel multi-objective road pricing model for economic, environmental and health sustainability. Transp. Res. Procedia **3**, 393–402 (2014)
1335. L. Wang, P. Xu, The watermelon algorithm for the bilevel integer linear programming problem. SIAM J. Optim. **27**(3), 1403–1430 (2017)
1336. M. Wang, R. Zhang, X. Zhu, A bi-level programming approach to the decision problems in a vendor-buyer eco-friendly supply chain. Comput. Ind. Eng. **105**, 299–312 (2017)
1337. Q. Wang, S. Wang, Bilevel programs with multiple potential reactions. J. Syst. Sci. Syst. Eng. **3**(3), (1994)
1338. S. Wang, F.A. Lootsma, A hierarchical optimization model of resource allocation. Optimization **28**, 351–365 (1994)
1339. S. Wang, Q. Meng, H. Yang, Global optimization methods for the discrete network design problem. Transp. Res. B Methodol. **50**, 42–60 (2013)
1340. S. Wang, Q. Wang, S. Romano-Rodriquez, Optimality conditions and an algorithm for linear-quadratic bilevel programming. Optimization **31**, 127–139 (1994)
1341. S.-Y. Wang, Q. Wang, L.C. Uria, A stability theorem in nonlinear bilevel programming. Questiió: Quaderns d'Estadística, Sistemes, Informatica i Investigació Operativa **20**(2), 215–222 (1996)
1342. X. Wang, P.M. Pardalos, A modified active set algorithm for transportation discrete network design bi-level problem. J. Global Optim. **67**(1–2), 325–342 (2017)
1343. X. Wang, Y. Wang, Y. Cui, An energy-aware bi-level optimization model for multi-job scheduling problems under cloud computing. Soft Comput. **20**(1), 303–317 (2016)
1344. X. Wang, Y. Wang, Y. Cui, A new multi-objective bi-level programming model for energy and locality aware multi-job scheduling in cloud computing. Future Gener. Comput. Syst. **36**, 91–101 (2014)
1345. Y. Wang, Y. Dvorkin, R. Fernandez-Blanco, B. Xu, T. Qiu, D. Kirschen, Look-ahead bidding strategy for energy storage. IEEE Trans. Sustainable Energy **8**(3), 1106–1117 (2017)
1346. Y. Wang, Y.-C. Jiao, H. Li, An evolutionary algorithm for solving nonlinear bilevel programming based on a new constraint-handling scheme. IEEE Trans. Syst. Man Cybern. Part C Appl. Rev. **35**(2), 221–232 (2005)

1347. Wang, H. Li, C. Dang, A new evolutionary algorithm for a class of nonlinear bilevel programming problems and its global convergence. INFORMS J. Comput. **23**(4), 618–629 (2011)
1348. Y. Wang, S. Liu, B. Zeng, Capacity expansion planning of wind power generation in a market environment with topology control (2017). arXiv preprint:1701.03172
1349. Y.B. Wang, D. Liu, X.C. Cao, Z.Y. Yang, J.F. Song, D.Y. Chen, S.K. Sun, Agricultural water rights trading and virtual water export compensation coupling model: a case study of an irrigation district in China. Agric. Water Manage. **180**, 99–106 (2017)
1350. Z.-W. Wang, H. Nagasawa, N. Nishiyama, An algorithm for a multiobjective, multilevel linear programming. J. Oper. Res. Soc. Jpn. **39**(2), 176–187 (1996)
1351. R. Wangkeeree, P. Yimmuang, Existence and algorithms for the bilevel new generalized mixed equilibrium problems in Banach spaces. Appl. Math. Comput. **219**(6), 3022–3038 (2012)
1352. J.D. Weber, T.J. Overbye, A two-level optimization problem for analysis of market bidding strategies, in *Proceedings of the Power Engineering Society Summer Meeting, 1999*, vol. 2 (IEEE, New York, 1999), pp. 682–687
1353. H.M. Wee, M.C. Lee, P.C. Yang, R.L. Chung, Bi-level vendor—buyer strategies for a time-varying product price. Appl. Math. Comput. **219**(18), 9670–9680 (2013)
1354. M. Weibelzahl, A. Märtz, Optimal storage and transmission investments in a bilevel electricity market model. Annal. Oper. Res. **287**(2), 911–940 (2020)
1355. U. Wen, Mathematical methods for multilevel linear programming, Ph.D. thesis (Department of Industrial Engineering, State University of New York, Buffalo, 1981)
1356. U. Wen, The "Kth-Best" algorithm for multilevel programming, Technical Report (Department of Operations Research, State University of New York, Buffalo, 1981)
1357. U. Wen, A solution procedure for the resource control problem in two-level hierarchical decision processes. J. Chin. Inst. Eng. **6**, 91–97 (1983)
1358. U. Wen, W. Bialas, The hybrid algorithm for solving the three-level linear programming problem. Comput. Oper. Res. **13**, 367–377 (1986)
1359. U. Wen, S. Hsu, A note on a linear bilevel programming algorithm based on bicriteria programming. Comput. Oper. Res. **16**, 79–83 (1989)
1360. U. Wen, S. Hsu, Linear bi-level programming problems—a review. J. Oper. Res. Soc. **42**, 125–133 (1991)
1361. U. Wen, S. Hsu, Efficient solutions for the linear bilevel programming problem. Eur. J. Oper. Res. **62**, 354–362 (1992)
1362. U. Wen, S.-F. Lin, Finding an efficient solution to linear bilevel programming problem: an effective approach. J. Global Optim. **8**, 295–306 (1996)
1363. U. Wen, Y. Yang, Algorithms for solving the mixed integer two-level linear programming problem. Comput. Oper. Res. **17**, 133–142 (1990)
1364. U.P. Wen, A.D. Huang, A simple tabu search method to solve the mixed-integer linear bilevel programming problem. Eur. J. Oper. Res. **88**, 563–571 (1996)
1365. A. Werner, Bilevel stochastic programming problems: Analysis and application to telecommunications, Ph.D. thesis (Section of Investment, Finance and Accounting, Department of Industrial Economics and Technology Management, Norwegian University of Science and Technology, Trondheim, 2005)
1366. D.J. White, Multilevel programming, rational reaction sets, and efficient solutions. J. Optim. Theory Appl. **87**, 727–746 (1995)
1367. D.J. White, Penalty function approach to linear trilevel programming. J. Optim. Theory Appl. **93**, 183–197 (1997)
1368. D.J. White, G. Anandalingam, A penalty function approach for solving bi-level linear programs. J. Global Optim. **3**, 397–419 (1993)
1369. G. Whittaker, R. Färe, S. Grosskopf, B. Barnhart, M.B. Bostian, G. Muller-Warrant, S. Griffith, Spatial targeting of agri-environmental policy using bilevel evolutionary optimization. Omega **66**, 15–27 (2017)

1370. W. Wiesemann, A. Tsoukalas, P.-M. Kleniati, B. Rustem, Pessimistic bilevel optimization. SIAM J. Optim. **23**, 353–380 (2013)
1371. R. Winter, *Zwei-Ebenen-Optimierung mit Stetigem Knapsack-Problem in der unteren Ebene: Optimistischer und pessimistischer Zugang* (Bachelorarbeit, TU Bergakademie Freiberg, 2010)
1372. A.T. Woldemariam, S.M. Kassa, Systematic evolutionary algorithm for general multilevel Stackelberg problems with bounded decision variables (seamsp). Ann. Oper. Res. **229**(1), 771–790 (2015)
1373. R.K. Wood, Deterministic network interdiction. Math. Comput. Model. **17**(2), 1–18 (1993)
1374. R.K. Wood, Bilevel network interdiction models: Formulations and solutions, *Wiley Encyclopedia of Operations Research and Management Science* ed. by J.J. Cochran, L.A. Cox, P. Keskinocak, J.P. Kharoufeh, J.C. Smith (Wiley, New York, 2010)
1375. C. Wu, Y. Ji, Resource allocation in multiple product design projects: a bi-level programming approach. Int. J. Control Autom. **9**, 271–280 (2016)
1376. S. Wu, Y. Chen, P. Marcotte, A cutting plane method for linear bilevel programming. Syst. Sci. Math. Sci. **11**, 125–133 (1998)
1377. W.-H. Wu, C.-Y. Chien, Y.-H. Wu, H.-H. Wu, J.-M. Lai, P.M.-H. Chang, C.-Y.F. Huang, F.-S. Wang, Inferring oncoenzymes in a genome-scale metabolic network for hepatocytes using bilevel optimization framework. J. Taiwan Inst. Chem. Eng. **91**, 97–104 (2018)
1378. Y. Xiang, L. Wang, A game-theoretic study of load redistribution attack and defense in power systems. Electr. Power Syst. Res. **151**, 12–25 (2017)
1379. W. Xiao, G. Du, Y. Zhang, X. Liu, Coordinated optimization of low-carbon product family and its manufacturing process design by a bilevel game-theoretic model. J. Cleaner Prod. **184**, 754–773 (2018)
1380. F. Xie, M.M. Butt, Z. Li, A feasible flow-based iterative algorithm for the two-level hierarchical time minimization transportation problem. Comput. Oper. Res. **86**, 124–139 (2017)
1381. H. Xiong, M. Chen, Y. Lin, N. Lv, X. Yan, K. Xu, C. Wu, Bi-level programming based contra flow optimization for evacuation events. Kybernetes **39**(8), 1227–1234 (2010)
1382. C. Xu, T. Chen, Incentive strategies with many followers. Acta Autom. Sin. **17**, 577–581 (1991, in Chinese)
1383. G. Xu, Y. Li, Steady-state optimization of biochemical systems by bi-level programming. Comput. Chem. Eng. **106**, 286–296 (2017)
1384. H. Xu, An MPCC approach for stochastic Stackelberg–Nash–Cournot equilibrium. Optimization **54**(1), 27–57 (2005)
1385. J. Xu, J. Gang, Multi-objective bilevel construction material transportation scheduling in large-scale construction projects under a fuzzy random environment. Transp. Plann. Technol. **36**(4), 352–376 (2013)
1386. J. Xu, Z. Li, Z. Tao, Bi-level decision making in random phenomenon, in *Random-Like Bi-level Decision Making* (Springer, Berlin, 2016), pp. 77–197
1387. J. Xu, Z. Li, Z. Tao, Foundations of random-like bi-level decision making, in *Random-Like Bi-level Decision Making* (Springer, Berlin, 2016), pp. 1–75
1388. J. Xu, Z. Li, Z. Tao, *Random-like bi-level decision making*. Lecture Notes in Economics and Mathematical Systems, vol. 688 (Springer, Berlin, 2016)
1389. J. Xu, Y. Tu, Z. Zeng, Bilevel optimization of regional water resources allocation problem under fuzzy random environment. J. Water Resour. Plann. Manage. **139**(3), 246–264 (2012)
1390. M. Xu, J.J. Ye, A smoothing augmented Lagrangian method for solving simple bilevel programs. Comput. Optim. Appl. **59**(1–2), 353–377 (2014)
1391. M. Xu, J.J. Ye, L. Zhang, Smoothing augmented Lagrangian method for nonsmooth constrained optimization problems. J. Global Optim. **62**(4), 675–694 (2015)
1392. M. Xu, J.J. Ye, L. Zhang, Smoothing SQP methods for solving degenerate nonsmooth constrained optimization problems with applications to bilevel programs. SIAM J. Optim. **25**(3), 1388–1410 (2015)

1393. M.H. Xu, M. Li, C.C. Yang, Neural networks for a class of bi-level variational inequalities. J. Global Optim. **44**(4), 535–552 (2009)
1394. P. Xu, Three essays on bilevel optimization algorithms and applications, Ph.D. thesis (Iowa State University, New York, 2012)
1395. P. Xu, L. Wang, An exact algorithm for the bilevel mixed integer linear programming problem under three simplifying assumptions. Comput. Oper. Res. **41**, 309–318 (2014)
1396. X. Xu, Z. Meng, R. Shen, A tri-level programming model based on conditional value-at-risk for three-stage supply chain management. Comput. Ind. Eng. **66**(2), 470–475 (2013)
1397. Z.K. Xu, Deriving the properties of linear bilevel programming via a penalty function approach. J. Optim. Theory Appl. **103**, 441–456 (1999)
1398. M. Yamagishi, I. Yamada, Nonexpansiveness of a linearized augmented Lagrangian operator for hierarchical convex optimization. Inverse Prob. **33**(4), 35 (2017)
1399. Y. Yamamoto, Optimization over the efficient set: overview. J. Global Optim. **22**(1–4), 285–317 (2002)
1400. H. Yan, W.H. Lam, Optimal road tolls under conditions of queueing and congestion. Transp. Res. A **30A**, 319–332 (1996)
1401. X. Yan, An augmented Lagrangian-based parallel splitting method for a one-leader-two-follower game. J. Ind. Manag. Optim. **12**(3), 879–890 (2016)
1402. X. Yan, R. Wen, A new parallel splitting augmented Lagrangian-based method for a Stackelberg game. J. Inequalities Appl. **2016**(1), 1–14 (2016)
1403. D. Yang, J. Jiao, Y. Ji, G. Du, P. Helo, A. Valente, Joint optimization for coordinated configuration of product families and supply chains by a leader-follower Stackelberg game. Eur. J. Oper. Res. **246**(1), 263–280 (2015)
1404. H. Yang, M.G.H. Bell, Transportation bilevel programming problems: recent methodological advances. Transp. Res. Part B **35**, 1–4 (2001)
1405. H. Yang, T. Sasaki, Y. Iida, Estimation of origin-destination matrices from link traffic counts on congested networks. Transp. Res. B Methodol. **26**(6), 417–434 (1992)
1406. H. Yang, S. Yagar, Traffic assignment and signal control in saturated road networks. Transp. Res. Part A Policy Pract. **29**(2), 125–139 (1995)
1407. L. Yang, R. Mahadevan, W.R. Cluett, A bilevel optimization algorithm to identify enzymatic capacity constraints in metabolic networks. Comput.Chem. Eng. **32**(9), 2072–2085 (2008)
1408. Q. Yang, A note on constrained qualification for bilevel programming. J. Math. Res. Exposition **19**, 359–366 (1999)
1409. İ. Yanıkoğlu, D. Kuhn, Decision rule bounds for two-stage stochastic bilevel programs. SIAM J. Optim. **28**(1), 198–222 (2018)
1410. D.-Q. Yao, J.J. Liu, Competitive pricing of mixed retail and e-tail distribution channels. Omega **33**(3), 235–247 (2005)
1411. Y. Yao, T. Edmunds, D. Papageorgiou, R. Alvarez, Trilevel optimization in power network defense. IEEE Trans. Syst. Man Cybern. Part C Appl. Rev. **37**(4), 712–718 (2007)
1412. J.J. Ye, Necessary conditions for bilevel dynamic optimization problems, *Proceedings of the 33rd IEEE Conference on Decision and Control, 1994*, vol. 1 (IEEE, New York, 1994), pp. 507–512
1413. J.J. Ye, Necessary conditions for bilevel dynamic optimization problems. SIAM J. Control Optim. **33**(4), 1208–1223 (1995)
1414. J.J. Ye, Necessary optimality conditions for bilevel dynamic problems, *Proceedings of the 36th IEEE Conference on Decision and Control, 1997*, vol. 2 (IEEE, New York, 1997), pp. 1405–1410
1415. J.J. Ye, Optimal strategies for bilevel dynamic problems. SIAM J. Control Optim. **35**, 512–531 (1997)
1416. J.J. Ye, Nondifferentiable multiplier rules for optimization and bilevel optimization problems. SIAM J. Optim. **15**, 252–274 (2004)
1417. J.J. Ye, Constraint qualifications and KKT conditions for bilevel programming problems. Math. Oper. Res. **31**, 811–824 (2006)

1418. J.J. Ye, Necessary optimality conditions for multiobjective bilevel programs. Math. Oper. Res. **36**(1), 165–184 (2011)
1419. J.J. Ye, X.Y. Ye, Necessary optimality conditions for optimization problems with variational inequality constraints. Math. Oper. Res. **22**(4), 977–997 (1997)
1420. J.J. Ye, D. Zhu, New necessary optimality conditions for bilevel programs by combining the MPEC and value function approaches. SIAM J. Optim. **20**(4), 1885–1905 (2010)
1421. J.J. Ye, D. Zhu, Q. Zhu, Generalized bilevel programming problems, Technical Report DMS-646-IR (University of Victoria, Department of Mathematics and Statistics, Victoria, 1993)
1422. J.J. Ye, D.L. Zhu, Optimality conditions for bilevel programming problems. Optimization **33**, 9–27 (1995)
1423. J.J. Ye, D.L. Zhu, A note on optimality conditions for bilevel programming problems. Optimization **39**, 361–366 (1997)
1424. J.J. Ye, D.L. Zhu, Q.J. Zhu, Exact penalization and necessary optimality conditions for generalized bilevel programming problems. SIAM J. Optim. **7**, 481–507 (1997)
1425. K. Yeh, M.J. Realff, J.H. Lee, C. Whittaker, Analysis and comparison of single period single level and bilevel programming representations of a pre-existing timberlands supply chain with a new biorefinery facility. Comput.Chem. Eng. **68**, 242–254 (2014)
1426. K. Yeh, C. Whittaker, M.J. Realff, J.H. Lee, Two stage stochastic bilevel programming model of a pre-established timberlands supply chain with biorefinery investment interests. Comput.Chem. Eng. **73**, 141–153 (2015)
1427. A. Yezza, First-order necessary optimality conditions for general bilevel programming problems. J. Optim. Theory Appl. **89**, 189–219 (1996)
1428. W. Yi, L. Nozick, R. Davidson, B. Blanton, B. Colle, Optimization of the issuance of evacuation orders under evolving hurricane conditions. Transp. Res. B Methodol. **95**, 285–304 (2017)
1429. P.-Y. Yin, Multilevel minimum cross entropy threshold selection based on particle swarm optimization. Appl. Math. Comput. **184**(2), 503–513 (2007)
1430. Y. Yin, Genetic-algorithms-based approach for bilevel programming models. J. Transp. Eng. **126**(2), 115–120 (2000)
1431. Y. Yin, Multiobjective bilevel optimization for transportation planning and management problems. J. Adv. Transp. **36**(1), 93–105 (2002)
1432. B. Yu, L. Kong, Y. Sun, B. Yao, Z. Gao, A bi-level programming for bus lane network design. Transp. Res. C Emerg. Technol. **55**, 310–327 (2015)
1433. J. Yu, H.L. Wang, An existence theorem for equilibrium points for multi-leader–follower games. Nonlinear Anal. Theory Methods Appl. **69**(5), 1775–1777 (2008)
1434. Y. Yu, F. Chu, H. Chen, A Stackelberg game and its improvement in a VMI system with a manufacturing vendor. Eur. J. Oper. Res. **192**(3), 929–948 (2009)
1435. D. Yue, J. Gao, B. Zeng, F. You, A projection-based reformulation and decomposition algorithm for global optimization of a class of mixed integer bilevel linear programs. J. Global Optim. **73**(1), 27–57 (2019)
1436. D. Yue, F. You, Projection-based reformulation and decomposition algorithm for a class of mixed-integer bilevel linear programs, in *Computer Aided Chemical Engineering*, ed. by K. Zdravko, B. Miloš, vol. 38 (Elsevier, Amsterdam, 2016), pp. 481–486
1437. D. Yue, F. You, Stackelberg-game-based modeling and optimization for supply chain design and operations: a mixed integer bilevel programming framework. Comput.Chem. Eng. **102**, 81–95 (2017)
1438. M. F. Zaman, S. M. Elsayed, T. Ray, R. A. Sarker, A co-evolutionary approach for optimal bidding strategy of multiple electricity suppliers, *Proceedings of the IEEE Congress on Evolutionary Computation (CEC), 2016* (IEEE, New York, 2016), pp. 3407–3715
1439. M.H. Zare, J.S. Borrero, B. Zeng, O.A. Prokopyev, A note on linearized reformulations for a class of bilevel linear integer problems. Ann. Oper. Res. **272**, 99–117 (2019)
1440. A. J. Zaslavski, Necessary optimality conditions for bilevel minimization problems. Nonlinear Anal. Theory Methods Appl. **75**(3), 1655–1678 (2012)

1441. P. Zeephongsekul, Stackelberg strategy solution for optimal software release policies. J. Optim. Theory Appl. **91**, 215–233 (1996)
1442. A.B. Zemkoho, Multicriteria approach to bilevel programming, Master's thesis (Universite de Yaounde I, Cameroon, 2007, in French)
1443. A.B. Zemkoho, Bilevel programming: Reformulations, regularity, and stationarity, Ph.D. thesis (TU Bergakademie Freiberg, Freiberg, 2012)
1444. A.B. Zemkoho, Solving ill-posed bilevel programs. Set-Valued Variational Anal. **24**(3), 423–448 (2016)
1445. B. Zeng, Y. An, Solving bilevel mixed integer program by reformulations and decomposition. Optimization online, pp. 1–34 (2014)
1446. D. Zhang, G.-H. Lin, Bilevel direct search method for leader–follower problems and application in health insurance. Comput. Oper. Res. **41**, 359–373 (2014)
1447. G. Zhang, J. Han, J. Lu, Fuzzy bi-level decision-making techniques: a survey. Int. J. Comput. Int. Syst. **9**(sup1), 25–34 (2016)
1448. G. Zhang, C. Jiang, X. Wang, B. Li, Risk assessment and bi-level optimization dispatch of virtual power plants considering renewable energy uncertainty, in *IEEJ Transactions on Electrical and Electronic Engineering*, **12**(4), 510–518 (2017)
1449. G. Zhang, J. Lu, The definition of optimal solution and an extended Kuhn-Tucker approach for fuzzy linear bilevel programming. IEEE Int. Inf. Bull. **6**(2), 1–7 (2005)
1450. G. Zhang, J. Lu, Fuzzy bilevel programming with multiple objectives and cooperative multiple followers. J. Global Optim. **47**(3), 403–419 (2010)
1451. G. Zhang, J. Lu, T. Dillon, An approximation branch-and-bound algorithm for fuzzy bilevel decision making problems, in *Proceedings of The 1st International Symposium Advances in Artificial Intelligence and Applications* (Citeseer, Poland, 2006)
1452. G. Zhang, J. Lu, T. Dillon, An extended branch-and-bound algorithm for fuzzy linear bilevel programming, in *Applied Artificial Intelligence: Proceedings of the 7th International FLINS Conference, Genova, Italy, 29-31 August 2006* (World Scientific, Singapore, 2006), pp. 291–298
1453. G. Zhang, J. Lu, T. Dillon, Decentralized multi-objective bilevel decision making with fuzzy demands. Knowledge-Based Syst. **20**(5), 495–507 (2007)
1454. G. Zhang, J. Lu, T. Dillon, Fuzzy linear bilevel optimization: Solution concepts, approaches and applications, in *Fuzzy Logic*, ed. by P.P. Wang, D. Ruan, E.E. Kerre. Studies in Fuzziness and Soft Computing, vol. 215 (Springer, Berlin, 2007), pp. 351–379
1455. G. Zhang, J. Lu, T. Dillon, Models and algorithm for fuzzy multi-objective multi-follower linear bilevel programming, *Proceedings of the IEEE International Fuzzy Systems Conference, 2007. FUZZ-IEEE 2007* (IEEE, New York, 2007), pp. 1–6
1456. G. Zhang, J. Lu, T. Dillon, Solution concepts and an approximation Kuhn–Tucker approach for fuzzy multiobjective linear bilevel programming, in *Pareto Optimality, Game Theory Equilibria*, ed. by P. Pardalos, A. Migdalas, L. Pitsoulis (Springer, Berlin, 2008), pp. 457–480
1457. G. Zhang, J. Lu, Y. Gao, An algorithm for fuzzy multi-objective multi-follower partial cooperative bilevel programming. J. Int. Fuzzy Syst. **19**(4, 5), 303–319 (2008)
1458. G. Zhang, J. Lu, Y. Gao, *Multi-level decision making: Models, methods and applications*, (Springer, Berlin, 2015)
1459. G. Zhang, J. Lu, J. Montero, Y. Zeng, Model, solution concept, and *k*th-best algorithm for linear trilevel programming. Inf. Sci. **180**(4), 481–492 (2010)
1460. G. Zhang, J. Lu, X. Zeng, Models and algorithms for fuzzy multi-objective multi-follower linear bilevel programming in a partial cooperative situation, in *Proceedings of the International Conference on Intelligent Systems and Knowledge Engineering 2007* (Atlantis Press, Amsterdam, 2007)
1461. G. Zhang, H. Sun, Y. Zheng, G. Xia, L. Feng, Q. Sun, Optimal discriminative projection for sparse representation-based classification via bilevel optimization, in *IEEE Transactions on Circuits and Systems for Video Technology* **30**(4), 1065–1077 (2019)

1462. G. Zhang, G. Zhang, Y. Gao, J. Lu, A bilevel optimization model and a PSO-based algorithm in day-ahead electricity markets, in *Proceedings of the IEEE International Conference on Systems, Man and Cybernetics, 2009. SMC 2009* (IEEE, New York, 2009), pp. 611–616
1463. G. Zhang, G. Zhang, Y. Gao, J. Lu, Competitive strategic bidding optimization in electricity markets using bilevel programming and swarm technique. IEEE Trans. Ind. Electron. **58**(6), 2138–2146 (2011)
1464. H. Zhang, Z. Gao, Bilevel programming model and solution method for mixed transportation network design problem. J. Syst. Sci. Complexity **22**(3), 446–459 (2009)
1465. J. Zhang, Approximating the two-level facility location problem via a quasi-greedy approach. Math. Program. **108**(1), 159–176 (2006)
1466. J. Zhang, Enhanced optimality conditions and new constraint qualifications for nonsmooth optimization problems, Ph.D. thesis (University of Victoria, Victoria, 2014)
1467. J. Zhang, X. Jia, J. Hu, K. Tan, Satellite multi-vehicle tracking under inconsistent detection conditions by bilevel k-shortest paths optimization, in *Proceedings of the 2018 Digital Image Computing: Techniques and Applications (DICTA)* (2018), pp. 1–8
1468. J. Zhang, Y. Qiu, M. Li, M. Xu, Sequential multi-objective optimization for lubrication system of gasoline engines with bilevel optimization structure. J. Mech. Des. **139**(2), 021405 (2017)
1469. J. Zhang, H. Wang, Y. Sun, A note on the optimality condition for a bilevel programming. J. Inequalities Appl. **2015**(1), 1–12 (2015)
1470. J. Zhang, C. Xu, Inverse optimization for linearly constrained convex separable programming problems. Eur. J. Oper. Res. **200**(3), 671–679 (2010)
1471. J. Zhang, L. Zhang, H. Huang, X. Wang, C. Gu, Z. He, A unified algorithm for virtual desktops placement in distributed cloud computing. Math. Prob. Eng. **2016** (2016)
1472. J.-Z. Zhang, D.-T. Zhu, A bilevel programming method for pipe network optimization. SIAM J. Optim. **6**, 838–857 (1996)
1473. L. Zhang, A fuzzy algorithm for solving a class of bi-level linear programming problem. Appl. Math. Inf. Sci. **8**(4), 1823 (2014)
1474. R. Zhang, Problems of hierarchical optimization in finite dimensions. SIAM J. Optim. **4**, 521–536 (1994)
1475. R. Zhang, Multistage bilevel programming problems. Optimization **52**, 605–616 (2003)
1476. T. Zhang, An elite particle swarm optimization algorithm based on quadratic approximations for high-dimension bilevel single objective programming problems. Int. J. Eng. Sci. Invention **7**(5), 90–95 (2018)
1477. T. Zhang, T. Hu, X. Guo, Z. Chen, Y. Zheng, Solving high dimensional bilevel multiobjective programming problem using a hybrid particle swarm optimization algorithm with crossover operator. Knowledge-Based Syst. **53**, 13–19 (2013)
1478. T. Zhang, T. Hu, Y. Zheng, X. Guo, An improved particle swarm optimization for solving bilevel multiobjective programming problem. J. Appl. Math. **2012**, 13 (2012)
1479. X. Zhang, D. Shi, Z. Wang, Z. Yu, X. Wang, D. Bian, K. Tomsovic, Bilevel optimization based transmission expansion planning considering phase shifting transformer, in *Proceedings of the Power Symposium (NAPS), 2017 North American* (IEEE, New York, 2017), pp. 1–6
1480. W. Zhao, R. Liu, D. Ngoduy, A bilevel programming model for autonomous intersection control and trajectory planning. Transportmetrica A Transp. Sci., 1–25 (2019). Online first
1481. X. Zhao, Z.-Y. Feng, Y. Li, A. Bernard, Evacuation network optimization model with lane-based reversal and routing. Math. Prob. Eng. **2016**, 13 (2016)
1482. Y. Zheng, T. Basar, Existence and derivation of optimal affine incentive schemes for Stackelberg games with partial information: a geometric approach. Int. J. Control **35**(6), 997–1011 (1982)
1483. Y. Zheng, D. Fang, Z. Wan, A solution approach to the weak linear bilevel programming problems. Optimization **65**(7), 1437–1449 (2016)
1484. Y. Zheng, G. Lei, X. Cao, A method for a ϵ-global optimal solution of linear bilevel programming. J. Math. Wuhan Univ. **33**(5), 941–945 (2013, in Chinese)

1485. Y. Zheng, J. Liu, Z. Wan, Interactive fuzzy decision making method for solving bilevel programming problem. Appl. Math. Model. **38**(13), 3136–3141 (2014)
1486. Y. Zheng, Z. Wan, A solution method for semivectorial bilevel programming problem via penalty method. J. Appl. Math. Comput. **37**(1–2), 207–219 (2011)
1487. Y. Zheng, Z. Wan, S. Jia, G. Wang, A new method for strong-weak linear bilevel programming problem. J. Ind. Manag. Optim. **11**(2), 529–547 (2015)
1488. Y. Zheng, Z. Wan, Y. Lü, A global convergent method for nonlinear bilevel programming problems. J. Syst. Sci. Math. Sci. **32**(5), 513–521 (2012, in Chinese)
1489. Y. Zheng, Z.-P. Wan, Z. Hao, An objective penalty function method for a class of nonlinear bilevel programming problems. J. Syst. Sci. Math. Sci. **33**(10), 1156–1163 (2013, in Chinese)
1490. Y. Zheng, Z.-P. Wan, K. Sun, T. Zhang, An exact penalty method for weak linear bilevel programming problem. J. Appl. Math. Comput. **42**(1–2), 41–49 (2013)
1491. Y. Zheng, Z.-P. Wan, G.-M. Wang, A fuzzy interactive method for a class of bilevel multiobjective programming problem. Expert Syst. Appl. **38**(8), 10384–10388 (2011)
1492. Y. Zheng, Z.-P. Wan, L.-Y. Yuan, Coordination problem of the principal-agent based on bilevel programming. Xitong Gongcheng Lilun yu Shijian/Syst. Eng. Theory Pract. **34**(1), 77–83 (2014)
1493. Y. Zheng, G. Zhang, J. Han, J. Lu, Pessimistic bilevel optimization model for risk-averse production-distribution planning. Inf. Sci. **372**, 677–689 (2016)
1494. Y. Zheng, G. Zhang, Z. Zhang, J. Lu, A reducibility method for the weak linear bilevel programming problems and a case study in principal-agent. Inf. Sci. **454-455**, 46–58 (2018)
1495. Y. Zheng, Z. Zhu, L. Yuan, Partially-shared pessimistic bilevel multi-follower programming: concept, algorithm, and application. J. Inequalities Appl. **2016**(1), 1–13 (2016)
1496. Y. Zheng, X. Zhuo, J. Chen, Maximum entropy approach for solving pessimistic bilevel programming problems. Wuhan Univer. J. Nat. Sci. **22**(1), 63–67 (2017)
1497. S. Zhou, A.B. Zemkoho, A. Tin, Bolib: Bilevel optimization library of test problems, Technical Report (University of Southampton, Southampton, 2018)
1498. S. Zhou, A.B. Zemkoho, A. Tin, Bolib: Bilevel optimization library of test problem version 2, *Bilevel optimization: advances and next challenges*, ed. by S. Dempe, A.B. Zemkoho (Springer, Berlin, 2020)
1499. Y. Zhou, S. Kwong, H. Guo, W. Gao, X. Wang, Bilevel optimization of block compressive sensing with perceptually nonlocal similarity. Inf. Sci. **360**, 1–20 (2016)
1500. X. Zhu, P. Guo, Approaches to four types of bilevel programming problems with nonconvex nonsmooth lower level programs and their applications to newsvendor problems. Math. Methods Oper. Res. **86**, 255–275 (2017)
1501. X. Zhu, P. Guo, Bilevel programming approaches to production planning for multiple products with short life cycles. 4OR—Q. J. Oper. Res. **18**, 151–175 (2020)
1502. X. Zhu, Q. Yu, X. Wang, A hybrid differential evolution algorithm for solving nonlinear bilevel programming with linear constraints, in *Proceedings of the 5th IEEE International Conference on Cognitive Informatics*, vol. 1 (IEEE, New York, 2006), pp. 126–131
1503. Z. Zhu, B. Yu, A modified homotopy method for solving the principal-agent bilevel programming problem. Comput. Appl. Math. **37**(1), 541–566 (2018)
1504. X. Zhuge, H. Jinnai, R.E. Dunin-Borkowski, V. Migunov, S. Bals, P. Cool, A.-J. Bons, K.J. Batenburg, Automated discrete electron tomography–towards routine high-fidelity reconstruction of nanomaterials. Ultramicroscopy **175**, 87–96 (2017)
1505. M. Zugno, J.M. Morales, P. Pinson, H. Madsen, A bilevel model for electricity retailers' participation in a demand response market environment. Energy Econ. **36**, 182–197 (2013)

The manufacturer's authorised representative in the EU is Springer Nature Customer Service Centre GmbH, Europaplatz 3, 69115 Heidelberg, Germany. If you have any concerns regarding our products, please contact ProductSafety@springernature.com

Printed and bound by CPI Group (UK) Ltd, Croydon, CR0 4YY
28/09/2026
02224061-0013